COPING WITH CLIMATE CHANGE

Proceedings Of The Second North American Conference
On Preparing For Climate Change: A Cooperative Approach

Convened By The Climate Institute
In Cooperation With Participating Agencies
And Universities Of The Canadian Climate Program

JOHN C. TOPPING, JR., EDITOR

Published By The Climate Institute
Washington, D.C.
June 1989

Climate Institute
316 Pennsylvania Avenue, SE
Suite 403
Washington, DC 20003
202/547-0104
Fax 202/547-0111
Telex 214 858 Climat UR

Library of Congress Catalog Card Number 89-62291
ISBN 0-9623610-0-3
Printed and bound in the United States of America

Cover Artwork From
"Failed Atmosphere" ©1988
Jerry W. Carter
Copies Of This Poster
Are Available At $15 Each
From The Artist At
Climate Art Fund
10602 Bucknell Drive
Silver Spring, MD 20902

CLIMATE INSTITUTE

CONFERENCE CO-SPONSORS

Geneva Steel
GE Foundation
The William Bingham Foundation
AT&T
American Gas Association
American Petroleum Institute
Ford Motor Company Fund
Motor Vehicle Manufacturers Association
Ebasco Services, Inc.
Texaco, Inc.
World Resources Institute
National Parks And Conservation Association
National Audubon Society
Atlantic City Electric Company
PPG Industries Foundation
American International Group
U.S. Environmental Protection Agency
U.S. Department Of Energy
National Climate Program Office
National Oceanic & Atmospheric Administration
U.S. Forest Service
National Aeronautics And Space Administration
National Science Foundation

TABLE OF CONTENTS

XX. DEVELOPING A FRAMEWORK FOR INTERNATIONAL CLIMATE COOPERATION

APPENDIX

FOREWORD

Public consciousness of the threat to humanity posed by global climate change has grown immeasurably over the past few years. In the six years since an EPA sponsored sea level rise conference first focused significant attention on the then obscure problem of the greenhouse effect, this issue has moved from being a concern largely of scientists to now being a subject of diplomatic discourse and great attention by heads of state.

As a scientist involved deeply in coastal implications of greenhouse effect induced sea level rise, I have long been conscious of the need for an institution to bridge the wide gulf between climate scientists and public and private sector decisionmakers. For the past three years the Climate Institute has sought to bridge this gap through a series of conferences and symposia as well as publications directed toward opinion leaders, planners and policy makers. Preparing for Climate Change, the Proceedings of the First North American Conference on Preparing for Climate Change, has proved an extremely useful reference for climate scholars and policy makers alike. These proceedings of the Second North American Conference with their specific focus on seven distinct regions within the Western Hemisphere should prove an essential reference work for those interested in potential implications of climate change.

Dr. Stephen Leatherman
Chairman
Climate Institute

INTRODUCTION

The Second North American Conference on Preparing for Climate Change may be the most ambitious assemblage of experts ever to assess impacts and response strategies to the twin challenges of greenhouse warming and stratospheric ozone depletion. Presentations were made by over 160 scientists, environmental leaders and policy makers from the Western Hemisphere, Europe and Asia in 38 sessions over a three day period.

Like the very successful First North American Conference on Preparing for Climate Change, this meeting drew extensively on the research performed for the Canadian Climate Program and the U.S. National Climate Program. The many studies completed for the U.S. Environmental Protection Agency's Effects and Stabilization reports to Congress enabled us to assemble some highly detailed regional impact panels.

Chapters in this volume correspond to the seven regional panels of the Second North American Conference, with discussions of implications of climate change for the Caribbean, the Arctic, California, the Southern United States, the Chesapeake Bay, Atlantic Canada and New England, and the Great Lakes.

This book also contains a policy overview of the climate challenge with contributions from U.S., Canadian, British and Caribbean governmental and corporate leaders. A chapter devoted to a scientific overview of climate change includes a skillful overview of the key scientific and policy issues involved in greenhouse warming, a seminal article on regional implications of climate change and the potential impacts of global warming on droughts and floods, and a panel discussion involving four of the world's leading stratospheric scientists. The next chapter includes penetrating discussions of climate feedbacks, oceans and atmospheric processes, and potential changes in weather circulation patterns. Critical scientific issues are also discussed in readily comprehensible terms in chapters on human health, agricultural, ecological and water resource implications of climate change. Subsequent chapters also include explorations of implications of climate change for coastal planning, urban planning and corporate planning. The three final chapters respectively explore prospects for reductions of stratospheric perturbants, energy implications of global warming, and mechanisms for engendering global cooperation to address the challenge of rapid climate change.

Over the past year a major priority of the Climate Institute has been the involvement of developing countries in climate impact and policy discussions. Embassies and United Nations missions of over 50 nations participated in two symposia which the Climate Institute convened in 1988 concerning implications of climate change for the Third World. The inclusion of a Caribbean regional panel and developing nation policy discussions in this conference should reinforce this effort. In 1989 the Climate Institute is sponsoring two major international conferences, both in cooperation with the United Nations Environment Programme. On May 3-5, a Conference on the Implications of Climate Change for Africa was held at Howard University in Washington, D.C. where it drew participants from over 20 nations. In December 1989, a World Conference on Preparing for Climate Change will be convened in Cairo, Egypt.

Each of these meetings and other more specialized past and projected meetings on implications of climate change for fisheries, wildlife, infrastructure, private sector investment, air quality management, the Chesapeake Bay and the Arctic have sought to foster broad-based sponsorship and participation. Development of an effective global response to the challenges posed by rapid climate change will require unprecedented cooperation among nations and various sectors of society. Joining in the sponsorship of the Second North American Conference on Preparing for Climate Change were six corporations, two corporate foundations, three trade associations, an independent foundation, two national environmental groups, two environmental research institutes, six U.S. government agencies and the Canadian Climate Program.

We anticipate that the proceedings produced from this conference will provide a valuable policy contribution to newly elected governments in the United States, Canada and Mexico, as well as to many individual decision makers throughout North America and the Caribbean.

John C. Topping, Jr.
President, Climate Institute and
Editor of the Proceedings

ACKNOWLEDGMENTS

This book is the most comprehensive single volume of articles yet assembled on climate change. For this we are deeply indebted to the approximately 150 authors, including many of the leading scientific and policy experts on climate change.

The Second North American Conference on Preparing for Climate Change which engendered these articles resulted from the initiative of Joseph A. Cannon, President of Geneva Steel. As Associate Administrator for Policy and Resource Management, Joe Cannon in 1982 commissioned the first EPA policy option studies concerning the greenhouse effect and in March 1983 presided over the first U.S. government conference concerning sea level rise. Shortly after becoming Assistant Administrator for Air and Radiation of EPA, Cannon initiated the risk assessments that laid the groundwork for the Montreal Protocol to Protect the Ozone Layer. Cannon's Chairmanship of the Conference and Geneva Steel's early contribution were instrumental in enabling the Climate Institute to build a broad base of sponsors.

Although we are grateful to all of the sponsors listed on the title page, three sponsors were especially crucial to the conference's success. The U.S. National Climate Program which helped spearhead the October 1987 First North American Conference on Preparing for Climate Change coordinated funding from a number of federal agencies, and we are thankful for the personal efforts of Alan Hecht and Bill Bolhofer. A major funder of the conference and a sponsor of many of the policy studies which made this conference so successful was EPA's Office of Policy Analysis. We appreciate the strong support from Dick Morgenstern, Dennis Tirpak, Joel Smith, Dan Lashof and Jim Titus of that office. We also acknowledge the pivotal support of the U.S. Department of Energy and especially Ted Williams.

Crucial to the success of this conference were the efforts of Climate Institute staff, especially Carl Cole and Nancy Wilson, and Cathie Hartnett, Cindy Kunz and Sharon Fischlowitz of Hartnett & Associates, the Institute's conference consultant.

For the production of this volume we are especially indebted to Jim Dyer and Dallas Hutchinson as well as Nancy Wilson, Editor of _Climate Alert_, and Carl Cole, the Climate Institute's Director of Administration.

We are also grateful to Jerry Carter whose masterpiece signifying the threat to our planet posed by a deteriorating atmosphere provided a memorable conference poster. This work appears on the cover of these Proceedings.

John C. Topping, Jr.
President, Climate Institute and
Editor of the Proceedings

REMARKS OF

Sir Crispin Tickell
Ambassador of the United Kingdom to the United Nations

It is an honour for me, but perhaps a curiosity to you, that an Ambassador should be speaking to you tonight. Why, you may ask, should the subject of your conference have anything to do with Ambassadors and become entangled in the wiles of diplomacy? Even more eccentric, why should this Ambassador be a member of the Board of the Climate Institute which has so admirably organized this conference?

There are, I think, four main reasons why problems of climate and preparing for the effects of climate change should be on the diplomatic as well as scientific agenda. First, diplomats can help in alerting international public opinion to the nature and scope of what may befall us. Second, they have a role in persuading world leaders--presidents, prime ministers, ministers of the environment and others--of the need for action. Third, they have practical knowledge of how to formulate policies within governments and to put those policies into effect; and last they can create the framework and draw up the instruments for comprehensive international treatment of essentially global problems. I shall have a word to say about each of these points.

Public opinion is of course crucial. Governments, especially elected governments, respond as much to public opinion as to any internal scientific or managerial advice. But if governments respond to public opinion, they also influence and to some extent lead it. Some 15 years ago, when I first began to study the problems of climate, I found that few believed that any serious climatic change was likely in anything less than hundreds of years. Of course, there were variations within broadly defined limits, and for those living on the margins between the tropics and the desert, or between temperate areas and the Arctic such changes could be cruel and disruptive. Likewise, few believed that the activities of our species could have much effect on the operations of the weather machine, which was seen, certainly by most economists, as one of the constants in human affairs. The reaction when some, including me, suggested otherwise was similar to that of the Victorian lady when told of the theory of evolution and her own likely descent from the apes. She said, "I hope such ideas are mistaken; but if not, I trust they will not become generally known."

In fact, public understanding of these issues has changed radically in the last 15 years and even more, perhaps, in the last 15 months. Most educated people have now heard of the greenhouse gases, of the role of the ozone layer in shielding life from ultraviolet radiation, of the effects of acid deposition down-wind of industry. Such opinion is sometimes influenced in the right direction for the wrong reasons. Thus, an exceptionally hot summer in North America, several years drought in sub-Saharan Africa, cyclones and flooding in Bangladesh, the dislocations produced by El Nino (or his sister La Nina) have been used to show that major damaging change is already upon us. I think that most here would agree that these events do not provide proof of such change. All could fall within the limits of natural variability. But let us be glad that the underlying problems of climatic change are now on the front pages of newspapers rather than on the back pages of learned journals.

In the process of increasing public consciousness, people within governments have played as big a role as those outside. I have in mind not only those close to the microphones but also officially funded research scientists, particularly in your country and mine. They have steadily built up the scientific case in the knowledge that the opening position of any government must be one of scepticism. Governments are elected for short periods and are understandably averse from taking expensive action to cope with long term problems. In my experience most governments ask the following questions: Can you prove that changes are taking place? What are their effects likely to be? Are they going to be good or bad? Is there anything which could or should be done about them?

We all know how difficult it is to demonstrate chains of cause and effect. Our familiar friend chaos creeps in at every link. But sometimes it is necessary to act before the links are all put together. A good example is the action which was taken last year on the ozone layer, enshrined in the Montreal Protocol, to limit the production and use of chlorofluorocarbons. What was then agreed with immense difficulty has already come to be seen as inadequate. It is for that reason that my government has decided to call a major international conference on chlorofluorocarbons and the ozone layer in London in March 1989. The ozone issue is relatively narrow and for that reason manageable. But it has enormous implications: for example, it leads straight into the linked questions of industrial development in poor countries, of technology transfer to convey information about substitutes, and of sanctions against possible offenders. Yet knowledge of ozone chemistry is still incomplete. Surprises may be awaiting us.

The Montreal Protocol is also a useful precedent for dealing with the much wider problem of global warming with its still greater implications for virtually every aspect of modern society. But if governments have been slow in the past to recognize these global problems, they have recently changed their minds. In this process the Brundtland Commission Report on Environment and Development had an appeal to a much wider audience than ever before. There was likewise a host of more specialized conferences and reports from the Global 2000 paper prepared under the Carter Administration to the Toronto Conference of a few months ago. My own Prime Minister made a memorable speech on the subject to the Royal Society in London on 27 September this year when she drew particular attention to three changes in atmospheric chemistry--increase in the greenhouse gases, thinning of the ozone layer, and acid deposition--which called for urgent action. In particular, she recognized the unprecedented rapidity of change and the role of human activities in producing it.

The main theme of this conference with its admirably wide agenda is on preparing for climatic change: how to formulate the right policies and how to carry them out. But I think most of us are still at the preparatory stage of trying to show that change is real and that its effects will reach far into our lives. Few have yet got their minds round the problems of whether we should simply try to adapt to change, whether to mitigate it, whether to correct it, or whether to consider a mixture of all three. Nor is it easy to see where to start.

After all, the problem is not simply that of the increasing quantities of atmospheric carbon dioxide through consumption of fossil fuel or burning the tropical rain forest. These phenomena, like the increase in atmospheric methane, and in chlorofluorocarbons and other chemicals arise from the development of industrial society in the last 250 years, from the desire of all people to enjoy its benefits, the corresponding increase in the population of human beings and their domestic animals, and the particular technologies which have been used to satisfy their energy requirements. Any attempt to manage or regulate the problem of global warming would thus go to the vitals of the present economic system, the imbalances within it, and the differing responsibilities of the countries concerned.

No wonder so many governments have quailed at the thought of doing anything about it. But I believe, and I imagine that most others here will too, that governments and peoples alike will have to do something about it if they are not to destroy themselves in the way that other animal species have destroyed themselves in the long biological history of our planet.

Adaptation to change, mitigation of the effects of change, correction of change, or all three, are the stuff of this meeting. But I notice that one point is absent from your agenda: the effects on population stability both within countries and regions and between them. In 1988 we have already seen that for a combination of natural and man-reasons the number of refugees and displaced people worldwide has greatly increased. The figures for Africa are truly alarming. Frontiers between rich countries and poor are already far from impermeable as your own long border with Mexico well shows. Yet if global warming is really upon us, then these problems will be greatly magnified, and future generations may come to look on the present as a time of relative stability and prosperity.

Last, I come to the international aspects. Already there is a panoply of institutions. The World Meteorological Organization has a long and honorable history, as has the International Council of Scientific Unions. The United Nations Environment Programme is more recent. Since 1979, there has been a World Climate Programme with four sub-programmes, including one on the impact of change. But without disrespect to existing institutions, and without attributing blame, their activities have not so far caught the general imagination. This year the Maltese, who had the credit for initiating the conference on the Law of the Sea, have new credit for putting an item on climate on the agenda of the United Nations General Assembly. Debates followed and the resulting resolution was passed unanimously. As is right, it remitted recommendations for action to the WMO and the UNEP to make proper use of the new Inter-Governmental Panel on Climatic Change, which had its first meeting in Geneva last month. The international community is now seized of the problem and has the machinery with which to work on it.

There are many here and elsewhere who would like to look forward to a Law of the Atmosphere on the same lines as the Law of the Sea. To them I counsel caution. The Law of the Sea conference was not a total success, and many governments including those of the United States and Britain, could not accept all the results. It was Winston Churchill who said that you should never look further ahead than you can see. For the moment it would surely be better for international work to begin without encumbering the debate with the apparatus of draft conventions and treaty making. Instead, we should adopt a step by step approach, designed to bring public opinion and governments along a kind of three-legged progress, until the need for international management or even regulation becomes manifest.

There are three simple thoughts on which I conclude. First, global warming yields no winners nor losers. The world is full of surprises. We do not know what is going to happen. Next, in an overcrowded world we are all likely to be victims. Last, the problem for the world is less the effects of change than the speed with which it is already upon us. Last month my Prime Minister said, "No generation has a freehold on this earth. All we have is a life tenancy with a repairing lease." Let us work to meet the terms of that lease in full.

Policy Keynote Speech of Ambassador Stephen Lewis December 7, 1988

Mr. Chairman, Mr. Grady and distinguished participants. I wanted especially to acknowledge Mr. Grady because I have spent my entire adult life palpitating at the prospect of acknowledging the presence of a president or president-elect, and Mr. Grady is probably as close as I will ever get. I may say at the outset that I want to distinguish my credentials by telling you that I am not an outdoorsman because outdoors to me has always implied activity and I am congenitally opposed to exertion of almost any kind, although I have to say that sleeping under the moon in Wyoming is the kind of somnambulant activity to which I might become addicted.

I am enormously gratified to be here. I want to begin if I may with a disclaimer. Chairing a conference on the international changing atmosphere is enormously gratifying but obviously it confers no particular legitimacy or expertise. And I therefore stand before you faintly bogus, incipiently fraudulent, in terms of the subject matter. Fortunately, one has in this room a surfeit of experts, a positive profusion of people who divined the arcane mysteries of this specialty, and therefore, it is unnecessary for me to attempt to embellish science with which you are all collectively familiar.

But I think that I can therefore say what is obvious to all of us, even those who are lay people like myself: the phenomenon of climate change is now well and truly documented. Warming trends obviously menace future international security on the basis of the work, often profound and searching, which has been done. Madame Brundtland, the Prime Minister of Norway, when she appeared before the Toronto Conference on the Changing Atmosphere in June of this year, talked of environment and climate as phenomena second only to nuclear war in the possibilities of apocalyptic consequences. I hesitate to raise things in an apocalyptic way but this group, beyond all other groups, will recognize the potential for catastrophe if we aren't urgently and readily mobilized. And it seems to me that only a combination of perversity, delinquency and myopia would forestall the world from taking the preventive measures that are necessary.

And in that context, the context of taking preventive measures and bringing collaborative activity, I'd like to make a few observations, and I think I can appropriately begin this way. As Bob Grady told you, I was fortunate enough to be sitting in the hotel room and watched General

Secretary Gorbachev from beginning to end. And I thought it was an enormously encouraging performance. I spent four years watching speakers at the podium of the General Assembly until only this last August, and I thought it was an enormously encouraging performance in a variety of ways. Intrinsically, it was vastly different and vastly preferable from the shoe banging spasms of a Khrushchev to the more urgently argued positions of a Gorbachev. But number one, it maintained the reversal of Soviet foreign policy which we have seen at play for some considerable time now. Number two, it is obviously in its own way a surge of a quest for international peace and security because on behalf of the Soviet Union he enjoins others to collaborate. And number three, it confirms the revival of international cooperation through the United Nations which is experiencing quite an astonishing metamorphosis at this point in time, quite a remarkable renaissance in international legitimacy.

And the more I think that nations in the world dealing with issues like climate change see the United Nations as an international vehicle through which nations can collaborate to seek positive and useful change, that is all to the good. And what, of course, was said this morning was merely the next step in a litany which began more than a year ago, and let me quickly remind you of it. I think, if memory serves me, it was September 16, 1987 when there appeared on the front page of Pravda an article under the byline of Mikhail Gorbachev in which he set out for the first time the particular disposition to glasnost and the willingness suddenly to embrace international organizations which the Soviets have largely inclined to disavow and to manipulate hitherto, to give those international organizations a new primacy in this world.

Then we move to October 5, 1987 where I was fortunate enough to witness one of the most vivid and dramatic moments I watched in my brief life sojourn at the United Nations. And that was the presentation of the Brundtland Commission Report on Environment and Development. And it was almost palpable to feel the sense of excitement that coursed through the General Assembly auditorium when Madame Brundtland set out what now emerges as a blueprint for environmental change for the next generation. And those who rose to speak and those who commented on it were heads of government and heads of state and environmental ministers who were clearly and passionately committed to a new environmental ethic, as Bob Grady spoke from the President-Elect of the United States. One week later, on October 12, 1987, the Director General of the World Health Organization mounted the podium of the United Nations to speak about the Global Program of Action to counter the pandemic of AIDS. And similarly, there was evidence throughout the General Assembly that the organization was beginning to understand that these international ingredients have drawn together, these issues which know no boundaries and have to be dealt with in that kind of way or the world is necessarily traduced into self defeat, sometimes even self-immolation.

Later in 1987, the Soviets confirmed that they were going to withdraw their troops from Afghanistan. Then we had the blessed phenomenon of a medium and short range arms accord on missiles between the United States and the Soviet Union, reducing the overall nuclear armament by a small fraction but a qualitative fraction. Then as we inched into 1988, we had the continuing peace negotiations between Iran and Iraq. And when that appeared to have become a fait accompli, there issued from the United Nations and the international establishment a series of hopeful prospects which collectively take one's breath away. The possibility of a plebiscite in the Western Sahara to end the squalid war between Morocco and the Independence Movement known as the Polisario; the progress that apparently is being made in Southern Africa and Angola and Namibia; the fact that the adversaries in Cyprus are dealing even now with a constitution which was drafted under the aegis of the United Nations; the possibility that if there is peace in Central America one day, economic reconstruction can be pursued through an international body; the possibility of Vietnam withdrawing from Cambodia (Kampuchea), then more and more nations sign the Ozone Protocol out of Montreal; and in the middle of this year as has been alluded to and, as everyone knows, the sudden preoccupying emergence of climate change as the centerpiece of environmental primacy for this world.

It was a real turning point. I remember reading some of the really fine clarity and lucidity to which Stephen Schneider is given, as we talked of the greenhouse facts of the warming trends internationally, and how in the middle of 1988 it had all come together for a variety of reasons all over the world. And that progress of events which really had their genesis in activities no more than about a year or so ago means that the work that you are doing collectively -- a conference of this kind sponsored by the Climate Institute -- is coming at an enormously propitious moment in time when the world is more susceptible to listen; when the possibilities of collaboration are real rather than self indulgent; where one has intuitive glimmers of optimism rather than the occasional moments of angst and despair.

And so the Toronto Conference on the Changing Atmosphere in June of 1988, if you will allow my excessive chauvinism for just a minute, was a particularly useful moment in time and a reflection of this growing, accelerating sense of internationalism. And inevitably you get together this group -- at that point, of two or three hundred scientists and some itinerant policymakers -- and you have to draft a program of action, an action plan which seems to me to be as durable in December as it was in June, although in June it was the product of sometimes tense and sometimes awkward behind the scenes negotiation. And you all know the ingredients of the plan. Again, the answers to the conundrums are not beyond the capacity of human beings to divine. They are, so far, beyond our capacity to implement, but the general outlines of what has to be done were indicated in Toronto, as I have no doubt that they will be indicated in Washington.

Number one, the sine qua non is a reduction in the emissions of carbon whether that is achieved through energy efficiency or through the pursuit of alternative energy supplies, alternative renewable energy supplies which don't rely on fossil fuels. And in the Toronto Conference where, those of you are not familiar with the details, the delegates sought a 20% reduction in carbon dioxide emissions measured in 1988 levels by the year 2005. Half of it achieved by efficiency, half of it achieved through renewable energy, all of it on the road to the stabilization of atmospheric emissions of carbon at roughly 50% of existing levels. Secondly, there was a poignant wish to have a World Atmosphere Fund, which monies would be raised from a surcharge on the purchase of fossil fuels and the use of fossil fuels. Number three, there would be massive reforestation efforts and very considerable efforts to prevent deforestation. Number four, that there should be a global atmosphere convention which might be put in place as early as 1992 but should certainly be pursued with fidelity and vigor in the intervening period. Number five, that there had to be a dramatic increase in research and development in all of the areas which spoke to limiting the consequences of climatic change. And number six, that the ratifications of the Ozone Protocol should continue apace and, as was indicated, that we phase out completely CFCs by the year 2000. Now, it is an easy recitation, but collectively, it amounts to a dramatic and significant shift in public policy all around the world.

And I was interested when I was reading through this very, very fine compendium, The State of the World. What a pleasure it is when people come together as they do at the Worldwatch Institute and put in place a report of such astonishing insight and clarity. I was interested when I read the chapter called, "Reclaiming the Future," to see how neatly it was put. Let me remind you all and share it with you. "Putting the world on a sustainable development path will not be easy given the environmental degradation and the economic confusion that now prevail. Modest increases in energy efficiency investments or family planning budgets will not suffice. Getting on such a path depends on a wholesale reordering of priorities, a fundamental restructuring of the global economy and a quantum leap in international cooperation on a scale that occurred after World War II. Unless the desire to ensure a sustainable future becomes a central concern of national governments, the continuing deterioration of the economy's natural support systems will eventually overwhelm efforts to improve the human condition. A sustainable future requires that a series of interlocking issues be dealt with simultaneously. Stabilizing population will prove difficult until poverty is reduced. It may be impossible to avoid a mass extinction of species as long as the Third World is burdened with debt. Perhaps most important, the resources needed to arrest the physical deterioration of the planet may not be available unless the international arms race can be reversed."

I thought that was as simple and eloquent a statement of the proposition as one could find, and therein lies the tale. Nothing we have done before corresponds to the challenge of what we must do now. Not even in the responses to the oil shocks of the 1970 and a very considerable progress that was made in the fields of energy conservation and the public policy shifts. Nothing we have done over the last twenty-five years corresponds to the imperatives which now obtain. As the Worldwatch Institute points out, despite that truth, yet not a single government on the face of the earth has yet adopted an energy policy that takes climate change into account. And I would add that not a single government or organization so far as I know, with the exception of the Institute itself, has managed to forecast the prospective costs of shifting policy and where perhaps the money might come from. And not a single government has contemplated the fundamental economic restructuring internationally which would cope with or counter the consequences of climate change.

We really are on the edge of the cliff looking into the precipice. There is agitated concern. There is no universal sense of urgency. Now I suppose one might argue that if the industrial countries were relatively self-contained, if this was a compartmentalized world, we might be able to make the indigenous policy alterations ourselves. We'd do it slowly. There would be dislocating economic and social consequences but we'd probably get around to it. But, of course, everybody in this room understands what the United Nations' and the internationalists' understanding is, that the rest of the world is indispensable to the process and the rest of the world, especially the developing countries cannot cope with the implications.

They simply cannot handle it. What was fascinating about that day in October '87, when Madame Brundtland introduced the World Commission on Environment and Development Report, was the response that came from Robert Mugabe, the President of Zimbabwe and head of the Non-Aligned Movement, and a fellow of enormous intelligence, let me tell you. I come from a country that shares the Commonwealth with Zimbabwe. But Robert Mugabe said with directness and candor which was really pleasing and quite simple. He said, "Look, you can't talk to us in the developing world about matters of the environment unless you are also prepared to address the concomitant imperative which is poverty. Because when you are talking about environment to the developing world, you are talking about poverty, and it is inappropriate to impose on the developing world the costs of compensating for the depredations of the developed world. We're not prepared to accept that. We're not prepared to stall whatever modest marginal growth we may have simply to serve to redress what the developed world has done."

So if you're talking about a serious international assault on what may be the devastating consequences of climate change or an environmental degradation, then you have to deal with the realities of the developing world and so far as the developing world can see, those things are not being dealt with. It was said in Toronto rather more softly and gently by Minister Salim, the Environmental Minister of Indonesia. It was shared in the negotiating behind the scenes by all of the representatives of the developing countries who wanted very much to be a part of the process but who looked askance at the western preoccupation with classic western environmental problems. And not understanding that, unless the realities of poverty and the uprooted in this inherent nature of developing economies is dealt with, then one can never have an adequate response.

I spend a lot of my time moving back and forth from Africa. In the forty-four countries of subSaharan Africa, you have this continuing phenomenon of per capita income declining, per capita consumption declining, population growth rates increasing. According to the World Bank and IMF, between thirty and thirty-five of those countries are engaged in wrenching structural adjustment programs internally in order to put their economies in shape and to move from the swamp of famine into long-term economic recovery. But they can't make it, and they can't make it because of the external constraints. They can't make it because their economies are stagnating. They can't make it because debt and debt-service obligations cripple the recovery. The debt-servicing obligations in the vulnerable economies of subSaharan Africa alone by 1995, not the debt, just the debt-servicing obligations, will mount as high potentially as forty-five billion dollars a year. It was three to five billion in the early years of this decade.

How does one ask countries to deal with the phenomenon of the consequences of climate change? To deal with reforestation? To deal with soil preservation? To deal with desertification? How does one ask them to do it when they are fundamentally and collaboratively penurious?

The third constraint, of course, is the plummeting commodity prices. SubSaharan Africa lost nineteen billion dollars or 29% of export earnings in the one year 1986 alone, and it dropped again in 1987. And one has, if you will forgive this slight rhetorical extremity or extravagance, one has what now amounts to the obscenity internationally of more than thirty billion dollars a year coming out of the developing countries and into the developed world instead of the flow being the other way as it was just ten years ago. One will never be able to deal with the consequences and the reality of these massive environmental shifts which are prophecy unless, in the developing world, there is the kind of support which allows them to make the economic adjustments. And not all of the creative and innovative measures which are adopted in the Western world, the Eastern world and the developed world generally will be sufficient to contain the consequences unless the developing world is engaged. So how does one work at the balance? Well, the balance is inordinately expensive. The Worldwatch Institute has, what for me was the first time I had seen, a table of this kind which gives rough estimates of additional expenditures to achieve sustainable development from 1992 to the year 2000, and the categories that are laid out are comprehensible and intelligent. They are protecting topsoil on crop land, reforesting the earth, slowing population growth, raising energy efficiency, developing renewable energy, retiring Third World debt.

There you have it, a simple panoply of policy that doesn't take prophetic insight to enumerate. And how much do they suggest will be required between 1990 and the year 2000 to achieve those ends? One point four trillion dollars, and in fact, it's an underestimate. The one place that their calculations, I say deferentially, are underestimated are in the retirement of Third World debt. My calculations tell me that what will be required to avoid what we fear to be the likely consequences of climatic change -- which are engaging all of your collective activities at this conference -- is roughly 1.7 trillion dollars over that ten-year period.

Well, where does the money come from? If you will forgive me, it comes from only one conceivable place and that's the money that we're now spending on the arms race, because there is no other source available. If a society is spending over a trillion dollars a year on the arms race, and over a decade you're going to require almost two trillion dollars in order to redress the depredations which we have imposed on the environment, then the only place it can come from is that reservoir of public expenditure, which means that the link between disarmament and development must become real. Indeed, the link must become a triad: disarmament, development, environment. And what is so hopeful about all of that is that, in the middle of 1987 under the auspices of the United Nations, there was an international conference on the relationship between disarmament and development which every single country in the world -- it makes me sad to add this -- save the United States, agreed to. And all of us fashioned a consensus document which demonstrated that, as we handled these questions of pressing international security and as we handled the phenomenon of disarmament. If the relationship between the President-elect and Mr. Gorbachev are consolidated, and if the world continues to move towards reductions in strategic weapons, conventional weapons, chemical bacteriological and radiological weapons, then the money that is freed has to go in some measure to development and environment rather than merely being used to satisfy the priorities of domestic economies.

But there's something else involved here that I want to put to you as strongly as possible, as scientists in this room assembled. It is truly important that, given this sense which you all have of the issue, that you move from analysis to advocacy. That is the true measure of a scientific community which is mobilized in defense of a cause. It is not without precedent, I remind you. I remind you that numbers of scientists all over the world and certainly in the United States, the United Kingdom and Canada, numbers of scientists who understood the full horror of the potential use of atomic weaponry -- many of whom had participated in the Manhattan Project and knew something about the consequences of the building of atomic weapons -- gradually over the years in the 1950s, they formed a group in solidarity under the auspices of the Bulletin of the Atomic Scientists, and fought vigorously, intelligently, indefatigably to get arms control policies in place and to shift away from the insanity of the arms race. Let me remind you as well of the physicians-- the physicians who suddenly decided some years ago that the greatest single public health hazard in this world would be a nuclear war. So they formed the International Physicians for the Prevention of Nuclear War. East and West collaborated, lobbied tenaciously and relentlessly and undoubtedly had some effect on the policies of the Eastern bloc -- certainly I think had some effect on the policies of the NATO countries -- in understanding that the arms race had to be stalled and that one had to inch its way -- the world's way -- back to rationality.

What we need now in a similar fashion for the environment is a grand coalition -- of scientists and environmentalists and non-governmental organizations and the policymakers who care to be involved -- to save this earth and humankind. I recognize it's not particularly your profession to be in the frontlines of advocacy. You're scientists, but there does come a moment in life when the scientific knowledge must be mobilized into the advocate's activities.

You know better than anyone else what the implications are for ecological integrity and ecological diversity when the forests of Madagascar disappear as they do. You understand better than anyone else the consequences for the world as the Brazilian forests are under assault. You understand better than most the ominous warnings that lurk in the succession of natural tragedies that have been visited on the country of Bangladesh, and the moral imperative inherent in dealing with the consequences, the human consequences, of that succession of tragedies. You understand better than most the march of the deserts in the Sahel, and the way in which these deserts continue to encroach on human survival. You grasp all of the implications -- the way most of mortal kind does not grasp them -- of the consequences of the heat waves for the American and Canadian Midwest. You chronicle the consequences of the holes in the ozone layer, and that implication for human health and well-being. You understand the imperative policy options -- whether they're timing or they're large; whether they're light bulbs or photovoltaic cells. This room is the repository of that wisdom and that knowledge.

And so my appeal today to you is that you combine science and advocacy, that you become both analysts and protagonists. By all means, don't give up the case. That's the nature of the professional imperative. But then move from the dispassionate observation to the passionate intervention. And do it with the collaboration of as many groups as possible because one is talking as one did with the scientists who opposed nuclear war. One is talking about the preservation of the planet. I don't pretend that you're the last best hope for human kind, but perhaps, collectively, you're the strongest voice for mobilizing change. We surely haven't come this far in human civilization to see it atrophy before our very eyes. I salute you and I throw the gauntlet to you. Thank you.

* Ambassador Lewis was Canada's Permanent Representative to the United Nations until August 1988. He is currently serving as Special Advisor on Africa to the Secretary General of the United Nations. Ambassador Lewis' speech followed a statement at the luncheon on behalf of President-elect George Bush by Robert Grady.

Concluding Address by Senator Albert Gore, Jr.
At Luncheon December 8, 1988*

Thank you for the introduction. But I want you to know I'm grateful for it. So many people have used that corny old story -- I wish my mother and father could have heard that. My father would have enjoyed it, my mother would have believed it. I think there's probably no more appropriate time to use that even though it has been used before. But I do appreciate it very much.

I am Al Gore. I used to be the next President of the United States of America. And I have been really looking forward to this conference. And I want to thank John Topping and the Climate Institute and congratulate all of those who have had a part in organizing this landmark set of conversations here that will be remembered as a turning point when awareness of the problems confronting the global climate took another quantum leap. And when the discussion turned for the first time to the details of what the solutions might look like. I do congratulate you and I do thank you for giving me a chance to participate here.

As Tom mentioned in over-generous terms, I have been trying to bring public attention to this issue for quite sometime and, like you, I was gratified that the miserable summer of 1988 focused the attention of the nation on this problem as it has never been focused before. Other environmental events have elevated awareness all over the world of what many now see as a crisis confronting the global ecosystem. The greenhouse effect is simultaneously the single most important and serious environmental problem we have ever faced and a defining metaphor to represent the ecological crisis of which it is a part: the loss of forestland; widespread famine and drought; the loss of species at a rate a thousand times greater than the last comparable period of species loss 65 million years ago; a population explosion which in some ways drives the other elements of the crisis. It took a million years for the world to arrive at a population of two billion people, and, in the course of a single human lifetime, it is moving from two billion to ten billion. We're already at five billion. The loss of topsoil, the loss of stratospheric ozone, the accumulation of greenhouse gases, all follow the same pattern -- a long period of stability throughout prehistory, through historic times, with the beginning of an upturn with the Industrial Revolution -- and then a sudden, dramatic, unprecedented acceleration during this century and particularly during the last half of the twentieth century.

Where are the comparable increases in leadership, information, compassion, cooperation and those qualities that will be essential to confronting the underlying question which is posed by the greenhouse effect, the nuclear arms race and the general ecological crisis we now confront? That question is, do we have the capacity as human beings to change in a single generation habits and attitudes and political patterns that have accumulated slowly through the millennium? We used to have a debate about appropriate technology. Many of you remember the discussions of whether or not a nuclear powerplant was an appropriate technology for a developing country. That serves for me as a metaphor of the theological question involved here. Are we appropriate as the species having dominion over the earth? In other words, can we find the capacity to control the destructive patterns which have now brought us to the point we're at today?

I just returned two weeks ago from Antarctica having gone because I learned during these eight years of hearings on the greenhouse effect that all the models predict a much more rapid warming at the poles than at the equator. Having learned as well that Antarctica plays a more important role in driving the world's climate system than any part of the earth and arguably more important than anything other than the sun and the rotation of the earth and the movement of the earth around the sun. I found interesting discussions there about the ozone hole. And you've had presentations describing for you the findings that as predicted the hole was shallower this year due to the change in the circulatory patterns that comes every twenty seven years. Going back a hundred and sixty thousand years, as far as they can go with this particular form of measurement, they've never found anything approaching the levels of CO_2 or methane or nitrous oxide, with less CFCs of course. And as they pulled out the ice cores, I was struck by the fact that they could point to the early 1970s and say, "You see right here is where the US Congress passed the Clean Air Act." They can read them the way foresters read tree rings. And an action by the Congress of the United States was so clearly evident that, going back a decade earlier, they said, "Here's where the world stopped nuclear testing."

And moving to the 1980s, of course, here we see a rapid accumulation of greenhouse gases which have been going on for sometime. But even to the naked eye, the so-called core eye, hoar, had a different pattern in the 1980s. The scientists are not prepared to publish conclusions but are instead exhaustively examining alternative hypotheses. But to an untrained eye, the pattern formed by partial melting and recrystallization seems to be obvious just looking at it.

The scientists also spoke of three potential catastrophes that could eventually come from a warming at the pole and particularly from a warming in Antarctica. Several of them have been discussed here at some length. I was surprised to learn that the cold water surrounding Antarctica absorbed more carbon dioxide than all the rain forests. It is as if the world has two lungs -- the forests and the Southern ocean. And the mechanisms by which the Southern ocean absorbs the CO_2, pulling it from the atmosphere, are extremely sensitive to temperature. The high salt content, the cold temperature, the intense biological activity, the turbulence that is present, all play a role in ways that are not yet fully understood, but temperature is one of the key variables. If, therefore, the temperature increases disproportionately in those very waters, what will the effect be on the mechanisms by which CO_2 is absorbed? The importance of that question is increased by the very fact that the oceans contain fifty times as much CO_2 as the atmosphere, and so relatively small changes in those mechanisms should have an important effect.

Second, as has also been discussed here, the West Antarctic ice sheet is believed to have broken up at least once and perhaps several times during pre-history -- during other climatic shifts that were not anthropogenic in nature and therefore took place over longer periods of time. The consequences for sea levels as you have heard would of course be catastrophic. And third, since Antarctica plays such a key role in creating the weather patterns of the world, dramatic changes in the temperature there could have consequences for the patterns by which cold is redistributed northward. In other words, the temperature at the equator and the temperature at the poles is a key equation in defining the way in which the weather system moves heat through the poles and moves coldness to the equator. That difference has been relatively stable for a long period of time. If there is a one degree centigrade increase at the equator and a ten degree centigrade increase at the poles, the analysis of the change cannot be confined to the linear effects of the warming itself. If the relationship is changed, then the pattern defined by that relationship could also change in ways the scientists describe as non-linear. In other words, the equilibrium of the system could change from one equilibrium state to another equilibrium state with wind currents and ocean currents and jet streams and the distribution of moisture and the temperature bands all being affected as the system searches for a new equilibrium.

If, as predicted, the potential catastrophe associated with the West Antarctic ice sheet is way off -- meaning two hundred to hundred and fifty years -- if the worst of sea level increases are farther off than the next few generations, then the initial impact which will affect our lives will come from the changes in these patterns by which temperature and moisture are redistributed. The real question we face is whether or not the political equilibrium can be changed to a new pattern before the climate system's equilibrium loses its current pattern. Because just as heat and cold and wind and ocean currents make up a weather system, national interest short-term calculations, read, "A whole manner of motivations good and bad," make up the world's political system. We must have leadership to change the way in which the world copes with the problem of the greenhouse effect.

There are obstacles to any political action. Let me enumerate them. First, this problem sounds unrealistic. It sounds like the plot for a bad science fiction movie. You're telling me it's a political problem? It's outside our experience. It's beyond what we refer to as common sense, and so we put it in the place in our minds that we reserve for things like Antarctica. And we tag it with the same mental labels. It's remote, alien, hard to get to, too unforgiving to stay there for long, hopelessly distorted by the maps of the world we're familiar with. And yet, it's real. And so in order to act, we must hurdle that political obstacle with conferences like this one and with actions by nations like the United States offering leadership.

I would like to see President-elect Bush appoint an ambassador-at-large to coordinate our activities on the greenhouse effect with those of other nations -- to seek out his or her counterpart in every other country now addressing this problem and work on new versions on the emerging agenda for international action.

The second obstacle to such action is that it has to be international in nature. We do not have sufficient precedents for working together on problems that will require difficult sacrifices by the peoples of the nations involved. And yet it is essential that we build on such precedents as we do have, such as the Montreal Protocol, such as the Antarctic Treaty of 1959, such as the Nuclear-Test Ban Treaty and Non Proliferation Treaty. With leadership, it is possible.

The third obstacle is that the solutions which now become apparent on the horizon are obviously so difficult that they may be beyond our ability to implement. And if there's a good chance that they're too difficult, then we might be wasting our efforts -- beating our heads against the wall by even trying to deal with the solutions so unthinkably difficult. I want to come back to that one in a moment. That's perhaps the toughest obstacle.

The next obstacle is the advice from some heard less and less these days -- thankfully -- that since the solutions are so difficult perhaps we should simply concentrate on adapting to the change which we can now predict. Some adaptation will have to occur. We know that a degree of warming has already been locked in. But ladies and gentlemen, the counsel to choose adaptation as the principal response to this challenge is the counsel to surrender and embodies a blindness to the degree of harm which will come from this problem unless we act.

The next obstacle is the similarly ill-advised suggestion that there are winners and losers and some countries may actually benefit. No country will benefit. No country will benefit -- even if some country had a marginal improvement in their ability to grow crops or the number of days defined as pleasant in the calendar. The political convulsions they would experience -- the illegal immigration; the consequences for their place in the world economy; the security threats; the economic challenges -- would all combine to completely negate any misguided calculation that they could be characterized as a winner.

The next obstacle is one you're dealing with: continued uncertainty about some key scientific evidence that we need to have solidified. And here it's important to be crystal clear about what's uncertain and what's not. There's no longer any uncertainty about the fact that we have a greenhouse effect, that it is occurring and that it is almost certainly going to bring consequences that will be catastrophic unless we change the patterns causing them. What is uncertain: how soon will the worst impacts occur; over what period of time will these mechanisms run; what regions will have which impacts; what role will the cloud system play in possibly mitigating some features of the phenomena or what is the chance of feedback mechanisms that could make the problem worse or of lesser significance. There are all kinds of uncertainties, but the central facts are now sufficiently clear to warrant action. There is the obstacle of a lack of awareness all over the world about how serious this problem is.

What can be done? First, we should start by raising awareness not only in the scientific communities but among the public. I have long proposed an international year of the greenhouse effect; the International Geosphere Biosphere Year accomplishes the purpose of focusing research and increasing knowledge in these areas. I believe it can usefully be supplemented by a highly visible international effort along the lines of the International Geophysical Year of 1957 - 58 to not only improve the research but also to involve the peoples of the world in the discussions of exactly what this problem is all about. We should simultaneously eliminate uncertainty and increase awareness and, as we improve the research, we should focus on problems that the scientific community has clearly identified. Here in the United States, we need a state-of-the-art super computer dedicated to climatic models with adequate time available to the leading scientists to improve the models and add new data. It's outrageous that we have to factor in the oceans as a second step when we do the analysis of what the greenhouse effect is likely to cause, just to use one example. There is a long list of items which should be improved there.

As we proceed to raise awareness and eliminate uncertainty, we should also begin moving on concrete actions to deal with the causes of the problem. I'd put these actions into two categories. First, things which should be done for other reasons anyway, and second, difficult actions which must be taken but which now appear to be unimaginably hard. First of all, in the category of things which should be done for any other reasons anyway -- first on the list is the elimination of chlorofluorocarbons. The Montreal Protocol was an achievement. I salute Ambassador Benedick and his colleagues. But with all due respect, a 35% reduction by the end of the century -- a 50% reduction from an increased base -- a total of 35% reduction from the current levels is far less than what is needed. A 90% reduction is needed just to stabilize the damage being done to the ozone layer and to stabilize the contribution to the greenhouse effect from CFCs. We know we should eliminate them for other reasons anyway. We should get on with that task.

Secondly, deforestation: we are losing an amount of forestland equivalent to one football field's worth every second, one state of Tennessee's worth every year. That should be stopped for other reasons anyway. It will be difficult, of course. The debt-for-nature swaps which have been widely proposed will be very difficult to implement. Quite frankly, as someone in public service and politics who has worked both with the environmental issues and with international finances, I could tell you that there are no mechanisms at all for creating a marriage between those two problems. Some tentative mechanisms have been created by the World Bank, but the large scale movements that are now required are going to be extremely difficult to put into place. But we must put them into place, and we should establish global goals for reforestation in every single country of the world. We can regenerate the capacity of the earth to remove the carbon dioxide from the atmosphere at least to an extent. I know that it only fixes it temporarily in the biosphere but it buys us time for the more difficult steps which can come later.

The more difficult steps include, first and foremost, a reduction in CO_2 emissions. This is really the cutting edge. This is the first problem in the category too difficult to realistically enact right now. But we must begin doing it nonetheless with improvements in efficiency and with discussions at the summit level -- not just between the United States and the Soviet Union, but in the discussions between United States and the OECD countries and in every world forum about an idea that I think is going to have to eventually be implemented. And that is a set of global goals for CO_2 emission reduction in every nation, with many of the advanced nations committing themselves to something like a CO_2 tax with a CO_2 credit against that tax for measures that reduce the emissions and avoid the tax. All the talk about a gasoline tax or a scrubber technology -- all of that in my view should be re-examined in the context of a CO_2 disincentive that will measure the CO_2 emissions regardless of the fuel.

Next, we have got to attack the problem of methane emissions, whether from landfills or particular forms of agriculture or the other sources that are not yet fully identified. We must have global goals for reductions in methane emissions. As you move down the list of causes to the smaller actors, we should similarly develop goals for reducing them as well. If we have success in the first steps, we may build our confidence and expand the limit of what we define as possible. I believe that we have the capacity to face up to this problem and to solve it. But we shouldn't underestimate how difficult it is. The winds of change in the world are approaching hurricane force. We can clearly see the need for change in our political system's response to this problem. In closing, I congratulate all of you on your contribution to improving the understanding of this problem, and I pledge to you my continued efforts to try to translate your knowledge and conviction into sensible solutions for the single biggest environmental problem we've every faced.

Thank you very much.

* Senator Gore was introduced by Thomas Grumbly, President of Clean Sites, Inc. and a member of the Climate Institute Board. Grumbly, who once worked in Congress for Gore, described his pioneering efforts in drawing Congressional attention to climate change.

OPENING SPEECH OF
JOSEPH A. CANNON, PRESIDENT OF GENEVA STEEL

I am delighted to welcome you to the Second North American Conference on Preparing for Climate Change: A Cooperative Approach. This meeting which I have the honor of chairing is probably the most comprehensive gathering ever assembled on our planet to discuss potential impacts of greenhouse warming and stratospheric ozone depletion and to explore response strategies to address those effects. Over the next three days this conference will hear over 160 experts from around the globe discuss implications of climate change for natural systems, wildlife, fisheries, water resources, coastal areas, air quality, urban planning, energy strategies and a multitude of other concerns.

The deliberations of this conference are likely to be a valuable input to governmental and private sector decisionmakers throughout North America. This conference is occurring at a time of political transition in the three largest nations of North America. National elections were held in November in both the United States and Canada and on December 1, Mexico inaugurated a new President after the most closely contested election in its history.

This unprecedented confluence of political transitions in the largest democracies of the North American continent coincides with an unusual stirring of interest in the possibility of fundamental change in our global climate. Just a few weeks ago representatives of over thirty nations met in Geneva, Switzerland under the auspices of the United Nations Environment Programme and the World Meteorological Organization to develop a systematic international process to address the prospect of significant global warming. Within this newly constituted Intergovernmental Panel on Climate Change the United States was given the lead in developing response strategies while the Soviets were given a lead in assessing impacts.

This October we witnessed some dramatic developments in ensuring a global response to stratospheric ozone depletion. Ratification of the Montreal Protocol by the European Community will ensure that the Protocol, the most significant international environmental agreement yet achieved, will take effect on January 1, 1989. Also, this October scientists and policy makers from around the world met at the Hague to lay the basis ultimately for a further tightening of the Montreal Protocol to protect the ozone layer.

This momentum is likely to continue over the next year. One of the few areas in an often divisive U.S. presidential campaign on which both major party candidates were fully agreed is the need to use the power of the White House to accelerate action on such international environmental challenges as greenhouse warming and stratospheric ozone depletion. President-elect Bush has stated that he will convene a conference on the global environment during his first year in office to address just such issues.

It is essential that this growth of interest among governments be matched by commensurate initiatives by the private sector and among the scientific and public interest communities. Some of you may have been puzzled when first looking at our conference program to learn that a steel company president was serving as the chairman of an international conference on preparing for the greenhouse effect. To some people that might seem incongruous, a little like Frank Perdue addressing a vegetarian gathering.

In fact, successful global initiatives to address the twin environmental challenges of greenhouse warming and stratospheric ozone depletion will require an unprecedented mobilization of private sector ingenuity and commitment. The sub-titles to both the 1987 and 1988 North American Conferences on Preparing for Climate Change have been "A Cooperative Approach." This underscores the reality that any effective response must involve cooperation among nations across the globe and among many often differing sectors of human society.

The response of the industrial sector is quite critical. From that sector will arise virtually all of the inventions that could reduce emissions of greenhouse gases and stratospheric perturbants. Energy conservation is beneficial to the bottom line of industrial energy users--such conservation will also restrain the growth of greenhouse gas concentrations.

Moreover, industry initiatives can also do much to enhance the prospects for successful international negotiations. In 1984 when I was Assistant Administrator for Air and Radiation of the U.S. Environmental Protection Agency, our office launched the analysis that was ultimately to result in the strong U.S. position in concert with Canada and the Scandinavian countries which led to the Montreal Protocol.

At the outset the manufacturers and users of chlorofluorocarbons were extremely skeptical about this U.S. initiative, and they did not hesitate to convey these views to EPA and the Congress.

Yet by the summer of 1986 the Alliance for Responsible CFC Policy, the principal U.S. CFC industry lobby, startled everyone by coming out for an international agreement to curb the growth of CFC emissions. This dramatically rescrambled the international political equation, undercutting the position of European industry groups which had encouraged their governments to resist any significant international controls. Similarly, the announcement this March by DuPont, the world's leading CFC producer, of its intention to phase out production of these chemicals, virtually guaranteed that the timetable of the Montreal Protocol for reduction of stratospheric perturbants will be accelerated. It is possible that the voluntary initiative of a Connecticut utility this year to finance planting of tropical forests to offset emissions from a new fossil fuel plant could have a similar galvanizing effect on future public policy.

I am heartened by the extensive industry participation in this meeting. This conference and its predecessor last year have drawn a wide array of sponsors and participants from the corporate sector, the environmental community, the scientific and research community, state, provincial and national government agencies in the U.S. and Canada, and the diplomatic community.

This success in embodying the cooperative approach envisioned by the Institute's first Chairman, Paul Pritchard, President of the National Parks and Conservation Association, has come much faster than any of us dreamed imaginable when the Climate Institute was established two and a half years ago.

The Climate Institute was founded in July 1986 by a number of us who had been senior environmental and policy officials in administrations of both U.S. major parties. All of us shared a strong concern that greenhouse warming and stratospheric ozone depletion were environmental challenges that could drastically alter the quality of life on our planet. Yet we sensed that most policy makers in countries across the globe viewed these challenges as problems to be addressed by our grandchildren.

The Climate Institute set out to serve as a bridge institution between scientists and scholars studying climate change, and those in the public and private sector who could fashion responses to mitigate the adverse effects of rapid climate change.

In the two and a half years since its inception the Climate Institute has moved from a U.S. to an international organization and has added some of the world's most respected climate scholars to its Board. These include Stephen Leatherman, who heads the Laboratory for Coastal Research at the University of Maryland; Stephen Schneider, distinguished scientist and author at the National Center for Atmospheric Research; Jim McCulloch, who recently retired as Director General of the Canadian Climate Centre; and Martin Parry of the University of Birmingham in England, widely regarded as the world's leading expert on climate and agriculture. Recently, we were gratified when one of the world's most respected diplomats, Sir Crispin Tickell, Permanent Representative from the United Kingdom to the United Nations, joined the Institute's Board. Ambassador Tickell, who has written a widely acclaimed book on <u>Climate Change and World Affairs</u>, will chair tonight's Awards Dinner.

Our Board members have played an integral role in the conferences and symposia of the Institute. This has been true alike of the scientists, senior corporate officials, and environmental leaders who comprise the Institute's Board.

A top priority of the Institute for the past year has been to work closely with decisionmakers from developing countries to assess implications of climate change for the Third World.

Two very successful symposia were held this year on implications of climate change for Third World nations. Diplomats from embassies and U.N. missions of over 50 nations participated in these two meetings in Washington and New York, respectively. We are gratified at the wide turnout for the Second North American Conference on Preparing for Climate Change from diplomats from North and Central America, the Caribbean, Europe and Asia. Appropriately, one of the honorees at tonight's Awards Banquet is Ambassador Richard Benedick, who was not only instrumental in the negotiation of the Montreal Protocol but who also chaired the Climate Institute's two climate change symposia for the diplomatic community.

Enormous as are the ecological and economic implications of significant climate change in Canada, the United States, Europe and Japan, the potential human toll is far greater in the Third World.

A one meter relative rise in sea level, a reasonable prospect by the middle of the next century, could displace as much as a quarter of the population of Bangladesh and a sixth of the population of Egypt and could jeopardize the very existence of some small island nations such as the Maldives and Kiribati.

In March 1983 when I headed the U.S. EPA's Office of Policy and Resource Management, we convened the first major meeting to assess implications of climate change for coastal areas. Thanks to the brilliant work of such EPA experts as John Hoffman and Jim Titus and outside scientists such as Steve Leatherman, this Sea Level Rise Conference helped to focus some interest among researchers and policy makers on the tremendous vulnerability of the earth's coastal regions to even modest fluctuations in sea level.

The vulnerability of coastal regions in the Third World to sea level rise will be a major focus of three upcoming international conferences being convened by the Climate Institute in cooperation with the U.N. Environment Programme.

The first of these will be a Conference on the Implications of Climate Change for Africa to be held May 3-5, 1989 at Howard University in Washington, D.C. A number of embassies from both sub-Saharan Africa and North Africa are working closely with us to ensure the success of this meeting. Implications of climate change and sea level rise for the Nile River Delta and potential impacts of climate change on African agriculture will be a major focus of this meeting.

In December 1989 the Climate Institute and UNEP are organizing a World Conference on Preparing for Climate Change in Cairo, Egypt. This conference, to be held on the Nile River, will highlight the vulnerability of much of the population of the Third World to even modest shifts in sea level. We also hope to focus on the vulnerability of the antiquities of that great nation to the ravages of air pollution and climate change. Fully aware of these threats, the government of Egypt is working to ensure the success of this meeting, and Suzanne Mubarak, First Lady of Egypt, is Honorary Chairman of this meeting.

A few months after the Cairo World Conference we expect in the spring of 1990 to convene a Conference on the Implications of Climate Change for Island Nations at the East West Center of the University of Hawaii. We hope to draw key policy makers from island nations whose very existence may be threatened by a several meter rise in the earth's sea level.

Any successful effort to build a global consensus for addressing climate change will require a detailed understanding of likely regional and localized impacts.

Thanks to the outstanding research and analysis performed over the last four years by the Canadian Climate Program and the multitude of outstanding climate impact studies commissioned by the U.S. Environmental Protection Agency for its two upcoming reports to Congress, we have a reasonable picture of the regional implications of climate change within North America.

This conference which is now beginning will provide the most detailed discussions ever held on the implications of climate change on a single continent. These deliberations will, I believe, prove useful to environmental and energy policy makers and newly elected governments in the U.S., Canada and Mexico. In addition, the ravages of Hurricane Gilbert have underscored the vulnerability of the entire Caribbean region to shifts in climate. We are delighted at the extensive participation from the Caribbean region.

The rapid development of sophisticated climate impact research on North America and the Caribbean will, I hope, set a standard for the essential impact studies in other regions of the world. With such an understanding of what is at risk, leaders of nations across the world can act for the global interest by acting for their own national interests.

We may thus build the global consensus necessary to make possible the international actions that will be discussed in the second and third days of this conference.

The first day of this conference will be devoted to a thorough review of scientific issues including stratospheric ozone depletion, changes in weather circulation patterns, climate feedbacks, ocean-climate interactions, and agricultural and public health implications. Before yielding this rostrum to my fellow Board Member and our Scientific Keynoter, Steve Schneider, I want to thank you for joining us for this watershed conference. We believe that the policy discussions of this conference will build on the very solid contributions made by this June's Toronto International Conference on the Changing Atmosphere. The conference proceedings which will be published by spring should provide a blueprint for policy makers and planners throughout North America and the Caribbean.

Remarks of

Robert Grady
Associate Director
U.S. Office of Management and Budget

Today, December 7th, is a very special day for those of us who worked on the Bush campaign. As you know, the president-elect has tried to reach out broadly in conducting his transition. He met last week with the heads of several major environmental organizations.

He patched up his differences with Michael Dukakis, Jesse Jackson, and Bob Dole. And today, he's asked us to set up a meeting with the guy who said that Pearl Harbor Day was December 7th.

In all seriousness, I want to thank you for giving me the chance to say some words on behalf of the president-elect's transition team. I will try to be brief--because I remember Adlai Stevenson's comment that the best after-lunch speech he ever heard was five words: "Waiter, I'll take the check."

Washington is quite a place to pick to conduct a conference on global climate change. It was an earlier American president, James Buchanan, who said that "Washington is no place for a civilized man to spend the summer."

This past summer, the American people noticed something was awry not only in the hothouse of Washington but seemingly across the land. The unusually hot summer wasn't necessary the product of long-term global climate change, of course, but it did serve to heighten the public's awareness--at least in this country of phenomena about which you experts have known for many years.

Not only were people talking about a warming of the earth's atmosphere, our national parks were burning. The legacy of decades of pollution of our oceans began showing up with increasing frequency on America's beaches. 1988, it seems, was the year the earth spoke back.

As alarming as the earth's message to all of us seems to be, this presents an enormous opportunity for those who care about its continuing health and safety. I'm here today, on behalf of the president-elect, to ask that you, the scientific experts who understand these problems, and we, the politicos, forge an alliance to seize the opportunity.

The opportunity is best described by the fact that--for the first time in our own recent electoral history, anyway--the environment became a major issue in this year's presidential campaign. The people are listening. And there is a consensus for action.

Now, during the campaign, the vice president made a simple statement: "I am an environmentalist," he said. And he meant it.

So before I go any further, let me formally pass along greetings from the president-elect:

As Mr. Topping may have told you, he very much wanted to come today, and he had hoped to make at least some short remarks about the subject of global warming and the environment generally.

Unfortunately, he had one small luncheon engagement today that he didn't think he could cancel--and that, of course, is with General Secretary Gorbachev in New York. So he hoped that you might understand.

If it is any consolation, I know that the president-elect has spoken at some length during these past few months of his intention to make the environment--and the steps that nations can take together to protect it--a subject of discussion at future international meetings. And, by the way, he has also said that he would like to make conservation education a regular feature of the education we provide the next generation--and they are going to face major challenges, and he believes that the key to a solution is greater understanding of the exact nature of the challenges.

In any event, the president-elect did ask me to communicate his regrets. He asked me to extend congratulations to last night's award winners: Dr. Rowland, Dr. Mitchell, and Ambassador Benedick. But perhaps most importantly, his commitment.

He pledged during the campaign to convene, at the White House, a Global Conference on the Environment--as a way of using the influence of his office to begin working toward an <u>international</u> solution to some of the problems you have been discussing in your meeting this week. His own view is that that is one useful role for American leadership--and his experience, on such matters as the Montreal Protocol which Ambassador Benedick played such an important role in crafting, indicates that international cooperation is indeed possible.

My own hope is that many of you will participate in developing some of the recommendations and ideas that come out of that conference--now is not too early to begin preparing for what promises to be an incredibly thought-provoking event.

He pledged during the campaign to reduce the American contribution to environmental problems that have global impacts:

acid rain, pollution of the oceans, emission of greenhouse gases and CFCs, and others.

And most importantly, he pledged to bring a new ethic to the office of the presidency: a conservation ethic.

Last spring, the World Commission on the Environment and Development issued a study that I'm sure many of you have seen. It was called "Our Common Future." And it began with this sentence: "The earth is one, but the world is not."

That sums up the most fundamental challenge before us--to bring a world with diverse cultural traditions, economic needs, and political philosophies together in a common mission of protecting the planet.

That we are in this together has been underlined by a series of transnational environmental incidents, from Chernobyl to Basel to the export of hazardous wastes.

But the quote also points up the opportunity presented by the election of a president as well versed in foreign affairs as George Bush. I sincerely believe he will use his foreign policy expertise, informed by the conservation ethic of which I just spoke, to work toward a global consensus on solving environmental problems.

The agenda is becoming clear: we must work together in reducing the emissions of carbon dioxide and other greenhouse gases. We must plan to phase out emissions of CFCs that destroy the stratospheric ozone layer. We must stem the alarming net loss of forests around the world.

I am not saying that these agenda items are easily tackled--clearly, it will be difficult as nations around the world pursue the economic development they have every right to expect or at least hope for.

What I am saying is that the president-elect is aware of these problems, and he is anxious to have the United States play a leadership role in addressing them. He spoke in the campaign of the need to ensure, for example, that environmental considerations are incorporated to a greater extent, in the design of development programs and lending practices both by agencies of the U.S. government and by multilateral organizations like the World Bank.

I also know, because I have spoken to him about it, that secretary-designate Baker is aware of and deeply concerned about these problems. One of his top aides has been working hard at developing proposals for debt-for-nature swaps, as one example.

These two leaders are outdoorsmen. They spent their summer holiday camping out under the August moon in the mountains of

Wyoming. It might sound like a trivial anecdote--but I think it bodes well for their appreciation of the marvels of this planet and their commitment to environmental protection.

The trick is to begin now to address problems which may seem to some to be far off in the future--to educate the public on the urgency of the need.

Neil Armstrong, an American astronaut who was the first man to walk on the moon, once said that "Science has not yet mastered prophecy. We predict too much for the next year, and yet far too little for the next ten."

Perhaps that's where the bully pulpit of the presidency comes in--to convince Americans, and peole around the world, to look ahead to our common future--because the risks are too great to ignore it.

I feel a little funny as a political advisor addressing a room full of scientists in their area of expertise. But I recall a definition that I once heard: "Science is the ascertainment of facts and the refusal to regard facts as permanent."

So from the political side, let me just say that we will continue to seek the mandate for change. We will work with you to make sure that it is positive. And we will pray with you that the steps we take today will leave our planet in better shape tomorrow.

US-USSR COOPERATION IN CLIMATE STUDIES

Alan D. Hecht
National Climate Program Office
11400 Rockville Pike
Rockville, Maryland 20852

US-USSR scientific cooperation in the field of climate studies has been ongoing since 1974 when the US-USSR bilateral Agreement on Environmental Protection was first signed. Under this broad Agreement, one working group (Working Group VIII) was organized to study the influence of the environment on climate.

WG VIII has grown significantly since 1974 and now includes research on both climate change and atmospheric chemistry, including ozone depletion. This Agreement provides a major channel by which joint publications, experiments, exchange of scientists and policy discussion have occurred.

The activities of this Agreement were highlighted in the Reagan-Gorbachev summit communique of 1987 when both leaders endorsed continued cooperation and called for the preparation of a "special report on future climates." The preparation of this report has become a major activity of this Agreement. The report will draw on the cooperative work undertaken by scientists within this program and will focus on comparing empirical and modeling methodologies for projecting future climate changes.

While the WG is administered by the National Climate Program Office, participating agencies include NOAA, NASA, NSF, DOE, EPA, and USGS. Universities and private industry are also represented.

Principal activities of WG VIII fall within one of the three major general goals of the Agreement:

(1) Organize research at the frontier of atmospheric sciences, such as the study of methane release in high latitudes.

(2) Pursue research that is mutually beneficial, such as joint experiments, joint oceanographic cruises and exchange of selected data sets.

(3) Contribute to international understanding of climate change and ozone depletion such as preceded the agreement signed in Montreal to limit the production of chlorofluorocarbons.

Specific projects included in the 1989 Protocol of activities include the following:

1. Complete the joint report on future climates. This report is intended to focus on theoretical and empirical evidence of climate change (including unpublished data from the USSR) and models and scenarios for future climate change. Soviet scientists have placed considerable emphasis on the use of paleoclimatic reconstructions as an analogue for future climate conditions. The report will deal with the methodology for these reconstructions and will analyze the value of such reconstructions as models for the future. This report is expected to be completed by the end of 1989.

2. Conduct joint measurements of ozone depletion in the Arctic. A joint experiment has recently been conducted in the Soviet Franz Josef Islands located approximately 82 degrees north in the Arctic Circle. The Heiss Island Project (HIP) took place from January to March with a goal to conduct both remote (ground based) and in situ (from balloons) measurements of the depletion of ozone and its relation to the formation of polar stratospheric clouds (PSCs). These clouds result when temperatures decrease below -80 degrees centigrade within the polar vortex during the Arctic winter. Heiss Island is ideally suited for this research since the vortex is generally centered over this region and the coldest temperatures necessary for PSCs are found here. This important experiment was coordinated with NASA/NOAA flights, based out of Stavanger, Norway, which investigated ozone depletion processes within the Arctic vortex. Both teams of investigators found evidence of ozone depletion.

3. Initiate methane sampling program. This year will mark the beginning of a three year effort to document methane releases in high latitude areas. Several federal agencies have prepared a research plan that was presented to Soviet scientists in December 1988. The plan has been jointly accepted and will be implemented in three steps. In 1989, American scientists, led by project leader Richard Gammon, NOAA, will visit the USSR to survey sites for field studies of methane. At the same time, methane samples will be collected from a number of sites and analyzed for both concentration and isotopic composition. In subsequent years a series of joint observations and experiments are planned which will aim at documenting the level of methane increases from various biological regions and in measuring their isotopic composition. In 1991, a combined study of sources and sinks of methane will be prepared.

4. Assemble paleoclimate data base for past warm climate intervals. Considerable effort will be made in 1989 to compile and document past climate changes in the USSR. Results of these studies provide an important data base for validating model simulations of past climate conditions. For example, during the period 6000-7000 years ago, model simulations suggest large areas of central Russia were significantly warmer than today. Adequate

documentation of these projections is crucial in validating the models.

5. <u>Dust and Aerosol Experiment in Central Asia</u>. A major field effort is planned to study the role of dust and aerosols on climate. A series of observations will be organized from Dushanbe and will focus on how dust from the deserts of China and Central Asia affect the radiation balance of the earth.

In addition to the above activities, a number of scientific exchanges and joint projects will take place in 1989. Plans will be developed for future joint cruises in the Pacific Ocean and on Lake Baikal. A joint monograph on aerosols will also be published. In total, over 60 activities will take place this year.

THE MONTREAL OZONE PROTOCOL: LESSONS FOR GLOBAL WARMING *

by

Richard Elliot Benedick

Ambassador, U.S. Department of State
Senior Fellow, The Conservation Foundation

A little over one year ago, representatives of countries from every region of the world reached an agreement unique in the annals of international diplomacy -- an accord which many observers had believed would be impossible to achieve. President Reagan described the treaty as "the result of an extraordinary process...of international diplomacy...[and] a monumental achievement."

The "Montreal Protocol on Substances that Deplete the Ozone Layer" establishes international controls on certain chemicals that can destroy the stratospheric ozone which protects life on earth from harmful radiation, and which can also change the global climate. By their action, the signatories at Montreal sounded a death knell for an important part of the international chemical industry, with implications for billions of dollars in investment and hundreds of thousands of jobs in such related sectors as food, plastics, transportation, electronics, cosmetics, fire prevention, and health care. The negotiators weighed the social and economic costs of replacing substances which contribute in many ways to modern standards of living against hypothetical dangers based on analysis at the frontiers of modern science -- and all this was <u>before</u> there was measurable evidence either of ozone depletion, or of actual damages from increased radiation or from climate change.

At Montreal, nations agreed for the first time on a worldwide regime for specified reductions of potentially damaging chemicals. This was not a response to an environmental disaster, such as Chernobyl or Seveso, but rather <u>preventive</u> action on a global scale. Moreover, the treaty did not take the timid path of controlling through "best available technology," a traditional accommodation to economic interests. Rather, it boldly established firm target dates for emissions reductions, with full knowledge that the technologies for accomplishing these goals did not yet exist.

---

* These remarks are excerpted from a forthcoming book, <u>The Ozone Protocol: A New Global Diplomacy</u>, to be jointly published by The Conservation Foundation and Georgetown University's Institute for the Study of Diplomacy.

The Montreal Protocol was a landmark because it symbolized a fundamental change both in the kind of problems facing the modern world, and in the way the international community can approach these problems. This new generation of issues reflects the interconnectedness of life and its natural support systems on this small planet, where localized activities can have global consequences, and where dangers are slow in developing, long-term in their effects, and not readily reversible. The concept is not obvious: a perfume spray in Paris helps to destroy an invisible gas 6 to 30 miles above the earth, and thereby contributes to deaths from skin cancer and extinction of species half a world distant and several generations in the future.

While international law is relatively well equipped for dealing with traditional transboundary environmental problems, the ozone issue represented uncharted territory in its worldwide scope, scientific uncertainties, and costs and risks extending far beyond normal policymaking time horizons. The negotiators confronted a threat which could affect every nation and all life on earth. The consequences were potentially disastrous, yet they could not be observed or predicted with certitude. The Montreal Protocol is thus a global prototype for decisionmaking under uncertainty: international consensus was forged on a balance of probabilities, where the risks of waiting for more complete evidence were finally deemed to be too great.

More than a year later, the events at Montreal have, ironically, acquired an air of inevitability. It all seems easy in retrospect. Some activists have even complained that the treaty was too little and too late. But memories are short: for the greater part of this period, and even after the actual negotiations began, many governments still had doubts over such fundamental questions as the possible degree of future damage to stratospheric ozone, the extent to which industrial products were responsible, the prospective growth of demand for these products, the significance of any adverse effects from ozone layer depletion, and how long before critical harm might occur.

A unique international process of scientific, technical, and economic analysis and assessment, reinforced by extensive informational and diplomatic initiatives of the United Nations Environmental Programme (UNEP) and the United States government, played an essential role in developing the consensus for concerted international action. The ozone accord broke new ground in its reconciliation of complicated scientific, economic, and political factors, in its handling of long-term risks, in its innovative provisions, and in the negotiating process itself.

Greenhouse warming is an even more complex issue than protecting the ozone layer, with many more contributing factors, more wide-ranging and uncertain consequences, and more economically painful choices. Nevertheless, despite the difficulties, global cooperation will be essential for any effective efforts to limit the magnitude and rate of temperature rise, and to adapt to the effects of climate change.

I would propose to you that the ozone protocol may well serve as a prototype for new diplomatic approaches to emerging global issues such as climate change. There was no single prime cause for the success at Montreal. Rather, it was a combination of key factors and events that made the agreement possible. Analysis of these elements offers insights into a possible methodology for dealing on an international level with climate change.

Firstly, the ozone history demonstrates the importance of building scientific consensus, by mobilizing the best possible scientists and the most advanced technological resources in a cooperative international effort. The development of a commonly accepted body of data and analysis, and the narrowing of the ranges of uncertainty, were instrumental in facilitating a political consensus among negotiating parties initially far apart in their positions.

In this process, close collaboration between scientists and government policymakers is crucial. This synergy contributed to the irresistible logic of the American position on ozone, and greatly strengthened the persuasiveness of US negotiators. The US government provided substantial financial resources for the necessary scientific research, and US policymakers paid attention to the results. The US government's negotiating stance demonstrated to other countries that it was prepared to accept considerable near-term inconvenience for the sake of a future good.

Secondly, in order to mobilize the political will of nations, public opinion must be adequately informed. Here again, individual scientists and national academies have a substantial role, but their findings must be translated and disseminated. International organizations such as UNEP and WMO, through publications and other activities, undertook major educational efforts on the ozone issue. Individual governments -- the US, Canada, Germany, the Nordic states -- were also particularly active in informing their own and other countries' publics. Legislative hearings can be important for airing scientific opinion and analyzing policy alternatives: the US Congress held a number of public hearings on ozone and climate change during 1987 and 1988, and the German Bundestag convened a special commission on atmospheric issues which received well-publicized testimony from many scientists in recent months.

Nongovernmental organizations can also make a considerable educational contribution, as well as promote research and legislation. The media -- particularly press and television -- obviously play a vital role in bringing issues before the public and thereby stimulating political interest. However, the temptation to overstate the case in order to capture public attention needs to be resisted. Exaggerated claims have a way of backfiring and providing ammunition to those who want to obstruct action. The case for ozone protection was built step-by-step, and generally avoided invoking apocalypse. Credibility is well worth preserving, even though it may require patience.

Thirdly, the ozone protocol process itself offers instructive insights for approaching other global issues. The idea of disaggregating a complex problem is valuable: climate change, for example, has so many aspects that it is impossible to deal with everything at once. The innovative fact-finding process which led to the Montreal agreement -- informal scientific and economic workshops which preceded formal negotiations -- is relevant and replicable.

The concept of an initial framework agreement, similar to the 1985 Vienna Convention on Protecting the Ozone Layer, is a useful model: it permits governments to agree in principle that a problem exists, and to launch coordinated scientific research to develop further data as a guide to policy. The next step -- corresponding to the Montreal treaty -- would be individual implementing protocols on specific aspects of the problem. It is worth noting that the ozone accord is itself an example of a partial solution to global warming, since CFCs may contribute 20 percent or more of the heat-trapping effect.

Fourthly, the mediating function of an international organization can be critical. UNEP's catalytic role in the events leading up to the ozone agreement has obvious implications for the future. In the informative and consensus-building stage, during the negotiations themselves, and later in the protocol implementation phase, UNEP was and will be indispensable. A great deal of the credit for the treaty should go to the personal efforts of the executive director of UNEP, Dr. Mostafa Tolba, an Egyptian scientist who -- I am very pleased to announce -- has just been re-elected for another four-year term. Tolba's strong presence was a major factor, commanding respect from all sides for his commitment and his sensitivity to national interests, particularly in the Third World. UNEP went far beyond a traditional secretariat function: it was a leader in mobilizing data and informing world public opinion and, through Tolba, it was a driving force in achieving the eventual consensus. It was UNEP, encouraging Third World governments which might otherwise have only marginal interests to participate in the process, that made the protocol truly global in scope. UNEP provided an objective international forum, free of the irrelevant and time-consuming debates on extraneous political issues which have often marred the work of other UN bodies. It was, in short, the very model of how a UN agency should operate in a complex international negotiation.

Fifthly, an individual country's commitment and policies can have a profound influence on the course of an international negotiation. The many scientific and diplomatic initiatives of the United States -- reinforced by actions of the US Congress, environmental groups, and industry -- were crucial in achieving the Montreal accord. Within the European Community, the ascendancy of Germany toward the end of the process was a significant factor.

A _sixth_ and final lesson from Montreal derives from the protocol itself: a dynamic and flexible instrument with many innovative features. Based on periodic scientific, economic and technical assessments, the treaty can be adapted to evolving conditions. There are even provisions for emergency meetings of parties in case of unexpected and fast-breaking developments. The protocol is not a static solution, it is an ongoing process.

Many of the treaty's provisions represent creative resolutions of complicated equity and technical problems, which can point the way for future protocols: the trigger mechanism for entry into force, the fixed target dates for the reductions, the process for reopening the timetable and reduction goals, the sensible transitional provisions for developing countries, the two-stage voting to reflect large stakeholders' interests, the treatment of trade and nonparties.

Montreal was not a radical treaty: it tried fairly to distribute economic burdens, it was sensitive to special situations, -- and for all of this, it should prove to be a lasting and precedent-setting model for international cooperation.

In conclusion, we have learned that this planet is more vulnerable than most of us had thought. Science is showing how activities of modern industrial societies, driven by consumer demands and by burgeoning Third World populations, can alter fragile natural balances which are not necessarily self-correcting. The Antarctic ozone hole conveys a philosophical warning that the atmosphere, upon which all life depends, is capable of surprises: there is a potential for large and unexpected, rather than incremental change. We can no longer pretend that nothing is happening, or that the planet will somehow automatically adjust itself to the billions of tons of man-made pollutants to which it is being subjected.

Mostafa Tolba has described the Montreal Protocol as "the beginning of a new era of environmental statesmanship." The ozone treaty reflected a realization that nations must work together in the face of global threats, that if some major actors do not participate, the efforts of others will be vitiated. In the realm of international relations, there will always be uncertainties -- political, economic, scientific, psychological. The protocol's greatest significance may be its demonstration that the international community is capable of undertaking complicated cooperative actions in the real world of ambiguity and imperfect knowledge. The Montreal Protocol can be a hopeful paradigm of an evolving global diplomacy, one wherein sovereign nations find ways to accept common responsibility for stewardship of the planet and for the security of generations to come.

Second North American Conference
December 1988

UNEP's Role in Addressing Climate Change

Dr. Noel Brown, Director of the United Nations Environment Programme's Liaison Office to the United Nations

I would like to begin by expressing my appreciation to Ambassador Benedick for the very kind tribute he has paid to my Executive Director, Dr. Tolba, and more directly to the United Nations Environment Programme. I believe the Montreal process is real and solid. I believe we in UNEP have learned a great deal from that process. I might add that it was ten years in the making which suggests that when you're dealing with complex problems such as climate change and atmospheric degradation, you need enduring and predictable institutions. I think it's a very great tribute to organizations such as the United Nations Environment Programme that we could stay in the course for that period.

We now are geared up for the second phase of climate change and hopefully that experience will stand us in good stead. At any rate, ladies and gentlemen, UNEP welcomes the opportunity to participate in the Second North American Conference on Climate Change with its emphasis on the cooperative approach and its search for broad cooperative strategies in various arenas including the Americas. Your encouragement of inter-American cooperation in this area is not only timely, but an urgent necessity given the importance of the region, its vast resource base, unique ecosystems such as Amazonia and the American Arctic. I was struck this morning listening to the rather vast area of cooperation between the United States and the Soviet Union since 1972 on critical environmental matters and this is as it should be. After all, these are leading powers both in science and in economics and in technology. But also one would hope that similar patterns of cooperation might occur among the states of the Americas. And that is why we welcome the focus of this session on inter-American cooperation.

We believe the sessions were also timely and very much in keeping with the most recent opinions of the United Nations in this complex and vital area. Let me explain. Yesterday, the United Nations General Assembly adopted by consensus a resolution protecting climate for present and future generations. This resolution was submitted by Malta and for the first time introduced the subject of climate change to full scale debate in the United Nations. Now we've heard a good deal this morning about the work of UNEP and WMO and other parts of UN systems, but for the first time, the question of climate change experienced a full scale debate at the United Nations itself. Initially, the act ran into difficulty as many delegations have problems with the concept of climate as part of the common heritage of mankind, as for example, the question of the oceans. Malta, when it introduced the act initially, opted for a concept comparable to that of the oceans trying to declare climate the common heritage of mankind. But even these difficulties were truly instructive as they obliged governments to wrestle with the concept with a view to finding the common ground. And the item also provided a forum for the developing countries to participate fully in the debate and to shape an outcome that will be consonant with your interest and concerns. And I might add that this is a very important consideration because, as you know, much of the discussion so far is really among the developed and industrialized countries. The developing countries have a sort of secondary, if not marginal role. The United Nations provide a unique forum whereby the developing countries now are able to be part of the process, to be players. It's providing an opportunity for what we might call a global learning process and stimulate it or provide the new stimuli to consensus building that is likely to accelerate and expand over time. For one thing, the idea is now firmly inscribed on the agenda of the United Nations and initially, at least, the Secretary General will be obliged to report annually on developments that are taking place, thereby keeping the subject under global review in this universal forum.

The climate change has also become a new vocabulary of concern for the United Nation's community as did environment some years ago and sustainable development last year. Secondly, the resolution sought to mobilize action in all sectors of the international community, responsible action -- the kind that we heard of so frequently referred to this morning. After all, the problem of climate change affects humanity as a whole and therefore, cannot be suffered by one or another government but must be dealt with in international forum. What is significant here, is that the resolution also called upon governments to subscribe urgently to the various international agreements on climate change and ozone depletion. It called particularly upon governments that have not done so already to ratify the Montreal Protocol, and this is something that I believe might accelerate that process.

To date, when I checked with the United Nations yesterday, I understand that we've had some sixteen ratifications: Mexico, United States, Norway, Sweden, Canada,

New Zealand, Egypt, Uganda, Ukrainian SSR, Japan, Luxembourg, Portugal, Byelo Russian SSR, Nigeria, Kenya and the Soviet Union. The sixteen were expecting the European Community to ratify the convention in a matter of days. So far, we expect at least eight more ratifications by the middle of December. If that does take place, then I believe we will be well on the road to entering the force in January as the treaty of the Protocol suggested. This we believe would be a monumental step forward and an acceleration of a process that we could commend very clearly, the responsible response of the world community.

But the Protocol, being a resolution of the United Nations, also called upon the United Nations Environment Programme together with WMO, to upgrade climate change on their own agendas and to start a process now that would bring about a series of significant steps in the next 18 months. I believe our Chairman referred to some of these steps and I certainly would like to just refer to at least five that would call for a resolution. First, it called for the UNEP WMO to try to accelerate the state of knowledge of the signs of climate change with a view to bring about a global consensus on what we know, establishing a sound database for policymaking. Secondly, the resolution calls upon the states to embark on programs and studies on the social and economic impact of climate change including global warming. Thirdly, possible response strategies to delay or mitigate impacts of adverse climate change. Next, the identification and possible strengthening of relevant existing international legal instruments having a bearing on climate. And finally, elements for inclusion in a possible future international convention on climate change.

I think this last provision has been somewhat contentious within the United Nations community as there are some people who are now willing to press if not for a law of the atmosphere, for some type of specific convention on climate change with binding provisions. Unless the prior steps outlined have been taken, unless we have a clearer database and a clearer sense of policy, it could be somewhat foolhardy to press now for a convention. What the resolution is calling for is a framework convention, or the elements that would lead to a framework convention, the kind that we had in the Vienna Convention that was adopted by the United Nations community in 1985. From this, we could opt for a series of protocols wherein the initial standards or provisions might follow from a framework convention.

But the question now is for some type of exploration of the elements that would go into the establishment of that kind of framework convention. In general, there were two other provisions which I believe may be of interest to our group this morning. And that is a resolution of the United Nations, Earth governments, intergovernmental organizations, and non-governmental organizations and such institutions, that treats climate change as a priority issue and what it takes to promote specific cooperative, action-oriented programs and research, so as to increase understanding on all sources and causes of climate change including its regional aspects and specific timeframes as with the cause and effect relationships of human activities and climate, and to contribute, as appropriate with human and financial resources, the efforts to protect the global climate.

Secondly, it encourages convening of conferences on climate change, particularly global warming, at the national and regional and global levels in order to make the international community better aware of the importance of dealing effectively and in a timely fashion with all aspects of climate change resulting from certain human activities. There's something else that is unique about this particular resolution in that it's requesting the Secretary General to bring the present document to the attention of all governments as well as intergovernmental organizations and non-governmental organizations in consultation with the United Nations and well established scientific institutions.

What I find significant here is not only the broad scope of the resolution, but the concern of the United Nations community that it be very widely distributed. Public understanding, as Ambassador Benedick suggested this morning in his remarks, is key: if indeed we are to get the kind of policies that are required -- the result of long-term complex forecasting -- we'll need publics that are informed. We'll need publics that are capable of supporting decisions that will be required to deal with the question.

Now I believe that there are some opportunities-- some unique opportunities -- posed not only by the document of the United Nations to which I have just referred, but by the development within the world community that might help the Americas to move forward. I believe it is quite clear that within this region, we do have a leadership of United States and Canada in scientific and research capabilities. And I believe this leadership is vital to the process that we are embarking upon at this time. And secondly, I believe that there is considerable legislative ferment in the United

States. One, for example, is rather encouraged by the global warming prevention bill of 1988 that was presented by Congresswoman Schneider. The various hearings within the United States over the last several months all encourage that if there is not yet a national dialogue on the subject, there is considerable public attention being paid to it.

We are also encouraged by the development of transnational cooperation between the United States and such countries as Brazil. The recent reports this summer from the Goddard Space Center of the forest fires in Brazil, the result of a cooperative effort, lead us to believe that we now have the ingredients of broader levels of cooperation among two of the major states within the Americas. But most importantly, I believe that we have the solid institutional regional frameworks for expanded cooperation among the states of the Americas. For example, we have the Organization of American States that might want to look at the issue of ecological security for the region. We have the Pan American Health Organization in place that might want to look at some of the effects of global warming and climate change. We have the InterAmerican Development Bank that might want to look at the investments that must be undertaken at this time to relieve us of some of the problems at this time that have been forecast and projected. There is also the Caribbean Development Bank that may also embark on this question. There is the Economic Commission for Latin America in the Caribbean that might want to look at the question of climate change in development planning. There is the World Bank with its various focuses on issues of the Americas. And finally, there is a Caribbean Environment Programme that was the special institution created by the United Nations Environment Programme.

I believe that the central challenge now is the definition of climate change in the respective agendas of these institutions. UNEP has made a start in this direction when last July we convened in Paris a meeting of the major development agencies and United Nations' specialized agencies with a view to ascertaining how far these bodies might address the question of climate and what they might do about it. For example, what would the FAO do in its own program and planning for the effects of climate change in agriculture? Or take UNESCO which has now a mandate to protect the cultural heritage of mankind. We heard last night about the antiquities in Egypt. What would UNESCO do about some of these basic cultural forms that may have been threatened by climate change? We would like to believe that the United Nations community and, more especially, these institutions of the Americas, might want to gear up to address these questions and place them on their agendas. We in UNEP are prepared to encourage that process and are prepared to convene such meetings of the American institutions to address the question of climate change, to exchange information, and as it were, to find a common language among the states of the region.

We have made a start in at least in one direction -- a start the Caribbean Environment Programme is concerned with -- we have a study on the implications of climate change on the wider Caribbean. The report of the task force is about to be completed and we believe that this might contribute very greatly to both the science and the literature on the question of climate change in the region. In establishing that task force, we tried to use a common format and are attempting to use it in the various regional seas program, the same regional seas program of UNEP in the Mediterranean, in the Southwest Pacific, in the Southeast Pacific, in the East Asian seas as well as in the Caribbean. The charge to the groups that are working on this is first to examine the possible effects of sea level changes on coastal ecosystems dealing with deltas, estuaries, wetlands and coastal plains. Secondly, to examine the possible effects of temperature elevations on terrestrial and aquatic ecosystems, including the possible effects on economically important species. Third to examine possible effects of climate physiographic and ecological changes on the socioeconomic structures and activities. Fourth, to determine the areas of systems which appear to be most vulnerable to the above changes. And finally, to prepare regional reports assessing the general and specific potential problems of the regions which may be caused by expected sea level rise and temperature changes.

Ladies and gentlemen, the reports are still to be concluded, but initial impression suggests a picture ranging from mixed to ominous. From the data in hand, there is strong reason to believe that there is significant ecosystem stress, erosion of the river delta regions, that there would be likely to be very serious problems involved in terms of salt intrusion as a result of sea level changes, that there would be loss of some wetlands, and finally, that there would be more frequent and ferocious storms.

Those of us who come from countries like my own, Jamaica, that experienced Hurricane Gilbert this summer, can't help but be apprehensive about any climate change that might create additional problems of the kind that were experienced in September of this year. And yet, the scientists tell us that if the temperatures do change as they predicted, that the atmosphere is likely to become more energy retentive and that we can expect more frequent and ferocious storms. Consequently, the governments of the regions have need to be concerned and have need to see what needs to be done at this time.

Once the reports are in, I believe that UNEP will have a fairly comprehensive view on the various problems that are likely to occur if global warming does take place in the semi-enclosed seas of the world and the coastal regions that surround them. We also believe that one of the principal consequences of such studies might help governments to place the issues on their national agendas and might encourage them to start thinking about the quality responses that are required at this time. And this is a very, very tough problem for many developing countries of the region. We're talking about long-term complex forecasting of consequences, about lacking capital over time in the next 20, 30, 40, 50 years. And yet, in order to stem the tide, we may have to make investments in the near term, and there is a very serious dilemma in terms of how the budgets may be sliced.

Secondly, we in the developing parts of the world have a kind of schizophrenic approach to problems such as these. On the one hand, there's a part of the brain that is scientific and secular, that is prepared to accept and to review the data as presented by the scientific community. On the other hand, there is another part of our brain that is quite providential, that somehow, we would hope for the best and believe that if God does not take care of it, somehow, it might be taken care of itself. We need to find ways of bridging that gap, of putting both sides of the equation together. We need also to provide more case studies and concrete results. My problem is expanding the model, in the case of Uganda, in trying to demonstrate what a two-degree shift in temperature would mean to the coffee production of the country of Uganda. Suddenly, the government has begun to take notice and the issue's no longer remote or abstract. Suddenly, we now begin to feel that this is something that's of immediate consequence to the future livelihood of the country.

We can only hope that the General Assembly resolution will continue to accelerate this kind of attention and give weight and authority to the findings of the scientific community. We believe that a step will have been taken by the decision just adopted by the General Assembly. For one thing, we believe the issue's now been mobilized. We now have embarked on a global debate on the question of climate warming. Secondly, we now have the capacity to place the issue on national agendas. We believe that this is extremely important, that governments now see the need and the necessity of dealing with this problem at the national level. Certainly, I believe the UN debate will have generated pressure for the ratification of the Montreal Protocol and, more directly, for the adjustments that must be made in the control provisions. As you know, there is a very serious debate in the scientific community asking if a 50% reduction is adequate or if we should gear it up to 85%. The context is changing as a result of the United Nations debate and I believe that this might very well encourage governments to take a harder look at what these provisions might be.

I believe also, the debate will force us to re-examine the role of UNEP. Ambassador Benedick and Dr. Hecht paid special tribute to my own program but the question that we must ask, given the explosive acceleration in the environment field in the question of climate change, does UNEP have the necessary capacity to deal with the issue or has the issue outgrown our institutional capabilities? The Soviet Union thinks so. At the current session of the General Assembly, foreign Minister Shevardnadze advocated the transformation of the UNEP governing council into an ecological security council. They argued that perhaps the Trusteeship Council of the UN which is on a decline, might be transformed into an ecological security council and proposed a timetable for discussing the question as to how far we might need to upgrade this. I think there may be merit to the argument, although one is not sure of the outcome and the timing. If indeed, we upgraded the status of the UNEP to an organ of the United Nations, it may have the stature of Security Council, Economic and Social Council, International Court of Justice and even the General Assembly itself. This would elevate the status of environmental questions within the United Nations community and give a more rapid response to the problems that we are experiencing now because I'm sure that all of you know that one of the difficulties that we face is that we are an organization headquartered in Nairobi. And whereas communications and technology has bridged many gaps, there still are problems in the rapidity with which we can respond to the world community.

We are also very encouraged on another front by the national leadership that is emerging on a variety of levels. We mentioned the United States and Soviet Union cooperation, and also Ambassador Benedick paid tribute to my own program, and I'd like to pay tribute to the United States leadership in the ozone debates and negotiations. There is no question that without U.S. leadership of the kind that we experienced on the Montreal discussions, I doubt that we would have tipped the balance as we did at the time. And I think a great deal of tribute has to be paid not only to the skill of Ambassador Benedick, but to the United States in staying the course over that period of time. The U.S. therefore is out front on this one and we hope the USSR. Canada is also assuming a very vital leadership position, as is Malta. The United Kingdom has now offered to host a major global summit conference on the ozone question in March of this year. And this will be a rather unique opportunity for all the members of the world community to take a second look at the Montreal Protocol just before we have the first meeting of the contracting parties. I know that the British are now negotiating very closely with the Chinese and India who have not yet signed the Protocol. One can hope that this meeting will lead to such tentative and specific results.

And finally, industry is now beginning to respond. One of the unique features about the negotiations at Montreal -- and before that -- is the extent to which industry became a major player. In the past, United Nations simply invites industry as a guest and not a partner in the process. We are rather excited by the fact, that in the first round at least, industry was a major player. We're excited about the fact that they are responding now to the question of climate change. IBM recently made a gift to UNEP of some 6 million dollars worth of equipment to help us monitor more closely the problems of climate change. One can only hope that this type of response will be accelerated and expanded. My program is very keen on exploring with industry the role of bio-technology in helping to alleviate some of the projected consequences. One of the problems we face is the question of methane discharge as the result of rice cultivation. Is there a possibility for a dry rice? Can bio-technology help us invent a different kind of rice cultivation? We're looking to the private sector to help us deal with this problem.

Ladies and gentlemen, the picture at this stage is complex, but the outlook is promising and for us at the United Nations Programme encouraging. We have some confidence as a result of the Montreal exercise and the ozone negotiations. I believe we have the confidence of the world community in terms of the way we have proceeded in trying to base policy recommendations on hard evidence. We have some continuity because of recent direction of my Executive

Director, and hopefully, we will be able to lead the process with a steady hand. But finally, we are encouraged by the fact that for the first time, there is some good news to report, that the scientists are now learning how to talk to the politicians, and the politicians are listening. And many political leaders now seek what they call "a greener image." And perhaps the greenhouse effect may have one consequence that may be hopeful for us all and that is it might help the greening of the political process. Thank you very much.

AGENDA FOR THE NEXT DECADE: CANADIAN PERSPECTIVES

Dr. D. Kirk Dawson
Director General
Canadian Climate Centre

I would like to thank the Climate Institute for providing me with the opportunity to participate in this excellent conference. I have to admit to being a little overawed as a relative newcomer in addressing such a large number of experts. Nevertheless, I believe that, as the Head of Canada's Climate Centre, I am well positioned to tell you something about our perspective on climate change. Why do I feel I am well positioned? It is because the Canadian Climate Centre is a component of the Atmospheric Environment Service (AES) which deals with weather, ice and air quality services and research. Not only that, we are part of the Canadian Department of Environment which is responsible for, amongst other things, an organization the equivalent of your EPA. Put simply, it means there is a relatively close working relationship between climate specialists, atmospheric chemists and the policy experts responsible for regulation and controls--a distinct benefit when dealing with major atmospheric change issues such as climate change, ozone and also acid rain.

Let me start by emphasizing that Canada is taking the threat of climate change seriously, and I am consciously using the word threat since we do not believe there are any ultimate winners in the kind of world being envisaged under a doubling of the concentration of CO_2 or the equivalent greenhouse gases. Our agenda for action over the next decade, like that of most countries, will unquestionably be influenced by regional concerns, our national perspective on the opportunities for action, and assessments of what we already know and don't know about the background science. On the latter, considerable efforts have already been brought to bear in a number of major international meetings in the last few years. In particular, I would like to recognize the role of the meetings held in Villach in 1985 and 1987, in Bellagio in 1987, in Toronto in June of this year and Hamburg just last month. Each of these meetings built upon the results of the preceding meeting, and resulted in a series of recommendations at the Toronto meeting that could serve us well as a starting framework for our proposed agenda. In addition to these international meetings, there have been numerous conferences and workshops at the regional and national level which have helped to address the regional and sectorial concerns and contributed to the perspectives of the issue at the national level. The First North American Conference on Preparing for Climate Change, organized by the Climate Institute, played an important role in this aspect.

Based on the results of the preceding meetings, it appears quite evident that we need much greater understanding of the physical, chemical and biological processes involved in climate change and its impacts and that any agenda must emphasize an enhanced level of scientific research. However, we believe we need to pursue three additional thrusts at the same time. As the delegates at the Toronto Conference stated in their final conclusions, "It is imperative that we begin to act right now." We need to:

(1) develop strategies for adapting to the inevitable changes in climate to which we are already committed;

(2) develop strategies for limiting and controlling the extent of such change;

(3) better inform the public and the decision makers.

These objectives, I should re-emphasize, are not mutually exclusive.

I would like at this point to review the conclusions of the Toronto Conference. Among its many recommendations it advocated the following:

(a) Research

- devote increasing resources to research and monitoring efforts within the WCP, IGBP and HRGC, particularly with respect to understanding climate change at the regional scale and better determining the role of the ocean
- increase significantly the funding for R&D in renewable energies
- support the work of the IGPCC to conduct assessments of scientific results and initiate discussions on response

(b) Adaptation Strategies

- expand funding for technology transfer and cooperation in coastal zone protection/management
- develop/support technical cooperation projects to allow developing countries to participate in related programs

(c) Limiting Strategies

- ratify the Montreal Protocol
- develop energy policies to reduce CO_2 emissions by 20% of 1988 levels by 2005 through energy efficiency and alternative energy options
- initiate development of a comprehensive global convention on the protection of the atmosphere
- establish a World Atmosphere Fund to mobilize a substantial part of the resources needed to pursue related action plans
- reduce deforestation/promote afforestation

(d) Public Awareness

- label products for user assessment of environmental effects
- disseminate conclusions/recommendations to governments of all nations and to future meetings (It should be noted I didn't bring any copies with me. It is our hotest publication and currently out of stock.)
- increase funding of NGOS for public education/ awareness campaign
- allocate more support to educational institutions for increased emphasis on this issue.

The Canadian perspective is conditioned not only by the results of the Toronto meeting but also by the results of an active domestic research effort under the Canadian Climate Program, which is now entering its second decade, and its active participation in the World Climate Program. It adheres strongly to the position that we are both capable of and required to pursue concurrent and interlinked activities in all four of the major thrusts mentioned above. Canada is now in the process of developing an action plan to do precisely that. That plan is being developed under a number of guiding principles, including the following:

i. The framework for action must be consistent with the broader principles of sustainable development and global stewardship as outlined by the Brundtland Commission in their report to the UN.

ii. It should draw upon and utilize existing national and international partnerships, such as the World Climate Research Program, and institutions as much as possible and create new ones only where needed.

iii. The enhancement of our knowledge of the scientific processes and their ultimate impacts is essential for the development of appropriate policy responses.

iv. The agenda is a shared responsibility of all sectors of society.

v. Individuals and decision makers need information on which they may make informed choices.

vi. The development of policy options in both limitation and adaptation can and should be pursued at the same time as we further our scientific investigations.

Based upon these recommendations, guidelines and principles, I would propose the following four point framework as our agenda of the next decade, a framework which must be coordinated globally but implemented locally.

Firstly, while all aspects of present activities under WCP, and those being considered for the IGBP and HRGC are significant, particular emphasis must be placed on the understanding of the global climate system, and the sensitivity of society to changes within it. Major uncertainties that need to be addressed include the role of oceans (both as a reservoir for greenhouse gases and as a sink, source and transporter of heat), clouds, sea ice and the hydrological cycle.

Scenarios for the regional characteristics of climate change are of particular interest, since the effect of climate change on ecosystems and society are regionally specific. Impact studies will need to identify those natural and human systems most sensitive to climatic change and investigate these in much greater detail than present studies are capable of providing. Such studies also need to look carefully at the integrated effects of climate change and other assaults on ecosystems, including acid precipitation, increased UVB radiation and ambient air quality.

The existing WCP provides an excellent framework for such activities. Implementation of its programs will, however, depend on strong and adequately funded national programs. In Canada, a National Climate Program was established in 1979. Results to date have included a much improved integration of cross-sectoral and interagency climate research, the development of an atmosphere-ocean GCM (which is at this moment being used for its first 2xCO2 climate change experiment), the completion of approximately 20 studies into regional and sectoral impacts of possible future climates, the establishment of several climate research professorships at McGill and Dalhousie universities, and the institution of climate round tables at the provincial level. Similar programs have been established in other countries. However, these are still far too few, particularly in the developing regions of the world, and most, including that of Canada, are under-resourced. These programs must be strengthened and broadened to include consideration of the linkages with other atmospheric issues. All must rely on international cooperation and collaboration, preferably under existing frameworks.

Secondly, scientific research can tell us what may happen, with what probability and what might be the first order impacts on the ecosystem and society. It cannot, however, address the practical question of how to respond, and the related social questions of costs and values. We need to develop appropriate strategies to adapt to possible changes by addressing the question of uncertainty and including it in our assessment of risks and costs. Here I believe we need to borrow from the how-to books of economics and actuarial sciences. We need to compare the consequences of preparing ourselves for predicted climatic changes that may not happen with the consequences of failing to prepare for changes that do occur. The results will be case specific. In some areas of human activity we may find the cost of anticipatory action to be too large, given existing uncertainties, and we may wish to risk the consequences of inaction. In others, the consequence of inaction may be potentially so costly, or the cost of action sufficiently small, that it becomes beneficial to pursue action as an insurance policy. Such strategies, discussed more extensively in the results of the 1987 Bellagio meeting, help direct us to those areas where anticipatory action should be pursued immediately, while identifying those that will need better understanding and more accurate assessments before action becomes justified. Much more work is required in developing the methodologies for such cost-benefit assessments and a pro-active program of educating the appropriate decision makers in the need to respond. Past work of institutions and agencies such as IIASA, UNEP and the Beijer Institute have provided a good beginning. Future work needs to seek the assistance of other organizations such as OECD, IFIAS, ICSS and UNU and should emphasize the most vulnerable regions of the world and the efficient exchange of methods and information.

Thirdly, while much of the preceding discussion, with respect to dealing with uncertainty, applies to limitation strategies as well, the latter must rely much more on cooperative global strategies. The participation of nations such as the USSR, China and the USA are obviously pivotal and essential. Hence the emphasis in the next decade will need to be focused on international conventions and protocols, strengthening those in existence and adding new ones to fill the major gaps. This will need to be integrated within a larger comprehensive framework, as emphasized by the Toronto Conference. Such a framework will depend largely on the role of UN agencies, perhaps as a component on its initiatives on sustainable development but must be encouraged at major meetings such as that of the G-7 summits and OECD. An initial international meeting of legal experts has already been scheduled to take place in Ottawa in February 1989 to consider the feasibility of a comprehensive framework convention and The Netherlands is planning to host a follow-up high level political meeting in the Fall of 1989 to consider further action. Nationally, countries must study carefully the economic and social benefits and costs of limiting atmospheric effluents and begin to evaluate energy and other industrial policies, particularly at the time of major changes in direction, to ensure the most environmentally benign, and ultimately the most economic, options are selected.

Lastly, implementation of action policies will not be the responsibility of scientists but of decision makers in government, industry and society in general. Effective transfer of information is, therefore, of paramount importance. Ultimately, in most societies it is the general public that dictates the actions of the politician and the nature of consumer products manufactured by industry. The general public is also the primary contributor to the inefficiency, wastefulness and environmental insensitivity of our socio-economic activities. Hence public education through the media, educational institutions, environmental groups, community service organizations and governmental programs is a fundamental part of any action strategy. Yet this needs to be pursued with caution since the issue of climatic change is a concern that has materialized over many decades, remains clouded with questions and uncertainties, and will not lend itself to quick and easy fixes. Our education efforts must emphasize the need for long term and continuing efforts at both improving our understanding and implementing our responsive actions.

In summary, as was emphasized earlier, we believe that the issues associated with climate change must be coordinated globally. It was for this reason that Canada was pleased with the decision of WMO and UNEP to create an Intergovernmental Panel on Climate Change. This panel should provide a key mechanism for reaching agreement internationally on the state of our knowledge of climate change, its impacts and the policy options open to nations to both limit and adapt to the changes that we have already committed ourselves to with the emissions of greenhouse gases that have already taken place around the world. It is our intention, therefore, to actively support the work of the IPCC and, where possible, provide experts and input into its deliberations.

At the same time we will continue to build upon bilateral opportunities wherever and whenever they appear, including activities under our MOUs with The Netherlands and The Peoples Republic of China. In the area of bilateral activities we were also encouraged by the recommendations that came from the first

United States-Canada Symposium on the Impacts of Climate Change on the Great Lakes Basin, and we would be interested in discussing with our colleagues in the USA, the possibilities of a joint integrated study of the Great Lakes Basin as a pilot project of how to respond to climate change.

As always, the future is uncertain and the way to proceed unclear. However, with the cooperation that has been evident in recent meetings and with the support of our political leaders we will, I am sure, be able to avoid the worst of our possible futures.

THE GREENHOUSE EFFECT: REALITY OR MEDIA EVENT*

Stephen H. Schneider**
National Center for Atmospheric Research***
P.O. Box 3000
Boulder, CO 80307

In 1988, "The Environment" became a media event, rivaling politics and baseball for the big stories of the year. The covers of Time and Newsweek, as well as top news stories on local and national television and radio, dominated the air waves in the summer of 1988 with drought, heat, hurricane winds, fire, and smoke. Topping the list of problems that has revitalized the environmental movement in 1988 is a century-old theory: global warming from the so-called "greenhouse effect". Did the events of the summer of 1988 finally demonstrate that the crescendo of warnings of many scientists over the past decade were too long ignored? Or, rather, was it just a random event of a perverse and fickle Nature requiring no action from us other than cleaning up the damage and trying to be less vulnerable next time? In July alone, I probably had 100 phone calls from journalists, many of which asked, in effect, okay, you have been carrying a banner about global warming for the past 15 years. Are you finally ready to give us an "I told you so"? Before I let on how I answer that question, first we need some background on just what the greenhouse effect is, what kind of scientific consensus exists over its likelihood, what it might mean for the environment and society, and ultimately, what can or should we do about it.

The greenhouse effect, despite the controversy we hear about it, is actually one of the most well-accepted and well-documented theories in the atmospheric sciences. Mars, a planet with a thin, predominantly carbon dioxide atmosphere, has a mean temperature well below that of most deep freezers. Venus, on the other hand, with a very thick, largely carbon dioxide atmosphere, has a temperature hotter than most ovens. The Earth, of course, with a moderate amount of atmosphere, contains liquid water,

---

*Prepared for World Monitor.

**Any opinions, findings, conclusions, or recommendations expressed in this article are those of the author and do not necessarily reflect the views of the National Science Foundation.

***The National Center for Atmospheric Research is sponsored by the National Science Foundation.

equable temperatures, and abundant life. Mars is too cold, Venus is too hot, and the Earth is just right--what some planetary climatologists have called the "Goldilocks phenomena"--is well understood to be a result of the greenhouse effect.

A greenhouse can be a pleasant place to grow tropical plants in cold winter climates. It also can be a very nasty, incredibly uncomfortable hotbox, like an overheated car with windows closed that was parked in the sun on a summer's day. The metaphor works both ways. One diplomat friend has urged me for years to drop the "greenhouse" metaphor since it doesn't convey sufficient unpleasantness to the public to motivate action--call it the "global heat trap", he recommended. This image has only negative connotations, and maybe more people would then pay attention. The last time I heard that advice was in May 1988, just before the infamous Summer of '88.

The way a greenhouse works is that the glass is relatively transparent to sunlight, allowing it to heat the inside of the house. Then, the panes block the air on the inside from mixing with the cooler air outside the greenhouse. They also, but to a much lesser extent, form a trap inside the greenhouse for another form of energy, so-called infrared radiant energy. All objects with heat give off radiation with different amounts of energy and at different wavelengths. The very hot sun gives off high-energy, short wavelength radiation, whereas the cooler Earth's surface gives off longer wavelength, lower energy radiation. The gases in the Earth's atmosphere and Mars and Venus also, have this "greenhouse" property: they are relatively more transparent to incoming solar energy than they are to outgoing infrared energy. This tends to trap radiant heat near the Earth's surface with greater trapping occurring the more the greenhouse gases in the atmosphere. So far, there is little controversy, except perhaps whether we call it a greenhouse effect or a heat trap.

The controversy begins to build though, when we have to translate extra heating of the Earth's surface in to future scenarios of climate changes. To do that, we have to project how much more greenhouse gases will there be in the future than there is now. The primary culprit for the buildup of greenhouse gases in the atmosphere is the burning of fossil fuels: coal, oil, and gas. The hydrocarbon molecules contained in these fuels started out as dead, buried plant matter hundreds of millions of years ago. Burning them for heat simply means combining their hydrocarbon molecules with oxygen. An inevitable byproduct is carbon dioxide (CO_2). CO_2 is a very effective greenhouse gas. Right now, there is incontrovertible evidence that there is about 25% more carbon dioxide in the atmosphere than there was a century ago, the buildup coming largely from the burning of fossil fuels but also from deforestation. The latter contributes because trees themselves put carbon dioxide in the air when burned and because the elimination of forests reduces one mechanism for buffering the atmosphere by taking the CO_2 out.

How much carbon dioxide will be injected in the future depends upon three factors: technology, standard of living, and population size. In order to project the emissions of carbon dioxide, we need to project the emissions per technology (coal is the worst, natural gas half as bad, solar and nuclear virtually CO_2 free). We also need to project the standard of living, that is, the per capita consumption of each of these technologies. Finally, we need to forecast the total population size using the technologies. Total emissions is the product of each of these factors. Obviously, each of these factors is difficult to project, and thus, there is controversy over just how much of this gas might build up into the future. Most estimates suggest that CO_2 will double sometime around the middle of the next century. But CO_2 is only part of the greenhouse story; other gases such as chloroflurocarbons (the very culprits in stratospheric ozone reduction and the ozone hole), methane (produced by animals and in rice paddies), and a host of other minor "trace" gases. These other trace greenhouse gases, taken together, add about as much to the enhanced greenhouse effect as increases in CO_2. All told, our climatic theories suggest that the future will see global averaged temperature some 2 degrees - 10 degrees F warmer over the next century than at present.

How are these estimates of two to ten degree F temperature increases made? Given some scenario of increasing trace greenhouse gases, we turn next to the primary tool used by climatologists in making quantitative estimations of how much global warming will take place: mathematical models of climate run on the best available super computers. Since no laboratory experiment can be built that remotely captures the complexity of the Earth's climate system, scientists instead build mathematical models. Equations are written down to represent the basic physical laws that govern the motions of the atmosphere, oceans and ice. These include the conservation of mass, the conservation of momentum, the conservation of energy and thermodynamic laws about the state of gases. However, the actual equations cannot be solved exactly, so approximation techniques are developed which involve creating a discreet number of points or "grids" around the globe at which solutions to the equations are approximated. Everything that occurs on smaller scales than these roughly 500 by 500 kilometer grid boxes is not explicitly treated in the model. Rather, we have to invent statistical rules to associate, for example, the average relative humidity at the four corners of the grid box with the average amount of cloudiness within that grid box. Winds, temperature, precipitation, sunlight, relative humidity, etc. are all then predicted internally at each of the model's grid points. The model at my lab uses a global grid with 4 1/2 degree latitude by 7 1/2 degree longitude spacing and 9 vertical levels. In this example, we divide the surface of the Earth into grids of 1,920 squares, with 9 levels, which makes 17,280 boxes with all these weather variables calculated every 30 minutes. It takes 10 hours of time on a Cray Supercomputer to compute a year's weather. Although the detailed predictions of different models built and maintained at different institutions around the world agree roughly that a doubling of carbon dioxide would increase global temperatures something like 3 to 5 degrees C, they disagree markedly as to the specific regional distribution of climatic changes and their time evolution. Nevertheless, these models do a credible job of predicting the very large difference in temperature between winter and summer, Mars and Venus, some aspects of the ancient Earth, so that it is a good bet that their global scale temperature predictions are accurate to about 50%.

At a recent Senate Energy Committee hearing in which Colorado Senator Tim Wirth had introduced a bill to try to curb the emissions of these gases, especially CO_2, some pointed conflicts arose among witnesses from the Department of Energy and other scientists, including me. The DOE argued that there is so much uncertainty associated with the details of how the climate will change it would be folly to invest present resources to hedge against a problem whose dimensions are not clear. I countered that platitudes about scientific uncertainty had for too long been used as an excuse to avoid action. When I first started in climatology in 1971, I recall a government official telling us we had ten years more to study without risk. Then, in front of a Congressional hearing in 1981, a Department of Energy official said essentially the same thing. It is not surprising that we will hear this advice again. Indeed, how long one should study "before" we act, is not a scientific judgment, but a <u>value</u> judgment, weighing the costs of any present investments to slow down the future climate change versus the costs of that change descending on us unchecked. There is no "right" or "wrong" there, simply a philosophical difference over how to weigh environmental and financial costs.

At Wirth's hearing, New Jersey Senator Bill Bradley arrived and began to press the scientists on the panel for some very specific reasons why any constituents in his state, for example, should worry about a degree or two change in the average temperature, or even more importantly, why they should be willing to pay extra money for insulation or drive cars that are smaller and get better mileage. Finally, he honed in on a critical question, is there anything that we were sure about?

I responded that it is indeed true that we can't be certain that the very warm and dry summer of 1988 was directly a result of the greenhouse effect. Nor could we directly attribute the unusual summer heat in the southeast in 1986 to the greenhouse effect, nor the devastation from hot weather in the cornbelt in August 1983, nor the crop- and people-killing temperatures in the late spring and early summer of 1981 in Texas, Oklahoma, Arkansas, and Missouri. It is quite possible that any of these events was simply a random occurrence of a perverse nature. In fact, it is no easier to determine whether these kinds of events are connected to the observed one-half to one degree F of warming that we think the world has undergone in the past 100 years than it is to determine if three snake-eyes in a row prove a pair of dice to be loaded. The more rolls we make, the surer we are of the true odds. The longer we wait to establish scientific certainty that the greenhouse effect has descended, the greater the dose of climate change we and the rest of the living things on this planet will have to adapt to over the next several generations.

One could think of a summer's climate as a giant game of chance. The climatic dice have not simply 12 faces, but 1,200 in this metaphor, including warm ones and wet ones and dry ones and so forth. What most climate scientists think will happen is that the world will warm up, and should warm up dramatically in the next century. If so, we would be "erasing" some of the "colder faces" and replacing them with an increasing number of warmer ones. And, as a number of climate theories suggest, there could be drier ones, too, at least in the interior of midlatitude continents. Therefore, no honest scientist can claim that 1988 or any of these heat waves in the 1980's were absolutely, certainly attributable to the greenhouse effect. In fact, there are a number of suggestions that our unusual drought pattern was the result of an out-of-position jet stream caused by unusually cold water in the equatorial Pacific Ocean. However, even if this proves true, increased global temperatures increase evaporation of water from our farm fields and make extreme heat waves more likely.

How important is a degree or two of temperature change, Bradley asked. Let's go back 18,000 years to the height of the last ice age, when mile-high ice sheets covered most of Canada and went as far south as the Great Lakes across New England and into New York State and the northern half of Long Island. It ended about 10,000 years ago, giving way to the present interglacial period in which civilization grew and flourished in the warm conditions. It is often surprising to many people to find out that the ice age was only about 9 degrees F colder than the present Earth average temperature. It took some 10,000 years for the planet to recover from that ice age. That was an event that literally revamped the ecological face of the world, radically altering what could grow where, the world's sea levels rose by hundreds of feet, species went extinct, others evolved, etc. All of this took place with some average rate of a few degrees F temperature change per <u>thousand</u> years. But now, if we continue to burn fossil fuels and deforest at current or expanded rates favored by some politicians, corporations, or economists, then it is quite plausible that we will be changing the climate something like 2 degrees - 10 degrees F in a <u>century</u>. This is some 10 to 50 times faster than the average natural rates of change following the Earth's recovery from the last ice age. What is scary in that scenario is not that we have the detailed description of who "wins" and who "loses" and precisely what changes where, but that we are virtually certain that this magnitude and rate of change is bound to cause major surprises. Amusingly, one senator (not Bradley) at the Wirth hearing interrupted, "Could you tell us, Doctor, what kind of surprises you specifically have in mind?"

"Well, sir," I replied somewhat sheepishly, "a surprise is something you really can't predict in detail." However, we could speculate on some major changes in the distribution of forest species, unusual changes in ocean temperature patterns, shifts in growing season, new agricultural zones, major changes in the likelihood of forest fires--an event made emotionally tangible by this summer's disasters--and the possibility that tropical diseases could expand out of their present regions. Of course, most surprises are just that, unforeseen. Already we have an example of a nasty surprise from human pollution of the atmosphere: the Antarctic ozone hole.

I am sure the constituents of New Jersey (or Texas) would have no difficulty relating to damages caused by storms which passed by their coast from time to time, pushing water against the shore in what is known as a storm surge. In such surges sea level can rise 5, 10, or perhaps as much as 18 feet, temporarily inundating coastal areas, causing tens of millions (and sometimes billions) of dollars in damage. How far each storm surge penetrates inland, and therefore how much damage it does, depends on the height of the sea level when the storm hits. If the storm hits at high tide, it is much more likely to do damage than if it hits at low tide. Similarly, if the world average sea level increases by several feet--as many climatologists project is likely over the next 100 years--then the probability of any particular storm surge damaging some part of a coastline every, say, 25 years, might change to that level of damage occurring, say, every 15 years. This could totally alter permissible land uses. We know that sea level is now about 5 inches or so higher around the world than it was a century ago. It is currently rising roughly one-half inch per decade. This is consistent with the fact that the world has warmed up by something as much as 1 degree F in the last century--quite possibly, but not certainly, as a result of the buildup of greenhouse gases.

Another way that people can understand why a few degrees warming deserves our concern is to focus on the unpleasant issue of intense heat waves. 1988 saw such conditions in most sections of the U.S., as many will undoubtedly remember. Certainly it seems logical that if the average temperature of the world is rising that heat waves will be more frequent or more intense. Indeed, two of my colleagues and I calculated just how we might be loading the climatic dice. For example, in Washington, D.C., the probability of 5 or more days in a row in July with afternoon temperatures greater than 95 degrees F is now about 1 chance in 6--the odds of getting one face of an unloaded die. If the temperature increased by "only" 3 degrees F, and nothing else in the climate changed, then the odds of that very unpleasant heat wave go up to nearly 1 in 2--3 faces of the newly loaded die! In Des Moines, the odds for this go from 1 in 17 to 1 in 5. And in Dallas, the odds of 5 or more 100 degrees + F days in a row goes from 1 in 3 to 2 in 3 if July average temperatures go up by 3 degrees F. Certainly, one doesn't need a graduate course in atmospheric science to fathom those kinds of gambles.

Also, when it is hotter and drier, there is a greater chance of forest fires, particularly in the U.S. west. Moreover, adequate moisture is essential to the U.S. maintaining its position as the chief grain exporting nation in the world. Noted environmental scientist Roger Revelle has suggested that one plausible scenario for the United States from greenhouse warming is that we would lose our comparative advantage as an agricultural exporter, since the warming of high-latitude regions in Canada and the Soviet Union could very well open new lands while the yields of grain in our greenhouse-warmed plains could be lowered somewhat relative to yields with present climatic conditions.

Although climatic models differ in their detailed predictions of regional climate changes that might occur at different parts of the world, they all agree that major fluctuations in the accustomed patterns of growing seasons, precipitation, solar radiation and so forth will be experienced. They also suggest that warming of the oceans will force the sea water to expand, just as liquid in a thermometer expands when heated, thereby causing some coastal inundation. Holland is the most obvious place to expect that sea level rise could be devastating. However, the Dutch have already invested billions of dollars in dikes and are actually less vulnerable to sea level rise than many other places in the world, such as Indonesia, Bangladesh, Venice or Florida. The Dutch do, however, worry that they will have to deliberately flood some of their reclaimed land not because the ocean will breach the dikes but because the increased sea level will cause salt water incursion into their ground water. To prevent this may require them to flood some of their now agricultural or suburban land with Rhine River water. Climate models differ about what happens to monsoon dependent land such as Africa or India, but there are strong suggestions that unlike the centers of mid-latitude continents that might dry, monsoon rainfall in India might increase. At first blush that might seem like an advantage, since drought frequently raises havoc with this monsoon dependent area. On the other hand 1988 saw disastrous flooding in Bangladesh. Farm lands, animal habitat and residences were destroyed. Therefore, if India and Bangladesh, for example, made the investment of the hundreds of billions of dollars required to build massive public works projects to control floods and store water for late irrigation, an increase in the reliability or intensity of monsoon rainfall would indeed be a major benefit. But, if those climate changes occurred very rapidly before such investments were made then the advantage of more intense or reliable rainfall could well be the disaster of increased frequency of flooding. In essence, while the details of regional changes are still debated it seems quite likely that there will be substantial redistribution of climatic resources around the world. The more rapidly the change evolves, the more likely it is that the negative impacts of unaccustomed climate would dominate the potential benefits that might accrue if the changes occured more slowly and we then had enough advanced warning to implement anticipatory actions.

Therefore, we come back to the question I posed at the outset: can I say, "I told you so". I think the strongest answer is, "I told you so--almost". We need more warmer years to be absolutely sure scientifically. The problem scientists face in trying to communicate complex and controversial issues with policy implications is formidable. On the one hand, our loyalty to the scientific method requires that we tell "the truth, the whole truth, and nothing but the truth"--meaning all the caveats, if's, and's, and but's, etc. On the other hand, as human beings, we would like to see the world a better place, which to many of us means reducing the risk of unprecedentedly rapid climatic change. That means offer up scary scenarios, few caveats, and get lots of media coverage. This "double ethical bind" I frequently find myself in cannot be solved by any formula, as each scientist has to decide for him or herself what the right balance is between being effective and being honest. To me, the prospect of global warming has been sufficiently compelling that it deserved everyone's attention, even after admitting the uncertainties up front. That philosophy didn't work so well over the past 15 years, I must admit, at least until the Summer of 1988 made global warming appear credible. The irony is that a solid case of good physics had been given vastly too little credibility for 15 years, whereas one, essentially random, hot event in 1988 has given us, perhaps, too much credibility. Not that I want people to ignore the Summer of 1988, for it reminds us we are still vulnerable to Nature. It is also the kind of event that could become more common with global warming. Therefore, I think the heightened public consciousness about the greenhouse effect at the close of business in 1988 has brought us to about the right level of concern, but unfortunately for the wrong reasons. I hope the inevitable cold and wet anomalies that the climatic dice will roll us sometime soon won't be equally over-interpreted in the other direction.

Finally, let's turn to the question about whether we should act to do something about the prospect of unprecedentedly rapid environmental change. How do we know our forecasts are right? I recall a number of years ago discussing with a congressional committee that our mathematical computer models were projecting major changes in climate and that, to me, this called for government action on improving energy efficiency standards, among other concrete actions. How do you verify these models, I was asked. One way is to check their performance in a seasonal cycle test. Winter is something like 25 degrees F colder than summer across the northern hemisphere, I said, and the models do extremely well at reproducing this much larger climate change than anything we project in the next century from human pollution.

"You mean to tell me," one representative said after the hearing, "that you guys have spent a billion dollars of the taxpayers' money proving that the winter is cold and the summer is hot?"

"Yes, sir," I replied, "and we are very proud of it--for if we couldn't reproduce that very large seasonal climate signal in our models, then I wouldn't have the nerve to stand before you and suggest that there is a good likelihood of major climate change from human pollution."

In fact, this piece of circumstantial evidence, taken together with the fact that the world is one-half to one degree or so F warmer than it was a century ago (while at the same time there is 25% more CO_2), and weighing in the Goldilocks phenomena too, makes a strong circumstantial case for the plausibility of most climatologists' warnings. So what do we do about them is the obvious next question.

In my value system, a prudent society hedges against potentially dangerous future outcomes, just as a prudent person buys health insurance. Of course, if one spent all one's resources investing to protect against every conceivable future risk, there would be nothing left to live on today. Thus, we need a priority system for determining what are prudent investments to help us cope with the greenhouse effect. It is my opinion that the best kinds of policies to deal with the greenhouse effect fall in the category of "tie-in strategies". Quite simply, a tie-in strategy would involve making an investment in some activity that reduces the amount of greenhouse gases that otherwise would have gone into the atmosphere, but at the same time also provides other benefits having nothing to do with climate. Such high-leverage investments are the goals of efficient business enterprises. The most obvious category of tie-in strategy investments is the more efficient use of energy. Fuel not wasted produces no CO_2, as well as no acid rain, no negative health effects from acute air pollution, and no further dependence on foreign energy supplies. Also, the more energy efficient a manufacturing process, that is, the less energy it takes to manufacture some product, the cheaper it will be to sell that product, particularly if the price of energy increases as many expect it will over the next several decades as oil supplies diminish. Presently, Japan, Germany, Italy, and other economic competitors are more than twice as efficient as the United States in terms of the amount of energy it takes to make their manufactured products. Thus, energy efficiency not only reduces the magnitude of climate change, but also reduces acid rain, negative health effects of air pollution in cities, dependence on foreign energy supplies, and increases competitiveness over the long term of American products. The strategic investment of several tens of billions of dollars a year to make America more energy efficient and ultimately more economically competitive, seems a valuable policy regardless of whether environmental change materializes as forecast. With all the benefits taken together, the case for action now is compelling.

However, such policies often run into the ideological opposition of those who claim that such decisions should be made by private investment. Indeed, private investment has over the past ten years substantially improved the energy efficiency of American transportation, power production, space heating, appliances, etc. But there is still a long way to go and in the past few years the rate of gains in energy efficiency have radically dropped off. Detroit is now arguing, and the administration tending to capitulate, that we should lower automobile gas-mileage standards. This is exactly the wrong way to go! It is a view dominated by short-term considerations, next year's profit and loss, without adequate regard to the protection of long-term environmental resources such as a stable climate or the reliability of energy supplies in the future.

Another tie-in strategy deals with the development of non-fossil fuel energy supplies, such as solar photovoltaic generators, or a new generation of inherently safer nuclear power plants. While I have not personally been a fan of nuclear power--and certainly would not advocate easing the licensing of existing power plants of the design that failed at Chernobyl or Three Mile Island--there are many designs now on the drawing boards that could satisfy a number of the concerns that people have legitimately had about nuclear power. Environmentalists shouldn't dismiss this option outright simply because of the dread "N-word". But, nuclear advocates must first prove that their new designs are essentially free of meltdown risk and can be cost effective relative to the alternatives, including energy conservation.

The bottom line of the global warming, greenhouse effect issue is that we insult the environment at a faster rate than we understand the consequences. Simple prudence suggests that modifying the global climate at 10 to 50 times the average natural rates of change is not a planetary experiment that we should glibly allow, particularly since there are many measures available that could substantially slow down our impact on Earth and at the same time buy many other benefits. The United States alone cannot prevent ozone depletion or the greenhouse effect. But we will have no moral persuasion over the Chinese, for example, who are rapidly developing polluting coal resources, unless we ourselves lead the way with substantial increases in the efficiency of our energy uses. Furthermore, if we want developing countries to pollute less, we will have to help them to deploy alternative technologies that can provide economic development opportunities with less ultimate pollution. At the same time, we will get the benefit of less total global pollution and new potential markets for our wares.

In 1975, I attended a meeting called by the late anthropologist Margaret Mead which she entitled "The Atmosphere:

Endangered and Endangering". I wondered why an anthropologist was getting involved in such controversial questions in physics. "Quite simple," she told me, "the atmosphere is the last symbol of global interdependence we have. If we can't solve some of our problems in the face of threats to this global commons, then I can't be very optimistic about the future of the world." The weather of 1988 has lifted the debate about climate change from ivy-covered halls and stone and glass government offices into the public consciousness. I hope that a cold, wet winter or normal summer or two won't deter us from undertaking the important steps needed first to slow down and eventually reverse the radical changes we are inflicting on our own world. To be sure, there are many uncertainties yet to be resolved. But we purchase insurance as individuals and defense forces as a society on strategic grounds even though there is great uncertainty about our personal health or how our nation may need to defend itself. Protecting the planet should also be a strategic goal, but of humanity; slowing down our pollution of the atmosphere is international insurance against the uncomfortable risks of future nasty surprises. Despite the uncertainty over details, to delay action is to commit the Earth and its inhabitants to larger amounts of more rapidly occurring environmental change than if we act now. The dilemma rests, metaphorically, in our gazing into a very dirty crystal ball. The tough problem is how long we clean the glass before we act on what we think we see inside.

Regional Greenhouse Climate Effects

J. HANSEN, D. RIND, A. DELGENIO, A. LACIS, S. LEBEDEFF, M. PRATHER, R. RUEDY
NASA Goddard Institute for Space Studies, New York

T. KARL
NOAA National Climatic Data Center, Asheville, North Carolina

ABSTRACT

We discuss the impact of an increasing greenhouse effect on three aspects of regional climate: droughts, storms and temperature. A continuation of current growth rates of greenhouse gases causes an increase in the frequency and severity of droughts in our climate model simulations, with the greatest impacts in broad regions of the subtropics and middle latitudes. But the greenhouse effect enhances both ends of the hydrologic cycle in the model, i.e., there is an increased frequency of extreme wet situations, as well as increased drought. Model results are shown to imply that increased greenhouse warming will lead to more intense thunderstorms, that is, deeper thunderstorms with greater rainfall. Emanuel has shown that the model results also imply that the greenhouse warming leads to more destructive tropical cyclones. We present updated records of observed temperatures and show that the observations and model results, averaged over the globe and over the United States, are generally consistent. Finally, we quantify recent greenhouse climate forcings, showing, for example, that chlorofluorocarbons have grown to be 25 percent of current increases of the greenhouse effect.

The impacts of simulated climate changes on droughts, storms and temperature provide no evidence that there will be regional "winners" if greenhouse gases continue to increase rapidly.

1. Introduction

We were asked to discuss in this paper regional climate impacts due to an increasing greenhouse effect, with emphasis on North America and the Caribbean. The conventional wisdom is that it is not yet possible to obtain reliable conclusions about regional climate impacts, principally for two reasons. First, the representations of atmospheric and surface processes in current climate models are highly simplified, and the models show a very wide range in their predictions for climate change at any particular region. Second, none of the existing climate models simulates the ocean realistically, and changes in ocean currents could alter regional climate.

These are valid concerns, especially if attention is focused on climate change at some specific time and locale. However, an increasing greenhouse effect undoubtedly implies some broad changes in the nature of regional climate, which it may be possible to investigate with existing modeling capabilities. In section 2 we consider changes of a drought index, defined as the difference between atmospheric supply of moisture and atmospheric demand for moisture. In section 3 we consider the impact of greenhouse warming on atmospheric stability as it affects convective storm intensity on scales from thunderstorms to tropical cyclones. In section 4 we discuss possible impacts of rising temperature itself, and we illustrate global and United States temperature trends. In section 5 we illustrate the magnitude of the 1988 North American and Asian heat waves and discuss the possible relation with the greenhouse effect. In section 6 we quantify recent known greenhouse climate forcings. Finally, in section 7, we summarize our conclusions.

2. Drought

We define a drought index, D, which is a normalized measure of the difference between atmospheric supply of moisture and atmospheric demand for moisture, as follows:

$$D(\text{current month}) = 0.9\, D(\text{previous month}) + d/\sigma$$

$$d = (\text{precipitation} - \text{potential evaporation})_{\text{actual}} - (\text{precipitation} - \text{potential evaporation})_{\text{climatology}}$$

$$\sigma = \text{standard deviation of } d$$

Potential evaporation is the evaporation which occurs if water is available. The ratio d/σ is a dimensionless measure of the precipitation deficit (or excess, if positive) in the current month. The drought index, D, includes a memory of precipitation deficit over preceding months. Note that the drought index continues to yield negative values after evaporation ceases due to lack of available water. It thus provides an indication of stress on vegetation. It is also a relevant measure of reservoir water balance, since there is normally water available for evaporation from a reservoir.

The drought index we have defined is similar to, but simpler than, the Palmer Drought Index[1]. The Palmer Drought Index has many locally defined parameters, which would make it impractical for us to obtain global results. However, we have calculated the Palmer Drought Index for the United States and have verified that the results obtained using it show characteristics of future drought intensification similar to those illustrated below. The factor 0.9 in the definition of D, which implies a time scale for recovery (i.e., a memory) of 9–10 months, is the same as in the Palmer Drought Index. We have tested recovery times as short as 1–3 months, verifying that the results discussed below are qualitatively unchanged.

We have computed the drought index D for some of our computer climate simulations published elsewhere[2]. These simulations were carried out with our global climate model (GCM) which has a global sensitivity 4.2°C for doubled CO_2. Ocean heat transports were assumed to remain the same in the next few decades as estimates for the recent past. Other characteristics and qualifications for these climate simulations have been documented.[2,3]

The drought index obtained for our trace gas scenario A is shown in Figure 1 for June–July–August (Northern Hemisphere summer) of four specific years. The temperature anomalies for the same years, relative to the 100 year control run, are shown for comparison in Figure 2. Scenario A[2] has rapid growth of trace gas emissions, for example, 1.5% per year for CO_2 and 3% per year for CFC's; we describe this scenario as "business as usual", because it may be realistic if there are no controls on trace gas emissions. The color scheme in Figure 1 divides the drought index into categories according to the percent of time that a given drought index occurs in the 100 year control run of the climate model; the control run had 1958 atmospheric composition. The drought index is defined locally, that is, relative to the control run climate at each location. Thus, for example, dry conditions in a rainforest only indicate that it is dry relative to the mean for that location in the control run.

The exact patterns of the drought index and temperature are of course not intended to be forecasts for individual years, because the climate patterns fluctuate almost chaotically on a year to year basis. However, it is meaningful to search for overall trends in the results. In the 1990's there is a tendency for more extensive dry conditions than in the control run. By the 2020's there is no mistaking the great intensification of drought at almost all middle latitude and low latitude land areas. There is also an intensification of wet regions, especially at high latitudes and in the intertropical convergence zone, the latter being the region where the low latitude trade winds of the two hemispheres collide.

A summary of the drought intensification with time is shown in Figure 3 for scenario A, averaged over all land areas except Antarctica. In this figure we define the degrees of dryness intensification which occur 1%, 5% and 16% of the time in the control run as extreme drought, drought, and dry, respectively. Drought conditions, which occur 5% of the time in the control run, have increased to 10% in the 1990's. In scenario A drought conditions continue to increase rapidly, to about 25% in the 2020's and about 45% in the 2050's.

What processes in the model lead to such rapid drought intensification? Of course, the principal factor is the higher surface air temperature, which increases the potential evaporation. More detailed analysis is hampered by the fact that droughts occur in different places in different years and are interspersed with wet periods. Therefore we have sorted model diagnostics according to drought index, which allows us to examine how the drought characteristics change, without concern as to where the droughts occur. We find that the regions with more negative drought index ("dry" regions) tend to warm more than the "wet" regions as the greenhouse warming increases. Several characteristic differences between the dry and wet regions in the control run, specifically reduced rainfall, fewer low clouds and less spring soil moisture in the dry regions, all tend to be further enhanced as the greenhouse warming increases. No doubt low antecedent soil moisture is one factor which helps determine the location of droughts, in part through positive feedbacks such as reduced evaporation and cloud cover. But many other factors, such as atmospheric longwave patterns and ocean temperature distributions, can influence the location and timing of droughts. We emphasize that, even as droughts intensify with a growing greenhouse effect, all of the droughts continue to be "natural", in the sense that their location and timing can be related to antecedent land, atmosphere and ocean conditions.

The qualitative picture which emerges is an intensification of both dry and wet extreme conditions as global temperature increases. A similar result is found for rainfall variability by itself,[4] but the effect is stronger for the drought index because of the effect of warmer temperatures. The fundamental mechanism is increased heating of the surface. In dry regions, where little water is available for evaporation, the increased heating goes mainly into increasing the air temperature, which reduces low-level cloud and thus causes further heating. But over the oceans and land regions which happen to be wet, the added greenhouse heating increases evaporation rates, leading to more intense storms, as discussed below, and to increased rainfall and floods.

Although the increasing drought frequency which we obtain may seem extreme, the changes of the drought index would be even larger if we used the climate parameters obtained by the GFDL

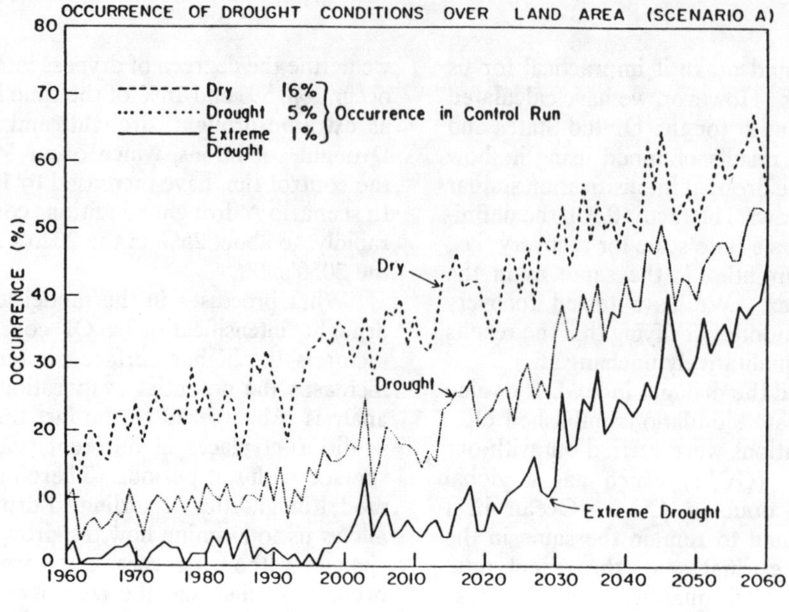

Figure 3. Drought occurrence as a function of time in scenario A. Results are averaged over all gridboxes which are more than 90% land, except that Antarctica is excluded.

(Geophysical Fluid Dynamics Laboratory) model of Manabe and Wetherald[5], because they obtain greater temperature increase and precipitation reduction at middle latitudes than we obtain with our model. The qualitative changes which we obtain in regions of increased drought, e.g., decreased low cloud cover and reduced spring soil moisture, are similar to the results which Manabe and Wetherald obtained for doubled CO_2 in North America and Asia, where their model developed strong drought conditions.

The intensification of both dry and wet extreme conditions is a plausible consequence of the increased surface heating and evaporation. The sense of this result is unlikely to depend upon precise simulation of regional climate patterns or on possible changes in ocean circulation. The magnitude of the effect does depend on regional climate feedbacks, such as decrease of low clouds with increasing drought intensity; this cloud feedback should be analyzed on the basis of global cloud observations. The results may also change somewhat as we improve the realism of the model, for example, by increasing the model's resolution[7,8] and improving the representations of ground hydrology and moist convection, which affect precipitation patterns. But it seems unlikely that such uncertainties will modify the sense of our result, that is, the intensification of both dry and wet extreme conditions.

3. Storms

Storms are generally not resolved by the coarse horizontal resolution of present global climate models. However, models can provide many climate diagnostic parameters which indicate how storm intensity is likely to change with increased greenhouse warming. Some of the specific quantities we have looked at are as follows:

a. Moist static energy

Moist static energy, the sum of sensible heat, latent heat and geopotential energy, is a useful indicator of the likelihood and penetration depth of moist convection. High values of moist static energy near the surface, relative to the air above, and high relative humidity favor deep penetrating convection (e.g., typical thunderstorms). Figure 4 shows the changes in the global mean vertical profile of moist static energy which occur in our transient scenario A and doubled CO_2 simulations. As the greenhouse effect grows, the maximum increase of moist static energy occurs near the surface, because of the higher absolute humidity associated with increased evaporation, and in the upper troposphere, because of the peak in greenhouse warming there. However, the surface moist static energy increase is 2 kJ/kg greater than that at higher altitudes, which represents a 20% enhancement of the lower tropospheric gradient of moist static energy, compared with current climate. Relative humidity changes at low levels are negative but very small in the climate simulations. The implication from these changes is that the warmer climate is prone to deeper, more penetrating convective events. A similar conclusion follows from the results obtained by Wetherald and Manabe[9] in a

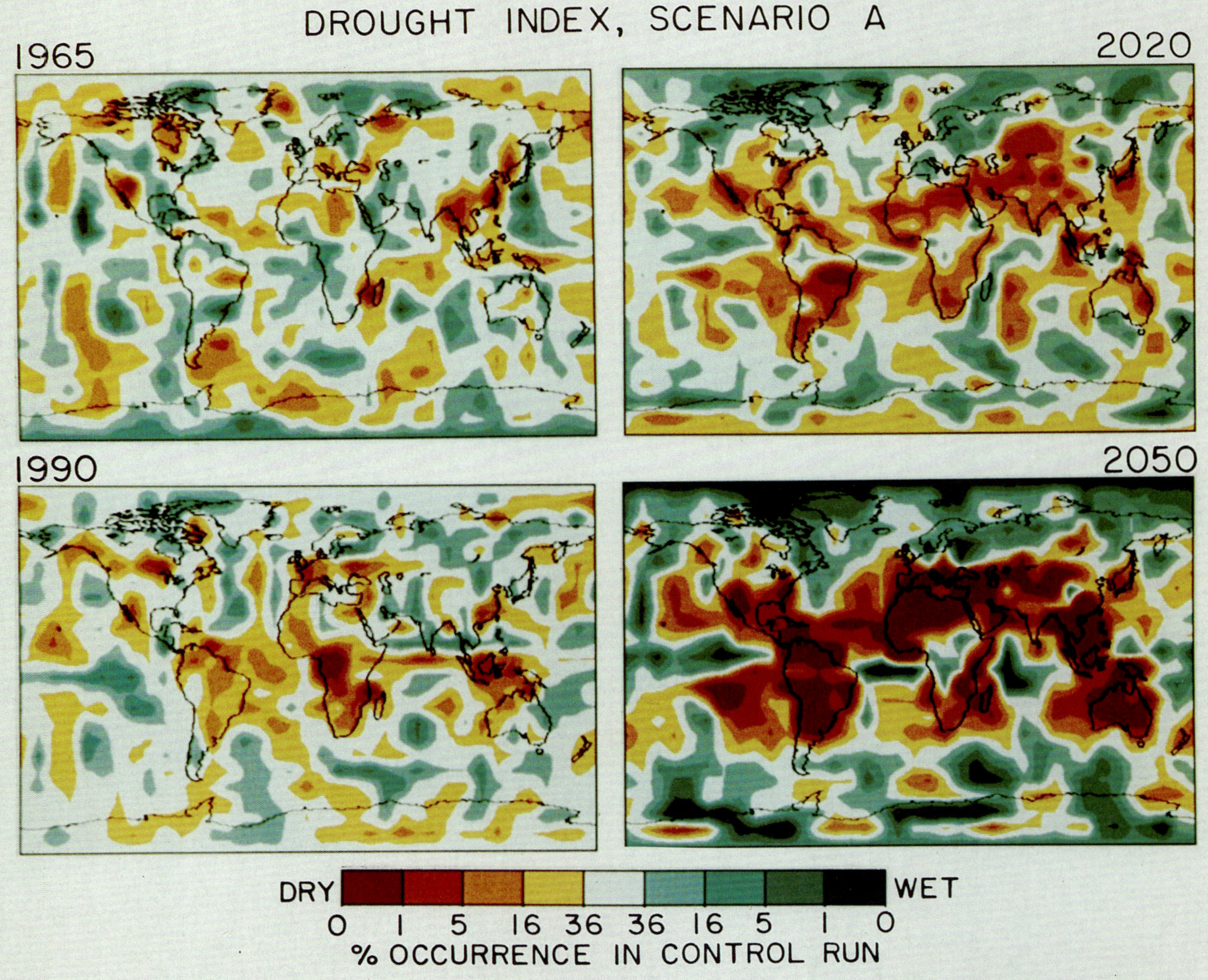

Fig. 1 Drought index for June-July-August in the GISS global climate model simulation for trace gas scenario A. The color scale is set by the frequency of occurrence of a given drought index in the 100 year control run which had 1958 atmospheric composition.

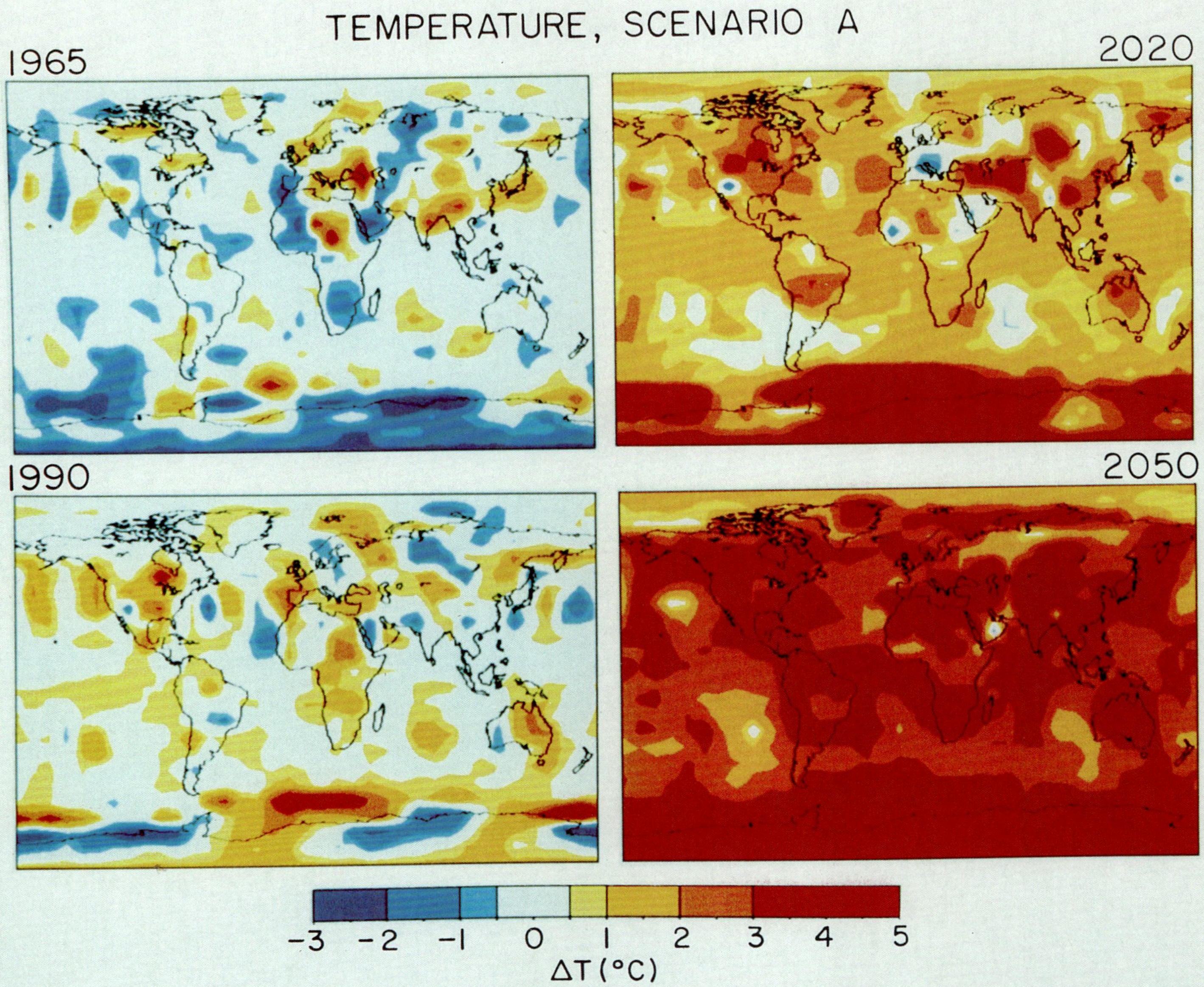

Fig. 2. Temperature change for June-July-August in the GISS global climate model simulation for trace gas scenario A, relative to the 100 year control run which had 1958 atmospheric composition.

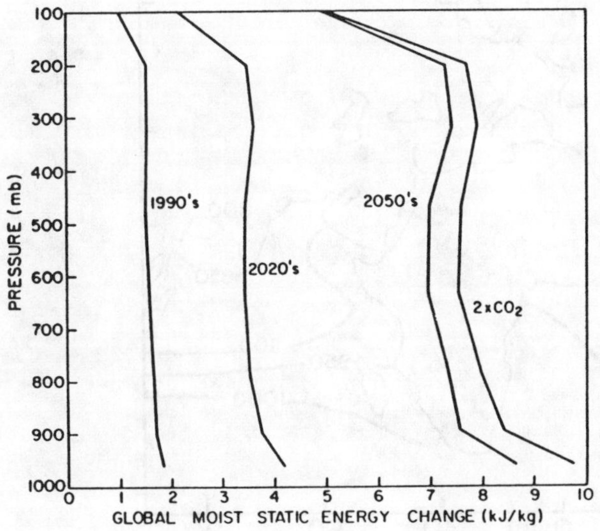

Figure 4. Change of global moist static energy in the doubled CO_2 and transient (scenario A) experiments with the GISS climate model.

GCM with a completely different cumulus parameterization. The tendency of deep convective cloud top heights to increase with increasing sea surface temperature in the tropical Pacific in the current climate[10] is also consistent with this conclusion.

b. Mass flux by moist convection

In our climate simulations the vertical mass exchange due to deep moist convection increases and the mass flux due to shallow convection decreases, consistent with the altered thermodynamic state discussed above. The global average depth of penetration by moist convection increases 20 mb of atmospheric pressure, from $\Delta p = 395$ mb to $\Delta p = 415$ mb, in the doubled CO_2 experiment, and to $\Delta p = 405$ mb by the 2050's in scenario A. The increases are largest near the equator and at middle latitudes. The height of these pressure surfaces increases by several hundred meters as a result of the warming.

c. Precipitation

The increased precipitation in the model, as the climate warms, is almost entirely in the form of moist (penetrating) convection. Changes in large scale (stratiform) rainfall are small, in fact slightly negative, especially at middle latitudes; the latter characteristic is probably a result of reduced synoptic-scale wave activity[11]. Atmospheric heating by moist convection, which is related to precipitation, increases 17% in the doubled CO_2 climate and by 10% by the 2050's in scenario A. These increases are caused by the higher absolute humidity (and latent heat content) of the warmer atmosphere and by the deeper penetration of moist convection, which allows a greater percentage of the latent heat to be released in condensation. Thus thunderstorms are more intense in the model, in the sense that they have higher cloud tops and produce more rainfall for a given mass flux.

Emanuel[6] used a simple Carnot cycle model to estimate the effect of greenhouse warming on the maximum intensity of tropical cyclones, based on the sea surface temperature changes in the doubled CO_2 experiment of the GISS model. Figure 5 shows the resulting minimum sustainable surface pressure which he obtained. With today's climate the minimum sustainable surface pressure is about 880 mb, but this decreases to about 800 mb for the doubled CO_2 climate. The corresponding maximum wind speed increases from about 175 mph to 220 mph. Since the kinetic energy increases with the square of the wind speed, Emanuel estimates that the destructive potential of hurricanes could increase by 40–50% with doubled CO_2.

These quantitative results were obtained for the ocean temperature warmings in the GISS model, which are as large as 4°C. Some other GCMs yield ocean warmings of only 2°C for doubled CO_2, which would imply wind speed increases half as large as those indicated here. Also, note that the maximum potential velocities are obtained in only a small percentage of hurricanes, and that the existing analyses do not permit prediction of the change in storm frequency. Nevertheless, it is obvious that the impact of greenhouse warming on tropical storms would have important implications for the Caribbean, Mexico and parts of the United States coast. And in addition to increased hurricane strength, it seems likely that higher ocean temperatures will lead to an expansion of the region which hurricanes frequent. For example, if the greenhouse warming continues unabated, hurricanes could become common along the entire east coast of the United States in the next century.

The picture that emerges from these diagnostics is an increased intensity of storms which are driven by latent heat of vaporization, both ordinary thunderstorms and mesoscale tropical storms. The basis for this is the increased evaporation and higher temperatures at low levels in the atmosphere, which yield more moist static energy at low levels and greater vertical penetration of moist convection. The magnitudes and detailed spatial and temporal patterns of changes in storms are sensitive to uncertainties in the parameterization of moist convection and other feedback processes in the model. For example, we cannot determine changes in the updraft speed or frequency of storms with the current version of the

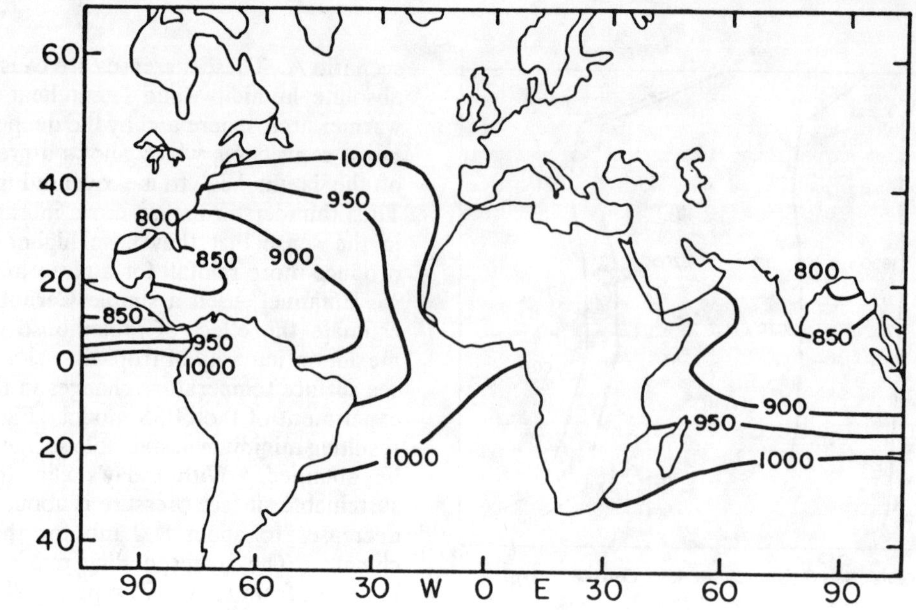

Figure 5. Minimum sustainable surface pressures in August as estimated by Emanuel[6] using sea surface temperatures from the doubled CO_2 experiment with the GISS model.

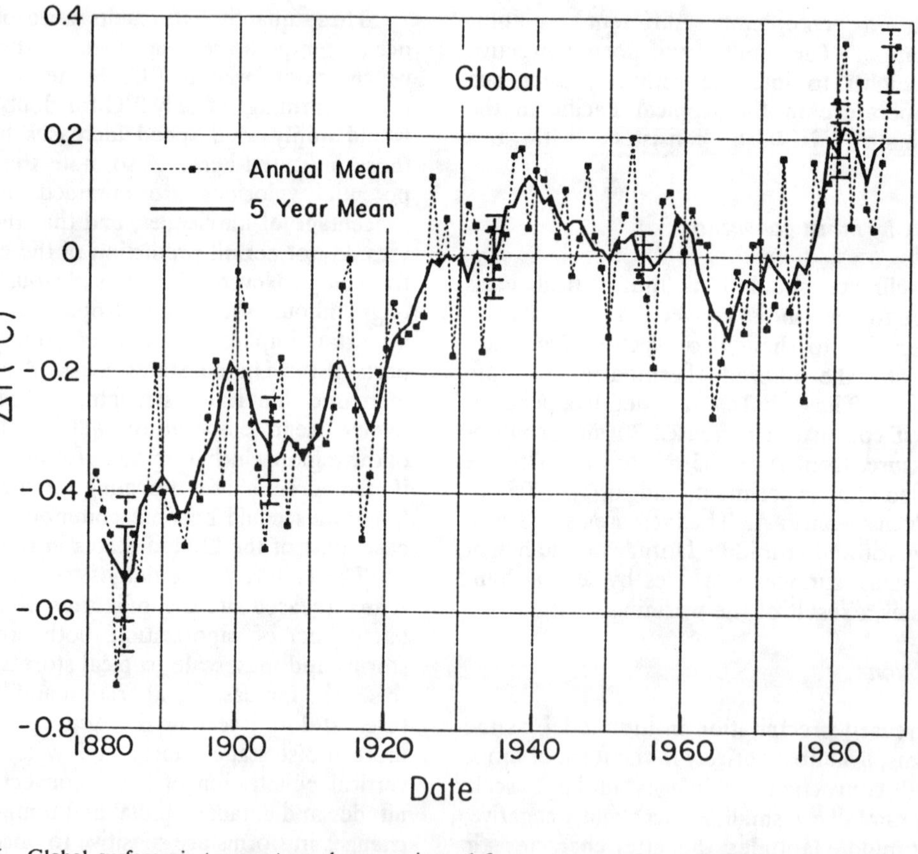

Figure 6. Global surface air temperature change estimated from meteorological station data. Uncertainty bars[13] account only for the incomplete spatial coverage of the stations. The error bar for 1988 is larger than that indicated for 1987, because of poorer station coverage and approximations in the near real time data used for the last several months of the year. Note also that no correction has been made in this figure for urban warming, which is estimated as 0.2°C for the century.

model, nor can we predict the nature of changes in special categories of storms which depend on local wind shear and mingling of different air masses (e.g., squall lines and tornadoes). However, the general nature of those changes we have described is determined largely by the Clausius–Clapeyron equation and is thus a straightforward consequence of fundamental moist thermodynamics.

4. Temperature

We examined elsewhere[2,12] temperature changes forecast by our global climate model for increasing greenhouse gases. In those papers we stressed the importance of a possible increase in the frequency of temperatures above some critical level. For example, we computed the average number of days per year that the simulated temperature exceeds certain limits in specific United States cities. Such quantities have an extremely large year to year variability, so they certainly will not increase smoothly as the world becomes warmer. However, it is meaningful to estimate how the probability of such extreme temperatures may change. For example, our climate model suggests that the probability of a hot summer in most of the United States may increase to 60–70% by the middle 1990's, as compared to 33% in the period 1950–1979 (Figure 6 of reference 2).

The simulated warming in our model is somewhat larger in Canada and smaller in Mexico and the Caribbean, as compared to the United States. But natural climate variability (fluctuations from year to year) also increases with increasing latitude, as illustrated elsewhere[13]. As a result, to a first approximation, the probability of a warm season relative to the local climatology is predicted to increase at a similar rate in these different regions[2].

The impact on the biosphere of increasing temperature will be dramatic at all latitudes, if the results computed for scenario A (rapid growth of trace gases[2]) are realistic. The poleward shift of isotherms by 50 to 75 km per decade in that scenario is faster than most plants and trees are thought to be capable of naturally migrating[14], and thus the warming could cause a decline of many species in North American forests. The productivity of crops which are sensitive to a run of consecutive hot days could suffer also, as indicated by calculations published elsewhere[2,15]. One impact in the Caribbean could be on coral reefs, since many coral populations are unable to survive if water temperatures rise above 30°C[16]. The expected increase in storm intensities, discussed in the section above, would be particularly important in the Caribbean.

Observations of current global temperature change are of special interest, because of the search for a long term warming trend attributable to the greenhouse effect. A preliminary update of the global temperature analysis of Hansen and Lebedeff[13], which uses MCDW (Monthly Climatic Data of the World) data available from NCAR, is shown in Figure 6. The last six months of 1988 are based on NOAA near real time data, adjusted for reporting biases as described elsewhere[17]. The use of these adjusted near real time data for six months affects the global temperature for the full year by at most a few hundredths of a degree. Note that Figure 6 has not been corrected for "urban" effects. As discussed by Hansen and Lebedeff[13] and below, approximately 0.2°C of the global warming in the past century in the MCDW data is estimated to result from urban growth effects.

Figure 6 indicates that 1988, within the error bar of measurement, was the warmest year in the history of instrumental records. Jones *et al.* (private communication) have recently reported that their analysis shows 1988 as the warmest year on record. The annual warmth for the globe in 1988 occurred despite rapid cooling at low latitudes between May and December, which was associated with an unusually strong negative phase of the El Niño cycle[18]. It will be particularly interesting to see whether this cooling of tropical surface air propagates to high latitudes and dominates global temperature trends over coming years. Some scientists have expressed the expectation that this negative El Niño will slow down the greenhouse warming by 30 to 35 years. Global temperature change does have some correlation with El Niños (Figure 3a of reference 17), especially the past two El Niño events, but the correspondence is far from overwhelming. It is a classical confrontation, like the tortoise vs. the hare: how long will it take a "small" global climate forcing to overcome the effects of a large negative El Niño fluctuation? If some recent climate simulations[2] are realistic, it will be at most only a few years before the global temperature records are raised further.

The global temperature record in Figure 6 is the average for all MCDW stations, urban and rural. We illustrated elsewhere[13] that if all stations associated with urban areas of population 100,000 or greater (about one third of the MCDW stations) are eliminated from this record, the global warming is reduced by 0.1°C. Based on studies of how the urban warming varies with population, we estimated that there may be an additional urban effect of about 0.1°C due to smaller cities. Thus we concluded that the warming over the period from the 1880's to the 1980's is reduced from 0.7°C to approximately 0.5°C, if urban effects are removed. Jones *et al.*[19] used an independent labor-intensive procedure to analyze urban effects, comparing nearby stations on a one-by-one basis to try to identify and partially correct for urban bias. After estimation of the remaining urban bias in their data, their result is consistent with ours in indicating a net global warming of approximately 0.5°C in the past century. The 0.5°C global warming is what we and others have used for empirical studies of the greenhouse effect and climate sensitivity[20].

The temperature trend in the United States can be examined in detail, based on comprehensive studies by Karl *et al.*[21,22]. They employ the recently completed Historical Climatology Network (HCN) data for 1219 stations, data which were meticulously scrutinized for biases resulting from such factors as station moves, time of observation changes and instrumental changes. By comparing urban temperature records with those of nearby rural stations, Karl *et al.*[21] obtained empirical relations between population and urban warming. The dependence of urban warming on population which they found is generally consistent with earlier studies, and thus does not modify our estimate of 0.5°C global warming in the past century.

Most of the stations in the Historical Climatology Network are located in sparsely populated regions, over 70% in areas with 1980 population below 10,000. Thus the urban warming which Karl *et al.* found in the raw HCN data was small, amounting to 0.06°C in this century. Here we use the urban-adjusted HCN record, that is, after removal of this urban warming, (1) to estimate the urban warming in the MCDW data for the United States, (2) to examine the United States temperature record for evidence of a trend, and (3) to compare with the results of our global climate model simulations. The HCN record has been updated to include 1985, 1986 and 1987.

Figure 7a compares the urban-adjusted HCN data for the contiguous United States with uncorrected data of Hansen and Lebedeff[13] based on MDCW stations. This comparison suggests that there is an urban warming bias of about 0.13–0.14°C/century in the MCDW data for the United States under the assumption that urban effects are the cause of the difference in the temperature trends. [The linear trend of the urban-adjusted HCN data of Karl *et al.* for the interval 1901–1987 is 0.26°C/century; the Hansen and Lebedeff[13] published data for the contiguous United States have a trend 0.39°C/century; if the Hansen and Lebedeff analysis is repeated using only stations within United States borders, the trend is 0.40°C/century.] Karl and Jones[23] estimated that the urban warming in the Hansen and Lebedeff data for the contiguous United States was close to 0.4°C/century; however, the assumed Hansen and Lebedeff temperature trend (provided by Hansen and Lebedeff) was incorrect, an error having been made in the integration over the contiguous United States. The correct comparison is that shown in Figure 7a. An urban warming of 0.4°C/century also would be inconsistent with the facts that: (1) the Hansen and Lebedeff[13] and Jones *et al.*[24] temperature trends are generally in close agreement (see, e.g., Figure 15 of Hansen and Lebedeff[13]), and (2) both Karl and Jones[23] and Jones *et al.*[24] estimate the urban warming in the Jones *et al.*[24] results to be about 0.1°C or less. These results indicate that urban warming for the United States is not larger than our estimate of 0.2°C for the full globe, despite the fact that per capita energy use in the United States is high and United States cities have experienced large vertical growth, relative to cities in the remainder of the world. Oke[25] and others have shown that, except perhaps for population and energy use, the most important factor in urban warming is the reduction of the skyview factor by vertical growth of the cities. Although indications are that urban effects do not qualitatively modify estimated global temperature trends, it is clearly important to carry out much more comprehensive analyses of the urban effects, as needed to provide optimum estimates of unbiased temperature change.

Figure 7b shows the HCN data for the contiguous United States (48 states) for the period 1901–1987. The linear trend of HCN data is 0.26°C/century. [The change from the trend reported by Karl and Jones[23] (0.16°C/84 years or 0.19°C/century) is due to the addition of data for 1985–87.] If ΔT were a linear function of time and the deviations from the straight line were normally distributed, it could be stated with 90% confidence that the slope of the temperature trend is positive, i.e., that there is a warming trend. But this confidence interval should not be taken literally because of inherent errors in these assumptions.

Figure 7c shows the Climatic Division (CD) data for the contiguous United States for 1901–1987, recently reported by Hanson *et al.*[26] The CD data are from 6,000 stations, including second order and cooperating stations. The small cooling bias on the CD data relative to the HCN data, about 0.1°C, may be a result of incomplete correction for time of observation bias in the CD data, which is known to introduce spurious cooling[21]. In any case, the HCN data, which have undergone elaborate station-by-station scrutiny, are the most reliable record of temperatures in the United States.

Figure 7d shows temperatures for the entire United States (50 states) for the period 1901–1987. This is the area-weighted average of the HCN data for 48 states and MCDW data for Alaska and Hawaii. The Alaska and Hawaii data are based on MCDW records for only stations associated with population centers of less than 5,000. The linear trend for United States temperatures is 0.33°C/century. The mathematical confidence (as defined above) that the slope

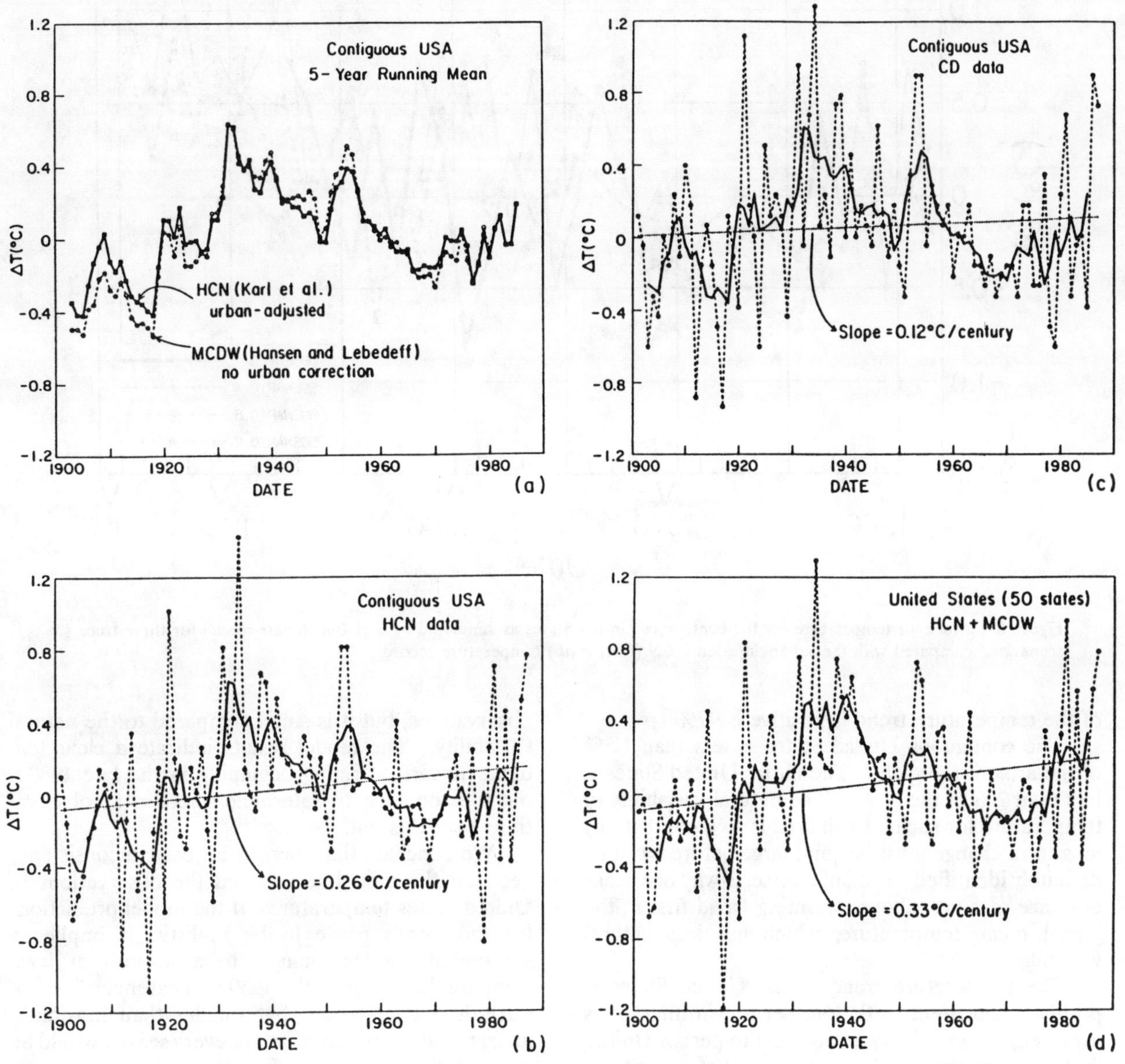

Figure 7. (a) Comparison of urban-adjusted Historical Climatology Network (HCN) temperatures for the contiguous United States and uncorrected Monthly Climatic Data of the World (MCDW) temperatures for the same region; both curves are 5-year running means. (b) Annual urban-adjusted HCN temperatures (Karl *et al.*) for the contiguous United States with linear trend. Data of Karl *et al.*[22] are updated with results for 1985, 1986 and 1987. (c) Annual Climatic Division (CD) temperatures (Hanson *et al.*[26]) for the contiguous United States with linear trend. (d) Temperature for the full United States (50 states). Contiguous United States data are from urban-adjusted HCN record; Alaska and Hawaii are based on MCDW stations associated with population centers of less than 5,000.

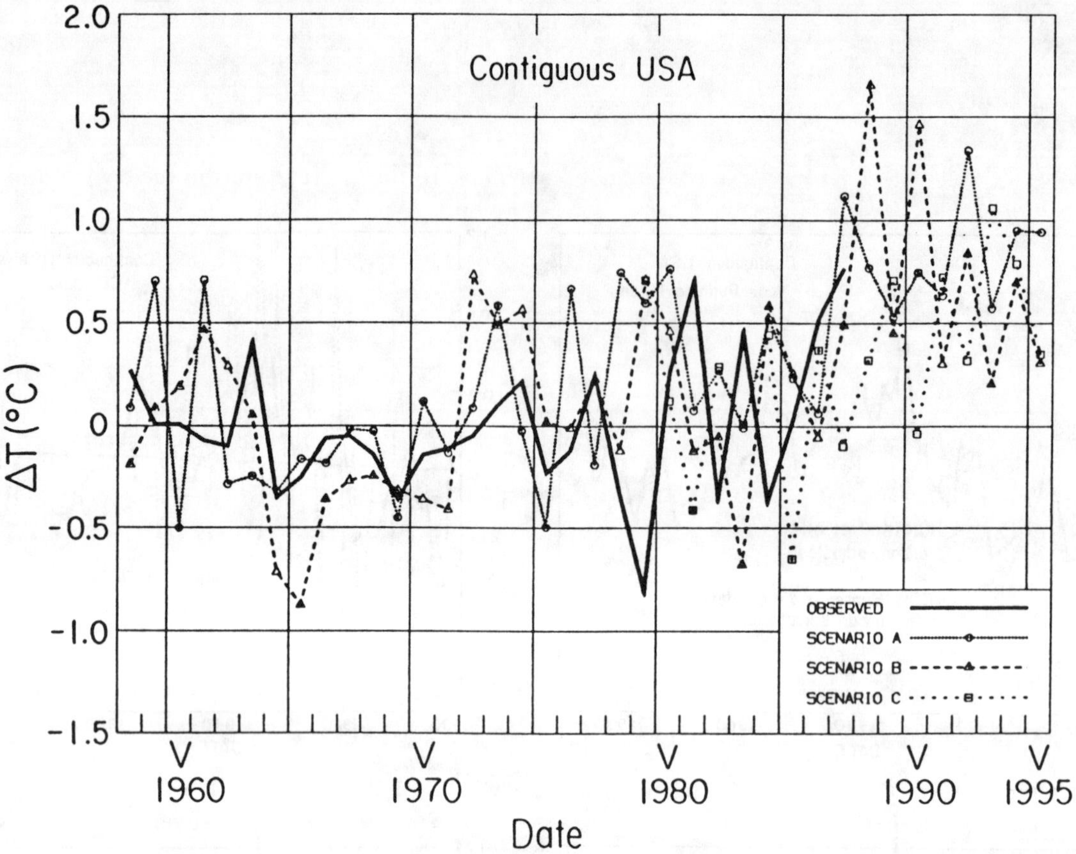

Figure 8. Surface air temperature for the contiguous United States as simulated by a global climate model for three trace gas scenarios,[2] compared with the urban-adjusted HCN (Karl *et al.*) temperature record.

of the temperature trend is positive is 97%.

The contiguous United States is less than 1.5% of the area of the globe. The entire United States is less than 2% of the globe. The natural variability of temperature for such a small area is so great that any long term change must be quite large before it can be definitely identified. That, of course, is why one seeks evidence of a greenhouse warming trend first in the global mean temperature, which has less natural variability.

The temperature trend in the United States is positive, consistent with long-term warming. As expected, the variability is too great to permit United States temperatures to be used as proof of a long term climate change. But, contrary to recent reports in the popular press, the United States temperatures do not provide a basis for questioning the reality of the global warming trend.

Finally, the HCN data is compared in Figure 8 with the surface air temperatures obtained for the United States in the three transient climate simulations with the GISS model[2]. There is warming at the end of the 30 year period in both the model and observations, but it is small compared to the natural variability. The model results indicate a clear tendency toward warming beginning in the late 1980's, but it is too early to determine whether the observations bear this out.

We conclude that there is no basic inconsistency between the model results and the observations of United States temperatures. If the model predictions for the 1990's prove to be realistic, it implies a substantial climate change, to a warmth at least comparable to that of the 1930's. That mean level of warmth remains somewhat smaller than maximum interannual fluctuations, so not every season would be warmer than normal. But it would represent a sufficient "loading" of the climate "dice" to be clearly noticeable.

5. Summer of '88

The summer of 1988 was warm and dry in much of the United States and Asia. The global context of recent temperature anomalies is given by Figures 10 and 11, which, respectively, show the temperature anomalies for the past four years and the four seasons of 1988. In the summer (June–July–August) of 1988 the mean temperature anomaly for the contiguous United States was about +1°C, but it was about +3°C in the hottest region near the United States–Canada border.

Is it possible that greenhouse warming played a significant role in the heat waves and droughts of 1988? Trenberth *et al.* state[27] "Any greenhouse gas effects may have slightly exacerbated these overall conditions during the 1988 drought, but they almost certainly were not a fundamental cause." Manabe, although he has been the principal proponent of the likelihood of increased drought with doubled carbon dioxide, stated[28] that, in view of the fact that global warming of 0.5°C in the past century was so small compared to the 4°C warming in his doubled carbon dioxide experiment, the greenhouse effect had "only a minor role, at most, compared to natural variability".

That rationale is reasonable. The natural variability of precipitation is even greater than the variability of temperature. On the other hand, the results illustrated in Figure 3 suggest that the greenhouse mechanism may be beginning to compete effectively with natural variability at about the present time, i.e., there begins to be a noticeable increase in the frequency of drought in the model. And the increasing greenhouse effect certainly does not have to cause changes which exceed the range of natural variability in order to have important consequences.

It should also be realized that some of the excursions of "natural variability" may be associated with global or large scale climate forcings. The great droughts of the 1930's came at a time of global warmth. That degree of global warmth could have arisen from purely internal fluctuations of the climate system. Our GCM produces similar long term variations in the absence of greenhouse forcing increases.[2] On the other hand, it is also possible that the warmth of the 1930's was related to global forcings, such as the net effect of the near absence of volcanic eruptions in the period 1910–1940, the growth of greenhouse gases in that period, and perhaps other factors such as change of solar irradiance. In any event, the Northern Hemisphere was very warm in the 1930's, and the present Northern Hemisphere temperature seems to be approaching a similar degree of warmth. This is another reason to be concerned about a possible relationship between large scale warming and drought.

We have examined the summer climate changes in our transient scenario A for the period 1986–1995, in comparison with the 100 year control run. For all gridboxes which are more than 90% land between the equator and 55°N (the region where the drought index increases most, see Figure 1) by the period 1986–1995 the mean summer warming is 0.7°C. For the same region and decade the temperature in the driest regions (the 10% of the gridboxes with most negative drought index) has increased about 1°C relative to the driest regions in the control run, to a level 3°C warmer than the control run climatology. Thus the average calculated warming is similar in magnitude to the mean warming in the contiguous United States in the summer of 1988, and the calculated warming in dry regions is comparable to the observed warming in the 1988 drought region.

Although these model results provide no information on the pattern of drought in 1988, they suggest the possibility that the increasing greenhouse effect could already play a significant role in summer heat. We emphasize that the timing and geographical distribution of any specific drought is determined primarily by short-term meteorological fluctuations and antecedent land, ocean and atmospheric conditions. The occurrence or intensity of a single drought can not be used to identify the role of the greenhouse effect in the drought. At the same time, determination of meteorological factors involved in any drought can not be used to disprove the role of the greenhouse effect; such a meteorological analysis provides no information as to whether the greenhouse effect is increasing the frequency and severity of droughts.

We are analyzing the results of our transient climate simulations in greater detail, but there are severe limitations on the information which can be extracted from the model with its present resolution and representation of physical processes. Thus our main effort is now in constructing the next version of the climate model, which will include higher resolution and improved representations of ground hydrology, moist convection and other processes. This should make it possible to do a more thorough analysis of the greenhouse role in summer heat waves and drought.

6. Greenhouse climate forcings

It is possible to make a reasonably accurate comparison of the global radiative forcings due to most of the greenhouse gases that man is presently adding to the earth's atmosphere. Table 1 lists trends of known greenhouse gases during the 1980's and the calculated climate forcing, ΔT_o, due to each gas. ΔT_o is the surface temperature change at equilibrium ($t \rightarrow \infty$) with no climate feedbacks included, computed with a one-dimensional radiative-convective climate model. Uncertainties of decadal trace gas abundance changes are typically less than 10% for the major species. Uncertainties of individual infrared absorption coefficients (Table 1) are larger than that, but averaged over all the greenhouse gases the uncertainty

Table 1. Global mean radiative forcing of the climate system (ΔT_o) due to estimated changes of several trace gases during the 1980's. CFC growth rates are based on UNEP 1989 compilation (a) and on Ramanathan *et al.*[30] estimates (b). Absorption coefficients used to compute ΔT_o were obtained from published laboratory measurements (P), unpublished laboratory measurements (U), and estimates scaled from other published data (E).

	ΔT_o (°C)	Growth Rate (%/year)	Estimated Abundance 1980	Estimated Abundance 1990	Remarks
CCl_2F_2 (F12)	.0143	4.5	297	468	a, P
CCl_3F (F11)	.0067	4.6	173	275	a, P
CCl_2FCClF_2 (F113)	.0042	12.1	15	52	a, P
$CHClF_2$ (F22)	.0032	7.1	58	110	a, U
CF_4 (F14)	.0009	2.5	70	90	b, P
$CClF_2CClF_2$ (F114)	.0004	6.7	4	7	a, P
CCl_4	.0004	1.0	95	105	a, P
$CClF_3$ (F13)	.0004	4.3	7	11	b, E
CF_3CF_2Cl (F115)	.0003	8.5	2	5	a, E
CH_3CCl_3	.0003	4.7	140	225	a, E
CCH_2CCl_2	.0002	3.8	30	44	b, P
CH_2ClCH_2Cl	.0002	2.4	30	38	b, E
C_2F_6 (F116)	.0001	3.2	4	5.5	b, E
$CHCl_3$	.0001	2.2	10	12.5	b, P

in the net climate forcing due to inaccuracies of the absorption coefficient data is probably of the order of 10%.

As summarized in Figure 9, CO_2 continues to be the dominant greenhouse gas in the 1980's, accounting for more than half of the net greenhouse effect due to those gases for which measurements are available. However, CFCs now account for one quarter of the current growth of the greenhouse effect. Until constraints were placed on CFC production in the early and middle 1970's, the growth rates of emissions of the principal CFC's were more than 10 percent per year. If CFC growth had continued unchecked, by this time the greenhouse effect of increasing CFCs would probably have exceeded that of CO_2.

There is one potentially important greenhouse gas, ozone, for which present measurements are inadequate to allow reliable greenhouse computations. There is evidence for stratospheric ozone loss and tropospheric ozone increases; both of these changes would lead to surface heating[29]. But ozone molecules in the upper troposphere are the most effective in influencing surface temperature. Very few measurements are available for that altitude range, and those measurements which do exist suggest an ozone decrease and thus a surface cooling,[29] especially at high latitudes. However, the net greenhouse effect of present changes in ozone is impossible to judge reliably without better observations.

7. Summary

Our climate simulations indicate that an increasing greenhouse effect causes an intensification of the extremes of the hydrologic cycle: (1) greater frequency (or areal coverage) and intensity of drought, and (2) more intense wet and stormy conditions. These general conclusions are not per se dependent on the accuracy of the climate model for specific regions, because the analysis avoids the need to predict exactly where the changes occur.

We find no evidence of regional climate "winners" with an increasing greenhouse effect. Droughts increase in the model at essentially all low latitude and middle latitude land areas, where almost all the world's population is located. Although annual precipitation increases at most locations in response to the global warming, the added rainfall occurs in intense (moist convective) events, not as gentle large scale rainfall, implying the likelihood of an increased frequency of flooding. Furthermore, our model results imply that an increased greenhouse effect will lead to

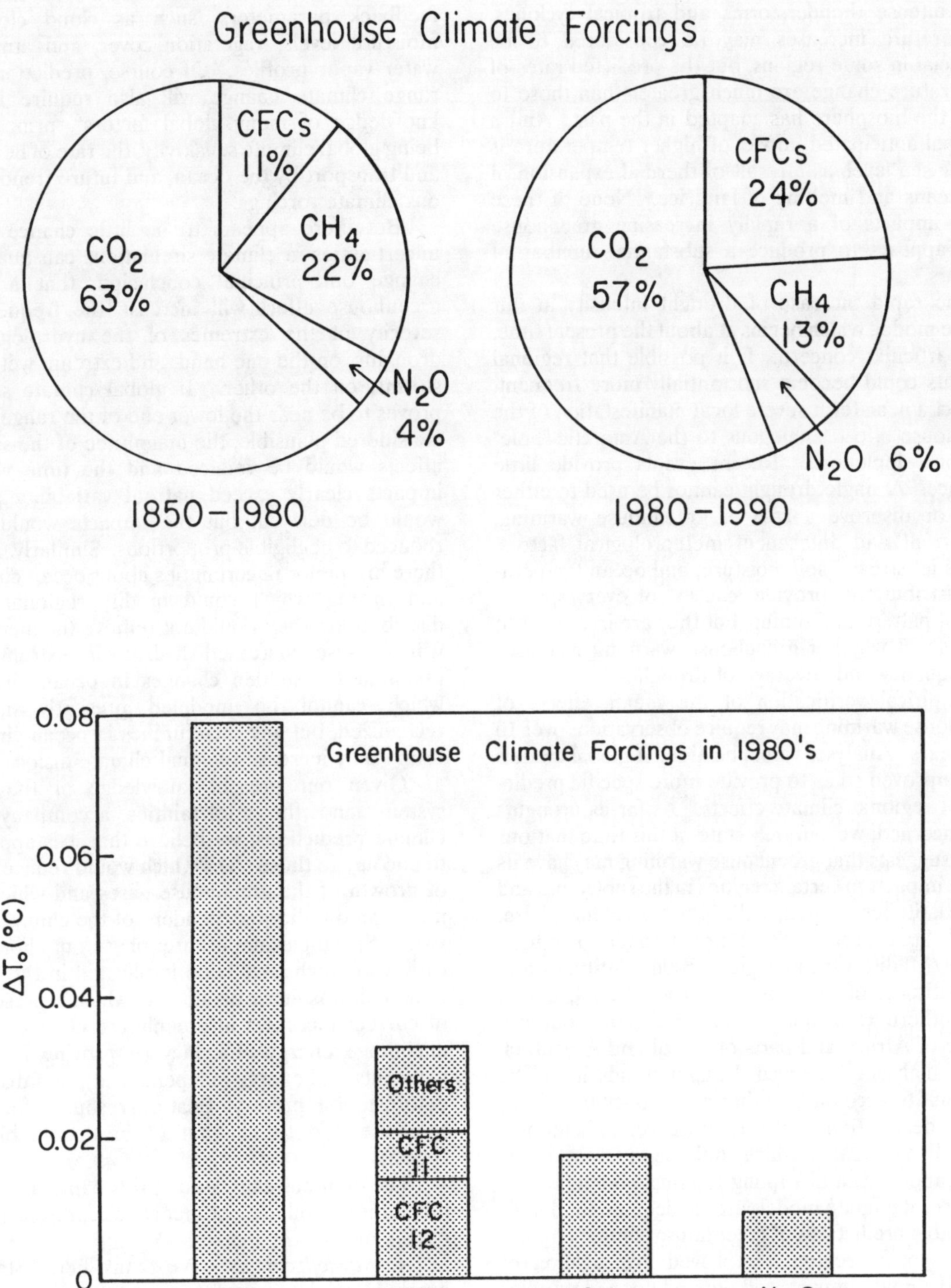

Figure 9. Global mean radiative forcing of the climate system due to estimated changes of trace gases (see Table 1). (a) Relative contributions of greenhouse gases to the total greenhouse climate forcing for two time periods. It is assumed that the 1850 abundances were: CO_2 (285 ppm), CH_4 (0.8 ppm), N_2O (0.285 ppm), CFCs(0). (b) Radiative forcings in the 1980's. The climate system response to this forcing involves many feedback processes, some of which are poorly understood; current global climate models suggest that the global temperature response at equilibrium is about 2–3 times larger than the global radiative forcing.

more intense thunderstorms and tropical cyclones. Temperature increases may be considered to be beneficial in some regions, but the predicted rates of temperature change are much greater than those to which the biosphere has adapted in the past. And a principal anticipated impact of higher temperature is a rising sea level, as a result of thermal expansion of the oceans and melting of land ice. None of these major impacts of a rapidly increasing greenhouse effect appears to produce a substantial number of "winners."

The rapid increase of drought intensity in our climate model, which begins at about the present time, is of particular concern. Is it possible that regional droughts could become substantially more frequent, in effect a near-term severe local manifestation of the greenhouse effect, analogous to the Antarctic "hole" of ozone depletion? Recent events provide little guidance. A single drought cannot be used to either prove or disprove a role of greenhouse warming. Concurrent and antecedent meteorological factors, such as jet stream, soil moisture, and ocean temperature distributions, provide "causes" of every specific drought pattern and timing, but they are irrelevant to the issue of whether greenhouse warming increases the frequency and coverage of drought.

Empirical verification of the major effects of greenhouse warming may require observations over 10 or 20 years. Analyses would be aided if global models were improved so as to provide more specific predictions of regional climate effects. As far as droughts are concerned, we can only state at this time that our model suggests that greenhouse warming may have its biggest impacts in certain regions in the subtropics and middle latitudes, such as, in the Northern Hemisphere: United States/Mexico/southern Canada, southern Europe/Mediterranean region, middle latitudes and lower latitudes of Asia, and the African Sahel; and, in the Southern Hemisphere: Australia, the southern quarter of Africa, and parts of Brazil and Argentina. But in analyzing observed drought trends it will be important to account for other anthropogenic effects, such as destruction of vegetation cover in semi-arid regions like the Sahel, which could be as important or more important in disrupting regional climate.

Current climate models are inadequate for detailed, reliable predictions of greenhouse climate impacts in any specific region. Improved investigation of regional climate impacts will depend upon (1) studies based on models with higher spatial resolution, and, especially, more realistic representation of key aspects of the "physics," such as moist convection, clouds, ground hydrology, and vegetation effects, and (2) global observations which permit analysis of key feedback parameters, such as cloud clover, soil moisture levels, vegetation cover, and atmospheric water vapor profiles. Of course, prediction of long range climate change will also require improved knowledge of many global factors[2], principal ones being global climate sensitivity, the rate of heat uptake and transport by the ocean, and future trends of various climate forcings.

But there appears to be little chance that the uncertainties in climate simulations can qualitatively change our principal conclusion, that a growing greenhouse effect will increase the frequency and severity of the extremes of the hydrologic cycle: droughts, on the one hand, and extreme wetness and storms, on the other. If global climate sensitivity proves to be near the lower end of the range which is considered plausible, the magnitude of the simulated effects would be reduced and the time when the impacts clearly exceed natural variability probably would be delayed, but the impacts would not be reduced to negligible proportions. Similarly, although there are major uncertainties about ocean circulation and mixing which could modify regional climate distributions, they would not remove the mechanisms which cause increased hydrologic extremes. The potential for sudden changes in ocean circulation, which cannot be modeled presently, must be recognized, but any such lurches in ocean circulation would only increase regional climate dislocations.

Given our present knowledge of the climate system, and the uncertainties accompanying any climate predictions, we believe that it is appropriate to encourage those steps which would reduce the rate of growth of the greenhouse gases and which would make good policy independent of the climate change issue. Specific examples are: phase out chorofluorocarbons (which have been implicated in the destruction of stratospheric ozone, and which represent 25% of current increases in greenhouse climate forcing), encourage energy efficiency (improving balance of payments and energy independence), and discourage deforestation (preserving natural resources for sustainable use and the habitat of invaluable biological species).

The opinions expressed in this paper are those of the authors, and are not meant to represent policy of NASA or NOAA.

Acknowledgements. We thank Peter Stone and Mark Handel for useful comments on this paper, Patrice Palmer for helping produce the color figures, Jose Mendoza and Lilly DelValle for drafting other figures, and Elizabeth Devine for desktop typesetting. This work has been supported by the NASA Climate Program and EPA grant R812962–01–0.

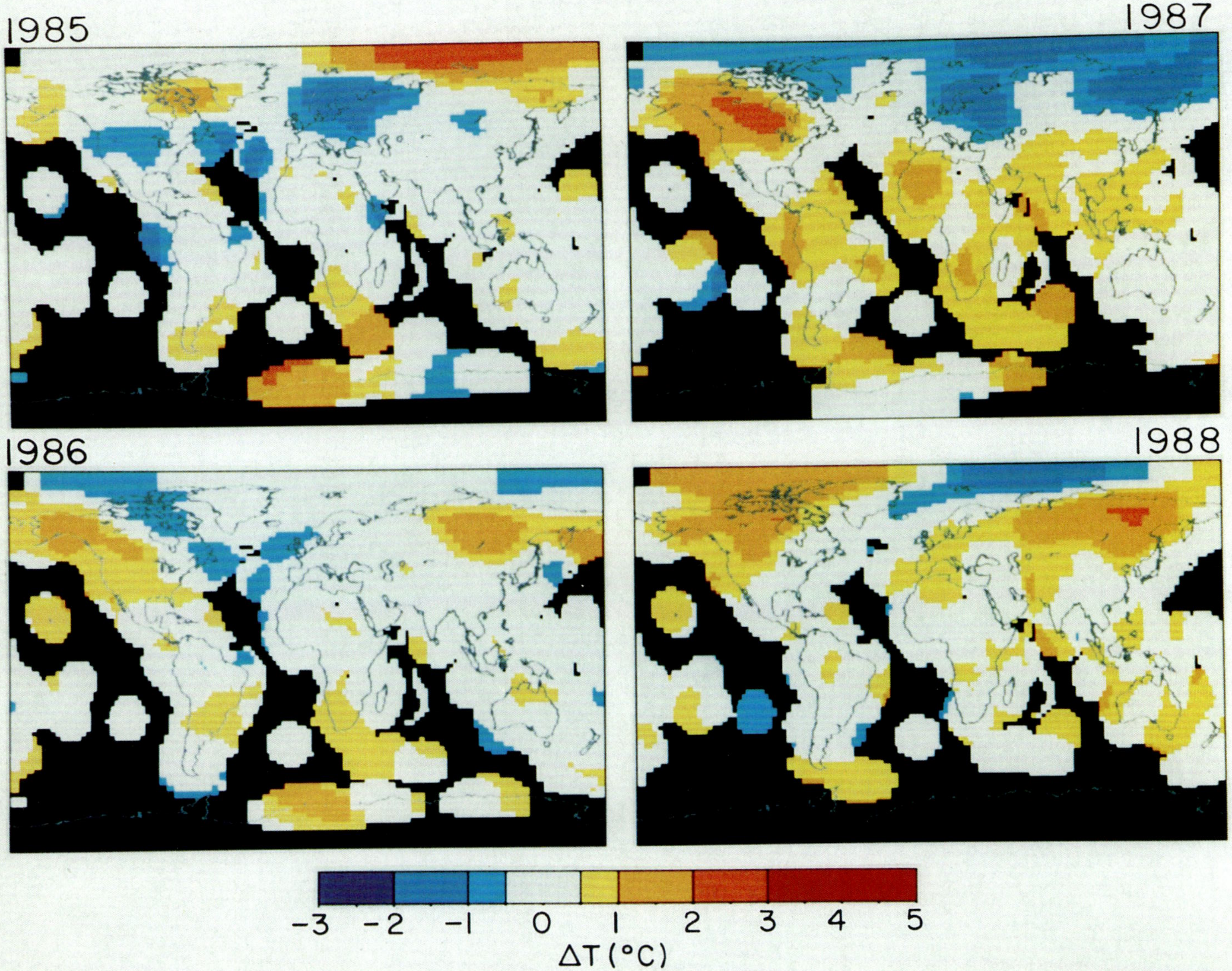

Fig. 10. Annual mean surface air temperature in 1985, 1986, 1987 and 1988. The zero point for each location is the 1951-1980 mean.

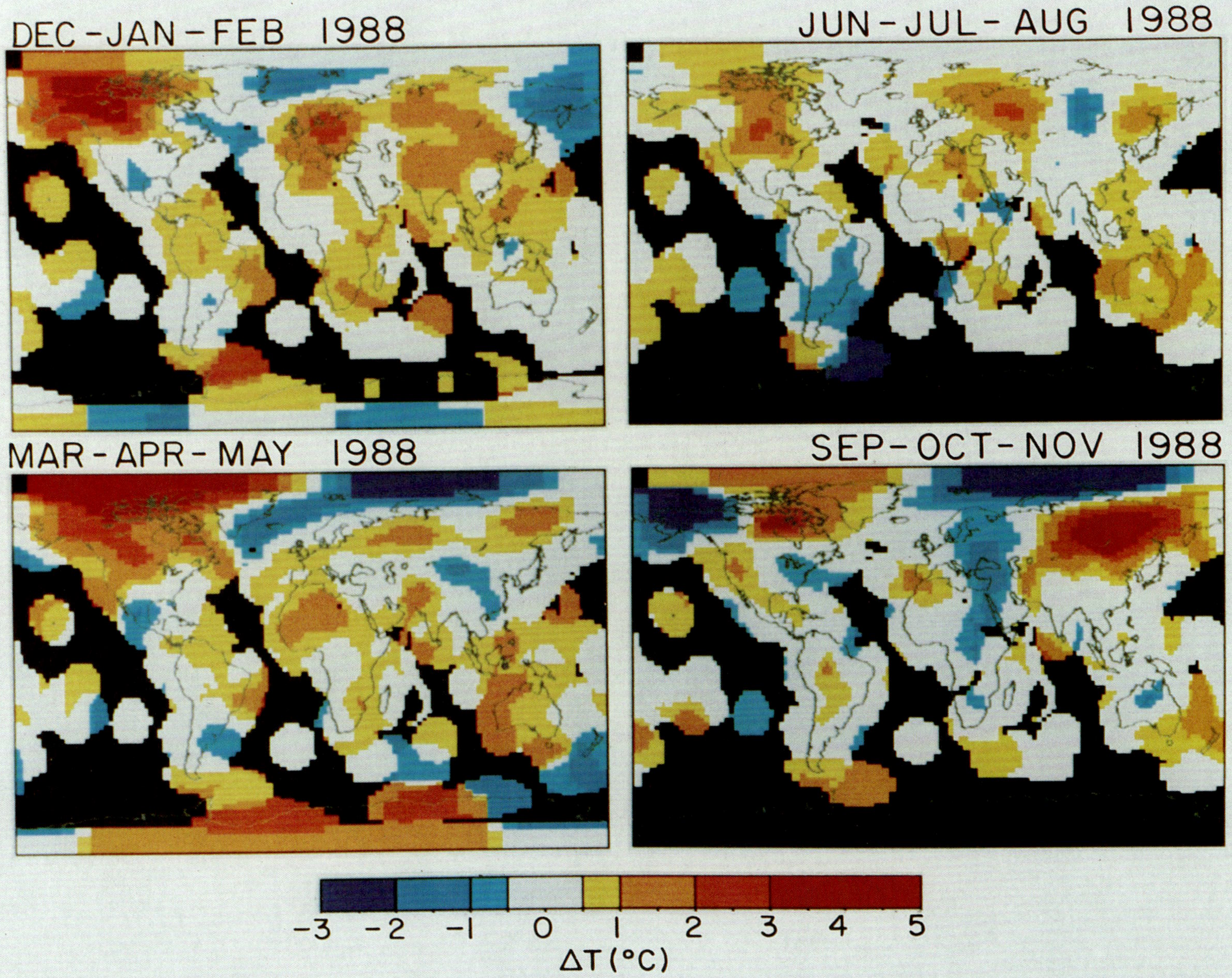

Fig. 11. Seasonal mean surface air temperature for the four seasons from December 1987 through November 1988. The zero point for each period and location is the 1951-1980 mean.

REFERENCES

1. Alley, W. M., *J. Clim. Appl. Meteorol.*, **23**, 1100, 1984.
2. Hansen, J., I. Fung, A. Lacis, D. Rind, S. Lebedeff, R. Ruedy, G. Russell, and P. Stone, *J. Geophys. Res.*, **93**, 9341, 1988.
3. Hansen, J., G. Russell, D. Rind, P. Stone, A. Lacis, S. Lebedeff, R. Ruedy and L. Travis, *Mon. Wea. Rev.*, **111**, 609, 1983.
4. Rind, D., Goldberg, R., and Ruedy, R., *Climatic Change*, in press, 1989.
5. Manabe, S. and R. T. Wetherald, *J. Atmos. Sci.*, **44**, 1211, 1987.
6. Emanuel, K. A., *Nature*, **326**, 483, 1987.
7. Rind, D., *J. Geophys. Res.*, **93**, 5385, 1988.
8. Rind, D., *J. Climate*, **1**, 965, 1988.
9. Wetherald, R. T., and S. Manabe, *J. Atmos. Sci.*, **45**, 1397, 1988.
10. Fu, R., A. D. DelGenio, and W. B. Rossow, *J. Climate*, 1989 (in preparation).
11. Rind, D., *J. Atmos. Sci.*, **43**, 324, 1986.
12. Hansen, J., I. Fung, A. Lacis, S. Lebedeff, D. Rind, R. Ruedy, G. Russell and P. Stone, *First North American Conference on Preparing for Climate Change*, October 27–29, 1987, Government Institutes, Inc., Washington, D. C., 516 pp.
13. Hansen, J. and S. Lebedeff, *J. Geophys. Res.*, **92**, 13,345, 1987.
14. Davis, M. B., *Climatic Change*, 1989 (in press).
15. Mearns, L. O., R. W. Katz and S. H. Schneider, *J. Clim. Appl. Meteorol.*, **23**, 1601, 1984.
16. Roberts, L., *Science*, **238**, 1228, 1987.
17. Hansen, J. and S. Lebedeff, *Geophys. Res. Lett.*, **15**, 323, 1988.
18. Ropelewski, C. F., *J. Climate*, **1**, 1153, 1988.
19. Jones, P. D., T. M. L. Wigley and P. B. Wright, *Nature*, **322**, 430, 1986.
20. Hansen, J., A. Lacis, D. Rind, G. Russell, P. Stone, I. Fung, R. Ruedy and J. Lerner, *Climate Processes and Climate Sensitivity*, American Geophysical Union, Washington, 368 pp., 1984.
21. Karl, T. R., H. F. Diaz and G. Kukla, *J. Climate*, **1**, 1099, 1988.
22. Karl, T. R., R. G. Baldwin and M. G. Burgin, NOAA Historical Climatology Series 14–5, National Climatic Data Center, Asheville, N.C., 1988.
23. Karl, T. R. and P. D. Jones, *Bull. Amer. Meteorol. Soc.*, **70**, 265, 1989.
24. Jones, P. D., P. M. Kelly and C. M. Goodess, *J. Climate*, **2**, 1989.
25. Oke, T. R., *J. Climatol.*, **1**, 237, 1981.
26. Hanson, K., G. A. Maul and T. R. Karl, *J. Geophys. Res.*, **16**, 49, 1989.
27. Trenberth, K. E., G. W. Branstator and P. A. Arkin, *Science*, **242**, 1640, 1988.
28. Manabe, S., Drought and greenhouse warming: what is the relation?, Strategic Planning Seminar, sponsored by National Climate Program Office, Washington, D.C., October 18, 1988.
29. Lacis, A. A., Wuebbles, D. J., and Logan, J. A., Radiative forcings of global climate by changes in the vertical distribution of ozone, *J. Geophys. Res.*, submitted, 1989.
30. Ramanathan, V., Cicerone, R. J., Singh, H. B., and Kiehl, J. T., *J. Geophys. Res.*, **90**, 5547, 1985.

Luncheon Panel on Stratospheric Ozone Depletion
December 6, 1988

Panelist: Dr. Robert Watson, NASA;
Dr. F. Sherwood Rowland, University of California at Irvine
Dr. Michael McElroy, Harvard University;
and Alex Chisholm, Atmospheric Environment Service, Canada

Watson: Ozone has been steadily depleting since probably the mid-1970s in Antarctica. The observations we have show that by 1987, there was a significant depletion of ozone over the Antarctic continent and in the region surrounding the Antarctic continent amounting to more than a 50% decrease in the October monthly mean in less than one decade. A major aircraft campaign co-sponsored by NASA, NOAA, NFS and industry, strongly supported by the international science community, did a series of measurements from Punta Arenas, Chile. Here what we've mainly noted was very high levels of chlorine monoxide and a very highly disturbed atmosphere.

In our first statement when we left Chile, we basically stated that the prime reason for the Antarctic ozone hole was manmade chlorine, strongly modulated in a year by the local meteorology. Now at that particular press conference, we strongly downplayed the role of bromine compounds. We basically said they were less than ten percent of the total contribution to the chemical destruction of ozone, and it was primarily a pure chlorine situation. Dr. McElroy would like to spend a few minutes probably clarifying that situation whereas, I believe now, we should look upon bromine as a much more important contributor to the ozone hole in Antarctica.

By 1988 as I showed, the ozone hole was nowhere near as low this year as in the last few years. That qualitatively does not surprise anybody. We've seen year to year modulation of the amount of ozone in Antarctica, and the fact that the ozone hole was significantly less depleted this year should not come as any surprise. The key issue is that while you have chlorine in the atmosphere at levels exceeding two parts per billion, there is always the chance of a major destruction of ozone in the Antarctic region depending on the meteorological conditions of that year.

Now this comes to a very key policy issue and it's very simply, what will the Montreal Protocol do or what will the Montreal Protocol not do. At present, there are three parts to a billion of chlorine in the earth's atmosphere. Even with a fully ratified Montreal Protocol where we can show gases such as fluorocarbons 11, 12, 113, 114 and 115 and some limited control on the halons, even with that fully implemented by all nations of the world, including nations such as China and India who have not signed the Montreal Protocol, the chlorine loading in the atmosphere will continue to increase and it will increase at least to the order of six parts per billion if not greater over the next few decades. If the theory is correct, that chlorine and bromine are implicated in the ozone hole in Antarctica, then it is quite clear that if the halogen concentration atmosphere is going to double in the next few decades, the Antarctica ozone hole is here forever.

How could we get to a situation where the Antarctica ozone hole were to disappear or be repaired? I believe it would require taking the chlorine-loading back to the levels of the late 1960s or early 1970s, which is on the order of two parts per billion. To reach two parts per billion, it would require on the order of a 95% phaseout of all fully halogenated compounds by all nations of the world. It would require probably at least a freeze of methyl chloroform which is not yet controlled in the Montreal Protocol. We would need to understand why carbon tetrachloride is still increasing in the atmosphere at one percent a year and find a way to stop that increase in carbon tetrachloride. And we would have to look seriously at all possible substitutes insofar as many of the substitutes proposed for chlorofluorocarbons do contain chlorine. That means a very careful examination of the ozone-depleting potential and the greenhouse-warming potential of all potential substitutes.

So in my opinion, the fact that the ozone hole in Antarctica in 1988 was not as deep as in the previous few years does not mean to say that anything about our understanding that chlorine can destroy ozone in Antarctica is flawed. It is quite consistent with year to year variability that we have come to expect. It also means that the Antarctic ozone is not protected under the Montreal Protocol and would require significantly stronger regulations in the future. And now I'd like to pass across to some words from Sherry Rowland.

Rowland: What I wanted to say is a little bit about the measurements of ozone in the Northern Hemisphere based on the Dobson Stations. The Dobson spectrometer was first used for measurements of ozone in a limited way about sixty years ago, in the 1920s. Two stations were set up at that time making measurements on a regular basis, and a regular basis means that they try to make a measurement everyday and do succeed on most days. And so you could end up with a daily and a monthly and a yearly record of ozone for the time period that these stations have been in operation.

There was a very big jump in the number of such stations in the time period of 1957 and 1958 during the International Geophysical Year. That was the time at which the Halley Bay Station, for instance, of the British Antarctic Survey, was established on the coast of Antarctica. That's the time for a number of stations being established around the world. The United States' contributions really started primarily around 1963 with the establishment of the current stations where measurements are made.

And so what we have on the basis of these measurements is a few stations going back in time. The station with the longest continuous record is the one in Arosa in Switzerland with a record that goes back to 1931. There are a fair number of stations since 1957 with continuous records. Many of the stations that were started during the IGY were then discontinued so that the number of stations in operation then is a much larger number than the ones that have continuous records up to the present time.

And in the NASA Ozone Trends Panel, what we did was to examine all of the Dobson Stations that have been reporting to the Ozone Data for the World, the official reporting. The official reports are sent to Toronto, Canada, and the data are published. And what the NASA Ozone Trends Panel did was to go over these data to try to see if there were objective ways of deciding which stations were providing good results and which ones were not. And one of the very useful ways of testing these stations is to intercompare them through the measurements on the same day with the overhead satellite. As Bob told you this morning, the satellite measurements themselves have been changing with time because of the degradation of the diffuser plate. But on any given day, the satellite is not degrading and it is giving reasonable comparison measurements as it overflies each of the Dobson Stations. And so the combination of the satellite and the Dobson Ground Stations together have allowed us to select the better Dobson Stations and to calibrate the satellite. And the combination has provided us with a much better view of what's happened to ozone over the last twenty-five years or so than we could get from either of these systems by itself. The satellite itself, of course, has only been giving data since 1978, within the limitations that I have described. But the procedure of the NASA Ozone Trends Panel was to examine the records at individual stations. And there are 18 good Dobson Stations between 35 and 65 degrees north. If you look at those records as they are published in the Ozone Data for the World, what you'll find is that during the winter months, all 18 of those stations have shown a loss of ozone over the time period of the last two solar cycles. That is 22 years. What we did was to look at the data starting in 1963, or in 1965, and ending in 1986. The argument behind this is that the solar cycle is certainly believed to have an effect on ozone and therefore, it would be very useful to have measurements that extended over at least two solar cycles. And so from 1965 through 1986 provides 22 years and two solar cycles. It also provides about ten of the quasi-biennial oscillations which are known to influence the amount of ozone that you can find over individual stations. The quasi-biennial actually has to do with the direction of windflow in the stratosphere over the equator. But this direction propagates -- this wind direction propagates both temperatures and ozone in both directions towards the poles. And so at individual stations, you can see the correlation with the equatorial wind direction, and by taking a 22-year run of data, you cover enough of the QBOs that you can assume that any long-term trend will be taken out, and of course you have to think about the seasonal effect which is well known. Well, as I said, if you look at the 18 stations between 35 degrees north and 65 degrees north, all of them showed a loss of ozone in the winter, and that's the fundamental discovery that the Ozone Trends Panel outlined.

We also combined these data into latitude band averages. That is not just trying to find out what the change was at an individual location, but what the average change was for each of the latitude bands. And as Bob reported, in the latitude bands from 30 degrees to 64 degrees north, all of them have shown losses of ozone in the winter, and that's to be expected because all 18 stations showed losses of ozone in the winter. In the summer, of those 18 stations: 5 of them seemed to indicate increases in ozone; 3 of them no change; and 10 of them losses. And that ends up saying that is not clear what's happening to the ozone in the summer. On the average, there seems to be a little bit less ozone than there was in the time period at the end of 1986, than there was in the 1960s, but it is not as clear a signal as the signal which occurs in the winter months which for our purposes were December, January, February and March.

I think that one of the important messages from the NASA Ozone Trends Panel is that one needs to examine the losses of ozone on a month by month basis. And when you do that, what you find is that there has been a loss in the winter, suggesting that whether the mechanism of loss that's being observed is influenced by the season and whether that's the temperature or something about the location, remains to be established. But the long-term trends taken with 22 years or more of data all show that there has been a loss of ozone in the winter in the Northern Hemisphere. And with that, I'll turn it over to Mike McElroy.

McElroy: What I wanted to do was to say a few words about the -- first of all -- about the mechanism for ozone loss in Antarctica. And if time permits, also a few words about how this might extrapolate to other latitudes. It was very clear from the measurements on the aircraft expedition a year and a half ago that levels of CLO and BRO, were very high in the Antarctica environment corresponding to the region where ozone disappeared. There's been a great deal of confusion, as Bob remarked, about the relative contribution from the two chemical schemes which are still viable as sinks for ozone in this environment. One scheme involves the formation of a CLO diamer as suggested by Mario Molina and co-workers. The second scheme involves the presence of a very small concentration of BRO at the five parts per trillion level interacting with chlorine to do the same thing, namely to catalyze the reaction with itself. Now with the new controversy with measurements, a problem has arisen due to some misconceptions propagated through the scientific media -- places where these data had really been published. But in any event, when you correct those errors, our best view from the assessment's empirical estimate of relative contributions of the CLO/CLO reaction and the CLO/BRO reaction to ozone loss would put them close to 50/50.

We believe that both schemes for ozone loss are important. In 1987, the bromine-catalyzed scheme accounted for between 30 and 45% of the total ozone lost. In 1988, the bromine scheme was relatively more important and that simply has to do with the dependence of the chlorine/chlorine scheme on the second power of the concentration of CLO. I think most people would agree that both schemes are important, chlorine and bromine. Both contribute to the loss of ozone in Antarctica, and the bromine scheme is going to have to be taken very carefully over other latitudes where the concentration of CLO is lower.

Now the more serious side of the question is how do we explain the interannual variability in ozone loss which was so dramatic between 1987 and 1988? The best view we have so far is that the chemistry will be very sensitive to the residual abundance of nitric acid. And these curves are simply intended to show you that by varying the residual level of nitric acid at the end of the Antarctic winter, you can get very large differences in the predicted loss of ozone over the spring.

Then the question backs off to what is it that controls the residual level of nitric acid? We've learned over the last few years that condensation chemistry is really a critical element in understanding what goes on in Antarctica, and we've learned a great deal about how it actually goes. The stratosphere everywhere contains a small number of sulfate aerosols. As the temperature goes down, the sulfate aerosols take on water. And as the temperature drops below about 195 degrees Kelvin, depending on where you are, then you form nitric acid trihydrate. And what that curve is intended to show you is that as the temperature goes from 196 degrees down to 195 degrees, down to 190 degrees, you essentially lose all of the nitric acid in the stratosphere. It precipitates. It falls out without any significant loss of water. And therein lies a problem, because the observations which are shown here indicate that there are occasions where you see significant concentrations of nitric acid and significant depletion of water vapor. How do you explain that?

Well, the explanation for that turns out to be very simple. As you cool the stratosphere down from 190 degrees, there's no nitric acid left. Water vapor condenses and falls. As the temperature goes down to 180 degrees, you can bring the water vapor mixing ratio down to 1 ppm. Those dashed lines simply indicate what happens if after denitrifying -- dehydrating the stratosphere, you admix air from outside, then depending on what temperature you admix at, you will set up a set of data which will lie along one of those dashed curves. So 1987 Antarctic data indicate temperatures that during the winter go as low as 180 degrees and indicate that the admixture of air from the outside was as small as 10% on occasions. That, we believe, is why the levels of nitric acid were so low in 1987. The difference in 1988 is not that the temperature did not get low. It did. Temperatures in fact in early winter of 1988 were as low as they were in 1987. At that time, you would have lost all the nitric acid and all the water vapor from the Antarctic stratosphere 1988 as efficiently as 1987. The difference is that in the month of July and early August, there was sufficient exchange with the environment outside to replenish both the nitric acid and the water vapor and, as a consequence, the chemistry-destroying ozone was less efficient. So inter annual variability in Antarctic ozone, we would now say, is attributable largely to wave activity in the Southern hemisphere interrupting the polar, vortex-exchanging mass primarily in July and that would be associated with changes in nitric acid and associated changes in the chlorine/bromine catalytic chemistry-destroying ozone.

What about other latitudes? One of the schemes that is rather intriguing-- one of the points that is rather intriguing about this curve which Bob Watson showed earlier -- is that if you look at the stage there shown by the triangles and the squares as a function of altitude, the remarkable fact is that the apparent loss of ozone is as large at 25 kilometers. As it is at 40 kilometers on a fractional basis, and on a real basis, the low altitude loss is going to contribute significantly more to the total cone because that's where most of the ozone is. So the question arises: What might be going on down at those altitudes? Now it's possible, a portion of that low altitude loss represents export from the Arctic and Antarctic. Alex Chisholm will talk about some very exciting new information from the Arctic, but I'm struck also by the fact that you see very strong evidence for dehydration of the stratosphere in tropical regions. The temperature gets as low as 190 degrees there, normally associated with the rising Hadley cell. Some of my data show clear indications of dehydration, dehydration as extreme as it is in the Antarctic. And there will be assuredly denitrification in that environment; it must take place. That sets the stage for gas phase chlorine/bromine catalyzed loss of ozone. And I think we have to look very seriously at the possibility of low altitude ozone loss in the tropics as well as in the polar regions. Obviously, loss in the tropics would be a lot more serious because the level of ozone is less there anyway. Okay. I'll pass it on to Alex Chisholm.

Chisholm: Thank you very much. Today, I'm going to talk a little bit about a structure which we have discovered in the Arctic which we think is very comparable to the Antarctic ozone hole. We're not convinced that we would want to call it an Arctic ozone hole. An Arctic ozone crater is probably more appropriate. This started with observations made by Atmospheric Environmental Service personnel at our Willis Station at Alert in the Northwest Territories which is about 82 degrees North. During the spring of 1986, we were there for another purpose and released ozone sonds which actually showed depleted profiles of ozone very much like the ones that were shown this morning from McMurdo in the Antarctic.

I think that for comparison, what I would like to do is show you the first viewgraph which is the ozone structure over the Antarctic. This is in October 1982. This graph comes from data from the TOMS Satellite. It's fudged a little bit. Fifty Dobson units of ozone have been added everywhere because in actual fact in the Arctic, there is more ozone. This will give you a better comparison of colors. The ozone hole is shown in blue and purple in the upper part and there is sort of a ridge that is outlined in red. Could we have the second one, Bob. This is the kind of structure that we have observed in the Arctic. You can see the outline of Greenland in sort of a 4 o'clock position and the ozone hole or ozone crater lies over the northern part of the USSR and up into the region of Spitzbergen Island. This is a fairly preferred location for this type of a structure, and what I want to point out is that the ozone crater is smaller in area. It's approximately half the size of continental USA rather than roughly the size of the USA, as is the case with the Antarctic ozone hole. It lasts for a shorter period of time. It's rather variable. It seems that at its maximum in January, as you can see, it's displaced quite a long distance from the North Pole, at about 65 to 70 degrees North latitude. In terms of its height, the depletion of the ozone layer occurs between approximately 10-16 kilometers in the Arctic as against a height of 13 to 23 kilometers in the antarctic.

The last slide I think will help to illustrate. We saw this morning tracks of the Antarctic ozone hole and the minimum in it. This is a rather crude representation. This is data taken from the 15th of January each year, 1978 through 1988, and you can see the ozone values dropping from approximately 330-340 units down to something just about 200 units on the end of 1988 coming up. And as was the case with the Antarctic circumstance there is a Quasi-Biennial Oscillation modulation on this as well. So that in summary, we think that there is something quite similar occurring in the Arctic in terms of the ozone layer. What it is related to, of course, is a big question and there's very much the hope that the Arctic aircraft expedition out of Norway this winter run by NASA will help to sort out the chemistry. Thank you very much.

Watson: I'm not sure that there's any time to go any further. I think the key thing to point out is that there have also been observations to date in the Arctic region that do suggest unusual chemistry. For example, Susan Solomon from the NOAA Labs saw elevated levels of chlorine dioxide, clearly an indication that it could have been elevated levels of CLO and BRO, the two key radicals in a chlorine and bromine system. When Jim Anderson flew his instrument on the ER2, in the Alaska area and across to the Spitzbergen last year, he was going to interesting regions of elevated concentrations with CLO. So at the moment there is strong indication that the ozone in the Arctic region in winter is decreased, a phenomenon that can't be allowed for by known natural variability. There are indications that both the chlorine system and much earlier -- it was shown quite convincingly -- that the amount of oxide of nitrogen, NO_2, is significantly decreased in the Arctic region. So we have evidence of unusual chlorine, unusual nitrogen chemistry. The campaign that Alex just alluded to hopefully will try and provide more answers. A very large team of scientists is not just enough. NASA/NOAA -- a large number of scientists from the university community including Mike McElroy as a theoretician -- will spend six weeks in Stavanger, Norway starting around December 26th or December 27th. And hopefully, we'll try and find out some clues as to why the ozone is indeed decreasing in the Northern Hemisphere, and whether this can be squarely blamed on the CFCs or whether or not it's just some unknown natural variability. I think Sherry would like to show one more view foil.

Rowland: Alex Chisholm mentioned that the low ozone that they were seeing tended to be in Northern, in Soviet Union and towards Spitzbergen, and there has been a Dobson Station as well running at Leningrad over this time period and I have a graph here of their data results which show very strongly a winter time loss. And if you look, for instance, at the units they are of the order of 40 Dobson-unit loss which is a 12% loss for ozone in the winter over Leningrad. So there is ground-based information that tends to look very similar to that from the satellite. There has been a heavy loss in certain places over Western Europe, but then particularly in the Soviet Unlion.

Watson: One comment I could make is I know that there are scientists in the Soviet Union and the Government who are equally interested in this problem. While the NASA/NOAA European Expedition will be from Stavanger, Norway, the Soviets will indeed be conducting their own experiments to find out what's happening in the Arctic this year. They will do some ground based observations and they will use two aircraft; one is a subsonic aircraft with a lighter on board, very similar to what we're doing. They're also hoping to use one of their supersonic transports to actually penetrate high into the lowest stratosphere up to 18-19 kilometers -- just like we will with the ER2. So while we are doing some experiments from Norway, the Soviets will indeed be doing their own experiments -- I do not know the location of it -- somewhere in the Soviet Union. So there's strong interest not only in North America and Europe, but strong interest also in the Soviet Union. McElroy: -- Just a comment that the key element in setting the stage for loss of ozone in the Arctic is likely to be the same as what it was in the Antarctic, namely the need to get rid of oxides of nitrogen. And I just want to comment that the time to precipitate nitric acid -- if it's precipitated in its trihydrate form at the relatively warm temperatures for that condense -- is in the order of a week or so. So you need to have temperatures that are below 197 or 198 degrees Kelvin for about a week. If on the other hand, you hit the frost point of water, about 190 degrees, then the particles grow to sizes in excess of 10 microns, and they fall like snowballs, and the time involved there is very short. All you need is a few hours to a day or so at 190 degrees and you can clean the nitric acid out of that region of the atmosphere. Once gone, you have a great chance for the chemistry to then run just as it does in Antarctica so long as you have enough trihydrate particle surfaces area left for the heterogenous chemistry to continue.

<u>Chisholm</u>: As an old cloud physicist, that really gives me a good bit of a jolt to think that clouds are once back in business again. The thought of being hit on the head by a ten micron particle in the Arctic is something that leaves me quite cold though.

<u>Watson</u>: And I think with that poor joke, we should finish.

The text of the executive summary of the press statement prepared by the scientists who participated in the Arctic Stratospheric Expedition follows on the next page.

AIRBORNE ARCTIC STRATOSPHERIC EXPEDITION

PRELIMINARY FINDINGS

February 17, 1989

This press statement has been prepared by the scientists who went to Sola Airfield near Stavanger, Norway to study the Arctic winter stratosphere. **The findings given here are preliminary.** Under normal circumstances, scientists studying such a complex scientific issue would take up to a year or more to disclose their initial findings. However, the issue of global ozone depletion due to human activities is one of justifiable public concern, and hence the public and policymakers should be kept abreast of the current scientific thinking. It is in this spirit that this provisional picture of the processes controlling ozone at high latitudes in winter is given. **This document cannot be referenced as a scientific publication.**

EXECUTIVE SUMMARY

The key findings of the 1989 Airborne Arctic Stratospheric Expedition are as follows:

- The chemical composition of the Arctic polar stratosphere was found to be highly perturbed.

- Specifically, in comparison with the abundances calculated using theoretical models that do not include chemical reactions occurring on the surfaces of Polar Stratospheric Clouds (PSC's)

  - the observed abundance of the ozone-depleting chlorine monoxide radical (ClO) was elevated by up to a factor of 50 in the Arctic lower stratosphere,

  - the abundance of chlorine dioxide (OClO) was increased by up to a factor of 50, consistent with increased abundances of ClO in the presence of bromine monoxide radicals (BrO),

  - the abundance of the major inactive chlorine reservoir, hydrochloric acid (HCl), was considerably reduced, indicating that there had been a conversion from the inactive to the active (ozone-depleting) chemical forms of chlorine, and

  - the abundances of nitric oxide (NO) and nitrogen dioxide (NO_2) were significantly reduced, thereby preventing the reconversion of ClO into an inactive form of chlorine.

- The measurements of the column abundances of several of the above compounds indicate that these perturbations occur over a wide range of altitudes in the stratosphere.

- The above observations add up to a consistent picture that is similar to the one drawn from the Antarctic data in 1987 and are believed to be the result of chemical reactions occurring on the PSC's that form at the cold temperatures.

- The chemical changes were associated with the presence of PSC's and are consistent with the picture of their role in (i) shifting the partitioning between the active and inactive chlorine compounds, (ii) removing the nitrogen oxides (NO + NO_2), and (iii) removing and repartitioning NO_y (the sum of all nitrogen-containing reservoir and radical species).

- No unequivocal signature of photochemical loss of Arctic ozone was identified before the end of this mission. However, by the end of this mission, a considerable portion of the vortex air was primed for ozone destruction.

- It is difficult to predict the degree of Arctic ozone depletion that will actually occur because it depends on the rate at which normal partitioning of chlorine species is reestablished. This rate is uncertain:
  - the level of exposure to sunlight is uncertain,
  - the degree of denitrification of the Arctic stratosphere (indicated by the amount of HNO_3) has not been adequately quantified with respect to altitude, and
  - the rate at which the Arctic air mixes with air containing ambient levels of the oxides of nitrogen is difficult to predict reliably.

- The relevant meteorological conditions for the formation of PSC's in the Arctic polar region were not unusual this year. Hence, the phenomena observed are expected to occur in most years.

- Until a comprehensive analysis of the data from this mission is accomplished and model calculations are performed, it is not clear whether the high latitude winter ozone decreases reported by the International Ozone Trends Panel are due to the perturbed Arctic chemistry.

- The observed PSC-induced buildup of reactive chlorine compounds represents an additional ozone-depleting process that was not included in the stratospheric ozone assessment models used as a basis for the Montreal Protocol.

The preliminary results of this mission have substantially increased confidence in the scientific understanding of the PSC-induced, chlorine catalyzed, ozone depletion phenomenon. As a result, it is clear that enhancements of chemically active chlorine compounds do indeed occur in both the Arctic and Antarctic stratosphere.

CLIMATE IMPACTS OF METHANE CLATHRATES

Gordon J. MacDonald*

Introduction. During periods of continental glaciation, concentrations of atmospheric carbon dioxide and methane were two-thirds and one-half lower, respectively, than the values that prevailed during interglacial and postglacial periods. These compositional changes represent the transfer of about 200 Gt of carbon into and out of the atmosphere. The decrease in atmospheric carbon took place gradually over 100,000 years, while the buildup of carbon dioxide concentration required only a few thousand years. Changes in the levels of these two infrared-absorbing gases provide a strong positive feedback to the alterations of earth's radiation budget associated with the earth's orbital fluctuations (Milankovitch theory).

Conventional representations of the carbon cycle identify three major reservoirs of carbon—oceans, fossil fuels, and soils. I suggest that methane clathrates form a fourth major reservoir; one that is sensitive to surface disturbances of temperature and pressure—brought about by ice cap growth and dissipation—and the accompanying fall and rise of sea level. Time constants for the movement of carbon to and from the clathrate reservoir are fixed by the thermal properties of sediments. Release occurs over thousands of years and trapping over many tens of thousands of years. The large mass of carbon stored in clathrates, coupled with the time scales for carbon transfer, suggests that clathrates played a significant role in mediating climate change during the ice ages.

Properties of Methane Clathrates. Clathrates are ice-like compounds in which methane and other gases are caged by water molecules. Clathrates resemble wet snow or ice in physical appearance and can form two distinct structures, the smaller of which traps methane. The small clathrate unit structure contains 46 water molecules, with up to eight molecules of methane, and leads to the formula $CH_4 \cdot 5.75\, H_2O$. The laboratory density of a fully filled methane clathrate is 0.91 g cm^{-3}, so one cubic meter of methane clathrate contains 170.7 m^3 of methane gas at standard conditions (STP). If only 90 percent of the sites are filled with methane, 156 m^3 of methane gas are contained within a cubic meter of clathrate. In terms of mass, a cubic meter of fully filled clathrate contains 122 kg of methane, 91.4 kg of carbon, and 789 kg of water; with a 90-percent filling of the gas sites, a cubic meter of clathrate contains 83.3 kg of carbon. When less than about 80 percent of the gas sites are saturated, the methane clathrate is no longer a stable phase (Davidson, 1983).

Thermodynamic Stability of Methane Clathrate. The phase diagram shown in figure 1 outlines the region of stability for methane clathrate (Vysniauskas and Bishnoi, 1983). At low temperature and high pressure, in the presence of methane and water, methane clathrate is the stable phase. An important feature of clathrate is its ability to exist over a range of temperatures and pressures at which pure ice is not stable. Methane clathrate is stable at pressures greater than 26 bars, and temperatures above 0° C.

Impurities in the chemical system will shift the clathrate phase boundary. Sodium chloride reduces water vapor pressure, moving the phase boundary toward lower pressures. The addition of larger gas molecules—carbon dioxide, ethane, propane, and hydrogen sulfide—tends to stabilize clathrates and to shift the equilibrium curve to higher temperatures. In natural sediments, the effects of the salinity in pore waters approximately cancel the effects of larger molecules (Kvenvolden and McMenamin, 1980), except in relatively

*Gordon J. MacDonald is Vice President and Chief Scientist of The MITRE Corporation, 7525 Colshire Drive, McLean, VA 22102. Copies of the full text of the paper, together with references, can be obtained from the author.

rare instances of high ethane and propane contents. Because of these canceling effects, the curve for a pure methane-water system provides a reasonable estimate for the stability region of naturally occurring methane clathrate.

Natural Occurrences of Methane Clathrates. The first confirmed samples of natural methane clathrate were retrieved in pressurized core barrels from a Prudhoe Bay well on March 15, 1972. Well logs confirmed the presence of several clathrate zones (Kvenvolden and McMenamin, 1980). Although strong gas shows in the drilling mud suggested free gas in the formation, anomalies in both the sonic and resistivity logs indicated ice or ice-like materials. The density log also indicated ice rather than free gas. Equivalent logging indications had been used by the Soviets in developing the Messoyakha field.

The first direct observations of gas clathrate in oceanic sediments were made by Yefremova and Zhizhchenko (1975) in shallow core samples from the Black Sea. The recovered microcrystalline aggregates of clathrates decomposed rapidly at the surface, releasing mainly methane and carbon dioxide. An almost pure, 1.05-m-long core sample of clathrate was unexpectedly recovered from Site 570 of the Deep Sea Drilling Project in 1982 (Kvenvolden et al., 1984). Wireline logging showed that the sample came from a solid interval of clathrate 3 to 4 m thick. The 1982 sample was the first natural clathrate taken at sea that was preserved for shore-based studies. Gas pressure measurements during decomposition indicated that the clathrate was very nearly filled to the formula $CH_4 \cdot 5.75\ H_2O$. The released gas was 99.4 percent methane, $\sim$ 0.2 percent ethane, and $\sim$ 0.4 percent CO_2. Deep Sea Drilling Project cores recovered dispersed clathrates from the Blake Outer Ridge of the Atlantic (DSDP Leg 76, Brooks et al., 1983). Clathrates were also found to be common in the slope sediments of the Middle America Trench off the shores of Mexico, Guatemala, and Costa Rica (Shipley and Didyk, 1982; Harrison and Curiale, 1982). Recently, clathrates have been found in sediments offshore Peru (Kvenvolden, 1988).

In the Gulf of Mexico, clathrates are found as 0.5- to 50-mm nodules, interspersed in layers 1 to 10 mm thick, or as solid masses greater than 15 cm thick. Eight samples taken on the Louisiana slopes are associated with oil-stained sediments, containing up to 7 percent extractable oil. Three of the recovered samples differed from other natural clathrates in having a low ratio of methane to ethane plus butane, $C_1/(C_2 + C_3)$, of 1.9 to 4.4 (Brooks et al., 1986). A further feature of the clathrates associated with oil sediments is a crystal structure large enough to enclose propane and other similarly sized molecules (Handa, 1988).

As indicated above, relatively few hydrate samples have been recovered. The scarcity of samples can be attributed to the difficulty in recovering and preserving material that rapidly decomposes when brought to the surface. Only two specific attempts to recover clathrates have been made (Kvenvolden and Barnard, 1983; Claypool et al., 1985). Yet clathrates have been found in widely differing geologic settings: Arctic permafrost, the tectonically active Middle America Trench, a sedimentary ridge off a continental passive margin (Blake Plateau), the very rapid depositional environment of the Gulf of Mexico, and finally, the stable environment of the Black Sea. The sediments enclosing clathrates vary from coarse sands on the North Slope to fine-grained hemipelogic muds and volcanic ash in the Middle America Trench. Clathrates are most commonly found as small crystals dispersed within the sediments or as larger nodules ($\sim$ 2 cm). Occasionally, clathrates occur in massive layers, with only a small admixture of contaminating sediments. Such a wide range of geologic environments suggests that where methane and water are available, clathrates may be ubiquitous, provided conditions in the sediments are within the methane clathrate stability region (see fig. 1).

Well data from Arctic regions has provided abundant indirect evidence for the presence of clathrates. Two exploratory wells drilled in the permafrost of the Mackenzie Delta showed sharp jumps in the methane content of their drilling muds after penetrating sand layers. Although the sands were very porous, their permeability was extremely low; also, the sonic log indicated a high velocity and high resistivity. Bily and Dick (1974) interpreted the anomalous observations to result from clathrate-plugged interstices. Similar data indicate the existence of clathrates in the Arctic Archipelago of Canada (Hitchon, 1974) and the Viluy gas field in Yakutia, Soviet Union (Makogon et al., 1971).

The most commonly used indicator of clathrate deposits in marine sediments is the appearance of an anomalous seismic reflection that mimics the ocean bottom in its appearance (Shipley et al., 1979). Bottom-simulating reflectors (BSRs) are best recognized on seismic records where they cut across other reflectors produced by sedimentary structure, etc. The reflector, arising from the impedance mismatch between the overlying, high-velocity clathrate and the underlying, lower-velocity sediment, is heightened if the underlying sediment is water- or gas-saturated. Such reflectors are usually characterized by both a large reflection coefficient and a polarity reversal.

Figure 2 shows the geographical location of all reported clathrate deposits, either inferred or actually sampled. Thus far, more samples have been taken from continental slopes than from the Arctic permafrost. There are no known occurrences of hydrates in Antarctica, but the sediments underlying the ice sheet should all be within the clathrate stability region (MacDonald, 1983).

Estimates of Total Carbon Stored in Methane Clathrates. In order to estimate the amount of carbon currently stored in clathrates, both the volume of the clathrate zone and the fraction of that volume occupied by clathrate must be estimated. Determining the volume of the clathrate stability zone depends to first order on the temperature gradient and to second order on the composition of pore waters. Geothermal gradients are well enough known to determine the volume of the clathrate stability zone to a factor of two, but the remainder of the task, determining the fraction of the zone filled by clathrate, must remain speculative. Factors essential to making estimations, such as porosity, carbon abundance, etc., are poorly determined or simply not known.

Cherskiy et al. (1985) have compiled geothermal data for the permafrost regions of Siberia. These data can be used to roughly estimate the areal extent, thickness, and volume of the clathrate stability zone, if one assumes that the effects of a deviation in the composition of pore waters compensate for the presence of gases other than methane in the clathrate. Table 1 lists estimated volumes of the clathrate stability zone. For those regions in which the zone contains Quaternary sediments (Timon-Pechora Province and

Table 1

Estimate of Carbon Stored in Clathrates in Regions of Permafrost in Siberia

Region	Area of Hydrate Stability (km^2)	Volume of Clathrate Zone of Stability (m^3)	Estimated Carbon Stored in Clathrates (Gt)
Timon-Pechora Province	6.7×10^4	2.66×10^{13}	10
West Siberian Platform	1.1×10^6	3.3×10^{14}	120
East Siberian Craton	1.8×10^6	8.1×10^{14}	150
Northeast Siberia	6.1×10^5	3.7×10^{14}	70

the West Siberian Platform), a porosity of 0.4 is assumed. In regions where older sediments are found in the clathrate stability zone, porosity is taken to be 0.2. The fraction of the available clathrate volume—porosity times total volume—that is actually occupied by clathrates is taken to be 1 percent. This percentage is based on North Slope wells, in which about 10 percent of the clathrate stability zone is actually filled by clathrates (Collett and Ehlig-Economides, 1983), and on the Messoyakha and Viluy fields (Makogon, 1978). However, because these fields are in known regions of methane accumulation, an average occupancy of 10 percent is deemed high. Instead, an arbitrary 1 percent figure has been used in constructing table 1. Based on a 1 percent filling, 350 Gt of carbon are stored as clathrate in the permafrost regions of Siberia, and 50 Gt in the North American Arctic, giving a world total of about 400 Gt of carbon for permafrost regions. As indicated, this estimate is highly speculative, but given the very large volume of near-surface sediments that lie within the clathrate stability zone, an estimate somewhat higher than 400 Gt may actually be more appropriate.

Over the sediment-rich continental shelves, the great variation in geothermal gradient produces a zone of clathrate stability ranging in thickness from 200 to 1,200 m (see fig. 3). In order to obtain an estimate of carbon stored in oceanic clathrates, an average value of 0.5 km is used. This gives a total volume of sediments in the clathrate stability zone of 31.3×10^6 km^3. If sediment porosity is taken as 0.4, then 12.5×10^6 km^3 are available for clathrate formation. Only a fraction of the total sediments will have a carbon content high enough to generate the amount of methane required to form clathrate. About 10 percent of the world's sea floor sediments contain concentrations of organic carbon greater than 1 percent (Premuzic et al, 1982; Budyko et al., 1987). If 10 percent of the available pore space of carbon-rich sediments (1 percent of total sedimentary pore space) within the clathrate stability zone actually contain clathrate, the total volume of clathrate is 1.25×10^5 km^3. The total volume corresponds to a carbon mass of 11,000 Gt for 90-percent-filled methane clathrate. This estimate does not include clathrate in abyssal plains sediments at depths greater than 3,000 m.

Ice Age Changes in Clathrates. Over much of the deep ocean, bottom water temperatures are about −1 to 0° C, the same temperatures that prevailed during peak glaciation. The stability of temperatures at the ocean floor indicates that oceanic sediments in deep water did not undergo major thermal perturbation during the ice ages. However, the land surface and sediments on shelves, particularly those in high-latitude regions, did experience major shifts in thermal conditions.

The thermal properties of sediments in cold regions have been carefully discussed by Lachenbruch et al. (1987). Thermal conductivity within sediments depends on porosity, conductivity of grain material, and whether the rock is frozen or thawed. Because of the higher conductivity of ice with respect to water, frozen sediments show a distinctly higher conductivity than do water-saturated sediments. However, even in areas where the average temperature is less than 0° C, fine-grained, clay-rich sediments may contain water rather than ice, due to poorly understood surface effects (Lachenbruch et al., 1982).

A rise of sea level results in rapid warming of low-lying, exposed-shelf, permafrost regions. The sharp change in surface temperature, from periglacial cold to the relative warmth of ocean bottom waters, leads to a downward-traveling temperature wave that can release methane from any underlying clathrate.

The evaluation of subsurface temperature for the flooding of a permafrost section that has an initial surface temperature of −10° C is shown in figure 4. The initial equilibrium temperature gradient is 31.25° C/km with a thermal conductivity of 1.76 Wm^{-1} °C^{-1}. These conditions correspond approximately to those that hold today around Point Barrow, Alaska (Lachenbruch et al., 1982). The initial top of the clathrate stability zone, 230 m down, is reached after 150 years, but the top of the stability zone is reduced by only 10 m after 500 years. However, after 2,000 years, the entire 400 m-thick stability zone is removed. The most rapid rate of release of methane occurs during the time interval between 1,000 and 2,000 years, as the geothermal gradient and phase boundary become nearly parallel.

The case of high-conductivity permafrost overlying lower-conductivity, water-saturated sediments is illustrated in figure 5. The solution that is valid in the upper layer is extended to the second homogeneous layer by requiring continuity in temperature and heat flow at the boundary of the frozen and thawed sediments. The initial clathrate zone is 1,000 m thick as a result of the assumed initial surface temperature of −15° C. Erosion of the top of the clathrate stability zone proceeds slowly until about 10,000 years. By 14,000 years, after the step warming of the surface, the clathrate stability zone vanishes.

The examples illustrated in figures 4 and 5 show that a sudden warming, resulting from inundation by the sea or the formation of a large lake, causes the elimination of the clathrate stability zone on a time scale of 2,000 to 14,000 years. Low initial surface temperatures and high-conductivity sediments favor the longer time scale.

Figure 6 illustrates the long time scale involved in the formation of a thick clathrate stability zone. The initial surface is taken to have a temperature of −1° C (maintained by the sea or lake) and is underlain by low-conductivity sediments. A stepwise drop in temperature to −15° C corresponds to a drop in sea level, which exposes the surface to cold periglacial conditions. The geothermal gradient drops below the clathrate stability zone after 2,000 years, but after 10,000 years, the thickness of the clathrate stability zone is just 230 m and will reach 480 m only after 60,000 years.

Sea level changes on the order of 100 m accompanied the formation and melting of continental ice sheets. The resulting change in pressure at depth had a major impact on the volume of the clathrate stability zone in ocean bottom sediments. The effect of changing sea level on the thickness of the clathrate stability zone is a function of both the ocean depth and subsurface temperature gradient, as is illustrated in figure 7. The time scale for the release of carbon from clathrates depends on the rate of sea level drop and on the thermal properties of the overlying sediments.

As pressure drops, clathrates dissociate, absorbing energy. The cooling of the dissociating clathrate will set up a temperature gradient between the clathrate and the surrounding sediments, initially drawing heat from sediments both below and above. At all ocean depths greater than about 250 m, dissociating clathrate can draw heat from overlying bottom waters. The dissociation of a 20-m layer of clathrate would take roughly 1,300 years. Because this time is short compared to the time scale for sea level drop, the latter would control the rate of methane release.

The clathrate zone would erode more rapidly in the upper layer; at the bottom of the zone, the large thermal inertia of the underlying sediments will cause the time scale for dissociation to be longer. Further, it is likely that released gases from bottom erosion will be trapped in place by the seal of the overlying clathrates.

The volume of the clathrate stability zone is greatly enlarged by continental glaciation. The amount of carbon stored under ice caps will depend on the carbon content and methane generation in the underlying rocks. If estimates of current continental clathrates are appropriate, then during the peak of glaciation, some hundreds of gigatons of carbon were stored at depth under ice. However, owing to the absence of relevant data, any estimate of the amount of carbon stored in this fashion must remain purely speculative.

When ice caps melt, underlying clathrates destabilize. If surface temperature drops from the 0° C assumed for the bottom of the ice cap to between −10 and −15° C (the temperature in the vicinity of the glacier), the clathrate would again stabilize on a time scale of thousands of years. The release of carbon is controlled by the balance between pressure destabilization and thermal stabilization as temperature drops in the thinning ice cap.

Release of Carbon by Future Greenhouse Warming. The release of methane from clathrates due to direct surface warming is a process that would require thousands of years, due to the slow downward diffusion of the surface thermal perturbation (Lachenbruch, personal communication; Kvenvolden, 1988).

Typical estimates of the global average warming for a doubling of the 1880 concentrations of greenhouse gases, which is expected to occur by the year 2050, range from 2 to 4° C. At higher latitudes, temperature increases may be twice the average global increase (Ramanathan, 1988). In a detailed analysis of subsurface temperature records in wells drilled on the North Slope, Lachenbruch and Marshall (1986) found evidence of a widespread secular warming of 2 to 4° C over the past century.

The time history of the development of the subsurface temperature field in the case of an exponentially growing surface temperature—2° C after 100 years—and higher-conductivity permafrost is shown in figure 8. After 100 years, the temperature disturbance enters the clathrate zone of stability, and methane release begins. Yet after 200 years, only 3 m of the stability zone will have eroded, and the amount of methane released would be less than 1 percent of the total stored, if thickness is taken to be 500 m. If a total of 400 Gt of carbon are stored in permafrost clathrates, the postulated exponential warming would release about 3 Gt of carbon. This amount of carbon is of the same order as the 5.5 Gt currently introduced annually by the burning of fuels.

As is illustrated in figure 5, significant amounts of carbon can be released when thermal gradient parallels the clathrate stability curve. Beginning 18,000 years ago, the rising oceans flooded areas of permafrost overlying subsurface clathrates. The thermal time constants are such that the downward-moving thermal perturbation could rapidly begin eroding the clathrate zone of stability. This intriguing possibility has been noted by Clarke et al. (1986), who believe that some 150 plumes seen by NOAA satellites near Bennett Island in the Soviet far Arctic are produced by methane released from clathrates. The upward-moving methane expands in the atmosphere, and, on cooling, forms ice nuclei to cause the plumes. Detailed examination of the reported plumes could provide important data with respect to the influence of decomposing clathrates on atmospheric chemistry.

Conclusions. Limited data from ocean drilling and seismic exploration leads to the conclusion that large quantities of carbon ($\sim$ 10,000 Gt) are stored in clathrates in carbon-rich sediments on continental shelves and in ocean basins. Much smaller quantities ($\sim$ 400 Gt) of carbon are currently locked in clathrates under permafrost regions of the world. The thermodynamic stability of clathrate limits the depth at which clathrate will be found to about 1,000 to 1,500 m. The nearness of the clathrate stability zone to the surface implies that changes in temperature and pressure at the surface will result in the destabilization or stabilization of clathrates over times on the order of several hundred to many thousands of years. The time scales are determined by the thermal properties of near-surface sediments.

Both the sensitivity of clathrates to surface changes and the time scales involved suggest that during the ice ages, clathrates played a significant role in determining atmospheric composition, and, in this way, provided an important feedback mechanism that amplified other changes associated with the advance and retreat of glaciers. The possibility that large quantities of carbon can be stored in clathrates is a relatively recent discovery—a discovery whose consequences for past and future climate have not yet been fully explored.

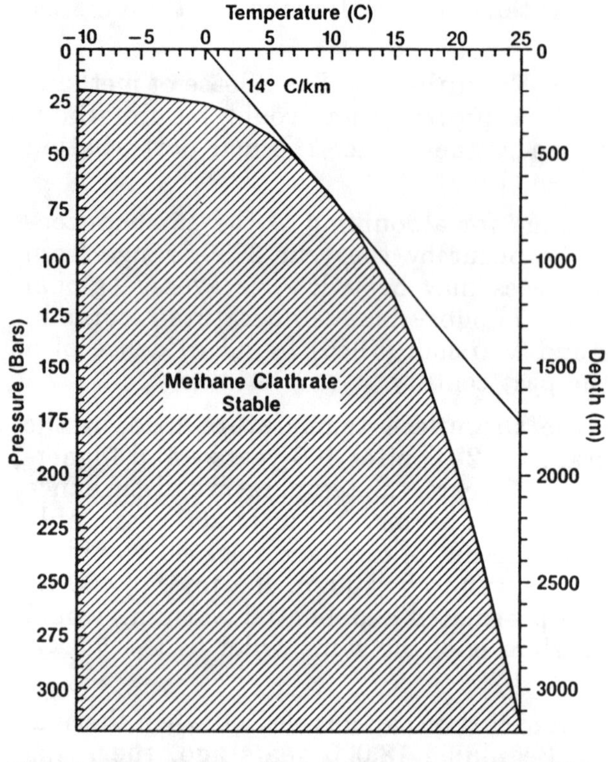

Figure 1. *Phase Diagram Showing the Boundary Between Free Methane Gas and Methane Clathrate for the H_2O-CH_4 System.*

The ice-water phase boundary would be a vertical line through 0°C. The depth scale is drawn assuming hydrostatic equilibrium for pore water. The limiting geothermal gradient of 14°C/km for the stability of methane clathrate outside permafrost regions is also indicated.

Figure 2. *Geographical Distribution of Methane Clathrate Occurrences.*

The map is updated from Kvenvolden and McMenamin (1980) and Kvenvolden (1988).

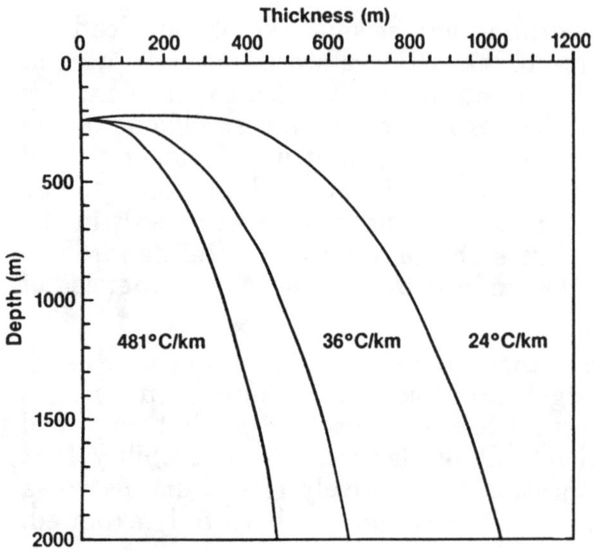

Figure 3. *Variation of Methane Clathrate Stability Zone as a Function of Ocean Depth and Thermal Gradient.*

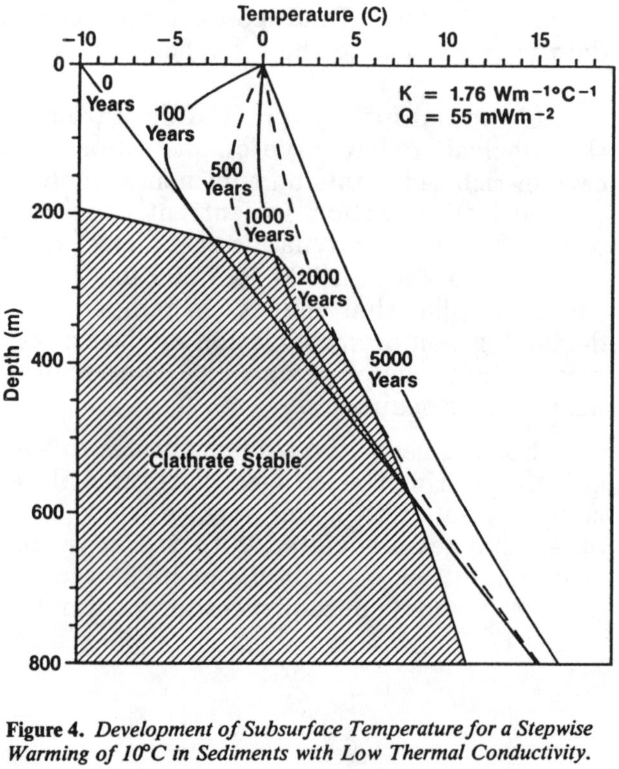

Figure 4. *Development of Subsurface Temperature for a Stepwise Warming of 10°C in Sediments with Low Thermal Conductivity.*

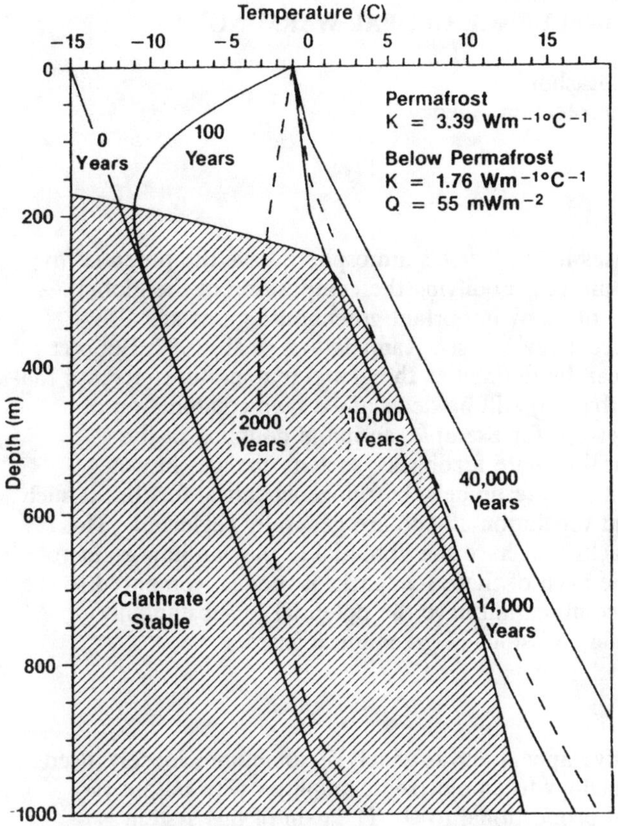

Figure 5. *Development of Subsurface Temperature Field Following an Initial Stepwise Warming of 14°C.*

The initial permafrost region extends to a depth of 920 m; the initial clathrate stability zone extends to a depth of 1200 m.

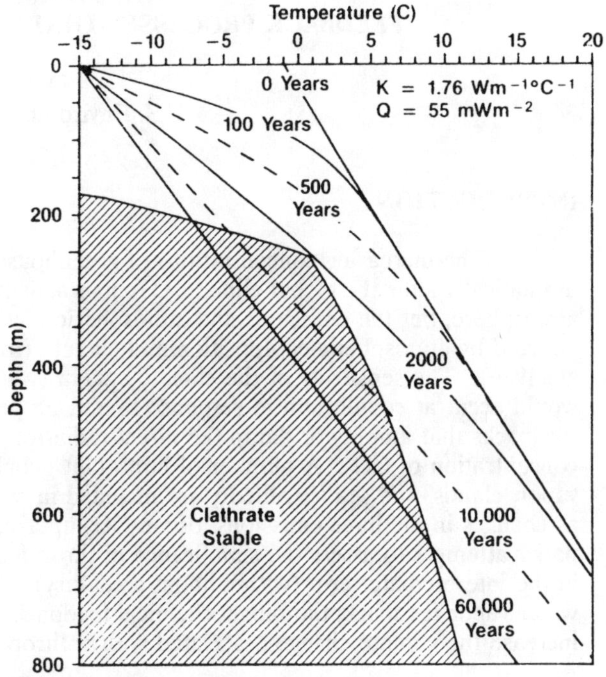

Figure 6. *Development of Clathrate Stability Zone as a Result of Surface Cooling.*

At the initial time the surface temperature drops by 14°C.

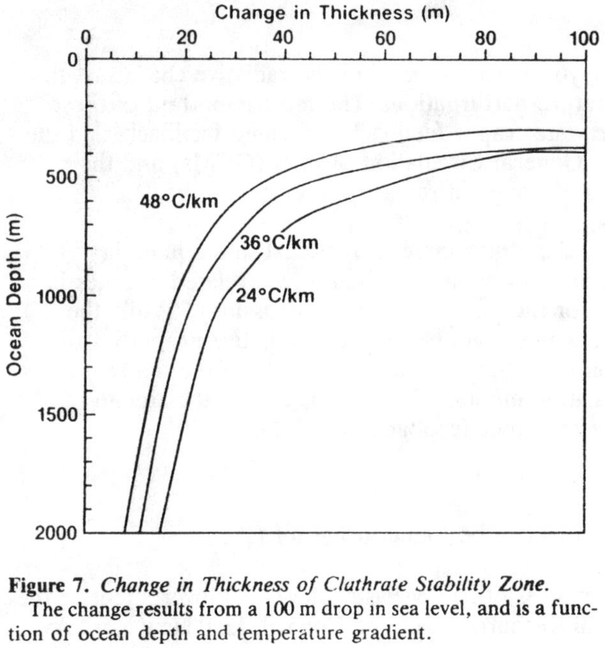

Figure 7. *Change in Thickness of Clathrate Stability Zone.*

The change results from a 100 m drop in sea level, and is a function of ocean depth and temperature gradient.

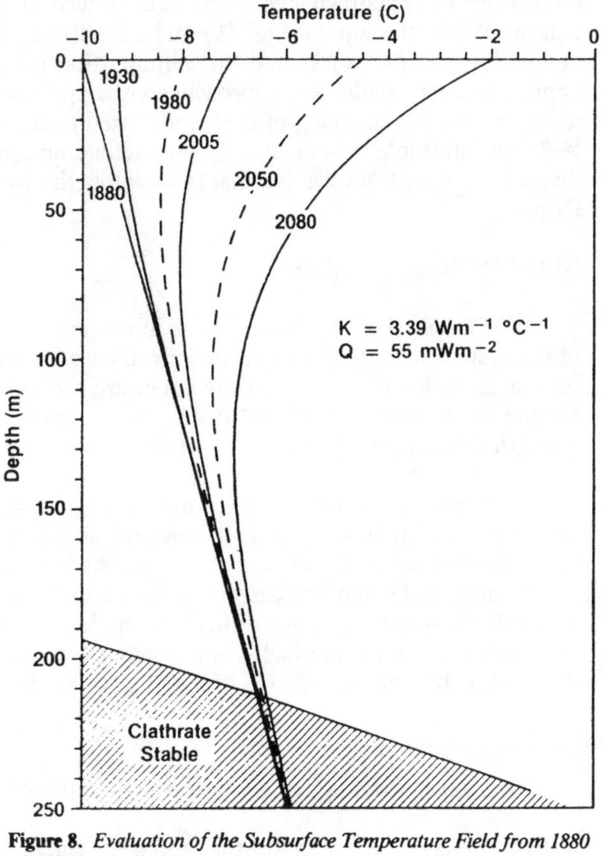

Figure 8. *Evaluation of the Subsurface Temperature Field from 1880 to 2080.*

The field is based on the assumption that the surface temperature increases according to $e^{\lambda t}-1$ with $\lambda = 0.011\ yr^{-1}$. The initial equilibrium temperature gradient is taken as $T = -10 + 0.01622\ z$.

THE DYNAMIC GREENHOUSE: FEEDBACK PROCESSES THAT CAN INFLUENCE GLOBAL WARMING*

Daniel A. Lashof**
U.S. Environmental Protection Agency

INTRODUCTION

The increasing concentrations of greenhouse gases in the Earth's atmosphere have the potential to dramatically alter the global climate. Human activity is directly modifying the composition of the global atmosphere, but the emissions and atmospheric lifetimes of many important greenhouse gases are also affected by atmospheric composition and climate (for recent reviews see Ramanathan et al., 1987; Bolin et al., 1986). The sensitivity of the climate system (which can be defined as the global temperature increase that would occur at equilibrium in response to a given perturbation) will be determined by a combination of feedbacks that amplify or damp the direct radiative effects of, for example, an initial doubling of the concentration of CO_2. A number of important geophysical climate feedbacks, such as changes in water vapor, clouds, and sea ice albedo, are included in current climate models, but biogeochemical feedbacks such as changes in methane emissions, ocean CO_2 uptake, and vegetation albedo are generally neglected. This paper attempts to assess the significance of those feedbacks that have the potential to play an important role in the intermediate-term (within about a century). While each of these feedbacks is modest compared to the water vapor feedback, the biogeochemical feedbacks in combination have the potential to substantially increase the climate change associated with anthropogenic emissions of greenhouse gases.

FEEDBACKS

The concept of feedback has considerable intuitive appeal and is widely, if sometimes loosely, used in discussions of climate change. The term is derived by analogy to an electronic amplifier with output (ω) determined by the input signal (ι) and a feedback signal proportional to ω. The gain of this system is defined by $g = (\omega - \iota)/\omega$, and the amplification is $\omega/\iota = 1/(1 - g)$. Note that as g approaches 1, ω approaches $+\infty$, while as g approaches $-\infty$, ω approaches 0. These simple relationships are emphasized to make precise the meaning of commonly used terms such as negative feedback ($g < 0$), positive feedback ($g > 0$), and unstable system ($g \to 1$). A useful consequence of the way gain is defined is that the gain for a linear system with several feedbacks is simply the sum of the gains associated with each individual feedback loop.

GEOPHYSICAL FEEDBACKS

Geophysical feedbacks in the atmosphere-ocean-cryosphere system alter the radiative characteristics of the system in response to an initial radiative or temperature perturbation. The most important of these feedbacks, at least on a short time scale, are probably the water vapor feedback, the cloud feedback, and the ice and snow feedback. These feedbacks are simulated in General Circulation Models (GCMs) and their strength determines the sensitivity of the climate system to a change in radiative forcing.

The **water vapor** feedback arises because warming the atmosphere increases absolute humidity (relative humidity is more-or-less constant) and water vapor is a potent greenhouse gas (related changes in the lapse rate are included with the water vapor feedback for the purposes of this discussion). While the significance of this feedback is not in doubt, its magnitude is uncertain because the radiative properties of atmospheric water vapor are difficult to model. Dickinson (1986) evaluated the strength of the major geophysical feedback mechanisms in various general circulation models. He obtains a central estimate of 0.39 and a 2σ range of 0.28-0.52 for the gain due to the water vapor feedback (Table 1).

---

*This paper summarizes a more complete discussion which can be found in Lashof (1989).

** The views expressed are the author's: They do not express official views of the U.S. Government or the Environmental Protection Agency. Current address: Natural Resourced Defense Council, 1350 New York Ave., NW, Washington D.C. 20005.

Changes in **ice and snow cover** contribute a positive feedback because warming the Earth reduces the planetary albedo and increases heat transfer from the ocean to the atmosphere by reducing the extent and persistence of sea-ice and snow cover. The importance of the albedo feedback is limited by vegetation and cloud masking of the surface reflectance. Dickinson (1986) obtains a central estimate of 0.12 and a range of 0.03-0.21 for the gain due to changes in ice and snow cover (Table 1).

The **cloud** feedback is probably the most complex and uncertain geophysical feedback in the climate system. In general, increases in the fractional cover or optical depth of low clouds would produce a negative feedback while increases in high (cirrus) clouds and increases in cloud altitude would produce a positive feedback. Hansen et al. (1984) fix cloud optical properties and find a substantial positive feedback (0.22). Roeckner et al. (1987) and Somerville and Remer (1984) argue that the liquid water content of clouds will increase with warming substantially altering their optical properties. In the model calculations of Roeckner et al. (1987) this produces a large negative feedback at the surface, more than offsetting the positive feedback from changes in cloud cover. Dickinson's analysis gives a central estimate of the gain due to changes in clouds of 0.09 and a range of -0.12 to 0.29 (Table 1).

It is the uncertainty in the net effect of the geophysical climate feedbacks that determines the quoted uncertainty in the equilibrium response of the climate system to a given perturbation, such as doubling CO_2. For this perturbation the no-feedback surface temperature increase would be 1.2-1.3°C (Ramanathan et al., 1987), while the resulting temperature change with feedbacks included is generally considered to be 1.5-4.5°C (National Research Council, 1979, 1983). Recent results, however, suggest that 1.5-5.5°C may be a more appropriate uncertainty range (Dickinson, 1986; Wilson and Mitchell, 1987). For the range 1.5-5.5°C due to doubling CO_2, the gain due to the geophysical feedback processes is 0.17-0.77 (Table 1).

BIOGEOCHEMICAL FEEDBACKS

Biogeochemical feedbacks are those that involve the response of the biosphere and components of the geosphere not considered in typical climate models. Included here, for example, are feedbacks that involve changes in the sources and sinks of greenhouse gases, and changes in surface properties such as albedo and transpiration that are mediated by the intermediate and long term response of land vegetation.

Release of Methane Hydrates

Potentially the most important biogeochemical feedback is the release of CH_4 from near-shore ocean sediments. Methane hydrates are formed when a methane molecule is included within a lattice of water molecules; there can be as much as one methane molecule for every six water molecules (Bell, 1982). The hydrate structure is stable under particular temperature and pressure conditions that can be found near the top of ocean sediment layers, and may be subject to destabilization as climate changes. Assuming that a 2°C global warming causes a 1°C temperature increase at the water-sediment interface I estimate a methane release rate of 220×10^{12} grams per year (Tg/y), based on Bell (1982) and Revelle (1983). I further assume that the fractional change in the mehtane mixing ratio at chemical equilibrium is 1.5 times the fractional change in methane flux due to chemical interactions that increase the lifetime of methane with increasing emissions (Thompson and Cicerone, 1986). Thus an increase in the flux of methane by 220 Tg/y would increase the methane concentration by 1.2 parts per million (ppm). Finally I assume that the radiative forcing due to changes in the methane mixing ration is enhanced by 70% to 0.17°C/ppm due to increased tropospheric ozone and stratospheric water vapor related to methane oxidation (Owens et al., 1985; Brasseur and De Rudder, 1987). Thus the gain from methane hudrate releases is given by (1.2 ppm)x(0.17°C/ppm)/2°C = 0.1. I subjectively assign a very wide uncertainty range of 0.01-0.2 to account for the large range in the estimates of total hydrates cited above, the possible oxidation of some methane in the water column, and the possibility that most of the hydrate zone is bathed in water that is thermally insulated from the atmosphere as argued by Kvenvolden (1988).

Tropospheric Chemistry

Climate-chemistry interactions have been reviewed recently by Ramanthan et al. (1987). In addition to the CH_4-CO-OH coupling and the stratospheric water vapor amplification discussed above, the chemical feedback with the largest climatic impact may be due to changes in OH resulting from changes in water vapor. A warmer Earth would have higher absolute humidity and thus more OH from the reaction $O(^1D)$ + H_2O ---> 2OH. This will, in turn, reduce CH_4 and tropospheric O_3 concentrations, resulting in a negative feedback on climatic change. Hameed and Cess (1983) find that climate-chemical feedback reduces tropospheric O_3 by 11% and CH_4 by 17% assuming that CH_4, CO, and NO_x fluxes remain constant. This produces a gain of -0.04 in thier model. This value depends strongly on the assumed background level of NO_x. In high-NO_x areas higher temperatures could increase O_3 formation, possibly reversing the sign of this feedback (Ramanathan et al., 1987). As current, and certainly future, NO_x distributions are highly uncertain, I assign a broad uncertainty bound to the gain from this feedback of -(0.01-0.06).

Oceanic Change

The oceans are the dominant factor in the Earth's thermal inertia to climate change as well as the dominant sink for anthropogenic CO_2 emissions. The ocean biota play an important role in carrying carbon (as organic debris) from the mixed layer to deeper portions of the ocean (see, for example, Sarmiento and Toggweiler, 1984). Thus, changes in ocean chemistry, biology, mixing, and large-scale circulation have the potential to substantially alter the rate of CO_2 accumulation in the atmosphere and the rate of global warming, and because the oceans are such an integral part of the climate system, significant changes in the oceans are likely to accompany a change in climate.

Ocean chemistry. The most straight-forward ocean feedback is on carbonate chemistry. As the ocean warms the solubility of CO_2 decreases and the carbonate equilibrium shifts toward carbonic acid; these effects combine to increase the partial pressure of CO_2 (pCO_2) in the ocean by 4-5%/°C for a fixed alkalinity and total carbon content. Because the total carbon content would only have to decrease by about one-tenth this amount to restore pCO_2 to its previous level, the impact of this feedback is to increase atmospheric CO_2 by about 1%/°C for a typical scenario according to calculations with and without coupling between temperature and carbonate chemistry, in a one-dimensional ocean CO_2 and heat uptake model. This results in a gain of 0.008 based on the temperature change realized in 2100.

Ocean Mixing. As heat penetrates from the mixed layer of the ocean into the thermocline the stratification of the ocean will increase and mixing can be expected to decrease, resulting in slower uptake of both CO_2 and heat. This feedback raises the surface temperature that can be expected in any given year for two reasons: First, the atmospheric CO_2 concentration will be higher because the oceans will take up less CO_2. Second, the realized temperature will be closer to the equilibrium temperature due to reduced heat transport into the deep ocean. The strength of this feedback is estimated by letting the mixing parameter in the model described above be proportional to $(\partial T/\partial z)^{-2}$, where $\partial T/\partial z$ is the temperature gradient at the base of the mixed layer. The effect is to decrease the mixing coefficient 30% by 2100. This results in a gain of 0.02.

Ocean biology and circulation. A more speculative, but potentially more significant, feedback involves a reorganization of the atmosphere-ocean circulation system as suggested by Broecker (1987). This possibility is suggested by the apparently very rapid changes in the CO_2 content of the atmosphere during glacial-interglacial transitions as revealed by ice-core measurements (eg., Jouzel et al., 1987). Only shifts in carbon cycling in the ocean are thought to be capable of producing such large, rapid, and sustained changes in atmospheric CO_2 (Keir, 1988; Sarmiento and Toggweiler, 1984; Seigenthaler and Wenk, 1984; Knox and McElroy, 1984). Because these changes in the ocean-atmosphere system may have been discontinuous (Broecker et al., 1985) the gain from this process in the future may be quite different from what it has been in the past; nonetheless, lacking coupled ocean-atmosphere GCMs, this is the only available basis for making an estimate. Assuming a glacial-interglacial temperature change of 5°C caused atmospheric CO_2 to increase by 80 ppm the gain from this feedback was 0.11. For the same relationship between temperature and CO_2 (16 ppm/°C) the gain would be lower in the future due to saturation of the CO_2 bands. If this feedback occurred when the CO_2 concentration was 500 ppm then the gain would be 0.06. Given the arbitrary assumptions required to make this estimate, I assign an uncertainty range of 0.0-0.1 for the gain from the ocean circulation/biology feedback.

A fundamentally different and potentially highly significant feedback involving ocean biology has been proposed by Charlson et al. (1987). Dimethyl sulfide (DMS) emitted by marine phytoplankton may act as cloud condensation nuclei (CCN) in remote marine locations where CCN are scarce, thereby affecting cloud optical properties. As noted above, changes in cloud optical properties can have a substantial influence on climate. Climate presumably affects biogenic DMS production, but the relationship is complex and poorly understood at this time (Charlson et al., 1987). DMS production is not directly related to primary productivity and interspecific differences in DMS production are large and may dominate direct climatic effects. While this mechanism was originally proposed as a potential negative feedback consistent with the Gaia Hypothesis, ice-core data indicate that non-sea-salt-sulfate aerosol levels were higher during the last glacial maximum, suggesting that biogenic DMS production may act instead as a positive feedback (Legrand et al., 1988).

Terrestrial Biota

The terrestrial biota interact with climate in a wide variety of important ways. The discussion here will emphasize the effects of large-scale reorganization of terrestrial ecosystems as well as the direct effects of temperature and CO_2 changes.

Vegetation albedo. Probably the most significant global feedback produced by the terrestrial biota, on a decades-to-centuries time scale is due to changes in surface albedo (reflectivity) as a result of changes in the distribution of terrestrial ecosystems. Dickinson and Hanson (1984) examined the differences in albedo between the CLIMAP vegetation reconstruction for 18,000 years before present and current vegetation and found that the planetary albedo was 0.0022 higher at the glacial maximum due to differences in mean annual vegetation albedo. A similar result was obtained by Hansen et al. (1984) using a prescriptive scheme to relate vegetation type to climate in GCM simulations for current and glacial times. This represents about a 1% increase relative to the current albedo of 0.3, and results in a gain of 0.06-0.08; this range was extended to 0.05-0.09 by Hansen et al. (1984) to account for uncertainties in the vegetation parameterization. Lacking other analysis, I assume that the low end of the range of gain calculated by Hansen et al. will apply in the future (namely, 0.05). This assumption is not unreasonable considering that the most important process--decreased albedo due to a poleward shift in the tundra-boreal forest boundary--should continue during the anticipated interglacial to super-interglacial transition. It is, however, also possible that increases in grassland and desert area, as a direct result of human activity or from climatic change, could increase albedo, partially compensating for decreased albedo at high latitudes. I therefore broaden the range assigned to this feedback to 0.0-0.09.

Carbon storage. Other significant feedbacks are related to the role of the terrestrial biosphere as a source and sink for CO_2 and CH_4. The carbon stored in live biomass and soils is roughly twice the amount in atmospheric CO_2, and global net primary production (NPP) by terrestrial plants absorbs about 10% of the carbon held in the atmosphere each year. On average this is nearly balanced by decay of organic matter, about 0.5-1% of which is anaerobic and thus produces CH_4 rather than CO_2. Small shifts in the balance between NPP and respiration, and/or changes in the fraction of NPP routed to CH_4 rather than CO_2, could therefore have a substantial impact on the overall greenhouse forcing, because CH_4 has a much larger greenhouse effect than CO_2 per molecule. Both NPP and respiration rates are largely determined by climate and NPP is directly affected by the CO_2 partial pressure of the atmosphere, thus the potential for a substantial feedback exists.

The relationship between climate and carbon storage is not straightforward. Various approaches can be used to estimate changes in carbon storage resulting from climate change, including life-zone classifaction systems (Emanuel et al., 1985a, b), statistical models (Lashof, 1987), and dynamic models (Solomon, 1986; Kholmaier, 1988; Woodwell, 1986). Collectively these approaches suggest that increased soil respiration in the boreal zone is likely to cause a release of CO_2 on the order of 0.25 PgC y^{-1} $°C^{-1}$. Stress on other ecosystems globally seems likely to contribute an additional flux of perhaps the same magnitude. Modeling this feedback by inserting a flux of 0.5 PgC y^{-1} $°C^{-1}$ into the carbon-cycle/temperature model described above yields a gain of 0.01; a range of 0-0.03 probably characterizes the uncertainty.

Carbon dioxide fertilization. Characterizing the global response of the biosphere to CO_2 fertilization is an equally thorny problem. While the short-term response of a number species has been measured, generalizing to long-term carbon storage at the ecosystem or global level is difficult and controversial (e.g., Gates, 1985; Houghton, 1987, 1988; Idso, 1988). In a review of global biospheric stimulation by CO_2, Gates (1985) concludes that the fractional change in NPP will probably be 0.25-0.5 times the fractional change in atmospheric CO_2. NPP and biomass are not directly related however, as is apparent from the difference between grassland and forest ecosystems, which may have similar annual productivity but vastly different carbon stocks. If it is assumed that forest biomass is limited by leaf area index and moisture transport (Cooper, 1988), then **maximum** stand biomass may not increase with increasing CO_2. On the other hand, open areas within a forest--gaps--would be filled in faster, and **average** stand biomass should increase. This hypothesis was investigated by Shugart (1984) by arbitrarily increasing the intrinsic growth rate of all species in a number of gap dynamic stand simulation models. With growth stimulations of 50-100% average biomass increased by 0.2-0.6 times the assumed increase in NPP. Using the center of both Shugart's and Gates' ranges implies that doubling CO_2 would lead to about a 15% increase in global biomass. Applying this relationship in the carbon cycle model described above assuming no lag between CO_2 increases and biomass increases, I find that CO_2 fertilization produces a gain of -0.013. This calculation does not take into account increases in soil carbon storage resulting from the higher rate of organic matter input. Kohlmaier (1988) found this to be an important contribution to the net flux to the biosphere from CO_2 fertilization. Considering Kohlmaier's result and the ranges given by Gates and Shugart, I adopt a central estimate for the fertilization feedback of -0.02 and a range of -0.01 to -0.04.

Methane emissions. The methane branch of the organic matter decay chain will also be perturbed by climatic change. Changes in water balance may be a critical factor, particularly in determining the land area subject to methanogenesis, as flooding seems to be the essential ingredient in producing the necessary anaerobic conditions. Unfortunately changes in wetland area cannot currently be predicted, so I assume here that area does not change and consider only the temperature effect on methanogenesis. For natural wetlands I further limit my attention to high-latitude bogs because coastal wetlands are not significant sources of methane and low-latitude wetlands tend to be carbon limited, suggesting that temperature change *per se* will not have a significant impact on their annual methane emissions. Methane emissions from bogs can be expected to increase as the result of two complementary factors: the length of the emissions season will increase because the period when the ground is frozen will decrease and the average emissions rate will increase because of the direct temperature response of microbes. The perturbation to bog emissions is estimated as the product of the season length and temperature effects summed over latitude bands (Lashof and Fung, unpublished). Emissions estimates are based on bog area from Matthews and Fung (1987) aggregated to the latitude bands of the GISS GCM, and values for the season length and temperature changes were derived from zonally averaged surface temperature values over land from GISS $2xCO_2$ and $1xCO_2$ results (Hansen et al., 1983).

Given the uncertainty that emerges from reviewing the literature on the relationship between methane emissions and temperature I have estimated the strength of this feedback using values for Q_{10} (the factor by which the methane emission rate increases for a 10°C temperature increase) between 1 (no temperature effect) and 4. It seems unlikely that the higher values reported in some of the literature would apply to annual average emission rates. Very high Q_{10} values probably reflect either the low-temperature start-up of methanogenesis (Q_{10} approaches infinity as the soil temperature approaches the freezing point) or seasonal variations involving covariance between temperature and other controlling variables (e.g. moisture and organic matter availability). My central estimate of the gain from this feedback is 0.01 (Table 1), based on $Q_{10}=3$, and the range (0.003-0.015) allows for the possibility that emissions from low-latitude wetlands could also increase somewhat. A similar estimate can be made for increased emissions from rice. My central estimate of 0.006 (Table 1) assumes that $Q_{10}=3$, current emissions are 100 Tg/y, and a global temperature increase of 4.2°C warms the major rice growing areas by 2.5°C. Assumptions based on the central estimate of Holzapfel-Pschorn and Seiler (1986)--$Q_{10}=4$, current emissions=120 Tg/y--yields a gain of 0.01, which is probably near the upper limit. I set the lower limit to 0, based on the possibility that emissions from rice paddies are carbon, rather than temperature limited.

Other Terrestrial Biotic Emissions

The biosphere plays an important role in emissions of various other atmospheric trace gases and these emissions are also likely to be influenced by climatic change. Although I will make no attempt to quantify these effects here, it is worthwhile to at least note some of the potentially important processes. For example, as much as half of nitrous oxide emissions are attributed to microbial processes in natural soils (Bolle et al., 1986). Emissions of N_2O tend to be episodic, depending strongly on the pattern of precipitation events in addition to temperature and soil properties (Sahrawat and Keeney, 1986). Thus it seems likely that climatic change would be accompanied by significant changes in N_2O emissions, although there may not be sufficient understanding of the microbiology to develop predictive models at present. The biosphere is also a key source of atmospheric non-methane hydrocarbons (NMHCs) which play an important role in global tropospheric chemistry; the oxidation of NMHCs generates a substantial share of global carbon monoxide, therefore influencing the concentration of OH and the lifetime of methane (Mooney et al., 1987; Thompson and Cicerone, 1986), and NMHCs participate in tropospheric O_3 formation. As much as 0.5-1% of photosynthate is lost as isoprene and terpene (Mooney et al., 1987). Lamb et al. (1987) found that biogenic NMHC emissions in the United States are about a factor of two greater than anthropogenic emissions. The ratio for the globe is probably greater. Emissions, at least for isoprene and α-pinene are exponentially related to temperature (Lamb et al., 1987; Mooney et al., 1987). The first-order impact of climatic change, then, is to increase NMHC emissions, producing a positive feedback through the CO-OH-CH_4 and tropospheric O_3 links. The actual impact when changes in ecosystem distribution are considered is uncertain, however, as different species have very different emissions (Lamb et al., 1987).

TABLE 1. ESTIMATED GAIN FROM CLIMATE AND BIOGEOCHEMICAL FEEDBACKS

FEEDBACK	GAIN	SOURCE
Geophysical		
Water Vapor[a]	0.39 (0.28-0.52)	
Ice and Snow	0.12 (0.03-0.21)	Dickinson (1986)
Clouds	0.09 (-0.12-0.29)	
Sub-Total[b]	0.64 (0.17-0.77)	
Biogeochemical		
Methane Hydrates	0.1 (0.01-0.2)	after Revelle (1983)
Tropospheric Chemistry	-.04 -(0.01-0.06)	Hameed and Cess (1983)
Ocean Chemistry	0.008	$\partial \ln(pCO_2)/\partial T = 4\%$
Ocean Eddy-Diffusion	0.02	$1/K \propto (\partial T/\partial Z)^2$
Ocean Biology & Circulation	0.06 (0.0-0.1)	after Sarmiento et al. (1984)
Vegetation Albedo	0.05 (0.0-0.09)	after Hansen et al. (1984), Dickinson and Hanson (1984)
Vegetation Respiration	0.01 (0.0-0.03)	Flux = 0.5 Pg y^{-1} $°C^{-1}$
CO_2 Fertilization	-.02 -(0.01-0.04)	15% biomass increase for $2xCO_2$
Methane from Wetlands	0.01 (0.003-0.015)	Lashof & Fung, unpublished
Methane from Rice	0.006 (0.0-0.01)	after Holzapfel-Pschorn & Seiler (1986)
Electricity Demand	0.001 (0.0-0.004)	after Linder and Inglis (1988)
Sub-Total[c]	0.16 (0.05-0.29)[d]	
TOTAL	**0.80 (0.32-0.98)**[d]	

NOTES:

[a] Includes the lapse rate feedback and other geophysical climate feedbacks not included elsewhere.

[b] Based on 1.5-5.5 K for doubling CO_2. The individual values do not sum to these values--see Dickinson (1986) for details.

[c] Based on selected biogeochemical feedbacks, considering which could occur together during the next century. See **Discussion** for details.

[d] Ranges are combined using a least-squares approach. That is, by letting $C - L = [\Sigma(c_i - l_i)^2]^{0.5}$, where C is the central estimate for the total, c_i is the central estimate for feedback i, L is the lower uncertainty bound for the total, and l_i is the lower uncertainty bound for feedback i. A similar calculation is performed for the upper uncertainty bounds.

DISCUSSION

It inappropriate to simply add all of the estimates displayed in Table 1 because the feedback processes may not be independent and may not occur at the same time. Hansen at al. (1984) have shown, for example, that the heat capacity of the oceans makes the response time of the system proportional to the square of the amplification due to geophysical climate feedbacks that act essentially instantaneously. Consideration of biogeochemical feedbacks introduces additional time-constants which may further delay the attainment of equilibrium. A "best estimate" of the overall impact of biogeochemical feedbacks over the next century therefore requires selectively combining the feedbacks in Table 1.

Combining the feedbacks that it appears could operate simultaneously over the next century: methane hydrates, 0.1; tropospheric chemistry, -0.04; ocean biology and circulation, 0.06; vegetation albedo, 0.025; wetlands, 0.01; and rice, 0.006, yields a "best estimate" of 0.16 for the gain from biogeochemical feedbacks. Adding this estimate to the gain from geophysical feedbacks based on the review by Dickinson (1986) gives a total gain of 0.80, which would increase the climate sensitivity to an initial doubling of CO_2 from 3.5°C to 6.3°C. Quadratically combining the error estimates gives a range of 0.05-0.29 for the biogeochemical feedbacks and 0.32 to 0.98 for the system as a whole, implying a climate sensitivity of 1.9 to >10°C.

Although the high end of the range calculated for the gain of the entire system implies an amplification factor of 50, this estimate cannot be taken seriously. Linearity, implicitly assumed in using gain to calculate amplification, is certainly not valid for temperature excursions much beyond the 2-5°C range used to calculate the gains. Saturation of the H_2O, CO_2 and CH_4 absorption bands, geographical limits on the ice, snow, and vegetation albedo feedbacks, and limits on the supply of carbon for the vegetation respiration and wetland feedbacks all tend to stabilize the system near the upper limit of sensitivity. Nonetheless, the analysis presented here shows, as suggested by Kellogg (1983), that biogeochemical feedbacks have the potential to make the climate system substantially more sensitive to perturbations than is generally assumed.

CONCLUSION

The largest feedbacks that will come into play during the next century are almost certainly the geophysical climate feedbacks (water vapor, clouds, ice and snow albedo). The other feedbacks discussed here are individually rather modest in comparison--each probably has a gain of $\leq$ 0.1, compared to 0.4 for the water vapor feedback. In a high-gain system, however, the total amplification is very sensitive to small additional gains. Thus, while there are many uncertainties, there is the potential for the biogeochemical feedbacks discussed here to substantially increase the overall sensitivity of the climate system. A climate sensitivity as great as 8° or even 10°C for an initial radiative forcing equivalent to doubling CO_2 cannot be ruled out. This analysis also gives an indication of the relative magnitude of various feedback processes, suggesting that the most important may be methane hydrates, ocean mixing and circulation, and vegetation albedo.

ACKNOWLEDGEMENTS

This work would not have been possible without the help of I. Fung, who provided important discussions, much of the computer code used to construct the ocean model, data used to calculate the wetlands feedback, and critical comments on a draft of the manuscript. Helpful comments were also provided by R. Dickinson, J. Hoffman, P. Martin, P. Tans, D. Tirpak, and two anonymous reviewers. An early version of this paper was presented at the Chapman Conference on the Gaia Hypothesis, San Diego, March 1988.

REFERENCES

Bell, P. 1982. Methane Hydrate and the Carbon Dioxide Question. In W. Clark, ed., *Carbon Dioxide Review: 1982.* Oxford University Press, New York. pp. 255-277.

Bolle, H., W. Seiler, and B. Bolin. 1986. Other Greenhouse Gases and Aerosols. In Bolin, B., B. Doos, J. Jager, and R. Warrick, eds. *The Greenhouse Effect, Climate Change, and Ecosystems.* John Wiley and Sons, Chichester. pp. 157-203.

Bolin, B., B. Doos, J. Jager, and R. Warrick, eds. 1986. *The Greenhouse Effect, Climate Change, and Ecosystems.* John Wiley and Sons, Chichester.

Brasseur, G. and A. De Rudder. 1987. The Potential Impact on Atmospheric Ozone and Temperature of Increasing Trace Gas Concentrations. *Journal of Geophysical Research* 92(D9):10903-10920.

Broecker, W. 1987. Unpleasant Surprises in the Greenhouse?. *Nature* 328:123-126.

Broecker, W., D. Peteet, and D. Rind. 1985. Does the Ocean-Atmosphere System Have More than One Stable Mode of Operation. *Nature* 315:21-26.

Charlson, R., J. Lovelock, M. Andreae, and S. Warren. 1987. Oceanic Phytoplankton, Atmospheric Sulphur, Cloud Albedo and Climate. *Nature* 326:655-661.

Cooper, C. 1988. Controls on Global Biomass. Paper presented at the Chapman Conference on Gaia Hypothesis. March 7-11, 1988.

Dickinson, R. 1986. The Climate System and Modelling of Future Climate. In B. Bolin, B. Doos, J. Jager, and R. Warrick, eds., *The Greenhouse Effect Climatic Change and Ecosystems.* John Wiley & Sons, Chichester. pp. 207-270.

Dickinson, R. and B. Hanson. 1984. Vegetation-Albedo Feedbacks. In J. Hansen and T. Takahashi, eds., *Climate Processes and Climate Sensitivity.* Geophysical Monograph 29, Maurice Ewing Volume 5. American Geophysical Union, Washington, D.C. p. 180-186.

Emanuel, W., H. Shugart, and M. Stevenson. 1985a. Climate Change and the Broad Scale Distribution of Terrestrial Ecosystem Complexes. *Climatic Change* 7:29-43.

Emanuel, W., H. Shugart, and M. Stevenson. 1985b. Response to Comment: Climatic Change and the Broad-Scale Distribution of Terrestrial Ecosystem Complexes. *Climatic Change* 7:457-460.

Gates, D. 1985. Global Biospheric Response to Increasing Atmospheric Carbon Dioxide Concentration. In B. Strain and J. Cure, eds., *Direct Effects of Increasing Carbon Dioxide on Vegetation.* Report DOE/ER-0238, U.S. Department of Energy, Washington, DC.

Hameed S., and R. Cess. 1983. Impact of a Global Warming on Biosperic Sources of Methane and its Climatic Consequences. *Tellus* 35B:1-7.

Hansen, J., A. Lacis, D. Rind, G. Russell, P. Stone, I. Fung, R. Ruedy, and J. Lerner. 1984. Climate Sensitivity: Analysis of Feedback Mechanisms. In J. Hansen and T. Takahashi, eds., *Climate Processes and Climate Sensitivity.* Geophysical Monograph 29, Maurice Ewing Volume 5. American Geophysical Union, Washington, D.C. pp. 130-163.

Hansen, J., G. Russell, D. Rind, P. Stone, A Lacis, S. Lebedeff, R. Ruedy and L. Travis. 1983. Efficient Three-Dimensional Global Models for Climate Studies: Model I and II. *Monthly Weather Review* 111:609-662.

Holzapfel-Pschorn, A. and W. Seiler. 1986. Methane Emission During a Cultivation Period from an Italian Rice Paddy. *Journal of Geophysical Research* 91:11803-11814.

Houghton, R. 1987. Biotic Changes Consistent With the Increased Seasonal Amplitude of Atmospheric CO_2 Concentrations. *Journal of Geophysical Research* 92:4223-4230.

Houghton, R. 1988. Reply to S. Idso, Comment on 'Biotic Changes Consistent With Increased Seasonal Amplitude of Atmospheric CO_2 Concentrations'. *Journal of Geophysical Research* 93:1747-1748.

Idso, S. 1988. Comment on 'Biotic Changes Consistent With Increased Seasonal Amplitude of Atmospheric CO_2 Concentrations' by R. A. Houghton. *Journal of Geophysical Research* 93:1745-1746.

Jouzel, J., C. Lorius, J. Petit, C. Genthon, N. Barkov, V. Kotlyakov and V. Petrov. 1987. Vostok Ice Core: A Continuous Isotope Temperature Record Over the Last Climate Cycle (160,000 Years). *Nature* 329:403-418.

Keir, R. 1988. On the Late Pleistocene Ocean Geochemnistry and Circulation. *Paleoceanography* 3:413-445.

Kellogg, W. 1983. Feedback Mechanisms in the Climate System Affecting Future Levels of Carbon Dioxide. *Journal of Geophysical Research* 88:1263-1269.

Knox, F. and M. McElroy. 1984. Changes in Atmospheric CO_2: Influence of The Marine Biota at High Latitude. *Journal of Geophysical Research* 89:4629-4637.

Kohlmaier, G. 1988. The CO_2 Response of the Biota-Soil System to a Future Global Temperature Change Induced by CO_2 and other Greenhouse Gases. Paper presented at the Chapman Conference on Gaia Hypothesis. March 7-11, 1988.

Kvenvolden, K. 1988. Methane Hydrates and Global Climate. *Global Biogeochemical Cycles.*

Lamb, B., A. Guenther, D. Gay, and H. Westberg. 1987. A National Inventory of Biogenic Hydrocarbon Emissions. *Atmospheric Environment* 21:1695-1705.

Lashof, D. 1987. *The Role of the Biosphere in the Global Carbon Cycle: Evaluation Through Biospheric Modeling and Atmospheric Measurement.* Ph.D. Dissertation, Energy and Resources Group, University of California, Berkeley (available from University Microfilms, Ann Arbor).

Lashof, D. 1989. The Dynamic Greenhouse: Feedback Processes that can Influence Future Concentrations of Greenhouse Gases and Global Climate. *Climatic Change,* in press.

Legrand, M. R., R. J. Delmas, and R. J. Charlson. 1988. Climate Forcing Implications from Vostok Ice-Core Sulphate Data. *Nature* 334:418-420.

Matthews, E. and I. Fung. 1987. Methane Emission From Natural Wetlands: Global Distribution, Area, and Environmental Characteristics of Sources. *Global Biogeochemical Cycles* 1:61-86.

Mooney, H., P. Vitousek, and P. Matson. 1987. Exchange of Material between Terrestrial Ecosystems and the Atmosphere. *Science* 238:926-932.

National Research Council. 1979. *Carbon Dioxide and Climate: A Scientific Assessment.* National Academy Press, Washington, D.C.

National Research Council. 1983. *Changing Climate.* National Academy Press, Washington, D.C. 496 pp.

Owens, A., C. Hales, D. Filkin, C. Miller, M. Steed, and J. Jesson. 1985. A Coupled One-Dimensional Radiative-Convective, Chemistry-Transport Model of the Atmosphere. 1. Model Structure and Steady State Perturbation Calculations. *Journal of Geophysical Research* 90(D1):2283-2311.

Ramanathan, V., L. Callis, R. Cess, J. Hansen, I. Isaksen, W. Kuhn, A. Lacis, F. Luther, J. Mahlman, R. Reck, and M. Schlesinger. 1987. Climate-Chemical Interactions and Effects of Changing Atmospheric Trace Gases. *Reviews of Geophysics* 25:1441-1482.

Revelle, R. 1983. Methane Hydrates in Continental Slope Sediments and Increasing Atmospheric Carbon Dioxide. In National Research Council, *Changing Climate.* National Academy Press, Washington, DC. pp.252-261.

Roeckner, E., U. Schlese, J. Biercamp and P. Loewe. 1987. Cloud Optical Depth Feedbacks and Climate Modelling. *Nature* 329:138-140.

Sahrawat, K., and Keeney. 1986. Nitrous Oxide Emission from Soils. *Advances in Science* 4:103-148.

Sarmiento, J., and J. Toggweiler. 1984. A New Model for the Role of the Oceans in Determining Atmospheric pCO_2. *Nature* 308:621-624.

Shugart, H. 1984. *A Theory of Forest Dynamics.* Springer-Verlag, New York, pp. 203-206.

Siegenthaler, U., and Th. Wenk. 1984. Rapid Atmospheric CO_2 Variations and Ocean Circulation. *Nature* 308:624-626.

Solomon, A. 1986. Transient Response of Forests to CO_2-induced Climate Change: Simulation Modeling Experiments in Eastern North America. *Oecologia* 68:567-579.

Somerville, R., and L. Remer. 1984. Cloud Optical Thickness Feedbacks in the CO_2 Climate Problem. *Journal of Geophysical Research* 89:9668-9672.

Thompson, A. and R. Cicerone. 1986. Possible Perturbations to Atmospheric CO, CH_4, and OH. *Journal of Geophysical Research* 91(D):10,853-10,864.

Wilson, C., and J. Mitchell. 1987. A $2xCO_2$ climate sensitivity experiment with a glbal climate model including a simple ocean. *Journal of Geophysical Research* 92:13315-13343.

Woodwell, G. 1986. Global Warming: And What We Can Do About It. *Amicus Journal* 8(Fall):8-12.

FUTURE CHANGES IN CLIMATE VARIABILITY

D. RIND
Goddard Space Flight Center
Institute for Space Studies
2880 Broadway, N.Y. 10025

For assessments of the impact of climate change on society, understanding how climate variability will change is an extremely important component. It is the extremes in temperature which are responsible for killing fruit trees in winter, or causing health problems in summer. It is the extreme hydrological conditions (floods/droughts) which affect water availability, lead to property damage or fires. While all of these effects are related to the mean conditions about which variability is occurring, the magnitude of the variability will amplify or diminish the importance of the mean condition changes.

A study was conducted of the changes in variability which arise in the GISS doubled CO_2 climate model simulation (Hansen et al., 1984) and the GISS transient climate change simulation (Hansen et al., 1988). Changes in future variability were assessed on three time scales: year-to-year (interannual) changes, day-to-day (daily) changes, and changes in the amplitude of the diurnal cycle. A full discussion of the results from this study is presented in Rind et al., (1989). Here we simply summarize the results, and put them in a broader perspective.

Temperature variability on all three time scales is generally reduced as the climate warms, although this is not true for all locations and all seasons. The reduction on the interannual and daily time scales is primarily because as climate warms, climate models indicate that the surface air temperature will warm more at higher latitudes. This "high-latitude amplification" of climate change results from high latitudes having greater atmospheric stability, so that warming is preferentially kept near the surface there. Furthermore, in the warmer climate, snow and sea ice are diminished at high latitudes, reducing the surface albedo and allowing more of the sun's energy to be absorbed.

Since high latitudes warm the most, the latitudinal temperature gradient is reduced. This has two effects: it reduces the temperature contrast of air masses which invade from low and high latitudes, and it reduces the intensity of extratropical storms, which advect different air masses into a region. Both effects reduce the variability in temperature that occurs; in the extreme case, if all the air masses were the same, or advection was nil, there would be little if any temperature variability at a particular site, except that associated with cloud cover/radiative changes. Since the latitudinal temperature contrast is not eliminated, only weakened, the reduction in variability does not occur everywhere, being experienced in about 2/3 of the model's extratropical grid boxes. The reductions are generally on the order of 10-15%, and are greatest in winter, the season for which the normal latitudinal temperature contrast is most severely reduced.

The diurnal temperature cycle is also reduced; for this time scale the effect is greatest in summer, and is associated with the increased greenhouse capacity of the atmosphere. With more CO_2 and water vapor, the loss of energy to space is inhibited, which limits nighttime cooling but does not noticeably impact daytime warming (as water vapor and CO_2 are poor absorbers of solar energy). Thus the day/night temperature contrast is reduced, again on the order of 10%, in most locations. The effect is greatest in summer when radiative processes become more important (relative to advection) and the model's cloud cover changes do not interfere.

In contrast to the reduced temperature variability, hydrologic variability tends to increase in the model as climate warms. This too has a straight-forward physical explanation. With the warmer surface air temperatures, more moisture is evaporated from the oceans. The hydrologic cycle (evaporation/precipitation) intensifies (by about 10% in the doubled CO_2 climate). The times when it rains it can thus rain more, and this produces greater contrasts with dry periods. The change in variability is thus strongly connected with the change in mean conditions, being of the same sign 3/4 of the time. Overall, 2/3 of the locations show increased hydrologic variability, with changes again on the order of 10%.

The results of this study were: temperature variability decreased and hydrologic variability increased as the climate warmed. What will this mean from a practical standpoint? The temperature variability reduction is likely to have little direct influence, since the mean temperatures are expected to be changing significantly from decade to decade. So while the variability about the mean may undergo some reduction, the fact that the mean itself is changing rapidly indicates that the temperatures experienced will be reaching new levels anyway.

The hydrologic variability increase is likely to be an underestimate of the actual effect. What was looked at specifically was precipitation. However, the effect on water resources also involves evaporation, which together with precipitation and runoff determine the available soil moisture. Evaporation increases as the climate warms, and potential evaporation (the atmosphere's demand for moisture) increases even more. The greater evaporation will reduce available moisture. In those locations where precipitation increases, the effect may be ameliorated, but where precipitation decreases, the effect will be amplified. Overall, then, the likelihood of droughts should increase substantially. This effect is quantified elsewhere in this issue (paper by Hansen et al.).

What effects have been left out that could contribute to changes in variability? The model does not include tropical storms, and recent analysis (Emanuel, 1987) indicates that the warmer ocean temperatures are likely to lead to an increase in the potential destructive power of hurricanes of 40-50%. This would augment the increased hydrologic variability discussed above.

Currently, interannual variability appears to be affected by changes in tropical ocean temperatures associated with El Nino events. While the model cannot make any specific forecast of how ocean dynamics will change, it is worth noting that the increased radiative forcing of the doubled CO_2 climate works to equalize temperatures in the eastern and western Pacific; the warming climate attempts to warm the cooler eastern Pacific more, for the heat put into the warmer western Pacific is largely used for increasing evaporation. It is this temperature contrast between the east and west Pacific that produces the most dramatic El Nino effects. Were El Ninos really to decrease with time, this would reduce interannual variability.

Other ocean circulation changes may have contributed to climate variability in the past. Specifically suggested in this regard are possible changes in North Atlantic Deep Water production (Broecker et al., 1985). Again, the model cannot comment upon this possibility; however, the largest changes in this effect seem to be associated with ice age initiation or ice melting effects, and it is not clear how the warming climate would force the system in this regard.

Finally, there has recently been some attention given to the possibility that significant interannual variability is related to solar cycle effects, as mediated by the phase of the quasi-biennial oscillation of zonal winds in the tropical lower stratosphere (Labitzke and van Loon, 1988). While this process is not well understood, and may ultimately turn out to be a statistical fluke, if it is real it implies that the climate system is much more sensitive to small amplitude perturbations than heretofore believed. The impact of the solar cycle on the received radiation is only on the order of 0.1%, and the doubled CO_2 climate is equivalent to a solar radiation increase of about 2%. The possibility thus exists that the real system may show a much larger change in variability than predicted by current models.

REFERENCES

Broecker, W.S., Peteet, D.M., and Rind, D., 1985: Does the ocean-atmosphere system have more than one stable mode of operation? Nature, 315, 21-25.

Emanuel, K.A., 1987: The dependence of hurricane intensity on climate. Nature, 326, 483-485.

Hansen, J., Lacis, A., Rind, D., Russell, G., Stone, P., Fung, I., Reudy, R., and Lerner, J., 1984: Climate sensitivity: analysis of feedback mechanisms. In Climate Processes and Climate Sensitivity, J.Hansen and T. Takahashi (eds.), American Geophysical Union, Washington, D.C. 130-163.

Hansen, J., Fung, I., Lacis, A., Lebedeff, S., Rind, D., Ruedy, R., Russell, G., and Stone, P., 1988: Global climate changes as forecast by the GISS 3-D model. J. Geophys. Res.,93, 5385-5412.

Labitzke, K. and van Loon, H., 1988: Recent work correlating the 11- year solar cycle with atmospheric elements grouped according to the phase of the quasi-biennial oscillation. Space Science Reviews, in press.

Rind, D., Goldberg, R., and Reudy, R., 1989: Change in climate variability in the 21st Century. Climatic Change, 13, in press.

IMPLICATIONS OF URBANIZATION FOR LOCAL AND REGIONAL TEMPERATURES IN THE UNITED STATES

Arthur Viterito
The George Washington University
Washington, D. C. 20052

It was known in medieval times that urban areas significantly modified their local climates. Particular attention was given to the deterioration of air quality and its unpleasant effects on personal comfort. By the 19th century, systematic investigation into the impacts of urbanization on local climates had begun and it is now apparent to the scientific community that virtually all atmospheric parameters are changed. A large body of empirical and theoretical evidence suggests that urban areas are warmer than their rural counterparts, have enhanced rainfall amounts, are generally less humid, and have reduced windspeeds. And yes, modern inquiry has proven what the medieval urbanite suspected: urban air quality is significantly poorer as many deleterious substances found in the urban environment are either absent from or are present in lowered concentrations in the surrounding countryside.

A great deal of attention has been given to the increased temperatures found in the urban environment. Oke has enumerated the causal mechanisms for this "heat island" phenomenon as (Oke, 1979):

1) Release of anthropogenic heat from building sides, roofs, stacks, chimneys, and automobiles;

2) Increased short wave radiation absorption due to urban canyon geometry;

3) Decreased net long wave (infra-red) loss due to the reduced sky view by the urban canyon;

4) Increased daytime heat storage and subsequent nocturnal release due to the thermal properties of building materials;

5) Increased sensible heat flux due to decreased evaporation from the replacement of vegetation with impervious surface materials;

6) Convergence of sensible heat in the urban core due to a reduction of wind speed;

7) Increased long and short wave absorption by particulates and other pollutants.

The amount of urban warming can be correlated with a number of urban parameters, most notable being the city's size, population, windspeeds, and amount of industrial activity. For analytical purposes, the urban population serves as an excellent predictor of heat island intensity. The author has recently sampled 30 U.S. metropolitan areas (see Table 1) for 1980 which range in size from 200,000 to 15,000,000 inhabitants and has found the average yearly urban vs. rural temperature difference to be 2.03 degrees Farenheit.

Table 1. Size and heat island intensity for 30 U.S. cities

City	1980 SMSA Population	Mean Yearly Urban/Rural Temp. Diff. (F)
New York, NY	15,590,300	5.4
Los Angeles, CA	9,479,440	3.3
Chicago, IL	6,779,800	2.5
Philadelphia, PA	4,112,930	3.0
Washington, DC	2,763,110	3.7
Boston, MA	2,678,760	2.7
Minneapolis, MN	1,787,560	3.8
San Diego, CA	1,704,350	3.0
Phoenix, AZ	1,409,280	2.0
Denver, CO	1,352,070	2.3
Portland, OR	1,026,140	1.8
Columbus, OH	833,648	2.5
Providence, RI	796,250	2.0
Norfolk, VA	770,784	1.4
Louisville, KY	761,002	2.3
Birmingham, AL	606,085	0.9
Rochester, NY	606,070	1.6
Tampa, FL	520,192	2.0
Nashville, TN	518,325	1.7
Hartford, CT	510,034	1.6
West Palm Beach, FL	487,044	1.6
Tulsa, OK	443,350	1.6
Albuquerque, NM	418,206	1.2
Austin, TX	379,560	0.6
Flint, MI	331,931	1.1
Mobile, AL	295,493	1.5
Little Rock, AR	295,133	0.3
Jackson, MS	265,051	1.0
Augusta, GA	251,250	1.2
Columbus, GA	214,591	1.5
Mean		2.03

Regression analysis shows that the intensity of urban warming is positively correlated with the logarithm of SMSA population. The form of the relationship is linear and is expressed as:

$$T\ (u-r) = -9.32 + 1.9\ LOG\ P, \quad (1)$$

where T(u-r) is the average yearly urban vs. rural temperature difference and P is the SMSA population. The analysis also yields a correlation coefficient of .86 (74% of the variance is explained), which is statistically significant.

The resulting prognosis suggests that population increases of .7% per year[1] will result in an urban temperature increase of .06 degrees Fahrenheit over the next 10 years. However, a number of cities in the "sun-belt" areas of the U.S. have been growing more rapidly than the national average. For example, the Phoenix metropolitan area grew 35% for the 1970 - 1980 period. At that rate, Phoenix can expect local temperatures to increase .29 degrees Fahrenheit per decade. For the relationship given in Equation 1, the decadal temperature increases which can be expected for a range of growth scenarios are summarized in Table 2.

Table 2. Urban temperature increases which can be expected for metropolitan populations exceeding 200,000.

% Annual Population Growth	Decadal Temperature Increase (F)
0.5	.041
1.0	.082
1.5	.123
2.0	.164
2.5	.204
3.0	.245
3.5	.285
4.0	.324

In addition to heat island intensification, continued population growth will result in the areal expansion of urbanized zones with corresponding "urbanization" of the temperature regimes in these newly developed areas. This escalating encroachment of urban development is clearly shown in the growing number of Standard Metropolitan Statistical Areas and in the number of counties that are now included in defined SMSAs (Table 3).

1 .7% is assumed based upon the national population growth projections of the U.S. Census Bureau.

Table 3. Metropolitan area characteristics since 1950 (Bogue,1985).

	1950	1960	1970	1980
No. of metropolitan areas	168	209	243	284
Percentage of population in:				
Metropolitan areas	56.1	63.0	68.6	74.8
Central cities	32.8	32.3	31.4	30.0
Surrounding areas	23.3	30.6	37.2	44.8
Nonmetropolitan areas	43.9	37.0	31.4	25.2
Number of counties or other subdivisions included in metropolitan areas	473	600	760	1,126

The author has shown that the "urban sprawl" process can result in temperature modifications on a regional scale (Viterito, 1988). In a study of the Baltimore-Washington corridor, it was shown that inter-urban population increase since 1950 resulted in the development of an urban "heat-corridor". That is, the heat islands of Baltimore and Washington are in the process of coalescing via steady temperature rise in the inter-urban continuum.

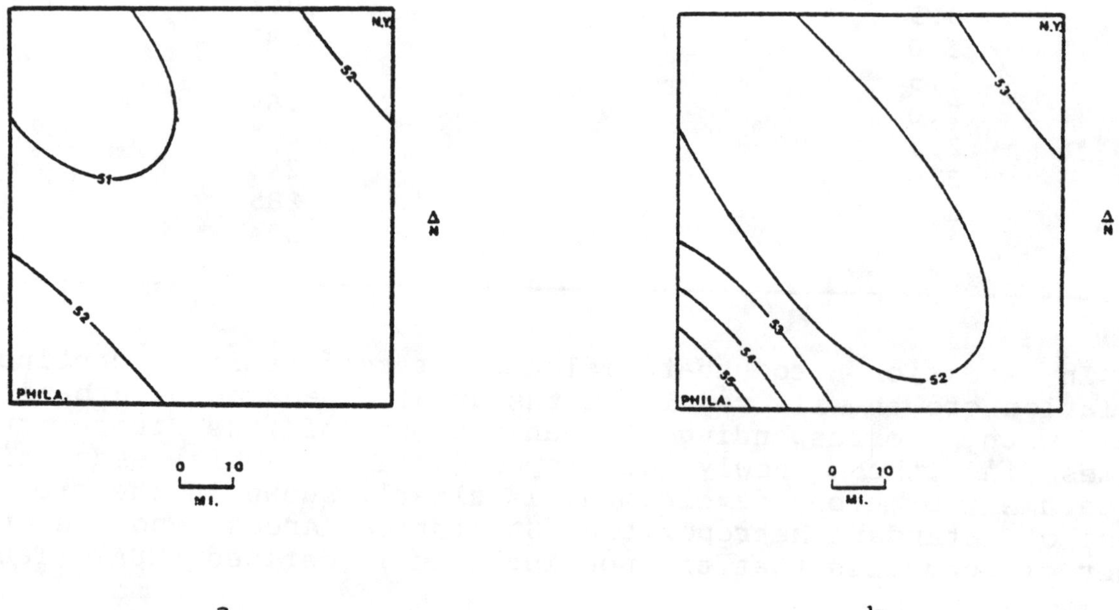

a b

Fig. 1. Quadratic trend surface of temperature for the Philadelphia - New York corridor: (a) 1950, (b) 1984.

A similar development can be seen in the New York-Philadelphia corridor (Fig. 1). The population in this region has increased from 3.8 million to 5.8 million during the 1950-1980 period. As evidenced by the trend surface analysis of the temperature fields for 1950 and 1984, there has been a small but significant increase of the areal extent of the New York and Philadelphia heat islands plus a steepened temperature gradient of the Philadelphia heat island. As this corridor becomes increasingly urbanized, its temperature field is certain to be modified towards the "heat corridor" configuration found in the Baltimore-Washington example.

A number of important points can be distilled from the preceding analysis. First is that many of the largest U.S. metropolitan areas have already experienced warming of a magnitude predicted by the global greenhouse warming scenarios. Any greenhouse warming that is realized is additive with respect to locally derived thermal increases. In the case of New York, a 3 degree greenhouse warming results in a total anthropogenic heat load of 8.4 degrees. For all cities, this compounded increase will lead to reduced heating requirements in winter, greater cooling requirements in summer, increased evapotranspiration, and higher heat stress. Also, increasingly larger areas will be affected as the conversion of rural land to urban fabric continues, thus promoting compounded warming on the regional scale. Finally, a number of researchers (e.g. Karl and Quayle, 1987, Viterito,1988) have recently raised important questions as to the representativeness of our land based temperature record. With increasing "urban contamination", a warming bias is introduced for a growing number of stations and at increasingly higher levels. Therefore, when defining any discernable temperature trend, caution must be exercised and the effects of urban growth accounted for.

REFERENCES

KARL, T. R. and QUAYLE, R. G.: 1987, "Climatic Change in Fact and in Theory: Are We Collecting The Facts?", The State Climatologist, II, 8-18.

BOGUE, D. J.: 1985, The Population of The United States, The Free Press, New York, 728 pp.

OKE, T. R.: 1979, Review of Urban Climatology 1973-1976, WMO Technical note No. 169, Geneva, 100 pp.

VITERITO, A.: 1988, "Changing Thermal Topography of The Baltimore-Washington Corridor: 1950-1979", Climatic Change, In Press.

POTENTIAL SHIFTS OF MONSOON PATTERNS
ASSOCIATED WITH CLIMATE WARMING

Eugene M. Rasmusson

1. NATURE OF MONSOONS

If the earth were a perfect sphere entirely covered by oceans, the climatological circulation patterns would be longitudinally uniform, varying only with latitude. The tropics would be characterized by a seasonally varying low-level cross-equatorial flow from the cool winter hemisphere toward the belt of highest sea surface temperatures (SST's)in the low-latitude summer hemisphere, i.e. toward the seasonal "thermal equator". Here the air would rise in towering convective clouds, releasing copious precipitation, then return via the upper troposphere to the winter hemisphere. Such a seasonally varying "Hadley Circulation" does indeed appear when the existing low-latitude flow is averaged around latitude circles, but contrary to the idealized picture, the observed circulation varies greatly with longitude. This variation reflects the presence of the continents, with their uneven topography and contrasting surface thermal characteristics. In particular, the heat capacity of the soil is small compared to that of the upper ocean mixed layer; thus the amplitude of the annual cycle in surface temperatures is far greater over land than over the oceans. This results in the seasonal reversal of the land/ocean temperature difference that drives the regional monsoon circulations.

The word "monsoon" is derived from the Arabic word "mausam" which means season. Ancient mariners applied this term to the seasonally alternating winds over the Arabian Sea. This summer-to-winter circulation reversal is a regional manifestation of the vast African-Asian monsoon system, which is a fundamental part of the global climate response to the annual cycle of incoming solar radiation.

Consider, for example, Northern Hemisphere summer conditions. The temperature difference between the relatively warm Afro-Asian land mass and the surrounding ocean areas results in lower surface pressure over the land, producing southerly to southwesterly inflow of moist air and extensive monsoonal precipitation over a vast region extending from West Africa eastward and southeastward into East Asia. In contrast, areas poleward of the northern limit of monsoon penetration, i.e. much of North Africa and central Asia, are hot and dry during the summer. The situation is reversed during the Northern Hemisphere winter, when relatively high pressure prevails over the colder continental areas. This results in a low-level southward outflow from the Asian continent toward the warmer waters and continents (Northern Australia and southeastern Africa) of the Southern Hemisphere tropics.

Similar but less pronounced monsoonal circulations occur over the low-latitude Americas, extending as far north as the southwest U.S. Monsoon regions exhibit a summer wet season, and most have a winter dry season, but beyond this common characteristic, regional monsoon precipitation regimes vary widely. For example, mountain ranges of even modest elevation produce dramatic regional variations in monsoon precipitation, sometimes even reversing the typical seasonal distribution. On the semi-arid margins, such as western India, Pakistan, and the west African Sahel, the rainy season is short and unreliable. In contrast, monsoon regions affected by tropical cyclones, such as Bangladesh, are subject to recurrent catastrophic flooding. Society tends to adjust to stable climatic conditions; the disruptive factors arise in connection with major departures from the norm. Above or below normal monsoon rainfall may be catastrophic or beneficial, depending on the area and manner in which it occurs.

The Afro-Asian monsoon both affects and is affected by other elements of the global general circulation. This fact was clearly recognized long ago by Sir Gilbert Walker(1924). Walker related year-to-year variations of the Indian monsoon to a global pattern of climate anomalies which he named the "Southern Oscillation". Almost half a century later, Jacob Bjerknes (1969) linked this atmospheric pattern of anomalies to pronounced year-to-year SST changes over the equatorial Pacific, thus identifying a remote link, or "teleconnection" between monsoon variability and tropical Pacific ocean temperatures. The complexity of these climate linkages vividly illustrates the need for a global view of both ocean and atmosphere variability when studying the processes associated with monsoon variability.

2. NATURAL VARIABILITY

The recurrent droughts and floods of the monsoon region are consequences of a high degree of natural monsoon variability. In fact, the monsoon circulations exhibit significant variability on a variety of time-scales, in response to a variety of causal factors, some identified and others unknown. Interannual and multi-millennia time-scales are of particular relevance to the question of global warming.

Orbitally-induced variations in insolation at the top of the atmosphere are the driving force behind much long-term climate change. Changes in the seasonal and latitudinal distribution of solar radiation are produced by changes in earth-sun geometry. Of particular importance to this discussion is the 22,000-year precession cycle which regulates the time of year when the earth-sun distance is a maximum or minimum and hence affects seasonality. Much of the multi-millennia time-scale variability that has altered the vegetation, ice volume, and SST over most of the globe during the past 18,000 years can be linked to the precession cycle (COHMAP members, 1988).

In its present orbit, the earth is nearest the sun during the Northern Hemisphere winter. Conditions were much the same 18,000 years ago, and reversed around 9,000 years ago, during the "Altithermal" period, when the earth was nearest the sun during the northern summer. This resulted in an amplified Northern Hemisphere annual cycle, since the hemisphere received around 8% more/less incoming radiation during summer/winter that it does today. Data and model results show an enhancement of the seasonally varying thermal contrast between land and sea, and stronger summer monsoons over Africa and Asia.

Turning next to the interannual time-scale, much of the monsoon region, notably India, Indonesia, northern Australia and China experience pronounced year-to-year variations in monsoon precipitation. Much of this variability has been linked to Walker's Southern Oscillation and the associated interannual variations in equatorial Pacific SST's; the so-called El Niño phenomenon. This El Niño/Southern Oscillation (ENSO) Cycle is a consequence of the large difference in SST between the warm water in the western equatorial Pacific and the cool equatorial upwelling water east of the dateline. The alternate warming and cooling in the central and eastern equatorial Pacific reflects the eastward expansion or westward contraction of the western Pacific warm pool. This leads, in turn, to major east-west shifts in the huge region of heavy precipitation that overlies the warm pool. The longitudinal displacement of this major atmospheric heat source, associated with condensation and the release of latent heat, has a pronounced effect on both the Pacific tradewind systems and the monsoon circulation, and also affects, to a lesser extent, the circulation in the temperate latitudes of both hemispheres.

Since the east-west SST gradient in the equatorial Pacific is at the heart of this oscillation, a change in ocean climatology could conceivably have a profound effect on the nature of monsoon variability. In the hypothetical case where SST became uniform across the entire equatorial Pacific, the basic source of the oscillation would be eliminated, thus removing a prime source of interannual monsoon variability.

3. EFFECTS OF GLOBAL WARMING

Two important questions arise regarding the effect of a global warming: (1) what changes would be expected in the prevailing climate of the monsoon region, and (2) what would be the effect on monsoon variability? Changes could occur as a consequence of changes in the strength or orientation of the monsoon circulation, and increases in atmospheric moisture content. In addressing these questions, it must be kept in mind that conditions during the transient phase of a warming may be quite different from those prevailing once a new equilibrium climate has been established. Having posed these questions, it must be acknowledged that there is as yet little to be offered in the way of definitive answers. Physical reasoning, while providing plausible hypotheses, is a highly uncertain exercise when applied to such a complex system, unless it can be supported by model results. Results from state-of-the-art general circulation models (GCM's) are available, and while very important, they must still be considered quite preliminary,, particularly with regard to the crucial hydrological processes.

It is tempting to view the Altithermal period of 6,000-12,000 years ago as an analog for a greenhouse warming. Northern Hemisphere summer continental temperatures were indeed higher during the Altithermal, but beyond this, the analog is less than totally comparable to a greenhouse warming scenario, as pointed out by Zhao and Kellogg(1988). The strengthened Northern Hemisphere Altithermal monsoons resulted from an amplification of the annual cycle, rather than changes in the mean annual heating. The enhancement of summertime monsoon rainfall must have been due to a larger land-sea temperature contrast coupled with a stronger moisture flow into the continental interior. The equilibrium land/sea temperature difference following a greenhouse warming would probably be less than that which prevailed during the Altithermal, since the oceans would also warm appreciably due to the year long warming. The Altithermal analog may be more appropriate for the transient phase, provided one assumes an ocean warming that lags the summertime continental heating. This hypothesis needs to be verified using improved ocean-atmosphere GCM's.

The most crucial questions of monsoon climate change center on changes in the hydrological cycle. Unfortunately, this is one of the least reliable aspects of current state-of-the-art GCM simulations. Recognizing this limitation, it is still instructive to compare model results to date. This has been done by Zhao and Kellogg(1988). They examined the performance of comparable models used by the Geophysical Fluid Dynamics Laboratory, Goddard Institute for Space Sciences, the National Center for Atmospheric Research, Oregon State University, and the United Kingdom Meteorological Office. They focused on a comparison of simulated soil moisture conditions. The largest scale features of the current climate were generally simulated quite well, although there were many significant differences in detail. For equilibrium conditions following a carbon dioxide doubling in the model atmospheres, a majority of the five models indicated an apparent intensification of the seasonally changing monsoon circulation over southern Asia, i.e. a change toward drier winters and wetter summers over most of India and Southeast Asia. Further north, the reverse was generally true. The authors warn, however, that these results must be viewed with great caution until they can be checked with improved climate models.

The question of possible changes in the nature of monsoon variability has as yet received little attention. As noted before, the ENSO cycle contributes a significant fraction of this variability. The sensitivity of the ENSO Cycle to climate change is an unresolved question. ENSO-like variability in simple coupled ocean-atmosphere models can be fundamentally changed by reasonable changes in model climate parameters. However, analysis of the frequency of El Niño occurrences, reconstructed from information derived from Peruvian colonial archives and historical narratives, indicates a relatively stable pattern of recurrence during the past 450 years, a period which includes the little ice age (Enfield, 1988). Tree ring data from the southwest U.S. appear to support this conclusion (Michaelson, 1988).

In regard to tropical storm activity, Emmanuel (1987) provides evidence that the maximum intensity of tropical storms is a function of SST. According to his hypothesis, a change of $1^{o}C$ in SST will change the minimum sustainable central pressure in hurricanes by 15-20 millibars, a very sizable change. If his reasoning proves correct, a significant increase in the intensity of tropical cyclones can be expected with even a modest warming of the tropical oceans.

4. FINAL REMARKS

On balance, GCM results to date give a tenuous indication that a global warming would result in a general increase in summer monsoon precipitation. Not all models agree, however, and in any event, the model deficiencies make these early results highly preliminary and far from conclusive. Simulations derived from models with an improved hydrologic cycle and a better ocean component are needed (1) to verify these results, (2) to resolve the regional pattern of monsoon change, (3) to determine the important transient response to increased greenhouse gasses, and (4) to determine the nature of monsoon variability after a warming. Experiments of this type are now in progress (S. Manabe, personal communication, 1988).

REFERENCES

COHMAP Members, 1988. Climatic changes of the last 18,000 years: Observations and Model simulations. Science, 241:1043-1052.

Bjerknes, J., 1969. Atmospheric teleconnections from the equatorial Pacific. Mon. Wea. Rev. 97:163-172.

Emmanuel, K. A., 1988. Toward a general theory of hurricanes. Am. Sci., 76:371-379.

Enfield, D. B., 1988. Is El Niño becoming more common? Oceanography Magazine (in press).

Michaelson, J. 1988. Long-period fluctuations in El Niño amplitude and frequency reconstructed from tree-rings. In: Interdisciplinary Aspects of Climate Variability in the Pacific and western Americas, D. H. Peterson, Ed., Am Geophys. Union Monograph (in press).

Walker, G. T. 1924. Correlation in seasonal variations of weather No. 9: A further study of world weather. Mem, India Meteorol. Dept. 24:275-332.

Zhao, Z. and W. W. Kellogg, 1988. Sensitivity of soil moisture to doubling of carbon dioxide in climate model experiments. Part II: the Asian monsoon region. J. Climate, 1:367-378.

UNDERSTANDING EL NINO AND OTHER LONG-TERM CLIMATE VARIABILITY OVER THE OCEANS

HENRY F. DIAZ

1. INTRODUCTION

An important part of the process associated with early detection of climate change and the development of plausible regional climate change scenarios involves a determination of the range of past patterns of climatic variability from the observational record and an understanding of its causal mechanisms. Understanding the role of the oceans in forcing interannual and longer time-scale climate variability is critical in this regard. Two types of phenomena are examined here. One is the El Nino/Southern Oscillation (ENSO) phenomena and the other deals with low frequency variability of sea surface temperature (SST), principally over the Atlantic and Pacific Oceans.

2. CLIMATE VARIABILITY and ENSO

The ENSO phenomena constitute the single largest source of interannual climatic variability in the earth's climate system. Its influence is felt most strongly in tropical areas principally over the Indo-Pacific region and the Pacific coast of South America, but strong responses (teleconnections) have been documented by many researchers for other regions of the globe, particularly for the extratropical North Pacific [1-6].

ENSO phenomena represent a large scale modulation of energy and mass exchanges between the oceans and the atmosphere as well as within each of these two domains. To understand ENSO and its effects on the climate, it is useful to first consider the "normal" or average conditions in the Pacific. Typically, two subtropical high pressure areas are centered over the eastern half of the ocean in both hemispheres, while a broad trough of low pressure is found near the equator in the western tropical Pacific. The resulting pressure gradients give rise to the Southeast Trades in the South Pacific and to the Northeast Trades in the North Pacific. These great wind systems converge a few degrees north of the equator to form the relatively narrow band of convective rainfall known as the Inter-Tropical Convergence Zone. Along the equator, east of about 180° the westward component of air flow induces upwelling of colder subsurface water and strong surface winds further depress the SST through evaporation. West of the dateline, however, surface winds are typically light and the occurrence of equatorial westerly winds help maintain a large reservoir of warm surface water that represents one of the major heat sources of the climate system. The energy stored there is made available to the atmosphere by the mechanisms of latent and sensible heat flux to be transported into high latitudes in order to help maintain the earth's heat balance.

The seesaw in sea level pressure between the eastern and western tropical Pacific is known as the Southern Oscillation and characterizes the atmospheric response associated with the ENSO phenomenon. The warming of the surface waters off the coast of South America and along the eastern equatorial Pacific, El Nino, characterizes the oceanographic response. Two indices of ENSO are now examined. The first, presented in Figure 1 is a time series of monthly departures from the long term mean of sea level pressure at Tahiti in

French Micronesia and Darwin, Australia (bottom graph). The second (top graph) is the normalized pressure difference between these two stations and constitutes an index of the Southern Oscillation (SOI). High (positive) index values correspond to times when the pressure gradient, and hence the trade winds, are stronger than normal and upwelling along the equator is enhanced resulting in lower SST. A low index or large negative SOI value corresponds to weaker pressure gradients and trade winds, reduced upwelling, anomalous ocean currents and higher than normal SST from the coast of Peru westward to vicinity of the dateline. The SST response in the eastern equatorial Pacific, known as the Nino 3 region [7] is illustrated in Fig. 2. The great ENSO event of 1982-83 is clearly evident, during which the SST exceeded four standard deviations above the mean.

The average SST anomalies associated with ENSO events during the past century are shown in Fig. 3 [8]. The figure represents the SST difference between the set of years experiencing an El Nino or warm ENSO event and the years experiencing cold conditions (sometimes referred to as La Nina). The greatest change occurs over the equatorial Pacific east of the dateline, with an opposite, but symmetrical SST response at middle latitudes of the Pacific Ocean in both hemispheres. The changes in SST and in the large scale wind field in the Pacific induce major changes in rainfall patterns. It is these changes and the interaction between the tropics and extratropics that lead to anomalous circulation patterns over the globe.

Figure 4 [8] shows the typical temperature anomaly pattern over the Americas during the months Dec.-Feb. corresponding to the mature phase of El Nino events. The pattern during La Nina is similar but of opposite sign. The changes in surface air temperature typically associated with ENSO events are of the order of 0.4 to 0.6^{o}C averaged over the global tropics [6]. The largest zonally-averaged anomalies occur some 6 to 14 months after the typical onset of the events, which is, on average, during the Northern Hemimsphere spring season (March-May). In mid-latitudes, the response varies, depending on the geographical region.

3. LOW FREQUENCY VARIABILITY

Analysis of the observational climate record reveals that the natural variability of climate, whether related to the ENSO or to other features of the system are normally superimposed on lower frequency changes. These usually take the form of decadal scale fluctuations, long-term trends and abrupt transitions from one quasi-stable regime to another. An example of this variability is shown in Fig. 5, which illustrates the evolution of SST anomalies averaged over 15-degree zonal bands across the entire Atlantic Ocean. Two features are worthy of note; one is a tendency for the North and South Atlantic to exhibit SST anomalies of opposite sign at any given time, the other is a suggestion that the anomalies may slowly propagate from one hemisphere to the other. By contrast, the Pacific Ocean tends to strongly reflect SST variability associated with the different ENSO phases, although larger spatial and longer temporal scales are also evident. Finally, it should be noted that temperature and precipitation trends around the globe are highly variable and exhibit a variety of temporal behavior, from slow drifts in the mean and/or variance to rapid changes over periods of only a few years [9].

It will be difficult to disentangle this very broad spectrum of climatic variability from the signals associated with the rise of atmospheric greenhouse gas concentrations. The uncertainties include significant shortcomings in the observational record associated with data sparseness as well as with data inhomogeneity. Furthermore, the spatial and temporal characteristics of climatic fluctuations associated with the transient evolution of climate from its present to its high greenhouse gas stage is unknown and may be different in any one location from its ultimate equilibrium state. These uncertainties suggest that some degree of caution is needed regarding the assessment of future regional climate change scenarios and the possible implementation of mitigative action.

4. REFERENCES

1. Bjerknes, J., 'Atmospheric teleconnections from the equatorial Pacific', Mon. Wea. Rev., 97, 163-172 (1969).
2. Horel, J.D., and J.M. Wallace, 'Planetary-scale atmospheric phenomena associated with the Southern Oscillation', Mon. Wea. Rev., 109, 813-829 (1981).
3. van Loon, H., and R.A. Madden, 'The Southern Oscillation. Part I: Global associations with pressure and temperature in northern winter', Mon. Wea. Rev., 109, 1150-1162 (1981).
4. Rasmusson, E. M., and J.M. Wallace, 'Meteorological aspects of the El Nino/Southern Oscillation', Science, 222, 1195-1202 (1983).
5. Ropelewski, C.F., and M.S. Halpert, 'North America precipitation and temperature patterns associated with the El Nino/Southern Oscillation (ENSO)', Mon. Wea. Rev., 114, 2352-2362 (1986).
6. Bradley, R.S., H.F. Diaz, G.N. Kiladis, and J.K. Eischeid, 'ENSO signal in continental temperature and precipitation records', Nature, 327, 497-501 (1987).
7. Rasmusson, E.M., and T.H. Carpenter, 'Variations in tropical sea surface temperature and surface wind fields associated with the Southern Oscillation/El Nino', Mon. Wea. Rev., 110, 354-384 (1982).
8. Kiladis, G.N., and H.F. Diaz, 'Global climatic anomalies associated with extremes of the Southern Oscillation' (1989, in press).
9. Elsaesser, H.W., M.C. MacCracken, J.J. Walton, and S.L. Grotch, 'Global climatic trends as revealed by the recorded data', Rev. Geophys., 24, 745-792 (1986).

Figure 1
(Source: NOAA/CAC)

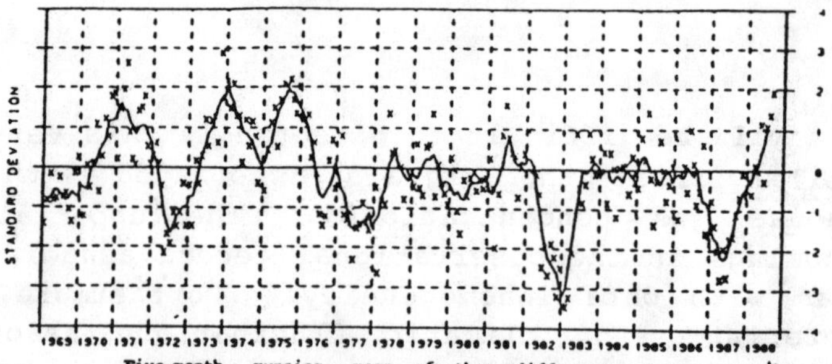

Five-month running mean of the difference between the standardized sea level pressure anomalies at Tahiti and Darwin (Tahiti-Darwin). Values are standardized by the standard deviation of the appropriate monthly mean. Crosses are individual monthly means.

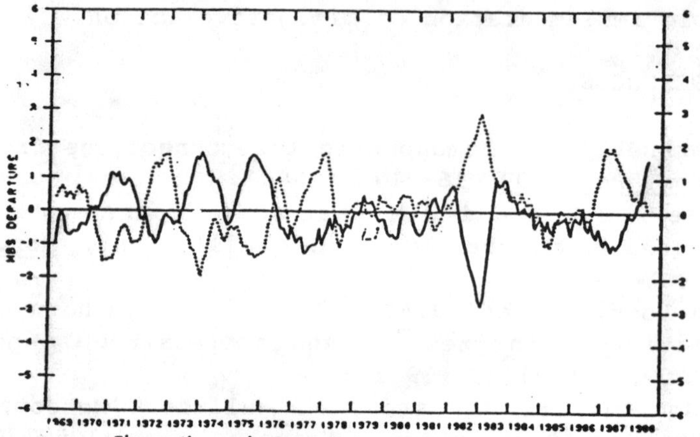

Five-month running mean of the sea level pressure anomalies at Darwin (dashed) and Tahiti (solid).

Figure 2

NORMALIZED SST DEPARTURES
NINO 3 (6°N-6°S, 150°W-90°W)
3 Month Running Means from 2° COADS and Ship Obs Data
Standard deviations and long term monthly means used in calculating departures were based on 1961 - 1980 data

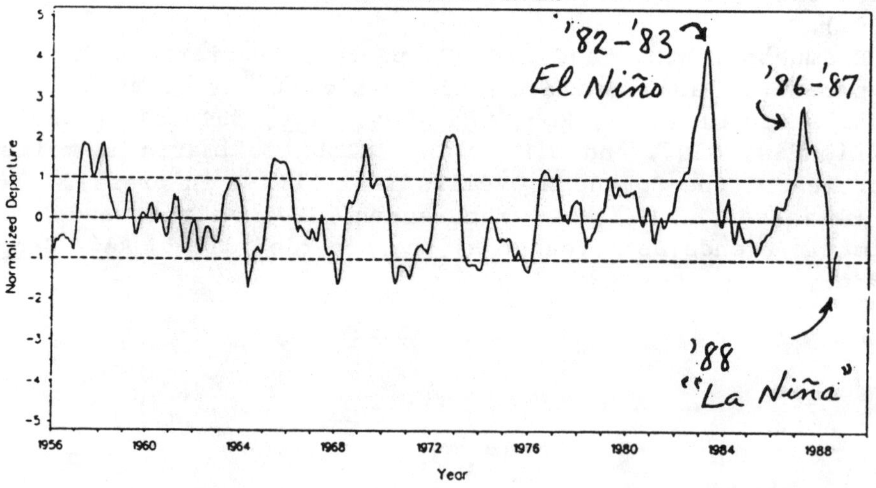

AVERAGE LONG TERM MEAN = 25.76 °C

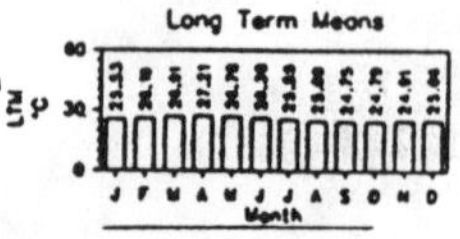

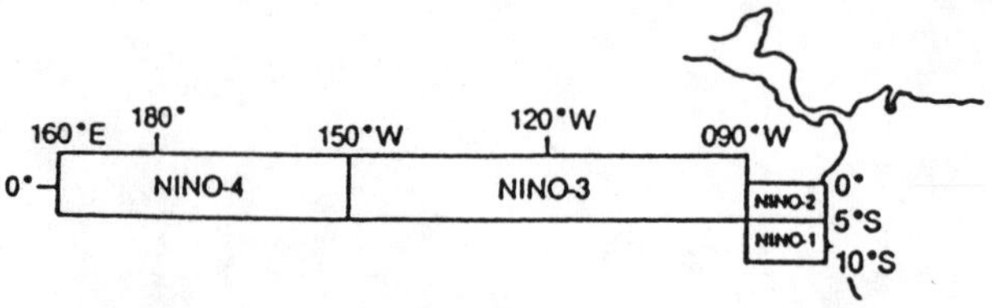

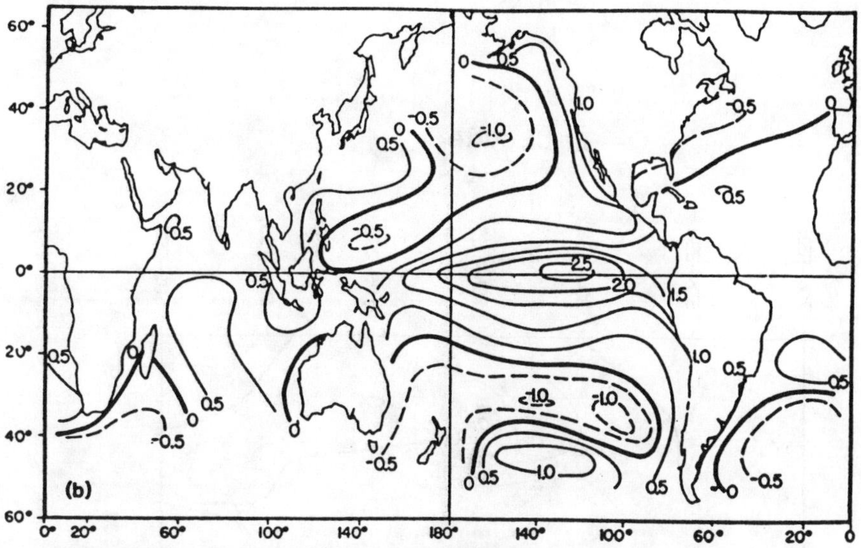

Figure 3: Average Sea Surface Temperature Difference During DJF (+1); Warm minus Cold Events (degree C). (See Ref. 8.)

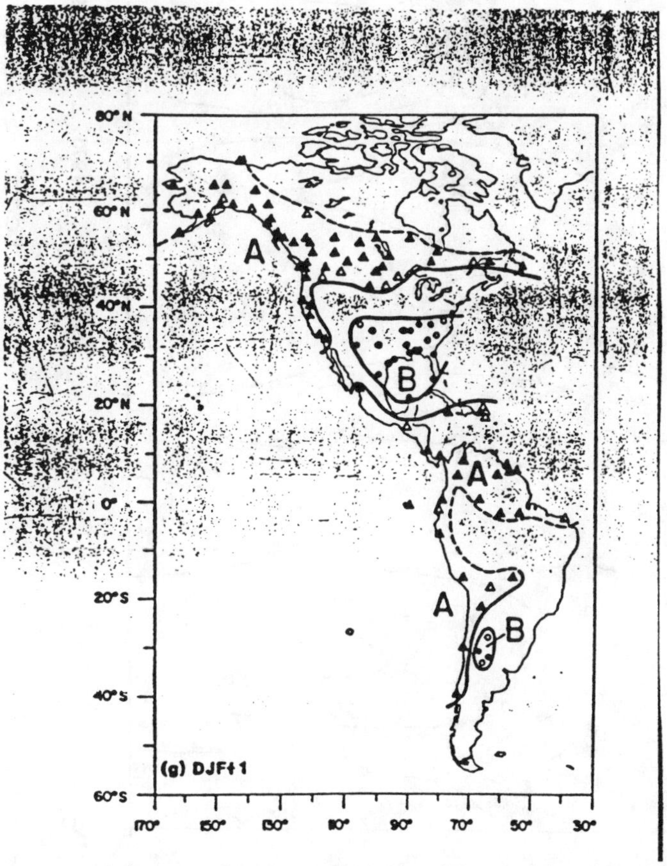

Figure 4: Location of Statistically Significant Surface Temperature Anomalies: Above (A) and Below (B). The Long-Term Mean During Warm ENSO Events for Dec-Feb (+1). (See Ref. 8.) Sign of Anomalies During Cold Events is Reversed.

Figure 5: ATLANTIC BASIN SST ANOMALIES (1950-86)

70W TO 20E

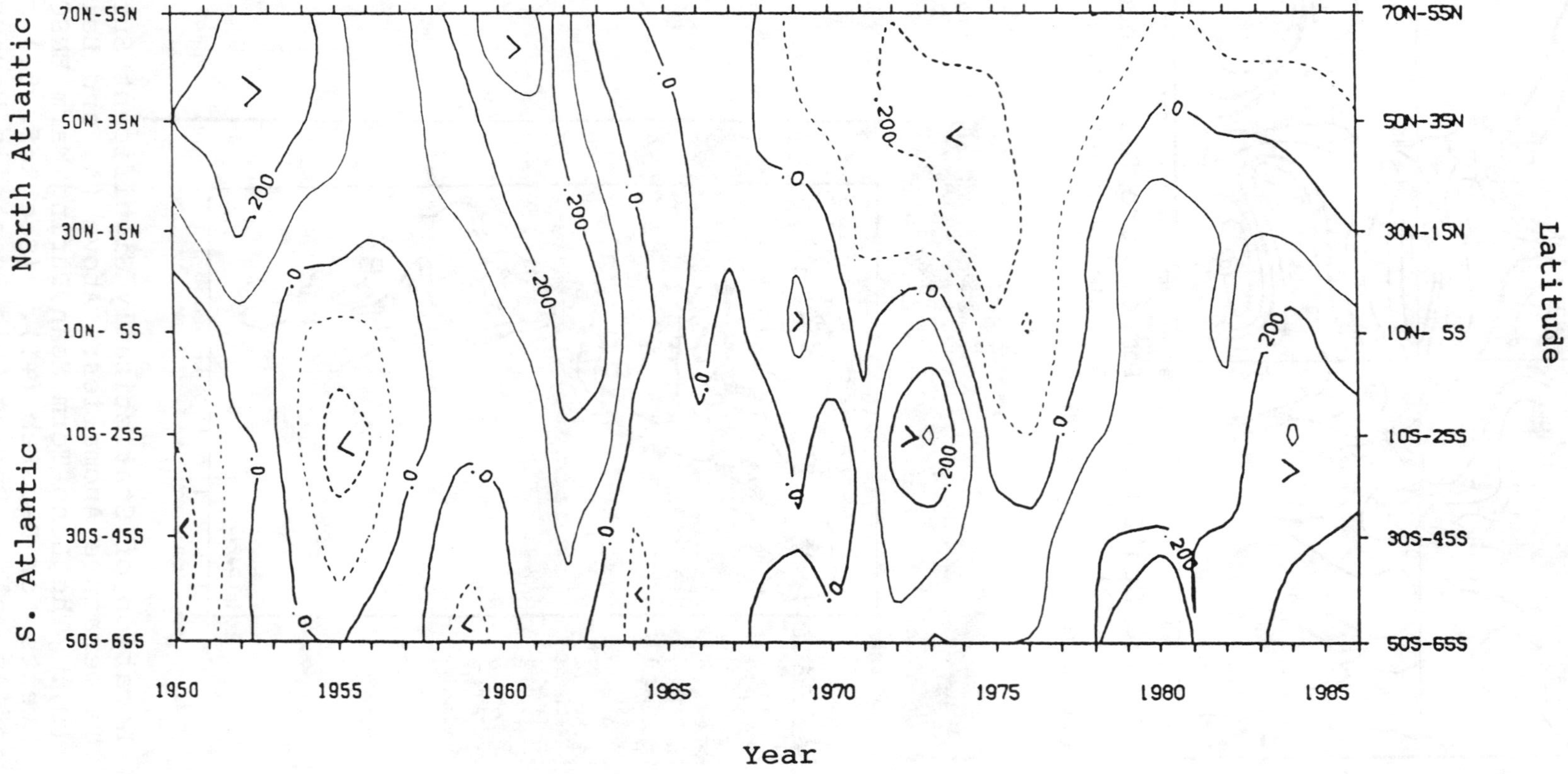

THE ATLANTIC OCEAN: UNSUNG PARTNER IN GLOBAL CLIMATE CHANGE

Robert L. Molinari
National Oceanic and Atmospheric Administration
Atlantic Oceanographic and Meteorological Laboratory
Miami, Florida 33149

INTRODUCTION

The rate at which greenhouse warming will occur is a critical unknown in mankind's ability to react to this climate change. Predictions based on models indicate global warming ranging from 1.5°C to 4°C in response to a doubling in greenhouse gases. However, whether this warming occurs in 50 years or 100 years is still not certain. Part of this uncertainty is related to an inability to predict the natural cycle of atmospheric temperatures. Hemispheric temperature changes approaching 1°C over a time span of 50 years have been observed in the past century. Since the natural and greenhouse effects are additive, the phase of the former cycle (i.e., are background temperatures rising or falling?) will determine, in part, how rapidly global warming proceeds.

To predict the effects of natural climate variability at periods ranging from interannual to decadal and longer an understanding of the role of the ocean in driving climate change is required. For instance, after intensive study, the Pacific Ocean has been shown to significantly impact global climate on interannual time scales through El Niño/Southern Oscillation (ENSO) events. Sea surface and subsurface temperature and surface current anomalies associated with El Niño have been shown to be strongly correlated with Southern Oscillation anomalies. The Southern Oscillation events, in turn, have been shown to disturb global climate. Present understanding of ENSO is such that predictions of the events and their implications appear feasible.

The Atlantic Ocean also seems to be a major player in establishing global climate change, particularly at periods of decades and longer. However, the present understanding of the Atlantic's role in climate change is much less developed than in the case of the Pacific. The following discussion presents an explicit summary of the relation between climate and the Atlantic and an implicit argument that our present knowledge is sufficient to warrant more intense research in the basin. This research is required to bring our understanding of the role of the Atlantic in global climate change to the level needed to predict Atlantic induced climate variability.

SCIENTIFIC BACKGROUND

Teleconnections have been defined as "a significant simultaneous correlation between temporal fluctuations in meteorological parameters at widely separated points on earth" (1). Through teleconnection patterns tropical atmospheric anomalies associated with ENSO events can be propagated to mid-latitudes. Teleconnection patterns have also been identified in the Atlantic which, as in the case of the Southern Oscillation, appear to be closely related to anomalies in near surface ocean conditions.

Early studies of teleconnections and air-sea interactions in the Atlantic were regional in nature. SST anomalies in the tropical Atlantic were observed to be correlated with rainfall anomalies in Northeastern Brazil (2). Severe droughts were associated with SST anomalies in the tropical Atlantic. These SST anomalies were, in turn, observed to occur during periods of anomalous near-surface temperature and current structure. For example, during early 1984, equatorial SST was higher than normal, the equatorial thermocline was deeper than normal and unusual eastward surface currents were observed south of the equator (3). Precipitation over northeast Brazil was anomalously high during this period consistent with earlier findings (2).

At more northerly latitudes, the North Atlantic Oscillation (NAO) is a teleconnection pattern with important climatic implications (4). The NAO is characterized by a "seesaw" both between the severity of winters in Greenland-Labrador and NW Europe and between sea level pressure in the vicinity of the Icelandic low (65°N) and a broad east-west high pressure area centered near 40°N, extending from the U.S. to the Mediterranean. A strong anomaly in SST has been associated with the NAO on interannual time-scales with largest anomalies observed south of Newfoundland, close to the Gulf Stream/Labrador Current boundary. The anomaly is characterized by lower temperatures to the north of the boundary and warmer temperatures to the south during times of increased NAO pressure gradient (referred to as a high NAO index). Recent studies have found the association between SST and European weather to be statistically significant and thus suitable for use as predictors in long-range forecasting models.

Changes in the NAO on interannual time scales are correlated also with atmospheric conditions over the continental United States. For instance, the NAO is strongly correlated with winter temperatures over the eastern U.S (6). Average anomalies in air temperature approach +2°C over most of the U.S. east of 95°W (the longitude of Houston) during periods of lower pressure in the Iceland Low and higher pressure to the south (i.e., a high index). Significant precipitation anomalies over the extreme eastern U.S. also are correlated with the intensity of the NAO.

Recent studies have shown that the NAO has a larger area of influence than merely the North Atlantic. There is also a significant node in sea level pressure variability in the northern Pacific Ocean. During periods of high NAO index, the Aleutian Low is stronger than normal. As might be expected, because of the changes in the intensity of the Aleutian Low, some statistically significant changes in surface air temperature over the western U.S. have been correlated with changes in NAO intensity (6). During times of high NAO index, higher temperatures are observed in the eastern and lower temperatures in the western U.S. Similarly, concurrent precipitation anomalies on the west coast are found, but these are not statistically significant.

Connections between NAO variability and atmospheric variability in the Atlantic tropics have also been found. The NAO has been correlated with rainfall patterns in central Africa (7). These changes in rainfall patterns have been attributed to shifts in the position of the Intertropical Convergence Zone, ITCZ (the boundary separating the northern and southern hemisphere Tradewinds). These tropical relations also appear in numerical models of the Atlantic which show that NAO-type variability could be associated with some of the earlier findings of teleconnections beteen tropical SST anomalies and Brazilian rainfall. Thus, the NAO is not merely a regional climate phenomenom but also has at least hemispheric and probably global implications, as does its Pacific counterpart, the Southern Oscillation.

Much of the interannual variability described above has been shown to be superimposed on longer time scales ranging from decades to several decades. For instance, a general increase in the NAO index was observed between the late 1890's and the early 1920's (5). Concurrently, the westerlies increased in intensity and were found over a larger latitudinal band. SST's south of Newfoundland increased by 2°C, with warming north of 40°N extending almost to Europe. This SST anomaly pattern is different than the dipole pattern associated with NAO variability on interannual time scales and the differences have been ascribed to the influence of ocean currents (5), as will be described in more detail shortly. Subsequently the NAO index decreased between 1922 and 1940 (the length of the available record).

Similar changes were found in a statistical analysis of a longer sea level pressure record (8). Two periods of change in the large-scale atmospheric circulation were observed, one in the early 1920's, and the second, during the early to mid-1950's, when the NAO increased. The decadal changes in the intensity of the NAO have been correlated with changes in northern hemisphere surface air temperature, drought in mid-America, sea level pressure and cyclone tracks on the east coast of the U.S., and the length of winter and summer seasons along the east coast. These changes can be summarized as follows: the NAO index was low between approximately 1925 and 1970; concurrently, (1) northern hemisphere surface air temperatures were above normal, (2) precipitation over mid-America was below average, (3) east coast sea level pressure was below average or average, (4) east coast cyclone tracks were generally located over the ocean rather than the continent, and (5) east coast summers were generally longer and winters shorter than normal. The period prior to 1925 was characterized by high NAO index and these conditions were in general reversed.

The atmospheric changes just described also have been qualitatively related to changes in the intensity of the circulation of the North Atlantic. At these decadal periods an increase in the intensity of the NAO is hypothesized to cause an increase in the intensity of the subtropical gyre in the north Atlantic and a decrease in the amount of warm water leaving this gyre to flow into the subpolar gyre. Conversely, a decrease in the NAO index results in a weaker subtropical gyre and increased advection of warm water from that gyre to the subpolar gyre. Prior to the 1930's, the period of high index, indirect observations of the North Atlantic circulation suggest that the subtropical gyre was indeed more intense than after the 1930's. As will be described in more detail additional study is required to define quantitatively, exact cause and effect relations between these oceanic and atmospheric features.

At decadal and longer periods, changes in oceanic heat flux play a major role in global climate change. The thermohaline component of ocean circulation (i.e., that component driven by global temperature and salinity differences rather than directly by wind) is an important part of the heat flux process. This circulation component can be thought of as representing the current structure in a vertical plane rather than flow in the horizontal plane, which is primarily wind driven. Recent numerical modelling studies show that changes in the intensity of the thermohaline circulation cause dramatic changes in the intensity of oceanic heat flux and global climate (9,10). In these models, the present climate state is characterized by a strong thermohaline circulation with a well developed deep circulation in the Atlantic, having the strongest flow along the western boundary in the form of a Deep Western Boundary Current (DWBC). However, when low salinity surface waters at northern latitudes of the Atlantic weakens or shuts off the thermohaline circulation, ice-age climate conditions are simulated in the model. Low salinity has been observed in the

Labrador Sea during times of reduced air-sea interaction in the region. Subsequent changes in the water mass properties of the DWBC have been observed on decadal time scales (11). Other evidence for changes in oceanic heat flux and North Atlantic atmospheric climate have been found in the paleoclimate record.

RECOMMENDATIONS FOR FUTURE STUDIES

Much of the evidence presented above is circumstantial and does not show conclusively that North Atlantic SST distributions and heat flux play a major role in driving global climate change (for instance, the SST features may merely be effects of the atmospheric anomalies). However, prior to the intensive studies of ENSO events much of the available knowledge was also primarily circumstantial and not conclusive. Definitive findings on the relative importance of the ocean in ENSO are now available because of the intense observational and modelling effort devoted to the problem. A similar effort is now deemed warranted to define the role of the Atlantic Ocean in global climate change.

The objective of such a study should be to develop the ability to predict decadal and longer climate changes. To achieve this objective, three goals must be met. A coupled ocean atmosphere general circulation model, GCM, must be developed and verified to perform the predictions. Ocean data must be collected to verify the model. Once verified, oceanographic data must be collected on a routine basis to initialize the predictive model runs. Considerable effort is being expended to develop coupled GCM's. Particular attention must be given to developing the methods needed to assimilate real time oceanographic data into these models as is done with meteorological data in atmospheric GCM's.

Two types of verification data are needed. Many of the important oceanographic processes are poorly parameterized (e.g., the flux of energy from the atmosphere to ocean, vertical and horizontal mixing processes, etc.) in oceanic GCM's. Process specific experiments are required to improve these parameterizations. These experiments represent the historical mode of operation for oceanographers, i.e., studies conducted by one or two investigators on short time and space scales. However, in addition, large-scale basin wide observations of ocean circulation, temperature and salinity structure are required for both model verification and initialization. This requires that oceanographers begin to operate in the data collection mode used by atmospheric scientists. Data must be collected on a global-scale with little concern for identifying particular data-sets with particular investigators. This approach has been particularly successful in the tropical Pacific where most recent advances in understanding of the ocean have been based on basin-wide data sets. The types of data required include temperature observations from merchant ships of opportunity, coastal sea level gauges, free drifting surface and subsurface buoys, and sea level and SST from satellite. Once in place these observing systems can also provide the data needed for model initialization.

REFERENCES

1. J. Wallace and D. Gutzler (1981) Mon. Wea. Rev., 109, 784-812.
2. A. Moura and J. Shukla (1981) J. Atm. Sci., 38, 2653-2675.
3. P. Hisard et al. (1986) Nature, 6076, 243-245.
4. G. Walker and E. Bliss (1932) Mem. Roy. Meteor. Soc., 4, 53-84.

5. J. Bjerknes (1964) Adv. in Geophys., 10, 1-82.
6. H. van Loon and J. Rogers (1978) Mon. Wea. Rev., 106, 296-310.
7. G. Meehl and H. van Loon (1979) Mon. Wea. Rev., 107, 1095-1106.
8. J. Kutzbach (1970) Mon. Wea. Rev., 98, 708-716.
9. F. Bryan (1986) Nature, 323, 301-304.
10. S. Manabe and R. Stouffer (1988) J. Climate, 1, 841-866.
11. R. Fine and R. Molinari (1988) Deep-Sea Res., 35, 1441-1450.

OCEANS: A DYNAMIC RESERVOIR FOR CARBON DIOXIDE

Taro Takahashi
Lamont-Doherty Geological Observatory,
Columbia University

INTRODUCTION

Even before the dawn of modern science, man recognized the moderating effects of the oceans on climate. This phenomenon is due mainly to two unique physical properties of water: its large heat capacity and its large latent heat of evaporation and condensation. An additional role of the oceans, more chemical in nature, has recently been recognized as an important factor for the future course of the Earth's climate change. The world ocean is a dynamic CO_2 reservoir and plays an important role in regulating the atmospheric CO_2 content, and hence the global climate. In this paper, the major processes governing the oceanic CO_2 reservoir capacity and their influence to the atmospheric CO_2 concentration will be discussed and a course for future scientific research will be recommended.

DISTRIBUTION OF CO_2 IN THE OCEANS

A large quantity of CO_2, as much as 50 times that present in the atmosphere, is dissolved in the global oceans. The amount of this oceanic CO_2 corresponds to about 20 times as much as that present in the land biosphere including both live and detrital material. Carbon dioxide molecules are exchanged between these major reservoirs and are cycled or stored in them via various physical, chemical and biological processes. Because of the large mass of the oceanic reservoir and its active interaction with the atmosphere, the oceans are considered to play a major role in regulating the atmospheric CO_2 concentration over a wide time scale, ranging from yearly to over many centuries. For example, the observation that the actual rate of atmospheric CO_2 increase is about a half that expected from the industrial CO_2 emission rate has been attributed generally to uptake opf CO_2 by the oceans (e.g. see Broecker et al., 1979, for review).

There are three chemical quantities which are pertinent to the discussion of the oceanic CO_2 reservoir: the total CO_2 concentration (TCO_2) in seawater, which represents the sum of all the CO_2 species dissolved in water including the undissociated carbonic acid (H_2CO_3), the bicarbonate ion (HCO_3^-) and the carbonate ion ($CO_3^=$); the alkalinity (ALK), which is a measure of excess cation charges needed to balance anions such as the hydroxyl, bicarbonate and carbonate ions; and the pCO2 , which is a measure of the thermodynamic driving potential for sea-air gas transfer. If the total CO_2 concentration in seawater is increased by uptake of atmospheric CO_2 and/or oxidation of organic debris, the seawater would become more acidic and hence have higher pCO_2. On the other hand, if it is decreased by photosynthesis, seawater would become more alkaline and have a lower pCO_2 value and hence tends to take up CO_2 from the atmosphere. The alkalinity in seawater is regulated mainly by the precipitation/dissolution of biogenic $CaCO_3$ shells. The dissolution of skeletal $CaCO_3$, for example, would increase the seawater alkalinity (two units of alkalinity are released by divalent Ca^{++} ions, where as only one unit of CO_2 is released by dissolution), and hence raise the pH (i.e. more alkaline) and lower the pCO_2. The acid produced in seawater due to the uptake of industrial CO_2 may be neutralized by the dissolution of sekeletal $CaCO_3$. This would prevent acidification of the oceans and hence allow the oceans to take up more CO_2 from the air. The value of pCO_2 which determines the direction and rate of the air-sea gas transfer is governed by the total CO_2 concentration, the alkalinity, and the temperature of seawater, all of which are, as in the examples mentioned above, changed by various biological, chemical and physical processes.

The distribution of the total CO_2 concentration (TCO_2) and the alkalinity (TALK), which have been obtained in various regions of the global oceans during the GEOSECS (Geochemical Ocean Sections Study) Program in 1972-1978, are shown in Figs. 1 and 2 (Takahashi et al., 1981-a; see Takahashi et al. (1981-b) for the listing of the regional mean values). The global mean values for temperature, salinity, alkalinity and total CO2 concentration have been estimated to be 3.9°C, 34.78 o/oo, 2372 uEq/kg and 2251 uM/kg respectively. Using the total volume of the ocean water of 1370×10^6 km^3 and the mean seawater density of 1.025 gm/cm^3, the total CO_2 content in the global oceans is estimated to be about 3.16×10^{18} moles (= 3.79×10^4 Gigatons as carbon or 1.39×10^5 Gigatons as CO_2). These mean alkalinity and total CO_2 concentration values for the whole oceans yield a pCO_2 value of 437 uatm at the mean deep water temperature of 1.5°C and 913 uatm at the mean surface water temperature of 19.2°C. This means that the deep ocean water is highly supersaturated with respect to the atmospheric pCO_2, about 340 uatm in 1988. If the atmosphere were brought to equilibrium with an hypothetical ocean chemically homogenized at the mean ocean temperature of 3.9°C, the atmospheric CO_2 concentration would increase by about 40%. If the temperature were increased to the mean surface water temperature of 19.2°C, the atmospheric CO_2 concen-tration would increase almost three-fold. The unique feature of the oceans is that a large body of deep water is highly supersatured with respect to atmospheric CO_2, and is capped with a thin layer (about 75 meters thick on the average) of warm and less dense seawater which prevents the rapid transfer of CO_2 from the deep ocean reservoir to the atmosphere. Understanding this "sequestering" process for the CO_2 in deep ocean water is one of the major questions central to the investigation of the oceanic CO_2.

PROCESSES CONTROLLING THE OCEANIC CO_2 RESERVOIR

Fig. 1 shows that the total CO_2 concentration in seawater increases rapidly with increasing depth from the surface to about 1000 meters, below which it changes slowly. This feature may be accounted for by the existence of a biological carbon pump, which transports carbon from the upper layers to the deep ocean via the gravitational settling of biogenic debris produced in the photic zone of the oceans. The debris is mostly oxidized to CO_2 during its sinking through the water column. The fact that the carbon content in deep ocean sediments is low (i.e. less than 1 % by weight) indicates that only a small portion of the biogenic debris reaches the sea floor. Fig. 2 also shows that the alkalinity increases with depth less rapidly than the total CO_2 concentration. In the Pacific and Indian Ocean profiles, it is seen that the alkalinity gradient is reduced at depths around 2000 meters, whereas the change in the total CO_2 gradient occurs at depths around 1000 meters. This suggests that, while the oxidation of organic debris (such as soft tissues) takes place at all depths in a water column (presumably more rapidly in warmer shallower waters), the dissolution of skeletal $CaCO_3$ occurs only in deep oceans below certain water depths. In fact, the solubility of $CaCO_3$ (i.e. minerals calcite and aragonite) increases rapidly with pressure: the solubility product doubles with a pressure increase of about 430 atm. (or about 4300 meters in water depth). While the seawater near the ocean surface is generally supersaturated several fold with respect to calcite, and it becomes undersaturated with respect to calcite (i.e. the shells of coccoliths and foraminifera) below about 4500 meters deep in the Atlantic Ocean and below a depth between 1000 and 3000 meters in the Pacific Ocean. The regional difference in undersaturation depths is attributed to the difference in the total CO_2 concentration and alkalinity in deep waters.

The density of seawater increases with decreasing temperature and with increasing salinity. In general, the ocean water is stablely stratified according to the density, and hence the vertical circulation rate is slow especially across the main thermocline down to about 1000 meters deep where the density gradient is steep. On the other hand, the deep ocean is ventilated more rapidly by the lateral motion of waters driven by the sinking of dense waters produced in high latitude oceans of both the northern and southern hemispheres. On the basis of the distribution of radiactive carbon-14 (which has a radioactive decay half-life of 5,700 years), the time scale for global ocean circulation has been estimated to be on the order of 1000 years. For this reason, the deep ocean can retain CO_2 transported to the deep water regime by the biological pump. However, a large regional difference exists in the rate of deep water circulation and the concentration of CO_2 in deep water reservoirs (see the North Atlantic (NA) and North Pacific (NP) curves in Fig. 2). This may be attributed to the much faster circulation or flushing rate of waters in the North Atlantic compared to those in the North Pacific. The deep Atlantic Ocean is ventilated mainly by two sources of waters flowing laterally in the opposite direction: the southward flowing deep waters which originated in the high northern latitude areas and the northward flowing deep waters which originated in the areas surrounding the Antarctic continent. The Antarctic waters are densest and hence are found at greater depths below the deep waters of northern origin. The deep Atlantic Ocean is thus flushed in a time scale of 100 years and hence has a relatively short time to receive the products of the biological pump. On the other hand, the deep North Pacific is ventilated only by the flow from the south since the Pacific does not extend far enough north to allow the surface water to become cold and dense enough to sink to abyssal depths. Thus, as indicated by much older carbon-14 ages of CO_2 dissolved in the North Pacific deep waters (Ostlund and Stuiver 1980), they are ventilated more slowly than the deep Atlantic waters, and hence receive greater quantities of the oxidation products of falling biogenic debris. In short, the holding capacity of the oceanic CO_2 reservoir depends mainly on the strength of the biological carbon pump and on the rate of ocean water circulation.

The strength of the biological carbon pump, however, is intimately coupled with the rate of ocean water circulation through the nutrient cycle. As the biogenic debris is decomposed during its descent through the water column, nutrient salts (i.e nitrate and phosphate) are also released into the surrounding water, and accordingly the concentrations of these salts increases with depth very similarly to the trend observed for the CO_2 concentration. The regenerated nutrient salts in deep water are transported back to the ocean surface by the vertical circulation of ocean water, and there they are utilized for photosynthesis. In the subtropical gyre regions of the oceans where the vertical circulation rate of water is greatly reduced due to the prevailing high degree of density stratification in the upper water column, the nutrient salt concentrations are often nearly zero, and the biological production, hence the strength of the carbon pump, is limited by the supply rate of nutrient salts. In contrast, in the oceanic regions where the nutrient salts are supplied to the photic zone by upwelling of nutrient-rich deep water, nutrient salts can not be all consumed by photosynthesis and are exported to the surrounding areas by surface currents. The equatorial belt of the Pacific, the coastal belts of Africa in the Atlantic and South America in the Pacific, and the high latitude areas of the southern and northern oceans are typically the areas where deep water upwelling occurs. Accordingly, the distribution of the biological pump is highly variable not only geographically but also with respect to time.

OCEANIC CO_2 RESERVOIR MODELS

Because of the complexity and the dynamic nature of the oceanic CO_2 reservoir as briefly discussed above, mathematical models are important and essential tools for gaining an improved understanding of the oceanic CO_2 system and its interaction with the atmosphere. Two fundamentally different classes of models have been developed: one is the box model and the other is based upon the general circulation model of the ocean. In the former, the oceans are divided into a number of oceanographically reasonable domains (or boxes) such as the surface mixed layer and deep water boxes, and the exchange rates of water between these boxes are assigned on the basis of observations, or estimated using constraints such as the distribution of tracers like carbon-14. Box models may be simple having a few number of boxes, or may be highly complex consisting of several hundred boxes. On the other hand, the general circulation model (GCM) for oceans is constructed on the basis of the laws of the physics governing the flow of fluids, and various biological and chemical processes relevant to the carbon cycle in the oceans are parameterized and superimposed on the dynamic circulation model. Because of the complexity of ocean GCM's with a modest spacial resolution (e.g. 5o x 5o) and prescribed sea surface boundary conditions (i.e. decoupled from the atmospheric GCM), a few tens of hours of computing time are required for one model calculation even though a state-of-the-art fast computer is used.

The ocean GCM results may be verified by comparing them with the observed distribution of various ocean tracers such as temperature, salinity, nutrient salts, anthropogenic chemicals and isotopes (e.g. CFC's, krypton-85 and the products of nuclear tests such as carbon-14 and tritium), and natural isotopes (e.g. carbon-14, argon-39 and helium-3). In Fig. 3, the distribution of the combined activity of the natural and bomb-produced carbon-14 observed in 1972 (Ostlund and Stuiver, 1980) in the western Pacific Ocean is compared with the model results computed by Toggweiler et al. (1989) at the Geophysical Fliud Dynamics Laboratory of NOAA at Princeton University. Although carbon-14 is taken up by the sea as CO2 and partially incoroporated into the biological pump, its concentration in seawater is primarily governed by the circulation of water and is not significantly affected by the biological transport due to the small proportion of biogenic CO_2 compared to the total CO_2 dissolved in seawater. High concentrations (positive numbers) observed in the upper layers represent mostly the effect of the nuclear-bomb carbon-14. The model results for the Pacific are in excellent agreement with the observations as seen in Fig. 3. On the other hand, their model results for the Atlantic tend to show the discrepancies in both the northern and southern high-latitude oceans, suggesting difficulties in modeling the deep water formation processes. Furthermore, an agreement between the model results and the observed disitribution of biological tracers such as dissolved oxygen and nutrient salts is yet to be attained. This calls for improvements in the formulation of the biological pump.

In conclusion, during the past decade, the size of observational data base for the distribution of CO_2, ocean circulation tracers and biological pump tracers has been steadily expanded. Oceanic CO_2 models including GCM's with the oceanic carbon cycle superimposed on it have been greatly improved during the same period. The next major challenge facing the scientific community would be the development of coupled atmospher-ocean GCM's and the acquisition of pertinent observational data for testing these models. Such an endevour requires a long term committment for financial support from the international community of governments and industries. A predictive capability thus developed for climatic changes would hopefully give the human community a tool for forseeing hazards lying ahead. Thus, the human community would be able to prepare for the obstacles on the road leading to and through the 21st century. It is a worthy investment for the welfare of the entire human community.

REFERENCES

Broecker, W.S., T. Takahashi, H.J. Simpson and T.-H. Peng (1979). Fate of fossil fuel carbon dioxide and the global carbon budget. Science, 206, 408-418.

Ostlund, H.G. and M. Stuiver (1980). GEOSECS Pacific radiocarbon, Radiocarbon, 22, 25-53.

Takahashi, T., W.S. Broecker and A.E. Bainbridge (1981-a). The alkalinity and total carbon dioxide concentration in the world oceans. In "Carbon Cycle Modelling", B. Bolin editor, SCOPE Vol. 16, J. Wiley & Sons, N.Y., 271-286.

Takahashi, T., W.S. Broecker and A.E. Bainbridge (1981-b).Supplement to the alkalinity and total carbon dioxide concentration in the world oceans. In "Carbon Cycle Modelling", B. Bolin editor, SCOPE Vol. 16, J. Wiley & Sons, N.Y., 159-199.

Toggweiler, J. R., K. Dixon and K. Bryan (1989). Simulations of radiocarbon in a coarse resolution world ocean model II: Distributions of bomb-produced ^{14}C. Jour. Geophys. Res., 94.

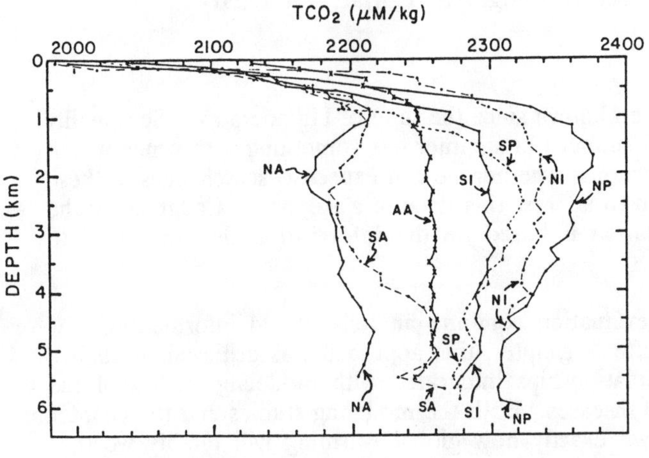

Fig. 1 - Distribution of the total CO_2 concentration in the oceans.

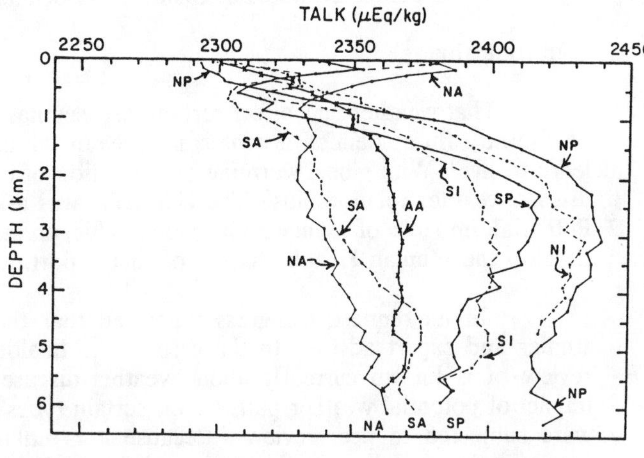

Fig. 2 - Distribution of the alkalinity in the oceans.

NA = North Atlantic
SA = South Atlantic
NP = North Pacific
SP = South Pacific
AA = Antarctic Ocean
NI = North Indian Ocean
SI = South Indian Ocean

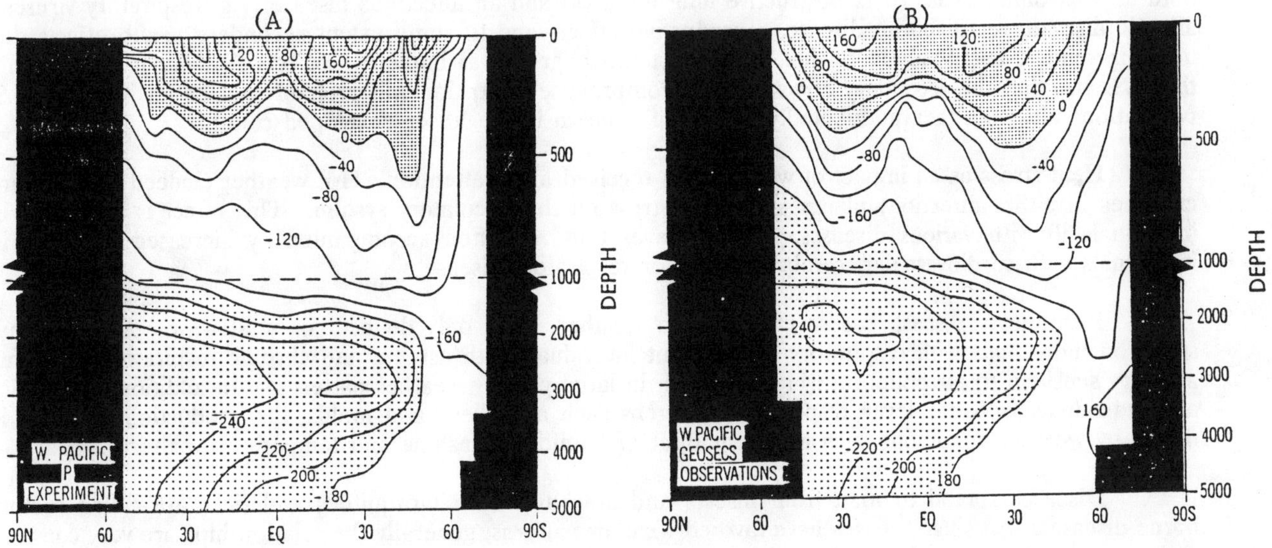

Fig. 3 - Meridional distribution of carbon-14 activity in the western Pacific Ocean. (A) Model results obtained by Toggweiler et al. (1989); (B) the observed distribution by Ostlund and Stuiver (1980). The carbon-14 activity is expressed in terms of parts per thousand (o/oo) difference from the standard wood sample representing 1973-74. The negative values indicate that the CO_2 dissolved in seawater is older than 1973-74, and -10 in the unit shown above corresponds to about 80 years in age.

OVERVIEW OF THE POTENTIAL EFFECTS OF CLIMATE CHANGE ON HUMAN HEALTH

Janice Longstreth, Clement Associates, 9300 Lee Highway, Fairfax, VA 22031

I. Introduction

That weather may affect certain diseases has been known since the time of Hippocrates[1]. Seasonality of disease - flu epidemics in winter, measles in fall and sunburn in summer - is something with which we all learn to live. With global warming one question of concern is whether we can expect to see changes in these disease-weather relationships. The U.S. EPA was asked to address this issue in a Report to Congress on the Potential Impacts of Climate Change[2]. This presentation is based on the information that was used to develop the Human Health chapter of that report.

In its request, Congress suggested that the evaluation combine analysis of old information, new studies and expert advice. In the case of the health effects chapter, this approach was achieved through a review of is known currently about weather-disease relationships integrated with modeling studies of the impact of potential weather patterns on certain types of diseases. Both the modeling studies and the chapter were subjected to peer review. Because it is unknown exactly how global warming will modify weather patterns, the modeling studies evaluated the probable impact of new weather conditions derived from a set of scenarios of future weather.

II. Overview of the health effects which are sensitive to weather

As a first step in the process of evaluating the potential impacts of global warming on human health, those health effects which showed a relationship to season, temperature or other weather-related factor were identified, either via review of the literature or via conversations with experts in the field.

Figure 1 shows a variety of the relationships identified. They include heat stress, respiratory disease both as a chronic disease (e.g., obstructive lung diseases) and an infectious disease (e.g. respiratory viruses), allergic disease, vector-borne disease, reproductive effects and to some extent secondary health effects due to compromised nutrition. Note that the Report to Congress was limited to effects that might occur within the U.S. and that the secondary effects due to compromised nutrition will probably not be an issue for U.S. populations although they undoubtedly will be of concern in the lesser developed countries.

Heat stress as an impact of weather has received much attention. Hot weather (indeed temperature extremes in either direction) places additional stress on the circulatory system. This effect is magnified in individuals ill with various diseases. Heat waves thus are often accompanied by increased mortality in individuals with cardiovascular, cerebrovascular or respiratory disease[3,4].

Respiratory disease has several links to weather. Not only do heat or cold waves put additional stress on individuals with respiratory disease, but individuals with asthma and chronic respiratory problems are very sensitive to air pollution which is driven in large part by weather patterns[5]. In addition, asthmatics and individuals with hay fever respond to allergens such as pollens and molds, the production of which is driven by weather conditions such as the amount of humidity, sunshine and temperature patterns.[6]

Diseases spread to man from insects and arachnids such mosquitoes and ticks are called vector-borne diseases. All such diseases have favored weather patterns; generally the relationships are very complex and depend not only on the response of the disease agent and its vector to temperature, humidity and light, but also on the response of the surrounding habitat and its animal inhabitants which serve as intermediate hosts for the vector[7].

The link between human reproductive effects and weather is far less solid than that between the other effects discussed above. However, in two recent studies of perinatal mortality and preterm birth [8,9], a summertime increase in these adverse effects was seen in two northern locations. One possible explanation is that perinatal infections are also increased during this time period. Further study will be required in order to determine if the observed seasonality is temperature related. If this were a temperature-related phenomenon, one might expect to see a latitude gradient in this effect in the U.S. with the warmer

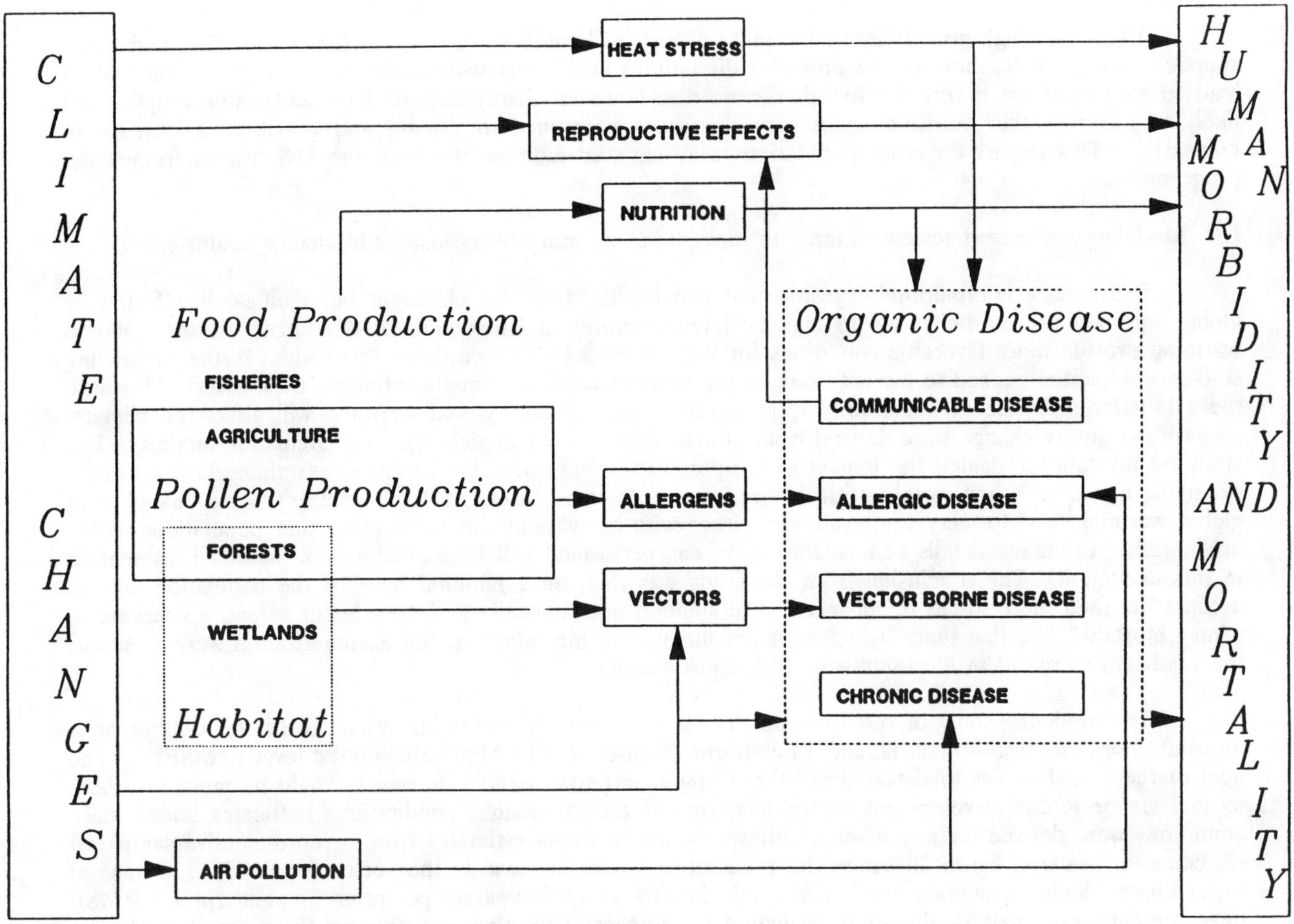

Schematic of How Climatic Changes Can Impact on Human Health

Figure 1 - Schematic of How Climate May Affect Human Health

southern states showing a higher incidence of these outcomes. The south does have a greater problem with infant mortality which heretofore has been ascribed to differences in socioeconomic status. Further study of this issue will be required in order to determine whether temperature contributes to perinatal mortality or preterm birth.

There is a high probability that climate change will cause disruptions in food production and water supplies. In the U.S., most of the potential disruptions envisioned will require some response but willnot lead to an impact on nutrition. In other countries, however, disruptions of food and water supplies are likely to affect human nutrition, leading to decreased resistance to disease and possibly an increase in epidemics. This impact on other populations may have an indirect effect on the U.S. via its foreign aid programs.

III. Modeling and expert review to indicate how global warming scenarios would change health effects.

Preliminary information suggested that two health effects which might be significantly affected by global warming are mortality due to increased temperatures and mosquito or tick-borne diseases (should warming provide more favorable conditions for these vectors.) To investigate these ideas further, modeling studies were commissioned to evaluate how future weather conditions might influence these effects. Because there is a large degree of uncertainty with regard to exactly how global warming will affect the climate, scenarios of future change were derived from global scale weather models (general circulation models). The study on mortality evaluated the impact of temperature on mortality for 15 cities, examining how mortality occurring above or below certain critical temperatures (termed threshold temperatures) would change with global warming[10]. Mortality was evaluated either with or without the assumption that populations would acclimatize. Complete details of how this study was performed will be the subject of another presentation in this document. The conclusion from the study was that, on a national basis, if the population fails to acclimatize, then there would be an increase in summertime mortality and, to a lesser extent, a decrease in winter mortality, i.e., that there would be a net increase in mortality. If full acclimatization were to occur, the predicted increases in mortality would be much smaller.

The modeling study of vector-borne disease evaluated the potential impact of future weather on a mosquito borne disease - malaria, and a tick-borne disease - Rocky Mountain spotted fever (RMSF)[11]. The goal of these studies was to determine if there might be parts of the U.S. which would be more favorable to malaria or RMSF development under scenarios of future weather conditions. Estimates under these conditions were derived for a number of cities. Figure 2a shows estimates from a representative sample of six cities for malaria; figure 2b shows the predictions for an increase in the vector for RMSF. In general the climate change scenarios used did result in any great increased potential for malaria or RMSF development. For malaria, the study evaluated the impact of weather not only on the vector but also on the parasite. In the case of Rocky Mountain spotted fever (RMSF), the approach taken was to evaluate whether future weather conditions would be more favorable for the development of the american dog tick - the vector for RMSF. (Previous information indicated that there is a fairly good correlation between the size of the dog tick population and the incidence of RMSF.) The models used in both these studies are based on existing information about the importance of weather parameters (e.g., rainfall, and temperature) on the development and life cycle of the vectors and, in the case of the malaria model, the parasite. One limitation of this study was that it held constant both the density of the intermediate host populations (field mice, birds, and/or deer) and the distribution of habitat between meadow and forest for both current as well as future weather conditions. Changes in weather could significantly affect both of these parameters. In an analysis of how changes in these two factors might affect the predictions, changes in the sizes of the host populations could cause the estimates to vary as much as 16 times.

IV. Conclusions/Implications

There are many interrelationships between weather and human health effects. As indicated in Figure 1, most of these are indirect. Thus weather patterns do not directly induce mortality (except perhaps in the case of lives lost during a hurricane) but rather it sets up a situation in which mortality is more likely, such as hot weather increasing the chance of heart attack. Some of the relationships are even more tenuous. Forinstance, poor weather may result in crop failure leading to famine and death due to starvation or decreased resistance to disease. In developed nations such as the United States, this category of weather related deaths does not seem that likely, but in the lesser developed countries, where populations exist on the

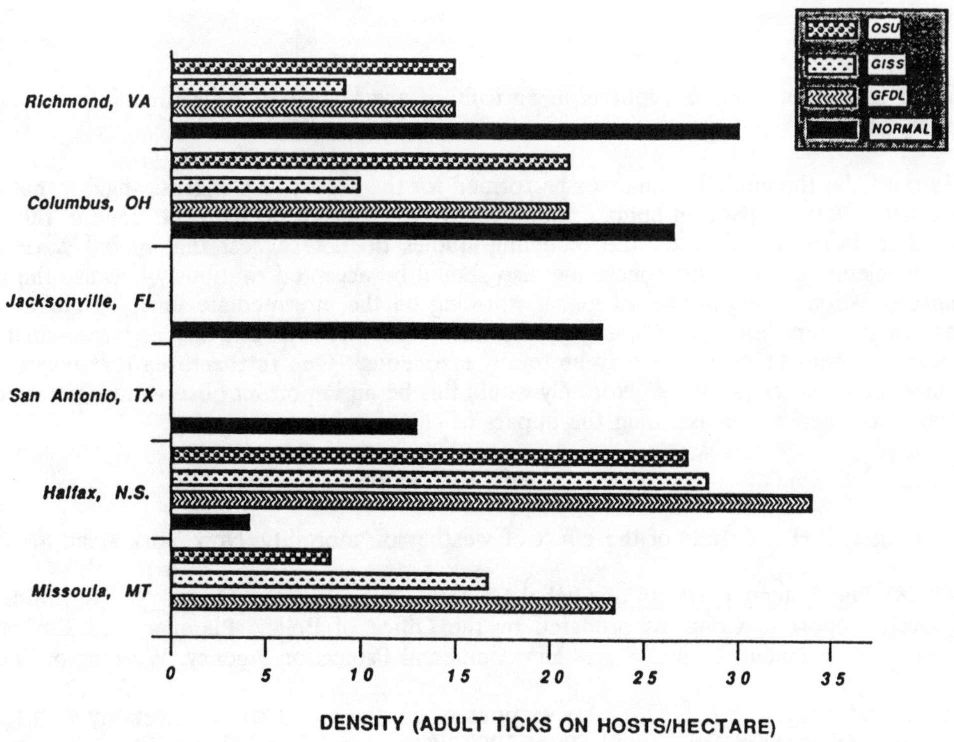

Figure 2A - Predicted Changes in Tick Density under Various Scenarios of Climate Change

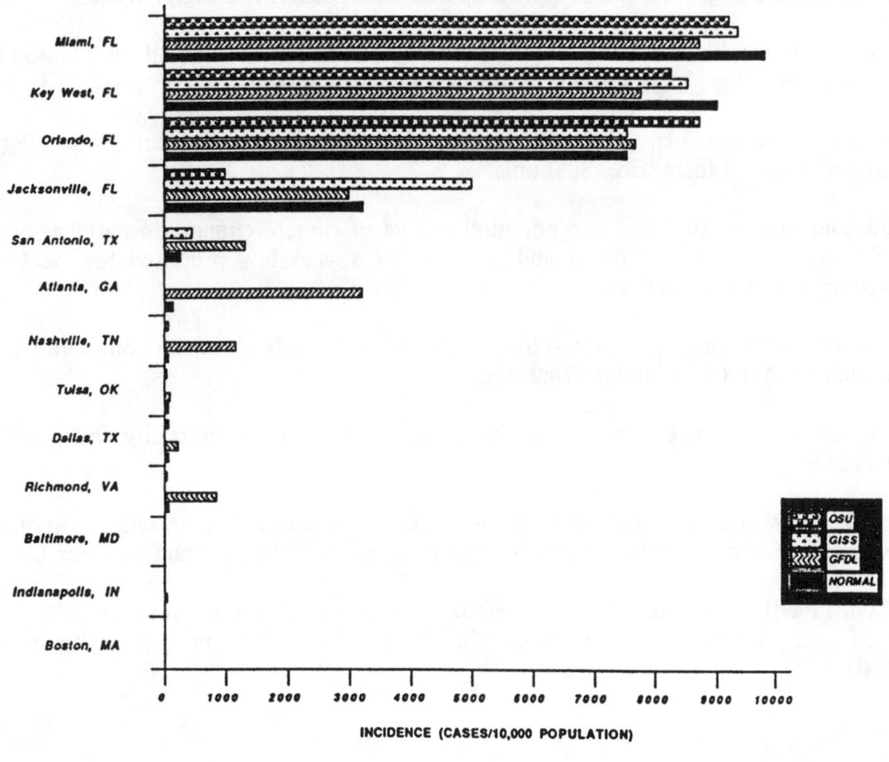

Figure 2B - Predicted Changes in Potential Malaria Incidence under Various Scenarios of Climate Change

borderline of starvation, such disruptions in agriculture are known to have terrible consequences in terms of human suffering.

In the U.S., the modeling analyses performed for the report to Congress suggest that global warming will have a net adverse impact on human mortality. More research needs to be done in this area, however. With regard to infectious diseases, the modeling studies do not suggest that global warming will lead to significant problems per se. This conclusion also should be accepted cautiously because the analysis did not include an estimation of the impact of global warming on the intermediate host. If global warming favors an increase in the population size/density of the mice, deer or birds that act as intermediate hosts for any of these vectors, then the estimates may be totally erroneous. One further area that needs research is that of infant mortality and temperature. Not only would this be an important observation, but such a relationship would have consequences for assessing the impact of climate change.

REFERENCES

1. Kutschenreuter, P.H. A study of the effect of weather on mortality. New York Acad Sci 22:126-138 1959

2. EPA (1988) The Potential Effects of Global Climate Change on the United States. Smith, J. and Tirpak D. (eds) Draft Report to Congress prepared by the Office of Policy, Planning and Evaluation, Office of Research and Development, United States Environmental Protection Agency, Washington, D.C.

3. Rogot, E. and Padgett, S.J. (1976) Associations of coronary and stroke mortaltiy with temperature and snowfall in selected areas of the United States 1962-1966. Am J Epi 103:565-575

4. White, M.R. and Hertz-Picciotto, I. (1984) Human health: analysis of climate related to health. In Characterization of Information Requirements for Studies of CO2 Effects: Water Resources Agriculture, Fisheries, Forests, and Human Health, MR White (ed) Department of Energy, DOE/ER/0236.

5. Grant, L.D. (1988) Health effects issues associated with regional and global air pollution problems. Draft document prepared for World Conference on the Changing Atmosphere Toronto

6. Lopez, M., and Salvaggio, J.E. (1983) Climate-weather-air pollution. chapter 54 in Allergy, Middleton, E and Reed CE (eds) C.V. Mosby Co., St. Louis

7. Wiseman, J. and Longstreth, J.D., The potential impact of climate change on patterns of infectious disease in the United States - Background paper and summary of a workshop prepared for the U.S. Environmental Protection Agency under contract no.

8. Cooperstock, M. and Wolfe, R.A. (1986) Seasonality of preterm birth in the collaborative perinatal project: demographic factors. Am J Epidemiol 124:234-41.

9. Keller, C.A. and Nugent, R.P. (1983) Seasonal patterns in perinatal mortality and preterm delivery. Am J Epidemiol 118:689-98.

10. Kalkstein, L.S. (1988) The impact of CO2 and trace gas-induced climate change upon human mortality. Report prepared for Office of Policy, Planning and Evaluation, EPA contract. number CR81430101.

11. Haile, D.G., (1988) Computer simulation of the effects of changes in weather patterns on vector-borne disease transmission. Report prepared for the Office of Policy, Planning and Evaluation Project Number: DW12932662-01-1

Climate change and parasitic diseases of man and domestic livestock in the United States.

Andrew Dobson, Biology Department,
University of Rochester, Ny 14627.

Introduction.

Current estimates suggest that parasitism of one form or another may be the most common form of life style in at least three of the five major phylogenetic kingdoms (May 1988; Toft 1986). These parasites cause enormous levels of disease, debilitation and human misery in the majority of the world's tropical countries. The data in Tables 1 and 2 contrast the impact of these diseases on human populations and *per capita* protein production from different parts of the world. Essentially, the diseases that result from infection by parasites are a major cause of the reduced *per capita* productivity of the human population in these areas. If the climate of the United States becomes significantly warmer, it seems likely that some of these pathogens will become more widespread, particularly in the South. In this paper, I briefly discuss some general properties of these parasites and indicate how climate change might effect the seasonal incidence of three important veterinary parasites and the possible introduction of several human parasites. A more detailed discussion of how climate change may effect the distribution of parasitic diseases is given in Dobson & Carper (1989).

Classification of parasite life cycles.

The enormous array of pathogens that infect humans and other animals may be conveniently divided on epidemiological grounds into <u>microparasites</u> and <u>macroparasites</u> (Anderson & May 1979; May and Anderson 1979). The former include the viruses, bacteria and fungi and are characterized by their ability to reproduce directly within individual hosts, their small size and relatively short duration of infection and the production of an immune response in infected and recovered individuals. Mathematical models examining the dynamics of these pathogens divide the host population into susceptible, infected and recovered classes. In contrast, the macroparasites (the parasitic helminths and arthropods) do not multiply directly within an infected individual, but instead produce infective stages which usually pass out of the host before transmission to another definitive host. Macroparasites tend to produce a limited immune response in infected hosts, they are relatively long-lived and usually visible to the naked eye. Mathematical models of the population dynamics of macroparasites have to consider the statistical distribution of parasites within the host population.

Direct and indirect life cycles.

A second division of parasite life histories distinguishes

between those species with monoxenic life cycles, and those with heteroxenic life cycles. The former produce infective stages which can directly infect another susceptible definitive host individual, the latter utilize a number of intermediate hosts or vectors in their transmission between definitive hosts. The evolution of complex heteroxenic life cycles permits parasite species to colonize hosts from a wide range of ephemeral and permanent environments, while also permitting them to exploit host populations at lower population densities than would be possible with simple direct transmission (Anderson 1988; Dobson 1988). However, parasites with heteroxenic life cycles are essentially constrained to those areas where the distributions of all the hosts in the life cycle overlap. Changes in the distribution of the these host species due to climatic changes, will therefore be very important in determining the areas where parasites may continue to persist and areas where parasites may be able to colonize new hosts.

Predictive models for parasites of domestic livestock.

Interactions between temperature and humidity seem to be of major important in constraining the geographical range of many of the monoxenic macroparasites that infect domestic livestock. A considerable body of data exists which examines the relationship between meteorological conditions and parasite outbreaks (Gordon 1948; Kates 1965; Levine 1963; Thomas 1974; Ollerenshaw 1974). Indeed the parasitologists of the 1950's and 1960's firmly believed that climate determined the distribution of a parasite species, while weather influenced the timing of disease outbreaks (Thomas 1974). This led to the development of large scale research programs designed to produce forecasts of when disease outbreaks were likely to occur in different areas and hence the best time to administer control measures.

Bioclimatographs.

One way of depicting the interaction between disease outbreaks and climate is through the use of bioclimatographs. The use of these diagrams for monitoring parasite outbreaks was originally suggested by Gordon (1948) in a study of the sheep nematode *Haemonchus contortus* (the barber's pole worm). Bioclimatographs are constructed by plotting the climatological conditions under which a parasite is able to exist and under which outbreaks occur onto a graph of mean monthly temperature and rainfall. When the mean weather data for each month of the year are added to this diagram it is possible to determine at which time of the year outbreaks of the parasite are likely to occur. Although the initial production of these diagrams requires a long term study of the parasite in any region, once the data are available that describe the conditions for both establishment and optimal development, then extrapolations may be made to other regions for which only the climate data are available.

Levine (1963) reviewed and extended the use of bioclimatographs to define and explain the distribution and seasonal incidence of a variety of gastro-intestinal parasites of sheep and cattle in the United States. Figure 1 illustrates bioclimatographs for three important veterinary nematodes for eight different parts of the United States. These figures may be used to determine changes in the seasonal abundance of these parasites in different parts of the U.S. due to climate changes. If we assume a 3-5 degree increase in temperature in the next 50-80 years, it seems likely that the seasonal distribution of *Trichostrongylus* and *Ostertagia* will certainly alter with a continuous period of infection throughout the winter at the edges of its range in Maryland and Oregon, a change in the main months of occurrence in the mid-West and perhaps a reduction in incidence in the South. In contrast, the transmission season for *Haemonchus contortus* (the Barber's pole worm) is likely to increase in all parts of its range in the United States.

It should be emphasized that many parasitologists have become frustrated with the use of bioclimatographs, primarily because they are based on data for mean temperature and rainfall and thus are usually only partially successful for <u>predicting</u> parasite outbreaks in any specific year. Similarly, derivation of bioclimatographs from laboratory determination of the parasite's minimum and optimum development constraints is not often possible, as the climate conditions experienced by the parasite larvae in the soil, are often different from those measured by the local weather station (Thomas 1974). However, because bioclimatographs reflect <u>climate</u> rather than weather, they may prove invaluable in determining both cahnges in seasonal patterns of incidence and whether long term climatic changes will permit specific parasites of domestic livestock to establish in regions where they are not at present a problem.

<u>Parasite of humans in Central America</u>.

The major threat to the human welfare in the United States as a result of climate change is likely to be the spread of parasitic diseases from Central America. The data in Table 1 illustrate that rates of parasitism in many Central and South American countries can reach as high as 30-50%. This contrasts sharply with incidence in the United States which rarely exceed 1-3%. Although the transmission rate of human parasites is aided by poor sanitation in many developing countries, climate also plays a significant role in determining rates of transmission. Over the next fifty years, the synergistic interactions between a warmer climate and an increasing human population is likely to considerably increase the importance of these pathogens in the southern United States.

Two parasitic diseases seem important in this context: Ascariasis and Chagas' disease. The former is caused by another monoxenic nematode, *Ascaris lumbricoides*, the latter is a microparasite transmitted between humans and its wildlife reservoirs by triatomine bugs, *Triatoma* spp. Between 10 and 50% of the population of people in Central America are infected with ascariasis, it is particularly prevalent in children (Crompton, 1988). Embryonation is temperature dependent, requiring 10-14 days at 30 ° C or 45-55 days at 17°C. A warmer moister climate in the Southern United States could lead to increased levels of infection by this parasite, particularly in poorer rural communities.

Chagas' disease is caused by the protozoan *Trypanosoma cruzi*, a distant relative of the pathogen that causes sleeping sickness in East and Central Africa. This parasite is present throughout much of Central and South America. The northern limits of its range seem constrained by the distribution of the triatomine bugs that act as vectors between humans and the range of wild species that act as reservoir hosts. In central and south America bats, opossums, raccoons, armadillos and some primates act as reservoirs. Wood rats, raccoons, opossums and domestic pets have all been recorded as hosts in the United States (Molyneux & Ashford, 1983). The first indigenous case of Chagas' disease in the United States was recorded in the 1950's (Woody & Woody, 1955). The major constraint on the spread of the disease may be the developmental time of insect vectors. These are temperature dependent, which would suggest a Northerly migration during the course of predicted climate changes. Acute forms of the disease may infect children while chronic forms are more common in adults where pathology results in cardiac damage, heart disfunction, and enlargement of the oesophagus and colon. The disease is one of the major causes of morbidity and mortality in South and Central America and there are no drugs available to treat the chronic stages (Molyneux & Ashford 1983). Vector control is hampered by a lack of susceptibility of the bugs to most insecticides and their increasing predilection for human housing as a habitat.

In the light of these comments, it remains ironic that the United States imports drugs from Central America while exporting arms to maintain security. From a human welfare perspective, it might be more rational to export drugs, and import students who can be trained as tropical biologists with the expertise to reduce the impact of parasitic diseases on human welfare in this area. This will have considerably more diplomatic impact on Central America's most significant threat to the US's social welfare than present US activities in this sphere. Yes, Colonel North, it may be time to hang the Uzi in the hall and study soil once more.

<u>Literature cited</u>.

Anderson, R.M. & May, R.M. (1979). Population biology of infectious diseases: Part I. **Nature 280**, 361-367.

Anderson, R.M. & May, R.M. (1985). Helminth infections of humans: mathematical models, population dynamics and control. **Advances** in **Parasitology 24**, 1-101.

Crompton, D.W.T. (1988). The prevalence of Ascariasis. **Parasitology Today 4**, 162-169.

Dobson, A.P. (1988). The population biology of parasite-induced changes in host behavior. **The Quarterly Review of Biology 63**, 139-165.

Dobson, A.P. & Carper, E.R. (1989). Global warming and potential changes in host-parasite and disease-vector relationships. In **'Consequences of Global Warming for Biodiversity'**, (Ed. R.Peters), Yale University Press.

Griffiths, R.B. (1978) The relevance of parasitology to human welfare today - veterinary aspects. In **'The Relevance of Parasitology to Human Welfare Today'** (Eds. A.E.R. Taylor & R.Muller), Symposia of the British Society for Parasitology 16, 41-66.

Kates, K.C. (1965). Ecological aspects of helminth transmission in domesticated animals. **American Zoologist 5**, 95-130.

Levine, N.D. (1963). Weather, climate and the bionomics of ruminant nematode larvae. **Advances in Veterinary Science 8**, 215-261.

May, R.M. (1988). How many species are there on earth? **Science (in press)**.

May, R.M. & Anderson, R.M. (1979). The population biology of infectious diseases. II. **Nature 280**, 455-461

Molyneux, D.H. & Ashford, R.W. (1983). **The Biology of *Trypanosoma* and *Leishmania*, Parasites of Man and Domestic Animals.** Taylor and Francis, London.

Ollerenshaw, C.B. (1974). Forecasting liver fluke disease. In **'The Effects of Meteorological Factors upon Parasites'** (Eds. A.E.R. Taylor & R. Muller), Symposia of the British Society for Parasitology 12, 33-52.

Peters, W. (1978) Medical aspects - comments and discussion II. In **'The Relevance of Parasitology to Human Welfare Today'** (Eds. A.E.R. Taylor & R.Muller), Symposia of the British Society for Parasitology 16, 25-40.

Toft, C.A. (1986). Parasites etc.. In **'Community Ecology'** (Eds. J.M.Diamond & T.J.Case), pp 445-463, Harper & Row, New York.

Woody, N.C. & Woody, H.B. (1955). American trypanosomiasis (Chagas' disease). First indigenous case in the United States. **J.Am.Med. Ass. 159**, 676-677.

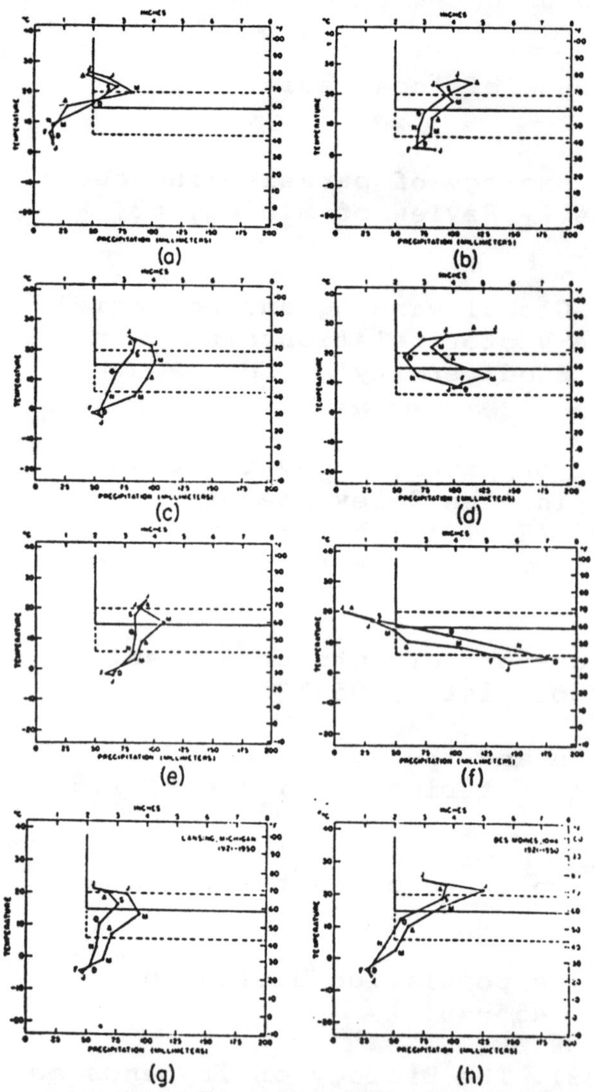

FIG. 2. Bioclimatographs of representative climates in relation to the epidemiology of ruminant gastrointestinal nematodes. Optimum conditions for pasture transmission of *Haemonchus:* 5 cm. or more total monthly precipitation and 15 to 37°C. mean monthly mean temperature. Optimum conditions for pasture transmission of *Trichostrongylus* and *Ostertagia:* 5 cm. or more total monthly precipitation and 6 to 20°C. mean monthly mean temperature. (a) Lubbock, Texas, 1921–1950. Bsk climate. (b) Beltsville, Maryland, 1921–1950, Cfa climate. (c) Urbana, Illinois, 1903–1954, Cfa climate. (d) Columbus, Georgia, 1921–1950. Cfa climate. (e) Williamsport, Pennsylvania, 1921–1950, Cfb climate. (f) Salem, Oregon, 1921–1950, Csb climate. (g) Des Moines, Iowa, 1921–1950, Dfa climate. (h) Lansing, Michigan, 1921–1950, Dfb climate.

TABLE 1.

Estimates of the numbers of helminthic infections in man. Figures (in millions) calculated from data of Le Riche (1967) which were based on those of Stoll (1947). There is little reason to believe that Stoll's basic premises have changed greatly in the last 30 years. Demographic data for 1975 are those published by the United Nations (1976).

	Tropics & Subtropics				Temperate Zones			
	Africa	Asia (excl. USSR)	C. & S. America	Oceania	N. America	Europe (excl. USSR)	USSR	World Total (1975)
Nematodes								
A. lumbricoides	159	931	104	1	5	39	30	1269
Hookworms	132	685	104	2	3	2	4	932
E. vermicularis	24	136	40	1	29	75	48	353
T. trichiura	76	433	94	1	1	41	41	687
T. spiralis	1		3		35	5	2	46
Others	9	49	21	< 1	1	1	3	85
W. bancrofti and *B. malayi*	59	300	22	2				383
Other filariae	178	57	39					274
Trematodes								
Schistosomes	168	88	15					271
Clonorchis group		36				< 1	1	38
Others		25						25
Cestodes								
T. saginata	32	11	2	< 1	< 1	1	29	77
H. nana	2	27	2		< 1	2	5	39
D. latum		2				3	10	15
Others	1	2					1	4
Total								
Population	401	2256	324	21	237	473	255	3967
Total Infections	841	2782	446	9	76	170	174	4498
% of Population Infected	210	123	138	43	32	36	68	113
Summary	4078 infections in 3002 = 136%				420 infections in 965 = 44%			

One of the six diseases included in the WHO·TDR Programme is leishmaniasis. This condition was almost excluded because statistical data indicated

THE RELEVANCE OF PARASITOLOGY TO HUMAN WELFARE TODAY — VETERINARY ASPECTS

R. B. GRIFFITHS

Food and Agriculture Organization of the United Nations, Rome, Italy

INTRODUCTION

I propose to address myself in the first place to the problems of the Third World. This is not to say that there are no problems remaining in the so-called developed countries but by and large they are equipped to deal with their important parasites. In most developing countries, however, there are many deficiencies. In a considerable number their problems have not been identified; and even where surveys have been made and where priorities for action have been established, field control efforts are often meagre because of the lack of money and trained manpower.

The need to intensify parasitic disease control in domestic animals is urgent not only from the human health standpoint because of the parasitic zoonoses but, from the viewpoint of human nutrition, the need is even more urgent since nobody will deny that, among the reasons for the low productivity of livestock and poultry in developing countries, parasitism is a highly important factor.

TABLE 2

Regional production of some livestock products per head of animal population. Source: Nestle (1974), cited by Smith (1976)

	Kg product/head of livestock population		
	Beef/Veal	Mutton Lamb	Pork
USA/Canada	93	11	98
Latin America	36	3	16
Africa	15	3	29
Near East	17	4	12
Far East	4	4	24

Tables 1 and 2 show the remarkably low levels of animal protein production and consumption in a large number of developing countries. Although these

POTENTIAL IMPACT OF GLOBAL WARMING: CHANGES IN MORTALITY FROM EXTREME HEAT AND COLD

Laurence S. Kalkstein
Center for Climatic Research
University of Delaware

The objective of this study is to evaluate changes in human mortality attributed to predicted changes in climate due to increased concentrations of CO_2 and other trace gases in the atmosphere. In addition, the historical relationships between weather and mortality are evaluated, and those weather variables most influential upon summer and winter mortality are identified. Finally, results from a preliminary study to determine the differential impact of weather and pollution on mortality in St. Louis are briefly discussed.

The evaluation covers 48 cities around the country, and daily mortality data for 11 summer and winter seasons are extracted and standardized in a manner that facilitates inter-city comparisons. The mortality totals are divided into various age and race categories, and separate evaluations are developed for all causes of death and those causes considered to be "weather-related."

The initial step involves the determination of historical weather/mortality relationships. Threshold temperatures, which represent the temperature beyond which mortality increases, are developed for each category within each city using an objective procedure (for a detailed discussion of all procedures used here, refer to Kalkstein, 1988 and Kalkstein and Davis, 1989). Once established, 12 weather variables are correlated with the mortality data for days beyond the threshold using an "all regression" multiple regression procedure which accounts for independent variable collinearity, and historical relationships are established.

Simple future weather scenarios are developed for 15 selected cities to provide rough future estimates of mortality trends. The historical algorithms for these cities produced from the regression procedures are utilized to provide "unacclimatized" estimates of mortality for each city assuming future climatic warming. "Analog cities," whose present weather duplicates a city's predicted weather in the future, are determined from a large pool of cities, and their historical mortality algorithms are used to predict "acclimatized" estimates of mortality for each of the 15 cities. "Partially acclimatized" estimates are produced as well, and are calculated by determining the mortality value midway between the acclimatized and unacclimatized estimates.

The following is a summary of results from the evaluation.

1. Threshold temperatures for summer and winter suggest that the impact of weather upon mortality is relative rather than absolute. The highest summer threshold temperatures were found in the South and Southwest (e.g. Dallas: 103°F, Phoenix: 112°F), while the lowest were found in the Pacific and Northeast regions of the country (e.g. Boston: 85°F, San

Francisco: 84°F). Thus, weather-induced increases in summer mortality were noted in cooler climates at much lower temperatures. During the winter, the lowest threshold temperatures were found in the coldest regions (e.g. Fargo: -9°F, Milwaukee: 8°F), while relatively high thresholds were uncovered for mild cities (e.g. Jacksonville: 50°F, San Francisco: 48°F).

2. During summer, the lag time between the weather mechanism and the associated mortality was short for all cities, never exceeding one day. In winter, the lag time was much greater, and in a number of cases, three day lag times were uncovered.

3. In both seasons, the age/race categories which appeared most sensitive to weather were total deaths and elderly deaths (greater than 65 years old). During summer, whites seemed to be somewhat more sensitive than non-whites (this varied on an inter-regional level), and in winter, non-whites seemed more vulnerable in southern cities. During both seasons, the smallest number of statistically significant weather/mortality models were found within the youngest age classes (ranging from 0-44 years old).

4. During summer, a much larger proportion of statistically significant weather/mortality models were found for "all causes of deaths" rather than "weather-related" deaths, indicating that mortality from a large variety of causes rises during extreme episodes. During winter the opposite was true, and the most significant relationships were found for "weather-related" deaths.

5. During summer a large number of the regression relationships exhibited R^2 values exceeding 0.250, and in a few cases the R^2 exceeded 0.500. During winter the opposite was true, and the most significant relationships were found for "weather-related" deaths.

6. For the total death models in summer, the most important independent variables influencing mortality were cooling degree hours (CDH, which measures the accumulation of degree hours above 90°F each day; this variable is a good indicator of the duration of warm conditions) and TIME (indicating that heat waves earlier in the season are more important than those late in the season). The importance of TIME implies that some degree of acclimatization might take place within a season. The relative significance of CDH and insignificance of maximum temperature suggest that the duration of the hot event, rather than its intensity, is most important.

7. Other weather variables appearing commonly in summer total death models were minimum dewpoint, afternoon visibility, and minimum temperature. In most cases it appeared that humid, uncomfortable conditions with warm nighttime temperatures contribute to heightened mortality, although in a minority of cities hot, dry conditions were most responsible.

8. Even stronger models were uncovered for the 65 and older age group, and the responsible weather variables were somewhat different. For example, maximum temperature was much more important here than in the total death models and CDH was less important. This suggests that the intensity of the heat wave is more important than duration for this age group. The impact of hot, dry weather was more noteworthy in the elderly models.

9. A regional comparison of summer models indicated a differential sensitivity to weather. The two most weather-sensitive areas were the Pacific and Continental North regions of the nation, areas noted for a lesser frequency of extremely high temperatures. The least weather-sensitive regions were in the South and Desert Southwest, where extreme heat is commonplace and human response is less dramatic.

10. During winter it appeared that overcast, damp days contributed to heightened mortality in many areas, regardless of temperature (but assuming that the temperature was below the threshold). Dewpoint, windspeed, and cloud cover seemed more influential in winter than summer, and surprisingly, windspeed was generally inversely related to mortality. This may imply that "wind chill" has a lesser impact on winter mortality than expected. Unlike the summer results, TIME was poorly represented in the winter models, indicating that the seasonal timing of offending weather is relatively unimportant.

11.Differential regional responses were difficult to uncover in winter, and weather variables which exerted the greatest influence upon mortality also demonstrated a lesser inter-regional variation than in summer. The greatest number of relationships with R^2 values exceeding 0.250 occurred in the Northeast.

12. Predictions of weather-induced mortality occurring during summer were attempted for 15 large cities, and it is estimated that over 1100 deaths (standardized to allow for inter-city comparison; refer to Lilienfield and Lilienfield, 1980 for standardization procedure) occur during an average summer season in the SMSAs of the 15 cities. St.Louis, New York City, and Philadelphia ranked first, second, and third, respectively, and each city averaged over 100 standard city deaths per summer. The seven highest ranking cities were all found in the Midwest or Northeast. Four of the five lowest ranking cities were found in the South, with New Orleans and Oklahoma City experiencing virtually no deaths attributed to weather in the summer.

13. Predicted future unacclimatized mortality in summer rose rapidly as the scenarios become warmer. The 7°F warming scenario yielded an estimated weather-induced mortality exceeding 9000 standardized deaths per year in the SMSAs of the 15 cities. Predicted future acclimatized mortality indicated that warming might have little impact on weather-induced mortality if acclimatization is complete. Estimates of future mortality assuming partial acclimatization (probably the most realistic scenario) indicated that sizable increases in mortality will occur as the weather becomes warmer (Table 1). Over 4500 added deaths will occur during an average summer in these 15 cities due to weather alone if the climate warms by 7°F and if people partially acclimatize to the warmth. This represents a fourfold increase over present weather-induced deaths.

14. Present-day estimates of winter mortality indicated that weather-induced deaths were much less important in winter than in summer. The results differed from those uncovered in summer regarding the relative impact of weather. Most of the estimated winter deaths occurred in regions with relatively severe winter climates, while the smallest number of deaths were found in mild weather cities. It appeared that the impact of

weather in winter is more absolute, while the impact of weather in summer tends to be relative.

TABLE 1. ESTIMATES OF FUTURE MORTALITY IN SUMMER ATTRIBUTED TO WEATHER USING WARMING SCENARIOS AND ASSUMING PARTICAL ACCLIMATIZATION[a]

City	0°F	1°F	2°F	4°F	5°F	7°F	Absolute Change from 0°F to7°F
Atlanta	33	38	73	148	132	208	175
Chicago	95	104	74	88	88	93	-2
Cincinnati	94	118	100	189	250	400	206
Dallas	0	39	54	79	99	167	167
Detroit	103	123	90	152	180	252	149
Kansas City	98	111	130	212	207	328	30
Los Angeles	46	56	66	82	123	270	224
Memphis	94	121	108	229	323	671	577
Minneapolis	86	102	71	105	216	167	81
New Orleans	0	0	0	0	0	0	0
New York	139	168	147	289	388	647	508
Oklahoma City	0	0	0	0	0	0	0
Philadelphia	118	83	120	220	297	472	354
St. Louis	183	139	199	275	495	833	650
San Francisco	29	36	40	49	38	64	35
TOTALS	1118	1238	1272	2117	2836	4572	3454

[a] Numbers represent the predicted increase in mortality during an average summer under each warming scenario.

15. Winter unacclimatized, partially acclimatized, and acclimatized predictions indicated that sharp drops in mortality are expected if the weather becomes warmer. The unacclimatized results differed from those uncovered in summer, when dramatic rises in mortality were predicted. The unacclimatized drop in winter may be related to fewer number of days below the threshold. Acclimatized mortality predictions approached 0 if a 7°F warming does occur.

16. There has been great difficulty in separating weather-induced mortality from pollution-induced mortality, as it is widely believed that weather situations which seem to be associated with high mortality are also associated with high pollution levels. At present we are in the midst of a pilot project to determine if weather-related mortality can be distinguished from pollution-related mortality. Using a synoptic climatological approach, air mass categories have been developed for St. Louis, and each day in the study period has been identified (refer to Kalkstein et al., 1987 for a discussion of synoptic methodology). Mean daily mortality has been computed for each air mass category, and it is apparent that one air mass type (which possesses particularly sultry weather conditions) has a 30 percent higher mean mortality than the others. This air mass category appears to possess average or below average readings of total suspended particulates, oxidants, ozone, nitrous oxides, and sulfur dioxides. Our very preliminary results for St. Louis suggest that weather, rather than pollution concentration, has a more important impact on the day-to-day variation in mortality. This topic obviously requires further investigation.

REFERENCES

Kalkstein, L.S. 1988. The impacts of predicted climate change on human mortality. _Publications in Climatology_ 41: 132 pp.

Kalkstein, L.S., and Davis, R.E. 1989. Weather and human mortality: an evaluation of demographic and interregional responses in the United States. _Annals of the Association of American Geographers_ 79: 44-64.

Kalkstein, L.S.; Tan, G.; and Skindlov, J. 1987. An evaluation of three clustering procedures for use in synoptic climatological classification. _Journal of Climate and Applied Meteorology_ 26: 717-30.

Lilienfield, A.M., and Lilienfield, D.E. 1980. _Foundations in epidemiology_. London: Oxford Press.

Climate Change as Natural Hazard: Three Perspectives from Natural Hazards Research Useful for Analyzing the Social Implications of Global Warming

William E. Riebsame

Natural Hazards Research and
Applications Information Center
University of Colorado
Boulder, CO 80309

Hazards researchers and managers have developed considerable understanding of social systems under environmental stress (Burton et al. 1978; Whyte 1986), and some aspects of this experience can be usefully applied to climate change (Heathcote 1985). Though the focus of hazards work has been on short-duration, extreme events rather than cumulative environmental changes, it provides at least three useful perspectives on the interaction of society and changes in the climate: (1) hazards theory explicitly analyzes the interaction of natural and social systems in an ecological framework; (2) hazards researchers recognize that people respond to the environment as perceived, rather than as defined by physical science; and (3) even slow, cumulative climate changes may be interpreted and responded to as discreet extreme events.

Hazards as Part of Human Ecology

The first useful perspective offered by hazards work is the conceptualization of hazards as occurring, by definition, only where and when natural extremes and social vulnerabilities interact, and its insistence on conjoint analysis of environmental and social systems. This "human ecological" perspective was codified in a model constructed by Kates (1971) to show hazards-society interaction as a cybernetic system dominated by negative feedbacks between vulnerability, impact, adjustment, and changing vulnerability and subsequent impact (Fig. 1). Kates later applied the ecological model to climate change (Fig 2), arguing that the impacts of climate were determined by both the nature of the fluctuation and the vulnerability or resilience of the social system (Kates 1985). Because vulnerability varies among societies and through time, so will the effects of climate changes.

Unfortunately, differential vulnerability to climate change (or differential resiliency to climate change) has thus far been addressed only with convention wisdom and intuition. There exists a widely shared belief that Third World societies are less able to cope with climate change because they have fewer resources to deploy in adapting. This proposition makes intuitive sense, but does little to explicate how any society will adapt, and where and when vulnerabilities will be

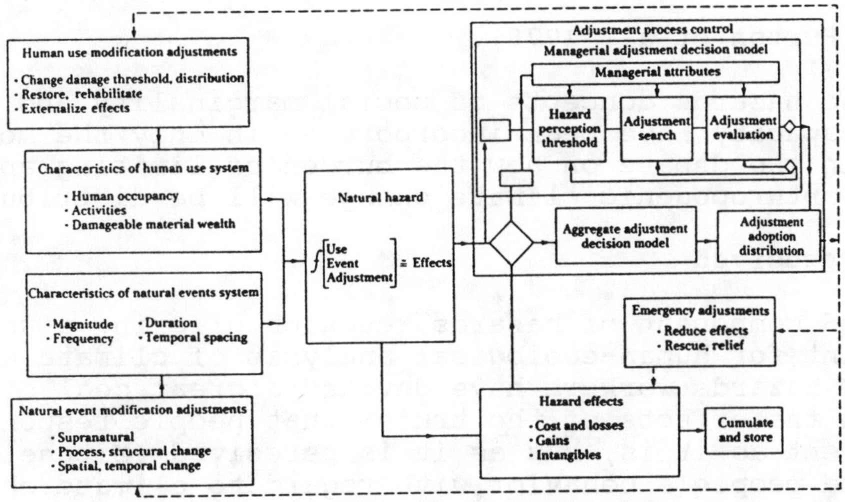

Fig. 1 Human-ecological model of natural hazards (Kates 1971)

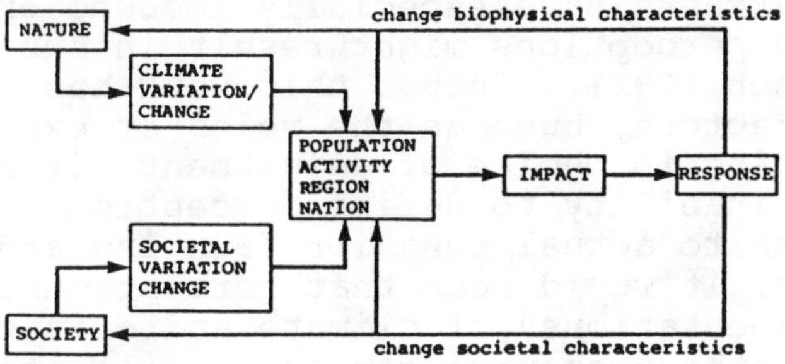

Fig. 2 Ecological model of climate and society (Kates 1985)

exacerbated or mitigated. It is worth noting that many traditional livelihood systems are quite resilient, and that it is usually in the transition from traditional to "developed" systems that social vulnerability emerges (Hewitt 1983).

Moreover, the fact that societal vulnerability to impacts changes over time is neglected in most impacts assessments. Only rudimentary efforts have been made to incorporate dynamic adaptation into impact projections (Parry et al. 1988), mainly because projecting social vulnerability is more difficult than simply projecting impacts on a static society; we have no widely accepted theories for predicting social change. Thus impacts projected even over several decades assume static societies. The threat may thus be over- or understated because social adaptation is certain to occur and to change vulnerability.

The human-ecological formulation of climate/society interaction (Fig. 2) incorporates a "black box" of social evolution that must be filled in by the researcher. It can encompass essentially any theory of how society actually works, subsuming the major paradigms including Marxist or materialist perspectives on differential social vulnerability along class lines (Watts 1983), structural perspectives of institutions and social constraints (Douglas and Wildavsky 1982), and the behavioralist perspective of perception, choice, and management of hazards (Burton et al. 1978).

Yet, the hazards concepts of social marginality and absorptive capacity have been incorporated in only the most facile way in the debate on how the burden of limiting and adapting to anthropogenic climate change will be distributed.

Hazards as Perceived

A second component of hazards research offering a useful starting point for human-ecological analysis of climate change is perception. Hazards workers have devoted a great deal of attention to the effects of the truism that people respond not to the environment as it is, but as it is perceived (Mitchell 1984). To understand people's behavior with regard to climate change then, one must first understand how they perceive it.

Hazards researchers have especially focused on how the filter of people's perceptions might result in sub-optimal adjustment (Saarinen 1982). Though this work has uncovered several relevant factors, such as the roles of experience and professional training in choice of adjustment, it also has been criticized for an inability to define perception unambiguously or to link perceptions to actual behavior (Bunting and Guelke 1979). Yet, within limits, it would seem that perceptions, or what might best be called "expectations" of climate and climate change among policy-makers and other resource managers, will affect how they respond and determine the range of options they consider. This perspective should be incorporated into climate change studies.

Hazards as Extreme Events

Besides being part of an interactive, human ecological web, and interpreted by human perceptions, hazards can also be seen as extreme geophysical events with certain characteristics that threaten social well-being. This definition is the one most likely to be transferred to the realm of climate change research because it is the easiest to operationalize (e.g., through measures of physical events). By definition, climate change will alter the nature of extreme events--in terms of frequency, magnitude, and spatial and temporal distribution. Surprisingly, however, most global change research to date focuses on changes in average conditions, partly because the higher moments of statistical distributions (e.g., variance and extremes) are difficult to measure and extrapolate.

Yet, it can be argued that social systems will respond not to slow changes in mean conditions, but to extreme events and to changes in the frequency with which critical thresholds are surpassed (Mearns et al. 1984). Thus, we will respond to worsened "storm surges" rather than gradual sea-level rise, to "droughts" on the Great Plains rather than cumulative drying, and to "heat waves" in cities rather than slowly warming summers.

The human tendency to perceive and respond to environmental changes as individual extreme events, as illustrated in Figure 3,

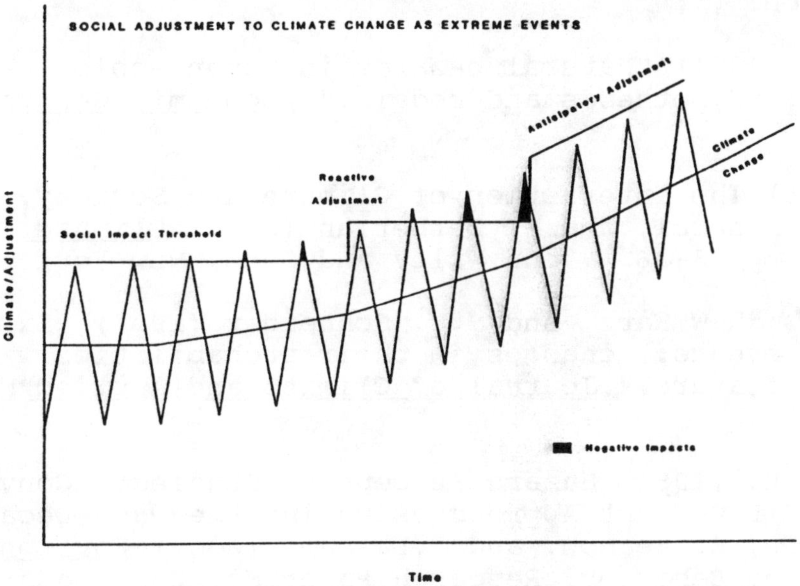

Fig. 3 Human adjustment to cumulative climate change as response to extreme events.

must be carefully analyzed and monitored as we adjust to climate change. It may be that "fixes" applied to the extremes of shifting distributions may not be sufficient or efficacious in response to the underlying cumulative trend. Crisis responses are not likely to be as efficient a way to adapt to climate change as more careful, integrated planning. This lesson is only slowly being learned in natural hazards mitigation (National Academy of Sciences 1987), and the persistent gap between scientific understanding of geophysical systems (we know a lot about how and where floods, earthquakes and other hazards occur), and our ability to reduce their negative impacts on social systems (losses keep increasing as society develops), is worth contemplating as we move forward with research to further define climate change.

References

Bunting, T.E. and L. Guelke (1979) "Behavioral and perception geography: a critical appraisal." Annals of the Association of American Geographers 69: 448-462.

Burton, I., R.W. Kates, and G.F. White (1978) The Environment as Hazard. Oxford University Press, New York.

Douglas, M. and A. Wildavsky (1982) Risk and Culture. University of California Press, Berkeley.

Heathcote, R.L. (1985) Extreme Event Analysis. In R.W. Kates, J.H. Ausubel, and M. Berberian (eds.) Climate Impact Assessment, pp. 369-401. John Wiley and Sons, New York.

Hewitt, K., ed. (1983) Interpretations of Calamity. Allen and Unwin, Boston.

Kates, R.W. (1971) "Natural hazards in human ecological perspective: Hypotheses and models." Economic Geography 47, 438-451.

______ (1985) The Interaction of Climate and Society. In R.W. Kates, J.H. Ausubel, and M. Berberian (eds.) Climate Impact Assessment, pp. 3-36. John Wiley and Sons, New York.

Mearns, L.O., R.W Katz, and S.H. Schneider (1984) "Extreme high-temperature events: changes in their probabilities with changes in mean temperature." Journal of Climate and Applied Meteorology 23: 1601-1613.

Mitchell, J.K. (1984) Hazard Perception Studies: Convergent Concerns and Divergent Approaches During the Past Decade. In T.F. Saarinen, D. Seamon, and J.L. Sell (eds.) Environmental Perception and Behavior. Research Paper No. 209, Department of Geography, University of Chicago.

National Academy of Sciences (1987) Confronting Natural Disasters. National Academy Press, Washington, DC,

Parry, M.L., T.R. Carter, and N.T. Konijn (eds.) The Impact of Climate Variations on Agriculture, Vols. 1 and 2. Kluwer Academic Publishers, Dordrecht, Holland.

Saarinen, T.F. (1982) Perspectives on Increasing Hazard Awareness. Monograph 35. Natural Hazards Research and Applications Information Center, University of Colorado, Boulder.

Watts, M. (1983) On The Poverty of Theory: Natural Hazards Research in Context. In K. Hewitt (ed.) Interpretations of Calamity, pp. 231- 262. Allen and Unwin, Boston.

Whyte, A.V.T. (1986) From Hazard Perception to Human Ecology. In R.W. Kates and I. Burton (eds.) Geography, Resources and Environment, Vol. II, pp. 240-271. University of Chicago Press.

OVERVIEW OF THE POTENTIAL HEALTH EFFECTS ASSOCIATED WITH OZONE DEPLETION

Janice Longstreth, Clement Associates, 9300 Lee Highway, Fairfax, VA 22031

I. Introduction

The stratospheric ozone layer acts as a protective shield, preventing much of the sun's ultraviolet radiation (UVR) from hitting the earth. This protection is somewhat selective for the shorter wavelengths of UVR are absorbed preferentially. The shortest wavelength region of UV is 200 to 290 nm and is termed UV-C; virtually all of it is absorbed by the ozone layer. UV-B, 290 to 320 nm, is only partially absorbed, and UV-A, 320 to 400 nm, is not absorbed at all.

With depletion of the ozone layer, more UV-B will reach the earth. Of the UV-B and UV-A that reaches the earth, UV-B has the most biologic activity per unit of energy. Irradiation with UV-B has been shown to cause cancer in animals and is also responsible for skin cancer and to some degree, cataracts. In addition, UV-B has an adverse effect on the immune system, suppressing the responses to foreign substances that enter the body via the skin. The consequences for this in terms of infectious disease are unknown, but data from several animal models of infectious diseases suggests that UV-B can affect the way animals recover from infections.

As part of an effort to assess the risk of stratospheric ozone depletion, EPA[1] evaluated the hazards associated with UV-B exposure and estimated the health risks associated with decreases in the ozone layer. This presentation is drawn from that risk assessment, a summary of which is given below.

II. Skin Cancer

There are three forms of skin cancer. Melanoma is a cancer of the pigment producing cells of the skin; it is the most dangerous form. It accounts for approximately 26,000 cases and about 7800 deaths, annually. Over the past 35 years, there has been a 200 percent increase in incidence and a 150 percent increase in mortality.[2] Non-melanoma skin cancer, which includes basal and squamous cell carcinoma, is more benign - so benign, in fact, that accurate records of incidence are not maintained. The last survey to evaluate incidence in a systematic fashion was performed in the 1977-1978 period. This survey found that between 400,000 and 500,000 individuals developed non-melanoma skin cancer annually; as compared to a similar survey performed in 1970-71, this represented a 15 to 20 percent increase in incidence rate[3].

A. Non-melanoma. The data linking sunlight and UV exposure to non-melanoma skin cancer is reasonably solid and consists of 1) the fact that those individuals who receive the most sun exposure and the lightest skin (i.e., having the least amount of protective pigment) are at the highest risk category, 2) the fact that those sites at greatest risk are the ones with the most exposure (e.g., face and hands), 3) the fact that individuals living closer to the equator (and therefore exposed to more UV) have a higher risk, and 4) the fact that UV induces non-melanoma skin cancer in animal models. On the basis of dose-response information drawn from the 1981 survey information, EPA[1] estimated that for every 1 percent increase in UV there would be between a 2 and 3 percent increase in non-melanoma skin cancer.

B. Melanoma. The data linking melanoma to sunlight exposure has been considered less clear-cut. Although individuals with lighter skin are at greater risk, it is the sites which receive intermittent exposure which show the greatest risk of melanoma. Individuals who cannot repair UV-induced DNA damage show a 2000-fold increase in risk of melanoma. However, those individuals who spend much of their time outdoors are at less risk than individuals who are exposed only on vacations and during occasional recreation exposure. Thus there was some concern as to a relationship between UV and melanoma. Recently the United States Environmental Protection Agency has completed a review of the available data and literature to assess the relationship of melanoma to UV-B exposure[4]. The salient conclusions of that review were that 1) the older literature, taken in conjunction with some very recent case-control epidemiologic studies, clearly indicated a role for sunlight in the etiology of melanoma, 2) the active portion of the solar spectrum was most likely to be in the UV-B region based on animal studies implicating UV-B in non-melanoma etiology and immunosuppression and on the sensitivity of XP patients, and 3) the dose-response relationships for incidence and mortality were estimated to be such that for every 10 percent increase in UV-B there would be between a 3 to 5 percent increase in melanoma incidence and between 2 and 10 percent increase in melanoma mortality. Very recent evidence confirms the role of UV-B in melanoma etiology, in that UV-B has now induced melanomas in two animal models, one in fish[5] and one in marsupials[6].

III.Cataracts.

Cataracts are opacities in the lens of the eye which impair vision. In the United States and other developed countries, cataract operations prevent most cataracts from causing blindness. Nevertheless, in the U.S., cataract remains the third leading cause of legal blindness[7]. Worldwide senile cataract is responsible for significant visural impairment in 30 to 45 million people - of these perhaps 12 to 15 million are blind. Current treatment rates are not keeping pace with current incidence rates - thus the problem will continue to grow[8].

The exact mechanism of the formation of cataract is still unknown. Epidemiological studies, laboratory animal studies, and biochemical analysis support the belief that some cataracts are etiologically related to exposure to ultraviolet radiation B (UV-B)[7]. Ultraviolet radiation A and other causes, e.g., nutritional deficiency, may also contribute to cataract formation. Recent epidemiological studies of occupationally exposed individuals have indicated that the incidence of posterior subcapsular and cortical cataracts are directly related to cumulative exposure to UV-B[9], thus changes in the amount of ambient UV-B radiation are likely to alter the incidence of cataracts. UV-B may also play a role in causing or exacerbating other eye disorders. Ozone modification that alters the amount of UV-B reaching the earth's surface is likely to change the prevalence (and incidence) of cataracts. The U.S. Environmental Protection Agency[1] has estimated, on the basis of epidemiologic data, that for every one percent decrease in stratospheric ozone, there will be between a 0.3 and 0.6 percent increase in cataracts.

IV. Immunologic Effects

Ultraviolet radiation (UVR) has been found to alter, both locally (in the skin) and systemically, the immune response (ImR) to antigens administered via the skin in humans and experimental animals.[10] The initial damage induced by UVR is inactivation of Langerhans cells - the principal antigen processing cell (APC) in the skin. APCs are required in the development of cell- or antibody-mediated ImRs[11]. In humans, loss of the Langerhans cells is accompanied by the migration into the skin of another type of APC which apparently has a predilection for interacting with a subset of T lymphocytes which suppress ImRs (suppressor T cells; Ts)[12]. The Ts which arise in the skin as a result of UVR treatment are specific for antigen and prevent the development of ImRs. In a tumor bearing animal, this immunosuppression may result in the outgrowth of tumor cells which in normal animals would have been destroyed by the immune system[13]. Although it is not certain that the damage to the Langerhans cells is entirely responsible for the immunosuppressive effect of UVR, it is clear that UV irradiation of skin reduces the ImR in that skin.[16] Even more important, it is also clear that it is the UV-B portion of the UV spectrum that is responsible for the depression of the ImR.[20]

One important facet of this UVR-induced immunosuppression is the role it plays in UV-B-induced carcinogenesis; it affects the host's ability to respond to tumor-specific antigens present on the UVR-induced tumor cells thereby increasing the risk for tumor development in animal models, probably by allowing the tumor to escape the normal immune surveillance mechanisms[14]. Thus the tumor cells are allowed to divide and establish a large and growing tumor in the host.

As indicated in Figure 1, the immunosuppressive effects of UVR may also have a deleterious effect on the ImR to those infectious diseases that enter through the skin, especially if the initial ImR to the agent takes place in the skin. As yet, little research has been done in this area. Recently published reports indicate, however, that UV irradiation during a first cutaneous infection with two very different organisms, the parasite Leishmania sp. and Herpes simplex virus, may result in an impairment of the ImR of the host to subsequent infections[15,16]. In the case of leishmaniasis, this could lead to the development of the more lethal form of the disease, visceral leishmaniasis. There is also some very preliminary evidence that other systems, e.g., the ImR to malaria, may also be affected by UV-B irradiation.[23]

The effects of UV-B and solar radiation on the human immune system have not been studied in sufficient detail to allow estimation of dose-response relationships for these effects. Qualitatively, it is known from animal studies that the doses of UV-B needed to induce immunosuppression are much lower than those

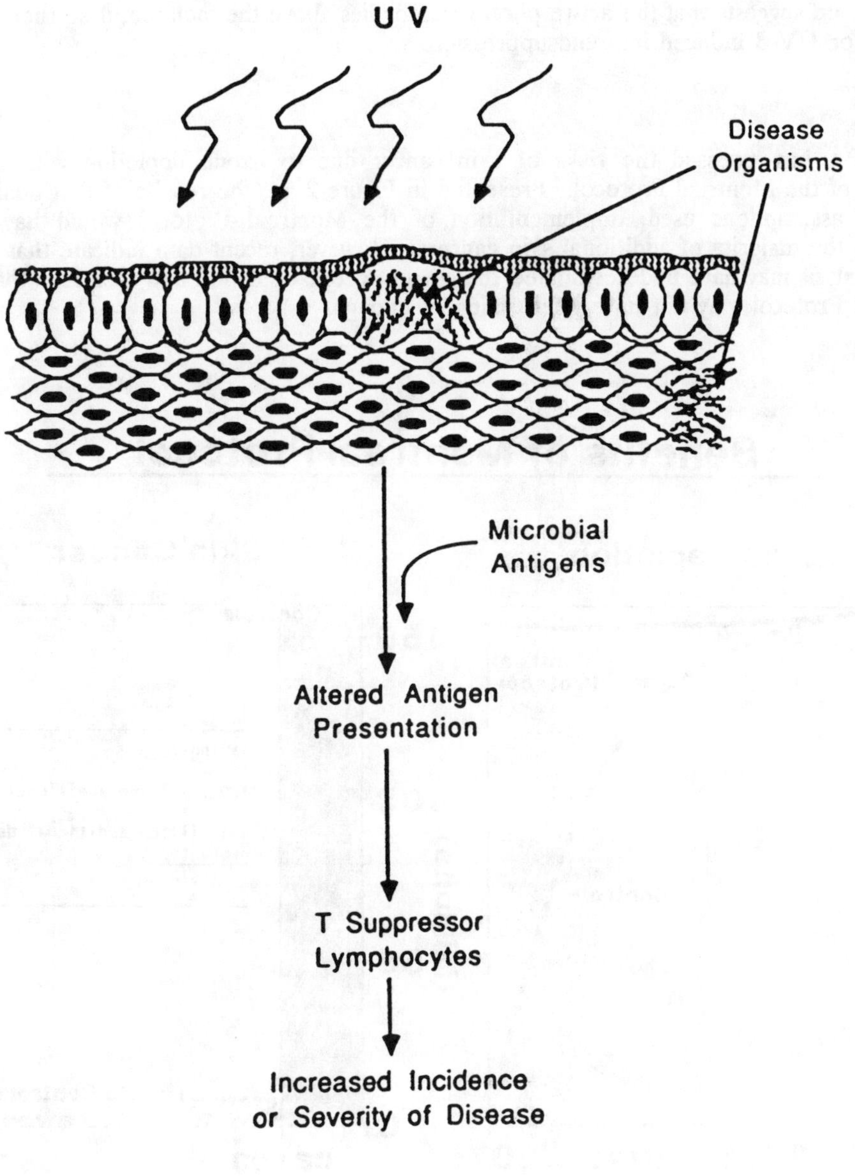

Figure 1 - Model for How UV May Affect Infectious Disease Processes

required for carcinogenesis. This may mean that exposure to low doses of UVR, even doses that do not cause a sunburn, may decrease the ability of the human immune system to provide an effective defense againstneoplastic skin cells or skin infections. In addition, one theory of the mechanism of the immunosuppression suggests that the active photoreceptor lies above the melanin; if so then all races may be equally at risk for UV-B induced immunosuppression.

V. Conclusions

The EPA has assessed the risks of skin cancer due to ozone depletion with and without the implementation of the Montreal Protocol. Presented in Figure 2 are the results of that analysis which show that under the assumptions used, implementation of the Montreal Protocol would have protected the population from the majority of additional skin cancers. However, recent data indicate that the assumptions used in EPA analysis may have underestimated the current ozone depletion and therefore that the conditions of the Montreal Protocol may not have been stringent enough.

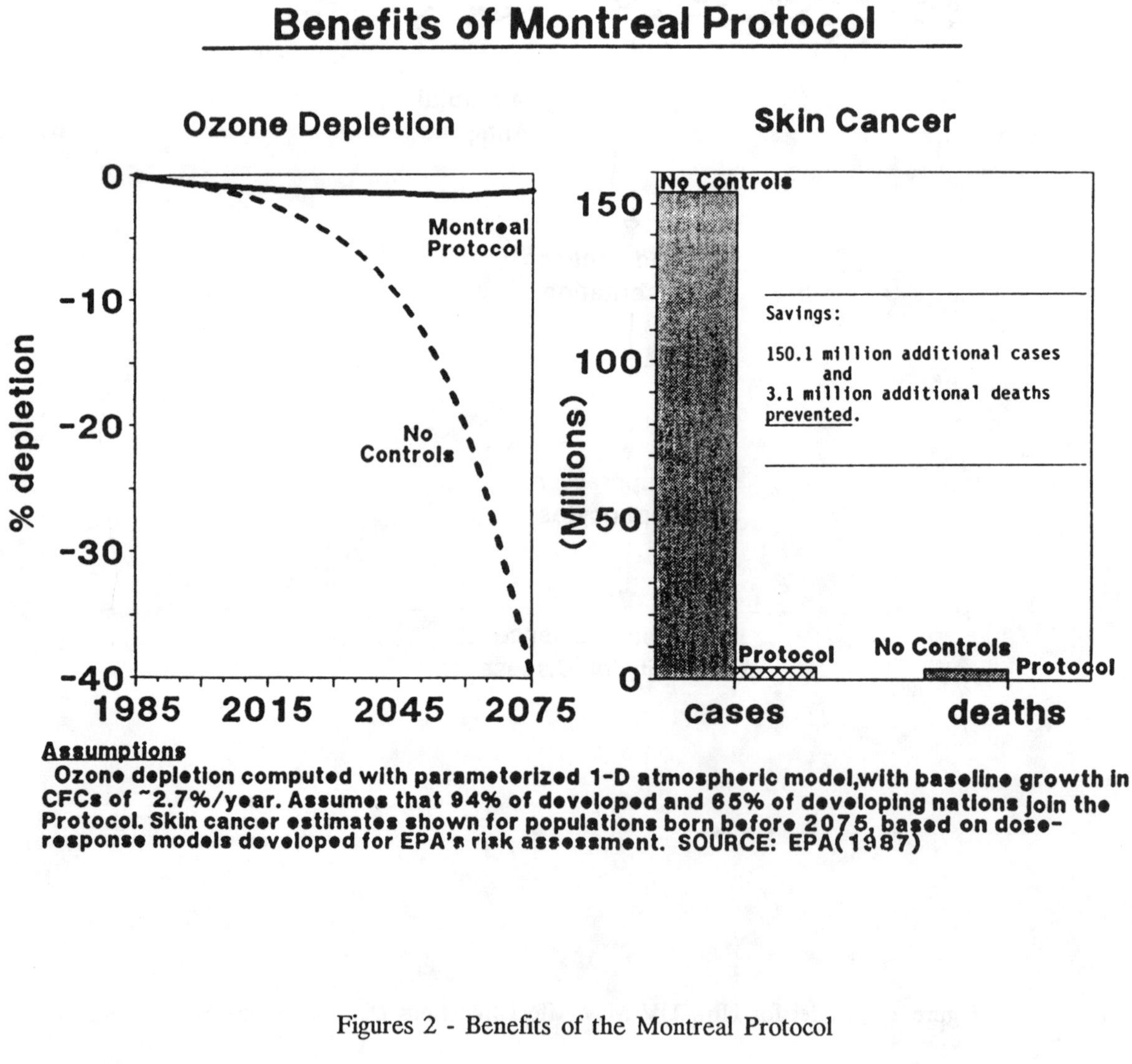

Figures 2 - Benefits of the Montreal Protocol

References

1. U.S. EPA, Assessing the Risks of Trace Gases that can Modify the Stratosphere Washington, D.C., Government Printing Office, EPA 400/1-87/001A-H, (1987)

2. NCI, Division of Cancer Prevention and Control, 1987 Annual Cancer Statistics Review, Including Cancer Trends: 1950-1985. National Cancer Institute (1988).

3. Scotto, J., T.R. Fears, and J.F. Fraumeni, Jr. Incidence of Nonmelanoma Skin Cancer in the United States. National Cancer Institute, U.S. Department of Health and Human Services, (NIH) 82-2433 (1981).

4. Longstreth, J.D. (ed) Ultraviolet Radiation and Melanoma With a Special Focus on Assessing the Risks of Stratospheric Ozone Depletion. EPA400/1-87/001D, Government Printing Office: Washington, D.C. (1987)

5. Woodhead, A.D., E. Grist, R.Schultz and R.B. Setlow. A fish model for estimating the wavelenghts effective in melanoma induction. 10th International congress on Photobiology, Programme and Abstracts p. 32, Jerusalem 1988

6. Ley, R.D. Monodelphis domestica: an animal model for studies in photodermatology including the induction of melanoma. 10th International Congrss on Photobiology, Programme and Abstracts, p. 68, Jerusalem. 1988

7. Pitts, D.G. et al. Optical Radiation and Cataracts, in Visual Health and Optical Radiation. M. Waxler and V. Hitchens (eds.) CRC Press Inc Boca Raton, Florida (1987)

8. Maitchouk, I.F. Trachoma and Cataract: Two WHO Targets. Int Nurs Rev 32:23-25 (1985)

9. Taylor, H.R., West, S.K., Rosenthal, F.S. et al. The effect of ultraviolet radiation on cataract formation. N Engl J Med 319:1411-5 (1988)

10. Kripke, M.L. Photoimmunology: the first decade. Curr Probl Derm 15: 164-175 (1986)

11. Stingl, L.A., Sauder, D.N., Iijima, M. et al. J Immunol 130:1586-1591 (1983)

12. Cooper, K.S. UV induced immunologic changes in humans, in Boerhaave Commissie voor Postacademisch Onderwijs in de Geneeskunde, Photodematologie, Leiden, the Netherlands (1988)

13. Kripke, M.L., and Fisher, M. S. Immunologic parameters of ultraviolet carcinogenesis. J Natl Cancer Inst 57:211-215 (1976)

14. Longstreth, J.D. UV-B Induced Immunosuppression and Infectious Disease. Report prepared for U.S. EPA, Office of Air and Radiation, Washington, D.C. October 1987

15. Giannini, S.H. and De Fabo, E.C.: Abrogation of skin lesions in cutaneous leishmaniasis by ultravilet B irradiation. In: Leishmaniasis: The First Centenary (1885-1985), New Strategies for Control. Hart, D.T. (ed.) NATO ASI Series A: Life Science, London: Plenum Publishing Company Limited, 1987

16. Perna, J.J., Mannix, M.L., Rooney, J.E. et al. Reactiviation of laten herpes simplex virus infection by ultraviolet light: A human model J Amer Acad Dermatol 17:473-8 (1987)

16.

STRATOSPHERIC OZONE DEPLETION, SUNLIGHT AND IMMUNE SUPPRESSION. A NEW CONNECTION. E. C. De Fabo, PhD & F. P. Noonan, PhD
Dept. of Dermatology, G. Washington U Medical Center, Washington, D.C.

My interest on the biological and health effects of stratospheric ozone depletion goes back nearly 12 years when as a photobiologist, I was project director for the U.S. Environmental Protection Agency on a national research program aimed at studying UVB (290-320 nm) effects on animals, plants and health should ozone depletion occur. At that time we were known as Biological and Climatic Effects Research (BACER) under the direction of Dr. Herb Wiser. Later we became the Stratospheric Modification Research Staff under the direction of Dr. Al Forziati. After more than 3 years in this position, I returned to the laboratory, even more interested in UVB effects on biological systems, particularly its health implications since nearly all of the ozone depletion models were indicating a real threat. In 1985-86, I returned to the EPA briefly as a photobiological consultant with John Hoffman's group, and was a principal author on the important and detailed melanoma report released last year. As we now know, those predictions of ozone depletion were correct. Whether significant depletions will occur over populated areas and to what extent is still largely unknown. My guess is that at some point we will begin to hear more about this. As an example, I mention the recent report of ozone depleted air over the southern coast of Australia (1) The specific biological effects of accompanying increases in UVB is also largely unknown and the amount of research funding for such studies has traditionally been low. I feel this research is even more relevant given what we now know about stratospheric ozone depletion.

Today, I wish to describe the results of our own research in photo-immunology particularly the discovery by Dr. Frances Noonan, my wife and the immunologist on the team, and myself of a unique photoreceptor (PR), found on human and other mammalian skin. This photoreceptor is capable of mediating interaction between sunlight and the immune system. This interaction is able to occur without and **I emphasize without,** direct irradiation on circulating blood cells meaning that whatever solar UVB is doing, it must be occurring in the outermost layers of skin. What is intriguing is that **solar UVB** radiation is directly converted, by this PR, into biochemical energy capable of interaction with the immune system. The resultant "end product" is the formation of a very specific type of immune cell, known as suppressor T cells. Our discovery of such an unusual immune-regulating skin photoreceptor is the first evidence of its kind.

Suppressor T cells function as "down regulators" of other immune cells called **effector cells.** Effector cells "go after" foreign antigens such as bacterial germs and viruses and destroy or eliminate them. Suppressor T cells keep "in balance" the activity of the effector cells much like two people on opposite ends of "teeter-totter" balance each other-when one is up, the other is down and vice versa. If these effector cells were left uncontrolled, indiscriminate attack on one's own cells might occur. Diseases due to uncontrolled immune attack on self antigens is what defines autoimmune disease.

It might help to define 4 other points: **UVB:** sunlight with wavelength limits between 290 and 320 nm. This waveband will increase if ozone depletion occurs. 290 nm is the approximate lower wavelength of sunlight able to penetrate to the earth's surface and 320 nm, is the approximate upper wavelength of sunlight expected to increase as the ozone layer gets thinner. Unfortunately, UVB is not precisely defined. Others may define it differently e.g. 280-315 nm. This is not a trivial difference and should be carefully noted when describing sunlamps as either "A" or "B" lamps, or in estimating UVB increases when describing ozone depletion. As we will see the biological "effectiveness" of different wavelengths of light can change dramatically over small ranges. **Photoreceptor:** a substance able to convert energy from one form to another, e.g. sunlight energy to biochemical energy. Such a substance is called a signal transducer. It may help to think of a photoreceptor as being like a TV antenna collecting invisible electromagnetic energy from the TV station and converting it to another visible form of energy i.e. a picture. Human and other mammalian skin is completely covered with the immune-regulating photoreceptor I am about to describe. **Antigen:** generally speaking, any foreign substance which interacts with your body such as viruses, bacteria, or skin sensitizing chemicals e.g. poison ivy. However,the term antigen is also used to describe **naturally** occurring proteins in your body. I shall use the term antigen in the sense of foreign substances interacting specifically with skin. Lastly, **immune system:** a mix of many types of blood cells. These may belong to the **humoral** or antibody-producing arm of the immune system or to the **cell-mediated** or **T-cell** arm. For our talk today we are concerned only with the T-cell type. In general, these can be the **attacking** cells (the effector or cytotoxic T cell) which "eliminate" foreign antigens or they can be the **down-regulating** suppressor T-cells which helps keep in balance the effector cells.

UVB radiation has been associated with skin cancer for many years (2). What is of recent origin was the observation by Fisher and Kripke (3), that skin tumors can be successfully transplanted only if the recipient of the transplanted tumor is given UV radiation before it is given the tumor. For example, mice given 1 hour exposure of UV radiation from an FS40 sunlamp, 3 times per week for about 6 months, will develop skin tumors. Transplanting a small piece of tumor into unirradiated, genetically identical mice, resulted in its being rejected within 4 weeks indicating these skin tumors were capable of stimulating an immune rejection response. However, when Fisher and Kripke **pre-irradiated** the recipients, using the same sunlamp but giving sub- carcinogenic doses (3 months), they found that the transplanted tumors would now be accepted and grow uncontrolled. In other words, the UV radiation from the sunlamp appeared to modify the animals immune system in such a way as to prevent tumor rejection. In subsequent experiments they showed it was an induction of suppressor T cells which allowed this to occur (4).

We proceeded to analyze the photobiological characteristics of this phenomenon. First we found that the percentage of animals which would reject a transplanted tumor depended on how much UV energy from the sunlamp they received i.e. by increasing the dose, a larger percentage of animals were able to grow tumors. Also, we found that we no longer needed to irradiate for a long time period. For example, as little as 1 hour a day for 3 days (dose fractionation) or one 3 hour exposure (constant dose) allowed on average about 30% of mice to successfully grow tumors. Another find was that if we lowered the intensity of the lamps and increased the time of exposure so that the total dose was the same (dose = intensity x time), we got about the same percentage of animals able to grow tumors for that dose. The conclusion here was that no matter how we treated the animals in terms of exposure to the sunlamp, the percentage of animals successfully growing tumors depended only on the total dose and the larger this total dose the greater the percentage of successfully transplanted tumors. At this point, we knew we had something different from straight-forward tumor induction since this type of reciprocity does not show up in photocarcinogenesis (5).

Another very important find revealed that by removing the UVB component of the FS40 sunlamp, the irradiated animals recovered their ability to reject a transplanted UV-induced tumor (6). This experiment definitively demon-strated it was UVB which modified the immune system of the irradiated animals allowing the transplanted skin tumors to grow out.

The next question was whether other immune effects might also be affected by UVB irradiation. Dr. Noonan joined our group to direct studies concerning the effect of UV radiation on the induction of delayed-type hypersensitivity (DTH) responses, immune responses regulated by T cells. The protocol was to shave the backs of two groups mice and cover their ears with black tape. We irradiated one of these two groups with the FS40 sunlamp for various time periods up to 3 hours. Following irradiation, we removed the ear tapes of both groups. Five days later a low concentration (1%) of contact sensitizer was placed on the abdomen of both groups and five days following this the same sensitizer was placed on both surfaces of each ear after the ear thickness had been measured with an engineer's micrometer. Twenty-four hours later the ears were again measured and the thickness recorded. Animals which had not been UV irradiated show approximately a 100% increase in ear thickness due to a DTH reaction. In brief, effector T-cells swarm to the site of the re-appearance of the sensitizing antigen, the ear (recall the sensitizer had been placed on the abdomen 5 days earlier) and cause a delayed hypersensitivity reaction (ear swelling). In contrast, animals which had been UV-irradiated before exposure to the sensitizing antigen, showed a greatly diminished response as indicated by a reduction in ear swelling. Using this protocol, experiments were conducted which addressed several questions: dose-response relationships; wavelengths involved; reciprocity of intensity and exposure; time delay required for maximum reduction in ear swelling, etc. Note these were exactly the same types of experiments I described earlier for the suppression of tumor rejection. Interestingly, our results indicated that except for dose, the photobiological characteristics involved in suppressing the DTH response to skin sensitizing antigens were very similar to those involved in suppression of tumor rejection (7). In general, we found that suppression of the DTH response was mediated by wavelengths in the UVB; was directly promotional to the total dose and independent of dose fractionation or intensity; and required a time delay of at least 72 h for maximum suppression as in the tumor system. Thus, suppression of the DTH response appeared similar to the tumor suppression system. As a direct result of the photoimmunological similarities of these two very disparate systems, we proposed a model which suggested a common mechanism between the two systems at the level of initial UVB absorption (8). In other words, a unique light-absorbing photoreceptor might be common to both systems.

To find such a photoreceptor, we completed what is called an action spectrum. Action spectroscopy is a powerful tool which can provide important information about the chemical nature of a photoreceptor (9). It is the absorption of photons by the photoreceptor, which starts a response following exposure to light. Usually such a photoreceptor acts as a signal transducer by converting light energy into biochemical energy which initiates a series of "dark" reactions leading to some observable change, e.g. in our system an immune suppression. By determining dose-response relationships for the individual wavelengths involved in initiating a light-driven reaction, action spectroscopy allows one to determine or at least approximate the in vivo absorption spectrum for the unknown photoreceptor.

To determine our in vivo action spectrum for immunosuppression (8) we used narrow-band wavelengths in non-overlapping steps of at least 5 nm, from 250 to 320 nm. The source used to determined the UV effect on the immune system was one I developed some years ago and is a unique interference-filter monochromator capable of producing narrow-band (2.5 nm) radiation in the UVB and UVC (< 290 nm) regions across a relatively wide field of exposure (~55 cm^2). This field size is large enough to irradiate the backs of three mice simultaneously (8). To complete the action spectrum we had to use about 1500 mice, requiring nearly 6000 ear measurements. It took 2 years to finish.

The optical core of the light source consists of 10 narrow-bandpass UV interference filters coupled to a 2.5 kW xenon arc. These filters block transmission of wavelengths outside the main bandpass to less than 0.1% from x-ray to the far-infrared. Dose response curves for the suppression of the DTH response to skin sensitizer were determined using each of the filters. From these dose-response curves, the amount of energy needed to induce 50% suppression for each wavelength was determined. From this value, the number of photons was calculated and its reciprocal plotted against wavelength to determine the configuration of the action spectrum. The shape of the action spectrum should correspond to the in vivo absorption spectrum of the unknown photoreceptor revealing its identity. Those wavelengths which are most effective at inducing immune suppression will require the least number of photons and will be the "peak" of the action spectrum. Normalizing this peak to 100 %, other wavelengths will fall at different positions on either side of it depending on their efficiency. Thus, if a discreet photoreceptor is responsible for absorbing UVB and inducing immune suppression, one should see a clear difference in the effectiveness of each of the wavelengths involved. Our detailed, in vivo action spectrum did show such a clear-cut wavelength dependence and strongly suggested the presence of a specific photoreceptor initiating interaction with the immune system.

The shape of our action spectrum showed that the most effective wavelengths at immunosuppression peaked at 270-275 nm with a gradual decline in effectiveness to about 3% at 320 nm. Since the peak wavelengths are most heavily absorbed in the epidermis approximately 10 times more efficiently than wavelengths of 320 nm, this observation implied the photoreceptor was superficially located. Following completion of the action spectrum, we analyzed a number of skin compounds which could absorb photons within the UVB range and whose absorption spectrum would match the action spectrum. We could find only one such compound and that was urocanic acid (UCA) a substance which makes up about 0.5% of the wet weight of the epidermis of human skin and is found almost exclusively in its outermost layer. UCA was first described in 1874 (**10**) and since that time no physiological role for it has ever been found. One role for it was proposed as a natural sunscreen, protecting DNA from UVB damage. However, this role was later questioned by other investigators (**11**). Nonetheless, it should be kept in mind that UCA will diminish the UVB flux on the skin simply because it is a good absorber within this range. To this extent some "protection" will be afforded to DNA. However, its primary role is, as our evidence shows, that of an immune-regulating photoreceptor. UCA also exists in two forms. The naturally occurring or "dark" form is known as the **trans isomer** while the form is takes after absorbing UVB is known as the **cis isomer** (12).

Because UCA was the only skin compound which accurately matched our detailed in vivo action spectrum, we put forward the hypothesis that the photoreceptor for UV-induced immunosuppression was trans-urocanic acid. Our hypothesis predicted that cis UCA would interact with the immune system to initiate immune suppression. Evidence supporting this conclusion has come from the following studies: removal of the stratum corneum where UCA is formed, before UV irradiation, prevented immunosuppression; mutant mice deficient in urocanic acid show little if any UVB-induced immune suppression to the CHS response; addition of purified cis isomer produces the same change in immune spleen cells, called antigen-presenting cells, important in the CHS reaction as UVB (**13**). Other investigators have shown: that UCA can induce suppressor T cells similar to those formed by UVB in a DTH response induced by Herpes virus I (**14**); that skin tumor yield (tumors per mouse) is greatly increased when murine skin is irradiated by UVB in the presence of trans UCA and these tumors were on average larger and more aggressive (**15**) and that production of InterLeukin-1, an important immune chemical, can be inhibited by UCA (**16**). All these results substantiate what our original model predicted (**8**) and clearly shows that urocanic acid can behave as a natural immunomodulator for cell-mediated reactions and suppressor cell induction, perhaps the first commonly occurring skin substance found to do so.

Interestingly, the doses of UV used experimentally to induce suppression can be realistically achieved by natural sunlight exposure. For example, we have calculated that a dose of UVB causing 50% systemic suppression of contact hypersensitivity in BALB/c mice can be received within a half day of sunlight exposure in June at Rockville, MD (**17**). Furthermore, using computer-generated sunlight intensities for any set of global coordinates (latitude and longitude) we can calculate that an exposure time of 30 min in summertime from latitudes 40 N to 40 S. (e.g. Canada to Australia) is capable of inducing immune suppression in mice. Thus low doses of immunosuppressive solar UVB can be easily obtained in either hemisphere. If this is so of what value might such a mechanism be to humans? Although uncertain, it may be to regulate against uncontrolled autoimmune attack against sun-damaged skin i.e. skin which has become antigenic by absorbing sunlight. Under normal sunlight exposure, this mechanism is probably acting in a protective fashion by producing the correct type of suppressor T cell to shut down such an autoimmune attack. Under stressed conditions, however, such as excessive sunbathing, or increased levels of UVB due to depletion of stratospheric ozone, the protective aspect could breakdown. For example, this mechanism might inadvertently end up "protecting" tumor cells along with "sun-damaged" cells by the formation of tumor-specific suppressor T-cells. In summary, evidence exists for a unique photoreceptor on mammalian skin, including human skin, capable of mediating interaction between sunlight and the immune system. This mechanism is probably protective preventing autoimmune attack against sun-damaged skin. If stressed by overexposure to UVB through excessive sunbathing, exposure to FS40 type sunlamps (suntanning salons), or stratospheric ozone depletion, inadvertent protection of skin tumors may occur allowing the tumor to grow unchecked. Not enough is yet known about the mechanism to adequately predict what other effects on the health of the skin, immune system or other systems could result as a consequence of stratospheric ozone depletion but the fact that we now know it exists suggests even more prudence in our daily exposure to sunlight.

References:

1. The day the ozone hole passed over Melbourne. 1988 New Scientist 17 December, p. 4

2. Roffo, A. H. (1934) cancer et soleil. Carcinomes et sarcomes provoques par l`action du soleil in toto. Bull. Assoc. Franc. EtudeCancer. 23:590

3. Fisher, M. S. and Kripke, M. L. (1977). Systemic aleration induced in mice by ultraviolet light irradiation and its relationship to ultraviolet carcinogenesis. Proc. Natl. Acad. Sci (USA). 74:1688.

4. Fisher, M. S. and Kripke, M. L. (1982). Suppressor T lymphocytes control the development of primary skin cancers in ultraviolet irradiated mice. Science (Wash., D. C.). 216:1133.

5. Forbes, P.D., R. E. Davies, and F. Urbach. (1978) Experimental ultraviolet photocarcinogenesis: wavelength interactions and time-dose relationships. Natl. Cancer Inst. Monogr. 50:31.

6. De Fabo, E. C. and Kripke, M. L. (1980). Wavelength dependence and dose-rate independence of UV radiation-induced immunologic unresponsiveness of mice to a UV-induced fibrosarcoma. Photochem. Photobiol. 32:183.

7. Noonan, F. P., De Fabo, E.C. and Kripke, M. L. (1981) Suppression of contact hypersensitivity by UV radiation and its relationship to UV induced suppression of tumor immunity. Photochem. Photobiol. 34:683.

8. De Fabo, E. C. and Noonan, F. P. (1983). Mechanism of immune suppression by ultraviolet radiation in vivo. I: Evidence for the existence of a unique photoreceptor in skin and its role in photoimmunology. J.Exp. Med. 158:84.

9. De Fabo, E.C. (1980). On the nature of the blue light receptor: still an open question. In: The Blue Light Syndrome (Ed. H. Senger). Springer-Verlag, Heidelberg, p. 187-197.

10. Jaffe, M. 1874 Ueber einen neuen Bestandtheil Des Hundeharns. Ber. Deut. Chem. Ges. 7:1669-1673

11. Scott, I. R. 1981 Factors controlling the expressed activity of histidine ammonia-lyase in the epidermis and the resulting accumulation of urocanic acid. Biochem. J. 194:829

12. Morrison, H., Avnir, D., Bernasconi, C. and Fagan, A. (1980). Z/E photoisomerization of urocanic acid. Photochem. Photobiol. 32:711.

13 Noonan, F.P., De Fabo, E.C. and Morrison, H. (1988). Cis-urocanic acid, a product formed by ultraviolet B irradiation of the skin, initiates an antigen presentation defect in splenic dendritic cells in vivo. J. Invest. Dermatol., 90:92.

14. Ross, J.A., Howie, S.E.M., Norval, M., and Maingay, J.P. (1987). Two phenotypically distinct T-cells are involved in UV-irradiated urocanic acid-induced suppression of the efferent DTH response to HSV-1 in vivo. J. Invest. Derm. 89:230, et al.,

15. Reeve, V. Greenoak, G., Canfield, P. Boehm-Wilcox, C, Gallagher C. Topical urocanic acid enhances UV-induced tumor yield and malignancy in the hairless mouse. Proc. Australasian Soc. Exp. Pathol. 19th Annual Meeting. Hobart, Australia

16. Rasanen, L., Jansen, C.T., Reunala, T., Morrison, H., Stereospecific inhibition of human epidermal cell IL-1 expression by cis Urocanic acid. Photodermatol. 4: 182-186.

17. De Fabo, E.C., Noonan, F.P. and Frederick, J. 1989. Biologically Effective Doses of Sunlight for Immune Suppression. J. Invest. Derm. 92:426A

IMPLICATION OF ULTRAVIOLET LIGHT IN SKIN CANCER AND EYE DISORDERS*

A. Kornhauser

Food and Drug Administration, Division of Toxicological Studies
Washington, D.C. 20204

This Conference deals with various aspects of the potential effects of climatic change. It couldn't have come at a better time, as a number of recent data strongly indicate a disturbing trend toward a significant global warming combined with widespread droughts and stratospheric ozone depletion. The observation that the 1980s are the warmest decade on record has resulted in increased public and governmental concern and prompted some urgent calls for action. The various scientific and policy approaches to this phenomenon are described elsewhere in this monograph. In this article, I will briefly discuss some aspects of the implication of a potential UVB increase, due to ozone depletion, to the incidence of skin cancer in humans. Due to space limitation, only a few pertinent references will be listed.

Light characteristics. The sun emits radiation over a wide range of wavelengths; the region 400-700 nm is known as the visible part of the sun spectrum, while ultraviolet (UV) covers wavelengths from 200 to 400 nm. Portions of the UV spectrum have distinctive features from both the physical and medical points of view. Figure 1 shows the accepted designations for the biologically important part of the UV spectrum. Wavelengths of less than 290 nm do not occur at the earth's surface because they are absorbed by the ozone layer in the stratosphere. As shown in this figure, ozone absorbs virtually all UVC and part of UVB. It is clear that any future ozone depletion would reduce the effectiveness of this shield, resulting in an increase of UVB energy reaching the earth and its population. The amplification factor has been calculated to be about 2, which means that a 1% decrease in ozone results in a 2% increase in UVB. The amount of UVB absorption depends significantly upon the angle of incidence of the sunlight to the ozone layer. Absorption increases considerably as the sun moves from its zenith, because the rays pass through the ozone layer at a greater angle, i.e., through a greater thickness of ozone. Therefore UVB varies significantly with seasons, times of the day, and latitude. On the other hand, UVA, which is not absorbed by ozone, is much less affected by these parameters.

In this paper we will concentrate on the health effects of UVB. Only about 1.5% of the sun's total energy at the earth's surface is found in the UVB region. It must be pointed out, however, that this narrow spectral segment is responsible for most biological and human health effects, including acute (sunburn), subacute (pigmentation), and chronic (aging, skin cancer, cataracts) events. It is also important to understand that UVB, when absorbed by cell or tissue constituents, has sufficient energy to produce chemical reactions, some of which are deleterious. On the molecular level, DNA and, to a lesser extent, proteins are the most sensitive UV targets. Absorption of UV results in an excited molecule; dissipation of this energy can result in adverse effects on the cell, which ultimately may lead to acute and/or chronic effects.

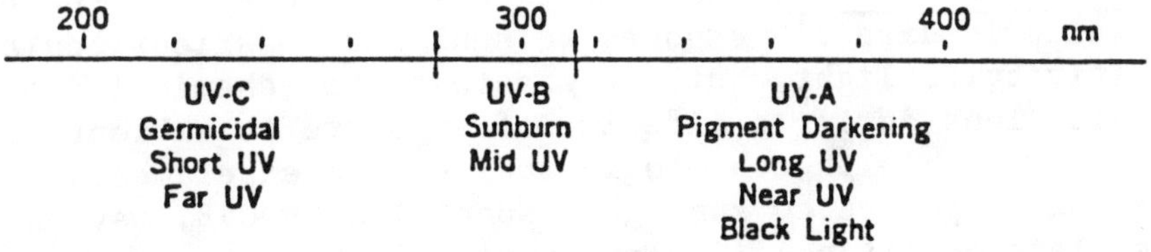

Figure 1. Biologically important regions of the UV spectrum.

Effects on skin cancer. Basal and squamous cell carcinomas of the skin (nonmelanoma skin cancers) will be discussed here. These are the most common malignant neoplasms occurring in the white population of the world. Currently, the annual incidence in the United States is estimated to be about half a million new cases, and the rates are increasing (Scotto, 1986). There is a general consensus that most of these malignancies are light-induced. Examination of the sun's role in the induction of human skin cancer does not lend itself to direct experimentation. However, extensive observations have strongly suggested the etiologic significance of light energy in the induction of these tumors. Skin cancer is most frequent in parts of the world where light-skinned people (Caucasians) are exposed to a large total amount of sunlight. Thus, the amount of ultraviolet exposure and the skin type and color are the most important factors. Numerous surveys have clearly established that the incidence of skin cancer is much higher in the southern than in the northern part of the United States. Further, it is established that skin cancers occur most frequently on the head, neck, arms, and hands--parts of the body exposed to sunlight.

Skin cancers can be produced experimentally on the skin of mice with repeated doses of UV radiation. The most effective spectral range for this process is the UVB region, which also produces erythema in humans. Though these and many additional arguments do not represent absolute proof, there is abundant epidemiologic evidence to support the role of sunlight in the etiology of nonmelanoma skin cancers.

Depletion of stratospheric ozone, recently reported by Scotto (1988), should result in an increase of UVB reaching the earth. As mentioned earlier, the incidence of skin cancers has risen significantly in the last decade. I must point out, however, that there is no proof to date that these two events are correlated in any way. As a matter of fact, ground-level monitoring at various geographic locations within the United States could not detect any UVB increase between 1974 and 1985 (Scotto, 1988). Therefore, the increase of skin cancers can be correlated mainly with a change of living habits, i.e., more people are spending more time exposed to the sun. These findings by no means indicate that we can relax; ozone depletion presents a serious threat to our health and environment. The fact that ozone depletion (if real) did not cause an increase in UVB at ground levels suggests that various meteorological, climatic, and environmental factors in the troposphere may play a greater role in absorbing UVB radiation than was previously suspected (Scotto, 1988). In any case, more interdisciplinary investigations are urgently needed to understand the factors that influence the UVB intensity in our environment.

Effects of UVB on the eye. Besides the skin, the eye is the only organ in the body which is directly exposed to sunlight. Various ocular tissues show sensitivity to light, but the greatest photochemical damage is induced to the ocular lens during a lifetime of exposure to ambient ultraviolet radiation. Available photobiologic and related experimental evidence indicates that exposure to sunlight, specifically UVB, may be a cause of cataracts (lens opacities) (Taylor et al., 1988). Cataracts occur more commonly in geographical areas with high solar irradiation; people spending a great deal of time outdoors are at higher risk. Recently, Taylor et al. (1988) undertook an epidemiologic study to assess the UV exposure among watermen at the Chesapeake Bay (Maryland) and the relationship of this occupation to cataract formation (Taylor, 1988). Their conclusion that there is an association between exposure to UVB and cataract formation supports the need for ocular protection from UVB. This study contributed significantly to our understanding of the role of UVB in cataractogenesis, a process which leads to a health disorder of major implications. More than one million cataract operations are performed annually in the United States. This number could significantly increase as we continue to face higher UVB levels as a result of ozone depletion. Prevention of photo damage to ocular tissues is the best approach to this problem. Ordinary sunglasses do not appreciably absorb in the UV region. Special lenses, recently introduced commercially effectively absorb in the UVB region and are highly recommended, particularly for individuals who spend a great deal of time outdoors in the sun.

Summary. Sunlight can be both beneficial and detrimental to humans. Photosynthesis, vitamin D synthesis, vision, and warmth are necessary for maintaining our lives. Sunburn, skin cancer, immune suppression and various phototoxic effects on skin and eyes are harmful; evolution of life, therefore, has required a continuous adaptation to light by simultaneous use of solar energy and protection against its detrimental effects. The danger to public health from sunlight arises mainly from rapid changes in our lifestyles, as well as from human-induced alterations of the environment which might change the quality and quantity of light reaching our habitats. Scotto (1986) presents epidemiological evidence linking solar radiation with nonmelanoma skin cancers. His analysis suggests that Caucasians in the United States have a 12-30% chance of developing these cancers within their lifetimes, even without the predicted ozone depletion. Although cure rates for nonmelanoma skin cancers are high (better than 95%), the extremely high incidence of this disease represents a major health and economic problem in this country and many parts of the world.

To prevent this serious health hazard, we must educate our population, primarily the segment at high risk (about 30% in the Unites States). These individuals should avoid excessive and repeated sun exposure, especially during peak hours (11 a.m.-2 p.m.). In addition, these subjects, particularly children, should avoid blistering sunburns induced by a single sun exposure, which can cause chronic disorders, including cancer, in later years. Sunscreens as well as appropriate clothing can provide an efficient optical filter against the harmful radiation. By increasing our understanding of light induced toxic reactions, we can improve our quality of life and live in harmony with our sun's rays.

REFERENCES

Scotto, J. 1986. Nonmelanoma skin cancer--UVB effects. In: Effects of Changes In Stratospheric Ozone and Global Climate, Volume 2: Stratospheric Ozone. J. G. Titus, Ed. Environmental Protection Agency, Washington, DC.

Scotto, J., Cotton, G., Urbach, F., Berger, D., and Fears, T. 1988. Biologically effective ultraviolet radiation: Surface measurements in the United States, 1974 to 1985. Science 239:762-764.

Taylor, H.R., West, S.K., Rosenthal, F.S., Munzoz, B., Newland, H.S., Abbey, H., and Emmett, E.A. 1988. Effect of ultraviolet radiation on cataract formation, N. Engl. J. Med. 319:1429-33.

---

*The views expressed in this paper are those of the author and do not necessarily reflect those of the Food and Drug Administration.

THE IMPACT OF CLIMATIC CHANGE ON AGRICULTURE

Martin L. Parry and Timothy R. Carter
Atmospheric Impacts Research Group
School of Geography
University of Birmingham
United Kingdom

INTRODUCTION

Changes in the composition of the earth's atmosphere will have direct effects on agriculture (for example, through the so-called "fertilization" effect of enhanced carbon dioxide levels on crop growth) as well as indirect climatic effects. Ideally, both effects would be considered together, but few such studies have been completed, and in this paper we focus on climatic effects alone. Climate is only one of a number of factors (both societal and environmental in origin) that may interact to affect agriculture. An ultimate goal must be to bring together all of these interactive components into a fully integrated strategy for impact assessment. However, this is not yet achievable as there are still important gaps in our knowledge (for example, the role of the oceans in regulating the earth's climate is not fully accounted for in global climate models). It has been common, therefore, to adopt a "partially integrated" approach, which considers some, though not all, of the interactions between climate and agriculture. Such an approach was adopted in a large international study of impacts on agriculture by the International Institute for Applied Systems Analysis (IIASA) and the United Nations Environment Programme (UNEP).[1] This paper provides a very brief summary of some methods and results of this study. A more detailed summary has been published elsewhere.[2]

A PARTIALLY INTEGRATED APPROACH

The partially integrated approach has two main elements. First, the assessment of effects of climatic variations is conducted using a hierarchy of models. At the top of the hierarchy are models or "scenarios" of climatic variation. These are used as inputs to biophysical models of first-order relationships (e.g., between certain climatic variables and crop yields). Outputs from these models are used as inputs to economic models of second-order relationships at the enterprise level. These consider the effects, for example, of changes in farm-level production on farm incomes, farm purchases, etc. As a further step, economic models of higher-order relationships (for instance, those between farm profitability and gross domestic product) can be employed to evaluate effects elsewhere in a region.

Second, two types of response to climate impacts can be considered: (a) adjustments at the enterprise level (which at the farm level might include changes of crops, increased irrigation, changes in fertilization, etc.), and (b) policy responses at the regional, national and international level. Many such responses are already built-in to agricultural systems (e.g., delaying planting because of inclement weather conditions, selecting alternative crops to cultivate, etc.), but these are not always incorporated in model-based impact assessments. Furthermore, if the magnitude of climatic change exceeds the bounds of those climatic variations for which present-day responses have been designed, other potential responses may also have to be considered.

CLIMATIC SCENARIOS

There are two main sources of information that can be used to specify a future climate. Firstly, general circulation models (GCMs) of the global atmosphere have been developed by several research groups around the world, which can simulate how a change in atmospheric composition may affect global climate. These are physically-based computer models, which produce estimates of different climatic elements (such as air temperature, precipitation rate and cloud cover) at different layers in the atmosphere and over a regular network of grid points distributed around the globe. Most model simulations to date have provided estimates of global climate for atmospheric conditions similar to those of today (the control case) and for conditions with atmospheric concentrations of carbon dioxide at double the control level (the 2 x CO_2 case). Some recent simulations have also been conducted to estimate the transient response to a gradual change in atmospheric composition.[3] These results are of interest because they provide an indication of the likely rate as well as the magnitude of climatic change.

Historical climatological data constitute a second, valuable source of information. They can be used both for characterising the present-day or "baseline" climate (e.g., adopting a standard 30-year climatological averaging period such as 1951-1980), and for indicating past occurrences of anomalous weather-years or periods of years that might serve as interesting analogues of possible future conditions under a changed climate.

IMPACT MODELS

The climatic scenarios provide input information for a range of first-order impact models, of which three main types can be distinguished. Agroclimatic indices combine into a single term those meteorological variables that most influence certain plant attributes. These measures are valuable indicators of large-scale agricultural potential, and can provide a useful background for more specific studies of crop response using other models.

Empirical-statistical models relate a sample of annual crop yield data to weather data statistically (though often with little explanation of the processes involved). These models are most effectively applied to situations where a strong relationship exists between crop productivity and a climatic variable. One of the major drawbacks of using these models to estimate effects of climatic change is the problem of extrapolating model relationships. The magnitude of climatic change inferred for a doubling of CO_2 is often well outside the range of climatic conditions under which the models have been developed. One method of increasing the range of model validity is to seek climatic analogue regions having a present-day climate similar to the climate projected for the study area, and to identify crop-climate relationships in these regions that can be incorporated in the statistical model.

Simulation models are developed based on an understanding of the relationships between the basic processes of plant growth and environmental factors. As such, after appropriate calibration, many of these models are potentially applicable to a wide range of environments. Extrapolation is thus less of a problem when using such models.

Models of second-order impacts are less easy to classify as there are so few examples of such models being used in climate impact assessments. For example, farm simulation models have been used to consider the effects of altered crop yields (estimated using a crop-climate simulation model) on income, purchases and profitability at the farm level in Saskatchewan, Canada. Subsequently, aggregate farm-level data have been used as inputs to input-output models to examine effects on the wider regional economy.[4] In regions such as the Soviet Union, where there is central planning of agriculture, land allocation models have been used to evaluate the optimal distribution of crops and other resources under different climatic scenarios.[5]

A SUMMARY OF POTENTIAL IMPACTS

Two types of impact were considered in the IIASA/UNEP study. There was firstly a sequential estimation of effects of climatic variations at different levels of the model hierarchy. These experiments were usually conducted assuming that agricultural practices and economic conditions are fixed at present-day levels. Subsequently, it was possible to investigate the efficacy of a number of adjustments by altering some of these assumptions to evaluate various options available to offset or mitigate the effects.

The magnitude of climatic change

There are large uncertainties associated with GCM estimates, and only a few general conclusions can be drawn from the various GCM projections of future climatic change. Naturally, these limit considerably our ability to project the likely impacts on agriculture. Four main findings seem to have emerged. First, that global mean annual temperatures are likely to rise in the future (for a CO_2-doubling) by between 1.5°C and 5.5°C; second, that this warming will be more pronounced at higher latitudes than at low latitudes; third, that the warming will probably be greater in winter than in summer; and fourth, that mid-continental areas may well become drier than at present.[6]

Changes in crop growth potential

It is important to emphasize that it is not the absolute magnitude of a climatic change that determines the nature of the effects as much as the change in climate relative to the exisiting (baseline) conditions. Thus, since agricultural potential in high latitude areas tends to decrease poleward, the same increases of temperature have greater relative effects on crop potential at higher than at lower latitudes. This latitudinal contrast is enhanced by the greater warming expected at high than at low latitudes.

The clearest implications for the growth of agricultural crops under a warmer climate are, first, a lengthening of the potential growing season and, second, a shortening of the required growing period. For example, for a rise in growing season temperature of about 3.5°C, the growing season in southern Saskatchewan is lengthened by 4-9 weeks, while the estimated maturation period for current varieties of spring wheat (the major crop grown in this area) is reduced by 4-14 days.[7]

Changes in crop yield

In regions where crops are constrained by low temperatures, estimates suggest that warmer conditions are beneficial to yields. For example, substantial increases in hay and pasture yields of more than 50% can be anticipated in Iceland under a 2 x CO_2 climate.[8] There are two major reservations to be noted, however. First, GCMs are unable to provide consistent or confident estimates of regional precipitation changes. Such information is critical for many regions such as Saskatchewan, where moisture as well as temperature is already a major constraint on crop yields.[9] A second reservation concerns the crops themselves. The estimated temperature increases under a CO_2 doubling are, in some cases, so great that present-day crop cultivars are ill-adapted to the longer and more intense growing season.

In many regions, the variability or stability of crop yields is as important a measure as the level of mean yields. At present, there is very little information about how climatic variability (i.e., diurnal, day-to-day and inter-annual variability) may change in the future. Thus, most impact experiments assume no change in absolute climatic variability relative to the present-day climate. Notwithstanding this assumption, a change in the mean climate can still be expected to lead to a change in the frequency of events that exceed certain thresholds of tolerance. For example, the frequency of high summer temperatures that are damaging for maize growth in the United States could increase markedly under a 2 x CO_2 climate even with no change in temperature variability around the mean.[10]

Higher-order effects of climatic variations on agriculture

A number of higher-order effects were identified in the IIASA/UNEP study. At the farm level, the effect of yield changes on farm profitability were examined in terms of net returns per hectare (assuming fixed input costs). In Finland, for example, such calculations imply substantial increases (varying in extent between different regions) in net returns to barley and oats under a 2 x CO_2 climate.[11] However, since there is heavy government support of agriculture in Finland, it is realistic to assume that subsidies would be adjusted to maintain an equitable regional distribution of income under a future climate.

Changes in farm profitability can also be expected to affect other non-agricultural sectors of the regional economy. Models which simulate such effects can provide reasonable estimates for scenarios of short-term climatic variations, assuming present-day relationships between sectors of the economy, but these become unreliable for scenarios of long-term climatic change. Impact experiments of this kind thus can give an indication of the sensitivity of the economy to climate, but not a prediction over the long term.

POTENTIAL ADJUSTMENTS

The types of results described so far assume no changes in technology, management, background economic conditions or policy. The IIASA/UNEP study tested a number of adjustments and policies to mitigate the impact of climatic change. Some of these are listed here.

* Changes in crop variety (e.g., to later-maturing varieties with higher thermal requirements)

* Changes in soil management (e.g., adjustments in fertilizer applications, altered use of fallow, etc.)

* Changes in irrigation scheduling (e.g., to adapt to the changed moisture and temperature regime)

* Changes in farm expenditure (e.g., to invest in storage facilities or equipment that might become cost-effective under a changed climate)

* Changes in regional land-use allocation (e.g., to exploit those crops that benefit from climatic change at the expense of crops that experience declines in productivity)

* Changes in national agricultural policy (e.g., to maintain national food security whilst avoiding over-supply, maintaining equitable regional farm incomes, etc.)

RESEARCH PRIORITIES

We have outlined a partially integrated approach for assessing the impacts of climatic change on agriculture, and have presented a few results obtained from the recent IIASA/UNEP study. These results are not predictions of future effects, but should be considered as measures of the present-day sensitivity of agriculture to climatic change. Such studies are still in their infancy, but several clear priorities for future research have emerged from them. These include requirements for:

* Improvements in the spatial resolution of estimates of future_climatic change, and more information about future changes in climatic variability

* Incorporation of other physical factors in simulations of first-order effects of climatic change (e.g., direct effects of CO_2 on plant growth, the occurrence of pests and diseases and their vectors)

* Development and use of a wider range of models for considering the higher-order effects of climatic change at the farm-, regional- and national-scales

* Broader consideration of the range of policy responses to effects of climatic change on agriculture.

REFERENCES

1. Parry, M.L., Carter, T.R. and Konijn, N.T. (eds) (1988). The Impact of Climatic Variations on Agriculture. Volume 1. Assessments in Cool Temperate and Cold Regions. Volume 2. Assessments in Semi-Arid Regions, Kluwer, Dordrecht, The Netherlands, pp. 11-95.

2. Parry, M.L. and Carter, T.R. (1989). An assessment of the effects of climatic change on agriculture, Climatic Change, forthcoming.

3. Hansen, J., Fung, I., Lacis, A., Rind, D., Lebedeff, S., Ruedy, R., Russell, G. and Stone, P. (1988). Global climate changes as forecast by Goddard Institute for Space Studies three-dimensional model. J. Geophys. Res., 93(D8), 9341-9364.

4. Williams, G.D.V., Fautley, R.A., Jones, K.H., Stewart, R.B. and Wheaton, E.E. (1988). Estimating effects of climatic change on agriculture in Saskatchewan, Canada. In M.L. Parry et al. (eds) op. cit., Volume 1, pp. 221-379.

5. Pitovranov, S.E., Iakimets, V., Kiselev, V.I. and Sirotenko, O.D. (1988). The effects of climatic variations on agriculture in the subarctic zone of the USSR. In M.L. Parry et al. (eds) op. cit., Volume 1, pp. 617-724.

6. Bolin, B., Döös, B.R., Jäger, J. and Warrick, R.A. (eds) (1986). The Greenhouse Effect, Climatic Change and Ecosystems, SCOPE 29, Wiley, Chichester, 542 pp.

7. Williams et al., op. cit.

8. Bergthórsson, P., Björnsson, H., Dyrmundsson, O., Gudmundsson, B., Helgadóttir, A. and Jónmundsson, J.V. (1988). The effect of climatic variations on agriculture in Iceland. In M.L. Parry et al. (eds) op. cit., Volume 1, pp. 383-509.

9. Williams et al., op. cit.

10. Mearns, L.O., Katz, R.W. and Schneider, S.H. (1984). Extreme high-temperature events: changes in their probabilities with changes in mean temperature, J. Clim. Appl. Meteor., 23, 1601-1613.

11. Kettunen, L., Mukula, J., Pohjonen, V., Rantanen, O. and Varjo, U. (1988).The effects of climatic variations on agriculture in Finland. In M. L. Parry et al. (eds) op. cit., Volume 1, pp. 513-614.

LIKELY EFFECTS OF CLIMATE CHANGE SCENARIOS ON AGRICULTURE OF THE USA

Leon Hartwell Allen, Jr., Robert M. Peart, James W. Jones,
R. Bruce Curry and Kenneth J. Boote

U.S. Dep. of Agriculture (L.H.A.), Agric. Eng. Dep. (R.M.P., J.W.J., R.B.C.),
Agronomy Dep. (K.J.B., L.H.A.), and
Fruit Crops Dep. (L.H.A.)
University of Florida
Gainesville, FL 32611-0621

Direct Effects of Carbon Dioxide on Plants

Photosynthesis generally shows a nonlinear increase with increasing carbon dioxide for plants having the C-3 biochemical pathway of photosynthesis. Therefore, Allen et al. (1) used a nonlinear model to predict photosynthetic, biomass, and seed yield responses of soybean to carbon dioxide relative to a concentration of 330 ppm. Table 1 shows that the rise of carbon dioxide from preindustrial values to present values could cause a 13% increase in seed yield. Furthermore, a doubling from 315 to 630 ppm could cause a 32% increase in seed yield, which is in agreement with a review by Kimball (2).

Although leaf stomatal opening tends to be smaller under elevated CO_2 conditions, transpirational water losses tend to show little difference because total leaf area increases and leaf temperature tends to rise slightly [about 2°C (3.6°F) for a doubling of carbon dioxide]. Water-use efficiency, the ratio of biomass accumulation to plant water use, increases with increasing concentration of carbon dioxide. However, almost all of the increase in water-use efficiency is due to an increase in photosynthesis, and only a small amount is due to decreases in transpiration.

Interactions of Carbon Dioxide, Temperature, and Water Stress

Few studies have been conducted on interactions of carbon dioxide and temperature as pointed out by Cure (3). Rice, corn, and soybean responded about 40% more to a doubling of carbon dioxide when grown under day/night temperatures of 28/23°C (82/73°F) than day/night temperatures of 23/20°C (73/68°F). These temperatures are not very high, but Idso et al. (4) reported that the biomass growth ratio for a 300 ppm increase in carbon dioxide increased somewhat linearly (0.087 per °C) with increasing mean temperature over the range of 19 to 34°C (66 to 93°F) for vegetative growth of several species of plants. Leaf photosynthetic rates are enhanced more by elevated carbon dioxide at high temperatures than at lower temperatures; therefore it would seem logical to expect increased growth rates when elevated carbon dioxide is accompanied by higher temperatures, at least up to some temperature threshold.

The temperature optimum of many temperate region species may be 25 to 30°C (77 to 86°F) at ambient carbon dioxide levels, but this optimum may increase 5 to 10°C (9 to 18°F) as carbon dioxide is increased. Species adapted to hotter and dryer climates have higher temperature optima for photosynthesis. The optimum temperature for photosynthesis and the upper temperature limit for photosynthesis for some species may increase up to 10°C from spring to summer and decrease in the fall. If this adaptability exists in crop plants, or if it can be introduced, then the implications of the impact of increasing temperature may be less severe.

The interactions of carbon dioxide and soil water availability (or water stress) is uncertain. Under high carbon dioxide, stress effects may be delayed for a few days in some crops because of the higher level of leaf starch buildup and lower stomatal conductance to water loss. This interaction with carbon dioxide may be highly dependent on other climate factors such as temperature.

Climate Change Scenarios

The climate change scenarios for a doubled carbon dioxide climate have been analyzed by Grotch (5) for four GCM's (CCM, GFDL, GISS, and OSU models). The June-July-August (JJA) median USA temperature increase was about 3.0, 5.6, 3.8, and 3.5°C (5.4, 10.1, 6.8, and 6.3°F) for the above models, respectively. The JJA changes in precipitation were +10, -25, +8, and +4% for the respective models. The GCM precipitation scenarios are more uncertain than temperature change scenarios and may vary considerably over regions. Another GCM from the United Kingdom Meteorological Office (UKMO) follows the pattern of the GFDL model with respect to the temperature and precipitation over the USA. All models predicted a warming with a doubled carbon dioxide climate. The highest temperatures of the GFDL (and the UKMO models) were associated with a reduced JJA precipitation.

Crop Responses to Climate Change Scenarios

Simulations of soybean (SOYGRO model) and corn (CERES-Maize) yields under doubled carbon dioxide climate change scenarios of the GISS model and the GFDL model were conducted for the Southeastern USA by Peart et al. (6) and Curry et al. (7) and for the Great Lakes Region by Ritchie et al. (8). Yields of wheat (CERES-Wheat) and corn were simulated for the Great Plains by Rosenzweig (9).

Yield simulations of soybean by Peart et al. (6) and Curry et al. (7) were conducted using a baseline of 30 years (1951-1980) of weather data at 19 sites in the Southeastern USA. To each year of baseline data at each site, temperatures were increased and precipitation changed proportionately based on the doubled carbon dioxide climate predictions of the two models. Simulations were conducted for the prevailing cropped soil type near each location of weather data. Current recommended cultivars and planting dates were used.

Simulations were run first based on climate changes only (temperature, rainfall, and solar radiation). Next, simulations were run based on irrigation water applied whenever the crop needed it. Finally, simulations were run with a carbon dioxide enhancement factor on crop photosynthesis (1.35 for soybean and 1.10 for corn). These enhancement factors were selected based on several sets of experimental data, but they do not include any temperature interactions with carbon dioxide on photosynthetic rates that may exist.

Table 2 shows the average climatic effects only (without direct carbon dioxide enrichment effects) for soybean. In the rainfed simulation, the GISS weather scenarios for a carbon dioxide doubling gave a 23% reduction in yield whereas the GFDL scenario gave a 71% reduction. Obviously, the reduction in summertime rainfall of the GFDL model led to drastic reductions in predicted yields. Under well-managed irrigation, soil water availability ceased to be a factor, and the average yields were decreased 18% and 19% for the GISS and GFDL scenarios respectively. High temperatures reduced yield by shortening the seed-filling period.

We cannot ignore the direct aerial fertilization effects of doubled carbon dioxide on leaf photosynthesis. Table 2 shows the average outcome of the SOYGRO model simulation with carbon dioxide fertilization plus climate effects for the rainfed and irrigation cases. Under rainfed cropping, the predicted yield increased 11% in the GISS model scenario, but decreased 52% under the GFDL scenario.

The carbon dioxide fertilization effect increased soybean yields by about 40% for the GISS model scenario both in the rainfed and in the irrigated simulations. This effect is somewhat larger than has been observed experimentally. Nevertheless, the biggest uncertainty in the prediction of crop yields for the Southeastern USA lies in the wide difference in the rainfall predicted by the two models for a doubling of carbon dioxide. Furthermore, under the GISS scenario, culture of soybean may become less competitive in this region compared to more northern latitudes.

Table 2 also shows that corn yields would be expected to decline only 8% in the GISS climate, but would decrease by 73% in the GFDL climate at 4 selected Southeastern USA locations. Irrigation increased predicted yields, but the GISS and GFDL models gave decreases of 18% and 27%, respectively, in comparison to the irrigated baseline weather crop. The yield reduction attributable to higher temperatures of the GFDL model was 10% with respect to the GISS model. Including the effects of carbon dioxide fertilization plus climate had little effect on the predicted yields of corn, a C-4 photosynthetic pathway plant.

Higher temperatures associated with the GISS and GFDL models led to a higher predicted irrigation water requirement. With combined direct carbon dioxide effects and climate effects the soybean irrigation demand was increased 33% under the GISS scenario and 134% under the GFDL scenario when averaged across all locations and years.

Ritchie et al. (8) found temperature to have a large effect on CERES-Maize and SOYGRO predicted yields of current crop cultivars in the Great Lakes Region based on GISS and GFDL scenarios. In the latitude band of 46-48° N they found a consistent increase in yields with warmer climate changes, whereas in the 38-46° N latitude region they found consistent decreases in yield. The positive effect of increased temperature at high latitudes is consistent with other earlier evaluations. Lower rainfall of the GFDL scenarios had an effect also, but irrigation apparently did not eliminate differences in predicted yield of soybean in the Great Lakes Region as it did in the Southeastern USA crop yield simulations. Soybean appeared to respond slightly more to direct carbon dioxide effects in the Great Lakes Region than in the Southeast.

Rosenzweig (9) also attributed a larger impact of temperature increase on yield reductions of wheat and corn in the Central and Southern Great Plains than to rainfall changes. The temperature increases shortened the duration of the crop life cycle. Her work implies greater demand for irrigation in this region and a possible shift of wheat production to the north.

Several reviews have been published describing cropping patterns that may shift northward and eastward under hotter and drier climatic change scenarios for North America. Agricultural productivity could be enhanced in Canada and Northern Europe, but decreased in the USA and Western Europe.

Agricultural Adjustment Options

Carbon dioxide fertilization effects can partially offset adverse climate, particularly for C-3 plants. More biomass and grain can be produced per unit water use and per unit solar radiation absorbed. However, elevated temperature can increase water use and offset partly the improved water use efficiency caused by elevated carbon dioxide.

Considerable disagreement exists among climate models on the range of predicted temperature change and the direction of regional precipitation. Therefore, the direction of agricultural adjustment options discussed below may not yet be on a firm footing.

Several options exist for agricultural adjustments to greenhouse effects. Germplasm improvements for drought-tolerance or heat-stress tolerance could be undertaken. We may need to investigate crop production responses in high temperature zones around the globe for both new management ideas and germplasm resources. Crop species could be substituted (e.g., pigeon pea for soybean). Certainly management adjustments are possible, such as earlier planting and possibly multiple-cropping. Crop production regions could migrate northward and eastward in North America. If overall crop production is curtailed by climatic change, then dietary adjustment may be necessary by committing more resources to human food rather than high quality animal feed. Meat supply could shift from feed lot cattle to grazing animals.

One obvious management possiblity is to increase supplemental irrigation in the relatively humid climates of the Eastern USA. This may require large scale feasibility studies and regional planning. Surface and aquifer water resources may need to be assessed for quantity and quality both under present and future hydrologic scenarios. Large scale delivery systems and interregional water transfers may be possible and desirable. The Mississippi River system could be more highly regulated as a water resource. Reduced streamflow under altered climate could lead to the need for a new infrastructure for transportation to substitute for barge traffic if streamflow decreases and/or water is withdrawn for other purposes. Competition could develop among potential users of available water.

Weather and Climate Variability

The real plague of agriculture is weather variabilty, which means rainfall variability. What production agriculture really needs is improved predictability of weather within the time span of a crop growing season, and improved predictability of season-to-season variability of weather and climate over the time span of a few years to a few decades. The difficulty of this challenge lies in getting a handle on cause and effect of variabilty rather than on response to one global perturbation (greenhouse effect gases).

Needed Tools For Future Predictions

Currently, there is a wide range of differences in prediction by the various GCMs. The differences that exist among GCMs in the seasonal and regional distribution of rainfall, especially, needs to be resolved or at least improved. (On the other hand, if there were not considerable diversity at this point in time of the GCM simulations of the earth's complex climate, I would be even more concerned.)

Although the SOYGRO model and CERES-Maize and CERES-Wheat models have been tested and widely used, other crop models for these same and other crops need to be adapted and used for prediction of direct carbon dioxide effects and the gamut of environmental interactions (e.g., Acock et al., 10). Specifically, better process-level components are needed for photosynthetic predictions with interactions of carbon dioxide and climatic factors. Background leaf and canopy studies of crop responses need to continue, including controlled environment studies of specific processes that are required for process-level crop models. Funded experimental and new crop model development programs such as the joint DOE-USDA effort throughout the 1980's are absolutely essential. These types of programs undergird each new round of climatic impact assessment on agriculture.

1. Allen, L.H., Jr., K.J. Boote, J.W. Jones, P.H. Jones, R.R. Valle, B. Acock, H.H. Rogers, and R.C. Dahlman. 1987. Response of vegetation to rising Carbon Dioxide: Photosynthesis, Biomass, and Seed Yield of Soybean. Global Biogeochemical Cycles 1:1:14.
2. Kimball, B.A. 1983. Carbon Dioxide and Agricultural Yield: An Assemblage and Analysis of 430 Prior Observations. Agron. J. 75:779-788.
3. Cure, J.D. 1985. Carbon Dioxide Doubling Responses: A Crop Survey. p. 99-116. In B.R. Strain and J.D. Cure (ed.) Direct Effects of Increasing Carbon Dioxide on Vegetation. U.S. Department of Energy, Carbon Dioxide Research Division, DOE/ER-0238, Washington, D.C.
4. Idso, S.B., B.A. Kimball, M.G. Anderson, and J.R. Mauney. 1987. Effects of Atmospheric CO_2 Enrichment on Plant Growth: The Interactive Role of Air Temperature. Agric. Ecosystems Environ. 20:1-10.
5. Grotch, S.L. 1988. Regional Intercomparisons of General Circulation Model Prediction and Historical Climate Data--TR 041. DOE/NBB-0084, U.S. Dep. of Energy, Carbon Dioxide Res. Div., Washington, D.C. 20545. 291 p.
6. Peart, R.M., J.W. Jones, R.B. Curry, K.J. Boote, and L.H. Allen, Jr. 1988. Impact of Climate Change on Crop Yield in the Southeastern USA: A Simulation Study. In J.B. Smith and D.A. Turpak, The Potential Effects of Global Climate Change on the United States. UESPA vol ___.
7. Curry, R.B., R.M. Peart, J.W. Jones, K.J. Boote, and L.H. Allen, Jr. 1988. Simulation as a Tool for Analyzing Crop Response to Climate Change. Paper No. 88-7512, Amer. Soc. of Agric. Engineers, St. Joseph, MI 49085-9659.
8. Ritchie, J.T., B.D. Baer, and T.Y. Chou. 1988. Effect of Global Climate Change on Agriculture: Great Lakes Region. In J.B. Smith and D.A. Turpak, The Potential Effects of Global Climate Change on the United States. USEPA vol. ___. (In Press).

9. Rosenzweig, C. 1988. Potential Effects of Climate Change on Agricultural Production in the Great Plains: A Simulation Study. In J.B. Smith and D.A. Turpak, The Potential Effects of Global Climate Change on the United States. USEPA vol. ___. (In Press).
10. Acock, B., V.R. Reddy, F.D. Whisler, D.N. Baker, J.M. McKinion, H.F. Hodgers, and K.J. Boote. 1983. The Soybean Crop Simulator, GLYCIM: Model Documentation. Responses of Vegetation to Carbon Dioxide, series no. 004. U.S. Dep. of Energy and U.S. Dep. of Agriculture Joint Program Progress Report of Research, Mississippi State, MS.

TABLE 1. SOYBEAN RESPONSES TO RISING CARBON DIOXIDE PREDICTED BY THE NONLINEAR, HYPERBOLIC, MODIFIED MICHAELIS-MENTEN MODEL. FROM ALLEN ET AL. (1).

ASSUMED PERIOD	CO_2 CHANGE	PHOTOSYN.	BIOMASS	SEED
years	ppm	-------percentage increase-------		
1800-1958	276-315	12%	10%	8%
1800-1986	276-345	20%	17%	13%
1958-2058	315-630	53%	43%	32%

TABLE 2. SOYBEAN YIELD SIMULATIONS (SOYGRO) AND CORN YIELD SIMULATIONS (CERES-Maize) IN BUSHELS PER ACRE UNDER TWO CLIMATE MODEL SCENARIOS WITH DOUBLED CARBON DIOXIDE (GISS AND GFDL) IN COMPARISON WITH BASE CLIMATE 1951-80) FOR THE SOUTHEASTERN USA. DERIVED FROM PEART ET AL. (6).

SOYBEAN YIELD SIMULATIONS					CORN YIELD SIMULATIONS				
BASE	GISS Model		GFDL Model		BASE	GISS Model		GFDL Model	
Yield	Yield	Dif.	Yield	Dif.	Yield	Yield	Dif.	Yield	Dif.
Climate Effects Only, Rainfed									
37	29	-23%	11	-71%	137	126	-8%	37	-73%
Climate Effects Only, Irrigated									
57	47	-18%	46	-19%	224	183	-18%	164	-27%
Carbon dioxide Fertilization plus Climate, Effects Rainfed									
37	41	+11%	18	-52%	137	130	-5%	35	-74%
Carbon dioxide Fertilization plus Climate Effects, Irrigated									
57	65	+13%	65	+14%	224	184	-18%	165	-26%

A SUMMARY OF CLIMATE CHANGE IMPACT STUDIES ON AGRICULTURE:

The U.S. Environmental Protection Agency's Report to Congress on The Potential Effects of Global Climate Change on the United States

Cynthia Rosenzweig

Department of Geography, Columbia University
NASA/Goddard Space Flight Center, Institute for Space Studies

Atmospheric scientists are projecting temperature increases by early in the next century to levels higher than any in at least the past 150,000 years. Associated alterations in rainfall and soil moisture are also expected, but these predictions are less well defined. Although the climate projections are uncertain, especially in regard to the rate of change and regional specificity, their magnitude creates the need for a broad consideration of the impacts of potential climate change on U.S. agriculture.

Recognizing this need, Congress sent a formal request to the U.S. Environmental Protection Agency (EPA) in 1986 to undertake a study of the health and environmental effects of climate change, including the impacts on agriculture. The EPA's Report to Congress on The Potential Effects of Global Climate Change on the United States is currently in draft form and undergoing revision after review by the EPA's Scientific Advisory Board. This paper describes the structure, review process, caveats, findings and implications of the agricultural studies done for the EPA report.

The studies completed for the EPA report are a first step in linking predictions of climate change to models of agricultural systems. While the studies represent the most comprehensive analysis, to date, of the potential impacts of climate change on U.S. agriculture, the findings are only initial projections of the sensitivity of various aspects of American agriculture to the estimated climatic changes, rather than predictions of what will occur. Much work remains to define more clearly the ranges of potential impacts on all aspects of the agricultural system, and to engender flexibility in society's actions in response to these potential changes.

Structure and Review Process

Agricultural scientists from universities and federal agencies completed studies in the following research areas: 1) crop growth and yield; 2) regional and national agricultural economics; 3) demand for irrigation water; 4) water quality; 5) pest-plant interactions; 6) direct effects of CO_2 on crop growth and yield; 7) impacts of extreme events; 8) potential farm-level adjustments; 9) livestock pests and diseases; and 10) agricultural policy.

In order to carry out the Congressional request for the

report, EPA developed a set of climate change scenarios based on global climate model (GCM) output for doubled levels of atmospheric CO_2. The results of several different climate models were selected in order to explore a broad range of future climates. These scenarios were used in models of agricultural systems to determine the potential effects of climate change. Some of the model studies were linked: simulated regional changes in crop yields and hydrology were used in a national agricultural economics model to estimate regional and national changes in crop production, land use, and demand for irrigation. Such a coordinated analytical framework is necessary in order to account for the effects of market forces on the agricultural system, and to evaluate the adequacy of the nation's resource base for agricultural production.

The review process included a review panel on the agricultural studies while work was still in progress, three independent peer reviews of each study, and the EPA Scientific Advisory Board public review of the summary report, chapter by chapter. In addition to these formal reviews, the draft report has been circulated to many federal agencies for comment. Suggestions from all of these sources are being incorporated into the formal interagency review draft of the report.

Sources of Uncertainty

There are, of course, many sources of uncertainty. The climate scenarios did not include changes in interannual variability, such as changes in the frequencies of drought or storms, even though these could affect crop yields significantly. The climate models have not yet been validated to project these types of changes. The effects of changes in levels of tropospheric and stratospheric ozone on crops were not considered, nor were changes in levels of pest infestations. The crop models, while among the best now available for large-area studies, are simplistic, and may not provide correct estimates for the extreme climatic conditions created by the climate change scenarios. The crop models can only approximate the physiological effects of CO_2 on crop growth and water use at this time. Finally, the range of possible alternative cropping strategies was not fully explored in the studies, such as substitution of different crop species, nor were potential changes in foreign demand for U.S. agricultural commodities.

Findings and Implications

Given global climate change as predicted by the global climate models, the main conclusions that may be drawn from the EPA studies are: (1) Significant regional shifts in agricultural production are likely; (2) U.S. food supply appears to be sufficient to meet domestic needs, but export levels could decrease; (3) demand for water for irrigation is estimated to rise in many regions of the country, and (4) the extent to which the direct effects of CO_2 will be beneficial is uncertain. A major caveat is needed at this point: <u>If droughts occur more frequently under changed climate, effects on agriculture may be more severe</u>.

Regional changes. If the current understanding of global climate warming is qualitatively correct, agricultural land in more southern latitudes may go out of production, while agricultural acreage is likely to expand in the north. All climate change scenarios tested show that the southern areas of the United States become less productive relative to northern areas. This is primarily because the high temperatures predicted for climate change will stress crop production more in southern areas than in northern areas where crops are currently limited by lower temperatures and shorter growing seasons. While the studies did take availability of agricultural soils into account, sustainability of expanded agricultural production in northern areas was identified as an important area for future research.

Other potential changes likely to occur include changes in infestation of crop pests, altered ranges of livestock pests and diseases, and higher levels of heat stress on livestock. For example, a northward extension of crop pest ranges and increased pest populations may reduce yields and affect livestock. Warmer temperatures may result in northward extension of the range of diseases and pests that now afflict livestock in the south, and could make conditions more favorable for introduction of new livestock diseases into the southern United States. These effects are all driven in large part by high temperatures.

Food supply and exports. The EPA studies imply that the projected climate change does not threaten food supply for the nation, even under more extreme climate scenarios. However, if droughts occur more frequently under climate change, this may not be the case. While production capacity of U.S. agriculture appears to be adequate to meet domestic needs, model projections that held demand for exports constant show that U.S. exports may decline significantly. It is important to note that climate change may expand or limit agricultural production in other areas of the world, and that changes in U.S. agriculture will take place in this global context. The overall result could be changes in the patterns of global food trade, in particular the demand for U.S. exports.

Demand for water for irrigation. Demand for water for irrigation is likely to rise with warmer temperatures. Increased demand for irrigation occurs for a number of reasons. First, crops tend to need more water. Second, farmers are likely to respond by irrigating in order to ensure stable yields in areas which become drier. Third, higher crop prices, caused by the inelastic demand for agricultural commodities, would make the costs of irrigation feasible for more farmers.

The EPA studies did not consider changes in competing demands for water (e.g., for industrial or residential use), which also may be altered in a warmer climate. Costs of developing irrigation systems in new locations and increasing capacity in currently irrigated areas may be high, and actual implementation of expanded irrigation systems depends on the economics of each local situation. Furthermore, rising demand for irrigation water could place additional stresses on water systems already being overdrawn, such as the Ogallala Aquifer in the Great Plains.

Direct effects of carbon doxide. Regional changes in U.S. agriculture are predicted even when beneficial physiological effects of increased carbon dioxide on crop growth are taken into account. While the beneficial effects of CO_2 may mitigate or actually overcome negative climate effects on yields in some areas, agricultural productivity shifts occur in any case. Thus, an optimistic estimation of yield increases caused by higher levels of CO_2 does not preclude significant changes in regional agriculture in the future.

Adjustments and Rural Communities

Regional changes in agriculture have important implications for rural communities. Farmers can make some adjustments to less severe climate change by planting and harvesting earlier and substituting better-adapted crop varieties and species. More severe climate change will likely require major adaptations, including expansion of irrigation infrastructure, and changes in farming systems (e.g., cropping to ranching). However, the industry of agriculture is composed of many people besides farmers, including farm equipment manufacturers, fertilizer and seed suppliers, and rural bankers, to name but a few. As production areas shift, climate change effects will reverberate through this larger community and are likely to result in structural changes in local economies such as relocation of markets and transportation networks. At its most extreme, climate change could cause dislocation of rural communities through farm abandonment.

Environmental Concerns

Natural environments may also suffer because of changes in agriculture. Shifts in regional production patterns imply changes in uses of natural resources. In more northward regions where pressure for expansion of agricultural production may grow, natural environments may be affected because of the potential for increased soil erosion, ground and surface water pollution (especially on sandy soils), and loss of wildlife habitats. In many regions, levels of chemical pesticide usage may change in response to changes in infestations of both crop and livestock pests.

Research and Policy

A strong research effort must be maintained in order to ensure technological improvements in agriculture and limit vulnerability to climate change, including continued development of crops which are drought and heat tolerant, and species acclimated to the projected changed climate regimes. Biotechnology may be able to contribute in this regard. One of the most crucial areas for further research is the projection of potential climate change effects at the international level, and the implications of the projected shifts in global agriculture for U.S. trade. These efforts must be accompanied by continued improvement in our ability to estimate regional climate changes.

It is also appropriate to examine current agricultural policies, such as the U.S. commodity programs, in light of the need for flexibility in adjusting to potential future warming. Do these and other farm policies encourage farmers to adapt as eadily as possible to changing climate conditions? Of course, given the large uncertainties about climate change, policies designed to help U.S. agriculture prepare should also make sense for today. For example, drought relief planning could be improved, thereby helping today's farmers and also providing insurance against potential future droughts. Other policies to examine in the light of climate change are land-use and water-resource management and water quality regulations. Finally, the implications of potential long-term changes in the levels of U.S. crop exports for the U.S. balance of trade should be considered by relevant policymakers.

Conclusions

While many critical uncertainties remain about the magnitude and timing of impacts, it appears that climate change is likely to affect agriculture and rural communities in the United States significantly in the coming century. If the higher temperatures that are predicted by global climate models occur, changes in regional crop yields, crop irrigation requirements, heat stress on livestock, and infestations of agricultural pests and diseases, may ensue. Model results showed regional changes in U.S. agriculture even when beneficial physiological effects of increased carbon dioxide on crop growth are taken into account.

The initial analysis of climate change impacts on U.S. agriculture provided by the EPA studies indicates that comparative advantage among farming enterprises and regions, and hence patterns of agriculture are likely to shift within the United States. While the domestic supply of agricultural commodities should be sufficient (if drought frequency does not increase), the impacts of climate change on agriculture in its larger role as a renewable resource base that provides industrial and domestic products beyond food and fiber have yet to be considered. Concern for climate change and U.S. agriculture, taken in the broadest sense as a major renewable resource, a primary user of the nation's land and water, a network of productive rural communities, and a strong influence on the environment, is amply justified. Furthermore, the need to study climate change effects in a global context, particularly in relation to the role of the United States as a reliable supplier of agricultural export commodities, becomes a necessity.

LIKELY IMPACT OF CLIMATE CHANGE ON CANADIAN AGRICULTURE

Dr. Barry Smit
University of Guelph

INTRODUCTION

Agriculture is an important sector in the Canadian economy and society, and given Canada's northern location, climate is a major constraint on agricultural activity. As a consequence, Canadian agriculture is often seen as particularly sensitive to climate change. This paper reviews implications of climate change for Canadian agriculture, highlights the issues of uncertainty and adaptability, and draws some conclusions for public policy.

SYNOPSIS OF STUDIES

Only a handful of studies have explicitly examined the impacts of climate change on Canadian agriculture. Most of the studies are based on a scenario where carbon dioxide and other greenhouse gases are assumed to be double the pre-industrial levels (the 2 X CO_2 equivalent). This scenario is used not because it is more likely to occur than any other scenario, but because it provides a useful benchmark from which to estimate the sensitivity of agriculture. Most of the studies use detailed climate-crop models to estimate the effects of greenhouse warming on crop yields and hence on production. To date much of this research in Canada has been conducted for the Prairies (Arthur, 1988; Stewart, 1988; Williams et al., 1988) and Ontario (Land Evaluation Group [LEG], 1985 and 1986), but other studies have looked at such impacts as the effects of sea level rise on Atlantic Canada (Lane and Associates Limited, 1988), and the implications for Canada of changes in global agricultural production opportunities (LEG, 1987).

GREENHOUSE AND AGRO-CLIMATE

The global climate change models are broadly consistent in their predictions of climatic conditions for Canada. Generally, under a greenhouse climate increased temperatures, longer growing seasons, increased thermal energy, and possibly increased precipitation can be expected. Overall this suggests an improved climate for agriculture in Canada. However, the increased solar energy also increases evapotranspiration, resulting in reductions in the availability of moisture for plants. Moisture supply is already an important constraint on agriculture in many areas, and under climate warming becomes critical. Of course, effects of these changes vary regionally with the moisture holding capacity of soils and other local conditions.

To visualize the changed conditions for Canada imagine an overall northward movement of agro-climatic environments. Warmer and drier conditions might be expected in the Southern Prairies, along with a shifting of the United States' Corn and Winter Wheat Belts into Canada, with increased opportunities occurring in northern areas where currently the growing season is a major limitation on agriculture.

IMPLICATIONS FOR REGIONAL PRODUCTION

In British Columbia agricultural production is probably most sensitive to climate change in the dry interior areas which would experience increased moisture shortages. More research has been undertaken for the Canadian Prairies, an area especially important for wheat and other grains. Under a greenhouse climate improved opportunities can be expected for many crops such as winter wheat, which requires a longer growing season, favourable seeding conditions, and an environment which allows winter survival. It is generally expected that wheat production opportunities would be enhanced. The ability to benefit from these changed climatic conditions varies within the Prairies, particularly with respect to the moisture regime. Manitoba is generally the least deficient in moisture and hence improved grain yields are widely expected in that province. Elsewhere prospects for expanded grain production are limited by modest supplies of moisture. The areas most sensitive to climate change are those already considered marginal. For example, in Northern Saskatchewan, Manitoba and Alberta agriculture is limited by the short growing season. With climate change, agricultural prospects will likely improve in these areas, but only modestly because of soil limitations. In contrast, the Southern Prairies are currently considered marginal because of their aridity. Under climate change droughts could occur more frequently further threatening the viability of agriculture in those regions.

For the provinces of Ontario and Quebec in Central Canada, the regions surrounding the Great Lakes and St. Lawrence Valley represent important agricultural production areas for corn, soybeans, grains, forages and horticultural crops. Under a greenhouse climate increased opportunities in the northern parts of these provinces might be expected, but the thin soils of the glaciated shield areas severely limit the prospects for expanded agriculture. In the southern areas, particularly the agricultural heartland of Southwestern Ontario, many crops are likely to benefit from the changed climate especially where there is ample moisture available. However studies have shown that major crops like corn and soybeans are sensitive to moisture stress in these areas, resulting in reduced yields under a greenhouse climate. On the otherhand, for fruit and vegetable crops, which are sensitive to winter extremes and the length of the frost-free season, a greenhouse climate is likely to expand opportunities. Fruits and vegetables which are currently limited to a viable range on the north shores of Lake Erie and the Niagara Peninsula may become viable over significantly larger parts of the province. Overall, if agriculture does not adapt to the changed climatic regime significant production losses can be expected under a 2 X CO_2 climate.

In the Atlantic Provinces ample moisture is available, so there are opportunities to benefit from the enhanced growing season length and heat supply. However, a one metre rise in sea level because of climatic change, has significant implications for the rich farming areas of Prince Edward Island and the Saint John River Valley of New Brunswick. These areas would become susceptible to increased flooding, erosion, elevation of the water table, perhaps salt water intrusion and other problems associated with excess moisture.

Throughout most of the regions of Canada climate change is expected to bring with it both improved opportunities and increased limitations. The opportunities generally relate to a longer growing season and increased thermal energy. The limitations generally relate to moisture supply: not enough moisture in many parts of Western and Central Canada, and too much in the low-lying areas of Atlantic Canada.

INTERNATIONAL COMPARISONS

Some observers conclude that Canadian agriculture would be a beneficiary of global warming. However the international implications for Canada would depend on the climate-induced changes and opportunities in Canada compared with those occurring elsewhere in the world. For example, an increase in wheat production opportunities in Canada would not necessarily be advantageous if opportunities were also enhanced in other major wheat producing areas of the world, especially when many countries are having difficulty selling what they currently produce. A recent critique of the international comparative position of Canadian agriculture showed that production prospects for wheat and other small grains may be enhanced in Canada and the Soviet Union and diminished elsewhere. This might translate to a somewhat improved comparative position for marketing of these products for Canada. Similarly with corn, the Canadian comparative position is expected to improve, perhaps resulting in a reduction in imports from the American Midwest. For some these crude analyses provide sufficient evidence to conclude climate change does not provide a major problem for Canadian agriculture. Agriculture is seen as very adaptable with Canadian agriculture a likely winner in the 'Climate Change Sweepstakes'. The remainder of this paper suggests caution should be exercised before drawing such conclusions.

VARIABILITY AND RISK

Most impact studies relate to changes in the so-called normal or average climatic conditions. Of course, under a 2 X CO_2 climate, just as under the present climatic regime, there is considerable variability from year to year around these norms. Climate-related problems in agriculture are invariably linked with these deviations from average conditions. For example, crises in the agricultural sector can be associated with dry years or excessively wet years, or frost coming later in the spring than usual. These deviations in climatic conditions from the average are realities of climate and have certain probabilities of occurring.

Some limited analyses of the implications of variability around the 2 X CO_2 climatic conditions have been conducted. Models of representative farms, run under scenarios with deviations in precipitation, show that while prospects may be enhanced under the average conditions, significant losses occur with departures from these averages, especially in dry years. In both Ontario and the Prairies agriculture is particularly sensitive to years which are drier than the norm. Agriculture is already sensitive to droughty conditions, as evidenced very vividly in the 1988 production year. Under climate change such dry years are likely to occur more frequently.

FARM MANAGEMENT OPTIONS

The first generation of agricultural impact studies, quite understandably, assumed little or no adjustment on the part of farm operators. Of course, if those in agriculture were aware that climatic conditions were changing then it is likely that certain adjustment strategies would be considered and implemented. For example, many opportunities to benefit from the increased temperature and growing season length exist, such as using higher yielding hybrids, or fall-sown crops, or introducing entirely new crops to particular localities. Another management option is reducing the risk associated with moisture deficiency by development of supplemental moisture supplies such as irrigation. Enhanced thermal conditions may allow the introduction of higher value horticultural crops which in turn might justify the capital expenditure involved with irrigation systems. However this increased need for a water source under climatic change may occur when the water supply is already taxed by demands from other sectors.

Other management options include development of hybrids resistent to moisture stress or which can mature in the shorter growing season when moisture is available. Buildings may need to be modified to avoid livestock losses under particularly hot conditions. For example, when advised that there was a reasonable probability of 1988 climatic conditions reoccurring sometime over the next ten years, a broiler operator concluded that an upgrading of his ventilation system was warranted to avoid the risk of a repeat of massive bird losses. Another management option is diversification, to reduce the susceptibility of farming to extreme events or the deviations from average conditions which are bound to occur.

ADAPTATION

Theoretically, agriculture is an activity which can adapt quite readily to gradual changes in environmental conditions. Many forms of agriculture operate on an annual cycle which allows for a change in the type of production from one year to another. Of course there are limitations to this since some forms of agriculture operate over more than a one-year cycle, and investments in capital infrastructure must also be considered. Despite this theoretical adaptability, in practice agriculture is very slow to adjust to environmental conditions. This can be attributed to several factors. One is behavourial, relating to human nature. People tend to believe that a bad year, for example a drought, is a rare event even when its known probability may be two years in five. Operators attempt to maximize returns assuming the normal or average year; they tend not to minimize risks recognizing the full range of conditions which are likely. Crop failures which occur in years which deviate from the average, and the resulting economic problems, increase pressure on farmers to take risks in subsequent years assuming that 'lightning will not strike the same place twice'.

Another factor limiting the adaptability of agriculture is the political and economic environment within which it operates. Pressure by lending institutions to maximize short-term returns encourages planning for a normal or good year. Furthermore when crop failures occur governments feel obliged to provide relief to farmers in the form of disaster aid; not aid to change the system of agriculture to one which is more compatible with the realities of environment. Thus the political economy tends to reinforce a cyclic inertia in agriculture. Many technological developments also limit the adaptability of agriculture. For example crop breeding results in varieties which produce higher yields, but over a narrower range of conditions, and which have much lower yields outside that range (Oram, 1985).

The consequence of all these factors is a system of agriculture which does well in an average year, but less well under conditions which deviate from that average. Agriculture is not very resilient to current variability in climate, although the probabilities of these variations from average conditions are known. How adaptable will agriculture be to unknown changes resulting from the greenhouse effect, and to variations in climate which will persist around those changed average conditions?

UNCERTAINTY

Climate impact studies have tended to assume a certain scenario of climate change and then assess the implications of that climate for the activity in question. Throughout this exercise there are many uncertainties which should temper any conclusions drawn about the implications of climate change. Uncertainties exist about the change in climate itself. While there is general agreement that the greenhouse effect is in place and that warmer global conditions can be expected, beyond that there is tremendous conjecture. The speed of change is not known, nor are the implications for agro-climatic conditions particularly at the regional level. And there is a real uncertainty about the pattern of change; the transient effects. Will there be a gradual and smooth change in climate or at certain thresholds will there be sudden changes? Most of the studies have focussed on a 2 X CO_2 scenario. This is analytically convenient but what of changes in climate below a 2 X CO_2 equivalent, and what about concentrations higher than this level?

There is also uncertainty regarding the direct effect of changes in CO_2 concentrations on plant growth. For some crops CO_2 enrichment acts like a fertilizer. However it might also change the moisture requirements of crops and may alter the sensitivity of plants to temperature. The impacts of other chemical characteristics of the environment such as high and low level concentrations of ozone are uncertain. Knowledge of the biological response of plants and ecosystems to climate change is based on very limited laboratory experiments. The implications of the timing of flowering, and changes in weed and insect infestation is very poorly understood. Uncertainty also exists regarding other conditions which influence agriculture, such as markets, trade and inputs. And as noted above there is considerable uncertainty about the human response to these changed environmental conditions.

To summarize the impacts of climate change on agriculture: the knowledge base is meager, the uncertainty is great, the potential benefits are modest, and the potential risks are significant.

IMPLICATIONS FOR GOVERNMENTS

Under climate change some gradual changes in regional patterns of agricultural production can be expected, either by conscious decision of those involved in agriculture and land use, or by repeated losses, failure, bankruptcy and new ownership. Both of these processes have repercussions for regional economies including the processing, input supply and trade sectors. Thus, impacts of climate change on agriculture are carried through to regional economies and social systems.

What are the implications for governments? The first is that governments should work to **prevent** further changes to the global atmosphere. Even for Canadian agriculture the risk:benefit ratio supports this conclusion, one which also makes sense for many other reasons. The second implication for governments is that agencies should **plan** for climate change so that a sector such as agriculture is not 'surprised' as conditions change, for example as droughts become more frequent. If it is assumed that such conditions are rare events then agricultural losses, family and community hardship, and unnecessary stress on regional and national economies are guaranteed. Planning for variability in climate is needed under current conditions; it becomes even more urgent under a scenario of climate change.

ACKNOWLEDGEMENTS

The support of the Atmospheric Environment Service of Environment Canada, the Social Sciences and Humanities Research Council of Canada, and the Ontario Ministry of Agriculture and Food is gratefully acknowledged. The paper benefitted from suggestions by Deborah Bond.

REFERENCES

Arthur, L. 1988. The implications of climate change for agriculture in the prairie provinces, _Climate Change Digest_, CCD 88-01. Downsview: Canadian Climate Centre.

Land Evaluation Group (LEG). 1985. _Socio-economic assessment of the implications of climatic change for food production in Ontario_, Publication No. LEG-22. Guelph: University School of Rural Planning and Development (USRP&D), University of Guelph.

LEG. 1986. _Implications of climatic change and variability for Ontario's agri-food sector_, Publication No. LEG-26. Guelph: University School of Rural Planning and Development, University of Guelph.

LEG. 1987. _Implications of climatic warming for Canada's comparative position in agricultural production and trade_, Publication No. LEG-27. Guelph: University School of Rural Planning and Development, University of Guelph.

Lane, P. and Associates Limited. 1988. Preliminary study of the possible impacts of a one metre rise in sea level at Charlottetown, Prince Edward Island, _Climate Change Digest_, CCD 88-02. Downsview: Canadian Climate Centre.

Oram, P. 1985. Sensitivity of agricultural production to climatic change, _Climatic Change_, 7 (1), pp. 129-152.

Stewart, R. 1988. Climate change - Implications for agricultural productivity on the Canadian Prairies. In _Preparing for climate change: Proceedings of the First North American Conference on preparing for climate change: A cooperative approach_. Rockville: Government Institutes, Inc., pp. 409-419.

Williams, G., R. Fautley, K. Jones, R. Stewart and E. Wheaton. 1988. Estimating effects of climatic change on agriculture in Saskatchewan, Canada. In M. Parry, T. Carter and N. Konijn (eds.), _The impact of climatic variations on agriculture. Volume I: Assessments in cool temperate and cold regions_, pp. 219-379.

HOW INCREASED SOLAR ULTRAVIOLET-B RADIATION MAY IMPACT AGRICULTURAL PRODUCTIVITY

by

Alan H. Teramura and Joe H. Sullivan
Department of Botany
University of Maryland
College Park, Maryland 20742 U.S.A.

Introduction

In March 1988, following an 18-month review involving over 100 of the world's leading atmospheric scientists, the International Ozone Trends Panel released conclusive findings that significant depletions in global stratospheric ozone have occurred. The panel cited a depletion of between 1.7 and 3.0% from 1969 to 1986 at latitudes between 30 and 64 degrees North. The panel found that this decrease is conclusively linked to atmospheric chlorine, and is in addition to the natural variation in ozone levels (NASA, 1988). At this time it is impossible to project future losses of ozone which will be associated with the inevitable increases in stratospheric chlorine and biogenically produced gases. Therefore, despite good prognosis for the cessation of ozone depletion based on the Montreal Protocol, it is premature to consider the threat of depletion to be ended.

Depletion of the ozone layer is of concern because the stratospheric ozone column is the primary attenuator of solar ultraviolet-B radiation (UV-B region, between 290 and 320 nm). A decrease in this ozone column would lead to increases in UV-B reaching the earth's surface. Though it represents only a small fraction of the total solar electromagnetic spectrum, UV-B is readily absorbed by important macromolecules such as proteins and nucleic acids, and thus has a disproportionately large photobiological effect. Therefore, it is not surprising that both plant and animal life are greatly affected by increases in UV-B radiation penetrating to the earth's surface.

There exists tremendous variability in the sensitivity of plants to UV-B radiation (see Teramura 1986 for a recent review) Some species show sensitivity to current ambient levels of UV-B radiation while others are apparently unaffected by rather massive UV enhancements. This issue is complicated further by reports of equally large response differences among cultivars within a species. Approximately two-thirds of some 300 species and cultivars tested appear to be susceptible to damage from increased UV-B radiation. However, the vast majority of the species tested have been annual agricultural species, which account for only 9% of global net primary productivity (Whittaker, 1975). Only a handfull of studies have evaluated the potential effects of UV-B radiation on forests or other natural ecosystems. Therefore the potential impacts of increases in UV-B radiation upon these ecosystems remain unknown.

UV-B Effects on Crops

During the past decade, only 10 field studies have examined the effects of UV radiation on the yield of some 22 crop species. Six of these were conducted over only a single growing season and two others were conducted over 2 years. Thus, only scant information exists for the annual variation which occurs in field studies. In several of these studies UV irradiation included the germicidal UV-C (between 200 and 280 nm) waveband. Since this shorter wavelength radiation would not naturally occur on the earth's surface even with massive ozone depletion, the conclusions of these studies may not be applicable to natural field situations. Nonetheless, overall yield was affected by UV-B radiation in about half of these studies. Another important feature, the quality of crop yield, has only been quantitatively examined in a few species including tomato, potato and sugar beet.

The effectiveness of UV-B irradiation on plant growth and productivity varies seasonally and is affected by microclimate and soil fertility. For example, under water stress or mineral deficiency, soybeans are less susceptible to UV-B radiation, but under low levels of visible radiation, sensitivity increases (Teramura 1983). Thus field validation studies conducted over several growing seasons are crucial in any UV-B impact assessment of agricultural productivity.

This paper will concentrate on the results of the only multi-year field study completed to date on the effects of UV-B radiation on crop yield and productivity.

In 1981, two soybean (Glycine max (L) Merr.) cultivars were chosen for study based upon preliminary greenhouse trials for UV sensitivity and planted into the field. Based upon overall growth performance, cultivar Essex was found to be sensitive while cultivar Williams was tolerant to UV-B radiation. Field experiments were conducted during May through October of 1981 to 1986 at the Agricultural Research Center, USDA, Beltsville, Maryland, U.S.A.

Supplemental UV-B radiation was supplied by filtered Westinghouse FS-40 sunlamps suspended above the plants. Lamps were filtered either with 0.13 mm thick cellulose acetate (transmission down to 290 nm) for supplemental UV-B radiation or 0.13 mm Mylar Type S plastic films (absorbs all radiation below 320 nm) as a control. The radiation filtered through the cellulose acetate supplied a weighted daily supplemental irradiance of either 3.0 or 5.1 effective kJ m^{-2} UV-B_{BE} using the generalized plant response action spectrum (Caldwell, 1971) normalized to 300 nm. Therefore, plants beneath these lamps received supplemental doses in addition to ambient levels of UV-B radiation. These increased levels of UV-B radiation (supplemental + ambient) were similar to those which would be received at College Park, Maryland, U.S.A. (39^{0}N) with anticipated 16 and 25% stratospheric ozone reductions during a cloudless day on the summer solstice (Green et al., 1980). The weighted irradiance of Mylar filtered lamps was 0, so plants beneath these lamps received only ambient levels of UV-B (8.5 effective kJ m^{-2} UV-B_{BE} on the summer solstice). Different UV-B doses were obtained by varying the distance between the lamps and the top of the plants. Complete irradiation protocols can be found in Teramura et al. (1980).

The results of this 6-year field study demonstrate intraspecific differences in UV-B sensitivity in soybean yield and quality (Table 1). However, the expression of these sensitivity differences to UV-B radiation was altered by other prevailing microclimatic factors. For example, under a simulated 25% ozone depletion, Essex had reductions in overall yield of 19-25% during 4 of the 6 years. The two years in which yield was not reduced, 1983 and 1984, were characterized as hot and dry with prolonged periods of drought. Parallel field studies have shown that the effects of UV-B radiation can be masked by drought-induced growth reduction.

Figure 1 gives an example of the potential effects of UV-B on yield in Essex soybean. Figure 1A shows the sources of yield losses in U.S. soybean under current levels of stratospheric ozone and Figure 1B shows the relative yield loss due to a 25% ozone depletion. If UV-induced yield reductions were 20%, then effective yield (i.e. harvest available for consumption) could be further reduced (from 56 to 36%). In contrast, yield increased

Table 1. Summary of UV-B radiation effects on soybean yield. A 25% ozone depletion was simulated over College Park, Maryland U.S.A.

Year	% Change in Yield	
	Essex	Williams
1981	-25	+22
1982	-23	+14
1983*	+6	-11
1984*	-7	+10
1985	-20	+4
1986	-19	+6

*Years with prolonged drought

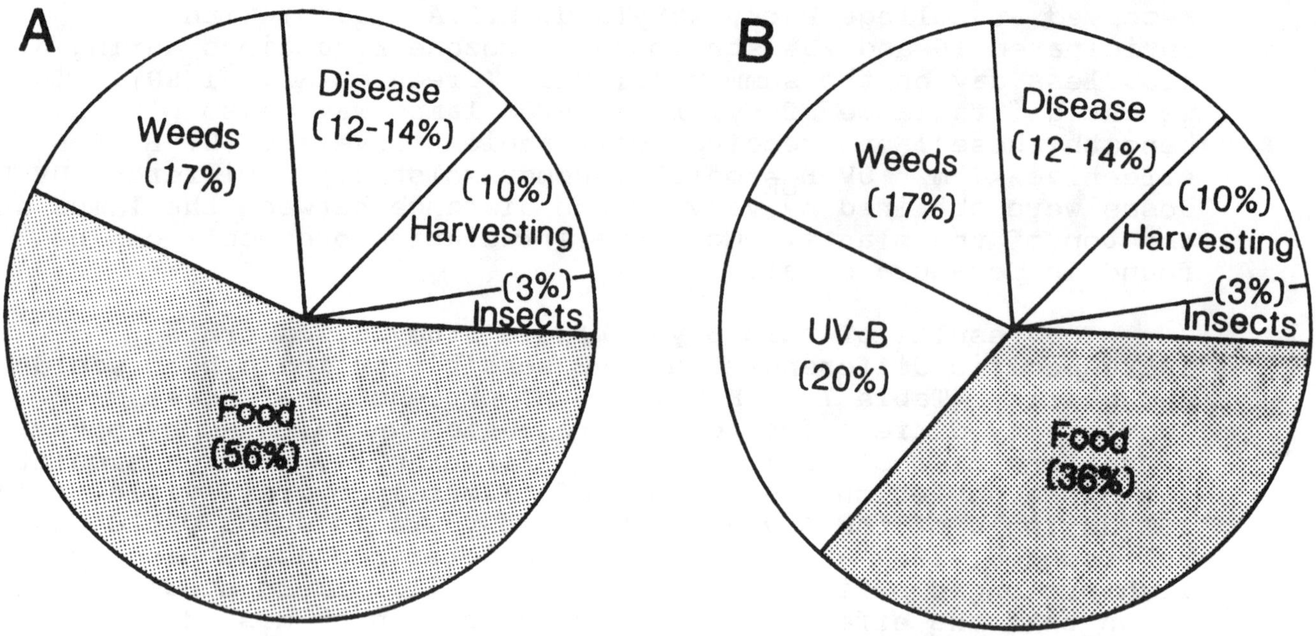

Figure 1. Sources of yield loss in U.S. Soybean under current levels of stratospheric ozone (A) and predicted from a 25% ozone depletion (B).

from 5 to 22% in 5 of the 6 years for the UV resistant cultivar Williams. Based upon previous studies (Biggs et al., 1981; Teramura and Murali, 1986) approximately two-thirds of the 50 soybean cultivars previously studied were found to be sensitive to UV-B radiation, while one-third were tolerant. This large intraspecific variation suggests that natural adaptations for tolerance to UV-B radiation exist in our modern soybean gene pool and therefore the potential for breeding and/or molecular biological techniques to improve tolerance also exists. However, no experimental exist on crop improvement with respect to UV-B tolerance at this time.

An assessment of the potential effects of increases in solar UV-B on plant productivity is also difficult to make due to the complex interactions between UV-B and other microclimatic factors. This is particularly true since parallel increases in greenhouse gases such as CO_2, methane, N_2O, etc., will lead to global climate changes including warming temperatures, changes in precipitation frequency and amounts, increases in atmospheric CO_2 concentrations, etc. However, the results of this study offer some insight by allowing the examination of the effectiveness of UV-B radiation under a range of environmental conditions. The number of precipitation events, air temperature, the number of days of low irradiance, and UV-B radiation all interact to affect crop yield. Although these various interactions are complex, linear models can be used to predict the combined effects of these factors on crop yield (Teramura and Sullivan, 1988). Models for both soybean cultivars tested in the six-year study include factors of temperature and the number of precipitation events in the season. The model for Essex, the UV-susceptible cultivar, includes total UV dose while the model for the resistant Williams cultivar does not. Both models are capable of predicting crop yield within 95% confidence intervals for the five-year period (1986 was excluded from analysis because artificial irrigation was used). Through the use of such models, it may be possible to more realistically assess the effects of increased levels of UV-B in concert with climatic changes than by a simple UV dose-response relationship. Such simple dose response relationships generally do not adequately explain field observations because of the complexity of interactions with other environmental variables. The interaction between UV-B radiation and potential climate changes needs further evaluation. Future changes in climate in concert with increasing levels of solar UV-B radiation reaching the earth could profoundly influence the productivity of soybean and other crops.

<u>Literature Cited</u>

Biggs, R.H., S.V. Kossuth and A.H. Teramura (1981) Response of 19 cultivars of soybeans to ultraviolet-B irradiance. Physiol. Plant. 53:19-26.

Caldwell, M.M. (1971) Solar UV irradiation and the growth and development of higher plants. In Photophysiol. VI (Edited by A.C. Giese), pp. 131-177.

Green, A.E.S., K.R. Cross and L.A. Smith (1980) Improved analytical characterization of ultraviolet skylight. Photochem. Photobiol. 31:59-65.

National Aeronautics and Space Administration (NASA) (1988) Executive Summary of the Ozone Trends Panel, NASA, Washington, D.C.

Teramura, A.H. (1986) Current risks and uncertainties of stratospheric ozone depletion upon plants. Risk Assessment for Environmental Protection Agency. 99 pp + appendix.

Teramura, A.H. (1983) Effects of ultraviolet-B irradiances on soybean. Plant Physiol. 65:483-488.

Teramura, A.H., R.H. Biggs and S. Kossuth (1980) Effects of ultraviolet-B irradiances on soybeans. II. Interaction between ultraviolet-B and photosynthetically active radiation on net photosynthesis, dark respiration, and transpiration. Plant Physiol. 65:483-488.

Teramura, A.H. and N.S. Murali (1986) Intraspecific differences in growth and yield of soybean exposed to ultraviolet-B radiation under greenhouse and field conditions. Environmental and Experimental Botany 26:89-95.

Teramura, A.H. and J.H. Sullivan (1988) Effects of ultraviolet-B radiation on soybean yield and seed quality. Env. Pol. 53(1-4): 466-468.

Whittaker, R.H. (1975) Communities and Ecosystems. MacMillan Co., New York.

Reducing Earth's Greenhouse CO_2 Through Shifting Staples Production To Woody Plants

Philip A. Rutter, MS
President, The American Chestnut Foundation
Vice President, Northern Nut Growers Association
Director, Badgersett Research Farm
RR1, Box 118, Canton, MN 55922

Abstract

Two systems of intensive food and fiber co-production using woody plants are proposed. Potential effects on atmospheric CO_2 are discussed, and could be highly significant if such systems were widely implemented. A maximum carbon fixation rate of 1.82×10^{13} g/ 10^6 hectares/ year is calculated, more than triple the average for maize. If all United States croplands planted to maize or soybeans in 1986 (55.5 million hectares) were planted to such woody crops, at least 1.01×10^{15} g carbon/year would be fixed, a large fraction of which could be sequestered for long periods of time or substituted for fossil fuels.

This paper considers a possibility which has so far been discussed little or not at all. Yet it is an option which might do much to ameliorate the increase of CO_2 in our atmosphere. It deserves careful examination.

I propose that we consider gradually shifting our agricultural system away from its current reliance on annual plants, and instead increasingly rely for foodstuff staples production on woody perennial plants. It may actually be possible to replace maize, rice and wheat with nuts and other kinds of fruits grown on woody plants. What I am suggesting is something substantially different from the commonly discussed "tree crops" concept of J. Russell Smith, which is concerned with basically traditional gathering of tree fruits; different from horticulture, which deals mostly with "luxury" crops; and different from "agroforestry" which calls for growing trees and crops together. The concept proposed here I call "woody agriculture": the intensive production of protein, carbohydrates and oils from highly domesticated woody perennial plants. Please note that I do not say "trees"— while one of the systems discussed here would use trees, others do not.

Why bother to contemplate such a sweeping change? Because a look at the distribution of solar energy available for photosynthesis each year clearly shows that annual crops cannot capture even 50% of it, whereas woody plants develop leaves very early, and are capable of capturing light and CO_2 throughout the growing season, even in cool weather (see figure).

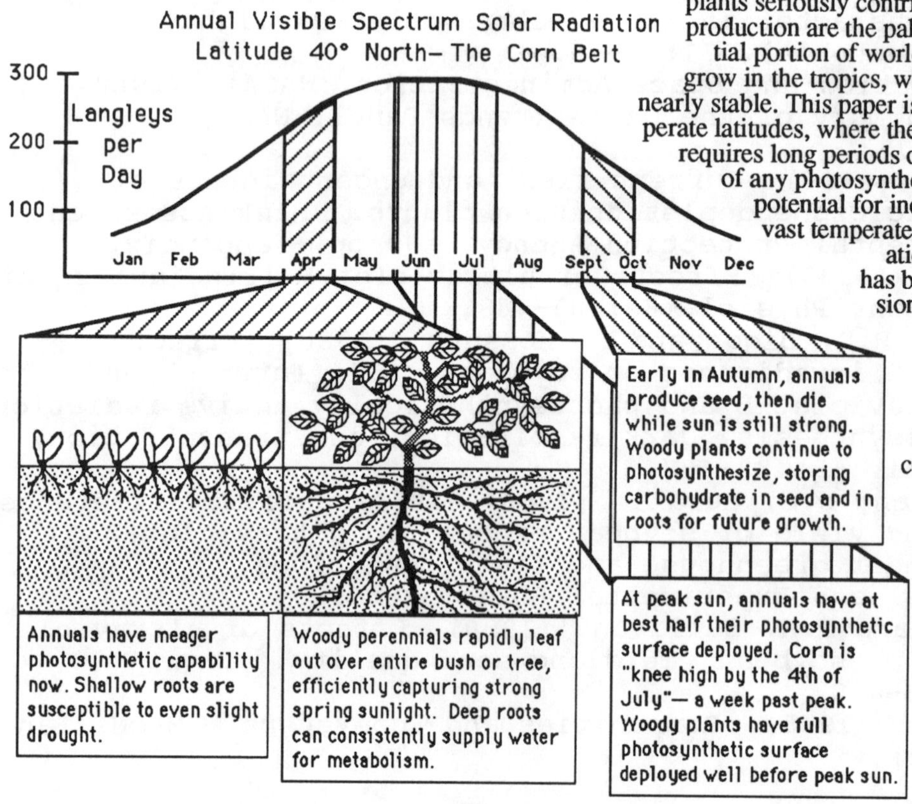

Our dependence on annual plants was inherited from our remote ancestors. There were excellent reasons why primitive peoples first beginning to rely on agriculture should utilize annual crops, (eg. the ability to harvest a crop one year after a migration to a new home site) but I contend those reasons have become obsolete. It may in fact be greatly to our benefit to begin a shift away from annuals and look to woody plants for the bulk of world food production. Currently, the only woody plants seriously contributing to international staples production are the palms, responsible for a substantial portion of world edible oils. Palms, however, grow in the tropics, where light and temperature are nearly stable. This paper is limited to discussion of temperate latitudes, where the culture of annual crops often requires long periods of time when fields are devoid of any photosynthetic potential, and because the potential for increased carbon fixation by the vast temperate land areas devoted for generations to annual crop production has been largely ignored in discussions of the global carbon budget *(14,15)*.

This paper seeks to convince the reader of the following points: that the benefits of woody agriculture could be immense; that woody agriculture could make a substantial contribution to control of atmospheric CO_2; and that all of the systems components for a woody agriculture either exist today or could be developed with current technology. Specific examples of woody crop systems that might be rapidly developed as staples are described below.

The quantitative estimates of the impact of extensive woody agriculture on atmospheric CO_2 offered here are simplistic, and can serve only to indicate the general magnitude of the effects. There are so many different factors interacting that complete estimates will require the attention of a specialist in mathematical models, which I am not. Several important beneficial effects are not estimated here.

Benefits Of Woody Agriculture

The benefits of using woody plants for agricultural purposes are many, and the reasons for shifting from annuals are compelling.

• Woody plants are capable of photosynthesis over a much longer portion of the temperate growing season. They do not, in general, photosynthesize over that entire time as efficiently as some crop plants can during the peak growing season, but the capability for photosynthesis over so much more of the growing season is very significant.

In addition to the phenomena described in the figures, even when deciduous trees have dropped their leaves the green twigs conduct photosynthesis whenever temperatures permit; contributing significant amounts of carbon fixation. As a measure of the potential photosynthesis: in April, when maize fields are bare, a wild oak forest in Minnesota had 10,300 grams/hectare of chlorophyll, in the oak twigs. By comparison, at its maximum in August, a field of maize contained only 13,000 g/ha, while the oak twigs and leaves then held 24,000 g/ha. In November, with the maize at 0 again, the oak twigs contained 7,000 g/ha *(3)*. Double-crop systems for annuals, such as the southern "corn belt" practice of winter wheat-soybeans, will do better than single-crop maize, but will still entail critical months where the photosynthetic potential of the field is very small.

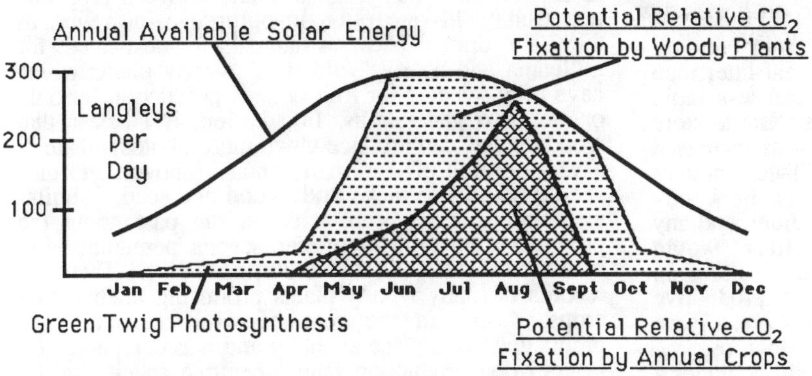

Woody plants, with their rapid early leaf deployment, multiple leaf layers, and longer growing season, capture significantly more solar energy than traditional annual crops. This means more, potentially much more, CO_2 fixed.

• Woody plants are by their nature more effective traps for light energy. Their more extensive and complex canopy structure, the early leafing, all lead to more thorough capture of available light. In a healthy forest, very little light penetrates to the ground; in a maize field, there are large amounts of unutilized light hitting the ground for most of the growing season.

• Deep rooted woody plants are able to sustain photosynthesis through moderate dry spells in ways that the initially shallow rooted annuals are unable to— when a young annual has to suspend photosynthesis because of a lack of water, established woody perennials will tap deeper water supplies and continue to fix CO_2.

As a result of the above three factors, temperate woody plants do lock much more carbon into biomass each year than temperate annuals. Illustrating the potential of woody plants, experimental stands of hybrid poplars bred for harvestable biomass have produced as much as 27.8 Mg(=10^6g)/hectare/year of dry, above-ground biomass *(9)* ; whereas the comparable figures for maize, including seed, average only 10-11 Mg/ha/year *(10)*. I believe it is justifiable to use a maximum attained figure for temperate woody biomass, which depended on optimized growing conditions, as the technology and genetics of such systems are relatively young, and are still improving.

Converted to carbon {0.5 X woody plant biomass, 0.43 X maize*(15)*}, this would be approximately 14 Mg carbon/ha/year for woody plants, and 4.5 Mg for maize; counting above-ground material only.

• Most of the carbon fixed by annual crops is cycled back to the atmosphere as CO_2 or methane within a year. Much (perhaps half) of the carbon fixed by woody plants, however, would be removed from the global carbon budget for a variable but relatively long time; decades or sometimes hundreds of years.

• Much of the below-ground biomass generated by deep rooted woody plants would be carbon removed from the yearly global carbon cycle for a very long time span. Unlike annuals, where most of the root mass is in the top layer of soil, tilled after the growing season, and rapidly converted to methane or CO_2, woody roots penetrate much deeper; 5 to 15 meters being common. Carbon used for constructing woody root systems would be locked out of the atmosphere as long as the plant remained alive, and after death of the plant, decaying root systems 5 meters deep would return very little carbon to the surface. Woody plant root systems vary greatly, but never contain less than 20% of the dry weight of the above-ground system, including leaves *(10)*. 30%, when fine rootlets are included, may be a conservative average. This would mean an additional 4.2 Mg/ha/year of carbon fixed.

Speculatively, it seems likely that some of the carbon used to form deep roots will ultimately be entirely removed from the biological carbon cycle. When deep roots die, their decomposition may result in the formation of carbonate ions, which would then readily migrate in groundwater. This carbon would then become part of the geological carbon cycle, perhaps to be deposited far from the original tree as calcite crystals, and locking the carbon out of the atmosphere on a geologic time scale.

• Some of the above-ground harvestable wood biomass produced as a by-product of staple food production would be used for energy generation to partially replace fossil fuels. In this way we would be recycling atmospheric CO_2 instead of constantly adding fossil carbon, slowing the build-up of greenhouse CO_2.

• Fossil fuels currently used in agriculture for yearly seed bed preparation and cultivation would be greatly reduced. With woody plants, it is probable that fields would need to be replanted only every 5-15 years, or even in some cases once in 100 years. It is possible fossil fuel use for maintaining woody agriculture plantings could be reduced by as much as one half from present requirements.

• The benefits of trees as carbon sinks are widely recognized, but at present trees are usually only considered for planting on marginal, ie. less productive, lands *(2)*. The ability to produce staple crops from woody perennials would enable us to put photosynthetic cover on our most productive and fertile lands; lands that are currently mere naked soil for much of the growing season. The amount of land benefiting from perennial cover could be greatly extended, and carbon capture very greatly increased.

In 1986 in the United States, the top ten maize producing states planted 23,836,000 hectares of maize; the top ten soybean producing states planted 19,862,000 hectares of soybeans; for a total of 43.7 million clean-cultivated hectares for just those two crops (fewer hectares were planted in 1987 and 1988, in an effort to reduce crop surpluses)*(8)*. At a rate based on the above calculations of theoretical potential, intensively cultivated woody plants would fix 1.4 X 10^{13} g carbon/10^6 ha/year above-ground and 0.42 X 10^{13} g carbon/10^6 ha/year in roots, or, for those 43.7 million hectares, 0.795 X 10^{15} g carbon/year. If all United States croplands planted to maize or soybeans in 1986 (55.5 million hectares) were planted to such woody crops, they would fix 1.01 X 10^{15} g carbon/year.

• Woody plants can produce far more leaf litter than annuals, and decreased tillage would mean less rapid oxidation in the upper soil layer: there is room to store substantial amounts of carbon in the soil as increased organic matter. Current cropping and tillage practices often deplete the organic content of soils. In woody agriculture, much of the abundant leaves, litter, and any woody material not harvested for fiber would eventually be incorporated in the soil, resulting in enriched, more water retentive, and more productive soils; and CO_2 removed from the atmosphere. (Some undisturbed forest soils typically have low organic contents, but woody agriculture soils would be under a completely different development regime, and might be more comparable to the highly organic prairie soils.)

How much carbon would it take to raise the organic content of the top 15 cm of the US cornbelt by 1%? The US Soil and Conservation Service estimates the weight of 15 cm of soil, at a bulk density of 1.3 (a reasonable average), to be approximately 2.24 X 10^9 grams/hectare. For the same ten maize and soybean producing states mentioned above, a 1% increase in soil organic matter for the top 15 cm of the 43.7 million hectares planted to those crops would require 98 X 10^{15} g of organic material. Using a conversion factor of 1.7 g of "organic matter" to 1 g of carbon, this is 57.6 X 10^{15} g carbon (1.3 X 10^{15} g carbon/ 10^6 hectare).

Increasing soil organic matter is a slow process, but considering the possibilities of adding more than just 1% organic matter to some soils, in woody agriculture fields around the world and in many climates, it is clear that this is a non-trivial storage potential. Carbon in organic material of temperate forest soils can have a "residency" of more than 100 years*(15)*. The soil could be a significant and highly desirable sink for some of the present atmospheric carbon surplus, providing systematic ways can be found to get the carbon into it. Policies encouraging woody agriculture would be one way.

Other Benefits

Besides improving the greenhouse CO_2 balance, there are other environmental benefits that would accompany the use of woody plants. The fact that the soil is not tilled on an annual basis would of course lead to greatly decreased soil erosion, both by rain splash and wind. Perennial cover holds water far better than annual systems by trapping moisture in the form of snow and fog, holding rainfall better, and allowing better penetration into the soil.

Woody agriculture would be far more sustainable than systems using annuals. Besides the above benefits to soil and water, the deep roots bring otherwise unavailable minerals to the surface for the enhanced nutrition of later generations- the leaf litter enriches and renews the soil in ways annuals do not.

Standard objections

Woody plants are seldom considered as having any serious potential for intensive agriculture because there are a series of unstated or unexamined assumptions about them which may in fact not be valid.

It is assumed that because woody plants put so much energy into wood, they cannot be as productive of seed as annuals. The comparisons made between crop and tree fruit yields are often inappropriate, however, as they are drawn between annuals domesticated for millennia and nearly "wild type" woody plants which have been selected for fruit or seed production for only one or two generations. In addition, it is clear that woody plants can produce much more photosynthate in a year per unit of land than annuals (above), and thus should be able to make both wood and seed. {Similar assumptions have been made in the past about the productive capabilities of herbaceous perennials, but measured yields belie the assumptions *(12)*.} There are records of many woody plants producing phenomenal crops *(4, 5)*. In the case of wild chestnut, crop production takes place annually and is accompanied by faster wood production than other tree species in the same forest *(6)*.

It is assumed that woody plants cannot be bred fast enough for staple crop needs because wild type woody plants usually require several years for each generation, sometimes up to 20 years. Precocity in woody plants, however, is not difficult to find. I have personally bred chestnut hybrids that have flowered 2-3 months after germination of the seed, producing useful amounts of pollen *(7)*. With an intensive effort similar to that now made for annuals, woody plants could certainly be bred fast enough to respond to disease challenges.

It is assumed that the usually high cost of clonal woody plantings could not be justified by the rather low prices received for staples. But the advent of plants from tissue culture, and the increasing potential for "artificial seeds" makes it likely that with economies of scale planting stock for woody agriculture could be very reasonably priced. Tissue culture also makes it possible to increase desirable genotypes fast enough for large scale needs. Hybrid poplar stands are currently established using unrooted cuttings *(9)*.

Requirements

For any suggested alternative form of agriculture to be worth considering, it must meet several specific requirements.

•It must be sufficiently recognizable to current farmers so that they can adopt it.

•It must be mechanizable. There is not enough hand labor in the entire world to pick the US maize crop. There must be ways to mechanize planting, care, and harvest.

•The mechanization requirement means that the crops themselves must be standardized— they must grow and ripen uniformly. For woody plants, this means there must be a way to produce and plant huge quantities of clonally propagated plantlets.

•There must be enough genetic malleability to the species being domesticated so that useful varieties can be identified and improved.

•There must be ways to very rapidly breed or otherwise produce new varieties resistant to newly evolved diseases.

•Production of a salable crop must occur within a very few years of planting, at most 3.

•The crop or its products must be stable in storage.

•The productivity, quality, and versatility of the new crops must equal existing crops.

Woody plants can almost certainly meet each of these requirements.

Specific examples

The paragraphs below show how, using the specific examples of chestnuts and hazelnuts, these requirements can be met, and why the usual objections may not apply to systems that are nearly available to us today.

I want to emphasize that though chestnuts and hazelnuts are discussed here, the kinds of agricultural systems being proposed are by no means limited to those two species: I use them for examples because I work with them, and because they may be close to actual utility. There are many other woody species which could with extensive but straightforward breeding be domesticated to the point where they would be suited for woody agriculture, in many latitudes and for many purposes *(4, 5, 6)*.

Could chestnuts and hazelnuts fit into current needs and markets? Yes. Chestnuts are comparable to maize for protein and feed value, being lower in oil but with higher quality protein— the limiting amino acid is isoleucine. When dried like maize, the nutritional value is stable. They are excellent animal feed and there are numerous traditional human culinary uses, ranging from soup to bread. Hazelnuts have good protein also, but have an oil content high enough to make them subject to rancidity in long term storage. Processing might involve pressing out the edible oil, and marketing the stabilized dried cake for both human and animal use. Both chestnuts and hazels also have considerable, though largely uninvestigated, potential as industrial feedstocks for products ranging from ethanol to plastics.

• The first specific proposed crop system involves harvesting chestnuts or hazelnuts on an alternate year basis, taking both nuts and a wood "biomass" crop every other year from established root systems. Envision a field, previously planted to maize, now planted solidly to a highly productive chestnut cultivar, plants being spaced about 1 m X 1 m. After a mechanized planting, the plants should take no more than 3 years to bear their first crop. At that time, they would be between 1.5 m and 2 m tall. In October, when the nuts are ripe but before they drop, such a field could be mechanically harvested with a combine, very much in the fashion of maize. A combine would strip nuts, husks, and some leaves from the branches, thresh out the nuts and blow the shattered husks back on the field.

A month later, after the woody stems and branches have gone completely dormant, the same field is harvested again; this time with a machine that cuts all the woody stems right down to the ground. This wood is chipped by the harvester, and sold as a biomass or fiber crop. The main stems of such plants are likely to be 3 to 6 cm in diameter, yielding chips large enough for chipboard production. The wood might also be extracted for salable chemical compounds (tannic acid is a distinct possibility) and ultimately put to use as fuel or pulp.

The year after harvest, the plants re-sprout from the established root system, growing 1 to 2 m high. Both chestnut and hazel seem to have the ability to re-sprout in this fashion for many years. <u>The second year after being cut to the ground, they bear nuts once more, and are harvested again.</u>

The field could be laid out in alternate strips, so that adjacent rows are in successive years of the rotation. This would increase protection for the soil and maximize edge effects for the bearing strips.

Benefits of cutting the wood to the ground every other year include:

Less concern about damage done to the wood by the nut harvesting machines, since the wood will shortly be harvested also.

Plants being harvested will always be the same size, simplifying machine handling.

Removal of old wood should reduce the need to control some diseases.

All pruning is eliminated.

A second salable crop is generated.

Marketing options are increased for the grower.

Could plants capable of such performance be found, or created? To a very limited extent, I have already grown a few multiple-species chestnut hybrids on my farm in Minnesota which have performed in just this way, bearing almost a kilogram of fresh nuts per root system *(7)*. Hybrid hazelnuts appear close to the same potential. It even seems possible that cultivars might be developed that would bear nuts each year, on new sprouts; this has already been accomplished with raspberries.

• A second type of crop system would use more conventional orchard technology, except that breeding would take the place of labor intensive pruning practices. Trees would be bred to take the approximate form of Lombardy poplars. Denser planting in early years would give way to wider spacing as the trees gained height, with an accompanying harvest of wood. With proper spacing to allow light penetration, such plantings would be very efficient collectors of solar energy, and even trees growing to heights of 50-60 feet would be able to fruit over their entire surface. The soil surface between trees might be planted to a tailored, shallow rooted perennial legume, which besides stabilizing the soil would fix some nitrogen for the crop system and also fix carbon when sufficient sunlight penetrates the overstory. Harvest would be by sweeping up fallen nuts or shaking them out of the trees; machines for both those tasks already exist. Large wild chestnut trees have already demonstrated the ability to bear heavily over the entire crown, year after year, so long as they receive full sun.

Advantages of this system are that:

Replanting might need to be done only once in 50-100 years.

If the system were used widely, it would allow very large amounts of carbon to be tied up for long periods of time.

The large stems of such trees would have multiple marketing possibilities, giving the farmer increased flexibility with his crop.

Pruning is genetically eliminated.

It is very similar to current orchard systems, making it more readily adopted.

Conclusions

Many other woody agriculture methods can be proposed. While the future of any such system is speculative, the potential ability of woody agriculture to remove CO_2 could be very great, and could constitute a major contribution to the eventual control of greenhouse gases. Not only is the present maximum carbon fixation rate of woody plants more than triple that of maize {1.82×10^{13} g carbon/10^6 ha/year above and below ground for woody plants*(9)*, *vs.* 4.8×10^{12} g carbon/10^6 ha/year for total maize field, including weeds*(10)* } but a large fraction of the carbon fixed in woody agriculture would not be immediately returned to the atmosphere, whereas most carbon fixed in annual agriculture is returned as CO_2 or methane within a year of fixation *(15)*. With some $1{,}500 \times 10^6$ ha under cultivation world wide *(15)* it is easy to see the potential impacts of woody agriculture could be tremendous.

How much carbon could the topsoil of temperate regions hold as organic matter from increased leaf litter? Soil will absorb 1.3×10^{15} g carbon/ 10^6 hectare for a 1% increase in organic content of a 15 cm soil layer.

These are simplistic initial estimates. For several important effects I have not yet prepared numerical estimates; the attention of a mathematical models specialist is needed. How much less fossil fuel would it take to grow crops that only had to be planted once every 5-10-100 years, instead of every year, or twice a year? How much fossil fuel could the woody biomass byproducts replace? How many forests could be left standing because fiber was being supplied by agriculture?

It should not be necessary to give up high crop productivity to achieve the desirable effects on atmospheric CO_2. The demonstrated ability of temperate woody plants to fix more than triple the carbon per year that maize can is an excellent indication that woody plants can be developed which would allocate a portion of that photosynthate to human-utilizable seed, in quantities at least equal to the present production of annual plants. Measured yields of woody plant fruit and seed, from systems and trees which in my opinion are primitive compared to the visible potential, already approach yields attained by annual crops *(4, 5, 6)*. It is entirely reasonable to believe that well developed woody agriculture systems would not be less productive than traditional agriculture, and an optimist might find reason to believe woody agriculture could actually be more productive than current practices.

Although this paper has been limited to considering temperate applications of woody agriculture, similar systems could certainly be developed for the tropics as well, and might be expected to utilize the extra heat and solar energy very efficiently. A well tailored tropical woody agriculture system might go a long way towards stabilizing the environment in those areas, once rain forest, now being so rapidly degraded by the need for constant tillage. Perhaps such systems could make more realistic the hope for "sustainable development".

No large woody agriculture system is going to appear soon; none will appear at all without increased appreciation of the benefits and a greatly enlarged research effort. Replacement of any substantial portion of annual crops could only be a gradual development; but the potential benefits are immense.

Woody agriculture is tenable today. The system component requiring the most development is the plants themselves, but there is no reason why suitable varieties cannot be bred. The Native Americans gave us the most compelling example of what transformations are possible through domestication. Starting with the wild grass teosinte, they created maize, an accomplishment modern workers have not equalled. Nothing similar has ever yet been attempted with woody plants, but the prospects are excellent. Woody plants are genetically rich, consistently containing much more variation than annual plants*(11)*.

Perhaps most importantly, woody agriculture represents a new option. And we are facing a future where *Homo sapiens* will assuredly need options.

Afterword and Acknowledgements: There are many considerations involved in the theory, rationale, and implementation of woody agriculture which cannot be discussed here because of space limitations. Those interested are encouraged to contact the author. I wish to express my deep gratitude to the following persons for their comments and suggestions for this paper: K. Davies, D. Egloff, R. Jaynes, R. Knutson, D. Lawrence, G. Miller, M. Widrlechner.

Bibliography

1. Neyra, Carlos A. 1985. Biochemical Basis of Plant Breeding, Vol. 1, Carbon Metabolism. CRC Press, Boca Raton, FL. 153 pgs.

2. Booth, W. 1988. Johnny Appleseed and the Greenhouse. *Science* , 7 October, pp. 19-20.

3. Ovington, J.D., & D.B. Lawrence. 1967. Comparative Chlorophyll & Energy Studies of Prairie, Savanna, Oakwood and Maize Field Ecosystems. *Ecology,* Vol. 48, #4, pp. 515-524.

4. Davies, Karl M. Jr. 1984. A Systematic Approach for

Indicating Potential New Perennial Crops for the Northeast. M.P.S. Special Project. Cornell University, Ithaca, New York.

5. Smith, J. R. 1953. Tree crops, a permanent agriculture. Devin Adair Co., NY.

6. —. 1980. Tree crops for energy co-production on farms. Symposium, US. Dept. of Energy, and Solar Energy Research Institute. National Technical Information Service. 259 pgs.

7. Rutter, P.A. 1987. Badgersett Research Farm— Plantings, Projects, and Goals. *Annual Report of the Northern Nut Growers Assoc.* pp.173-186

8. USDA National Agricultural Statistics Service, Annual Crop Summary, January, 1989

9. Heilman, P.E. & R.F. Stettler. 1985. Genetic Variation and Productivity of *Populus trichocarpa* T.&G. and its Hybrids. II. Biomass Production in a 4-year Plantation. *Canadian J. of Forest Research*, Vol. 15, #2, pp. 384-388.

10. Ovington, J.D., D.Heitkamp, & D.B. Lawrence. 1963. Plant Biomass & Productivity of Prairie, Savanna, Oakwood and Maize Field Ecosystems in Central Minnesota. *Ecology*, Vol. 44, #1. pp. 52-63.

11. Ledig, F. Thomas. 1986. Heterozygosity, Heterosis, & Fitness in Outbreeding Plants. *in:* Conservation Biology; the Science of Scarcity and Diversity, ed. Michael Soulé. Sinauer Assoc. Inc., Sunderland MA. pp. 77-104

12. Jackson, W. & M. Bender. 1984. Investigations into Perennial Polyculture. *in* Meeting the Expectations of the Land, eds. W. Jackson, W. Berry, B. Colman. North Point Press, San Francisco.

13. Edwards, G. & Walker, D. 1983. C_3, C_4; mechanisms, and cellular and environmental regulation, of photosynthesis. University of California Press. 511 pgs.

14. Detwiler, R.P., & C.A.S. Hall. 1988. Tropical Forests and the Global Carbon Cycle. *Science*, Vol. 239, 1 January, pp. 42-47.

15. Ajtay, G.L., P. Ketner, & P. Duvigneaud. 1977. Terrestrial Primary Production and Phytomass. *in* The Global Carbon Cycle. Bolin,B. *et al.* eds. John Wiley& Sons. New York. pp. 129-181.

FACTORS AFFECTING BIOLOGICAL METHANE PRODUCTION

C. C. Delwiche
Department of Land, Air & Water Resources, University of California, Davis, CA 95616.

Although the atmospheric concentration of methane is low, less than 2 parts per million (ppm), compared with Carbon dioxide at about 320 ppm, it plays a significant role in the atmospheric heat exchange processes. Much of it comes from biological sources, some from fossil fuel burning or venting, and some from combustion. Enormous (but poorly estimated) quantities of this gas are stored as gas hydrates or "clathrate" methane, bound in the crystal lattice of permafrost or in deep ocean sediments under high pressures.

In this type of gas hydrate storage, the quantity of methane contained per unit volume far exceeds storage in "solution" or as a gas. One liter of water (55 moles), when converted to ice, could hold as much as 164 liters of methane (at standard temperature and pressure) as the hydrate combination (Davidson et al., 1978). This theoretical limit probably is seldom reached, but the storage capability it implies is impressive.

Most atmospheric methane, whether of "fossil" or more immediate sources, probably originated in biological reactions. Diagenesis in deep sediments undoubtedly generates some methane, but the line between biological and non-biological sources is uncertain. Our concern here is with the possible contribution of agricultural activities to the budget of methane, and it is in the agricultural system, almost entirely biological, that we will examine the process. The underlying question is the extent to which changing agricultural practices (largely increased rice production) may be responsible for the increase in atmospheric methane that has been observed over the past few dacades.

In the absence of oxygen, some organisms can gain useful energy in the alteration of organic compounds, either by "fermentation" in which part of a molecule is oxidized and part reduced, or by other reactions in which some compound other than oxygen serves as "electron acceptor" for the oxidation of another compound with a yield of energy. In the biological formation of methane (methanogenesis), a process carried on by some of the most primitive and most remarkable microorganisms we know, there is the added complication that the cooperation of more than one species is involved, a process known as interspecies hydrogen transfer. Thus ethanol can be broken down hydrolytically to yield acetate ion and two moles of hydrogen according to the following:

1) $$CH_3CH_2OH + H_2O \dashrightarrow CH_3COO^- + H^+ + 2H_{2(g)}$$

At pH 7 this reaction yields sufficient energy to be useful for a microorganism until the hydrogen gas reaches a partial pressure of about 10^{-3} atmospheres, after which it is inhibited by the presence of the hydrogen.

The acetate ion can be broken down to methane and carbon dioxide by the following:

2) $CH_3COO^- + H^+ \text{--->} CH_4 + CO_2$

This reaction also is energy-yielding at pH 7, provided the concentrations of methane and carbon dioxide do not get too high.

But the hydrogen from reaction 1) above and the carbon dioxide of reaction 2) can be handed off to another methanogen that carries out the following:

3) $CO_2 + 4\ H_2 \text{---->} CH_4 + 2\ H_2O$

also with an energy yield.

The combination of these three reaction types, together with the alcoholic fermentation that preceded them, serves to break down a carbohydrate source completely to methane, carbon dioxide and water although not stoichiometrically, so other products can be expected to be formed.

Other organic compounds, including poorly soluble materials such as fats and waxes, chitin, and compounds with some complex ring structures, are not broken down effectively under anoxic conditions. Lignin, a major constituent of wood and plant cell walls in general, is not subject to complete breakdown, and for this reason, woody materials buried in anoxic muds can persist for millenia.

Lignin is a complex polymer serving to bind together the linear elements (cellulose) of the plant cell wall, giving woody tissues their characteristic rigidity and strength, serving a function much like the polymer binding glass fibers in a fishing rod. It is made up of repeating aromatic rings joined together with three-carbon chains and with methoxy substituents on the rings. Lignin is not water soluble, and it is extracted from the wood (as for example, in paper manufacture) only by the use of harsh solvents yielding degradation products of the parent molecule, among which are vanallin, isohemipinic acid and other methoxy-substituted aromatics. It is probable, but not proven, that methoxy groups from lignin make significant contribution to the methane yield in the microbial degradation of organic materials, leaving behind a more recalcitrant modified lignin surface.

With this brief overview of some of the chemistry involved in biological methane formation, it is now possible to look more carefully at a specific agricultural practice that is known to produce methane, sometimes in large quantities. The formation of methane (swamp gas) in water-logged environments has long been recognized, and the process has been examined in some detail,particularly in rice paddies, over the past several years by a number of workers. The results of some of these investigations are summarized in table 1.

Description	kg CH_4 per ha per year	Reference
Paddy Rice	540	Holzapfel-Pschorn & Siler (1986)
Paddy Rice	250	Cicerone, et al. (1983)
Paddy Rice	420	Cicerone & Shetter (1981)
Fresh Water Lake	470	" " "
Salt Marsh	57	DeLaune et al. (1983)
Brackish Marsh	970	" " " "
Fresh Marsh	2130	" " " "
Salt Marsh	531	King & Wiebe, (1977)
Tidal Estuary	210	Chanton & Martens, (1988)
Marine Sediment	733	Martens et al., (1986)

Table 1. Some values from the literature for methane production in various water-logged environments (recalculated or extrapolated to annual values where appropriate).

Rice does appear to be a major methane source under some conditions. Based upon their observations, Holzapfel-Pschorn and Seiler estimate an annual global production of between 70 and 170 Tg (Tg = 10^{12} g) methane per year for 1979 conditions, perhaps as much as one third of the total annual production of this gas. The available figures, of which those in table 1 are only a sample, emphasize that one can not pick an "average" figure for methane production under rice or any of the other environments, and to give a global estimate may be misleading.

Our own field investigations have yielded values for methane production varying over a wide range as influenced by a number of external factors. In order to evaluate the significance of these factors, we have conducted increasingly controlled studies, finally resorting to an "artificial" soil made up of sponge rock (pearlite) held in position by coarse sand, with all of the mineral elements coming from shelf reagents. On the basis of these studies and the work of others, it has been possible to make the following generalizations:

1. Methane formed under rice is from two sources:
   a) Organic matter there before the crop was planted or added organic matter.
   b) Organic matter having its origin in the present crop, probably mostly senescent root material.

2. The added or other exogenous organic matter is the major methane source, and can range from negligible to 1000 or more Kg carbon per hectare per year.

3. The maximum methane production from this source observed in our laboratory studies has accounted for approximately 5 percent of the added organic carbon.

4. Soil texture and water penetration rates influnce methane production.

5. An initial burst of methane production, shortly after the field is flooded, probably comes from methoxy groups of lignin and other sources that require only a modest lowering of oxidation potential.

Extrapolating from the limited data we have, one can ask the extent to which an increase in the area of rice production world-wide has increased the release of methane from this source, and thereby contributed to the greenhouse gas burden of the atmosphere. There is no quantitative answer available right now, but for several reasons, the answer probably is that there has been no large effect.

In the first place, rice culture practices vary greatly from one locale to another and have changed greatly in the past few decades. In many cases inorganic fertilizers have been substituted for the traditional organic sources. In these situations, methane production probably is less than it was when large organic additions were used.

When otherwise well areated soils are converted to paddy rice production and the field is water-logged for a significant part of the year, there probably is a general accrual of organic matter in the soil, with correspondingly less carbon released to the atmosphere as carbon dioxide. Methane is more effective as a greenhouse gas than carbon dioxide because of its much lower concentration and the lessened likelihood that its absorption bands are saturated. For this reason it is not correct to say that it does not matter whether carbon gets into the atmosphere as one molecule or the other. But since there is an overall greater storage of carbon in the soil under the conditions that make for methane production, the net greenhouse effect is not easily assessed. The uncertainties are great enough, and the stakes high enough, however, that a careful evaluation of the contribution of this and other agricultural practices to the total of greenhouse gases is needed.

The general rise in atmospheric levels of this gas has not been satisfactorily explained. Some of the reactions for methane production are carbon-dioxide dependent. One would expect that if the system is sufficiently tightly coupled, methane production would track carbon dioxide concentration. From a biogeochemical point of view, portions of the carbon cycle are exceedingly tightly coupled, and so this is not an unreasonable possibility.

With a general global warming, it would appear that the venting of fossil methane trapped as methane hydrates in permanently frozen organic soils might become a large methane source that would overshadow the contribution from rice fields. Which of these sources contribute most to the observed present increase is not clear.

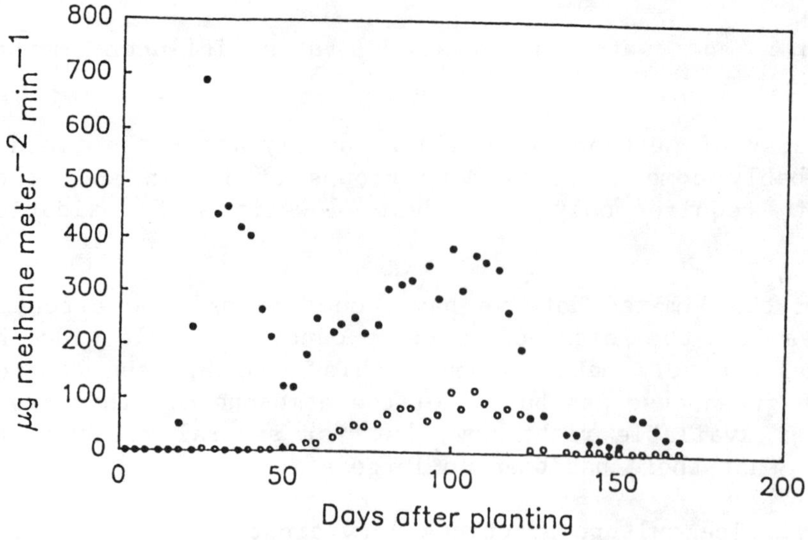

Figure 1. Effect of added organic matter on methane production: (●) = Ground rice straw added at rate of 8000 Kg. per hectare. (○) no added OM. Data from Delwiche and Cicerone, unpublished.

REFERENCES:

Chanton, J.P., and C.S. Martens, (1988), Seasonal Variations in Ebullitive Flux and Carbon Isotopic Composition of Methane in a Tidal Freshwater Estuary, Global Biogeochem. Cycles, 2: 289-298.

Cicerone, R.J., and J.D. Shetter, (1981), Sources of Atmospheric Methane: Measurements in Rice Paddies and a Discussion, J. Geophys. Res., 86:7203-7209.

Cicerone, R.J., and J.D. Shetter, (1983), Seasonal Variation of Methane Flux From a California Rice Paddy, J. Geophys. Res, 88:11022-11024.

Davidson, D.W., M.K. El-Defrawy, M.O. Fuglem, and A.S. Judge, (1978), Natural Gas Hydrates in Northern Canada, in Proceedings, Third International Conference on Permafrost, 1978, National Research Council of Canada, Ottawa, vol. 1, pp. 938-943.

DeLaune, R.D, C.J. Smith, and W.H. Patrick Jr., (1983), Methane Release From Gulf Coast Wetlands, Tellus, 35B:8-15.

Holzapfel-Pschorn, A., and W. Seiler, (1986), Methane Emission During a Cultivation Period From an Italian Rice Paddy, J. Geophys. Res., 91:11803-11814.

King, G.M., and W.J. Wiebe, (1977), Methane Release From Soils of a Georgia Salt Marsh, Geochem. et Cosmochim. Acta, 42:343-348.

Martens, C.S., N.E. Blair, C.D. Green, and D.J. Des Marais, (1986), Seasonal Variations in the Stable Carbon Isotopic Signature of Biogenic Methane in a Coastal Sediment, Science, 233:1300-1304.

GREENHOUSE GASES RELEASED TO THE ATMOSPHERE FROM DEFORESTATION FOR FARMLAND

R. A. Houghton

The conversion of forests to agricultural lands is estimated to have released between 0.4 and 2.5 x 10^{15} g carbon, as CO_2, to the atmosphere in 1980 (Houghton et al. 1985, Detwiler and Hall 1988). This release was approximately 10% to 50% of the annual emissions of CO_2 from combustion of fossil fuels. Most of the biotic release was from tropical regions. Outside the tropics, deforestation was approximately in balance with reforestation, and the net flux of carbon in these regions is believed to have been close to zero in 1980 (Melillo et al. 1988). The reason for the release of carbon following deforestation is that forests hold 20 to 100 times more carbon per unit area than agricultural lands. The carbon held in trees and in the organic matter of soil is oxidized following deforestation and released to the atmosphere. Part of this oxidation occurs rapidly as trees are burned; part of it occurs over years to decades as dead plant material, wood products, and soil organic matter decay.

The global release of carbon from changes in forest area cannot be measured directly. It is calculated from two types of data: data on the rates and types of land-use change, for example, deforestation, and data on the carbon stocks of terrestrial ecosystems.

<u>Rates of deforestation</u>. Only two studies have estimated the rate of deforestation for the tropics (Myers 1980 and FAO/UNEP 1981). When constant definitions are applied to these two studies, their findings differ by less than 10% over the entire tropics, although for individual countries the differences are sometimes greater than 100% (Houghton et al. 1985). They differ most in their estimates of change in the area of the fallows of shifting cultivation. The FAO/UNEP study reported the area in fallow forests to be increasing. Myers reported a substantial annual reduction (Houghton et al. 1985). The difference between the two estimates, and indeed the accuracy of either estimate, could be determined by analysis of existing satellite imagery. No systematic monitoring of changes in the area of forests using satellites has yet been initiated, however.

<u>Carbon in the vegetation and soils of tropical forests and the fate of that carbon following deforestation.</u> Estimates of the amount of carbon held in the vegetation of tropical forests differ by a factor of two. The higher estimate is based on the direct and destructive sampling of biomass in forest ecosystems. The lower estimate is based on the volumes of the growing stocks of wood (FAO/UNEP 1981). When the data on volumes are converted to units of biomass, the results are about half those determined from direct measurement. This discrepancy accounts for a large portion of the current range of estimates of the biotic flux.

The conversion of forests to agriculture generally leads to a reduction in the amount of carbon held in vegetation and soil. The rate of reduction depends on the ecosystems undergoing change. The loss of carbon from land is almost entirely as CO_2. Losses through erosion or as dissolved organic matter following disturbance are small relative to the loss as CO_2. Little of the carbon is believed to be released as methane. Even at the highest methane yields reported for termites, less than 1% of the total release would be as methane. The sources of information and data used to calculate the rates of release and accumulation of carbon on land following changes in land use have been documented elsewhere (Houghton et al. 1985, 1987).

Estimates after 1980. Recent estimates of the release of carbon from deforestation ($0.4 - 2.5 \times 10^{15}$ g/yr) are for the year 1980. No estimates have been calculated for the years after 1980, because there has been no comprehensive analysis of tropical deforestation since then. Rates of deforestation were projected to 1985 by the FAO/UNEP survey (1981), but the projections appear low for the only large region investigated since 1980, the Brazilian Amazon Basin. The FAO/UNEP study projected an average rate of deforestation for all Brazil of 1.9×10^6 ha yr^{-1} for the period 1980 to 1985. A recent estimate by Setzer and Pereira (1988) for the Brazilian Amazon alone found the rate in 1987 to be about 8×10^6 ha yr^{-1}, more than four times higher than the FAO/UNEP projection.

Whether the large increases measured in the Brazilian Amazon are typical of other parts of the tropics is not known. If they are, the current net release of carbon could be as much as four times higher than the release in 1980, or between 2 and 10×10^{15} g annually. It seems unlikely that deforestation in the rest of the tropics is keeping pace with Brazil. Current rates of change in the area of forests are so poorly known, however, that high rates cannot be ruled out. The ignorance is remarkable given the importance of deforestation to climatic change and the demonstrated ability of existing satellite data to measure the change in forest area.

Projections to the Year 2100. In order to estimate how significant the biotic flux might be in the future, projections of tropical deforestation and its associated release of carbon were computed to the year 2100. Many such projections have been made for the emissions of carbon from fossil fuels. However, the assumption of a logistic curve, appropriate for the extraction of fossil fuel from increasingly inaccessible deposits, is probably inappropriate for projecting rates of deforestation. Forests are generally deforested at their edges, and as the area of a forest declines, the ratio of edge to area increases, as will the number of farmers and ranchers at these edges.

Three projections to the year 2100 are described below. In the first projection (linear), the change in the average annual rate of deforestation between 1980 and 1985 (FAO/UNEP 1981) was assumed to continue to the year 2100. Under this assumption, the total rate of deforestation increased from 11.3 X 10^6 ha yr^{-1} in 1980 to 15.8 X 10^6 ha yr^{-1} in 2079, when the available area of closed forests in Latin America was eliminated. Using the higher estimates of biomass (see above), the annual net flux of carbon in this projection remained between 2.4 and 2.8 x 10^{15} g until the Latin American forests disappeared (Fig. 1). The total net flux over the period 1980 to 2100 was 288 x 10^{15} g C, more than twice the release that occurred in the 130-yr period preceding 1980. With the lower estimates of biomass, the annual net flux remained near 1.0 x 10^{15} g, and the total 120-yr flux was 122 x 10^{15} g C, a value about equal to the release over the previous 120-yr period. The comparison between the pre- and post-1980 periods is not strictly comparable because the projections for the future did not include changes outside the tropics. Reconstruction of changes between 1850 and 1980 included the temperate zones as well as the tropics.

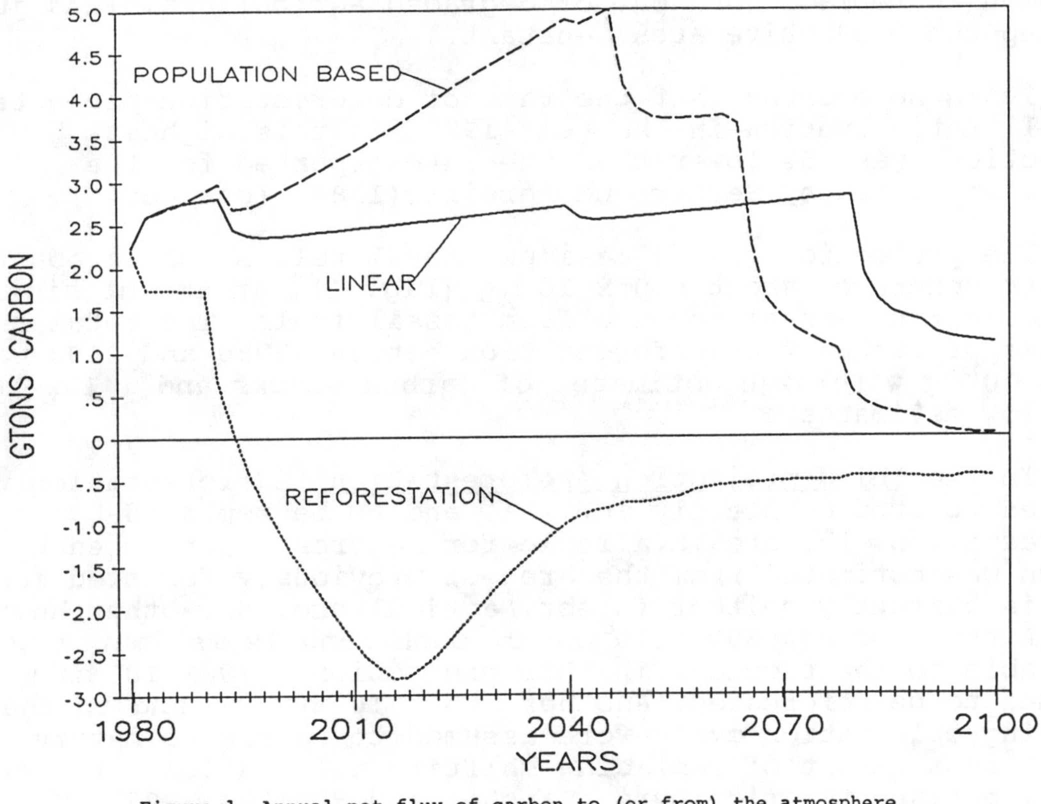

Figure 1. Annual net flux of carbon to (or from) the atmosphere as a result of projected rates of deforestation (or reforestation). GTONS = 10^{15} g.

In the second projection (population growth), deforestation was assumed to be a function of the rate of growth of populations and was projected, between 1981 and 2100, on the basis of population projections by the U.S. Bureau of the Census and the World Bank. These projections show a global population of 13.5 X 10^9 in 2100. Rates of deforestation in each of the three tropical regions were increased or decreased as rates of population growth (%/yr) increased or decreased relative to 1980. The rate of deforestation increased to about 29.7 X 10^6 ha yr^{-1} in 2045, when the area of forests in Asia was exhausted. Within the next 30 years all tropical forests had been eliminated. Thus, under this projection, the world's tropical forests were gone in less than 100 years.

The rationale for this projection was that expanding numbers of people will require an expansion of agricultural resources, at least some fraction of which will be provided by an increased area of crops and pastures. Advances in biotechnology and social mechanisms of food distribution may, of course, alter this projection. On the other hand, much of the deforestation in past decades has been by people outside economic markets, people who can purchase neither food nor technology, but who must use any land available for subsistence. Furthermore, some deforestation has been to replace worn out or degraded agricultural land just to keep the productive area constant.

It is noteworthy that the rate of deforestation projected for all Latin America in the year 1987, in this highest projection, was 25% lower than the rate reported for the Brazilian Amazon by Setzer and Pereira (1988) (see above).

The projection gave a maximum annual release of carbon to the atmosphere of about 5.0 x 10^{15} g (Fig. 1), an amount similar to the current annual release from fossil fuels. The total net release of carbon from deforestation between 1980 and 2100 was 334 x 10^{15} g with high estimates of carbon stocks and 141 x 10^{15} g with low estimates.

In the <u>third projection (reforestation)</u>, deforestation was assumed to stop completely in 1991, and to be replaced by massive reforestation. The area available for reforestation in each region was estimated from the area of previously forested land that is currently neither forest, agriculture, nor other human use. A total of 500-600 X 10^6 ha of such land is estimated to be available in the tropics. In this projection, 520 X 10^6 ha were assumed to be reforested. Another 365 X 10^6 ha of land in the shifting cultivation cycle were assumed to regrow to mature forest as a result of replacing shifting cultivation with low-input, permanent agriculture (Sanchez and Benites 1987). The total area reforested over the 110-year period 1990-2100 was thus 885 X 10^6 ha. The area is approximately 38% of the area of tropical forests (including fallows) in 1985 (FAO/UNEP 1981). This projection assumed no expansion of agricultural area into forests after 1990 and, hence, requires a substantial increase in yields to supply food for expanding populations.

The rates of reforestation were arbitrarily chosen, but the annual withdrawal of carbon from the atmosphere reached values as high as 2.8 x 10^{15} g with high estimates of biomass (Fig. 1). The total flux of carbon, including both the release from deforestation and the accumulation from reforestation over the 120-yr period, was a net accumulation of carbon on land of about 100 x 10^{15} g for high estimates of biomass, and 40 x 10^{15} g for low estimates.

In summary, future policies can lead either to the complete elimination of tropical forests in the next century, with a release of as much as 330 x 10^{15} g C in addition to that released from fossil fuels, or to an expansion of forest area and a withdrawal of as much as 100 x 10^{15} g C from the atmosphere. The latter alternative would contribute significantly to stabilization of greenhouse gases in the atmosphere.

REFERENCES

Detwiler, R.P. and C.A.S. Hall. 1988. Tropical forests and the global carbon cycle. Science 239:42-47.

FAO/UNEP. 1981. Tropical Forest Resources Assessment Project. Forest Resources of Tropical Asia, Forest Resources of Tropical America, and Tropical Forest Resources of Tropical Africa. FAO, Rome.

Houghton, R. A., R. D. Boone, J. M. Melillo, C. A. Palm, G. M. Woodwell, N. Myers, B. Moore and D. L. Skole. 1985. Net flux of CO_2 from tropical forests in 1980. Nature 316:617-620.

Houghton, R. A., R. D. Boone, J. R. Fruci, J. E. Hobbie, J. M. Melillo, C. A. Palm, B. J. Peterson, G. R. Shaver, G. M. Woodwell, B. Moore, D. L. Skole, and N. Myers. 1987. The flux of carbon from terrestrial ecosystems to the atmosphere in 1980 due to changes in land use: Geographic distribution of the global flux. Tellus 39B:122-139.

Melillo, J.M., J.R. Fruci, R.A. Houghton, B. Moore, and D.L. Skole. 1988. Land-use change in the Soviet Union between 1850 and 1980: causes of a net release of CO_2 to the atmosphere. Tellus 40B:116-128.

Myers, N. 1980. Conversion of Tropical Moist Forests. National Academy of Sciences Press, Washington, D.C.

Sanchez, P.A., and J.R. Benites. 1987. Low-input cropping for acid soils of the humid tropics. Science 238:1521-1527.

Setzer, A.W., and M.C. Pereira. 1988. Amazon biomass burnings in 1987 and their tropospheric emissions. Manuscript.

THE RELATIONSHIP OF GLOBAL CLIMATE CHANGE TO OTHER AIR QUALITY ISSUES

J. Christopher Bernabo
Science and Policy Associates, Inc
Landmark Building - Suite 400
1333 H Street N.W.
Washington, D.C. 20005

INTRODUCTION

Climate is the primary environmental variable influencing natural systems and human activities. Changes in the extremes, variability, and mean conditions of climate will affect a series of existing air quality concerns including acid deposition, attaintment of National Ambient Air Quality Standards (NAAQS) particularly for ground-level ozone, and the depletion of stratospheric ozone.

These individual issues are linked by a complex set of relationships between emissions, atmospheric processes, and environmental consequences. Policies to address any one of these issues could influence the other problems because of the physical and societal linkages that exist among them. These linkages must be better understood in order to develop sound approaches to this ensemble of related issues.

EMISSION RELATIONSHIPS

The emissions of many air pollutants are climate sensitive. The amounts of SO_2, NO_x, CO and VOCs are influenced by climatic conditions in the source regions. For example, if a cold winter increases the combustion of heating oil in a region, the related emissions would rise. Similarly, hotter summer temperatures cause greater demand for air conditioning and thus electricity; if coal is used to produce the electricity then SO_2 and NO_x emissions could increase during that season.

Transportation emissions (NO_x, CO, VOCs) also are climate dependent due to the affects of weather on combustion efficiency and air conditioning usage. VOCs released from fuel tanks, solvent use, and chemical plants are more volatile in higher temperatures; therefore, warmer conditions can cause elevated levels of VOCs. In fact, the majority of emission sources and processes are affected by weather conditions, and thus are modulated by any changes in climate.

In addition to these traditional air pollutants (SO_2, NO_x, CO, VOCs), the emissions of several major greenhouse gases are temperature dependent. For example, greater use of fossil fuels in response to warmer climate could result in more CO_2 emissions.

CFCs emissions not only can destroy stratospheric ozone but also they are potent greenhouse gases. Depletion of strato-spheric ozone will increase the penetration of UV-B radiation and could further add to tropospheric warming.

Natural sources of emissions also are sensitive to climatic conditions. The rate of CH_4 released from permafrost regions, nitrogen emitted by soil microbes, and VOCs derived from vegetation are all temperature dependent.

ATMOSPHERIC PROCESSES LINKAGES

Climatic change alters the basic weather features that control the transport, chemical transformation, and eventual deposition of all air pollutants. A region's climate is defined by the average condition of variables such as temperature, precipitation, humidity, cloudiness, and winds. When the Earth's climate changes, it results in shifting patterns, extremes, frequencies, and mean conditions of the regional weather.

Stratospheric ozone depletion and tropospheric ozone pollution are linked through a series of complex atmospheric processes. A decrease in stratospheric ozone will result in an increase in UV-B radiation reaching the Earth's surface. This increase in UV-B irradiance will enhance such photochemical reactions as the production of tropospheric ozone and hydrogen peroxide, which also plays an important role in the formation of acid deposition. In addition, warming-induced emission increases of ozone precursors (NO_x, VOCs) from natural and man-made sources could further exacerbate the problem of ambient air quality.

The influence of prevailing weather on the patterns of air pollution can even obscure the distribution of sources. As shown in Figure 1, the annual pattern of ozone concentrations for the United States can be very different between two consecutive years. This condition does not reflect a shift in emissions, but rather was the direct result of different weather patterns.

Overall, climatic changes will have significant effects on pollution levels because of the different types and frequencies of weather that will prevail in various regions. These complex atmospheric relationships are just beginning to be understood.

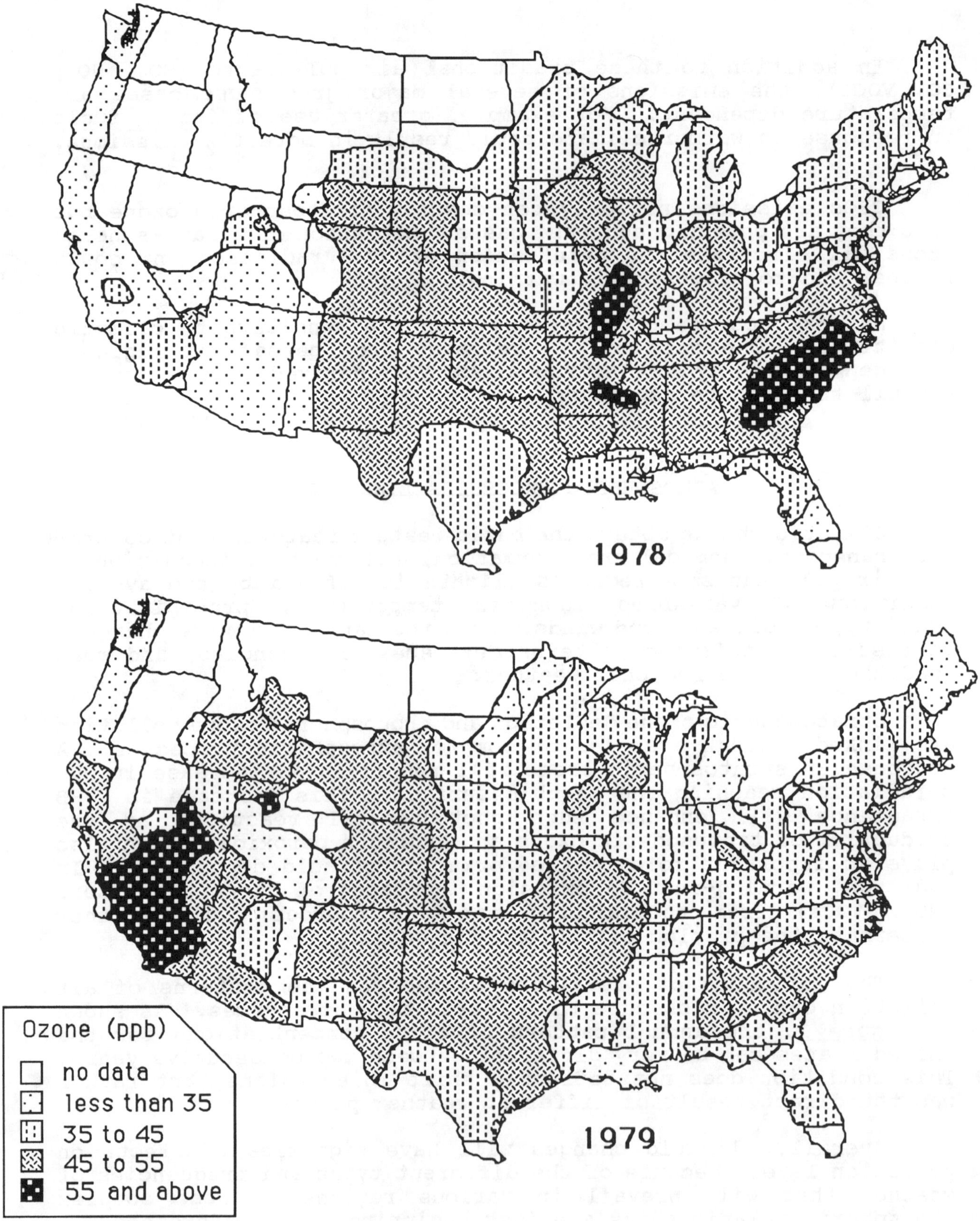

FIGURE 1. Effects of meteorology on pollution patterns: contrasting years of ground level ozone concentrations in the United States.

ENVIRONMENTAL EFFECTS RELATIONSHIPS

The effects of various air pollutants on human health, vegetation, aquatic systems, and man-made building materials are climate dependent to different degrees. Synergistic effects also may exist between the responses of different pollutants under various temperature and moisture conditions. For instance, the level of ozone that causes damage to a leaf is related to the amount of heat or moisture stress at the time of exposure.

The sensitivity of certain materials to acid deposition and other air pollutants is influenced by the surface temperature and time of wetting. UV-B radiation also plays an important role in the degradation of certain materials.

To study the effects of pollutants on sensitive systems, researchers try to hold major climate-related variables constant. Understanding the complex interactions of climate- and pollution-induced stresses will require more sophisticated experimentation and modeling. Simple extrapolation of pollution effects assuming a fixed climate will not suffice in the future.

FUTURE AIR QUALITY

Over the next century climate will change due to some combination of natural and anthropogenic causes. Man's activities are increasing the concentrations of greenhouse gases and that may produce a warming in mean global temperatures. However, the timing, magnitude, and environmental consequences of the anticipated global warming still are uncertain.

Global climate changes will influence man-made and natural emissions, atmospheric processes, and environmental as well as societal systems. Climate acts to modulate other air and water quality issues, each of which will be affected to varying degrees by changing conditions in the atmosphere. Acid deposition, attainment of national pollution standards for ground-level ozone, and depletion of stratospheric ozone each are linked to climate through temperature, precipitation, humidity, solar radiation, cloudiness, etc.

Past assumptions that climate is a long-term constant are incorrect and will lead to unreliable predictions of future pollution levels. Numerous air pollution - climate relationships exist and could result in complex and unexpected interactions in the future. A better understanding of the physical and societal linkages between these air quality issues are needed before the next generation of more holistic pollution control strategies can be developed.

HOW MUCH CERTAINTY IS ENOUGH ?

The question of whether we have enough scientific certainty to undertake specific societal actions to address the issue is a value judgement not a technical conclusion [Bernabo, 1986; Schneider, 1989]. There is no end point in science called regulate. The policy debate on what to do, how and when, therefore must include broad participation from governments, industries, environmental groups, international organizations and the public.

Scientists play a crucial role in improving the information base for policy decisions by contributing their objective knowledge to the policy debate. If they choose to go beyond science and advocate policy actions, it should be understood that any policy recommendations made reflect their personal values in interpreting the technical information.

Science alone does not provide the answers to policymakers' questions because science is mute on the values that guide the decisions societies make. Social, economic and political factors dictate how societies choose to respond to environmental issues.

People, including scientists, can honestly disagree on what the significance of an issue is for society and how we should respond. Ultimately, our legislators have the responsibility for representing our societal values and arbitrating between the inevitably competing interests.

CITATIONS

Bernabo, J.C. (1986) " Science and Policy: Notes from a Former Congressional Fellow", EOS, Transactions of the American Geophysical Union 67, 82.

Schneider, S.H. (1989) "The Greenhouse Effect: Science and Policy", Science 243, 771-781.

WORLD RESOURCES INSTITUTE
A CENTER FOR POLICY RESEARCH

1709 New York Avenue, N.W., Washington, D.C. 20006, Telephone 202-638-6300

Linkages Between Climate Protection and Air Quality Strategies

Dr. William R. Moomaw, Director
Climate, Energy and Pollution Program

There is a sense that we are being overwhelmed by an overabundance of local, regional and global problems. The traditional policy response of waiting for each to become sufficiently serious before it is addressed on an individual basis is no longer a viable approach.

Because the atmospheric problems of global climate change induced by an accelerating greenhouse effect, urban and regional air pollution, acid deposition, and stratospheric ozone depletion are interlinked both scientifically and economically, it is necessary that proposed solutions recognize this fact and be designed appropriately. Unless this is done, we are not only likely to exacerbate one problem as we attempt to solve another, we may also miss cost effective options which only become apparent when an integrated policy analysis is undertaken.

It is possible to begin with any one of the atmospheric issues as a starting point and examine its relationship to each of the others. Because it is the most all encompassing, I have chosen in this paper to utilize global climate change as the central issue in this analysis and to relate the other problems to it.

Looked at from the perspective of the infrared heat trapping gases involved, about 50 percent of the increase in global warming arises from carbon dioxide, approximately 18 percent each from methane and chlorofluorocarbons (CFCs), perhaps 8 percent from tropospheric ozone, 6 percent from nitrous oxide and a small contribution from other sources.

Policy is not, however, based upon chemistry, but rather on economic sectors. When seen in this light, three quarters of the carbon dioxide comes from the energy sector (the remainder is principally from deforestation and agricultural practices), but additional energy sector activities contribute methane, ozone and nitrous oxide to raise the energy sector contribution to approximately 57 percent according to an analysis carried out by the U.S. Environmental Protection Agency.

It is readily apparent that combustion of fossil fuels is also primarily responsible for acid deposition and the urban and regional smog that adversely affects human health and damages crops and forests. What is not so immediately obvious are some of the strategies that might simultaneously address these problems.

Tropospheric ozone arises from atmospheric photochemistry. The ingredients are volatile organic compounds (VOCs) and oxides of nitrogen (NOx) that arise during combustion in vehicles, electric generating stations and to a lesser extent from industrial and domestic heat production. NOx is also a major precursor of acid deposition. From a global warming perspective, ozone may account for only 8 percent of the problem, but there are several considerations that suggest it may be more important than that.

First, there exist feedback mechanisms that link increased global warming to higher ozone production. This ozone increase is expected to occur partly because of increased electricity production to meet greater demand for air conditioning which can be expected to generate more NOx and other air pollutants. More important is the strong correlation between ozone production and higher temperatures. It is expected that the number of heat waves is likely to increase sharply in many urban areas even for a relatively small rise in global average temperatures creating ideal conditions for greater smog and ozone formation.

Second, unlike other contributors to global warming, once formed, ozone has an atmospheric lifetime of only hours to days rather than decades to a century. Therefore eliminating the precursors of ozone, brings about the simultaneous and immediate reduction in a greenhouse gas and a source of air pollution and acid rain. In other words, a conventional clean air strategy can have benefits for climate change as well. Not all clean air strategies will however achieve these simultaneous benefits. For example, the proposal to replace gasoline with methanol to reduce conventional air pollutants will produce large quantities of extra carbon dioxide if that methanol is produced from coal.

A similar argument can be made regarding the importance of considering methane controls in multiple contexts. The natural atmospheric sink for methane is its reaction with hydroxyl radical. Because large quantities of a non-greenhouse gas, carbon monoxide, that come largely from incomplete combustion in automobiles, compete for the highly reactive hydroxyl radicals, concentrations of methane are higher than they would otherwise be. Once again, an appropriate, conventional air pollution control strategy, this time for carbon monoxide, could address both local air pollution, climate change, and to a lesser degree stratospheric ozone depletion.

A final argument that favors linking policies for global climate change and conventional air pollution is that a framework of laws currently exists to address the latter, while it may take some time to develop a similar body of law for climate change. There is already a major impetus to strengthen clean air standards particularly in non-attainment areas of the United States such as Southern California and in Arizona and Colorado. Citizens are far more likely to demand that their local air quality be improved immediately and galvanize their political leaders to do something about it than they are to mobilize to address a long term issue such as climate change. Hence it may be possible to begin slowing global warming sooner using this strategy than to wait for national greenhouse policies to be developed and international agreements to be completed.

While the examples considered here suggest the potential for simultaneously addressing more than one atmospheric issue at once, many other multiple benefit solutions are conceivable. Several possible strategies are listed in the following table.

STRATEGIES FOR ADDRESSING GLOBAL WARMING

Near-term Policies

1. Energy efficiency - Acid deposition, air pollution, energy security, competitiveness, jobs.
2. Rapid CFC phase-out - Stratospheric ozone
3. Switch to natural gas - Acid deposition
4. Control CO and NOx - Air pollution
5. Reduce methane release - Air pollution, stratospheric ozone
6. Utility regulatory reform - Air pollution, acid deposition
7. Reforest - Air pollution, social benefits
8. Population stabilization - All areas

Long-term Policies

1. Develop renewable energy sources - All areas
2. Reexamine the role of the next generation of nuclear power - All areas
3. Retrofit existing dams for hydro power before developing new projects - All areas

THE INTERACTION OF AIR POLLUTION PROGRAMS AND GLOBAL CLIMATE CHANGE

Jerry Emison, Director
Office of Air Quality Planning and Standards
U.S. Environmental Protection Agency

It is a pleasure to have an opportunity to discuss the air pollution program in this country and the interaction that exists between air quality management and global climate issues. The relationship is significant and far-reaching with the potential for benefiting both problems. The air program has been forced to adopt dramatic changes over the past two decades and we welcome this new challenge.

Planning and implementing the air pollution program in this country centers around the concept of air quality management (AQM). This is a highly analytical and flexible process that involves selection of specific goals from an identified array, evaluation of alternative mitigation strategies, field implementation, measuring progress and formal reassessment, and it has served us well. Despite significant growth in population, urbanization, vehicle miles travelled and change in the industrial base over the past 20 years, we have made noticeable air quality gain. For example, average levels of particulate matter in ambient air are down by 10%; SO2 by 35%; lead by nearly 90%.

The concept of AQM and its wide acceptance in air pollution surprises many people not associated closely with the air program. There has been a tendency to categorize new pollution as a somewhat simple engineering problem involving application of control technology to obvious sources. Actually, AQM from the start has been forced to work well beyond the physical characteristics of the problem and to deal with broad societal programs and goals. There are many compelling examples of this. One of the early and major challenges in the Clean Air Act of 1970 was to evaluate industrial growth (new point sources) to prevent significant deterioration. Then came transportation control measures, inclusion of urban growth in plans to maintain the ambient air quality standards, exhaust inspection programs for individually owned automobiles, often in conjunction with safety inspection, application of economic sanctions involving new construction permits and highway funds and more recently a focus on consumer products for control of toxics and ozone. The interplay of environment and economic growth, land use, transportation systems and energy management is not new territory for air pollution.

I noted earlier that we welcome the new challenge represented by new issues such as climate, stratospheric ozone depletion, acid deposition, and critical loadings of NOx. Actually, the air program needs the broad-based public interest and concern being given to these emerging problems and the new thinking and initiative being shown. The cost of conventional air pollution is skyrocketing. Control for volatile organics often reaches $2,000 a ton, and proposals in the Los Angeles basin call for $8,000-10,000 per ton; total national expenditures for air pollution are estimated at $30 billion per year, over $300 million is spent each year simply to administer State and local air agencies. Recent emphasis on fine particles, acid aerosols, visibility, ozone non-attainment, risk assessment and toxics are complex and resource-intensive. Narrow, single problem analyses and solutions just will not be accepted with problems of this magnitude.

Correspondingly, the billions of dollars this country has invested in the air quality management system over the past two decades has much to offer to the attack on emerging problems. A few examples:

- National monitoring networks and sophisticated data systems are in place and well understood throughout the country.

- An array of dispersion and atmospheric chemical models are in use with field personnel available.

- A well-established and broad-based research program totaling greater than $125M annually is underway including health and welfare effects, engineering, atmospheric reactions, analytical methods and risk assessment. This effort is enhanced by a variety of scientific advisory bodies and peer review groups.

- A strong infrastructure has been developed including a 10,000 member professional society, graduate programs at more than a dozen universities and good dialogue with a variety of interested and knowledgeable citizens' groups, environmental organizations and trade associations.

- Comprehensive and technically trained field organizations implement the program. These represent 20 years of experience in air monitoring, source emissions, standard setting and enforcement and a total of over 8,000 people in ten EPA Regional Offices and agencies in all States and over 100 cities.

We have discussed briefly the opportunity presented by the marriage between the more conventional and established air effort and the attack on emerging issues such as climate modification. However, a cautionary note must be mentioned. Literally thousands of legally binding air emission regulations are applicable to many of the same sources being discussed in strategies to mitigate these emerging problems. The slate is not clean. Existing standards frequently are based on well-accepted health and welfare mandates that are independent of the current considerations such as climate or acid deposition. These regulations are well understood and established and must be integrated fully into any mitigation strategies.

I have tried to make several points here today:

1. The air quality management program in this country has a long history in interacting effectively with broad societal programs and goals.

2. As the cost and impact of conventional air pollution efforts increases, we must align ourselves with related programs.

3. The nation's investment in air pollution infrastructure can contribute much to the attack on these emerging issues.

4. Existing air pollution regulations must be integrated fully into these several programs.

Fortunately, I believe that the need and opportunities are being recognized and that real interaction is occurring. Some examples: dialogue on the Clean Air Act amendments is very broad-based with much emphasis on energy conservation; we are reviewing the form of all air regulations relative to pollution prevention, energy efficiency and climate impacts; an Ozone Futures Study is underway looking at the potential impact of climate and stratospheric ozone on future emissions and atmospheric reactions; State and local air agencies have established a Global and Regional Air Quality Effects Committee; pollution prevention is a new ethic throughout the Agency designed to enhance consideration of all environmental goals in the implementation of specific media programs.

It is clear that current activities can make a difference. We are talking about the same limited air resources. It is vital that we recognize our mutual interests and our technological and philosophical similarities and find ways to optimize and integrate the entire atmospheric effort.

LIKELY EFFECTS OF GLOBAL CLIMATE CHANGE ON FISH ASSOCIATIONS OF THE GREAT LAKES

Henry A. Regier[1], John J. Magnuson[2], Brian J. Shuter[3], David K. Hill[2], John A. Holmes[1], J. Donald Meisner[1]

[1] Department of Zoology, University of Toronto
Toronto, Ontario M5S 1A1

[2] Center for Limnology, University of Wisconsin Madison
Madison, Wisconsin 53706

[3] Ontario Ministry of Natural Resources
Maple, Ontario L0J 1E0

ABSTRACT

In the Great Lakes Basin, climate warming will likely lead to a major expansion of waters that are quite warm in summer, to the advantage of the centrarchid family (black basses, sunfish, etc.) and the percichthyid family (white bass, white perch, etc.). There will also be a major expansion of waters that attain the cool range during midsummer, especially in the northerly half of the Basin, to the advantage of the percids (walleye, yellow perch, etc.). In many parts of the Great Lakes, nearshore and stream waters will become too warm in summer for the salmonids (lake trout, brook trout, Pacific salmons, lake whitefish, lake herring, etc.), but optimal temperatures will appear earlier in spring and extend later in fall.

If other environmental factors are optimal, and if the fish species can freely rearrange themselves geographically and hydrographically, then climate warming should lead to an increase in overall production of preferred species. But a number of ecological mechanisms have been identified that may involve some degrading synergism between warming and other factors of the habitat. Some pessimism in this respect would now be prudent. Attention should now be directed to the development of appropriate adaptive fishery strategies to climate warming of Great Lakes waters.

A. INTRODUCTION

At the 1987 Conference of The Climate Institute we reported (Regier et al. 1988) on likely impacts of climate warming on Great Lakes fish based on early studies (e.g. Meisner et al. 1987, 1988). We and others have made further scientific progress since then mostly because of the following initiatives:

- joint ecological research (Magnuson et al. 1988) in which we were aided by hydrological research (Blumberg and DiToro 1989, McCormick 1989) all supported by the U.S. Environmental Protection Agency (see also Smith 1988);
- a Symposium on Climate Change and Fisheries convened under the auspices of the American Fisheries Society in Toronto in September 1988 (Hill and Magnuson 1989, Holmes 1989, Meisner 1989, Regier et al. 1989, etc.);
- a Symposium on Climate Impact on the Great Lakes organized by Dr Stanley Changnon of the Illinois Natural History Survey for the U.S. Environmental Protection Agency and the Canada Atmospheric Environment Service and convened near Chicago in September 1988 (Changnon 1988, Magnuson and Regier 1988).

In the present paper we report on our current understanding of some of the more important impacts that climate warming will likely have on Great Lakes fish. We cannot be very confident about our predictions for two reasons: the atmospheric and hydrologic experts cannot yet be accurate and precise about changes to be expected with respect to physical phenomena that are of great importance to fish; and, as fish ecologists, our own understanding of how changes in such physical factors will ultimately affect fish is neither accurate nor precise. But we judge that our overall understanding is sufficient to provide some useful information to long-term planners.

In our studies we (see Magnuson et al. 1988) have used climate warming scenarios developed by Oregon State University (OSU), the General Fluid Dynamics Laboratory (GFDL) and the Goddard Institute of Space Studies (GISS). These were interpreted for us with respect to offshore waters of Lake Michigan by McCormick (1989) and Lake Erie by Blumberg and DiToro (1989). With respect to the issues addressed in the present paper the ecological consequences of differences between existing scenarios and their hydrological interpretations generally did not differ greatly. This lack of major disagreement may be due to the fact that our overall "model" is as yet very approximate.

Atmospheric and hydrological modellers have not yet developed scenarios for "wind and humidity" that are as "reliable" as the scenarios for temperature. The seasonal wind regime in particular is important with respect to the establishment of vertical (e.g. thermocline) and horizontal (e.g. thermal bar) structures and currents of the water mass. Spatial distributions of various fish species are strongly influenced by these thermal phenomena (Magnuson et al. 1979, Christie and Regier 1988), and especially by extent and frequency of extreme events.

With respect to the temperature regimes, as influenced by climate change, of the waters of the coastal zone and upland streams we derived our own scenarios of water temperatures from air temperatures of the GISS scenario, described by Smith (1988) and our own simulations that relate water temperature regimes to air temperature regimes for non-stratified shallow waters (Magnuson et al. 1988).

Much of all of this is pioneering work based on the results of recent scientific advances. At the present stage, our findings are not to be interpreted as definitive. But they are based on much research experience: altogether the authors have devoted over 50 person-years to the study of the "thermal ecology of fish", broadly defined.

B. DESCRIPTION OF ECOLOGICAL SYSTEM

The Great Lakes lie within a relatively small watershed Basin (Fig. 1) that has long been the industrial heartland of North America and is now home to some 37 million people. In a world in which fresh water is becoming a scarce resource our wealth is reflected in the fact that these lakes contain about one sixth of all the surface fresh water of the globe, not including the glaciers and groundwaters. This chain of lakes drains through the St. Lawrence River into the Gulf of St. Lawrence where the freshwater outflow plays a role in the reproduction and recruitment of the large herring fisheries of the Gulf.

By 1950, the southern third of the Basin had become a vast ecological slum because of careless agricultural, forestry, industrial and urban development in preceding decades. A series of studies then led to binational conventions and agreements to rehabilitate the fisheries and remediate the bad practices that degraded water quality. Many billions of dollars have been spent in these efforts by the various governments at different levels. Some programs have been relatively successful: control of the predaceous sea lamprey, correction of over fishing, reduction in loadings of old-fashioned industrial wastes and of plant nutrients from sewage, and discontinuation of the use of persistent pesticides. Prospects are improving that loadings of hazardous contaminants and acidic substances will progressively be contained and reduced. How will climate change interact with other cultural stresses, -- will some be reactivated?

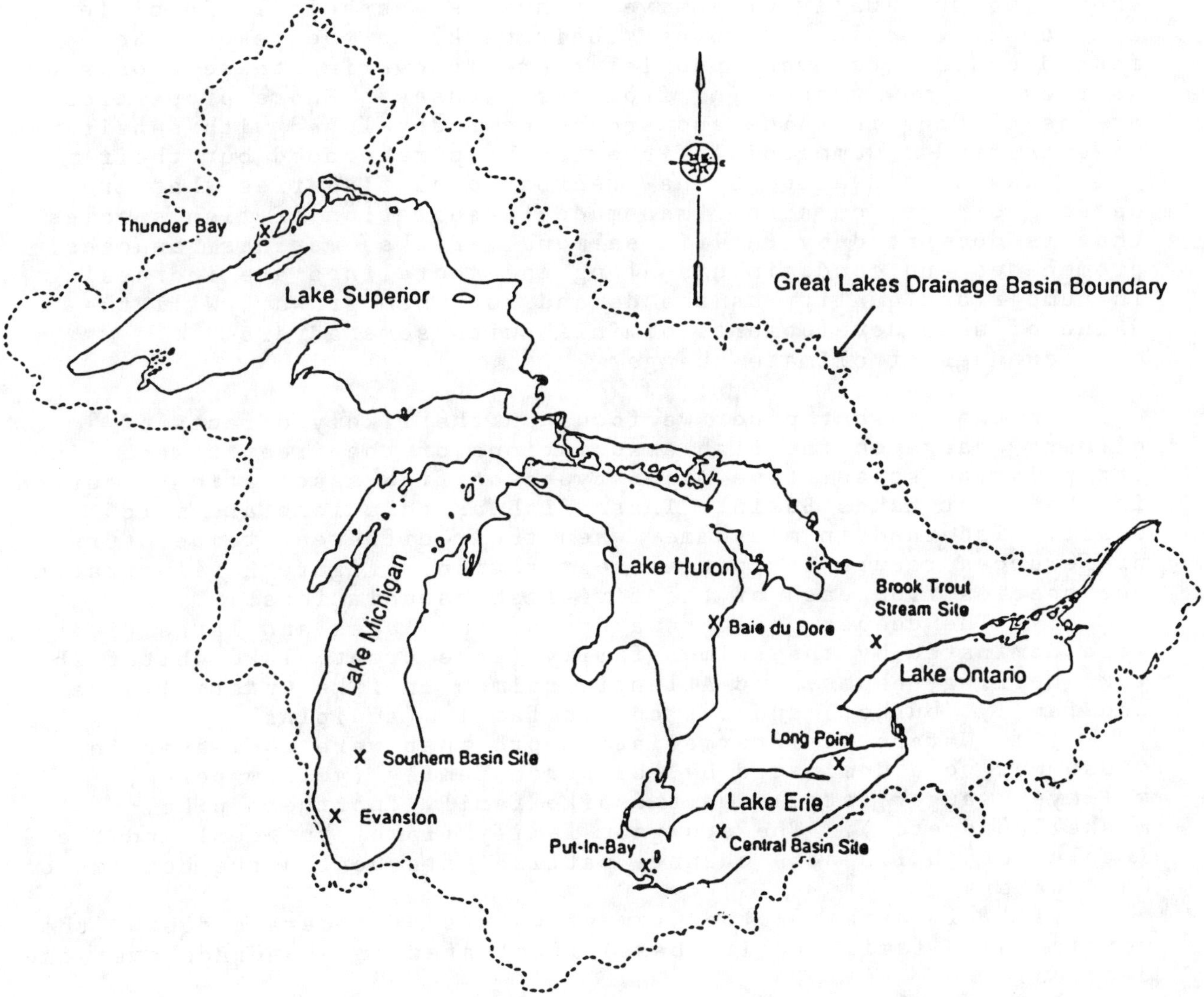

Figure 1. The chain of Great Lakes and its drainage basin boundary. The study by Magnuson et al. (1988), on which the present paper is based, included analytical examination and quantitative assessments of likely impacts of climate change at each site identified with an X.

By 1950, the ecological quality of the lakes, streams and shores was generally offensive to humans, especially in cities and towns. People had turned their backs to the lakes. Some four decades later many people are rediscovering these ecosystems as they are now recovering from past abuses. Shore properties are escalating in value and are being redeveloped with sensitive uses in mind. Commercial fisheries have rebounded but their value has been dwarfed by new recreational fisheries that are based, in part, on a new "man made" association of fish species that is dominated by Pacific salmons. Parks, marinas, beaches, promenades and condominiums along the shorelines are increasing in number and quality as the demand for them grows. Will the value of such developments diminish with some adverse consequences of climate change?

In the present paper we focus on the likely effects of climate change on the fish associations of the Great Lakes. In its pristine state, five major types of fish association occurred in the Great Lakes Basin. These fish associations are more easily discerned in midsummer when their different temperature preferences result in spatial segregation. Figure 2 illustrates one species from each of the five fish associations.

i) The deeper, colder waters of the lakes and large rivers were dominated by the salmon family (lake trout, lake whitefish, lake herring, chubs, and Atlantic salmon in Lake Ontario), the cod family (burbot) and the cottid family (sculpins).

ii) Waters of intermediate depth that were cool even in midsummer were dominated by the perch family (yellow perch, walleye, sauger, etc.) and the pike family (northern pike, muskellunge,etc.). The sturgeon family (lake sturgeon) and the ictalurids (bullheads, channel catfish) frequented the bottoms of cool waters.

iii) In a few of the warmest waters of moderate depth, the percichthyid family (white bass) fluctuated in abundance over the decades.

iv) Shallow, warm inshore waters, especially of bays and streams, were the home of the sunfish family (sunfishes, crappies, smallmouth bass, largemouth bass, etc.) and the cyprinid family (minnows).

v) In the forested coldwater tributary streams, again the salmon family (brook trout) and cottid family (sculpins) predominated.

At the outset, it seems likely that each of these five types of associations will react somewhat differently to climate warming in a particular locale. In summer, coldwater habitats may shrink toward the north in the lakes and upstream into the headwaters; coolwater habitats may be shifted offshore, upstream, and northwards; warmwater habitats may expand from the south northward and from the protected bays into the lake proper.

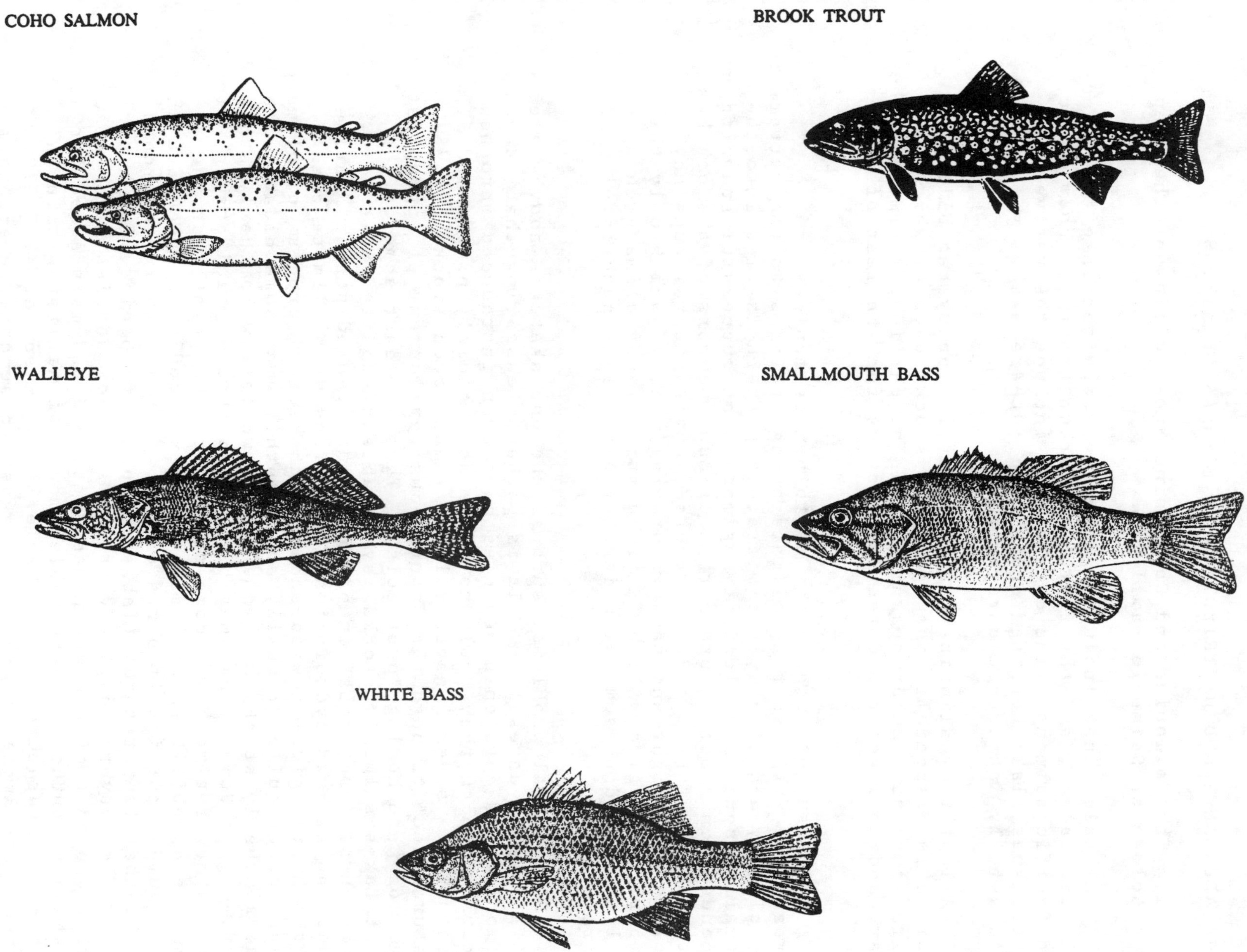

Figure 2. Representative fish species mentioned in the text. Diagrams were taken from Scott and Crossman (1973).

C. ABUNDANCE AND DISTRIBUTION OF FISH ASSOCIATIONS

1. The Salmonine and Coregonine Sub-Families of the Deep, Coldwater Salmonid Family

The salmonines include the native deepwater lake trout and Atlantic salmon, and the introduced brown trout, rainbow trout, and Pacific salmons. The coregonines include the native whitefishes, lake herring, chubs, etc.; there are as yet no successfully introduced coregonines.

All of the salmonines have the evolutionary capability to radiate into specially adapted local stocks each of which migrates to particular spawning areas at particular times of the year, and to feeding and resting grounds for the rest of the year.

One issue of uncertainty as to how salmonines and coregonines will react to climate change relates to the matter of how closely and tightly the migratory pattern is tied into the normal seasonal hydrological regime. That successful spawning is dependent on close co-ordination of many members of a stock has often been inferred by researchers. Environmental cues may include both light and temperature regimes, of which only the temperature regime will be altered strongly by climate change. Will this lead to some disorientation and less successful reproduction?

Behavioural problems, such as the one sketched above, are still largely unresolved. It is perhaps reassuring that many salmonine and coregonine species have been introduced into new waters to which they could not have been closely programmed. Eventually - several decades later - many such stocks become "naturalized" through the spontaneous development of appropriate migratory and homing "protocols". Thus at least some of the Great Lakes salmonid stocks may be able to modify their behavioural "protocols" in order to survive and perhaps thrive under the altered hydrological conditions with climate change. If the hydrological consequences of the actual process of climate change occur quite gradually and smoothly over some decades, then some southerly stocks may wane slowly while new northerly stocks may emerge. Some stocks may be extinguished in habitats in which the thermal regime has become altered to the point that the adaptive capabilities of a stock are exceeded. We speculate that this would occur if all parts of the water mass were to exceed the "lethal temperature" (Table 1) for that species, for a period of at least several days a year. A stock or species might also fail if no part of the water mass attained and maintained for at least some months an appreciable amount of habitat within the "optimal temperature range" (Table 1) (Magnuson et al. 1979, Christie and Regier 1988). We suggest these considerations more

Table 1. Temperature relations[1] of some fishes of the fish association of the Great Lakes.

Group (habitat)	Temperature (°C)		Thermal Niche[2] (C°)	
	Lethal	Optimum	Narrow	Broad
Lake Salmonids (deep water)				
lake trout[3]	24	10	8 - 12	5 - 15
lake whitefish	26	12	10 - 14	7 - 17
coho salmo	25	15	13 - 17	10 - 20
Percids (medium depth)				
walleye	31	20	18 - 22	15 - 25
sauger	30	22	20 - 24	17 - 27
yellow perch	31	23	21 - 25	18 - 28
Percichthyids (medium depth)				
white bass	36	29	27 - 31	24 - 34
white perch	36	32	30 - 34	27 - 37
Centrarchids (shallow water)				
smallmouth bass	35	28	26 - 30	23 - 33
largemouth bass	36	30	28 - 32	25 - 35
pumpkinseed	35	32	30 - 34	27 - 37
Stream Salmonids (upland)				
brook trout	25	15	13 - 17	10 - 20
brown trout	26	17	15 - 19	12 - 22
rainbow trout	26	19	17 - 21	14 - 24

[1] Data from Coutant 1977, Magnuson et al. 1989 and Wismer and Christie 1987.

[2] Thermal niche is defined as narrow: optimum ± 2.0 C°; and broad: optimum ± 5.0 C°.

[3] Common names follow Robins et al. 1980.

as hypotheses than as inferences, at this stage of our research studies.

With respect to habitat of the preferred temperature for feeding by salmonines and coregonines, all the lakes (except perhaps Lake Erie) will continue to contain large amounts of optimal habitat for large parts of the year. But the relative amount of optimal habitat may be smaller during summer in the more southerly lakes because much of the water column may be too warm. It will likely be larger during early spring and late fall than it is at present. If so, then the likelihood of an occurrence of a "summer squeeze" (Coutant 1987a,b) might be of crucial importance with respect to the persistence of a stock of salmonines or coregonines.

A "summer squeeze" has in the past already occurred, apparently, in Lake Erie (Regier and Hartman 1973). Various coregonine species historically spent some time, at a juvenile stage of their life cycle, in the cold bottom waters of the Central Basin. When these bottom waters were stripped of their oxygen, due to excessive enrichment of Lake Erie with sewage and agricultural fertilizers, the coregonines disappeared. We judged that such eutrophication was not solely to blame for the demise of the coregonines in Lake Erie, but did play a synergistic role with other stresses (Regier and Hartman 1973).

Billions of dollars have been spent to reduce the fertility of Lake Erie so that eventually the cold bottom waters will again have sufficient oxygen so that preferred species like the coregonines can thrive there. Climate change may pose a threat to such recovery, under several possible conditions, e.g. the warmer temperatures of surface water might stimulate ecological production and other limnological processes that might again lead to anoxia of the bottom waters.

The non-native sea lamprey (a petromyzontid) is now a scourge in the Great Lakes in that it preys on valued salmonines and coregonines. In the adult stage this fish prefers cold deeper waters of lakes, but in the young larval stage it prefers cool waters of streams, rivers and shallow nearshore water of the lakes (Holmes 1989). Fisheries agencies now spend some 6 million dollars annually to control the lamprey's numbers.

Current evidence suggests that some streams in the coldest region of northeastern Lake Superior watershed may now be marginal habitat for sea lamprey larvae; with climate warming these streams may become optimal habitat with respect to temperature. This would lead to increased harm to valued salmonids in these waters and to increased cost of sea lamprey control. Currently some streams at the southerly edge of the Lake Erie watershed may occasionally reach temperatures beyond the narrow thermal niche of sea lamprey larvae. Here further warming of waters in summer may render these streams inappropriate for lamprey production. Climate warming might also work to the disadvantage of Lake Erie's salmonids (see above). In the absence of salmonids, the favoured prey of sea lamprey, there would be no great motivation to control the sea lamprey. Reduction both of sea lamprey and salmonids in Lake Erie should reduce the need for sea lamprey control in Lake Erie.

2. The Percids of Cool Waters of Medium Depth.

Percids -- and the other three associations to be discussed below -- can, in winter, tolerate the cold waters in which the salmonids spend the winter. In summer percids prefer warmer waters than do the salmonids (see Table 1), and the following two associations prefer even warmer waters. Again, it is the temperature preferred in summer which determines whether we designate a species (or family) as cold, cool, or warm.

In the Great Lakes Basin, the percids of special interest to humans include the walleye, sauger and yellow perch. There are many species of small percids -- the darters -- that serve as prey to various preferred species. Generally the darters frequent the bottoms of shallow waters and prefer cool waters in summer.

Large percids of the Great Lakes thrive in relatively enriched waters of shallow to medium depth that are cool in summer. They tend to be abundant in the large enriched bays of the Great Lakes, especially in the warmer southerly parts of Basin. Currently the limnological regime, including the thermal aspects, of the Western Basin of Lake Erie is nearly optimal for walleye. Similarly the surface layer (epilimnion) of the Central Basin of Lake Erie is nearly optimal for yellow perch. Both of these stocks are the "resources" for some highly valuable fisheries.

It may be that climate change may warm Lake Erie sufficiently that much of its surface waters would become somewhat too warm for the large percids and more favourable for the percichthyid association (see below). This might also occur with some of the shallower parts of large bays, e.g. south-western Saginaw Bay of Lake Huron, southern Green Bay of Lake Michigan and western Bay of Quinte of Lake Ontario. In all of these the percichthyids might successfully dominate the percids, though likely not to the point of extinction of any percid stocks. Percids generally are more tolerant of the degrading effects of human interventions than are salmonids, -- in part because the habitats of percids are unlikely to suffer catastrophic anoxia, as with the isolated colder, deeper waters preferred by salmonids. Also the percids will move to somewhat deeper cooler (but not "cold") waters if the surface temperatures become too warm in summer.

With climate change, some of the moderately enriched bays of the more northerly waters of the Great Lakes Basin will likely experience increased abundance of percids such as walleye, yellow perch and darters. Percids may also become more abundant along open shores of the lakes, at least where moderate fertility and some turbidity exist. There will be large parts of the exposed shore of Lake Superior where temperatures, fertility and turbidity will not reach levels preferred by percids. These waters will remain primarily the habitat of salmonids, though the salmonids may not frequent the surface water for some weeks in summer.

An exotic percid, the ruffe, has recently been introduced into the Duluth area of Lake Superior, apparently through the release of ballast water from an ocean-going vessel. This percid is somewhat smaller than the yellow perch, and is not therefore likely to be valued highly by fishermen. It has a bad reputation in its native northern Europe for feeding on fish eggs, including those of some valued coregonines. If current attempts at eradication are unsuccessful, then the ruffe may expand into the cool-water habitats of all the lakes. With climate warming all the shallower parts of the upper lakes may become appropriate habitat for ruffe, where it may be a pest.

3. The Percichthyids of Warm Waters of Medium Depth.

The native white bass has occurred, in fluctuating and occasionally major abundance, in warmer, more southerly parts of the Great Lakes. The white perch has gradually invaded these waters in recent decades from waters south of the Basin by using canals (Johnson and Evans 1989). Two other percichthyid species thrive in more southerly waters -- yellow bass and striped bass -- and these could and perhaps would invade the lakes, or be introduced by fisheries interests, if the waters were to become warmer due to climate change.

The percichthyids thrive in somewhat enriched or mesotrophic or eutrophic waters, as with the percids. Their feeding preferences are quite similar, hence they compete for some food items. With climate change the percichthyids may thrive, at the expense of percids, in such southerly waters in which the percids are currently very abundant, as sketched in the preceding section.

4. The Centrarchids of Warm Shallow Waters.

The centrarchid family includes the black basses (largemouth and smallmouth), the crappies, rock bass and sunfishes such as pumpkinseed and bluegill. These species thrive in shallow waters that are structurally complex due to rocks, logs, docks and large aquatic plants. Currently in the Great Lakes they are most abundant in the warmer southerly bays and rivers, along shore and on shallow nearshore reefs. They are quite sensitive to environmental pollution and other forms of habitat degradation, -- more so than the percids and percichthyids but resembling the salmonids in this respect.

A variety of centrarchid species have their northerly limits of distribution just to the south of the Great Lakes (Mandrak 1988). Some of these, including the spotted bass, are candidates for invasion or introduction into southerly Great Lakes waters with climate change. None of the centrarchid species are now viewed as pests, but some of the smaller sunfishes have utilitarian value only as forage for the larger, more preferred centrarchids in summer and percids at times other than midsummer.

Currently the largemouth and smallmouth bass thrive even in the warmest waters of Lake Erie, but they may migrate to somewhat cooler and deeper waters in summer. With climate change the spotted bass from south of the Basin may appear in such waters. Meanwhile largemouth and smallmouth bass will likely thrive in waters further north, including some protected but unpolluted bays of Lake Superior (Magnuson et al. 1988).

These proposed changes in the distribution of the various black bass species will presumably be accompanied by changes in various sunfish species (Mandrak 1988). Very little attention has been focussed as yet on likely effects of climate change on the distributions of small species of centrarchids.

5. The Salmonids of Cold Streams.

The native brook trout, the introduced brown trout, and the young of rainbow trout, Atlantic salmon and Pacific salmon thrive in the colder waters of Great Lakes streams, rivers and nearshore zone that have not been degraded by humans. Various kinds of human interventions cause such waters to become warmer in summer thus making them unfavourable for salmonids. Of course other forms of degradation also put the sensitive salmonids at a disadvantage with respect to other species of organisms of lower value to most humans.

In northern streams within the Great Lakes Basin, that are well shaded and not degraded, the full length of the stream may now be habitable even at the height of summer. More southerly, unshaded parts of streams generally become too warm for salmonids. In such streams salmonids persist in shaded upland reaches of streams that are continuously fed by cold water from aquifers. In the Great Lakes Basin the aquifers of most importance to stream salmonids occur in moraines in which many stream headwaters originate. Aquifers, at depths of 10 to 30 metres from the surface, generally have near-constant temperatures throughout the year. An aquifer's temperature is usually within 1 or 2 C^{o} of the average annual air temperature (Meisner et al. 1988). At the southern edge of the Great Lakes Basin the aquifer-fed springs that feed Ohio streams that flow into Lake Erie now have a temperature of about $12^{o}C$. The sources of Ontario streams that flow through Toronto now register about $9^{o}C$. Such cold springs, if they flow abundantly, serve to maintain low water temperatures for some distance downstream, and these cold waters serve as refuges for stream salmonids in the warmest part of summer (Meisner 1989).

With climate warming the aquifer temperatures will gradually increase in approximate proportion to the increase in average annual air temperature. Also the stream waters will warm more rapidly due to higher air temperatures in summer. Thus the amount of cold water habitat in summer will shrink and with it, presumably, the productivity of these waters for stream salmonids because of "summer squeeze". These problems may be exacerbated if the recharge of aquifers is decreased, if more water from upland aquifers is pumped for human uses, if shading trees and vegetation are removed from the edge of streams, if the streams are dammed, etc.

The fish species that will likely benefit from a warming of headwater streams are small percids and minnows (cyprinids). The latter have modest utilitarian value as bait fish for the capture by anglers of large percids and centrarchids.

D. THERMAL ECOLOGICAL ZONES, AN OVERVIEW

In the preceding section we have sketched five associations of fish species as related to habitat temperatures in midsummer. In each case we referred primarily to one family of fish species as indicative of current conditions and what we might expect with climate warming. Clearly we have simplified reality for this exposition, but we do not think that we have distorted it. In this section we sketch an overview as a kind of summary of what has already been presented above.

First consider the warmest part of the Great Lakes as they exist currently, i.e. Lake Erie. Consider the Eastern Basin of Lake Erie with its aquifer-fed tributaries, larger rivers, protected bays, and its offshore deep and thermally-stratified water. Currently this Eastern Basin has appropriate thermal habitat for all the five fish associations sketched above. The salmonid habitats, both with respect to headwater aquifers and deep bottom waters, have been degraded somewhat through pollution and other human abuses, but are still appropriately cold in midsummer.

In Lake Erie's Eastern Basin the thermal zonation in midsummer may be sketched most readily by focussing first on the imaginary line in the coastal zone that separates the land-dominated part of the Basin from the water-dominated part. In the warmest part of summer the warmest waters may be found near this imaginary line. Where the waters are protected from the stronger currents and larger waves of the rivers and the offshore lake waters, centrarchids abound, at least in waters that are not degraded by humans. Still in the protected bays in the warmest surface waters and at some distance from the shore the percichthyids now thrive. In somewhat cooler waters of the more open parts of the coastal zone, down to moderate depth, the larger percids are abundant in midsummer. Smaller percids are found in the cool reaches of tributaries. The cool waters bounding the coastal zone are in turn bounded by the cold spring-fed headwaters in the morainic headlands and the cold deeper waters of the thermally stratified Eastern Basin. In both of these salmonid zones the temperature is still suitably cold in summer but the habitat has been degraded to some extent by other human practices so as to make much of this habitat of poor quality for the sensitive salmonids.

Currently there is a kind of symmetry in the fish associations, as sketched above. The centrarchids thrive in quite shallow habitats along the boundary between the land- and water-dominated parts of the Eastern Basin. Next to them in the more open surface waters of protected bays and estuaries are found the percichthyids. The warm centrarchid and percichthyied waters are bounded by cool-water zones with percids both in upstream and offshore directions, which in turn are bounded by cold-water zones of salmonids. Whether one progresses upslope on the land side or downslope on the lake side or from the more protected bays thence along shore to the open coast, we note a similar progression from centrarchids and percichthyids to percids and salmonids, in midsummer. In spring and/or fall the spawning distributions overlap to a far greater extent, but they tend to separate again during the winter. The most intense competition for food generally occurs with warm temperatures during summer, and this may explain the niche-separation, as sketched above, that occurs in summer (Magnuson et al. 1979).

With climate warming the amount of habitat that is thermally optimal for centrarchids and percichthyids will likely expand laterally along shore and somewhat further offshore and upstream in the larger tributaries. In such waters the warming will likely favour them over the percids. The summer habitat of the latter will in turn likely retract away from the warmest waters of the Basin and extend further offshore into somewhat deeper waters. In streams the habitat for small percids will likely expand further upstream and contract from the warmest waters of the coastal zone. Finally the habitat for salmonids will likely contract both into the deeper offshore waters and into the more shaded spring-fed upland streams, again in summer.

For contrast, consider the state of the Thunder Bay sub-basin of northern Lake Superior, which is very roughly comparable to the Eastern Basin of Lake Erie with respect to geography and hydrography. In the Thunder Bay sub-basin the amount of centrarchid habitat is now relatively small but will expand somewhat, in the streams and their lakes as well as in Lake Superior waters, with climate warming (Magnuson et al. 1988). There are currently very few percichthyids but small populations may appear with climate warming. Large percids occur in some abundance but spottily, -- these populations will likely expand notably, unless the fertility of the waters is reduced markedly through efforts against cultural eutrophication. The accidentally introduced ruffe may also invade these waters and thrive in them, perhaps at the expense of some salmonines and coregonines. The salmonids will likely be excluded in summer from larger areas of warm waters in the protected parts of the coastal zone. But this does not necessarily imply that the overall production of salmonids will fall, -- see below. On the whole, with climate warming, Thunder Bay waters will likely support modest populations of centrarchids and percichthyids, and moderately strong populations of percids in the coastal zone, with continuing strong populations of salmonids further upstream and offshore,-- at least if the populations of the exotic sea lamprey and ruffe are limited to relatively low abundance.

We have sketched the likely consequences of climate warming in one of the warmest and one of the coldest parts of the Great Lakes Basin. Intermediate between these are bays such as Green Bay, Saginaw Bay and Georgian Bay. Here we can expect thriving populations of all five associations, with some adjustments of relative abundances consistent with those sketched above. Of course we assume that our efforts to reverse the cultural degradation of such waters will continue to be successful.

E. FISH PRODUCTION RATES AS A FUNCTION OF HABITAT TEMPERATURE AND OTHER ENVIRONMENTAL FACTORS

In the preceeding text we have focussed on the likely effects of climate warming on the geographic and hydrographic distributions of some associations of valued fish species. Now we turn to some effects that climate warming will have on physiological and ecological rates that contribute to the harvestable production of fish species.

Fish are cold-blooded or ectothermic, i.e. their body temperatures are nearly equal to the seasonally-varying temperature of their aquatic habitat. Through behavioural means they generally "know enough" to migrate to waters that they find favourable. Thus if waters of suitable ecological quality are accessible to them, then they will likely be found in a fairly narrow range of temperature that has been called their "thermal niche" or "optimal temperature range" -- see the discussion in Section C above, and see Table 1.

It has long been known that a variety of physiological and developmental processes of ectothermic organisms are strongly influenced by temperature. In situations where the habitat is benign in that all the necessities of life are present in abundance and any harmful conditions are only of minor importance, the mathematical expression for a first order chemical reaction has often been used to characterize physiological or ontogenetic process, at least with the younger stages of an organism's life. It happens that conventional, intensive fisheries tend to constrain an aquatic ecosystem to a stressed state in which much of the ecological productive process is sustained by younger organisms throughout the ecological network (Rapport et al. 1985, Regier et al. 1988). Thus the first order chemical reaction should provide an approximate model for various physiological and ecological phenomena of exploited, fish-dominated aquatic ecosystems such as the Great Lakes (Regier et al. 1989). In general terms this expression is

$$\frac{dx}{dt} \cdot \frac{1}{x} = k,$$

$$\text{or} \qquad x_t = x_o e^{k(t_1 - t_0)}$$

Here x is the amount of biomass, t is chronological time and k is the specific rate constant.

With ectotherms at a juvenile or "log-phase" stage of their maturation process in benign circumstances, it is known that the rate process is approximated by the Arrhenius expression for chemical reactions:

$$k = Ae^{-\frac{E}{R} \cdot \frac{1}{T}},$$

$$\text{or} \qquad \ln k = \ln A - \frac{E}{R} \cdot \frac{1}{T} .$$

Here T is absolute temperature (^oK) and A, E and R are constants as explicated by the theories of quantum mechanics and thermodynamics.

Regier et al. (1989) have shown that taken together these two approximate models provide close fits to data on a variety of rate processes with fish and other aquatic ectotherms. Examples are shown in Table 2.

Table 2. Representative regression equations relating some physiological and ontogenetic processes of fish to absolute temperature (°K), where $\log_e$ rate = a + b(1/T). All regression slopes are significant ($P < 0.001$). The equations shown here are a subset of the 37 regressions reported by Regier et al. (1989). Common names for fish follow Robins et al. (1980).

Species	Regression a	Regression b	N	r^2	Temperature range (C)	Source[a]
Egg Incubation Period						
Brook trout[b]	-32.58	10328	41	0.9892	1.6-14.8	1
Growth Rate						
Carp[c]	25.23	-7772	52	0.5643	15.6-26.4	2
Developmental Period						
Walleye[d]	-9.063	3659	20	0.7467	-9.4-20.5	3
Fish Yield						
Fish yield[e]	29.74	-8115	43	0.6876	-5.2-25.6	4

[a] Sources of data: 1 - Embody (1934); Knyazev (1987); 3 - Colby and Nepszy (1981); 4 - Schlesinger and Regier (1982).
[b] Duration of the incubation period (in days) from fertilization to 50% hatching.
[c] Data are specific rate of weight increase in carp fingerlings.
[d] Age at which 50% of the female walleye in a given population are sexually mature estimated by us from probit paper using Colby and Nepszy's (1981) data.
[e] Data represent the residuals from the regression of mean annual fish yield (kg/ha yr) of all commercially important species combined. Residuals were calculated from a multiple regression as follows: Residual = $\log_e$observed - $\log_e$expected where $\log_e$expectd is $\log_y$ield - 0.4898($\log_e MEI_{25}$). The MEI_{25} term is a measure of lake fertility.

In preceding sections of this paper we predicted that various fish species would alter their seasonal lateral and vertical spatial distributions so as to find favourable thermal habitat. The lakes already contain valued native and non-native fish species that favour a variety of thermal niches that together span the entire range of water temperatures to be expected with climate warming. In addition species adapted to somewhat warmer waters further south are not now effectively isolated from Great Lakes waters and can find their ways into those waters if they become thermally more favourable (Mandrak 1988).

All of the considerations above are consistent with the hypothesis that fish production in total should increase with climate warming. That this may occur is supported by comparative studies of ecosystem productive processes in waters of different latitudes and of different average temperatures (e.g. fish yield equation in Table 2). But other considerations cast some doubt on this optimistic hypothesis.

The entire ecological production process in the Great Lakes Basin rests ultimately on the photosynthetic process of plants. Where all other factors are in benign ranges, the rate of photosynthesis depends both on the intensity of illumination and on the temperature. Plant species also have their preferred "thermal niches" as with fish, and there is usually a seasonal progression in the maximal production of different species, with each species producing maximally when the habitat temperature is approximately optimal, other factors remaining benign. Plants also have preferred "illumination niches", so maximal production may depend on the co-occurrence of optimal thermal and illumination conditions.

With climate warming the temperature will increase, but direct radiation of the Basin by the sun will not increase. Whether the photosynthetic process, and also net primary production, will increase with climate warming, as expected from the models presented above, will depend on whether other factors such as nutrient supply become available in sufficient abundance to support the increased production.

Highly-enriched waters in the more developed parts of the Great Lakes Basin have been subjected to excessively high levels of nutrients during this century. Strong efforts have been made in recent years to reduce the nutrient loadings so as to foster recovery of more desirable ecosystemic states with suppression of primary production. Acceptable levels have been defined with respect to current climate conditions, and no attention has yet been directed to what would be optimal loading with climate warming.

Some hydrographic basins of the Great Lakes may be vulnerable to adverse synergistic interactions between climate warming and other human influences. The Central Basin of Lake Erie might be such a case (Magnuson et al. 1988, Blumberg and DiToro 1989). In contrast some quite sterile (oligotrophic) colder waters further north in the Great Lakes Basin may benefit from synergism between climate warming and moderate cultural eutrophication, if other factors remain benign.

Wetlands of the Great Lakes are strongly dependent on the occurrence of fluctuations in water levels (Stephenson 1989, Whillans 1989). These coastal ecosystems appear to benefit on balance from variabilities in water level at three scales of spatial and temporal resolution: small, short-term; medium, seasonal; large, decadal. If climate change will alter the variability regimes of one or more of these, then coastal wetlands will be affected. Individual wetlands may shrink or expand or shift spatially, and the vegetation communities may undergo changes in composition. The major impact on the fish associations will likely relate to local changes in abundance of centrarchids, relative to percichthyids and percids.

Every kind of human influence on the Great Lakes, as practiced in the past, has had its degrading consequences. Billions of dollars have been spent to remediate such degrading practice and to foster recovery of desirable features of the ecosystem through rehabilitation. If climate warming is already inevitable, we would be lucky indeed if the effects of climate warming on the Great Lakes aquatic ecosystem would in general be considered desirable by humans. From what we now perceive, it would be prudent to infer that some of the consequences will be desirable and some will be undesirable. Based on past experience we should expect that nasty surprises will occur.

Major new management strategies will presumably be needed to benefit from desirable consequences and to meliorate or mitigate the undesirable consequences. Our understanding is as yet insufficient to provide guidance on what those strategies might be.

F. REFERENCES CITED

Blumberg, A.F. and D.M. DiToro. 1989. The effects of climate warming on Lake Erie water quality. Trans. Am. Fish. Soc. (submitted).

Changnon, S.A. 1989. Summary of Symposium on Impacts of Climate Change on the Great Lakes Basin, Chicago, Illinois, September 1988. Illinois Natural History Survey, Champaign, Illinois.

Christie, G.C., and H.A. Regier. 1988. Measures of optimal thermal habitat and their relationship to yields for four commercial fish species. Can. J. Fish. Aquat. Sci. 45:301-314.

Colby, P.J. and S.J. Nepszy. 1981. Variation among stocks of walleye (Stizostedion vitreum vitreum): management implications. Can. J. Fish. Aquat. Sci. 38: 1814-1831.

Coutant, C.C. 1977. Compilation of temperature preference data. J. Fish. Res. Bd. Can. 34:739-745.

Coutant, C.C. 1987a. Thermal preference: when does an asset become a liability? Environmental Biology of Fishes 18: 161-172.

Coutant, C.C. 1987b. Poor reproductive success of striped bass from a reservoir with reduced summer habitat. Trans. Am. Fish. Soc. 116: 154-160.

Embody, G.C. 1934. Relation of temperature to the incubation periods of eggs of four species of trout. Transactions of the American Fisheries Society 64: 281-292.

Hill, D.K. and J.J. Magnuson. 1989. Potential effects of climate warming on the growth and prey consumption of Great Lakes fishes. Trans. Am. Fish. Soc. (submitted).

Holmes, J.A. 1989. An ecological early warning system of climate change in the Great Lakes. Trans. Am. Fish. Soc.(submitted).

Johnson, T.B. and D.O. Evans. 1989. The effect of winter severity on recruitment of white perch in the Laurentian Great Lakes, with implications of climate warming. Trans. Am. Fish. Soc. (submitted).

Knyazev, I.V. 1987. Use of the van't Hoff coefficient to analyze the growth of carp fingerlings. Hydrobiological Journal 23: 91-94.

Magnuson, J. J., L. B. Crowder and P. A. Medvick. 1979. Temperature as an ecological resource. American Zoologist 19:331-343.

Magnuson, J. J., J. D. Meisner, and D. K. Hill. 1989. Potential changes in the thermal habitat of Great Lakes fishes following global climate warming. submitted to Trans. Am. Fish. Soc.

Magnuson, J.J. and H.A. Regier. 1989. Effects of climate change on Great Lakes fish, In Changnon, S.A. (Ed). Proceedings of

Symposium on Impacts of Climate Change on the Great Lakes Basin, Chicago, Illinois, September 1988. Illinois Natural History Survey, Champaign, Illinois.

Magnuson, J. J., H. A. Regier, D. K. Hill, J. A. Holmes, J. D. Meisner, and B. J. Shuter. 1988. Potential responses of Great Lakes Fishes and their habitat to global climate warming. U.S. Environmental Protection Agency, CR-814644-01-0, Office of Policy Analyses, Washington, DC

Mandrak, N. 1989. Geographic distribution of fish species in the Great Lakes Region: an assessment of the effects of CO_2 climate change. J. Great Lakes Res. (in press).

McCormick, M.J. 1989. Potential climatic changes to the Lake Michigan thermal structure. Trans. Am. Fish. Soc. (submitted).

Meisner, J. D., J. L. Goodier, H. A. Regier, B. J. Shuter, and W.J. Christie. 1987. An assessment of the effects of climate warming on Great Lakes basin fishes. Journal of Great Lakes Research 13(3):340-352.

Meisner, J. D., J. S. Rosenfeld, and H. A. Regier. 1987. The role of groundwater on the impact of climate warming on stream salmonines. Fisheries 13(3):2-8.

Meisner, J. D. 1989. Simulating reductions in summer brook trout habitat of two southern Ontario streams due to climatic warming. submitted to Trans. Am. Fish. Soc.

Rapport, D.J., H.A. Regier, and T.C. Hutchinson. 1985. Ecosystem behavior under stress. American Naturalist 125: 617-640.

Regier, H.A. and W.L. Hartman. 1973. Lake Erie's fish community: 150 years of cultural stresses. Science 180: 1248-1255.

Regier, H.A., J.A. Holmes and J.D. Meisner. 1988. Likely impact of climate change on fisheries and wetlands, with emphasis on the Great Lakes. pp. 313-327, In J.C. Topping, Jr. (Ed). Preparing for climate change, Proceedings of the First North American Conference on Preparing for Climate Change: A Cooperative Approach, October 27-29, 1987, Washington, D.C. Government Institutes Inc. Rockville, Maryland. 516 pp.

Regier, H.A., R.L. Welcomme, R.J. Steedman and H.F. Henderson. 1988. Rehabilitation of Degraded River ecosystems. Canadian Special Publication of Fisheries and Aquatic Sciences, 104 (in press).

Regier, H.A., J.A. Holmes and D. Pauly. 1989. The influence of temperature on ectothermic aquatic living systems: a summary of empiric data. Trans. Am. Fish. Soc. (submitted).

Robins, C.R., R.M. Bailey, R.M.,Bond, C.E., Brooker, J.R., Lachner, E.A., Lee, R.N.,and Scott, W.B. 1980. A List of Common and Scientific Names of Fishes from United States and Canada. 4th edition, Amer. Fish. Soc., Spec. Publ. No. 12, Bethesda, MD.

Schlesinger, D.A.,and H.A. Regier, H.A. 1982. Climatic and morphoedaphic indices of fish yields from natural lakes. Trans. Amer. Fish. Soc.111:141-150.

Scott, W.B. and E.J. Crossman. 1973. Freshwater fishes of Canada. Fisheries Research Board of Canada, Bulletin 191. 969 pp.

Smith, J.B. 1988. Methodology. Chapter 3, In J.B. Smith and D.A. Tirpak (Eds). The potential effects of global climate change on the United States. Draft Report to Congress. Volume 1: Regional Studies.

Stephenson, T.D. 1989. Significance of Toronto area wetlands for fish. In Proceedings of Ontario Wetlands: Inertia or Momentum?, 20-21 October, 1988. Federation of Ontario Naturalists, Don Mills, Ontario.

Whillans, T.H. 1989. Successional response to water levels in a marsh fish community of Lake Erie. Trans. Am. Fish. Soc. (submitted).

Wismer, D. A., and A. E. Christie. 1987. Temperature relationships of Great Lakes Fishes: a data compilation. Great Lakes Fishery Commission, Special Publication No. 87-3, Ann Arbor, Michigan.

POTENTIAL IMPACT OF STRATOSPHERIC OZONE DEPLETION ON MARINE ECOSYSTEMS

Robert C. Worrest[1], Hermann Gucinski[2], and John T. Hardy[3]

[1]U.S. Environmental Protection Agency, Office of Research and Development, Washington, DC
[2]NSI Technology Services Inc., Corvallis, Oregon
[3]Department of General Science, Oregon State University, Corvallis, Oregon

Introduction

As a result of stratospheric ozone depletion, ultraviolet-B radiation (290-320 nm; UV-B) reaching marine environments is likely to increase over the next few decades (Ozone Trends Panel 1988; Hoffman and Gibbs 1988). Available information suggests that UV-B radiation can have a variety of deleterious effects on marine ecosystems (Worrest 1986). The purpose of this report is to point to the current uncertainties in existing information and to speculate on the range of possible effects. We hope that controversial aspects will stimulate discussion and additional needed research.

Assessing the Dose of UV-B Radiation in the Sea

In the sea, the amount of radiation reaching any given depth depends upon the total amount reaching the sea surface, the angle of incidence, the degree of sea-surface roughness (which determines the amount reflected back into space), and the scattering and absorption within the water column. In clear ocean water UV-B radiation is reduced to 1% of the surface level at a depth of about 30 m, while in moderately productive coastal waters the 1% level may occur at only 1 or 2 m (Baker and Smith 1982).

Shorter wavelengths of UV-B radiation are generally more biologically damaging than longer wavelengths. Therefore, the daily biologically effective dose at depth must be estimated by normalizing each wavelength by an applicable biological action spectrum and integrating downward spectral irradiance over all angles of the sun during a day (Smith and Baker 1979).

However, current estimates of UV-B exposure at depth in the water column remain uncertain for several reasons. First, estimates of attenuation values at wavelengths shorter than 310 nm have a high degree of uncertainty because the total subsurface irradiance declines rapidly at these shorter wavelengths. Second, the choice of action spectra can change the estimate of biologically effective dose at any depth by an order of magnitude or more. Third, changing sea states or sky conditions introduce additional uncertainty in the open sea. Finally, all estimates of the dose of UV-B radiation received by marine plankton populations remain somewhat uncertain because of a paucity of data on their local movements and their vertical distribution in the water column.

Effects on Marine Productivity

Increased UV-B radiation can reduce the productivity and species diversity of marine phytoplankton (Worrest et al. 1978, 1981; Worrest 1983; Worrest 1986). Phytoplanktonic productivity in the upper few meters is already inhibited by ambient levels of UV-B radiation, and enhanced levels of UV-B radiation are likely to further decrease primary production (Lorenzen 1979; Maugh 1980; Smith and Baker 1980; Wolniakowski 1979; Worrest 1982).

A 16% ozone depletion in summer over temperate pelagic waters would result in about a 47% increase in DNA-action-spectra-weighted UV-B radiation reaching a depth of 1 m (Figure 1). Assuming continuous residence at that depth, primary productivity in about half the species of the fewer than ten species examined by Worrest et al. (1981) would be reduced 20% after a 5-day exposure to the increased radiation.

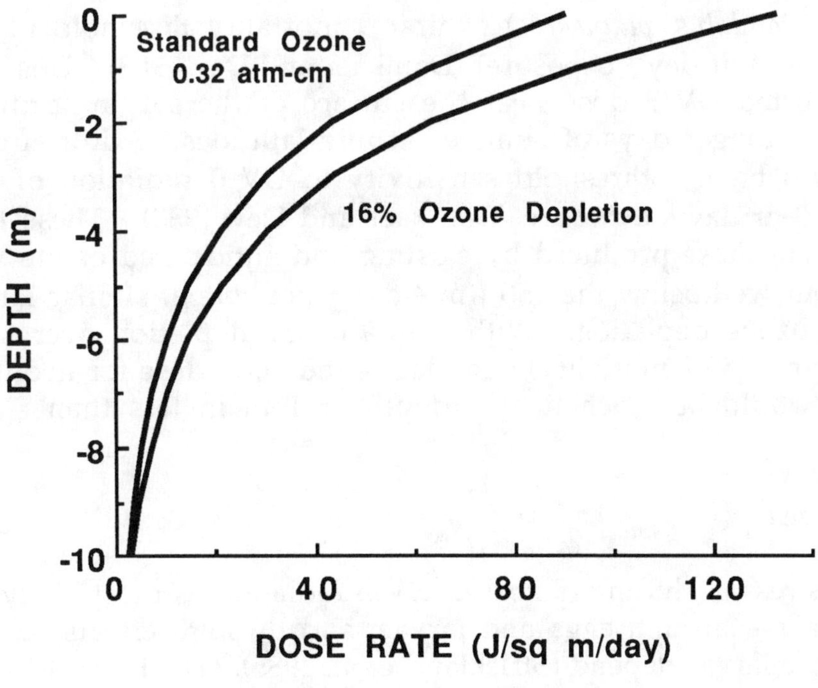

Figure 1. Penetration of biologically effective UV-B radiation (DNA weighted) into pelagic ocean water. Conditions are July 2, 40°N latitude, low dissolved and particulate material. Incident irradiation at sea surface based on Green et al. (1980).

Using models of Smith and Baker (1979) and Baker and Smith (1982), Worrest (1983) estimated that a 16% depletion in stratospheric ozone at temperate latitudes would result in a 5% decrease in marine primary production integrated over the entire euphotic zone. However, a large degree of uncertainty remains in extrapolating measured effects of UV-B radiation to effects on overall in-situ productivity.

Shallow water habitats prevent the escape of motile organisms to greater depth. Rooted aquatic vegetation, such as sea grasses, are known to have widely differing UV-B sensitivity (Trocine 1981). Also at risk are the coral reefs, which depend on the viability of the associated zooxanthellae (Jokiel 1980). These habitats may be at significantly greater risk than the open ocean.

Effects on Zooplankton

Marine invertebrates differ greatly in their sensitivity to UV-B radiation. For example, the copepod *Acartia sp.* suffers 50% mortality from UV-B radiation dose rates below those currently present at the sea surface. By contrast, some shrimp larvae (*Pandalus sp.*) tolerate dose rates at the sea surface greater than those forecast for a 16% ozone depletion.

Larval shrimp (*Pandalus platyceros*) exhibit a mortality threshold of 15 J m^{-2} day^{-1} (DNA weighted 3-hr day^{-1} exposure) (Damkaer and Dey 1983). This threshold is close to the reported UV-B dose near the surface of the sea in spring, but is exceeded during the longer days of summer at mid-latitudes. Adult euphausiids (*Thysanoessa raschii*) have a threshold sensitivity to UV-B radiation of 51 J m^{-2} day^{-1} for a similar 3-hr day^{-1} exposure (Damkaer and Dey 1983). These threshold levels are higher than those produced by existing and anticipated ozone levels for spring spawners, but well below the 150 J m^{-2} day^{-1} possible at similar latitudes in July even without ozone depletion. With a 16% ozone depletion over temperate pelagic waters, a lethal (50% mortality) cumulative radiation dose for about half the species examined would be reached at a depth of 1 m in less than 5 days (in summer).

Effects on Fisheries

Anthropogenic stress, including increased UV-B radiation, is most likely to affect fisheries in two ways—through eggs and larvae, and through effects on the food chain upon which the larvae depend (Strickland et al. 1985). The bulk of the world's marine harvest of fish, shellfish and crustaceans largely depends on species that have eggs and larvae that occur at or near the sea surface (Hardy 1982; U.S. Fish and Wildlife Service 1978) where they will be exposed to increasing levels of UV-B radiation.

Enhanced solar UV-B radiation directly reduces the growth and survival of larval fish (e.g., Hunter et al. 1979, 1981). Chapman and Hardy (1988) used the data from Hunter et al. (1979, 1981) to estimate that, for the Washington State coastal shelf in June, a 16% ozone reduction levels would result in increases in larval mortality at 0.5 m depth of 50, 82 and 100% for anchovy larvae of ages 2, 4 and 12 days, respectively. In Oregon, anchovy larvae occur coincident with high radiation levels between June and August with a peak in July (Richardson and Pearcy 1976). Since virtually all anchovy larvae in the northern California, Oregon and Washington shelf area occur within the upper 0.5 m, a 16% ozone reduction level could be expected to lead to large increases in larval mortality.

Evidence indicates that increased UV-B irradiance could also result in fishery losses through indirect effects on the planktonic food web. Several authors have suggested that fishery yield decreases in a power law fashion with decreases in primary production (Ryther 1969; Oglesby 1977; Nixon 1988). Thus, using the equation that fisheries yield increases as productivity raised to the 1.55 power (Nixon 1988), a 5% decrease in primary production (estimated for a 16% ozone depletion by Smith and Baker 1980 and Worrest 1983) will yield reductions in fish yield of approximately 6 to 9%. A 7% reduction in fish yield, if it occurred on a global basis, would then represent a loss of about 6 million tons of fish per year.

Indirect effects may also occur in the form of altered patterns of predation, competition, diversity, and trophic dynamics as species resistant to UV-B radiation replace sensitive species. The combined effects of direct (larval mortality) and indirect (food web) losses cannot as yet be predicted, nor have assessments been made of adaptive strategies or genetic selections that could minimize population or ecosystem effects.

Risks and Uncertainties

While current data suggest that predicted increases in UV-B radiation could have important negative effects in the marine environment, uncertainties regarding the magnitude of these effects remain large. Among them are problems of extrapolating laboratory findings to the open sea, and the nearly complete absence of data on long-term effects and ecosystem responses. Additional information is needed in several areas before more reliable assessments of risk are possible. Needed research includes:

- Accurate and appropriate biological action spectra for marine species
- Dose-response data on a greater variety of phytoplankton, zooplankton, ichthyoplankton and shallow water benthos than is now available
- Long term effects—if embryos or larvae are exposed to UV-B radiation, is the survival of the adult population affected (or their offspring)?
- Field studies that lead to better understanding and application of laboratory findings

- Detailed temporal and spatial distribution data for early life stages to determine exposure

- Effects on Antarctic and other ecosystems

- Data on the mechanisms and ranges of possible adaptation or genetic selection to increased UV-B radiation

- Effects on important marine biogeochemical cycles involving compounds such as methane, nitrous oxide, carbon monoxide, and organo-sulfides

- Effects of the sea surface microlayer in mediating air/sea exchange, and microbial and neuston interaction

Only when the results of such studies are available will an accurate assessment of the risks posed to the marine environment from increasing UV-B radiation be possible.

References

Baker, K.S. and R.C. Smith. 1982. Bio-optical classification and model of natural waters. *Limnol. Oceanogr.*, **27**, 500-509.

Chapman, J. and J.T. Hardy. 1988. Effects of middle ultraviolet radiation on marine fishes. Final Report Oregon State Univ. US EPA Coop. Agrmt. CR-812688.

Crawford, M. 1987. Landmark ozone treaty negotiated. *Science*, **237**, 1557.

Damkaer, D.M. and D.B. Dey. 1983. UV damage and photo-reactivation potentials of larval shrimp, *Pandalus platyceros*, and adult euphausiids, *Thysanoessa raschii*. *Oecologia*, **60**, 169-175.

Dey, D.B., D.M. Damkaer and G.A. Heron. 1988. UV-B dose/dose-rate responses of seasonally abundant copepods of Puget Sound. *Oecologia*, **76**, 321-329.

Döhler, G., R.C. Worrest, I. Biermann and J. Zink. 1987. Photosynthetic $^{14}CO_2$ fixation and [^{15}N]-ammonia assimilation during UV-B radiation of *Lithodesmium variabile*. *Physiol. Plant.*, **70**, 511-515.

Green, A.E.S., K.R. Cross and L.A. Smith. 1980. Improved analytic characterization of ultraviolet skylight. *Photochem. Photobiol.*, **31**, 59-65.

Hardy, J.T. 1982. The sea surface microlayer: biology, chemistry and anthropogenic enrichment. *Prog. Oceanogr.*, **11**, 307-328.

Hoffman, J.S. and M.J. Gibbs. 1988. Future concentrations of stratospheric chlorine and bromine. U.S. Environmental Protection Agency. Office of Air and Radiation. EPA 400/1-88/005.

Hunter, J.R., J.H. Taylor and H.G. Moser. 1979. Effect of ultraviolet radiation on eggs and larvae of the northern anchovy, *Engraulis mordax,* and the Pacific mackerel, *Scomber japonicus,* during the embryonic stage. *Photochem. Photobiol.,* **29**, 325-338.

Hunter, J.R., S.E. Kaupp and J.H. Taylor. 1981. Effects of solar and artificial ultraviolet-B radiation on larval Northern Anchovy, *Engraulis mordax. Photochem. Photobiol.,* **34**, 477-486.

Jokiel, P.L. 1980. Solar ultraviolet radiation and coral reef epifauna. *Science,* **207**, 1069-1071.

Jokiel, P.L. and R.H. York. 1984. Importance of ultraviolet radiation in photoinhibition of microalgae growth. *Limnol. Oceanogr.,* **29**, 192-199.

Lorenzen, C.J. 1979. Ultraviolet radiation and phytoplankton photosynthesis. *Limnol. Oceanogr.,* **24**, 1117-1120.

Murali, N.S. and A.H. Teramura. 1985. Effects of UV-B irradiance on soybean. VI. Influence of phosphorus nutrition on growth and flavonoid content. *Physiol. Plant.,* **63**, 713-716.

Maugh, T.H., II. 1980. Ozone depletion would have dire effects. *Science,* **207**, 394-395.

Nixon, S.W. 1988. Physical energy inputs and the comparative ecology of lake and marine ecosystems. *Limnol. Oceanogr.,* **33**, 1005-1025.

Oglesby, R.T. 1977. Relationships of fish yield to lake phyto-plankton standing crop, production and morphoedaphic factors. *J. Fish. Res. Board Can.,* **34**, 2271-2279.

Ozone Trends Panel. 1988. Executive summary. Coordinated by the National Aeronautics and Space Administration, Washington, DC, March 15, 1988.

Richardson, S.L. and W.G. Pearcy. 1977. Coastal and oceanic fish larvae in an area of upwelling off Yaquina Bay, Oregon. *Fish. Bull.,* **75**, 125-146.

Ryther, J.H. 1969. Photosynthesis and fish production in the sea. *Science,* **166**, 72-76.

Smith, S.V. 1981. Marine macrophytes as a global carbon sink. *Science,* **211**, 838-840.

Smith, R.C. and K.S. Baker. 1979. Penetration of UV-B and biologically effective dose-rates in natural waters. *Photochem. Photobiol.,* **29**, 311-323.

Smith, R.C. and K.S. Baker. 1980. Stratospheric ozone, middle ultraviolet radiation, and carbon-14 measurements of marine productivity. *Science*, **208**, 592-593.

Strickland, R.M., D.J. Grosse, A.I. Stubin, G.K. Ostrander and T.H. Sibley. 1985. Definition and characterization of data needs to describe the potential effects of increased atmospheric CO_2 on marine fisheries from the Northeast Pacific Ocean. Report to U.S. Department of Energy (DOE/NBB-0075), Contract No. W-7405-ENG-48. School of Fisheries, University of Washington, Seattle. 139 pp.

Trocine, R.P., J.D. Rice and G.N. Wells. 1981. Inhibition of seagrass photosynthesis by ultraviolet-B radiation. *Plant Physiol.*, **68**, 74-81.

U. S. Fish and Wildlife Service. 1978. Development of fishes of the mid- Atlantic Bight: An Atlas of Egg, Larval and Juvenile Stages. USFWS/UBS/- 78/12. Washington, GPO.

Wolniakowski, K.U. 1979. The physiological response of a marine phytoplankton species, *Dunaliella tertiolecta*, to mid-wavelength ultraviolet radiation. M.S. Thesis, Oregon State University, Corvallis, OR, 101 pp.

Worrest, R. C. 1982. Review of literature concerning the impact of UV-B radiation upon marine organisms. The Role of Solar Ultraviolet Radiation in Marine Ecosystems. J. Calkins, ed., Plenum Publishing, pp. 429-457.

Worrest, R.C. 1983. Impact of solar ultraviolet-B radiation (290-320 nm) upon marine microalgae. *Physiol. Plant.*, **58**, 428-434.

Worrest, R.C. 1986. The effect of solar UV-B radiation on aquatic systems: an overview. Effects of Changes in Stratospheric Ozone and Global Climate. Vol. 1. Overview, J.G. Titus, ed. U.S. Environmental Protection Agency and United Nations Environment Programme, pp. 175-191.

Worrest, R.C., H. Van Dyke and B. E. Thomson. 1978. Impact of enhanced simulated solar ultraviolet radiation upon a marine community. *Photochem. Photobiol.*, **17**, 471-478.

Worrest, R.C., B.E. Thomson and H. Van Dyke. 1981. Impact of UV-B radiation upon estuarine microcosms. *Photochem. Photobiol.*, **33**, 861-867.

A MECHANISM FOR GREENHOUSE-INDUCED COLLAPSE OF MAMMALIAN FAUNAS

Dewey M. McLean
Department of Geological Sciences
Virginia Polytechnic Institute and State University
Blacksburg, VA 24061

INTRODUCTION

Today, we face possible major CO_2-induced greenhouse warming. By the Hansen et al. Scenario A (1), Earth could warm by 4°C by the year 2050, comparable to the 5°C warming that ended the last ice age 10,000-12,000 years ago, but at a rate 10 to 40 times faster. Environmental heat, by damaging developing embryos, can kill on a global scale (2, 3, 4). In the warming at the end of the last ice age, most of Earth's large mammals suddenly became extinct; coeval dwarfing among the extinct species and survivors, and skeletal abnormalities, were common. Integration of heat-embryogenesis dysfunction with ice age climatology, and the fossil record, produces new climate-bio-evolution coupling that provides perspective on how mammals fit into the climatology of the modern world, and predictions on how a modern greenhouse might affect them.

In an evolutionary sense, mammals have lived mostly in cool glacial conditions, and only briefly in the hot interglacial intervals that end the ice ages. Today, we live in an interglacial world so hot that summer heat already kills embryos on a vast scale; we may have little thermobiological tolerance for severe greenhouse warming. A Scenario A greenhouse could trigger general collapse of mammalian faunas, and especially so in the middle latitudes.

AIR TEMPERATURE-EMBRYOGENESIS DYSFUNCTION COUPLING

That modern summer heat lowers fertility of mammals (including humans) is well documented. Seasonal drop in reproduction due to heat and humidity is one of the most serious problems of the dairy industry (5, 6), and especially in tropical and subtropical areas. Lowered conception rates due to effects of maternal heat stress upon developing embryos is the dominant factor. For example, the early bovine embryo is extremely sensitive to maternal heat stress, and thermal stress from 30 hours after the onset of estrus until day 7 of pregnancy increases the incidence of abnormal or retarded embryos (7).

Thatcher (8) notes that fertility is inversely related to the maximum environmental temperature the day after insemination and to uterine temperature at, and the day after, insemination. For lactating cattle, air temperatures above 21°C (70°F) cause core temperatures to rise sufficiently to reduce conception rates. For Florida cattle, as maximum air temperatures rise from 21.1 to 35°C (70 to 95°F), conception rates drop from 40 to 31%. Maximum temperatures from 30.7 to 32.7°C (87 to 90°F) exert a thermal stress associated with decreased fertility. Another study noted that as rectal temperatures increase by 1.0°C (1.8°F), pregnancy rates drop from 61 to 45% (9). In another, two groups of heifers were exposed, after breeding, to 32.2°C (90°F), and a constant 21.1°C (70°F); whereas none in the first maintained pregnancy, about half of the low temperature group did so (10).

Figure 1, after Badinga et al. (11), shows the negative relationship between conception rate and air temperature. Based on breeding records for over 12,000 Florida cattle, conception rates of lactating cows decreased sharply when the maximum air temperature on the day after insemination exceeds 30°C (86°F). Rates dropped from 52 to 32% as the maximum air temperature rose from 23.9°C (75°F) in March to 32.2°C (90°F) in July, and remained low in the hot months. A greenhouse, by increasing the number of hot days per year, would expand the time of lowered conception rates into the current cool prime breeding months.

For Virginia cattle, the optimum for cattle conception is from 10°C and 23°C (50 and 73° F); the drop in fertility when the day after insemination exceeds 23°C is typical of southern climates, and may reflect heat damage to embryos during early cleavage stages (12). For Arizona cattle, the cool month conception rate of 50% may drop to 20% in the hot months (13). Missouri herd winter conception rates of 50% drop to 0-20% for July and August (14).

For hogs 18 to 21°C (64 to 70°F) is optimal for productive performance (15); each Celsius degree above 21°C (70°F) reduces hog gain 35 to 60 grams below optimum. Sows mated during hot weather show reduced farrowing rates, due partly to reduced conception rates and to increased early embryo death.

Impact of a greenhouse may be further evaluated via examination of the "acceptable temperature ranges" (the average daily temperature of long-term exposure causing nominal losses in production of livestock) of farm animals (16). They are: dairy cattle (lactating, or within two weeks of breeding), 4 to 24°C (39 to 75°F); calves, 10 to 26°C (50 to 79°F); beef cattle, 4 to 26°C (39 to 79°F); sheep, 4 to 24°C (39 to 75°F); poultry over 10 days old, 13 to 27°C (55 to 81°F); and laying hens, 7 to 21°C (45 to 70°F). A greenhouse that would greatly increase the number of days/year above the acceptable temperature range would extract a severe toll among important farm animals. .

The Hansen et al. Scenario A model (1) estimates the number of days/year exceeding 35°C (95°F) by the year

2050. For Washington, DC, the number of +95°F days will increase from 6 today, to 49 by the year 2050; for New York City from 2.8 to 22; for Memphis, 18 to 82; and for Omaha, 13 to 50. The number of days exceeding 30°C (86°F), which cause sharp drop in conception rates for cattle (11), will expand catastrophically into the current prime cool breeding season. A Scenario A greenhouse would likely trigger general collapse of mammalian faunas. The mechanism by which it would operate is next developed.

High air temperature can decrease male fertility; however, seasonal decreases in fertility associated with heat stress are due primarily to the female (17). Figure 2 shows how maternal hyperthermia can trigger collapse of populations. In this causal-loop diagram, which is based on established principles, pairs of related variables are identified and linked by arrows. Independent variables are at the tail of the arrow, and dependent at the head. The arrows, which may be read as "causes," or "affects," express asymmetrical, irreversible, relationships. Next, causal pairs are linked to others. Closure indicates a feedback loop that is either negative (stabilizing) or positive (runaway expansion, or collapse). Discussion of the diagram follows, beginning in the upper left corner.

As evident from the above discussion, high environmental temperatures "cause" heat stress in female mammals. As a defense against heating, blood flow is shunted to peripheral tissues in order to dissipate heat to the environment (18); this action reduces blood flow to the uterus where embryos are developing (19, 20). Hyperthermia induced a 25-48% decrease in uterine blood flow (UBF) of ewes (21). UBF is a developing embryo's source of oxygen, water, nutrients, and hormones, and also carries damaging metabolic heat away from the embryo. Reduction of uterine blood flow can damage, or kill, developing embryos; a 46% reduction in UBF in heat stressed rabbits produced increased dead fetuses and runts (22).

Air temperature of about 21°C (70°F) cause the core temperature of lactating cattle to begin to rise. At air temperature of about 30°C (86°F), the female's core temperature can have risen by about 1.5°C (2.7°F). An embryo must be maintained at a nearly constant, optimum, temperature; a 1.0-1.5°C (1.8-2.7°F) rise above optimum will kill most embryos. High humidity lowers the air temperature at which core temperature begins to rise.

Figure 2 (middle left) shows that increased embryo death reduces population numbers which, in turn reduces embryo numbers, and so on, producing positive feedback that can trigger reduction of mammalian faunas, or the population collapse indicated in Figure 2 (lower left).

New information is emerging on how, specifically, maternal hyperthermia damages developing embryos. Embryo mortality in hyperthermic cows may be due to heat-induced alterations in synthesis of conceptus proteins involved in embryo development and maternal recognition of pregnancy (23); embryo mortality may result, in part, from failure of the embryo to produce biochemical signals at the proper time to prevent corpus luteum regression. Pigs and sheep respond similarly (24).

Heat stress can produce proportional dwarfs about three-quarters normal size; intrauterine growth retardation resulting from exposure of pregnant females to high temperatures during late gestation may be due to reduced uterine blood flow (25, 26). Hyperthermia can also produce skeletal abnormalities, the effects dependent on the stage of pregnancy at the time of exposure (27).

HUMAN CONCEPTION VERSUS TEMPERATURE

Human conception rates in tropical to warm temperate climate zones show cyclical tracking of the seasons. Larger seasonalities of conception occur in hotter climates. For the subtropical and warm temperate climate zones, maximum conception is in the cool months, and drops off in the hot months. For India and Pakistan, the winter-summer difference in conception rates is 45-50%; for Florida and Louisiana, 35%; and for the interior of North America (Missouri, Kansas, Wisconsin, etc.), and Austria, 15-20% (28). Control of indoor climate in developed countries, and the high humidity of undeveloped tropical regions are factors. For Hong Kong (29), the winter-summer difference is 30%; the conception curve resembles that for cattle (Figure 1), except that human conceptions drop sharply at about 22°C (72°F)

That mammalian faunas are reproductively habituated/adapted to climate zones is clearly seen among humans (28). Maximum conception rates for tropical (Ceylon, etc.) and subtropical (southern India, northern Australia, etc.) climate zones are at 27 and 25°C (81 and 77°F), whereas for the warm temperate (middle and southern United States, Mediterranean region, etc.), the maximum is at 16°C (61°F).

TROPICAL VERSUS MID-LATITUDE GREENHOUSE WARMING: IMPLICATIONS

Schlesinger and Mitchell (30) discuss several general circulation models that estimate surface air temperature change for doubled atmospheric CO_2: the GFDL (Wetherald and Manabe, 1986), GISS (Hansen et al., 1984), and the NCAR (Washington and Meehl, 1984) models. Generally, warming is minimum in the tropics in both summer and winter seasons, and ranges from 2-4°C (3.6-7.2°F). For North America, winter warming is greater than summer; the GFDL and GISS simulations show warming of 4°C (7.2°F) in the south to 10°C (18°F) in the north. The NCAR model simulates less than 4°C (7.2°F) over most of North America. Polar amplification of warming during a greenhouse may explain why, 10,000-12,000 years ago, tropical areas were lightly affected by extinctions, but middle and high latitudes severely so; North America lost nearly 70% of the genera of large mammals. Geopolitical ramifications emerge here: in a modern greenhouse, middle and high latitude mammalian faunas will face greater warming than will tropical, and are in greater danger.

GREENHOUSE WARMING AND MAMMALS: PREDICTIONS

Integration of heat-embryogenesis coupling with ice age climatology, and the fossil record, provides predictions on how rapid warming affects mammals (and birds, and reptiles)--past, present, and future. Future greenhouse warming will likely produce: (1) Heat-induced embryogenesis dysfunction that will reduce population numbers; large mammals, because of their relatively small surface/volume ratios which cause them to retain body heat to a greater extent than do small mammals, will suffer greater losses than small. (2) Dwarfing: intrauterine growth retardation, resulting from exposure to high environmental temperatures during late gestation, produces proportional dwarfs about three-fourths the size of adults. (3) Skeletal abnormalities: hyperthermia produces fetal malformations in some species, the effects dependent upon the stage of pregnancy. (4) Reduction of the middle latitude winter breeding season because of relatively greater winter warming associated with greenhouse warming. (5) Relative severity of thermal effects upon middle to high latitude mammalian faunas; because of the polar amplification of warming, the middle and high latitudes will heat up relatively greater than tropical regions, subjecting mammalian faunas to a relatively greater heat load to have to adapt to.

ICE AGE CLIMATOLOGY: HOW WE FIT IN

In the past two million years, Earth has experienced numerous ice ages, or glacial cycles (Figure 3) driven by variations in Earth's orbit, and by varying amounts of CO_2 in the atmosphere. During each ice age, continental ice sheets developed in the high-middle latitudes of the northern hemisphere. Each glacial cycle lasts about 100,000 years, and is terminated by rapid climatic warming known as interglacials which last about 10,000 years.

In an evolutionary sense (for the past two million years) mammals have lived mostly in cool 100,000 year glacial cycles, and only briefly in hot 10,000 year interglacials. Today, we live upon a thermal plateau in the hot interglacial climate that ended the last ice age (Figure 3); maximum warmth of this interglacial was about 6,000 years ago, 0.5-1.0°C (0.9-1.8°F) warmer than today. The question mark expresses the uncertainty of the future. If humans were not present on Earth, the climate would likely cool into a new ice age, according with mammalian evolutionary experience. However, our production of CO_2 may create greenhouse warming off the thermal scale of mammalian experience. Already, prior to any warming, the heat of this interglacial world kills embryos on a vast scale; we may have little thermobiological latitude to operate within before warming will begin to drive bioevolution irreversibly toward fulfillment of the above-cited predictions.

PREDICTIONS TESTED AGAINST THE PLEISTOCENE-HOLOCENE MAMMALIAN EXTINCTIONS

The global mammalian extinctions of 10,000-12,000 years ago contain more information coupling climate to bioevolution than any other extinction event in Earth history. Being so recent, modern physiological principles directly apply to the extinct mammals, and to the survivors, allowing isolation of climate-physiology couplings that show how modern mammals fit into the climatology of this interglacial world, and provides perspective on how a greenhouse might affect us. Modern mammals are but the survivors of those extinctions.

The coldest part of the last ice age, and maximum southward advance of the ice sheets, was 18,000-20,000 years ago; most mammals survived the cold and reduced living space and food supplies. Then, during warming that ended the ice age, while living space was expanding, and food supplies were increasing, were global scale mammalian extinctions and dwarfing, and skeletal abnormalities (Figure 3). From 15,000- 8,000 years ago, nearly 70% of North America's megafauna died out, with most perhaps disappearing about 12,000-10,000 years ago. Many genera may have dwindled significantly by 12,000 years ago, and many yet to be dated may not have survived beyond 12,000 years ago (31). This antedates the proposed 11,000 year-ago arrival of humans in North America (32).

The predictions linking rapid warming to embryogenesis dysfunction seem to work for the extinctions of 10,000-12,000 years ago (2, 3, 4): large mammals exceeding 44 kg (100 lbs) were most selected against, dwarfing occurred on a global scale, skeletal abnormalities occurred in some genera, and the extinctions were most severe in high and middle latitudes.

Not all ice age terminations produce extinctions; that of glacial cycle C about 125,000 years ago did not. Extinctions occur mostly at terminations of the most severe cold stages (33). The extreme cold stage of glacial cycle C was not as severe as that of cycle B 18,000-20,000 years ago; evidence is that in Illinois tulip trees and pines persisted throughout the former, but not the latter (34).

After climatic warming had reduced and fragmented mammalian faunas, human predation (35) could have been a factor in eliminating some species.

FUTURE CONSIDERATIONS

The keys to appraising the impact of rapid greenhouse warming upon mammalian faunas are these: (1) Today, mammals are generally habituated/adapted to the climate zones they live in. (2) During a major carbon cycle-induced disorganization of the biosphere, finding new optimum thermal niches may be difficult. (3) The world we live in is a hot interglacial one, prior to the onset of any greenhouse warming. (4) Because heat is already killing embryos on a vast scale, many species may already be near to their upper thermal limits. (5) Greenhouse warming at a rate 10 to 40 times natural would outpace the capacity of many species to reduce their surface/volume ratios sufficiently for females to maintain optimum uterine thermal environment for developing embryos. (6) In the middle latitudes, the cool months provide the prime breeding season, and the greater wintertime than summertime warming during a greenhouse would reduce breeding time in the middle latitudes. (7) Because of (6) and also polar amplification of greenhouse warming, middle-high latitude faunas face greater warming than do tropical faunas. (8) Emergence of a Hansen et al. (1) Scenario A greenhouse could trigger general collapse of mammalian faunas of the middle latitudes.

REFERENCES

1. Hansen, J., Fung, I., Lacis, A., Lebedeff, S., Rind, D., Ruedy, R., and Russell, G., 1988, J. Geophys. Res. 93:9341; **2.** McLean, D. M., 1981, Am. J. Sci. 281:1144; **3.** McLean, D. M., 1986, in The Quaternary of Virginia--a Symposium Volume, Virginia Division of Mineral Resources, Pub 75, McDonald, J. N. and Bird, S. O., Eds., 105; **4.** McLean, D. M., 1988, in Global Environmental Protection Act of 1988, Pub. 100-843, U. S. Senate, Washington, DC; **5.** Stott, G. H., Wiersma, F., and Woods, J. M., 1972, J. Am. Vet. Med. Assoc. 161:1340; **6.** Gwazdauskas, F. C., Wilcox, C. J., and Thatcher, W. W., 1975, J. Dairy Sci. 58:88; **7.** Putney, D. J., Drost, M., and Thatcher, W. W., 1988 a, Theriogenology 30:195; **8.** Thatcher, W. W., 1974, J. Dairy Sci. 57:360; **9.** Ulberg, L. C., and Burfening, P. J., 1967, J. Amin. Sci. 26:571; **10.** Dunlop, S. E, and Vincent, C. K., 1971, J. Anim. Sci. 32:1216; **11.** Badinga, L., Collier, R. J., Thatcher, W. W., and Wilcox, C. J., 1985, J. Dairy Sci., 68:78; **12.** Gwazdauskas, F. C., Lineweaver, J. A., and Vinson, W. E., 1981, J. Dairy Sci. 64:358; **13.** Monty, D. E., and Wolff, L. K., 1974, Am. J. Vet. Res. 35:1495; **14.** El-Rabeie, H. M., 1983, Effect of Missouri Climate on the Reproductive Performance of Holstein Dairy Cattle, M. Sci. thesis, University of Missouri, Columbia; **15.** Curtis, S. E., 1985, in Stress Physiology in Livestock, Vol. II, Yousef, M. K., Ed., CRC Press, Boca Raton, 59; **16.** Hahn, G. L., 1976, in Progress in Animal Biometeorology, v. 1, Johnson, H. D., Ed., Swets and Zeitlinger B. V., Amsterdam, 496; **17.** Stott, G. H., 1961, J. Dairy Sci. 44:1698; **18.** Rubsamen, K., and Hales, J. R. S., in Stress Physiology in Livestock, Vol. I, Yousef, M. K., Ed., CRC Press, Boca Raton, 143; **19.** Roman-Ponce, H., Thatcher, W. W., Caton, D., Barron, D. H., and Wilcox, C. J., 1978 a, J. Anim. Sci. 46:167; **20.** Roman-Ponce, H., Thatcher, W. W., Caton, D., Barron, D. H., and Wilcox, C. J., 1978 b, J. Anim. Sci. 46:175; **21.** Oakes, G. K., Walker, A. M., Ehrendranz, R. A., Cefalo, R. C., and Chez, R. A., 1976, J. Appl. Physiol. 41:197; **22.** Leduc, B., 1972, J. Physio. 225:339; **23.** Putney, D. J., Malayer, J. R., Gross, T. S., Thatcher, W. W., Hansen, P. J., and Drost, M., 1988 b, Biol. of Reprod. 39:717; **24.** Biggers, B. G., Geisert, R. D., Wetteman, R. P., and Buchanan, 1987, J. Anim. Sci. 64:1512; **25.** Brown, D. E., Harrison, P. C., Lewis, J. A., and Wallace, M. H., 1977, J. Anim. Sci. 44:442; **26.** Brown, D. E., and Harrison, P. C., 1981, J. Anim. Sci. 52:1114; **27.** Hafez, E. S. E., 1969, in Adaptation of Domestic Animals, Hafez, E. S. E., Ed., Lea and Febiger, Philadelphia, 61; **28.** Macfarlane, W. V., 1974, in Progress in Biometeorology, v. 1, Pt. 1B, Tromp, S. W., Swets and Zeitlinger B. V., Amsterdam, 557; **29.** Parkes, A. S., 1971, J. Biosoc. Sci. Suppl. 3:13; **30.** Schlesinger, M. E., and Mitchell, J. F. B., 1985, in Projecting the Climatic Effects of Increasing Carbon Dioxide, MacCracken, M. C., and Luther, F. M., Eds., United States Dept. of Energy, DOE/ER-0237, 81; **31.** Grayson, D. K., 1987, Quat. Res. 28:281; **32.** Martin, P. personal communication, 1988; **33.** Webb, S. D., 1984, in Quaternary Extinctions, Martin, P. S., and Klein, R. G., Eds., Univ. Arizona Press, Tucson, 189; **34.** King, J., personal communication, 1988; **35.** Martin, P., 1984, in Quaternary Extinctions, Martin, P. S., and Klein, R. G., Eds., Univ. Arizona Press, Tucson, 354.

ACKNOWLEDGMENTS

I extend my deepest appreciation to John C. Topping, Jr., President of the Climate Institute, for the invitation to participate in the exciting and timely Second North American Conference on Preparing for Climate Change: A Cooperative Approach, and to Dr. Frank Gwazdauskas, Professor of Dairy Science, VPI & SU, whose pioneering works on effects of climate upon reproduction, and counsel, have been invaluable to me.

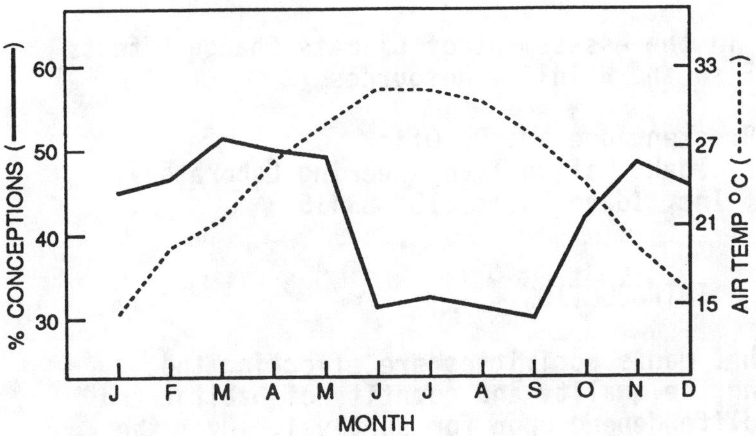

Figure 1. Breeding records of over 12,000 lactating Florida cattle, from Badinga et al. (11), show the negative relationship between cattle conception rates and air temperature. The sharp drop in conception occurs when the maximum air temperature on the day after insemination exceeds 30°C (86°F). As the maximum air temperature rose from 23.9°C (75°F) in March to 32.2°C (90°F) in July, the conception rate dropped from the cool month rate of 52% to the hot month 32%, and remained low during the hot months.

Figure 2. Causal-loop diagram. Uterine blood flow (UBF) is an embryo's source of critical resources, and carries damaging maternal metabolic heat from the embryo. Environmental heat reduces UBF, damaging, or killing, embryos. Increased embryo death, by reducing population numbers that, in turn, reduce embryo numbers, can produce the closed loop positive feedback shown on the middle left of the figure. Greenhouse warming of the magnitude and rate of the Hansen et al. Scenario A (1) could trigger the exponential collapse of mammalian faunas indicated in the lower left inset.

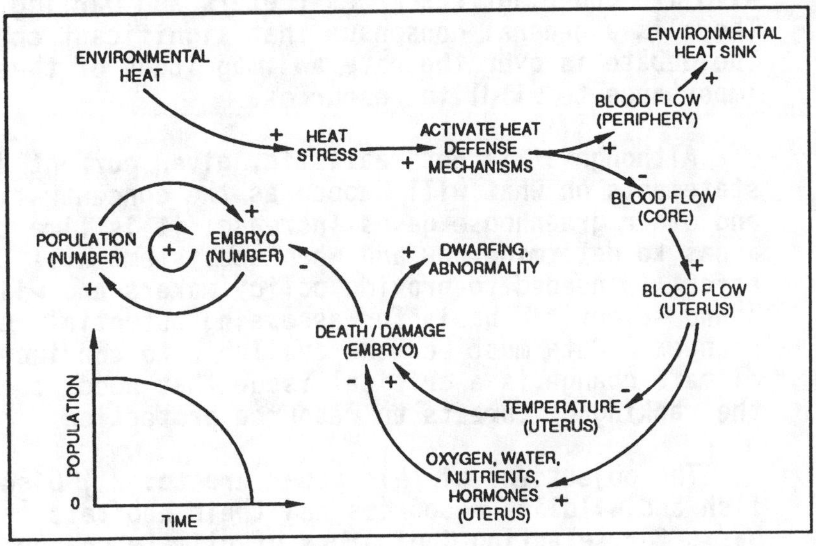

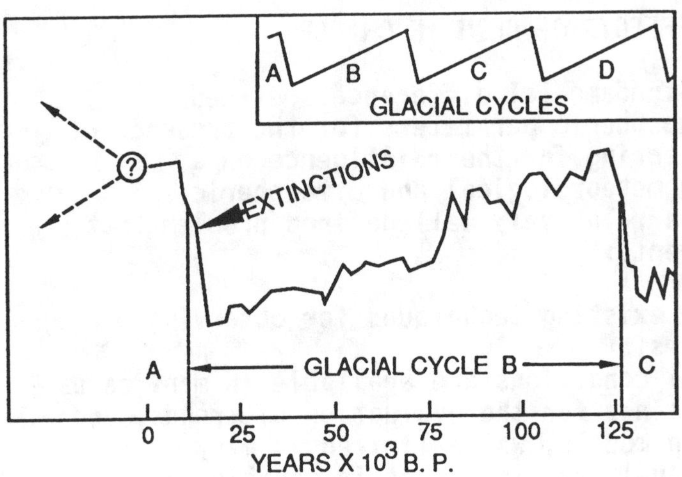

Figure 3. For the past two million years, Earth has experienced ice ages, or glacial cycles, each of about 100,000 years duration. Ice ages are terminated by rapid warming termed interglacials which last about 10,000 years; we live in an interglacial world. Most mammals survived the coldest stage of the last ice age 18,000-20,000 years ago, at which time ice sheets covering large areas of the high-mid latitudes of the Northern Hemisphere would have reduced living space and food resources. Then, from 12,000-10,000 years ago, as Earth was warming, and living space, and food supplies were expanding, global extinctions, dwarfing, and skeletal abnormalities occurred. Modern mammals are the survivors of those extinctions.

Monitoring Concepts Useful in the Assessment of Climate Change Effects on U.S. Fish and Wildlife Resources

R. P. Breckenridge, M. D. Otis
Scientific Specialists, Idaho National Engineering Laboratory
EG&G Idaho, Inc, Idaho Falls, ID 83415

INTRODUCTION

Most scientists agree that man's activities are effecting the earth's climate and impacting the quality and quantity of habitat and resources that fish and wildlife depend upon for survival. Over the past few years, there have been numerous conferences, workshops and scientific articles that have discussed and assessed potential effects of climate change on the biological diversity and management of fish, wildlife and plant resources (Peters and Darling, 1985; Pain, 1988). There is a general consensus that significant changes are occurring. The debate is over the rate and magnitude of these changes and their importance to wildlife resources.

Although it is not realistic, given current data, to make definitive statements on what will happen as the concentrations of carbon dioxide and other greenhouse gases increase; it is time to monitor sensitive areas to determine how and when ecosystems will change. This monitoring effort is needed to provide policy makers and wildlife managers with a sound scientific basis for assessing potential responses to these changes. Data must be made available to convince policy makers that climate change is a critical issue that needs to be given priority in the ranking of threats to resource protection.

The objectives of this paper are to: 1) Discuss the sensitivity of fish and wildlife resources and their habitats to climate changes as a basis for selecting indicators of climate change effects; 2) outline an anticipatory monitoring approach for trend assessment; and 3) identify regional and national challenges for management response to effected wildlife resources.

MONITORING THE EFFECTS OF CLIMATE CHANGE

It is important to note the fundamental differences between measuring meteorological and atmospheric parameters for the presence of global climatic changes and monitoring for their influence on a particular ecosystem. Measuring meteorological and atmospheric parameters of climate change is a relatively well defined problem that is based on three important elements:

1) changes can be measured using existing techniques for observing standard meteorological parameters,
2) data describing meteorological conditions are available in many cases in enough detail to develop baselines for the evaluation of trends, and
3) global climate and circulation models, while limited in many respects, provide a solid conceptual framework that is capable of generating testable hypotheses about the direction and magnitude of anticipated changes.

Monitoring for the effects of climate change on an ecosystem needs additional attention with respect to all three of these elements. First, reliable indicators of climatic effects are only starting to be identified. Some candidate ecosystem parameters are available that can be incorporated into a monitoring program, but most have been developed for other purposes and their value as measures of climate change is unproven. Second, for many of the candidate parameters, baseline data is lacking, making it difficult to interpret observed changes. Finally, the conceptual framework linking anticipated changes to measurable effects are often inadequately developed. The general principles are available but in this context are unproven.

The development of testable links between anticipated climatic changes and observable ecosystem parameters is essential to any monitoring effort. In some cases, adequate baseline or paleoecological data (McLean, 1988) are available to develop hypotheses that can be statistically tested. These data sets should be the focus of current activities by resource agencies and regulators to identify indicator species that are/or could be most sensitive. It has been many years since scientists first started sounding the alarm about ecosystem shifts due to increases in greenhouse gases. It is unclear what changes have occurred since then, but data do show increases in temperature and CO_2 concentrations (Detwiler and Hall, 1988; Anon, 1988.)

Any monitoring program, directed at detecting climate change or evaluating its effects, must recognize the long-term nature of the problem. Current predictions indicate the possibility of significant change over the next several decades. This time scale is short on a geological time scale, but is very long compared to the funding time scale for a typical environmental monitoring program.

INDICATORS OF CLIMATE CHANGE EFFECTS

A monitoring plan should be a cooperative effort that includes local, regional, national and global conditions. The plan should focus on the use of indicators that can be designed into a complete program that have potential for showing a detectable response within practical limits of time and effort. An ideal indicator should have the following features:

<u>Sensitive Ecosystems</u> - Given the variation in climatic conditions and in natural ecosystems, monitoring indicators should have a realistic chance of showing detectable effects. Any monitoring program that provides direction for management of wildlife resources should be aimed at key indicators of subsequent effects. Ecosystems that are not intensively managed will, in general, be more sensitive and probably subject to earlier change than managed systems. Ecosystems that are in regions expected to experience large climate changes are more likely candidates. Finally, ecosystems that are on the edge of sharp environmental gradients are likely to be more sensitive. Existing wildlife refuges, national and state parks, forests and other ecosystems that meet these criteria include, but are not limited to: 1) high latitude wetlands and terrestrial systems (tundra/permafrost);

2) high elevation lakes and terrestrial systems (montane/alpine); 3) coastal marshes (marine/estuarine); and 4) arid inland wetlands and terrestrial systems (desert islands/native grassland).

<u>Links to Climate Change</u> - To interpret results, the chosen indicators should be linked (either directly or as a surrogate) to specific climatic variables being measured as part of the monitoring plan. There are at least four broad categories of indicators that may be linked to climate change. These are: 1) changes in the abiotic environment related to water, its abundance, quality, and distribution; 2) spatial distributions of key species of plants and animals that directly measure shifts in habitat; 3) ecosystem structure parameters such as diversity indices and relative abundances of key species that indicate habitat shifts and ecosystem stress; and 4) ecosystem function parameters such as production/respiration ratios that measure changes in stress.

<u>Small Natural Variation</u> - Detecting a long-term signal in the presence of relatively short-term noise (natural variation) imposes severe constraints. The ideal indicator for monitoring purposes would have a small natural variation compared to the anticipated change due to climatic effects. Statistical techniques should be used for estimating the accuracy of measurement and duration of monitoring required to detect that change.

<u>Adequate Baseline Data</u> - An adequate baseline must be established if monitoring results are to be interpreted. It is important that data be obtained (if possible) on the climatic conditions that are linked to the environmental indicator, preferably measured at the same location over the same time period. The literature provides some good information for potentially useful indicators (Cooperrider et al., 1986).

<u>Measurements</u> - In the interests of maintaining a long-term monitoring program, indicators should be as simple, stable, and inexpensive as possible. Simple measurements require the minimum of complex analysis and interpretation. Stable measurements are not sensitive to small variations in technique or instrumentation among researchers. Basic ecosystem functions are the key parameters that should be investigated. Simple and cost-effective techniques are not necessarily "low tech", labor intensive approaches. Remote sensing by satellite, for example, can detect changes in vegetation distribution and a number of ecosystem function parameters quite reliably provided sufficient "ground truthing" is available.

<u>Frequency and Timing of Measurement</u> - The duration of several decades required for monitoring programs is inescapable. Frequency of measurements within that program duration are key elements of monitoring design. Indicators with strong seasonal dependence should be avoided for routine monitoring purposes unless the measurement is easily performed or the optimum timing can be reliably predicted.

Problems of natural variation and non-climatic influences can make interpretation difficult even for the best of indicators. Therefore, reliance should not be placed on a single measure of climate change effects. Monitoring a selected suite of indicators along with appropriate meteorological parameters offers the greatest chance of early detection of climate change effects and allows the greatest latitude for management responses.

A key aspect of putting a monitoring strategy together is to take advantage of the knowledge base that exists. Resources would be well spent if the initial data assessment stage identifies sensitive species that meet the criteria of many of the items discussed in this section. To pursue a monitoring program without considering this information and not addressing the functions of the ecosystem would probably doom a program to failure.

MANAGEMENT OPTIONS AND POLICY ISSUES

Resource managers, regulators, and policy makers need to look at the entire system being managed to assess individual and cumulative impacts. In many cases, focusing on only one issue jeopardizes the future health of the resource. The importance of advanced planning to ensure that all issues are evaluated adequately cannot be understated. This requires a strongly proactive stance. The proactive position allows for the planned maintenance and enhancement of a resource, the reactive position makes it difficult to address more than today's problems and leaves future protection of resources in question.

The ultimate challenge is whether resources can be managed with enough foresight to provide habitat to those resources that cannot adapt quickly enough to man's imposed changes. It is critical that scientists communicate with policy makers to ensure that proper data are collected to address future management and/or protection of the resource. Here are a few general concepts and ideas that could help in the development of strategies on resource management policy development.

1. Should policy and management decisions continue to protect current communities or should new communities be allowed to establish? These decisions include public and scientific images of particular species/populations/ communities, species importance relative to communities and ability of a community to withstand anticipated change. It may be that some species will have to be sacrificed to provide more funds for the whole. Some decision criteria and procedures will need to be established jointly by scientist and policy makers to address resource allocation.

2. Mitigation options need to be assessed to determine how effective they might be at addressing climatic change effects. If concepts like human movement of seed stock, developing migration paths and designing artificial habitat are to be considered options; then scientific approaches, cost/benefits policy and management issues must be addressed.

3. Resource managers should include a section within their short/long-term planning and impact documents for consideration of various potential climate change effects on their respective resources and how these impacts might affect management decisions/practices.

4. Land acquisition should give specific considerations to: 1) establishment of corridors for migration of displaced species; and 2) acquisition of new lands based on potential new ranges of various species/communities of concern.

5. Monitoring is a significant issue for management and policy makers' immediate consideration, programs need to be developed on a cooperative basis.

In summary, it is important to encourage federal, state, and private managers and policy makers to pool key scientists to monitor and plan now for how our resources will be managed decades from now. It will be this type of advanced planning and policy making that protects our Fish and Wildlife resources for future enjoyment.

ACKNOWLEDGEMENTS

The authors appreciate the review by Dr. G. B. Wiersma, Mr. R. C. Rope and Mr. J. E. Cornish.

This work was supported by the Environmental Protection Agency and the U.S. Department of Energy under Contract No. DE-AC07-76ID01570.

LITERATURE CITED

Anon., "Taking the Worlds Temperature", The Economist, July 9, 1988, p. 80.

Bailey, R. G., Description of the Ecoregions of the United States, U.S. Department of Agriculture, National Forest Service, Ogden, Utah.

Breckenridge, R. P., M. D. Otis, J. G. Cornish, R. C. Rope and R. E. Trout, "Challenges of Assessing Climate Change Effects on Fish and Wildlife Resources in the Western United States", 1988 Joint Conference of WAFWA/WDAFS, Albuquerque, NM, July 10-13, in press.

Cooperrider, A. Y., R. J. Boyd and H. R. Stuart, Inventory and Monitoring of Wildlife Habitat, U.S. Department of Interior, Bureau of Land Management, U.S. Government Printing Office, Washington, DC, 1986.

Detwiler, R. P. and C. A. Hall, "Tropical Forests and the Global Carbon Cycle", Science, 239, 1988, pp. 42-47.

McLean, D. M., "Climatic Warming and Mammalian Evolution/Extinction", Statement to U.S. Senate Subcommittee on Hazardous Wastes and Toxic Substances, September 14, 1988.

Pain, S., "No Escape from the Global Greenhouse", New Scientist, November 12, 1988, pp. 38-43.

Peters, R. L. and T. D. S. Darling, "The Greenhouse Effects and Natural Reserves", Bioscience 35:11, 1985, pp. 707-717.

Woodward, F. I., Climate and Plant Distribution, Cambridge Studies in Ecology, Cambridge University Press, New York, 1987.

CLIMATE CHANGE EFFECTS
ON
FISH, WILDLIFE AND OTHER DOI PROGRAMS

Indur M. Goklany*
Office of Program Analysis
Department of the Interior
18th and C Streets, NW
Washington, DC 20240

CLIMATE CHANGE AND THE DEPARTMENT OF THE INTERIOR

DOI is directly responsible for managing about a quarter of the U.S. land mass, and natural resources in the Outer Continental Shelf (OCS). It manages much of the water resources in the Western U.S. It also is responsible for the economic and social well being of American Indians, Alaska Natives and the peoples of the U.S. territories. If climate changes, DOI would be affected in a variety of ways:

- DOI is the primary manager of the nation's natural ecosystems. It also has special responsibilities in preserving native habitats and species, fish and wildlife, biodiversity, and wetlands. Most assessments indicate that the major adverse effect of climate change will be in precisely these areas. However, some studies also indicate that certain species of fish and wildlife could be more abundant in the event of climate change.

- DOI is the manager of much of the nation's energy resources, and it has a special stake in this nation's energy security. Response strategies could involve changing the emphasis on related programs.

- Because DOI is the largest landowner in the nation, any response strategy involving reforestation or afforestation can only succeed with active DOI participation.

- DOI manages several areas in or near coasts including Parks, historic structures, and Wildlife Refuges which could be affected by sea level rise.

- DOI manages about 80% of Alaska which will be profoundly affected by climate change. (Climate change is expected to be magnified as one goes northward).

*** Views expressed here are the author's and not necessarily the Department of the Interior's.**

- DOI is responsible for promoting the social and economic well being of several U.S. territories (e.g., Guam, American Samoa, Virgin Islands). These are all islands and will be affected by sea level rise.

- The Bureau of Indian Affairs will be affected because a good portion of the Indian lands are in the arid southwest, and because Indians have valuable water and energy holdings (especially coal).

- The Bureau of Reclamation has planned and developed water resource projects throughout the Western U.S., and it manages a substantial portion of these. It has responsibilities towards flood control, power generation, dam safety as well as water deliveries for use in agriculture, municipalities, industry, recreation and other in-stream uses. If climate changes, changes in water supply and demand will almost certainly occur at every location and in every river basin. This would certainly affect DOI's mission in this area.

- U.S. Geological Survey traditionally has had research responsibilities in the areas of water resources, biogeochemical cycles, paleoclimatology and data handling and management.

Table 1 indicates the various ways in which the different bureaus within DOI could be affected by climate change. Evidently, both the conservation and development sides of DOI's house will be affected by any climate change and actions designed to ameliorate the threat of climate change.

FRAMEWORK FOR RATIONAL POLICY ANALYSIS

While today there is broader recognition than in the past of the possible effects of climate change on DOI's functions and responsibilities, this has not resulted in any concrete actions or responses within the bureaus. Policy makers, given scarce resources and budgets, are unlikely to embark on major policy fixes without rational policy analysis of the climate change issue.

As a start, one needs to separate what is likely from what is possible. Inherently, anything that does not violate a law of nature (e.g., the third Law of Thermodynamics) is, possible. Thus dealing with merely what is possible is not a wise use of resources. Ideally, one would need to know the expected value of climate change effects (as a function of time). However, it is probably more practical to provide information on the direction, magnitude and geographical extent of effects and the rate at which these effects would likely be manifested. Such analyses must of necessity be at a regional or, possibly, smaller scale.

The importance of knowing about the rate of change of effects cannot be overemphasized because response strategies generally have human and financial costs associated with them. A measure implemented before it is necessary will take away human and financial capital from other worthwhile enterprises. In addition to numerous social, public health and environmental problems which need immediate attention, generation of society's wealth is one such enterprise -- such wealth will eventually make response measures more affordable. (This is especially important for developing nations where standards of living -- and levels of health and social services -- are low). Moreover, the longer one can afford to wait, the greater the likelihood that technological progress will reduce the costs and consequences of responses. On the other hand, if technological progress proves to be too little or too late, and the greater the effect to which one cannot adapt, the costlier will it be to postpone limitations or reductions in greenhouse gases. This indicates the need to reduce uncertainties regarding the direction, magnitude, geographical extent and rate of change of effects.

Once such information has been obtained, one must analyze response strategies in terms of their effectiveness, and their social, economic and environmental consequences. These must be compared with each other and with the consequences of doing nothing. In this context, effectiveness would be measured by (a) the increase in the adaptability to climate change, or (b) slowing down of the rate at which climate changes or more importantly, the rate at which adverse effects of climate change occurs.

The state-of-the-science, however, is such that rational analysis using the framework outlined above is not currently feasible. As the discussion below shows, estimating effects itself is a formidable task. Because of this, expensive measures to adapt to or limit climate change _per se_ are unlikely to be adopted. However, the possible adverse effects of climate change could provide the argument tipping the balance in favor of measures that almost make sense on their own and which -- incidentally -- would also diminish climate change or its effects.

EFFECTS CALCULATIONS

Figure 1 is a simplified wiring diagram showing how effects should be estimated. This diagram shows four basic elements (greenhouse gas concentrations, climatic measures and abiotic factors, effects on hydrosphere and biosphere, and human activities). The breakdown into four elements is somewhat arbitrary because human activities could be subsumed under the biosphere element. The Figure also indicates three basic iterative steps. First, based on natural processes and human activities, the rate at which greenhouse gas (GHG) concentrations will change must be estimated. This must be done using biogeochemical cycles (e.g., the carbon cycle, methane cycle). Second, the rate of change for relevant climatic measures and/or abiotic factors affecting the biosphere, hydrosphere and atmosphere must be calculated. This could be done using general circulation models (GCMs). Third, the rates at which effects occur in the biosphere and hydrosphere must be calculated using various "effects models". As the Figure shows, each element affects the other. Moreover, within each element, there are numerous sub-elements that are interconnected with each other.

Reality, however, is far short of Figure 1. Figure 2 shows how the most sophisticated of the effects studies to date have been done. Comparing the two Figures indicates that few of the interconnections have yet been factored into the calculations. Notably, little consideration has been given to (a) the connections between biosphere, hydrosphere and atmosphere, and (b) many of the climatic measures and abiotic factors that could affect the biosphere and hydrosphere (e.g., year-to-year variability, magnitude and frequency of extreme events). Moreover, even where these connections have been considered, within and between each element there are several unknowns and uncertainties associated with the science and modeling tools used today. For example, most of the crop, vegetation and forest models are designed for and based on today's atmosphere and simply may not be appropriate for future high CO_2 atmospheres. Similarly, current tools to estimate changes in greenhouse gas concentrations and climatic measures have substantial limitations.

Figure 2 also indicates that the more sophisticated present day assessments treat calculation of effects as essentially the last step in a 3-step calculation. Thus, confidence in effects is inherently low.

Table 2 indicates one group of experts' opinion of how long it would take for scientists to reach consensus on a number of important climatic measures using GCMs (AAAS, 1988). Clearly, consensus on effects is at least as far away as the 10-50 year time frames indicated on that Table.

SUGGESTED RESPONSE STRATEGIES

A number of response strategies have been suggested as methods of mitigating possible impacts of climate change on fish, wildlife and habitats. These include:

- Purchasing more land to improve odds of survival of native habitats, to establish artificial habitats, and to "restore" old or new habitats (EPA, 1988).
- Purchasing more land to establish migratory corridors (Peters, 1988).
- Factoring the possibility of climate change into the planning and management documents produced by the various land management agencies (e.g., Goklany, 1988).
- Factoring climate change into decisions to purchase additional lands (e.g., consider sea level rise when purchasing shoreline properties)(e.g., Goklany, 1988).

- Research to identify, document and study existing species to establish baseline and to improve odds of future survival.

- More active human management, e.g., to establish new habitat or help species migration (e.g., Goklany, 1988).

However, the low level of confidence in effects estimates and the lack of knowledge of impacts at the regional (or smaller) level probably precludes any near term implementation of costly options such as additional land purchases because questions such as which plot of land to buy or which migratory corridor to establish cannot be answered yet.

An alternative and more cost-effective option to land purchase would be to ensure that requirements for land and water uses which compete with natural ecosystems are reduced. Worldwide, these are agriculture, grazing, the quest for fuelwood and the expansion of human settlements. Dealing with these would also directly address many of the present causes of deforestation and loss of biodiversity. Such measures could include:

- Making agriculture more efficient in terms of land and water use, e.g., by focussing research on developing appropriate new cultivars.

- Reducing subsidies for agriculture and grazing.

- Increasing incentives for more dense human settlements while assuring environmental quality standards are met.

That such measures would be effective can be seen by the fact that wildlife in several parts of the Northeastern U.S. (e.g., New York and Connecticut) is more abundant now than at anytime since the turn of the century. Much of this has been attributed in large part, to the decrease in agricultural acreage (Kolbert, 1986; Hatch, 1987). For example, in 1910, New York State had 24 million acres of farmland; now it has less than 5 million acres. Because of this there has been a net increase in forest land even though over the same time period human settlements expanded from about half to 5 million acres. The sheer quantity of land freed up by displacing agriculture could not have been bought reasonably by direct purchases.

Not only would such measures allow more land to stay in a "natural" state, they would also reduce overall land prices making purchase of land for conservation more effective. Other benefits of such measures include:

- A probable increase in global CO_2 sink capacity.

- Less methane (CH_4) from grazing. The importance of reducing CH_4 cannot be overemphasized. See Table 3. Pound-per-pound it has a greater potential for greenhouse warming than carbon-dioxide (CO_2). One estimate indicates that it has already contributed about 27% of the total greenhouse warming since pre-industrial times (compared to 59% for carbon dioxide) (Dickinson and Cicerone, 1986). It's atmospheric concentration is now growing at a faster rate than CO_2's (1.1% per year vs. 0.4% per year) (Wuebbles and Edmonds, 1988). Also CO_2 can be beneficial to agriculture and vegetation because it can increase both the photosynthesis rate and drought resistance in many plants (see e.g., Warrick et al., 1986), whereas CH_4 has little or no redeeming value associated with it.

- More water for in-stream uses such as recreation or hydroelectric power.

SUMMARY

Many DOI programs and responsibilities would be significantly affected were climate to change or if measures to limit the build-up of greenhouse gases were adopted. Nevertheless, changes in natural resource management programs requiring substantial resource commitments are unlikely to be undertaken unless there are significant reductions in the uncertainties and gaps in knowledge regarding the direction, magnitude, geographical extent and timing of effects due to climate change. Such improvements in knowledge could take a few decades -- rather than a few years.

It is extremely important to reduce uncertainties regarding the timing of effects because responses will require human and financial resources to implement. The greater their social, economic and environmental costs, the more important it is to reduce uncertainties. Implementing responses before they are necessary deprives other worthwhile social, public health and environmental enterprises of much needed, but scarce, resources. This is especially true for developing countries where standards of living -- and levels of health and social services -- are low.

In the meantime, several low cost measures could be implemented. Perhaps the most effective of these would be those designed to reduce land and water uses competing with natural ecosystems and habitats. Such measures include supporting research or reducing barriers to (a) making agriculture more efficient in terms of land and water use, (b) eliminating subsidies for agriculture and grazing, and (c) increasing densities of human settlements.

Finally, it may be more cost-effective to reduce CH_4 than CO_2. Moreover, the latter may help agriculture and natural systems become more adaptable, while CH_4 has virtually no redeeming features.

REFERENCES

AAAS, 1988. Table provided by American Association for the Advancement of Science, Climate Project, 1414 H St., NW, Washington, DC, December 1988.

R.E. Dickinson and R.J. Cicerone, 1986. "Future global warming from atmospheric trace gases", Nature, 319, pp. 109-115.

EPA, 1988. "The Potential Effects of Global Climate Change on the United States", draft Report to Congress, prepared by U.S. Environmental Protection Agency, October, 1988.

I.M. Goklany, 1988. Remarks and Comments at Climate Institute Wildlife Symposium, Washington, DC, January 21-22, 1988.

D. Hatch, 1987. "Naturalists List the Beasts", New York Times, Section 23, p. 1, September 6, 1987.

E. Kolbert, 1986. "For Wildlife the Times are Very Good", New York Times, Section E, p.9, November 23, 1986.

R.L. Peters, 1987. "Effects of Global Warming on Biological Diversity: An Overview", Proc. of the First North American Conference on Preparing for Climate Change: A Cooperative Approach, October 27-29, 1987.

R.A. Warrick, R.M. Gifford with M.L. Parry, 1986. "CO_2, Climatic Change and Agriculture" in "The Greenhouse Effect, Climatic Change and Ecosystems", ed. B. Bolin et al., SCOPE 29, J. Wiley and Sons, pp. 393-474.

D.J. Wuebbles and J. H. Edmonds, 1988. "A Primer on Greenhouse Gases", prepared for U.S. Department of Energy, Office of Energy Research, Washington, DC 20545, March, 1988.

Table 1: Possible Effects of Climate Change on DOI Bureaus

	Sea level rise, Wetlands	Water resources, precip. changes	Vegetation, Crops	Energy Production	Bio-diversity	Integrity of parks, refuges, wilderness areas
NPS	x	x	x		x	x
FWS	x	x	x	x	x	x
GS	x	x				
BR		x	x			
BM				x		
BLM		x	x	x	x	x
MMS	x			x		
BIA		x	x	x		
TIA	x					
OSM				x		

NPS = National Park Service
FWS = Fish and Wildlife Services
GS = Geological Survey
BR = Bureau of Reclamation
BM = Bureau of Mines
BLM = Bureau of Land Management
MMS = Minerals Management Service
BIA = Bureau of Indian Affairs
TIA = Territorial and International Affairs
OSM = Office of Surface Mining Reclamation and Enforcement

FEARLESS PROJECTIONS

date: 7/22/88

TABLE 2

PHENOMENA	PROBABLE GLOBAL ANNUAL AVERAGE CHANGE	DISTRIBUTION OF CHANGE: REGIONAL AVERAGE	DISTRIBUTION OF CHANGE: CHANGE IN SEASONALITY	* INTERANNUAL VARIABILITY	SIGNIFICANT TRANSIENTS	CONFIDENCE OF PROJECTION: GLOBAL AVERAGE	CONFIDENCE OF PROJECTION: REGIONAL AVERAGE	ESTIMATED TIME FOR RESEARCH THAT LEADS TO CONCENSUS (YEARS)
TEMPERATURE	+2 to +5 C	-3 to +10 C	YES	DOWN?	YES	HIGH	MEDIUM	0 to 5
SEA LEVEL	+10 to +100 cm	(2)	NO	???	UNLIKELY	HIGH	MEDIUM	5 to 20
PRECIPITATION	+7 to +15 %	-20 to +20 %	YES	UP?	YES	HIGH	LOW	10 to 50
DIRECT SOLAR RADIATION	-10 to +10 %	-30 to +30 %	YES	???	POSSIBLE	LOW	LOW	10 to 50
EVAPOTRANSPIRATION	+5 to +10 %	-10 to +10 %	YES	???	POSSIBLE	HIGH	LOW	10 to 50
SOIL MOISTURE	???	-50 to +50 %	YES	???	YES	???	MEDIUM	10 to 50
RUNOFF	INCREASE	-50 to +50 %	YES	???	YES	MEDIUM	LOW	10 to 50
SEVERE STORMS	???	???	???	???	YES	???	???	10 to 50

Notes:

SOURCE: AAAS, 1988

(1) For an "equivalent doubling" of atmospheric CO2 from the preindustrial level. These are equilibrium values, neglecting transient delays and adjustments.

(*) Inferences based on preliminary results of David Rind (GISS) personal communication for Continental U.S.

(2) Increases in sea level at approximately the global rate except where local geological activity prevails.

(???) No basis for quantitative or qualitative forecast.

Table 3: Comparison of the Effects of Methane and Carbon Dioxide

	CO_2	CH_4
Greenhouse Warming Contribution to Date	59%	27%
Growth rate per Year	0.4%	1.1%
Greenhouse Warming as a Function of Concentration	logarithmic	linear
Direct Effects on Plants		
Fertilization	+	
Drought Resistance	+	
Resistance to other Stresses	+(?)	

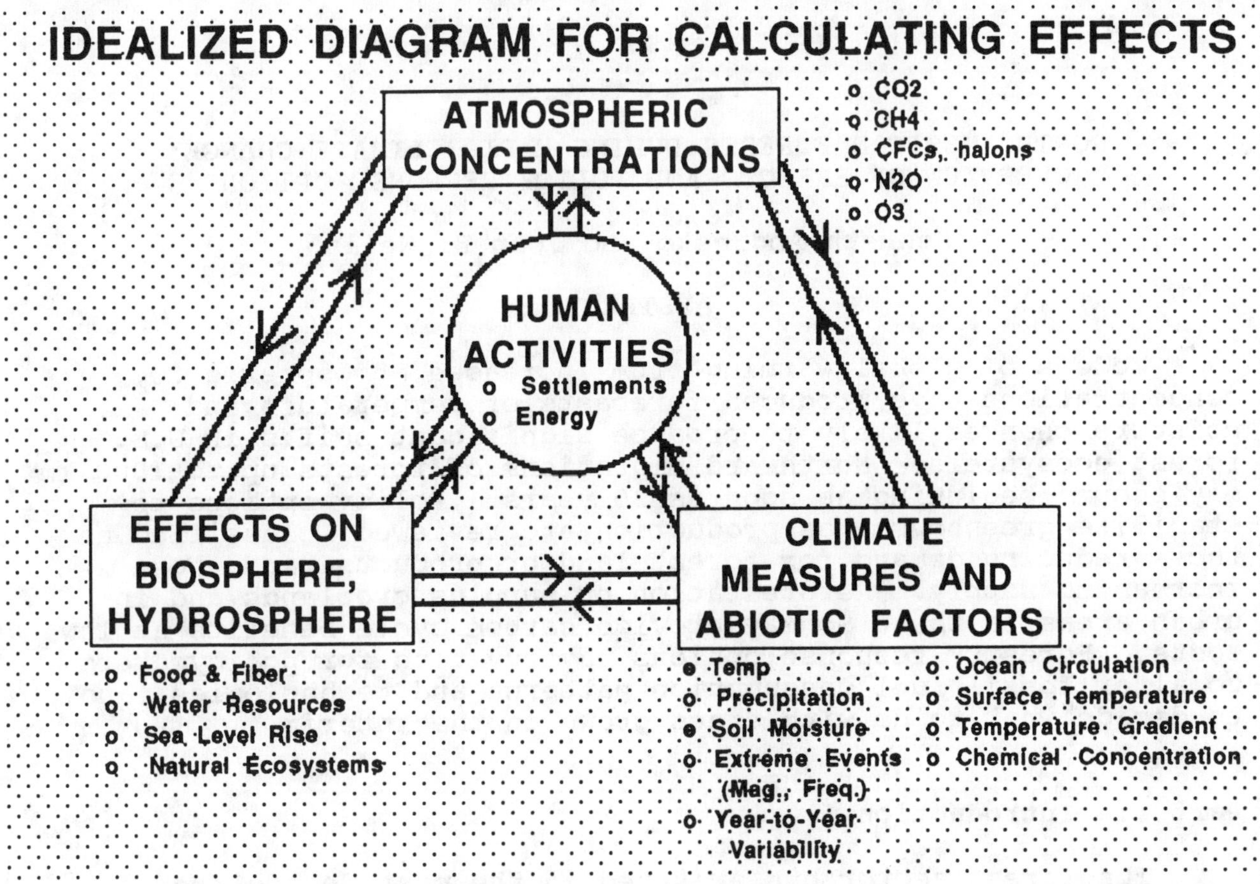

Figure 1

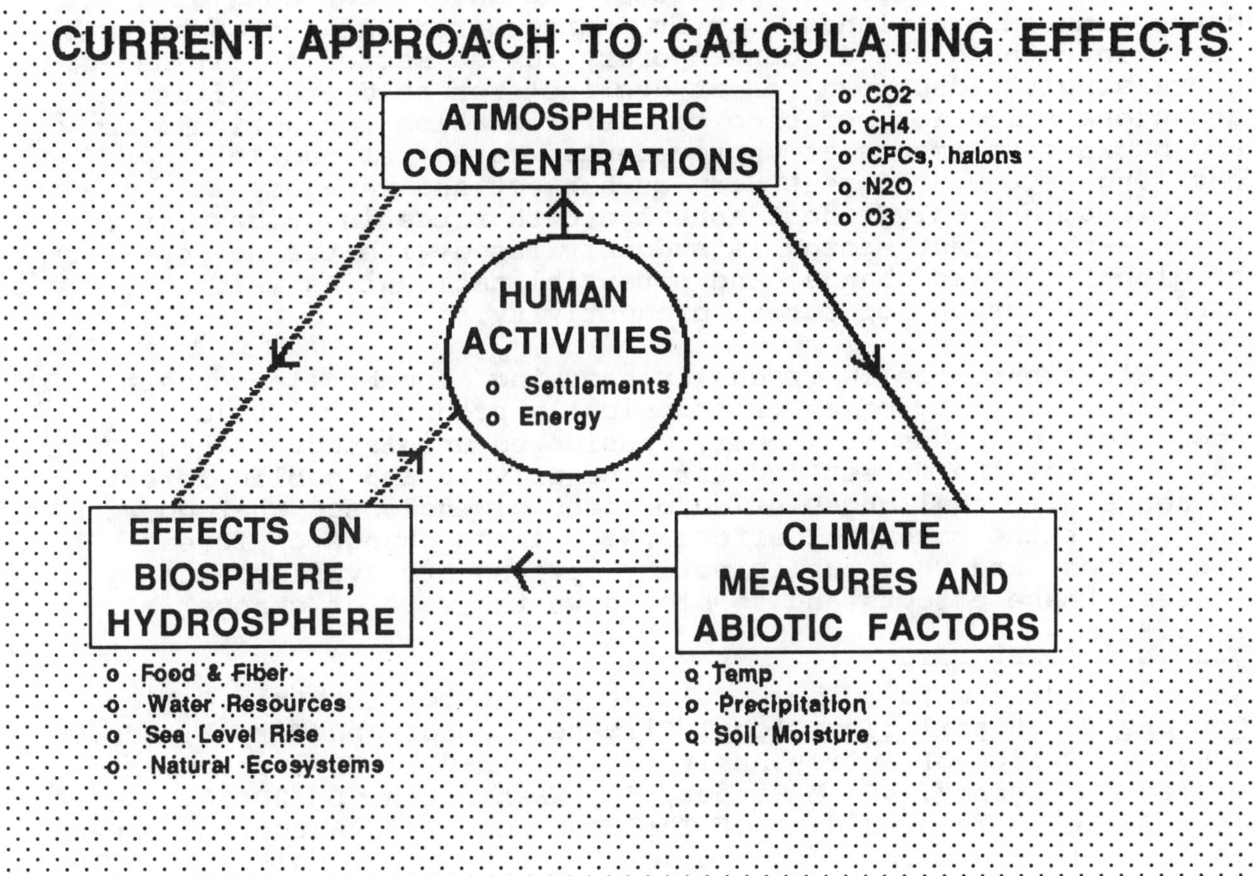

Figure 2

NORTH AMERICAN FORESTS DURING RAPID CLIMATE CHANGE: OVERVIEW OF EFFECTS AND POLICY RESPONSE OPTIONS

Kenneth Andrasko and John B. Wells

ABSTRACT

Models of global warming from increased atmospheric CO_2 concentrations have produced forecasts of temperature and moisture changes likely to produce significant shifts in U.S. forest ecosystems. Northward migrations of forests up to 1000 km may occur, beginning as soon as 30 years. Policy options to stabilize greenhouse gas production are reviewed, classified as those reducing demand for forest land or products, and those increasing supply. Reforestation on surplus croplands and in urban areas in the U.S. may provide carbon cycle benefits at low costs. Forestry response options may buy time for policymakers to alter fossil fuel consumption patterns and technologies, and to decide what to do with trees grown in the interim.

Part I: Introduction

The forest sector has awakened in the past decade from a century-long dream of bountiful stands available at low costs, healthy demand for forest products, and government subsidies for harvesting federal lands. As it emerges from the cocoon of this dream into the glare of contemporary forest sector realities--of increasingly widespread forest decline from acid precipitation and ozone pollution, of tropical deforestation for agriculture and human settlements at unprecedented rates, of public challenges to U.S. Forest Service harvest and management practices, and only modest annual growth rates in prices (in real terms)--the forest sector is suddenly faced with the new specter of global climate change, and potential national as well as global disruptions of forest productivity.

Climate change is gradually becoming an umbrella concept unifying a variety of anthropogenically produced climatic and ecological problems previously considered unrelated. As such, it may confront the forest industry and private and public forest managers with challenges of unforeseen dimensions. What do we know about the potential effects of climate change on forest ecosystems, and what public policy options are available to address these effects and to slow down the rate of change?

Kenneth Andrasko is author of Forestry chapter in EPA's Policy Options for Stabilizing Global Climate, Lashof and Tirpak, eds., 1989, in Office of Policy Analysis/SSB, PM-221, Environmental Protection Agency, 401 M Street, SW, Washington, DC 20460.
John B. Wells is president of The Bruce Co., and author of papers on forestry and climate change, at The Bruce Co., Suite 210, 1100 6th St., SW, Washington, D.C. 20024.

Forests cover about one-third of the earth's land, or 4 billion hectares, of which about 42% is in developed countries (mostly temperate) and 58% in developing countries (mostly tropical). Forest ecosystems store 20-100 times more carbon per unit area than croplands (Houghton, 1988), and play a critical role in reducing ambient CO_2 levels, by sequestering atmospheric carbon in the growth of woody biomass.

Anthropogenic alterations of forest ecosystems, however, now account for an increase in atmospheric CO_2 equal to about 20-30% of total emissions from combustion of fossil fuels, as carbon stored in vegetation and soils is released by clearing, fire, or decay (Houghton, 1988). Estimates of net carbon flux from deforestation for 1980 range from 0.4-2.6 Petagrams of carbon per year (Pg C/y), with 1-1.6 Pg a reasonable average, compared to 5 Pg C released from fossil fuel combustion (Detwiler and Hall, 1988).

Forests in temperate regions are essentially now in balance in terms of carbon cycling, with annual incremental growth rates roughly equal to timber harvest and deforestation for urban growth and other land uses. Consequently, temperate forests do not currently contribute significantly to increased atmospheric CO_2.

Temperate forests potentially could be expanded through widespread reforestation programs into former forest ranges, in order to reduce net carbon emissions. Trees newly planted in urban areas would respond to the greenhouse problem in two ways, first by reducing the need for air conditioning and hence electricity, and second by increasing the uptake of carbon (Akbari et al). Active forest management to maintain high amounts of standing biomass, to reduce tropical deforestation, and to aggressively reforest surplus agricultural or degraded lands, offers significant potential for slowing global atmospheric buildup of greenhouse gases (CO_2, N_2O, and CH_4) emitted by biomass burning and disturbed tropical soils.

Part II: Potential Effects of Climate Change on U.S. Forests

Studies conducted thus far on the potential impact of changes in temperature and water regime in a scenario of doubled CO_2 (twice current atmospheric CO2 concentrations, abbreviated as $2xCO_2$) suggest a wide range of effects on individual trees and forest communities (e.g., see Shands and Hoffman, 1987; Smith and Tirpak, 1989; Meo, 1987). Rising levels of CO_2 are likely to alter growth rates and productivity of trees and plants by affecting biochemical processes like photosynthesis, respiration, and water use efficiency that play an important role in determining the distribution of species and the productivity of forests. Forest community composition (i.e., the ratios of species within a forest type) is likely to be changed, as species more tolerant of water stress, warmer temperatures, pest infestations, fire, and more frequent extremes of wind, temperature, and drought or flood outcompete species at the southern edges of their ranges or unable to adapt to these alterations.

Most early studies of possible climate-induced perturbations of forests have used standard General Circulation Models (GCMs) of global climate (e.g., the GISS, GFDL, and OSU models; see Smith and Tirpak (1989) for comparison) to predict changes in regional temperature and moisture regimes under 2xCO_2, and their consequences for forest growth and distribution. These changes from elevated CO_2 are not likely to be uniform geographically or temporally; further, the various models do not offer consistent forecasts. The climate may become wetter and warmer in some regions, and drier and warmer in others. Temperature changes are expected to be most pronounced in higher latitudes, and less in the tropics. Regional climates will change more than global averages, but GCMs presently do not provide adequate resolution to allow reliable forecasting of effects on specific forests.

The effect of CO2 enrichment (also called "fertilization") on forest growth remains unclear, despite indications of enhanced growth. Laboratory studies of the effects of elevated CO_2 levelson plants have documented increased rates of photosynthesis, lowered plant water use requirements, increased carbon sequestering, and increased soil microbial activity fixing nitrogen for fertilizer, thereby stimulating growth (Hardy and Havelka, 1975; Drake et al, 1988). Kramer and Sionit (1987) have demonstrated that current levels of ambient CO2 are insufficient and actually limit tree growth; thus, increasing available CO2 is likely to stimulate photosynthesis. However, little work has been done <u>in situ</u> on forest communities over extended time frames. The net effect of CO2 enrichment combined with forest decline from climate change and air pollution remains uncertain.

An overview of potential effects of doubled CO2 on selected forest systems and species is presented in Table 1.

Part III: Response Options for Stabilizing Climate Change

Given the implications for forest resources, can we devise technical control options and public policies to reduce the buildup of greenhouse gases, and slow the rate of climate change?

The Environmental Protection Agency has prepared the most comprehensive assessment of control options and strategies to date, for all sectors--transportation, residential and commercial energy, agriculture, etc. (Lashof and Tirpak, 1989). Its report offers an overview of forestry control options, summarized below.

The forest sector presents challenges to foresters and policy makers devising technical response options not found in other sectors, because of the vast and highly differentiated acreage, low level of resource management and economic return, uncertain ownership, low levels of financial support for research

---

Table 1: Summary of Potential Effects of Rapid Warming on Forest Ecosystems in North America

Range Shifts:
* Southern ranges of many Eastern U.S. forest species may shift northward 200-1000 kilometers (km)

* Northern ranges of Eastern species could shift up to 700 km northward, although actual migration could be as low as 100 km, due to problems with seed dispersal and survival

* Forest health and survival in the long term may depend on how fast climate stabilizes, and whether large-scale forest declines result from pests, stress from air pollution, fires, and drought

* Forest composition (predominance of species) may change significantly, as species less water dependent and in the northern part of their range tend to become dominant

Changes in Productivity of Forests:
* Productivity declines of 46-100% could result along the southern edges of species ranges, depending on levels of soil moisture

* Productivity could increase along northern limits of ranges of some Eastern forests, as slow-growing conifers are replaced by hardwoods growing more rapidly

* Declines may begin in 30-80 years

Policy Implications of Climate Change for Forestry:
* Forest management agencies and institutions should begin to consider the effects of climate change and incorporate them into management and research plans

* Potential strategies available to maintain forest productivity and species diversity include introduction of subtropical species into Southern forests, massive reforestation on surplus croplands, and research on genetics and seedling survival

Source: Smith and Tirpak, 1989

---

and field trials, and poor quality of data concerning tropical forests.

Some time scale problems are unique to forestry--very slow cycling times, compared with agriculture (annual crop cycles) or political decision making (cycling at about 2-5 years). Rotation lengths (the time necessary for one cycle of a forest crop to be planted, grown, and harvested) in forestry are typically 25-80 years in temperate zones, and 8-50 years in the tropics.

Essentially, forestry climate response options are effective for a fixed period, although long, since trees eventually die and release carbon through decay. Forest options may buy time for policymakers to alter fossil fuel consumption patterns and technologies, and to decide how to use trees grown thus far.

The most frequently suggested forestry response measure is global or national reforestation of available lands with fast-growth plantations (e.g., Postel and Heise, 1987). Sedjo and Solomon (1988) have proposed that the current annual atmospheric increase in carbon (approximately 2.9 billion tons) could be sequestered for about 30 years in approximately 312 million hectares of plantation forests for a cost possibly as low as $125 billion, a large but not inconceivable sum. All such estimates thus far suffer from their application of high growth rates to vast tracts of highly differentiated site conditions. More detailed analyses are underway by EPA and other organizations.

Afforestation offers a stream of ecological and economic benefits worthwhile on their own merits: forestry products industry jobs, maintenance of biodiversity, watershed protection, nonpoint source water pollution reduction, recreation.

Forest strategies to slow global warming, each of which offers several technical options, are summarized in Table 2. All strategies could be pursued simultaneously. Some are better suited for industrialized country forests and policy conditions, and some are more appropriate for developing nations forest circumstances. Both biomass growth rate maximizing and total standing stock (volume) maximizing strategies are needed. Species characterized by high volumes of biomass (e.g., Douglas fir in the Northwest) usually grow slowly (e.g., 80-100 years to mature), and offer the most value from a carbon cycle perspective in industrialized country conditions that favor stable, long-term forest protection and intensive management of stands.

Developing countries are more likely to be able to sustain strategies involving high growth rate species, essential to restore degraded and desertifying lands, and to produce requisite fuelwood and timber (although stands of high-value hardwoods like mahogany and teak are also needed, to generate foreign exchange). Tropical nations face rapid forest depletion, high population growth rates, and limited institutional capability to guarantee long forest rotation times in the face of these realities. Lashof and Tirpak (1989) provide a description and comparison of the spectrum of response options, including cost estimates.

Several forest management options are available in North America. Sufficient new forest area growing at average U.S. forest rates is not available to offset our current annual production of 1.3 Pg C. Per capita annual carbon production for 237 million Americans is about 5 t C/capita (Marland, 1988), or about 6 ha/capita (15 ac/capita) at average growth rates for commercial forests in the U.S. (those producing greater than 1.4

Table 2: Forest Sector Strategies for Stabilizing Climate Change

A. Reduce Demand for Forest Land and Forest Products

1. Substitute sustainable, sedentary agricultural technologies for swidden (shifting) agriculture

2. Decrease consumption of forest for cash crops and development projects, through environmental planning and management

3. Improve the efficiency of biomass (fuelwood) combustion in cookstoves and industrial uses

4. Decrease the production of disposable forest products (e.g., paper, paper packaging) by substituting durable wood or other goods, and by recycling wood products

B. Increase Supply of Forested Land and Forest Products

1. Improve forest productivity on existing forests, through management and biotechnology on managed and plantation forests

2. Increase harvest efficiency, by harvesting more species with methods that reduce damage to standing trees and use all biomass

3. Establish plantations on surplus cropland and urban lands in industrialized temperate zones (U.S.), to produce high volumes of standing forest biomass and/or fast-growth species to fix carbon

4. Reforest degraded forests and establish plantations and agroforestry projects in the tropics, using both fast-growth and high-biomass species on short rotations for biomass and timber.

m^3/ha/y and not set aside in parks) totalled 195.3 million ha in 1977, with a net average growth rate of 3.15 m^3/ha/yr, or 0.82 t C/ha/yr. Thus 6.3 ha/person (15.7 ac) of forest to sequester 5 t C/yr for 237 million people requires 1.5 billion ha of average forest--a tract 50% larger than the U.S.'s 919 million ha area (Marland, 1988).

However, intensive forest management may offer a partial solution. The U.S. Forest Service (1982) estimates that if current commercial forests simply were fully stocked, then their net annual growth could be increased by about 65 percent, which would sequester an additional 103.5 million t C/y. USFS also estimates that forest owners already could gain 18 billion cubic feet of annual forest growth (327 million t of wood, or 0.16 Pg C) by embarking on currently economically feasible opportunities offering a 4% annual return on investment (Hagenstein, 1988).

Doubling commercial forest productivity--instead of forest area, as above--on current forest lands in the U.S. would offset an impressive 5 Pg C (not impossible, if simply fully stocking forests would produce 65% gains). Increased stocking, forest management, and harvest rates, and sequestering the timber produced in construction and paper products would be necessary to attain full carbon cycle benefits.

Intensive management techniques available to improve productivity include selection of species for site conditions through provenance trials with seeds, thinning and release cuttings, control of spacing among trees, weed and pest control, fire suppression, fertilization, irrigation, and planting genetically improved seedlings. All options requiring significant labor tend to be prohibitively expensive on large scales, although volunteer labor from youth or citizen groups might make these options more attractive.

The implications for carbon fixing are clear: a 30% rise in biomass production allows 30% less land in forest--or a 30% offtake of timber--to achieve the same results.

According to the Forest Service report for FY 1987, total U.S. tree planting by federal and state agencies covered about 3 million acres in 1987 (USFS, 1988). Additional enticements, on the order of $220-345/ha ($90-140/ac), would likely stimulate tree planting on hundreds of thousands of hectares. The Conservation Reserve Program (CRP) of USDA paid an average of $219/ha ($125/ha ($50/ac) in bid price, plus half of establishment costs at an average of $94/ha), to plant 648,000 ha in trees (1.6 million ac) from 1986 to mid-1988. Modification of the CRP has been proposed as the quickest, most cost effective way to stimulate tree planting at the scale necessary to partially offset CO_2 emissions (Dudek, 1988). For example, all new CO_2 emissions from fossil fuel electricity plants projected for 1987-96 (25,223 megawatts total, producing 166.7 million t CO_2) could be offset by planting about 9 million ha (22 million ac) of trees fixing 5 t C/y (Lashof and Tirpak, 1989).

A review of urban forests in 20 cities showed that for every four trees removed, only one tree is planted (Moll, 1987). An urban tree is about 15 times more valuable than a forest tree in terms of reducing CO_2 emissions. Trees break up "urban heat islands" by providing shade that reduces cooling loads (air conditioning) in warm weather through reduced building heat gain, and cuts heating loads in cool weather by slower evaporative cooling and increased wind shielding. Three strategically placed trees per house can cut home air conditioning energy needs by 10-50 percent (Akbari et al,. 1988). The American Forestry Association recently launched Project ReLeaf to plant 100 million trees, and estimates savings of 40 billion kilowatt-hours of energy from these new trees, which could absorb 18 million tons of CO_2 annually (Sampson, 1988).

Obstacles to Large-Scale Reforestation in the U.S.

Economic obstacles to widespread reforestation center on the high costs of site preparation, planting, forest management, and financial incentives to private landowners. Ecological drawbacks include low genetic variability in vast monocultural stands, and reduced resistance to pest infestations (e.g., gypsy moth, pine bark beetles) capable of widespread forest decline and mortality.

Utilization of the biomass produced on plantations in gas-fired turbines or otherwise to produce electricity would entail environmental side effects, including production of CO_2, CO, and N_2O; persistent smog and health effects in areas with thermal inversions or poor mixing and circulation; and potential impacts on water quality. Air pollution--acid precipitation, ozone and other photochemical oxidants--is already affecting the health of forests in the U.S., Europe, and China. Decline in at least 7 coniferous and 4 broadleaved European and 8 North American important tree species has been documented since 1979. Carbon storage is problematic. Wood simply could be stored in durable products (e.g., paneling, flooring, paper), but large volumes might severely disrupt trade markets and depress prices.

The long-term process of climate change is likely to complicate the task of actively expanding net forest area, a topic explored in Smith and Tirpak (1989). The natural rate of forest migration--about 100 to perhaps 400 km per 1000 years--appears unlikely to keep up with the likely rate of forest decline (Davis and Zabinski, in press; Shands and Hoffman, 1987; Smith and Tirpak, 1989).

Forests migrating north in response to rising temperatures will tend to encounter poorer soils, slow natural seed dispersal methods, stresses on ecosystems, competition from other land use sectors and ecosystems also responding to climate change, and either reduced or increased precipitation and water supply conditions, depending on the region. Climatic change therefore may affect the viability of reforestation strategies at the same time efforts are made to use forest as a mitigation measure.

Summary of Forestry Technical Control Options

Slowing tropical deforestation, and rapidly expanding tropical and temperate reforestation (especially in urban areas), may offer two of the most cost-effective policy responses to increasing CO_2 emissions. However, only rudimentary estimates of the feasibility, costs, and consequences of large-scale reforestation have been performed. Forest managers may need to begin to consider site characteristics as dynamic, rather than static, considerations in silvicultural decisions. Accelerated research in forest genetics and seed dispersal methods is warranted to develop viable seed dispersal methods.

REFERENCES

Akbari, H., J. Huang, P. Martien, L. Rainier, A. Rosenfeld, and H. Taba. 1988. The Impact of Summer Heat Islands on Cooling Energy Consumption and CO_2 Emissions. Paper presented at ACEEE Summer Study on Energy Efficiency in Buildings, Asilomar, Cal., August.

Davis, M. and C. Zabinski, in press. Rates of Dispersal of North American Trees: Implications for Response to Climatic Warming. Forthcoming in Proceedings of the Conference on Consequences of the Greenhouse Effect for Biological Diversity. Washington, D.C., October 4-6, 1988, World Wildlife Fund-U.S.

Detwiler, R. and C. Hall, 1988. Tropical Forests and the Global Carbon Cycle. Science 239:42-47.

Drake, B., P. Curtis, W. Arp, P. Leadley, J. Johnson, D. Whigham. 1988. Effects of Elevated CO_2 on Chesapeake Bay Wetlands. III. Progress report to Department of Energy, Carbon Dioxide Research Division, Office of Energy Research, Washington, D.C., 101 pp.

Dudek, D. 1988. Offsetting New CO_2 Emissions. Unpublished paper, Environmental Defense Fund, New York, September.

Hagenstein, P. 1988. Forests Over the Long Haul. Paper presented at Conference on Natural Resources for the 21st Century, sponsored by American Forestry Association, Washington, D.C., November 14-17, 1988.

Hardy, R. and U. Havelka. 1975. Nitrogen Fixation Research: A Key to World Food. Science 188:633-43.

Houghton, R. 1988. The Flux of CO_2 Between Atmosphere and Land as Result of Deforestation and Reforestation from 1850 to 2100. Unpublished paper prepared for Environmental Protection Agency.

Kramer, P. and N. Sionit. 1987. Effects of Increasing Carbon Dioxide Concentration on the Physiology and Growth of Forest Trees, in Shands, W. and J. Hoffman, 1987. The Greenhouse Effect, Climate Change, and U.S. Forests. The Conservation Foundation, Washington, D.C. 304 pp.

Lashof, D. and D. Tirpak. 1989. Policy Options for Stabilizing Global Climate. Environmental Protection Agency, Report to Congress, Washington, D.C., c. 800 pp.

Marland, G. 1988. The Prospect of Solving the CO_2 Problem Through Global Reforestation. U.S. Department of Energy, Office of Energy Research, Washington, D.C. DOE/NBB-0082, 66 pp.

Meo, M., ed. 1987. Proceedings of the Symposium on Climate Change in the Southern United States: Future Impacts and Present Policy Issues. Symposium held May 28-29, 1987, New Orleans, Louisiana, by Environmental Protection Agency, Washington, D.C.

Moll, G. The State of Our City Forests. American Forests, May/June.

Postel, S. and L. Heise. 1988. Reforesting the Earth, in L. Brown et al., State of the World 1988. Norton, New York. 237 pp.

Sampson, R. 1988. ReLeaf for Global Warming. American Forests, November/December:9-14.

Sedjo, R. and A. Solomon. in press. Climate and Forests. Paper prepared for Workshop on Controlling and Adapting to Greenhouse Forcing, held by Environmental Protection Agency and National Academy of Sciences, Washington, D.C., June 14-15, 1988. Forthcoming in N. Rosenberg, W. Easterly, P. Crosson, and J. Darmstaedter, eds. <u>Greenhouse Warming: Abatement and Adaptation.</u> Resources for the Future, Washington, D.C.

Shands, W. and J. Hoffman, 1987. <u>The Greenhouse Effect, Climate Change, and U.S. Forests</u>. The Conservation Foundation, Washington, D.C. 304 pp.

Smith, J. and D. Tirpak. 1989. <u>The Potential Effects of Global Climate Change on the United States</u>. Environmental Protection Agency, Report to Congress, Washington, D.C. c. 600 pp.

USFS (United States Forest Service, Department of Agriculture). 1982. <u>An Analysis of the Timber Situation in the United States, 1952-2030</u>. Forest Resource Report No. 23, Washington, D.C.

USFS (United States Forest Service, Department of Agriculture). 1988. 1987 U.S. Forest Planting Report. Washington, D.C., April 1988, 13 pp.

Climate Change and Forest Fires

Michael A. Fosberg
USDA Forest Service
Washington, DC

The summers of 1987 and 1988 brought a great deal of attention to the relationship between forest fires and weather. In 1987, an unusually high incidence of lightning started fires in California and Oregon resulted in 1500 fires and 370 thousand hectares burned. Continued drought in 1988 also brought fire to the Yellowstone area and with it, 730 thousand hectares burned. These are not the largest fires we have experienced, but certainly are ones which received much attention on television and in the press. With greenhouse warming also receiving much attention, it was inevitable that attempts to link the greenhouse effect to future forest fires, and what protection might be required.

To understand the relationship between fire and climate change, we need to understand fire behavior. Our knowledge of fire behavior integrates the amount and structure of the vegetation (the fuels) and weather. Two measures of fire behavior are of interest here, rate of propagation and flamelength or fire line intensity. These two measures tell us area burned, damage, and effort required to supress a fire.

Forests are composed of fuel elements of varying size, from conifer needles, small twigs and leaves to the large boles of trees. Each of these elements respond differently to weather. Fine dead fuels (twigs, etc.) respond rapidly to diurnal variations of relative humidity while large dead fuels may take several months to dry following winter rains. Living vegetation is also consumed in a fire. Vegetation under drought stress has lowered moisture content frequently contributing to fire propagation and intensity. The fine fuels have high surface area to volume ratios and determine the rate at which fires propagate. Large fuels contain the energy that will be released during fires and determine the intensity of a fire. The mixture of fuel sizes and the amount of fuel of each size determine the potential fire behavior of a given ecosystem. Grasslands contain no large fuels. In grasslands, fire can spread rapidly and typically, does not result in significant damage. Heavy timber ecosystems which have been cut and allowed to cure have large fuel elements. In the cured state, much of the fine fuels, the foliage, has fallen to the ground and become somewhat compacted. Fire spreads slowly through this complex, but can be very intense and substantial damage may result.

ECOSYSTEMS AND POTENTIAL FIRE BEHAVIOR

In this paper, I will use a simplified version of the U.S. National Fire Danger Rating System (Deeming et. al. 1972). This simplification results in potential fire behavior for a number of ecosystems and is based on measured structure and composition of those ecosystems as expressed through the size and amount of fuels. For both fire spread rate and flamelength, I have assigned a value of 100 to the maximum in each category. For example, for grass fuels, fuel A in table 1, the rate of spread is 100 while the flamelength has a value of 13. Chaparral brush fields, fuel B, have a rate of spread of 52 on this scale and a flamelength of 100. The remaining fuel types are:

C. Open overstory forest with grasses or other herbaceous plants as a ground fuel. Young conifer plantations, open ponderosa, sugar, longleaf, slash, and sand pines, as well as pinyon juniper stands characterize this class.

D. Ecosystems in which there is heavy loading of fuels 2 cm or less in diameter and in which living fuels burn readily. The low pocosins of the Atlantic states, and black spruce stands at high latitudes are represented by this model.

E. Hardwood and mixed conifer-hardwood stands during the dormant season, before leaf fall has been compacted.

F. Young brush fields which contain little or no dead materials. Laurel, mountain mahogany and young stands of chamise and manzanita are represented here.

G. Dense conifer stands where heavy buildup of downed timber has accumulated.

H. Closed short-needle conifer stands, hardwoods and mixed hardwood conifer stands after leaf fall has been compacted.

I. Clear cut timber where little material has been removed.

For determination of effects of climate change on forest fires, only the flamelength will be used because flamelength is directly correlated with suppression cost.

On the scale described in Table 1, 20 represents the limits of personnel working a fire line, 50 represents the limits of direct attack and 60 represents crowning fires where suppression forces are withdrawn for safety reasons. The Yellowstone fires ranked about 100 to 150 on this scale. The expected potential fire behavior for fuel type G was 55. The most intense fire in recent years, the Sundance fire of 1967 in Montana, also is fuel type G had a value of 200. This fire totally reduced the forest to black ash.

CHANGES IN FIRE SEVERITY POTENTIAL FROM CLIMATE CHANGE

A number of future ecosystems have been simulated from general circulation model runs for a doubled CO_2 environment. These range from superposition of ecosystems on climate shifts (Holdridge life zones) in which steady state ecosystems move with climate change, to use of sucessional or gap-phase models in which forests develop, go through their life cycle, and die. The first, the superposition approach, totally ignores the fact that species migrate at different rates (Davis 1981) and assumes that future forests will look just like today's, but just be somewhere else. It also ignores trauma in ecosystems - fire, insect and disease. The gap-phase models describe the processes of forest succession and allow forests to grow, individual trees to die and in turn be replaced by the same forest species, new tree, species or other vegetation depending on individual ability to compete for energy, nutrients, and water. The gap phase simulations are most interesting not only because they are a more realistic description of the ecosystem dynamics, but also because they allow an assessment of the transition period from the current climate to some future climate.

I use three examples in this paper to illustrate the effects of future climate on forest fire potential in a doubled CO_2 environment. These are from the Lake States, the Rocky Mountain West, and Tennessee.

For the Lake States, Botkin _et al_ (1988) determined that the current ecosystem of mixed hardwoods and conifers would become a mixed hardwood stand on wet sites and that the uplands would become a hardwood savannah. In the case of the wetlands the ecosystem change would not materially affect the fire potential remaining as a fuel type E through the change. The upland would experience an increase in potential fire severity transitioning from fuel type E (19) to fuel type C (41). The dead materials produced during the transition as the forest opened up, could increase the interim fire potential to a fuel type G (55). Recall here the changes required in fire suppression tactics required when this index exceeds 20 and 50. This particular scenario was constructed using the GISS model. If the same gap-phase model is run using the NCAR climate (Soloman and West, 1987) pines remain and there is no net change in fire potential.

For southeast Tennessee, Solomon and West (1987) predict the oak-hickory forests to prevail, a severity index of 19. Miller _et al_ (1987) predict loblolly pine plantations as a possible future forest, an index of 41. While this forest was predicted by the Miller superposition geographic shift of a steady state forest in a future climate, it is reasonable because loblolly pine is intensively managed and would probably have been intentionally introduced.

The Yellowstone area of the Rocky Mountains is expected to remain in lodgepole pine (Leverenz and Lev, 1987). The area which burned in 1988 would initially be represented by young trees (fuel type F) and over time return to the fuel type G of heavy timber with much dead material.

DIRECT EFFECTS OF CLIMATE CHANGE

There are few definitive studies of direct effects of climate change on fire frequency and severity. Direct effects are used here to define the changes in drought frequency, humidity, precipitation and other weather elements that determine day to day variation and interannual variability in fire behavior. Fried and Torn (1988) compared the changes in area burned under the current and a double CO_2 climate. They found that there would be a two-fold increase in modest sized fires (a few hundred hectares) and a three-fold increase in fires greater than 1000 hectares. Fried and Thorn based their studies on an area of the California Sierra Nevada in which the ecosystem is expected to remain unchanged in a future climate.

An analysis of drought from AD 372 to 1985 (Stahle, 1988) showed increased interannual variability when decadal variability was high. Stahle correlated tree rings with the Palmer Drought Index for a site in the Southeastern United States. Because the Palmer Drought Index is used to calculate the moisture content of live vegetation, it is an ideal analog for the moisture content of live vegetation, and therefore ideal analog for fire danger variability.

SUMMARY

The ecosystem models used to project future forests are particularly sensitive to the moisture projections of the general circulation models, as demonstrated by the Great Lakes Simulation. Using 9 broad discrete fuel types to describe the ecosystem decreased the sensitivity of the future fire severity projections somewhat and suggested that direct weather effects may be more important in comparing future problems with current problems. The one case examined in which the transient ecosystem was considered showed a temporary increase in the problem before reaching a value above the current dangers but less than the transient peak.

Little can be said about transient solutions in general because the ecosystem models do not currently address trauma, that is, fire, insect and disease. This should be a topic of high priority to include in the ecosystem models.

TABLE 1

FUEL TYPE	RELATIVE RATE OF SPREAD	RELATIVE FLAME LENGTH
A	100	13
B	52	100
C	54	41
D	23	45
E	11	19
F	8	7
G	8	55
H	5	21
I	19	85

References

Botkin, D.B., R.A. Nisbet, and T.E. Keynales; 1988:. Effects of Climate Change on Forests of the Great Lakes, Final Report to EPA, Univ. of Calif., Santa Barbara, 40 pp.

Davis, M.B; 1981: Quaternary history and the stabililty of forest communities, In West, D.C., H.H. Shugart, and D.B. Bothkim ,eds, Forest Succession: Concepts and Application, New York Springer Verlag, pp. 132-153.

Deeming, J. E., J. W. Lancaster, M. A. Fosberg, R. W. Furman and M. J. Schroeder; 1972: National Fire Danger Rating System, USDA.Forest Service Research Paper Rm-84, 165 pp.

Fried, J. S. and M. S. Torn; 1988: The Altered Climate Fire Model: Simulating the effects of climate change on the effectiveness of a wildland fire initial attack program (manuscript in preparation, Univ. of Calif., Berkeley).

Levenrenz, J. W. and D. J. Lev, 1987: Effects of Carbon Dioxide Induced Climate changes the natural ranges of six major commercial tree species in the western United States, Shands, W.E. and J.S. Hoffman, eds., In: The Greenhouse Effect, Climate Change and U.S. Forests, The Conservation Foundation, Washington DC pp. 123-156.

Miller, W. F.; P. M. Dougherty and G. L. Switzer. 1987: Effects of rising carbon dioxide and potential climate change on loblolly pine distribution, growth, survival and productivity in Shands, W. E. and J. S. Hoffman, eds., The Greenhouse Effect, Climate Change and U.S. Forest. The Conservation Foundation, Washington, DC, pp. 157-188.

Solomon, A. M. and D. C. West; 1987: Simulating Forest Ecosystem responses to expected climate change in eastern North America:Applications to Decision making in the Forest Industry Shands W. E. and J. S. Hoffman, eds., The Greenhoue Effect, Climate Change and U.S. Forests; The Conservation Foundation, Washington D.C. pp.189-218.

Stahle, D. W.; M. K. Cleaveland and J. G. Hehr; 1988: North Caorlina Climate Changes Reconstructed from Tree Rings.AD 372 to 1985. Science 240:1517-1519.

CLIMATE CHANGE AND THE CANADIAN FOREST

James B. Harrington
Forestry Canada
Petawawa National Forestry Institute
Chalk River, Ontario

Abstract

Industrial society, through the careless handling of waste products, is threatening to damage or destroy enormous areas of boreal forest. Potential damage occurs in two entirely different ways. First, sulfur and nitrogen oxides, create acid rain which leaches minerals from plant leaves and nutrients from the soil, thereby weakening trees and increasing their susceptibility to damage by drought, insects and disease. A byproduct of these emissions, ozone, reduces photosynthesis and still further reduces the tree's vitality.

Second, and of greater consequence, is the production of greenhouse gases CO_2, N_2O, CH_4, CFC's and O_3 which are predicted to warm the atmosphere to unprecedented levels. If present model predictions are correct, much of Canada's boreal forest west of the Hudson Bay will disappear to be replaced by grassland and brush. The consequences to Canada's economy would be devastating.

A plea is made to intensify research on the relationship of climate to trees, fire, insects and disease. Without such basic data, intelligent resource planning will not be possible.

Most Canadians harbor a deep, almost mystical, love of their wilderness including all the wild plants and creatures that inhabit it. This fact is clearly evident in a gallery of Canadian paintings where lovely jack pines, watchful loons and crimson maples are common themes. Those in Canada who are aware of environmental changes affecting their wilderness are deeply concerned. This paper deals with two problems, one new and one old: namely, the effect on Canada's forests of climate change, and the continuing effect of air pollution.

Now that free trade between the United States and Canada is a reality let me begin by discussing a free exchange that has been going on for some time. I am referring to the exchange of air across our border. During my talk, delivered at the Second North American Conference on Climate Change: A Cooperative Approach on December 7, 1988, I showed two slides taken from my deck in Deep River, Ontario. The pictures were taken on two days in the early fall, one when the wind was in the northwest and the other when the wind was in the south. The first slide, taken in clear Canadian air, shows the high land two kilometers away, across the Ottawa River in Quebec, where every tree stands out clearly, every branch etched against the sky. The second slide, taken when the wind was in the south, shows the trees and even the hills themselves obscured by sulphuric and nitric acid mist. Rain here commonly occurs with southerly winds. At these times, the pH of rain becomes acidic, falling about twice a month to as low as 3.5, the acidity of mild vinegar.

Some of our trees are sick, sugar maple being the most severely affected at present. Although we have not been able to attribute this sickness directly to air pollution because of the confounding effects of insect defoliation, drought and cultural practices, we are certain that air pollution is a contributing cause. Certainly, the presence of ozone, which suppresses photosynthesis by as much as 36% in hardwoods (Reich, 1987) can not be helping the maples. Cook _et al_ (1987) conclude that forest decline in northeastern United States is due either to climatic shocks or to air pollution. Acidic air pollution is a prime suspect (Linzon, 1985; McLaughlin _et al_, 1987). Johnson and Siccama (1983) find high acidic pollution loads in areas of red spruce decline but find that the effects of drought and air pollution cannot be readily separated. Despite the uncertainty of the research results, it is clear to me that reducing photosynthesis, leaching minerals from the leaves of trees and depleting the soil of nutrients must be contributing to the current forest decline in Canada (Hauks and Wright, 1986; Hinrichsen, 1987; McLaughlin, 1985).

The blight of air pollution has been worsened by two measures intended to improve air quality. High stacks and the cessation of leaf burning permitted increasing amounts of industrial and automotive pollution to be emitted while holding urban communities within EPA guidelines. These measures were promulgated, I suppose, with malice aforethought.

The United States is not the only industrial country poisoning its own air. In Greece and Italy one can rarely take a photograph of the landscape due to the pall of acid mist. French travel brochures recommend Pau as the best site for a view of the Pyranees. Yet, in summer, one cannot see the Pyranees from Pau. The high pollution levels and dying forests of northern Europe are well known.

In the 1940's the US sued the Trail, B.C. smelters and forced them to reduce SO_2 emissions (Hewson, 1945). More recently extreme pressure has been placed on Mexico to curb sulfur emissions from their new copper smelter just across the border. Why then is there such a reluctance on the part of the United States to control its own pollution? Now, with climate change occupying front stage center, we in Canada are concerned that the United States may be altogether distracted from controlling air pollution.

Now let me discuss the second problem and the major theme of this conference, the prospects for Canadian forestry under the impact of climate change. All the major global circulation models, for a doubling of CO_2 (or the equivalent considering all greenhouse gases) now show a mean global temperature increase of about 4 degrees centigrade. Positive feedback from snow and sea ice is expected to amplify the warming, particularly in high latitudes. The warming is predicted to reach 12 degrees centigrade in northern Canada in winter and as much as 8 degrees centigrade in south central Canada in summer. If these predictions are correct, they imply a devastating impact on forestry in Canada. Are they correct?

Many facts point to a need for caution in completely accepting the model predictions. They fail, for example, to accurately portray the present climate using a 1 x CO_2 atmosphere. Furthermore, the grid spacing of the major models is so great that the Rocky Mountains or the Hudson Bay are barely represented. The Hudson Bay, a major determinant of the climate in eastern Canada, is represented in the Goddard Institute for Space Studies, GISS, model by a single grid point. Among the models agreement on the magnitude of climate change is, in part, the result of manipulations by the modelers. For example, Mitchell (1983) adjusted his ocean temperatures upward by two degrees to bring them into alignment with other models. Finally, many feedback mechanisms, such as the variation of ice and snow albedo with temperature or the albedo of clouds, are not known with confidence. Considering the large possible errors in the climate models, it would be unwise, at present, to do more than speculate on impacts.

Despite reservations regarding the accuracy of the GCM's Canadian foresters are deeply concerned about potential impacts on our forest industry. The impact, for example, of the 50% decline in summer soil moisture predicted for Central Canada by Manabe and Wetherald (1986) would have devastating consequences. As you are probably aware, most of Canada's boreal forests are composed of even aged post-fire stands. Stringent fire control in the southern part of Western Ontario has reduced the return period of fire from the pre-control 50-70 years to approximately 500 years. Even in the north, where fire control is less rigorous, the return period is about 100 years (Harrington and Donnelly, 1978). A 50% soil moisture reduction in this fire prone area would increase the area burned by fire (Harrington _et al_, 1983; Flannigan and Harrington, 1988) and seriously affect regeneration.

Another potentially serious problem is blowdown by tornados and downbursts. At present these destroy about 10% as much forest as fire, roughly 0.02% annually in Ontario (Harrington and Newark, 1986). Higher surface temperatures combined with a higher tropopause and cooler stratosphere could be expected to increase the intensity of destructive thunderstorms. Most tree species in Canada are shallow rooted, and ill adapted to withstand strong winds.

Far the most serious potential consequence of climate change to forestry in Canada is the predicted retreat of the boreal forest in the face of increasing temperature (Bolin _et al_, 1986) and the inability of many boreal forest species to advance northward into the arctic tundra at a comparable rate (Davis _et al_, 1986; Ritchie, 1987). Wheaton _et al_ (1987) have produced an excellent assessment of the implications of climate change to forests in the prairie provinces and Northwest Territories. They show, using the results of the Geophysical Fluid Dynamics Lab, GFDL, and GISS models that the southern boundary of the boreal forest zone will shift northward 250 to 900 km whereas the northern boundary will shift only 80 to 700 km. Along the southern margin the boreal forest will quickly be replaced by grassland over a period of less than 100 years. At the northern boundary of the boreal forest, pioneer species will advance only 100 to 150 km a century (Davis _et al_, 1986). The present boreal forest, excluding forested barren land, extends over a latitudinal range of only about 600 km in this area and, therefore, will largely disappear if the present GCM predictions are correct. This scenario presents a frightening prospect for a country in which 10% of its population work in forest-related industries.

Perhaps there are mitigating circumstances. Perhaps fire can be controlled so that boreal species can survive until they are replaced by more southerly species. Perhaps the 50% decline in soil moisture predicted by the GFDL 2 x CO_2 scenario is incorrect. Perhaps cloud or oceanic phytoplankton feedback will be strongly negative. Perhaps the grid spacing is too coarse or vegetation-atmosphere interactions have been improperly modeled, or a shift in storm tracks will bring more precipitation to the boreal forest zone. At the present stage of modeling we cannot have absolute confidence in any of the models, especially at the regional scale.

We in Forestry Canada (formerly the Canadian Forestry Service) firmly believe that the climate will be warming. We are concerned about the potential effects on our forests and on many other aspects of Canadian life. But, as yet, the estimates of actual changes in temperature and precipitation are too uncertain for us to take action beyond the planning stage.

Our present knowledge of forest ecology and of the causes of climate change is about in the same position as was cancer research nearly twenty years ago. Vast expenditures were made in a vain attempt to conquer cancer at a time when the necessary basic research had not been done. Models of forest response to climate are important in clearly identifying critical knowledge gaps (Solomon, 1986). But modeling should not replace the basic research on which the models feed.

Accurate predictions of forest reaction to climate change will require us:

1) to determine the climatic limits to seed production, germination and seedling survival for all important species such as Black and Bliss (1980) have done for black spruce;
2) to determine the climatic ranges, rates of growth and genetic diversity of various species under conditions of changed thermal and photoperiod conditions;
3) to study the ecology of survival and dominance of northern species under changing climatic conditions;
4) to study the likely changes in soil chemistry and the effects of melting permafrost;
5) to continue studies of rates of migration;
6) to estimate changes in fire and blowdown;
7) to study the effect of changing climate on insects and disease;
8) to adequately monitor vegetation, soils, climate, insects and disease in the forested regions of Canada;
9) to develop a remote sensing technology to increase the efficiency of monitoring; and
10) to maintain an adequate archive accessible to all scientists.

Such a program will be necessary if we are to solve the enormously complex environmental problems attendant on climatic change. New resources in manpower and money will be essential if we are not to suffer the humilation of being caught unprepared for these unprecedented changes in our environment.

In entering this new and exciting arena let us, in our haste to solve new problems, not forget that life may not be worth much if we neglect an old and growing problem--air pollution.

REFERENCES

Black, R.A. and L.C. Bliss. 1980. Reproductive ecology of _Picea mariana_ (Mill.) BSP., at tree line near Inuvik, Northwest Territories, Canada. Ecological Monographs, 50, 331-354.

Bolin, B., B.R. Doos, J. Jager and R.A. Warrick. 1986. Scope 29: the greenhouse effect, climatic change and ecosystems. John Wiley and Sons, Chichester, endpapers.

Cook, E.R., A.H. Johnson and T.J. Blasing. 1987. Forest decline: modeling the effect of climate in tree rings. Tree Physiology, 3, 27-40.

Davis, M.B., K.D. Woods, S.L. Webb and R.B. Futzma. 1986. Dispersal versus climate: expansion of Fagus and Tsuga into the upper Great Lakes region. Vegetation, 67, 93-103.

Flannigan, M.D. and J.B. Harrington. 1988. A study of the relation of meteorological variables to monthly provincial area burned by wildfire in Canada (1953-80). J. Appl. Meteorol. 27, 441-452.

Harrington, J.B. and R.E. Donnelly. 1978. Fire Probabilities in Ontario's boreal forest. Proceedings of the Fifth Joint Conference on Fire and Forest Meteorology, Mar. 14-16, 1978, Atlantic City, N.J. American Meteorol. Soc. 1-4.

Harrington, J.B., M.D. Flannigan and C.E. Van Wagner. 1983. A study of the relation of components of the Fire Weather Index to monthly provincial area burned by wildfire in Canada 1953-80. Information Report PI-X-25, Petawawa National Forestry Institute, Chalk River, 65 pp.

Harrington, J.B. and M.J. Newark. 1986. The interaction of a tornado with rough terrain. Weather, 41, 310-318.

Hauks, M. and R.F. Wright. 1986. Regional pattern of acid deposition and forest decline along a cross section through Europe. Water, Air and Soil Pollution, 31, 463-474.

Hewson, E.W. 1945. The meteorological control of atmospheric pollution by heavy industry. Quart. J. Roy. Meteorol. Soc. 71, 266.

Hinrichsen, D. 1987. The forest decline enigma. BioScience, 37, 542-546.

Johnson, A.H. and T.G. Siccama. 1983. Acid deposition and forest decline. Environ. and Sci. Technol., 17, 294A-305A.

Linzon, S.N. 1985. Forest damage and acidic precipitation. E.B. Eddy Distinguished Lecture Series. Faculty of Forestry, Univ. of Toronto, Toronto, 1-30.

Manabe, S. and R.T. Wetherald. 1986. Reduction in summer soil wetness induced by an increase in atmospheric carbon dioxide. Science, 232, 626 et seq.

McLaughlin, S.B. 1985. Effects of air pollution on forests. J. Air Pollution Control Assoc. 35, 512-534.

McLaughlin, S.B., D.J. Downing, T.J. Blasing, E.R. Cook and H.S. Adams. 1987. An analysis of climate and competition as contributors to the decline of red spruce in high elevation Appalachian forests of the Eastern United States. Oecologia, 72, 487-501.

Reich, P.B. 1987. Quantifying plant response to ozone: a unifying theory. Tree Physiology, 3, 63-91.

Ritchie, J.C. 1987. Post-glacial vegetation of Canada. Cambridge Univ. Press.

Solomon, A.M. 1986. Transient response of forests to CO_2-induced climate change: simulation modeling experiments in Eastern North America. Oecologia, 68, 567-579.

Wheaton, E.E., T. Singh, R. Dempster, K.O. Higginbotham, J.P. Thorpe, G.C. Van Kooten and J.S. Taylor. 1987. An exploration and assessment of the implications of climatic change for the boreal forest and forestry economics of the prairie provinces and Northwest Territories: phase one. Saskatchewan Research Council, Technical Report No. 211, Saskatoon, Sask.

CLIMATE CHANGE AND U.S. FOREST MARKETS

James L. Regens[1], Frederick W. Cubbage[2], and Donald G. Hodges[3]

Atmospheric concentrations of greenhouse gases are likely to produce global-scale climatic changes over the next 50 to 80 years affecting important societal activities (Budyko and MacCracken, 1987). Substantial uncertainty surrounds the rate and magnitude of sub-global impacts, but existing studies suggest plausible impacts on growth patterns for managed forests (see Hodges et al, 1988). This study summarizes the results of such analysis for the dominant regions in the U.S., the South and Pacific Northwest, in order to identify policy responses that might ameliorate adverse implications of climate change for forestry.

PLAUSIBLE ECOSYSTEM AND FOREST MANAGEMENT IMPACTS

Limited data suggest plausible biophysical effects for entire forest regions due to the doubling of CO_2 and other trace gases (see Hodges et al., 1988). The potential impact on loblolly pine, the primary commercial tree species in the southern United States, and Douglas-fir, the major species in the Pacific Northwest, depends upon the composition of individual forest stands. Our study reveals that modeling responses to climate change by major commercial tree species involves an array of assumptions. These include projecting the frequency and distribution of severe climate events, determining the impact of climate change on forest insects and disease, estimating fire incidence, and linking the implications of those factors to forest management practices. For example, warmer weather might increase fire incidence but wetter winters might reduce forest fire problems, since most fires occur in the spring. And, if climatic changes occur fairly rapidly, forests might be more susceptible to insect and disease attack since generational succession occurs more quickly among pests than trees. As a result, to the extent that global climate change alters those ecosystems, forest management practices, especially stand establishment, stand manipulation and harvesting, will be affected.

The Southeastern United States. Increased summer temperatures and reduced precipitation will tend to modify forest management practices to enhance stand survival. For example, site preparation practices must minimize ecological damage to the more environmentally-sensitive sites typical of the

---

[1]Institute of Natural Resources, University of Georgia, Athens, GA 30602 (USA)

[2]School of Forest Resources, University of Georgia, Athens, GA 30602 (USA)

[3]Southern Forest Experiment Station, U.S. Forest Service, New Orleans, LA 70113 (USA)

Central Highlands because soil loss would reduce site productivity and moisture-holding capacity below their already low levels. A northward shift in the loblolly pine range would produce higher costs for stand establishment, maintenance, and harvest.[1] Assuming these additional costs were only $1 per acre per year, which seems extremely conservative, annual cost increases would exceed $100 million, in constant 1987 dollars.

The Pacific Northwest. The management focus for Douglas-fir also will change over the next 30-50 years but not primarily from climate change. Instead, new management regimes will emerge because of shifts from predominately old-growth to younger second-growth stands. Climate change, however, will be significant enough to create site-specific problems along the margins of the current natural range, especially water stress. Assuring adequate regeneration is likely to be the most serious problem in the marginal regions since seedling survival is largely dependent on favorable air and soil temperatures and rainfall. Planting methods already exist to regenerate such sites, but they may prove too expensive for the returns received. Moreover, those border regions cannot withstand intensive site-preparation practices which result in erosion or further loss of soil moisture.

IMPLICATIONS FOR THE U.S. FOREST INDUSTRY

Forestry traditionally has involved relatively long planning horizons and substantial uncertainty surrounding forest growth. The above discussion suggests that global climate change could alter significantly the resource management practices of U.S. forest products firms. Evaluating the impact of those effects on U.S. forest markets is essential for sound forest management.

To the extent the southern U.S. experiences altered precipitation patterns during the growing season, the resulting reduction in specific gravity for harvested timber would increase the overall volume of wood required by mills, decrease pulp yields, and reduce lumber quality and strength. Valuing product input changes that lack good biological dose-response functions is obviously tenuous. But increased wood fiber demand of even 1 or 2 percent would result in significant costs. And, individual pulp plant production would tend to decline slightly because less wood would be processed in the fixed digestor space. This is the case since digestor efficiency is, in part, a function of wood density with efficiency increasing as density increases so that yields would decrease compared to stated plant capacity. And, incremental expansions of existing mills must take into account chemical recovery capacity and energy balances, both of which are available only in lumpy units. Expanding plant capacity on a retrofit basis requires substantial redesign and large capital investments. Moreover, the retrofit option may not exist if processing plants relocate to acquire softwood supplies at competitive prices or due to sea level increases.

To the extent sea levels rise, as much as an additional 10 percent of total loblolly pine volume might be lost due to the flooding of the Coastal Plain. Table 1 illustrates the impact on total mill capacity from varying changes in sea level. Clearly, if severe enough, higher sea level gradients could force substantial mill relocation and/or increase costs for protecting low-lying coastal pulp mills.

TABLE 1

IMPACT OF VARYING INCREASES IN SEA LEVEL ON SOUTHERN AND NORTHWESTERN PULP AND PAPER MILLS

Elevation (m)	Mills Impacted (n)		Capacity (Tons/24 Hr)		Percentage Total Capacity	
	South	Northwest	South	Northwest	South	Northwest
3	16	1	16,720	475	14.6	1.7
5	18	2	18,945	925	16.6	3.3
7	23	6	23,305	5,435	20.4	19.4

Source: Adapted from McKeever (1987)

Note: Figures are cumulative. Percentage calculation based on total mill capacity of 114,315 tons/24 hours in the South and 27,980 tons/24 hours in Northwest.

Lumber yields should decline due to increases in the amount of earlywood harvested.[2] Such losses would involve physical damages from increased warpage and poorer drying of predominately earlywood lumber. Poorer quality lumber also may reduce market share for southern pine due to declining consumer preference for the product.

Massive migration of the southern pine resource northward might generate a corresponding shift in the industry's processing infrastructure as new plant construction occurs. Several factors would make such a shift costly. First, the southern pulp and paper industry is located predominately in the coastal areas which provide good access to water supplies and international ports. The southern mountains and Central Highlands lack readily available water supplies and well-developed transportation facilities which severely limits the physical desirability of those areas for plant locations. The high cost of building new greenfield pulp and paper mills also constrains relocation as long as existing capacity can be utilized. High entry costs for new mills (at least $500 million in current dollar terms) make such investments problematic, and even firms with adequate capital might lose competitive advantages to foreign producers having lower marginal costs. On the other hand, because sawmills require considerably less investment (approximately $10 to $20 million) and probably could shift with greater ease, the net impact should be less in the Pacific Northwest where the majority of plants are sawmills or other solidwood product producers.

Mills unable to relocate must pay more to transport logs or, eventually, cease business if costs become too great. This scenario is plausible given a recent study which concluded that southern forestry is likely to have stable supplies at best through the year 2030 (see U.S. Forest Service, 1988). Supply changes due to warming trends are certain to exacerbate any incipient problems. Marginal coastal mills typically would be affected by resource base shifts first. And, those remaining open would probably have to pay more for protection measures against rising sea levels.

The economic impacts of such shifts are obvious. Employment levels would decline in areas experiencing reduced plant output or closures. Table 2 reveals that such closures could have a significant impact in the South, where forest products manufacturing has been a major employer for the past 40 years. Direct wages formerly paid to employees of the firm and indirect wages for services and purchases made by the firm and its employees in the local community would be lost.

CONCLUSION

Estimating possible impacts of global climate change on U.S. forest markets is a significant research challenge. Data for economic valuation are scarce. As a result, the uncertainty surrounding economic consequences is substantial. The industry's capability to adapt to likely changes needs to be considered. Important factors include the impact of increased competition domestically for land and water resources and internationally for world market shares with foreign producers, as well as impacts on valuable non-market forest resources such as recreation and wildlife. Obviously, this creates problems for long-term forest management. Nonetheless, preliminary data suggest that prospective global warming, with its resultant changes in regional climatic conditions, will adversely impact forestry. At the very

TABLE 2

EMPLOYMENT IN MANUFACTURING LUMBER AND WOOD PRODUCTS AND PULP AND PAPER PRODUCTS, SOUTH AND NORTHWEST REGIONS

State	Thousands of Employees				Percentage Change (1972-1986)
	1972	1976	1982	1986	
SOUTH					
Alabama	38.4	39.9	38.0	50.4	31.2
Arkansas	29.7	28.5	27.4	31.5	6.1
Florida	31.6	28.9	30.9	41.2	30.3
Georgia	45.8	47.7	50.3	65.9	43.8
Louisiana	29.6	28.9	22.0	24.4	-17.6
Mississippi	28.4	26.3	25.5	30.8	8.4
North Carolina	45.2	49.4	49.7	58.4	29.2
Oklahoma	4.7	5.0	5.8	5.6	19.1
South Carolina	25.7	26.0	24.8	31.1	21.0
Tennessee	31.6	31.8	29.7	38.2	20.8
Texas	42.4	47.0	47.5	64.5	52.1
Virginia	34.4	34.3	32.8	38.4	11.6
TOTAL:	387.3	394.7	383.4	480.4	24.0
NORTHWEST					
California	90.3	104.2	83.7	99.0	9.6
Idaho	15.2	19.0	13.6	15.0	-1.4
Montana	9.2	9.3	6.8	8.1	-12.0
Oregon	84.7	87.6	65.2	72.9	-14.0
Washington	65.5	71.6	55,2	53.1	-19.0
TOTAL:	264.9	291.7	224.5	248.1	-6.4

Source: Adapted from U.S. Forest Service (1987); Warren (1987)

least, the loblolly pine range will tend to move northward creating potential environmental and economic problems associated with stand establishment, forest growth, transportion to mills, and perhaps even mill closings.

There may be compensatory responses. Biotechnology may offset partially such impacts by developing seedlings that are drought resistant and highly responsive to increased CO_2 levels. Slash pine may expand northward, filling a void left by the retreating loblolly stands, and because slash pine generally is straighter, it may prove to be a superior replacement. For southern forests, it may become warm and wet enough for very fast growing species such as Eucalyptus or Caribbean pine to prosper.[3] Similarly, projections suggest that the benefits from increased CO_2 and the upslope movement of the winter snowpack in the Cascades will outweigh the harmful effects of increased temperatures and longer summer droughts. Market forces and engineering innovations may foster adaptation by the forestry sector to new, more efficient processing technologies. Moreover, because climate change is global in nature, U.S. forestry may adjust comparatively better with such an outcome giving American firms a competitive advantage despite increased costs.

On the balance, however, economic prospects for U.S. forest markets are likely to be negative. For example, Binkley (1987) projects that the impact of CO_2-induced warming on global forestry will cause income from timber sales to decline 20 percent in the eastern U.S. and 26 percent in the West by 2030.

Individual forest products firms face two different types of questions (see Sandenburgh et al., 1987). The first involves resource acquisition and management decisions. Typically, companies are likely to delay altering site management practices until changes in current operations seem warranted. In contrast, resource acquisition represents a more immediate concern. Firms may find it desirable to consider potential climate change impacts when purchasing or leasing timberland. The second type of decision involves research and development. Determining appropriate responses to climate change is difficult given the scarcity of reliable information, particularly on a subglobal basis. Thus, firms may conduct or fund research focusing on climate change's impact on forestry. However, based on the long time horizons for impacts and the forest industry's historical reluctance to fund substantial research programs (see Hodges and Harris, 1988), a significant commitment to climate change research is doubtful.

Given the uncertainty currently surrounding industry-specific assessments of climate change on a subglobal basis, from a policymaking standpoint, the key question is whether sufficient evidence exists to warrant direct governmental action. The most obvious options available include increasing energy efficiency, reducing the rate of deforestation in the tropics, and controlling global chloroflourocarbon emissions. Moreover, if those scientific uncertainties are to be diminished, an ambitious research effort is needed to delineate empirically the plausible effects of climate change as well as the implication of alternative policies aimed at reducing its undesirable consequences.

NOTES

1. Miller (1983) asserts that the northward-shift of the range generally would lessen the overall loblolly pine site index due to a shift to poorer sites competition with existing agricultural uses.

2. Latewood is produced during the dry summer months; earlywood is produced during the wetter spring months. Miller (1983) estimated that decrease of 5 to 7 feet in site index could occur throughout the region, which would result in a 10% yield reduction on pine land.

3. Eucalyptus is prone to cell damage from rapid drops in temperature not just cold temperatures alone. Given the U.S.'s continental climate, such changes would probably still occur despite higher mean regional temperatures.

REFERENCES

Binkley, C.S. (1987) "A Case Study of the Effects of CO_2-Induced Climatic Warming on Forest Growth and the Forest Sector: Economic Effects on the World's Forest Sector, " in M. L. Parry, T. R. Carter, and N.T. Konjin, eds., The Impact of Climatic Variations on Agriculture (Dordrecht, Netherlands: Reidel).

Budyko, M.I. and M.C. MacCracken (1987) "US-USSR Meeting of Experts on Causes of Recent Climate Change," Bulletin of the American Meteorological Society 68(3):237-243.

Hodges, D.G., J.L. Regens, and F.W. Cubbage, (1988) "Evaluating Potential Economic Impacts of Global Climate Change on Forestry in the Southern United States, "Resource Management and Optimization 6(3):

Hodges, D.G. and T.G. Harris, Jr. (1988) "U.S. Forest Products Research: Trends and Outlooks," Forest Products Journal 38 (7/8):26-32.

McKeever, D.B. (1987) The United State Woodpulp Industry. Forest Products Laboratory Resource Bulletin FPL-RB-18 (Madison, WI: U.S. Forest Service).

Miller, W.F. (1983) "Alternative Futures for Resource Management," pp. 17-28 in Water--A Resource in Demand Southern Cooperative Series, Bulletin 288 (Starkville: Mississippi Agricultural and Forestry Experiment Station, Mississippi State University).

Sandenburgh, R., C. Taylor, and J.S. Hoffman (1987) "How Companies Can Respond to Rising Carbon Dioxide and Climate Change," in W. Shands and J. Hoffman, eds., The Greenhouse Effect, Climate Change, and Forest Management in the U.S. (Washington: Conservation Foundation).

U.S. Forest Service (1988) The South's Fourth Forest-Alternatives for the Future (Washington: U.S. Forest Service).

Warren, D.D. (1987) Production, Prices, Employment and Trade in Northwest Forest Industries, First Quarter, 1987. (Portland: U.S. Forest Service).

TROPICAL FORESTS AND CLIMATE

Norman Myers

As tropical forest cover is eliminated, there appears to be--much earlier understanding to the contrary--a number of significant repercussions for climatic regimes, whether at local, regional or global levels. These are manifested through disruptions of hydrological systems and precipitation patterns through changes in albedo and associated evapotranspiration rates, and through contributions to the greenhouse effect. Certain of these processes are well established scientifically, others are more speculative pending further research. Despite much uncertainty, we now know, through illustrative examples drawn from a variety of sites in the humid tropics, that deforestation can indeed lead--though does not necessarily lead--to significant consequences for climatic regions, notably in the way of disruption of evapotranspiration and rainfall patterns.

We urgently need to identify and define the ecological linkages at work, and analyse such physiobiotic mechanisms as may pertain. Further, we need as much quantification as is available on issues of climatic repercussions that stem from deforestation. What significant consequences could arise for major development sectors such as agriculture, energy and human settlements? What policy responses should be generated?

In particular, we need to focus on deforestation-derived buildup of carbon dioxide in the global atmosphere, insofar as this represents one of the principal causes of the greenhouse effect that is overtaking the Earth's climatic systems and that threatens salient sectors of economic development in both the developed and developing worlds. One management response appears to offer substantial scope to counter the buildup of carbon dioxide, even though it has been little addressed in systematic fashion: it is a massive tree-planting programme in the humid tropics. Tree plantations absorb carbon dioxide from the atmosphere, and the humid tropics with year-round warmth and moisture are by far the best place for fast-growing tree plantations. Reforestation on a suitable scale in the humid tropics--accompanied of course by measures to halt deforestation--could eventually serve to sequester carbon in amounts significant for our efforts to counter the greenhouse effect.

The basic calculations amount to this. One the one hand, fossil fuels emit around 5.2 billion metric tons (gigatons) of carbon dioxide into the global atmosphere each year, an amount that has remained more or less stable in recent years. In 1980 tropical forests emitted roughly 1.8 gigatons. While this was only one third of the fossil-fuel amount, deforestation has been increasing rapidly. Whereas in 1980 a total of 36,000 square miles of moist forest were destroyed throughout the tropics, in 1987 at least 33,000 square miles of forest were burned in Brazilian Amazonia alone. By the year 2020 the figure for tropical forest emissions of carbon could climb to almost 5 gigatons--after which it would decline sharply for the simple reason that there would not be much forest left to burn.

Of the carbon loading of the atmosphere, only half remains in the skies. The rest disappears into the oceans, or into some other sink. Thus the annual net increment in 1980 was somewhere around 3.5 gigatons. (Remember too that carbon dioxide emissions cause only about half of the greenhouse effect. The rest comes from buildup of other trace gases, such as methane, and nitrous oxide--which derive from a variety of sources, including an unassessed amount from tropical deforestation.)

Of course a grand-scale reforestation exercise in the humid tropics would be pointless unless it were accompanied by efforts to slow and even halt deforestation. India, Thailand and the Philippines among other countries talk increasingly about deforestation as a "national emergency." Suppose enough countries were to take vigorous action, this could reduce forest-burning emissions of carbon by more than half, or (say) one gigaton, meaning a net amount of half a gigaton. This would reduce the overall net increment of carbon dioxide to 3 gigatons per year.

How many trees would we need to plant? According to some exploratory calculations by this writer, and by Drs. George Woodwell and Richard Houghton of the Woods Hole Research Center, Mass., also Dr. Gregg Marland of the Oak Ridge National Laboratory and Ms. Sandra Postel of the Worldwatch Institute, a tropical plantation can absorb an annual average of around 4 tons of atmospheric carbon per acre, or more than 2500 tons per square mile. So we would have to think in terms of 400,000 square miles of plantations in order to soak up one gigaton of atmospheric carbon. To eliminate the 1980 net buildup of 3 gigatons would require trees covering 1.2 million square miles, an expanse equivalent roughly to all states east of the Mississippi.

Where can we find enough space in the humid tropics? This leads into a crucial calculation, so let's check a few further figures. Of 3.2 million square miles already deforested (roughly half the original extent of tropical forests), about 640,000 square miles lie in watersheds, and they urgently need reforestation to safeguard topsoil, river flows and the like. In addition, many tropical countries also need to plant trees in 220,000 square miles to expand their fuelwood stocks, plus 40,000 square miles for commercial timber supplies. There is some overlap between these three figures, so let's reckon that at least 800,000 square miles are required for immediate reforestation--an amount that tropical-forest governments agree on already. As for the further 400,000 square miles needed to make up the 1.2 million square miles, in lowland Southeast Asia at least 350,000 square miles of deforested lands are not given over to crop growing or other productive human activity. They have degenerated into poor-quality scrub and brush, or coarse grasslands, good for little apart from reforestation. All in all, then, finding sheer space should be no insuperable problem.

At an average tree-planting cost of $160 per acre, the exercise would require a budget of just over $120 billion, or $12 billion per year for a ten-year effort. Surely a good bet. Not only would tropical nations themselves benefit through restored watersheds and the like (India suffers an average of $1 billion of crop losses and other damages from flooding in the Ganges valleylands each year, due in major measure to deforestation in the Himalayan foothills.), the global community would be spared some greenhouse costs. A moderate sea level rise (just through heating of the ocean surface, not through melting of the ice caps) could levy coastal-erosion and other costs along the United States' shoreline, according to the latest EPA calculations, of $110 billion. To revamp the U.S. network of dams and irrigation systems in the wake of a greenhouse effect could cost anything up to $23 billion, while other agricultural costs in the United States could eventually prove at least as large if not much larger. Further costs related to sea level rise and disrupted agriculture in other parts of the world would surely turn out to be similarly great.

In short, tropical reforestation could turn out to be a super-sound investment all round.

MITIGATING CLIMATE CHANGE: STRATEGIES TO FINANCE RETENTION OF TROPICAL FORESTS

Dr. Ata Qureshi
Climate Institute, Washington, DC

During the 1980's six years have been recorded as the warmest since official records were kept. On the global average 1988 was the warmest year in more than a century. The world is experiencing heavy floods in some parts, droughts in others and heat increases in many places. Carbon dioxide (CO_2), methane (CH_4), oxides of nitrogen (NO_x), chlorofluorocarbons (CFC's) and other trace gases resulting from industrial, energy and agricultural activities are implicated in global climate change.

While use of fossil fuel is the main source of carbon release (5.2 gigatons annually) into the atmosphere, forest burning and tropical deforestation is the second largest source (over 1.8 gigatons annually). Over the past century or so, atmospheric CO_2 has risen 25% to about 350 ppm. If present trends continue, the increased CO_2 levels in the atmosphere could raise the Earth's temperature by 1.5 -4.5 degrees C by the year 2050.

Worldwide Deforestation as One of the Major Causes of Climate Change and Environmental Degradation

It has been estimated that 1.7 billion acres of forest cover have already been lost worldwide. Since 1860 forest clearance has contributed 90-180 billion tons of carbon to the atmosphere. The United Nations Food and Agriculture Organization (FAO) has estimated the worldwide deforestation as 11.3 million hectares within only one year, 1980 (Table 1).

Table 1

Region	Area Deforested in 1980 (hectares)
Tropical Asia	2,016,000
Tropical Africa	3,676,000
Tropical America	5,611,000
World Total	11,303,000

Source: FAO Forestry Paper No. 30, 1982

Fig. 1

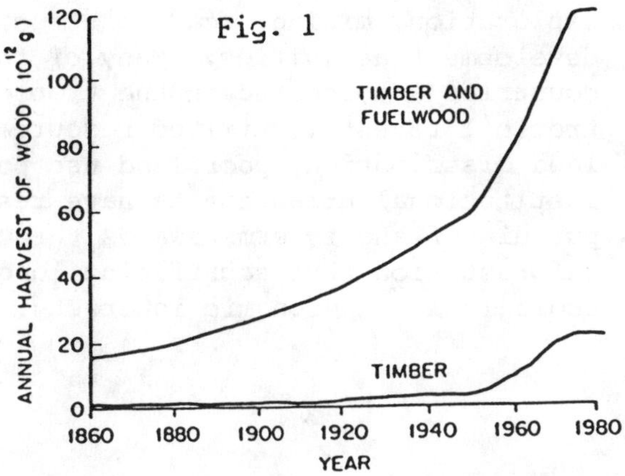

Dramatic increase in the quantity of wood harvested annually within the past century in Southeast Asia, as an example, is represented in Figure 1 above. In the Himalayas (Pakistan, Nepal, India) forest clearance has resulted in the loss of topsoil from the upper watersheds, reducing agricultural productivity of the land, depositing silt in the rivers, decreasing capacity of dam reservoirs and increasing flash flooding. Aerial photographs and satellite images show that small uninhabitable islands are being formed in the Indian Ocean and the

Bay of Bengal as a result of deforestation in the upper Himalaya mountains. Over 400 million people are affected by such misuse of watersheds (Figure 2).

Fig.2

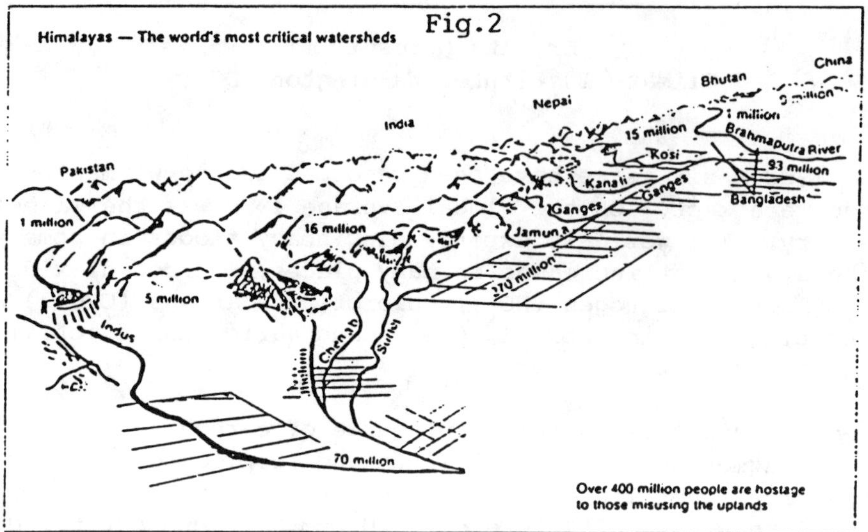

Fig. 3

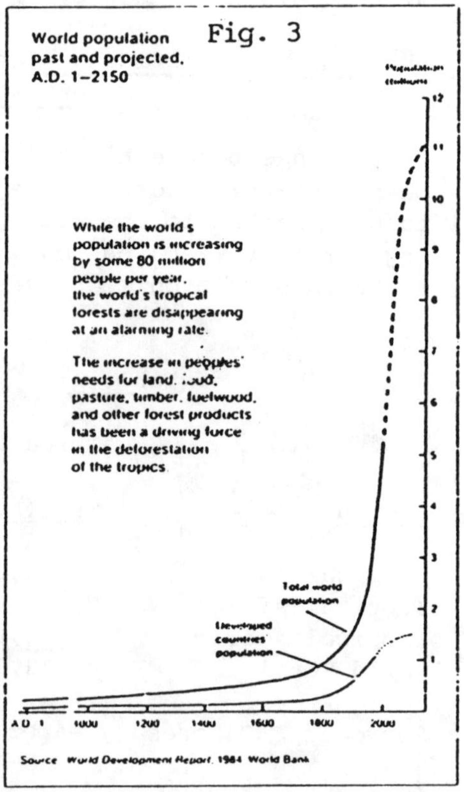

Why Are Tropical Forests Being Depleted?

Increased population pressures (Fig. 3) have been fueling this relentless destruction. Forests are cleared for food crops, ranch lands, timber, fuelwood, shifting cultivation, government sponsored colonization programs, and for planned and not so well planned projects of oil exploration, mining, road construction and other development activities. Many of the developing countries are located in the tropics with belts of tropical forests. Limited resources, inadequate land distribution, poor land use policies and weak institutional arrangements have resulted in the pursuit of short-term rewards through deforestation thus sacrificing long-term ecological and economic interests.

Strategies for Retention of Tropical Forests

A. Strategies at the Country Level

One of the strategies for tropical forest retention at the country level is to make the country's policy makers, policy implementers and the public more aware of the various short and long-term consequences of large scale deforestation. Most tropical soils are nutrient poor, and it is through the existing vegetation that the nutrient cycling occurs in the tropical forests. Once that balance is disturbed through large scale clear-cutting or other means of forest removal such as burning, soils may become impoverished or even irreversibly unproductive because of sun drying or through removal of topsoil under the torrential rain in the tropics. Then agriculture is usually no longer sustainable except by heavy inputs.

Although it is becoming increasingly clear that the value of cleared forest land is rarely as great as expected, little attention has been given in the past to developing analytical models that give full weight to the economic value of intact forests. Methods of figuring true economic worth of tropical forests would help decision makers know environmental and economic costs of deforestation, and it might lead to more equitable compensation by those who benefit from these forests. Producing such models might also be one of the most effective measures to be taken in defense of tropical forests and thus help governments able to plan rationally for the future. This also raises the question of putting a price tag on the genetic sources of the forest ecosystems of developing countries. The genetically unique and diverse nature of plants and animals are as much a national natural resource as the oil and minerals under the forest floor. Irreplaceable loss of genetic diversity, wildlife habitat, watershed function and value of tropical forests in influencing evapotranspiration, local rainfall and climate must be considered.

Decision makers should be presented with alternatives to tropical deforestation. Such alternatives could include intensification of agriculture in existing productive agricultural areas, raising firewood and charcoal production through energy plantations around metropolitan areas where demand for energy is usually high. Sensitive areas need to be protected. Afforestation of barren unculturable lands and restocking of areas suitable for industrial raw material production need to be carried out using environmentally sound harvesting and regeneration techniques. Research, education and forestry extension must be geared to meet sustained future needs of the society. Institutional infrastructure needs to be strengthened to handle such needs. To achieve these and other similar objectives, prioritization of programs and reallocation of budgets would be necessary.

B. Strategies at the Regional and International Level

Many of the countries where tropical forests are located are poor or otherwise have limited resources. Some of these countries are cutting their forests for short-term gains, for instance, to earn foreign exchange to pay debts or simply to get the money for domestic activities. Only a small portion of the budget is used for forest restoration or environmental conservation.

FAO's worldwide estimate for annual deforestation is 11.3 million hectares and only 1.1 million hectares are replanted. The ratio of deforestation to replanting in Asia is 5:1 and in Africa as wide as 25:1. Even through reallocation of budgets for forestry programs, the developing countries of the tropics would not be able to bring a balance between forest harvesting and restocking on their own. Climatic considerations by these countries are still in the inception stage. It is here that the role of regional and international organizations in helping these countries to protect and maintain tropical forests is of increasing importance. Such conservation and reforestation efforts would not only benefit that particular country itself but also the adjoining countries of the region through such benefits as protection of downstream siltation and regulation of river flows. There would also be global benefits such as climatic change amelioration, forest product trade enhancement, etc.

Hither-to-fore, the attention paid to the forestry sector programs by the regional and international development organizations has been rather small as compared to other sectors. Figure 4 shows average annual lending from 1980-84 by the major development banks including the World Bank, Inter-American Development Bank, Asian Development Bank and the African Development Bank. Of the total annual amount of 20.3 billion dollars lending, forestry sector lending was only 0.1 billion dollars or 1/200th of the total.

Fig. 4

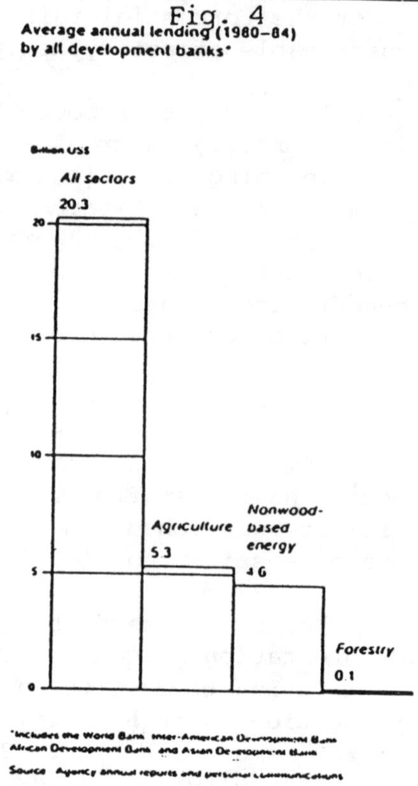

Bilateral agencies such as the U.S. AID, CIDA, and ODA, regional organizations such as ADB, ASEAN, and ESCAP; and international development organizations such as the World Bank, UN FAO, UNDP, and UNEP, as well as the private sector, need to consider increasing their share of efforts in protecting, maintaining and enhancing the tropical forests for a variety of benefits derived from them including the reversal of the greenhouse effect and global warming.

An Innovative Mechanism for Saving the Tropical Forest-- Debt for Nature Swaps

Third World debt is estimated at 1.3 trillion dollars and many developing debtor countries find it difficult to pay interest on their external debt let alone the loan principal. Out of this Third World debt problem has emerged an innovative program in 1987 to save some tropical forests. Debt for Nature Swaps, as they are commonly known, involve an exchange or cancellation of a foreign country's debt in return for the debtor's country's commitment to use a given amount of local currency funds to protect national parks, conservation of forests or other environmental activities. These swaps can occur when certain conditions exist:

1. debtor country's debt must be available through donation or purchase on the secondary market.
2. debtor country government must agree to the terms of the swap, and
3. a party other than the creditor's bank or debtor country must be interested in obtaining the debt.

Debt for Nature Swaps have been arranged in Ecuador (Figure 5), Bolivia (Figure 6), Costa Rica (Figure 7) and the Philippines. The non-governmental organizations purchase some of the debt at discount rate, then turn over the promissory note to the central bank of the host country which commits local currency to environmental protection. The mechanism of a swap is explained (Figure 8) in the case of Ecuador, where agreement was signed by the World Wildlife Fund and Nature Conservancy on April 5, 1989 to absorb $9 million of Ecuador's foreign debt at about 12 cents on the dollar. The Central Bank of Ecuador agreed to redeem the promissory note in local currency at 100 cents on the dollar. This is the largest of six such swaps. U.S. conservation groups have picked up a total of $13 million in foreign debts in earlier swaps--two in Costa Rica, and one each in Bolivia, the Philippines and Ecuador.

Although the impact of Debt for Nature Swaps might be minimal on the debt situation itself, funds provided by these swaps can go a long way in improving natural resource enhancement and conservation of tropical forests within the individual countries.

U.S. Response to Tropical Forests Retention

Response from the United States Congress towards protection of tropical forests has been overwhelming. A number of bills have been introduced during the 100th and 101st Congresses concerning tropical forests, the greenhouse effect and climate change. The Tropical Forest Protection Act (H.R. 3010) was introduced by Congressman John Porter and Congressman David Obey. Senator Robert Kasten and Senator Alfonse D'Amato introduced the Tropical Forest and Wetlands Protection Act (S. 1538). Both these bills have a primary focus on protection of tropical forests and wetlands study to determine areas unsuitable for agriculture and require determination by the U.S. Treasury Secretary on institutional capacity to protect such forest resources in concert with such organizations as the Multi-Development Banks (MDB's) such as the World Bank. Continuing Resolution H.R. 395 (public Law 100-200) specifically directs Department of Treasury to promote environmentally sustainable economic growth through MDB's. Though World Bank and IMF loans cannot be converted to grants, are not saleable on the secondary market, and are not reschedulable, the Treasury Department recommends a number of initiatives including debt for nature pilot project.

Several U.S. organizations and agencies such as the Environmental Protection Agency, the Forest Service of the U.S. Department of Agriculture, Smithsonian Institution, National Academy of Sciences, National Science Foundation, a number of universities and several environmental organizations have started to include programs on tropical forests as they relate to amelioration of climate change. In February 1989, EPA submitted a report to the U.S. Congress on policy options for stabilizing global climate including reducing demand on forest lands and increasing supply of forest products. The U.S. Forest Service which initially started climate change research activity in its Fire Research Unit is now also looking towards initiating components on tropical forest conservation and use of reforestation to mitigate climate change in its International Forestry Program.

Figure 5

DEBT-FOR-NATURE SWAPS - EXAMPLE

E C U A D O R

PURPOSE:	TO PROTECT PARTS OF AMAZON RAIN FOREST, ANDEAN HIGHLANDS, COASTAL SAVANNAH LANDS AND MANGROVE FORESTS
NGOs:	WORLD WILDLIFE FUND FUNDACION NATURA
GOVT. AGENCIES:	CENTRAL BANK OF ECUADOR
BANKS:	CITIBANK BANKERS TRUST
DEBT SWAP:	$1 MILLION (SOLD FOR $354,500)
DATE ARRANGED:	OCTOBER 1987

Source: Congr. Res. Svc.

Figure 6

DEBT-FOR-NATURE SWAPS - EXAMPLE

B O L I V I A

PURPOSE:	TO CREATE RESERVE SURROUNDING BENI BIOSPHERE 334200 ACRE TRACT
NGOs:	FRANK WEEDON FOUNDATION CONSERVATION INTERNATIONAL INTER-INSTITUTE TECHNOLOGY COMMISSION BOLIVIAN NATIONAL ACADEMY OF SCIENCES
GOVT. AGENCIES:	MINISTRY OF AGRICULTURE CENTRAL BANK
BANKS:	CITIBANK
DEBT SWAP:	$650,000 (SOLD FOR $100,000) $250,000 ADDITIONAL ENDOWMENT FROM US AID & BOLIVIAN GOVT.
DATE ARRANGED:	JULY 1987

Source: Congr. Res. Svc.

Figure 7

DEBT-FOR-NATURE SWAPS - EXAMPLE

C O S T A R I C A

PURPOSE:	GUANACASTE NATIONAL PARK IN SANTA ELENA--32500 ACRES AND OTHER CONSERVATION PROGRAMS
NGOs:	NATIONAL PARK FOUNDATION OF COSTA RICA WORLD WILDLIFE FUND THE NATURE CONSERVANCY
GOVT. AGENCIES:	MINISTRY OF NATURAL RESOURCES CENTRAL BANK
BANKS:	FLEET NATIONAL BANK, RHODE ISLAND, USA SWISS BANK CORPORATION COSTA RICA COOPERATIVE BANK
DEBT SWAP:	$5.4 MILLION $254,000 ADDITIONAL DONATION
DATE ARRANGED:	AUGUST 1987

Source: Congr. Res. Svc.

Fig. 8

MECHANISM FOR ECUADOR--WWF

DEBT-FOR-NATURE SWAP*

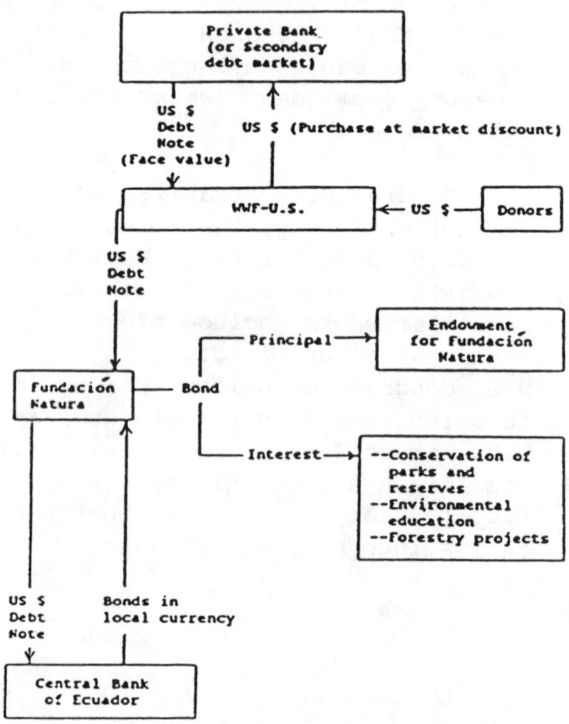

Summary and Conclusions

Almost 25% of the annual carbon emissions to the atmosphere is due to tropical deforestation. If future global warming is to be controlled, deforestation has to be curtailed. In order to reverse the greenhouse effect, existing forests need to be restocked and already cleared areas and unculturable waste lands afforested.

To provide for the increasing human population, intensification of agriculture, development of wood reserves, and sustainable land-use systems are needed. Policies of each country must be reshaped to allow for protection of climate. Research and educational institutions need to be strengthened. Both policy makers and the public must be motivated for concerted actions.

Reprioritization of government programs and increased funding for forestry development, climatic amelioration and environmental protection at the national, regional and international levels both by the public and private sectors would be necessary to protect this planet for ourselves and the future generations.

From "CLIMATE CHANGES AND U.S. WATER RESOURCES":*
A BRIEF SUMMARY AND THE RECOMMENDATIONS

Paul E. Waggoner, editor*

Summary

Characteristically, climate includes a large annual variability of temperature or precipitation around an average or mean, even if the mean is not changing. Thus we cannot tell whether or not the climate is changing by looking at a single year or even the sequence of a decade. Why then, do we suspect climate is changing?

Because the CO_2 in the atmosphere is increasing. During more than a century since the beginning of the Industrial Revolution it rose about 50 parts per million by volume (ppmv) to reach 316 in 1959, and since 1959 it has increased another 34 ppmv.

Solar energy reaches the earth's surface as short wave radiation, much of it light, and it readily passes through CO_2. The radiation leaving the earth and balancing the incoming short waves is, however, long waves, which are obstructed by CO_2. Thus more CO_2 means the terrestrial temperature must rise to push the balancing long waves through the atmosphere to outer space. A more detailed analysis takes into account the location of the continents and oceans and the circulation of the atmosphere in a large computer model of the atmosphere. It predicts an average global temperature rise of 2 to 5 degrees C (4 to 9 degrees F), an increase of 7 to 15% in precipitation, and a rise of 10 to 100 cm in sea level. These changes are predicted when the effects of a doubled CO_2 content are felt. The doubling is expected in the middle of the 21st Century unless present burning of carbon in coal petroleum and wood decreases. Less burning seems unlikely because we do not have other large sources of energy and because developing nations are bent on industrialization.

---

* From "Climate Changes and U.S. Water Resources," edited by Paul E. Waggoner, Report of AAAS Panel on Climatic Variability, Climate Change and the Planning and Management of U.S. Water Resources, John Wiley and Sons, New York, New York, 1989.

Wide geographic variations are expected. In some regions temperature and precipitation may actually decrease. Unfortunately, the model tells us little about what will happen at any specific location. Although it does predict that the interior of the large continents at mid-latitudes will be drier, it does not tell us what will happen in California or Iowa.

Even after a climatological prediction of temperature and precipitation are made, however, we still do not know whether water resources will meet our demands. Thus we begin with the present resources and vulnerabilities. We examine how evaporation is subtracted from present precipitation to leave runoff as water resources and how their variation is tempered by reservoirs and aquifers. These effects of CO_2 increase will be both positive and negative with precipitation decreasing in some places and increasing in others. More CO_2 will make plants grow faster and will temper their transpiration of water.

We also examined the effects on human activities of the changes in water resources caused by climate change.

Usually the probabilities of extremes like flood or drought will change more than averages.

In the West the effects of a change will be greatest simply because it is arid, making a small change a large relative change. Increased evaporation from warming will have a larger relative effect on water resources in the arid places.

Maintaining the present irrigated area in the West will be hard although increased efficiency may make it less difficult. In the East irrigated area is expected to increase several fold but still be small.

Everywhere seasonality will change as snow melts earlier, freezing is less persistent and the abundant water floods rivers and fills reservoirs at new seasons.

Along the coasts rising sea level will reduce fresh water in aquifers and push the tongue of salt water further up estuaries, jeopardizing intakes of fresh water.

Fluctuations are damped by larger watersheds, organizations encompassing larger regions and storage. Storage includes surface reservoirs and aquifers, which have the advantage of no evaporative loss.

Climate change heightens the need to review rules about water, especially ground water which is badly overdrawn. Transfers are already occurring as traders bring market forces to bear, even under present institutional arrangements, which are inherently conservative to protect joint interests.

The 80% of Americans who depend on urban water supplies will be supplied, but it will cost more.

Recommendations

Although climate change seems an irresistible force that even a nation cannot forestall, a nation certainly and even a locality or person can do something about water resources. Thus action can follow recommendations. We grouped our Recommendations in those for scientists, who must enlarge our understanding of the natural phenomena, their impacts on society and our alternatives to cope. Those for public bodies, which must defend the nation, set the rules of society's game and sometimes construct and operate public works. And those for private persons and enterprises, who being by far the largest group, must perform most of the acts and reap most of the consequences.

Scientists

PROSPECTS FOR CLIMATE CHANGE

Because everyone's decisions, actions and fate depend upon anticipating correctly the water resources in a changed climate and only scientists are likely to be equipped for the needed predictions, our first and fundamental Recommendation must concern them.

> **Scientists investigating climate change should make a special effort to improve predictions on the scales of time and space most relevant to the management of water resources, i.e., the scales of decades and the scales of large hydrologic surface and groundwater basins, setting bounds on the likely changes of averages, extremes, interannual variability and the rate of changes.**

PUTTING FORECASTS TO USE

Some changes are highly likely, e.g., warming; whereas other things will long remain uncertain, e.g., precise changes in runoff. Meanwhile, practical managers must decide upon actions now with consequences several decades hence, after some expected climatic changes have occurred. Thus scientists should inform practical managers continuously about what is certain, what are the bounds on the uncertain, and how soon change is likely.

Organizations of professional water resource managers, e.g., ASCE, AWRA, and WPCE, should meet formally and regularly with scientists concerned with water and climate change. In these meetings scientists will share with the practitioners what is known about climate change effects, what is likely to be known soon, and what will likely remain uncertain. In turn practitioners will share with scientists what hydrological knowledge they need to manage water resources reliably and efficiently.

FROM CLIMATE TO WATER RESOURCES

Even a perfect prediction of precipitation will not precisely predict water resources. First we must properly couple all the phenomena of evaporation, infiltration and storage examined in our chapter "From Climate to Water Resources." Further, as the chapters on climate prediction explain, the models of global circulation that predict climate change cannot be correct until the feedback loops between earth and atmosphere are correct. Thus a Recommendation to elucidate the feedback and improve the forecasting of hydrologic effects of climate change.

Diagnose the relations between climate factors and past and present streamflow variability at the scale of the major US water resources regions. Improve existing monitoring and data collection systems to support these diagnostic studies.

<u>Public bodies</u>

GUIDELINES FOR DECISION

Governments from the Federal to a myriad of municipalities and entities from legislatures and courts to regulatory agencies set the rules for who shall be protected, who shall own, and how water resources shall be managed. Public bodies build dams, canals and pipelines that store and transmit water. Among the climatic changes that governments and other public bodies are likely to encounter are rising temperatures, increasing evapotranspiration, earlier melting of snow-packs, new seasonal cycles of runoff, altered frequency of extreme events, and rising sea level. Thus a crucial Recommendation is directed to them.

Governments at all levels should re-evaluate legal, technical and economic procedures for managing water resources in the light of the climate changes that are

likely. These procedures include techniques for using water, criteria for design and operation, guidelines for choosing among large structures and alternatives, and regulations and methods for allocating water among users. To increase the flexibility of water systems, government guidelines should incorporate both the certainties of global climate change and the uncertainties about local details. Where professional societies draw guidelines, they should put in their manuals of practice the same certainties and uncertainties.

INSTITUTIONAL FLEXIBILITY

Climate change increases the value of flexible institutions for allocating water. Governments charter these institutions and set the rules for what they may and may not do. Governments must give due regard to equity and public good as in protecting the value of water in a stream as well as the water withdrawn. Or in protecting the rights of those like native peoples and farmers who may be damaged by reallocation of water.

Governments should permit and encourage flexible institutions, including markets, that can allocate water for its most beneficial use as climate changes, sometimes forestalling crises. Remembering their role as the protectors of the public good, governments should protect public values in water while at the same time increasing flexibility in allocating water.

ORGANIZING OVER "PROBLEM SHEDS"

Climate change will often cause problems that make integration, management and exchanges over regions advantageous. Although these regions may correspond to present boundaries of states or other well known regions, new regions may be required to match the problems and opportunities for solution. Hence our next Recommendation.

Governmental or quasi governmental entities should be encouraged to facilitate the planning and operation of water systems over areas proper to the solution of problems of climate change.

ENCOURAGING RESEARCH TO ANSWER PRESSING QUESTIONS

We employed engineering, agronomy, ecology, history, economics and knowledge of practical management in anticipating impacts and responses to climate and water-resource change. The relations between climate and water cut across the boundaries of the classic disciplines of science. Although separate research by different disciplines is not wasted, collaboration by scientists of diverse disciplines, sharing questions, knowledge and responsibilities is crucial. Since governments must solve problems to defend the nation or state they govern, a Recommendation about science is directed to them.

> **To foresee changes and effects of climate and water resources takes interdisciplinary research. Until institutional and personnel problems such as funding and professional recognition are dealt with, interdisciplinary research will be hard to sustain in government, universities or anywhere. Governments needing knowledge of future climate and water resources and other public entities, too, should encourage interdisciplinary research by their rules and money.**

<u>Private persons and enterprises</u>

INVESTING WISELY

Climate change will open opportunities to those who adapt. Climate change could increase the return from investment that copes first with variability and then with change.

> **During planning, managers should be alert for economical measures to increase flexibility and accommodate climatic variability, sea-level rise, and as we learn more about it, climate change. They should exploit opportunities to retain or increase flexibility of systems, especially since such measures may be fairly inexpensive if put in an original design.**

OPPORTUNITIES FOR WISE WATER USE

Conservation is an important means of adapting to the hydrologic impacts of climate change. Climate change could increase the return from innovation that fosters conservation.

Industrial and agricultural innovators should be alert for profitability of water conserving technologies increased by climate change. Then, through personal and community efforts the public can take early advantage, for example, of new crops and cultural practices and of new industrial techniques for extracting more benefit per gallon of water.

REMOVING VULNERABILITIES

All regions of the United States already have vulnerabilities to climate changes. Some regions exceed warning levels for all or most indicators. Many eastern and northwestern basins are vulnerable to increases in droughts or floods because of small storage volumes. Western basins are susceptible to decreased precipitation and runoff because of both high demand relative to available supply and great reliance on groundwater. Also western basins are already vulnerable to the variability of flows. Thus the Recommendation:

Decrease existing vulnerabilities, which climate change will make more harmful. In some places they can be decreased by arrangements for transferring water to new uses, in others by building new structures and in others by increasing the benefit of each gallon.

KEEPING COOL

The phrase "future shock" expresses the emotions of insecurity and confusion evoked by change. Uncertainty about what scientists agree upon with reasonable certainty adds to the confusion. Thus the final Recommendation:

Those reporting about climate change bear a special responsibility for accuracy, conveying the real complexities and uncertainties, and not oversimplifying. Scientists must make extra effort to explain clearly in conservative and understandable terms.

VULNERABILITIES OF UNITED STATES WATER SYSTEMS TO CLIMATIC CHANGE

Peter H. Gleick [1]

(prepared for the American Association for the Advancement of Science (AAAS) Climate and Water Panel Report on: Climatic Variability, Climate Change, and U.S. Water Resources, John Wiley and Sons, Inc., New York)

Climatic changes will affect the design, construction, and operation of water-supply systems. The long lifetimes of existing facilities means that they may be subjected to climatic conditions for which they were never designed. For new facilities, early planning for the hydrologic impact of climatic changes can prevent their failure when conditions change and thus save expensive redesign or reconstruction, although building for future climatic changes that never materialize could mean overbuilding facilities--an uneconomic prospect. The appropriate response is incorporating into new designs those climatic changes that we are confident will occur. At the same time, we must explore the vulnerabilities of the existing water-supply infrastructure to existing climatic variability, and identify other common-sense, inexpensive changes that make water supply more flexible and less vulnerable to a range of plausible futures. This paper discusses critical vulnerabilities of water-supply systems to climatic variability and change, and identifies measures for evaluating these vulnerabilities in the United States.

VULNERABILITY OF WATER-SUPPLY SYSTEMS TO EXISTING CLIMATIC VARIABILITY AND TO CLIMATIC CHANGE.

Water-supply systems include varied physical and economic mechanisms that provide water over a range of spatial and temporal scales. Among the most important components of such systems are physical structures, such as interbasin and intrabasin aqueducts, natural and artificial surface reservoirs, and sub-surface storage reservoirs called aquifers. Increasingly important, however, are the economic and institutional mechanisms that permit the transfer of the rights of water use from one region or user to another.

Insights into water resource vulnerabilities and

---

[1] Director of Global Environment Program, Pacific Institute for Studies in Development, Environment, and Security, Berkeley, California, 94709.

sensitivities can be gained by looking at the ways hydrologists measure the flexibility or resilience or reliability of the existing system. Water-resource systems are designed and operated for future events for which no exact time can be forecast (Linsley et al. 1982) and for which no perfect knowledge about conditions exists. Thus we rely extensively on the past hydrologic record. Occasional failures are undesired, but they are recognized to be a result of these normal uncertainties.

Systems are therefore built to be "resilient", and to operate under a range of conditions such that the present value of the cost of failure is low--presumably lower than the cost of preventing the failure by changing the original design. Similarly, systems should be "robust", i.e. insensitive to error, either random or otherwise. Most large hydrologic systems are designed to be resilient and robust, and in a stable climate these systems operate satisfactorily without any sophisticated or elaborate projections about future climate.

The normal uncertainties about climatic extremes are compounded under conditions of changing climate when we don't know precisely what hydrologic changes to anticipate. Yet knowing if streamflow or other hydrologic factors will exceed or fail to reach certain values is critical to the economic and social evaluation of a project and to estimates of the consequences of the failure of that project.

Vulnerabilities of water-supply systems to climate can be separated into three categories, listed in Table 1: meteorological and climatological vulnerabilities, hydrologic system and design vulnerabilities, and geographical and societal vulnerabilities. The variables listed are not meant to be all-inclusive, but indicate particular sensitivities.

Table 1
Vulnerabilities to Climatic Variability

	Important Variables [1]	Examples of Vulnerable Basins [2]
Meteorological and Climatological		
Timing	annual, seasonal, daily flow	California
Quantity	absolute and seasonal flow	Great Basin
System and Design		
Physical	storage volume, peak flows, maximum release rate	Tennessee
Operational	timing of flows, timing of demand	Mid-Atlantic
Legal or Institutional	distribution of water rights & options for transfer	Lower Colorado
Economic	water pricing, cost of storage or delivery	California
Geographical and Societal		
High Demand	level and timing of demand	Lower Colorado
Flood prone	peak flow, flood storage volume	Ohio
Water Quality	minimum flows, patterns of use	Mid-Atlantic
Rainfed Agriculture	timing of rainfall, evapotranspiration rates	Arkansas-White-Red
Hydroelectricity	seasonality of flows, reservoir storage volume	Pacific Northwest

Notes to Table 1.

1. U.S. Water Resources Council "water resource regions". "Sample Basins" are basins with physical, hydrological, or institutional characteristics that make them vulnerable to climatic variability or change.

INDICATORS AND MEASURES OF THE VULNERABILITY OF HYDROLOGIC SYSTEMS TO CLIMATE

I have identified generic types of vulnerabilities to climatic conditions and referred to specific instances or regions of note. The extent of actual vulnerabilities to climatic change depends on the variables in Table 1 and on regional details of future climatic changes. But how can such sensitivities be measured? One way is developing and evaluating specific quantitative indicators. Presented below are five such measures that evaluate different sorts of regional vulnerabilities.

While other quantitative measures can be developed, the ones here offer insights into some of the important vulnerabilities discussed earlier--for example, floods and droughts, groundwater overuse, and reliance on hydroelectricity. These measures have been evaluated on an annual basis for the 21 official U.S. water resource regions.

Many water problems are local, and aggregating river basins into the water resource regions can hide individual problems. The indicator ratios should therefore also be calculated for smaller regions. For example, California is vulnerable by four of the five measures--all but groundwater overdraft. But within California, some sub-basins are already greatly overdrawing groundwater, while in others demand is well below available supply. Similarly, although the values in Table 2 are calculated on an annual basis, it is often valuable to look at seasonal indicators as well. A region with sufficient annual supply may well find seasonal shortages to be a problem if there is a distinct bias in either seasonal supply or demand.

It is also important to note that the absolute numbers in Table 2 are less important than the relative vulnerabilities. The fact that the Tennessee Basin has a ratio of demand to supply of 0.01 does not mean that it is 9 times better off than the Lower Mississippi, where the ratio is 0.09. These indicators are most useful for identifying regions where special attention should be paid to particular water-resource problems.

1. <u>Regions with low storage volume relative to renewable supply:</u>

$$\frac{S}{Q}$$

An indicator of ability to withstand prolonged drought or severe flooding is the ratio of storage volume (S) to the quantity of supply (Q). With large storage compared to supply, short-term droughts are less likely to cause major water shortages. Conversely, where this ratio is small, climate-induced reductions in water or changes in the frequency or intensity of droughts and floods, will be strongly felt. Storage systems whose capacity is less than 60 percent of annual flow may be particularly vulnerable to wet and dry extremes. Those basins where S/Q is below 0.6 are therefore of special concern.

Two groups of regions stand out as being vulnerable to storage-related problems--the eastern portion of the country and the far west. Basins such as the Colorado and the Missouri with extensive storage are less vulnerable to flooding and drought. It must be noted, however, that even in basins with ratios of storage to supply close to the indicator, such as California, the operation of existing systems may be able to mitigate some of the worst impacts of extreme hydrological events. On the other hand, climatic changes may have a greater impact on such marginal areas.

2. <u>Regions with high consumptive use relative to available renewable supply.</u> $\frac{D}{Q}$

Where consumptive use (defined as use of water that excludes re-use, such as through evaporative loss or contamination) is high, or supply is low, annual or seasonal shortages are likely to be felt. The ratio of consumptive use or demand (D) to renewable supply or quantity (Q) is perhaps the best measure of the relation between these two variables. Where demand is a large fraction of total available water, shortages may be a problem. Climatic changes that alter the timing or magnitude of flows affect regions where D/Q is high. Szesztay (1970) considers water to be a critical factor for development in regions where D/Q is 0.20 or above.

Basins vulnerable to supply problems tend to be clustered in the western part of the United States, where concerns about supply are already very high. Climatic changes that affected the quantity of available water would have a most noticable effect in these regions.

3. <u>Regions with a high proportion of hydroelectricity relative to total electricity.</u>

$$\frac{H}{E}$$

In regions that depend on hydroelectricity, droughts affect total electricity availability and its average cost. Because hydroelectricity is typically the cheapest form of electricity to produce, less water usually increases consumption of fossil fuels, and thus the cost of electricity. The ratio of electricity produced by hydroelectric systems (H) to total electricity produced (E) indicates this vulnerability. Since no absolute measure exists, this study assumed that regions dependent on hydroelectricity for more than 25 percent of total demand on an annual basis merit special attention. This measure can also be used to identify seasonal vulnerabilities, where peak demand or peak hydroelectricity generation has a strong seasonal component.

According to Table 2, the western United States has the greatest reliance on hydroelectricity, with the Ohio, Missouri, and Tennessee Basins. A climatic change that reduced water availability in these regions would have a serious economic impact. At the same time, these regions would be best able to take advantage of increased runoff to reduce their dependence on other energy sources, such as fossil fuels that are more expensive or more environmentally damaging.

4. <u>Regions with large groundwater overdraft relative to total groundwater withdrawals.</u>

$$\frac{GO}{GW}$$

Many regions depend on groundwater for part of their freshwater needs. Some of these regions already pump groundwater faster than the natural recharge rates. Where the ratio of groundwater overdraft (GO) to total groundwater withdrawals (GW) is high, water availability is already a problem. These regions are especially vulnerable to climatic changes that decrease water availability. Indicator GO/GW warns regions principally or solely dependent on groundwater that they are vulnerable to droughts, groundwater contamination by saltwater intrusion, or population growth. As with the indicator for hydroelectricity, a region with GO/GW above 0.25 merits special attention.

The regions vulnerable to groundwater problems identified in Table 2 include most of the major agricultural areas of the central part of the country. Many of these regions overdraw groundwater because of the unreliability of surface water supplies and because of high demand relative to water availability. Other regions known to have local groundwater problems, such as California, are not highlighted here because the entire basin as a whole is not in overdraft. The point made above about the importance of doing this type of analysis at the local scale is especially relevant here.

5. Regions with variable streamflow.

$$\frac{Q_{05}}{Q_{95}}$$

Q_{05} is the flow quantity exceeded only five percent of the time; whereas Q_{95} is the quantity exceeded 95 percent of the time. Thus, the lower the ratio of these two variables, the lower the variability and the higher the dependability of flow in a basin (Stockton and Boggess 1979). High variability indicated by Q_{05}/Q_{95} suggests vulnerabilities to the extremes of flooding and drought. A median value of 3 was chosen here to suggest regions sensitive to variability. Problems with variability are most evident in the central and western regions of the United States.

Evaluation of Indicators

These five measures summarized in Table 2 for the 21 United States water resource regions warn of regional sensitivities and vulnerabilities to climatic variability and change. In the Missouri, the Great Basin, and California regions, four or five out of the five ratios are high; although the Missouri Basin and the Great Basin have relatively large storage, they also have high demand compared to supply, depend on hydroelectricity for over 25 percent of their total supply, overdraw groundwater, and have highly variable streamflow. Three other regions exceed three out of the five measures and the warning lamp is on in all regions.

Table 2

INDICATORS OF VULNERABILITY TO CLIMATIC CONDITIONS

MEASURE OF:	STORAGE	DEMAND	HYDRO	OVERDRAFT	VARIABILITY	
River Basin Region	$\frac{V}{Q}$ [1]	$\frac{D}{Q}$ [2]	$\frac{H}{E}$ [3]	$\frac{GO}{GW}$ [4]	$\frac{Q05}{Q95}$ [5]	Number Exceeding Limits [6]
New England	.15	.01	.03	.00	2.23	1
Mid-Atlantic	.10	.02	.03	.01	2.38	1
S. Atlantic-Gulf	.16	.03	.08	.06	2.93	1
Great Lakes	.08	.02	.12	.02	2.31	1
Ohio	.12	.02	.02	.00	2.42	1
Tennessee	.23	.01	.28	.00	1.84	2
U. Mississippi	.14	.03	.01	.00	2.89	1
L. Mississippi	.31	.09	.02	.09	3.75	2
Souris-R-R	.93	.07	.00	.00	6.33	1
Missouri	1.12	.29	.25	.25	4.22	4
Arkansas-W-R	0.45	.17	.10	.62	5.59	3
Texas-Gulf	0.61	.23	.01	.77	9.90	3
Rio Grande	1.89	.64	.09	.28	22.00	3
Upper Colorado	2.61	.33	.04	.00	4.00	2
Lower Colorado	4.22	.96	.27	.48	1.42	3
Great Basin	.35	.49	.25	.42	3.92	5
Pacific NW	.19	.04	.93	.08	1.92	2
California	.42	.29	.30	.12	4.48	4
Alaska	.00	<.01	.50	.00	1.46	2
Hawaii	.01	.05	.00	.00	2.71	1
Caribbean	.05	.06	.00	.05	4.44	2

(Table 2 continued next page.)

Table 2 (continued)

INDICATORS OF VULNERABILITY TO CLIMATIC CONDITIONS

Notes to Table 2.

1. Measure of storage. Ratio of maximum basin storage volume V to total basin annual mean renewable supply Q (1985). Regions with values below 0.6 have small relative reservoir storage volumes. Large reservoir storage volumes provide protection from floods and act as a buffer against shortages.

2. Measure of demand. Ratio of basin consumptive depletions D (including consumptive use, water transfers, evaporation, and groundwater overdraft) to total basin annual mean renewable supply Q (1985). Water is considered a decisive factor for economic development in regions with values above 0.20 (Szesztay 1970).

3. Measure of dependence on hydroelectricity. Ratio of electricity supplied by hydroelectric facilities H to total basin electricity production E (1975). Regions with values 0.25 or above have a high dependence on hydroelectricity.

4. Measure of groundwater vulnerability. Ratio of annual groundwater overdraft GO to total groundwater withdrawals GW (1975). Regions with values of 0.25 or above already have groundwater supply problems.

5. Measure of streamflow variability. Ratio of 5% exceedence flow to 95% exceedence flow. The 5% exceedence flow is the streamflow exceeded five percent of the time--a very high flow. The 95% exceedence flow is the streamflow exceeded 95 percent of the time--a very low flow. Values of 3 or above suggest high streamflow variability. Data are from 1975.

6. Overall Sensitivity. The number of measures exceeding the suggested limits in notes 1 to 5 are presented in this column. Thus, the Missouri, the Great Basin, and California appear to be particularly susceptible to climatic conditions, with four or five out of the five measures exceeded; the Lower Colorado, the Rio Grande, and Texas-Gulf exceed three out of the five measures. At least one problem appears in each of the 21 areas. More detailed regional or seasonal assessments will reveal other problem areas.

Data from U.S. Water Resources Council (1978) and U.S. Department of the Interior (1986).

SUMMARY AND CONCLUSIONS

Because water systems are naturally vulnerable to climatic variability and change, they are designed with a degree of resilience and robustness that permits them to operate during varied conditions. When failures do occur, they are caused by unusually severe weather, or occasionally, the failure of components of the system itself to perform as designed.

Under some circumstances, climatic changes caused by the greenhouse effect will increase the risk of failure or the cost of maintaining or building facilities. In other cases, climatic changes will make changes in existing facilities imperative, or require new facilities. Lacking precise predictions on expected regional hydrologic changes, indications about vulnerabilities and sensitivities can permit design or operation changes to reduce risks of failure.

Unless long-term climatic change can be distinguished from temporary variability, wrong decisions could be made. Perhaps the greatest difficulty facing resource managers is separating a permanent climatic change from climatic variability, since the appropriate responses differ. A short drought, for instance, is often mitigated by surplus water from outside the basin, or by temporary changes in water use. A long-term shortage, however, requires permanent changes in demand, perhaps by changing prices, or the physical structures themselves.

One of the most difficult climatic changes for most regions to handle would be increased variability--and thus more frequent extremes: higher peak floods, more persistent and severe droughts, greater uncertainty about the timing of the rainy season. Few such changes would be beneficial. Changes in variability cause the most costly societal responses, such as flood control, large water transfers, and storage.

Qualitative vulnerabilities of water resources to climatic change are caused by system design, operational inflexibility, and legal and institutional constraints. Each constraint reduces the flexibility of water systems to respond to climatic changes.

Finally, quantitative indicators of water-resource vulnerability and sensitivity to both climatic variability and changes were developed. The five examples include measures of variability, groundwater use, hydroelectric generation, demand, and reservoir storage capacities. They identify regions with special vulnerabilities and warn where responses are needed to make water resources more flexible and resilient to coming climatic changes. Every region of the United States is vulnerable to disruptions of water supply or changes in water quality due to climatic changes. Our challenge is to identify these vulnerabilities and react with foresight and ingenuity.

REFERENCES

Linsley, R.K., Jr., Kohler, M.A., and J.L.H. Paulhus. 1982. Hydrology for Engineers. Third Edition. (McGraw Hill Book Company, Inc., New York.) 508 pp.

Stockton, C.W. and W.R. Boggess. 1979. "Geohydrological Implications of Climate Change on Water Resources Development." U.S. Army Coastal Engineering Research Center, Fort Belvoir, Virginia. 206 pp.

Szesztay, K. 1970. The Hydrosphere and the Human Environment: Results of REsearch on Representative and Experimental Basins. Proceedings, Wellington Symposium, UNESCO Studies and Reports in Hydrology, #12. UNESCO Publications.

U.S. Department of the Interior. 1986. National Water Summary 1985: Hydrologic Events and Issues. United States Geological Survey. Water Supply Paper 2300. (U.S. Government Printing Office, Washington, D.C.)

U.S. Water Resources Council. 1978. The Nation's Water Resources: 1975-2000. Volume 2. Second National Water Assessment, U.S. Government Printing Office. Washington, D.C.

CLIMATE CHANGE, CO_2 FERTILIZATION AND EVAPOTRANSPIRATION

Norman J. Rosenberg[1], Mary S. McKenney[1] and Philippe Martin[2]

I. INTRODUCTION

The question of how climate change may alter ET is complex for several reasons. Greenhouse warming may alter not only temperature, but cloudiness, windiness, and humidity as well. All four factors determine the atmospheric demand for water vapor or its "drying power". In addition, changes in soil moisture caused by climatic change will affect the amount of water available to plant roots, and thus the amount that can be transpired. Also to be considered is the fact that one of the major greenhouse gases, carbon dioxide, can affect plant growth and resistance to the passage of water vapor into the atmosphere. Any analysis of the possible impacts of climate change on ET should examine all of these factors.

Most previous studies of the hydrologic consequences of climate change have considered ET as a function of temperature (e.g., Revelle and Waggoner, 1983; Gleick, 1987), or as a function of the effect of carbon dioxide on stomatal or canopy resistance (e.g., Idso and Brazel, 1984; Aston, 1984). We set out to examine how the various possible changes in climate and plant behavior described above can affect ET. Three ecosystems--a wheat field in Nebraska, a grassland in Kansas, and a forest in Tennessee were used as test cases by Martin et al (1989). Results of that analysis are described here and additional analysis done for an irrigated alfalfa field by Rosenberg et al. (1989b). Results of both studies and their implications for the response of evapotranspiration to climate change are reviewed here.

II. MODELS AND ASSUMPTIONS

A. The Energy Balance

Net radiation provides the energy that drives important processes at the earth's surface: heating the soil, heating the air and vegetation, fueling photosynthesis and evaporating water. We keep account of these simultaneous processes by the energy balance,

$$Rn + S + H + LE + P + M = 0 \qquad \text{[eq. 1]}$$

where (all terms are in units of W m^{-2}):

S is flux of heat into or out of the soil;
H is exchange of sensible heat between air and surface;
LE is latent heat flux (or energy consumption by evapotranspiration), where L is the latent heat of vaporization, and E the rate of evaporation;
P is energy consumption by photosynthesis;
M is miscellaneous storages and metabolic processes that consume or liberate small quantities of energy.

---

[1]Climate Resources Program, Resources for the Future, 1616 P Street, NW., Washington, D.C.

[2]National Center for Atmospheric Research, P.O. Box 3000, Boulder, CO. 80307

Note that both LE (expressed in units of energy) and ET (expressed in units of water) refer to the same quantity and we use these terms interchangably.

Some of the energy of net radiation is conveyed into the soil, but when vegetation is well watered most is consumed by evapotranspiration which also has the effect of cooling the vegetation and soil surface. A smaller portion of the net radiation is converted into sensible heat and warms the air. When, however, the plant cannot conduct water into the air as rapidly as the atmosphere demands, evapotranspiration decreases, foliage and soil become warmer, and more heat is convected and radiated away to heat the atmosphere.

B. The Resistance Model of Evapotranspiration

Equation 1 shows that the availability of energy, whether from net solar radiation of the field or forest or from sensible heat advection, puts an upper bound (the potential evapotranspiration, ETP) on the water that can be evaporated or transpired. Actual rates and amounts of evapotranspiration, however, are controlled by other atmospheric, plant and soil factors that are identified below.

A plethora of theoretical, mechanistic and statistical equations, formulae and models have been developed to predict ET (see e.g., Brutsaert, 1982; Rosenberg, et al., 1983; American Society of Agricultural Engineering, 1985). Here we use a single conceptual approach--one that permits the importance of the many climate and plant factors that influence ET to be evaluated within the same framework.

The equation of Monteith (1965) which introduced resistance terms into an earlier model of ETP by Penman (1948) allows consideration of cases of either optimal or limited water supply. The so-called Penman-Monteith model is given by:

$$LE = \frac{s\,(Rn + S) + \rho_a C_p \left(e_s - e_a\right)/r_a}{s + \gamma[(r_a + r_c)/r_a]} \qquad \text{[eq. 2]}$$

where:

e_s is the saturation vapor pressure;

e_a is the actual vapor pressure;

s is the slope of the saturation vapor pressure curve with respect to temperature;

r_a is the aerodynamic resistance;

r_c is the canopy resistance;

ρ_a is the density of air;

C_p is the specific heat of air at constant pressure;

γ is the psychrometric constant.

In addition to its completeness, the Penman-Monteith equation [eq.2] has other advantages (e.g., Saxton, 1975; Beven, 1979). Because its structure is based upon underlying physical principals, it is better suited for use in simulation than empirically based methods (McCuen, 1974). In addition, it has been successfully used to estimate evapotranspiration from both crops and forests (Rosenberg et al., 1983).

In order to use the Penman-Monteith (hereafter P-M) method to estimate the impacts of climatic change on ET we must first identify the kinds and magnitudes of climatic changes expected to result from greenhouse warming, and the kinds and ranges of plant responses to elevated CO_2 conditions.

C. Changes in Climate and Plant Response

The range of possible climatic changes due to greenhouse warming is the subject of considerable debate. For most of the climatic variables in [eq. 2], the magnitude and even the direction of expected change is highly location specific and subject to much uncertainty. Here we rely primarily upon the estimates of Schneider et al (1989) to define the range of changes examined in this study. Although changes in precipitation are not considered directly by the Penman-Monteith model, such effects are included via their impact on atmospheric humidity and stomatal resistance.

Changes in plant response (growth, stomatal regulation) can be due to climatic changes and the direct effects of higher CO_2 concentrations on photosynthesis and on stomatal regulation. The mechanisms controlling these responses were thoroughly reviewed by Rosenberg et al (1989a) in preparation for the simulation modeling reported below. The range and sign of the changes considered in the simulations are as follows:

Climatic factors	
Temperature	- 4 to 10°C
Net radiation	± 30%
Humidity (vapor pressure)	± 20%
Wind speed	± 20%
Plant factors	
Leaf area index	± 30%
Stomatal resistance	± 60%

III. METHODS

We used the P-M model [eq. 2] in conjunction with field observations from four distinctly different ecosystems to estimate how changes in the climatic and plant factors discussed above may cause ET rates to change. We first parameterized the model for each of the ecosystems, then tested its sensitivity to changes in individual factors. We then used the model to predict the net effect on ET of simultaneous changes in climate and plant parameters. Various combinations of change were examined within the likely ranges discussed earlier in this paper.

A. The Ecosystems

Field observations of ET rates and climatic conditions and periodic observations of plant characteristics were obtained from studies in four different ecosystems. The ecosystems include:

- Mixed forest, mostly deciduous (oak, hickory, some loblolly pine) at Oak Ridge, TN (35°57′N; 84°17′W; 365m above mean sea level, msl); August 2-9, 1984 (Verma et al, 1986).

- Tall grass prairie near Manhattan, KS (39°03′N; 96°32′W; 480 m above msl); July 29 - August 6, 1986 (Verma et al, 1988).

- Wheat at Mead, NE (40°51′N; 96°45′W; 359 m above msl); May 20 to June 24, 1985 (S. B. Verma, University of Nebraska, Lincoln, personal communication, 1987).

- Irrigated alfalfa also at Mead, NE; June 9-25, 1976 (Brakke et al, 1978).

Details of site characteristics, observation protocols, and climatic conditions on days of highest and lowest LE flux in the first three ecosystems are given in Martin et al. (1989) and for the alfalfa in Rosenberg et al. (1989b). The wheat and tall grass prairie were observed during years of greater than average rainfall, the forest during a normal year when moisture was ample, and the irrigated alfalfa during a drought year. ET was measured by eddy correlation in the grassland and forest, by the Bowen ratio-energy balance method in the wheat, and by a modified version of that method in alfalfa.

Climatic data included measurements of temperature, net radiation, humidity, and wind speed. Plant measurements included canopy height, leaf area index, and in the case of the wheat, stomatal resistance. It was necessary to estimate r_a and r_c parametrically since direct measurements of these factors were very limited in number. In all cases, r_c was computed as r_s/LAI.

IV. RESULTS

A. Sensitivity. Figures 1 and 2 for alfalfa show the percentage change in LE in response to a percentage change in each climate and plant factor for days of high and low latent heat flux. (Temperature is not plotted here, as a percent change in this term has no physical meaning.) On the low flux day, response is dominated by the climatic factors, R_n and e. This is true to a lesser extent on the high flux day. Response to r_s is also strong, particularly on the high flux day when strongly advective conditions were present. A non-linear response is apparent in this variable, and also in LAI. On both days, LE is relatively insensitive to changes in wind speed. Although temperature is not shown here, it should be noted that LE responds with an increase of approximately 6-7 percent for each 1°C change in temperature. Therefore a 3° C increase in temperature yields an 18-21% increase in LE. Sensitivity to temperature and to the other variables is determined partly by model structure, and partly by the values of all of the climatic and plant parameters.

The patterns of relative sensitivity shown here for alfalfa are fairly typical of those for the other ecosystems, although a wide range of responses was found among days within a given study, as well as among ecosystems. A 3° C increase in temperature caused LE to change, on average, by 20% in the wheat field, 17% in the grassland, and 24% in the forest. The impact of a 40% increase in stomatal resistance was about the same for the wheat and alfalfa (causing a 12 and 13% decrease in LE, on average), and slightly higher in the grassland and forest (-17% and -16%, respectively). On some days LE was more sensitive to changes in r_s than to any of the climatic variables. Wind speed was relatively insignificant in all ecosystems, and for that reason is not shown in Tables 1 or 2.

When multiple factors are varied simultaneously, the combined impact on ET is approximately additive in most cases. (See Table 1). For example, if a 3°C change in temperature increases LE by 20%, and a 10% increase in R_n causes 5% greater LE, the combined effect will be a 25% increase in LE. This is not strictly the case with r_s and LAI, both of which cause a somewhat non-linear response in LE over the range of change shown here.

Model sensitivity obviously does not tell the whole story about likely responses of ET to climate change. One must also consider the expected magnitude of change in each of the climatic and plant factors. Even a relatively insensitive parameter may be quite important if it changes dramatically with greenhouse warming. Likewise, very small changes in an influential parameter may be imperceptable in terms of LE.

B. LE Response to Scenarios of Climatic Change

Table 2 illustrates of the results of the simulations in which multiple factors were varied simultaneously to create scenarios of climatic change. Values for each of the four ecosystems are shown (average of all days). Both scenarios shown here include a 3°C rise in temperature. For Scenario I we impose a 10% increase in net radiation and a 10% decrease in humidity. Both of these changes accentuate the impact of a temperature increase in creating more drying conditions. In the second Climate Scenario, R_n and e change in a direction which mitigates the impact of increased temperature (ie. cloudier and more humid). We then impose upon these climate scenarios varying levels of change in the plant parameters. Stomatal resistance is varied from 20 to 60% to represent a probable increase in resistance in response to CO_2 and other factors. LAI is varied by +15% to represent the effects of CO_2 fertilization, and by -15% to include the possibility that factors such as increased water stress and increased leaf temperature will act to decrease growth.

As can be seen, increases in ET due to climate change alone (Scenario I, line 1) range from 29% in the grassland to 40% in the forest. The introduction of CO_2-induced changes in the plants, i.e., increased stomatal resistance and greater plant size, modifies the impact of climatic change. An increase in r_s of 60% reduces the climatic effect most in the forest (from 40% to 11%) and least in alfalfa (from 31 to 25%). If CO_2 fertilization occurs and LAI increases, a stomatal response can more than compensate the effects of increased leaf area (lines 7,9,11). If LAI decreases by 15%, the combined effect of a 60% increase in r_s is to nearly or completely offset the impact of climate on ET (line 12).

Under Climate Scenario II, the pattern of ET response to changes in plant factors is much the same as in Scenario I, but because evaporative demand is lower, ET is actually reduced in many cases below the control (or "no climate change") case.

IV: DISCUSSION AND CONCLUSIONS

These analyses illustrate the danger of assuming that global warming will increase evapotranspiration in all circumstances. It is necessary to consider not only temperature change but all other climate changes that may accompany it, as well as changes in plant growth and stomatal regulation caused by rising carbon dioxide. We have attempted here to define a range of possible ET responses, and to examine the potential for the effect of increased temperature to be offset or accentuated by other changes in climate or vegetation. However, the results of these simulations cannot be taken as the final word on the question of how and to what extent ET will be affected by greenhouse-induced climate change and CO_2-induced changes in plant growth and stomatal response. Some limitations of this study should be discussed.

First, our lack of firm knowledge regarding the magnitudes of change to expect in both climate and plants limits our ability to predict the ET response. General circulation models of the atmosphere (GCM's) have been used to make predictions of the types of climatic change which may occur. However a number of problems currently exist with these models (Washington and Meehl, 1984; Grotch, 1988), and their forecasts must be interpreted with care. In Martin et al. (1989) we applied climate change scenarios from three different GCM's to the wheat, forest and grassland ecosystems. Not surprisingly, LE predictions varied greatly depending upon the GCM used. It is noteworthy that in all cases predicted LE was lower when all climate changes were considered than when temperature changes alone were taken into account. This lends support to our cautionary note that it is important to consider the potential of other climatic changes to partially offset the influence of greenhouse warming. However, until more reliable predictions of climatic changes are available, our attempts to predict ET effects can be no more than an examination of a range of plausible changes.

Many unknowns also remain regarding the direct effect of carbon dioxide on plants. We cannot be completely confident that the simulated responses of ET to changes in stomatal resistance would actually occur in fields and forests exposed to a future CO_2-enriched atmosphere. We know from many experiments that when a climatic factor (e.g., net radiation, humidity) changes, changes in transpiration can be measured. However, CO_2-enrichment of the open air has not yet been accomplished for a long enough period to establish whether the stomatal closure effect actually occurs and whether, if it did, it would be transitory or persistent.

On the other hand CO_2 enrichment of the air in controlled environments does lead to an increase in leaf temperature which means that evapotranspiration is reduced to <u>some degree</u>. The picture is uncertain: however, changes in canopy resistance--a function of both stomatal resistance and leaf area (plant growth)--cannot be excluded from further consideration at this stage as a possibly important modifier of the effects of climatic change on ET.

Likewise, the extent to which the "CO_2 fertilization effect" - the increased growth that has been observed in greenhouse plants grown under conditions of elevated atmospheric CO_2 - will occur is a topic of debate. This effect has been demonstrated in controlled environments, but there is some question as to whether or not it will occur under field conditions and over extended periods of time for long-lived species (Solomon and West, 1987). An increase in the amplitude of the Mauna Loa record of atmospheric

CO_2 concentration is suggestive of a global biospheric response to CO_2 fertilization (Revelle and Kohlmaier, 1986), although this effect may be explained by other phenomena, as well. Although evidence is not yet conclusive, a growth response to CO_2 is certainly one possibility to be considered.

In assessing the utility of the modeling approach used here, we should also ask the following questions: Have all the relevant variables been considered? Are their functional relationships and interactions properly represented by our model? The Penman-Monteith model is one of the most complete and physically based of the commonly used methods for estimating ET. However some limitations should be noted. Since the model does not explicitly consider soil moisture effects on ET, changes in precipitation cannot be simulated directly, but only by varying r_s. Many unknowns remain concerning the net effect of changes in the many factors which may influence this variable, including vapor pressure deficit, air temperature, soil moisture, and CO_2 concentrations. Potential interactions and feedbacks between climate and plant response are not considered here, although these could potentially be included by modifying the estimates of the r_s parameter.

Paw U and Gao (1988) have pointed out that significant errors can occur in Penman-Monteith estimates of LE because of the model's use of a linearized form of the saturation vapor pressure function. However, they found that the P-M equation yielded sufficiently accurate estimates of ET when field data from a semi-arid climate were used. It was only when tested with an artificial range of more extreme values where surface and air temperatures diverged by more than 5°C that errors became apparent. In this study we assume that we are operating within the range over which P-M performs sufficiently well. However we cannot rule out the possibility that some of our simulation runs which consider large changes may reach beyond the limits of this range.

These simulations indicated that wind speed has relatively little impact on ET - a result that seems contrary to much experimental evidence. Particularly in the alfalfa study, where highly advective conditions were present, one would expect a larger response. The relatively poor fit between modeled and measured LE for this ecosystem (41% on the windiest day) may be indicative of problems with either the model itself, or with the parameterizations used here.

Another important issue is whether these analyses, especially those made in small fields, can be extrapolated to larger areas and regions. According to Jarvis and McNaughton (1986) transpiration in forests in weakly decoupled from stomatal behavior (their Omega factor, Ω, approaches zero) i.e. the role of physiology is most significant but the atmosphere exerts some control. In wheat and grassland, however, especially when well supplied with water, transpiration is more responsive to atmospheric demand (Ω approaches 1). Because of the nature of the calculations and parameterizations we used, such effects were not apparent in this study. Paw U and Gao (1988) challenge the validity of the Jarvis-McNaughton formulation and point out that, especially where Ω approaches 1, (Paw U and Gao figure 6) large errors in ET estimation can occur. For now we see no reason to assume that atmospheric factors would, indeed, overwhelm the transpiration suppressing effects of CO_2-induced stomatal closure as Jarvis and McNaughton have suggested they might.

The implications of these findings with respect to runoff and other hydrological phenomena are discussed in Rosenberg et al (1989a). With respect to another important aspect of the water resource issue--crop production--we conclude again that changes in ET predicted on the basis of increased temperature alone may be exaggerated. Increases in ET could either be moderated or augmented by other conceivable climatic changes. The effect of elevated CO_2 concentration on stomatal functioning could exert a moderating influence on ET that would probably not be counteracted entirely by an increase in plant size (leaf area index). In dryland agriculture changes in ET rate would have important effects on plant growth, development and yield by altering the timing, duration and severity of periods of water stress. In irrigation agriculture changes in ET will directly determine how much more or less water must be provided to grow the crops.

Acknowledgement

We are indebted to Dr. Bruce A. Kimball of the USDA Water Conservation Laboratory, Phoenix, AZ for much of the information concerning plant responses to CO_2 enrichment and direct climatic effects which appeared first in Rosenberg et al. (1989a). This research was conducted with support of the National Science Foundation as part of the American Association for the Advancement of Science study on Climate Variability and U.S. Water Resources.

REFERENCES

Amer. Soc. Agri. Engineers. 1985. Advances in Evapotranspiratio. ASAE Publ., 453 pp.

Aston, A.R. 1984. The effect of doubling atmospheric CO_2 on streamflow: a simulation. Journal of Hydrology 67: 273-280.

Beven, K. 1979. A sensitivity analysis of the Penman-Monteith actual evapotranspiration estimates. J. Hydrol. 44: 169-190.

Brakke, T.W., S.B. Verma, and N.J. Rosenberg. 1978. Local and regional components of sensible heat advection. J. Appl. Met. 17: 955-963.

Brutsaert, W. 1982. Evaporation Into the Atmosphere: Theory, History, and Applications. D.Reidel Pub. Co., Boston, 299 pp.

Gleick, P.H. 1987. Regional consequences of increases in atmospheric CO_2 and other trace gases. Climatic Change 10: 137-161.

Grotch, S.L. 1988. Regional intercomparisons of general circulation model predictions and historical climate data. U.S. Dept. of Energy, DOE/NBB-0084, Report TR041.

Idso, S.B. and A.J. Brazel. 1984. Rising atmospheric carbon dioxide concentration may increase streamflow. Nature 312: 51-53.

Jarvis, P.G. and K. B. McNaughton. 1986. Stomatal control of transpiration: scaling up from leaf to region. Advances in Ecological Research 15: 1-49.

Martin, Ph., N.J. Rosenberg and M.S. McKenney. 1989. Climatic change and evapotranspiration: simulation studies of a wheat field, a forest and a tall grass prairie. Climatic Change, in press.

McCuen, R.H. 1974. A sensitivity and error analysis of procedures used for estimating evaporation. Water Res. Bull. 10: 486-497.

Montieth, J.L. 1965. Evaporation and environment. Symp. Soc. Exper. Biol. 19: 203-234.

Penman, H.L. 1948. Natural evaporation from open water, bare soil and grass. Proc. Roy. Soc. London, Ser. A, 193: 120-145.

Paw U, K.T. and W. Gao. 1988. Applications of solutions to non-linear energy budget equations. Agric. For. Meteorol. 43: 121-145.

Revelle, R.R. and G. Kohlmaier. 1986. Increasing amplitude of seasonal CO_2 cycle. In: C.Rosenzweig and R. Dickinson (eds.), Climate-Vegetation Interactions. UCAR, Office for Interdisciplinary Earth Studies, Boulder, CO, Report OIES-2.

Revelle, R.R. and P.E. Waggoner. 1983. Effects of a carbon dioxide-induced change on water supply in the western United States. In: Changing Climate. National Academy of Sciences, National Academy Press, Washington, D.C.

Rosenberg, N. J., B. L. Blad, and S. B. Verma. 1983. Microclimate: The Biological Environment. John Wiley & Sons, New York, 495 pp.

Rosenberg, N.J., B.A. Kimball, Ph. Martin and C.F. Cooper. 1989a. Climate Change, CO_2 Enrichment and Evapotranspiration. In: P.E. Waggoner and R. Revelle (eds.), Climate and Water. John Wiley & Sons (in press).

Rosenberg, N. J., M. S. McKenney and Ph. Martin. 1989b. Evapotranspiration in a greenhouse-warmed world: A review and simulation. Agri. and Forest Meteorology (in press).

Saxton, K. E. 1975. Sensitivity analyses of the combination evapotranspiration equation. Agric. Meteorol. 15: 343-353.

Schneider, S., P. Gleick, and L. O. Mearns. 1989. Prospects for climate change. In: Climate and Water: Climate Change, Climatic Variability, and the Planning and Management of U.S. Water Resources. Wiley Press, New York (in press).

Solomon, A. M. and D. C. West. 1987. Simulating forest ecosystem responses to expected climate change in eastern North America: applications to decision making in the forest industry. In W.E. Shands (ed.), The Greenhouse Effect, Climate Change and U.S. Forests. Conservation Foundation, Washington, D.C.

Verma, S.B., D.D. Baldocchi, D.E. Anderson, D.R. Matt, and R.J. Clement. 1986. Eddy fluxes of CO_2, water vapor, and sensible heat over a deciduous forest. Boundary-Layer Meteorology 36: 71-91.

Verma, S.B., J.K. Kim, and R.J. Clement. 1988. Carbon dioxide, water vapor, and sensible heat exchanges of a tall grass prairie. Boundary-Layer Meteorology, submitted.

Washington, W.M. and G. A. Meehl. 1984. Seasonal cycle experiment on the climate sensibility due to a doubling of CO_2 with an atmospheric general circulation model coupled to a single mixed-layer ocean model. J. Geophys. Res. 89 (D6): 9475-9503.

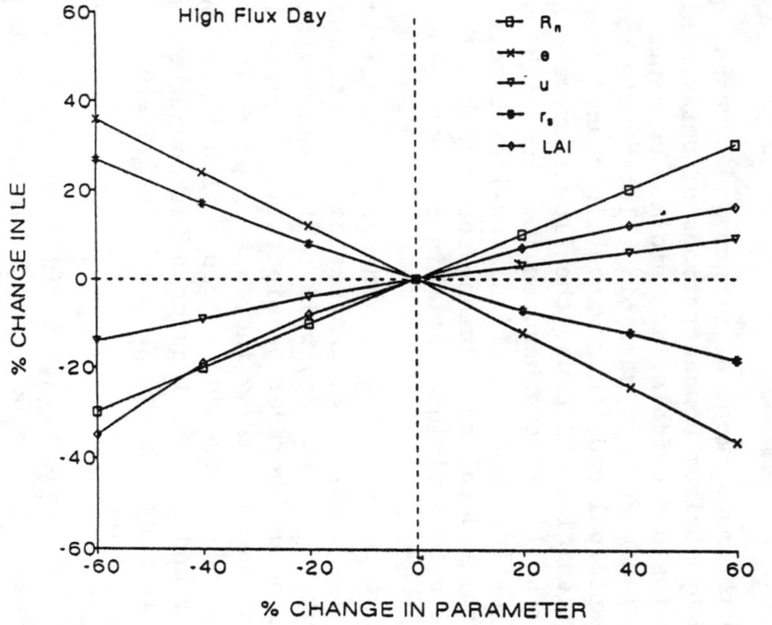

Figure 1

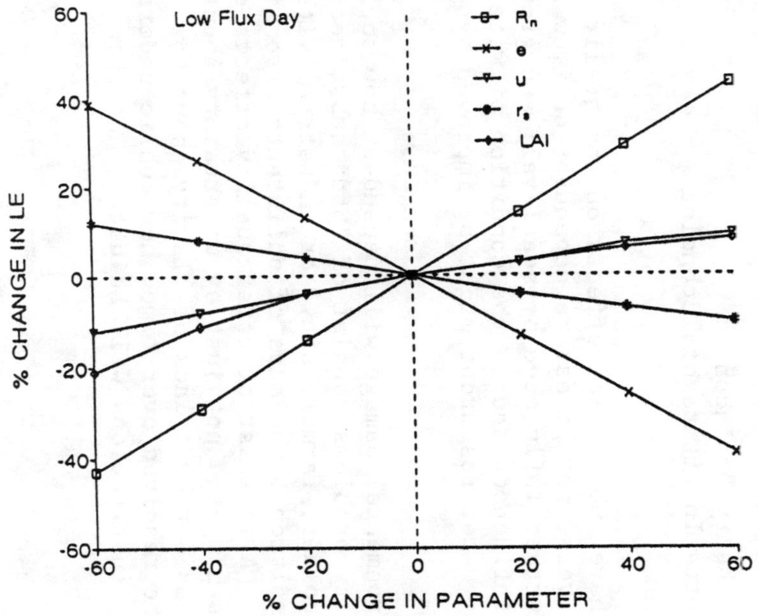

Figure 2

Figure 1. Sensitivity of latent heat flux in an irrigated alfalfa field to changes in net radiation, humidity, stomatal resistance, and leaf area index on June 25, 1969. Unperturbed LE flux was 555 W m^{-2}.

Figure 2. Sensitivity of latent heat flux in an irrigated alfalfa field to changes in net radiation, humidity, stomatal resistance, and leaf area index on June 13, 1969. Unperturbed LE flux was 371 W m^{-2}.

LIKELY EFFECTS OF CLIMATE ON WATER QUALITY

Henry D. Jacoby
Massachusetts Institute of Technology

Changes in climate will have substantial effects on the quality of the nation's waters. Streams, rivers, and lakes are dependent on flow and temperature; estuaries are further influenced by sea level; and groundwater is subject to many of these influences and to the condition of waters at the surface. The quality of some water resources might be improved by climate change; most will not.

The severity of the environmental damage will depend on how society responds. In the main, quality problems consist of "bad" substances put into the water by some householder, farmer, city, or industry. With shifts in climate and hydrologic conditions, the mass of pollutants also will change as the country adapts. A forecast of water quality effects of climate change, therefore, is hampered not just by uncertainty about physical and biological systems but by a lack of clarity about how scientific uncertainty will be resolved over time, how public understanding will grow, and how regulatory institutions will adjust.

1. A Catalog of Effects

The interaction of these physical systems and human institutions can be explored by first considering what might happen if there were no societal response. What if designated uses continued unchanged, with no adaptation in pollution control policy, even in the face of changes in temperature, precipitation, evapotranspiration, and the hydrologic cycle?

Rivers and Streams. In the management of rivers and streams, quality characteristics are defined by maximum concentrations of unfavorable substances like salinity or toxic chemicals, and minimum concentrations of a favorable one, dissolved oxygen. For any rate of pollutant discharge, which is the net of industrial, municipal and other waste minus any cleanup, in-stream quality depends on streamflow. Because the flow in rivers and streams may vary by a factor of 10 or more between low- and high-water periods, quality usually is defined for a critical low-flow condition. For any zone of a river or stream, federal regulations require that threshold concentrations of pollutants not be violated at this critical flow level.

One effect of a shift in climate will be to change these critical low flows. In some regions low flows may rise, increasing dilution and improving water quality. But in others they will fall, reducing stream quality. Deterioration will occur wherever precipitation decreases or where increased evapotranspiration overwhelms any increase in precipitation. Quality also will be reduced under increased precipitation if wet periods are shifted toward the winter and spring, leaving drier conditions in late summer when low flows occur in many regions of the United States.

Changes in flow would particularly affect the quality of streams in arid regions, where the problem frequently is salinity rather than municipal and industrial waste. An example is the Colorado River which each year delivers around 9 million tons of salt to the Gulf of California. Roughly half of this load would occur naturally because the river runs through an ancient inland sea which is rich in marine minerals. The other half results from irrigation, reservoir operations, and export of water from the basin.

To maintain the quality of the Colorado, a number of salinity control projects have been built or are planned, including control of irrigation losses and point sources of natural pollution (salt springs), and development of desalination plants. Even with planned new projects, increasing water use and anticipated diversions will lead to violation of salinity standards early in the next century. A climate-induced reduction of average flow of 10 to 20 percent would make the standards impossible to meet without much more stringent and expensive salinity control measures. A flow reduction of as much as 50 percent, as some studies contemplate, would make the lower basin waters unfit for many current uses.

Estuaries. Warming plus a reduction in critical low flow will create problems in estuaries similar to those in rivers and streams. In addition, penetration of salinity from the oceans will increase. A crucial aspect of estuary quality is the front or boundary between fresh and salt water, because salinity affects municipal and industrial uses and aquatic life. In most estuaries, the front moves up and down within some range (often several miles) in response to tides and the flow of incoming rivers. With a shift in climate, changes in freshwater runoff and increases in sea level will have a combined influence, and the salt front may move far upstream.

An example of the deterioration that may result is provided by the Delaware Estuary. The Delaware River drains portions of southern New York, eastern Pennsylvania and western New Jersey, and it forms a tidal estuary stretching from Trenton past Philadelphia, Camden and Wilmington to the sea. In the 1960s, the Basin experienced years of unusually low flow. The salt front moved up in the Estuary, creating significant problems for the water supply of Philadelphia and other cities and industries that draw directly from it. (Groundwater was degraded too, as discussed below.) If climate change made low-flow conditions more frequent or severe, and sea level rose, these quality problems would be magnified.

Lakes. Greenhouse-induced climate change will affect lake quality in three ways: changes in throughput and volume, higher water temperatures, and reduction in ice cover. In a stream receiving a fixed amount of a pollutant, a lowering of streamflow will increase pollutant concentration. The same process works in lakes if there is a reduction in inflow or increase in evaporation, only with a time lag. The time to a new steady state is a function of the volume of the lake.

Air temperature and wind influence quality through the dynamics of temperature stratification and seasonal overturn of waters. In the summer months, the temperature stratification of a lake creates a zone in its lower levels which is poorly mixed with more oxygen-rich upper regions. Warmer lake temperatures increase the rate of biological activity throughout a lake but most importantly in its lower regions and bottom sediments. Accompanying changes in temperature tend to reduce the volume of poorly-mixed water near the bottom, lowering the supply of oxygen to these biological processes, and potentially creating anaerobic conditions.

Higher temperatures would also reduce ice cover. For example, the maximum extent of ice cover in an average year is about 60 percent of the area of the Great Lakes. Ice-temperature correlations indicate that a 4° to $5^{o}C$ atmospheric warming could reduce this coverage to somewhere in the range of 0 to 15 percent. The environmental implications of such a change are not well understood, but there is evidence that biological activity increases under ice cover, and that some species of cold water fish are dependent on this environment.

Groundwater. Groundwater resources differ greatly in their vulnerability to climate change. "Confined" aquifers, which are overlain by impermeable material, have very limited recharge. The quality of this resource is decoupled from climate at the surface, at least for a few centuries. A second category includes "unconfined" aquifers in areas of high rainfall and low evaporation. Precipitation exceeds evaporation and transpiration in most years, groundwater is easily recharged under current climate, and it would continue so even after climate change. There are serious problems of contamination of these waters, usually as a result of careless disposal of municipal sanitary wastes, industrial chemicals, garbage, and radioactive substances, or of spills, leaks and the runoff of agricultural chemicals and road salt. To the degree that a particular aquifer is charged from rivers or lakes, and their quality changed, groundwater quality could be influenced, but climate is a minor aspect of the total problem.

The third category includes unconfined aquifers in dry regions, where there is a shifting annual balance between precipitation and evapotranspiration. Groundwater is recharged only in the wet years. If a region becomes dryer, less groundwater will be available for human use. If the dryer climate increases the demand for groundwater, the pressure on the resource will be magnified. For some regions of the country the effect on water quantity may be substantial. The implications for quality are likely less serious. As in humid regions, groundwater quality will deteriorate if quality deteriorates on the surface. These aquifers are recharged only during high runoff years, however, when pollutant concentrations are diluted, so the effect of climate on quality should be small.

Finally, deterioration may occur in aquifers bordering on the ocean or recharged from estuaries. In a coastal aquifer undisturbed by human withdrawals, a balance is maintained between the zones of salt and fresh groundwater. At the coastline, the freshwater table is at mean sea level. Inland, the water table is higher, the fresh water floating on top of the heavier salt water. If withdrawals exceed recharge, the water table is drawn down and salt water penetrates inland polluting wellfields near the beach. Then, if sea level rises, the coastline will move inland, reducing the area of the aquifer. If an aquifer rises only a few feet above mean sea level (say, on a low island) then its volume may be reduced substantially. Where the effects of rising sea level and pumping are combined, an aquifer can be destroyed for water supply.

When an estuary is underlain by an unconfined aquifer, the quality of the groundwater will reflect the salinity of the estuary waters. Again the Delaware Estuary is an example. In the 1960s there was damage to the Potomac-Raritan-Magothy aquifer system, which is the main source of water for municipalities and industries in southern New Jersey. In the drought the salt front moved far enough upstream to recharge the aquifer with saline water, degrading some wellfields. If climate change raises sea level and lowers the flow in incoming rivers, events like those of 1961-66 will become more common and troublesome.

Biological Diversity. Changes in water quality will bring a redefinition of areas where species can be sustained, and will determine whether some survive at all. For example, fish are sensitive to temperature and oxygen, and the geographical occurrence of various freshwater species will be altered even if previous standards for dissolved oxygen are maintained. In estuaries, aquatic life is peculiar to conditions of temperature and salinity; changes in temperature, flow, and sea level may alter local ecosystems substantially. Shellfish and other species may migrate upstream with an advancing salt front, to face conditions that may be better or worse than before.

Moreover, many forms of life, including insects, plants, algae and fungi, vertebrates, and invertebrates and microorganisms, are specific to local ecosystems, such as a particular lake or estuary, or a set of local streams. They are not widely dispersed nor can they easily migrate under changing conditions. Rapid changes in the temperature regime, combined with variation in dissolved oxygen, salinity, ice cover, seasonal flow, and area of wetlands may create stresses some cannot withstand. The potential loss of species, and the implications for local ecosystems, are poorly understood and far beyond the scope of this paper. Our ignorance is troublesome considering that the effects could be substantial.

2. Societal Response

Analysis of how these effects may be ameliorated requires a forecast of when individuals and institutions will know that action is required, and what they will do at that point. The effects above would occur if society takes no notice of the changes under way. At the other extreme are the results under perfect foresight. For example, if a reduction in critical low flows was a prospect nationally, technical treatment standards would likely be tightened for all dischargers. Were changes in national discharge standards inadequate to avoid the quality problems in particular streams, other measures now used for meeting state standards would be available. Restraints might be placed on new polluting activities in a basin; special treatment requirements might be imposed; or additional dams might be built to maintain critical low flows during drought. In short, the existing water quality management system would adjust, much as it would in the face of changes unrelated to shifts in climate.

For some streams the cost of additional cleanup might be too great. Downgrading would follow. However, with decades to plan and adjust, and with most corrections embodied in new investment rather than retrofit of old, adaptation would be cheaper than if done today. Much of the degradation associated with municipal and industrial waste would never occur. Other circumstances leave less room for optimism. To maintain the current quality standard in the Colorado River under a substantial reduction in flow would require removal of millions of tons of salt from the River, an unlikely event even if technically feasible.

Many of the same opportunities exist for correcting climate effects in estuaries and lakes. All the effluent control measures mentioned above could be applied. Problems of saltwater intrusion in estuaries could be ameliorated by additional storage or changes in the operation of existing works upstream. In the case of anoxic conditions in lakes, some of the climate-induced problem could be counteracted by increased efforts to control nutrient discharge. On the other hand many lake effects, such as those related to temperature alone, ice cover, and lake volume and level, would be beyond human intervention.

Groundwater also offers opportunities for adaptation, to maintain quality if not quantity. For all but coastal aquifers, climate-induced quality problems could be ameliorated by a cleanup of surface waters and improved control waste discharge on land. The potential is also great for management of coastal aquifers, mainly through reduced pumping.

3. The Future

Of course, the magnitude of future climate change and its hydrologic effects cannot be forecast accurately. The prospect for timely and efficient adaptation depends on how this uncertainty is resolved over time. The greater the uncertainty about the effects, and the longer and more complex the likely path to its resolution, the harder it will be for water authorities to know what to do.

For one thing, acquiring the knowledge to support action is complicated not only by physical lags in the system but by limitations on our ability to measure changes in the environment even after they have occurred. Long periods of data are needed to identify a change with any confidence because climatological and hydrologic systems are so highly variable. If regional rainfall, cloud cover and storm frequency were likely to undergo monotonic change, new observations would be expected to narrow the distribution of ultimate outcomes. We would progressively learn more, and develop tighter and tighter confidence intervals on forecasts. This may not happen, however. Because of likely differential warming of continents and oceans, transient conditions may be created which will introduce misleading signals along the way. A progressive reduction in the range of estimates need not occur. Over long periods, perhaps decades, uncertainty regarding in key climate variables (particularly at the regional level) may not be reduced much.

In this complex circumstance there are but a few prescriptions of what to do now. A premium is placed on flexibility in water quality management systems, to adapt to a wide variety of circumstances. Another early action should be careful study of possible big mistakes along the way. Some quality effects of changing climate cannot be fixed either now or later. Others can be fixed now, but they can be fixed later as well, as knowledge improves. Quality studies thus should concentrate on identifying those effects that might be avoided or corrected now, but will be irreversible later.

Finally, the most obvious way to prepare for the uncertain future is to control current pollution more effectively.

[A longer version of this paper is in Paul E. Waggoner (ed.), Climate Change and Water Resources, Report of AAAS Panel on Climatic Variability, Climate Change, and U.S. Water Resources, Wiley-Interscience, 1989.]

BEACH RESPONSE STRATEGIES TO ACCELERATED SEA-LEVEL RISE

by

Stephen P. Leatherman, Director
Laboratory for Coastal Research
University of Maryland
1175 Lefrak Hall
College Park, MD 20742

Introduction

A significant portion of the United States population lives within the coastal zone, with many buildings and facilities located at elevations less than 3 meters (10 feet) above sea level. These structures are presently subject to damage during storms, and this hazard has grown increasingly serious as sea levels have risen during the twentieth century. Greenhouse-induced warming is expected to raise water levels at historically unprecedented rates, resulting in increased beach erosion and flooding.

Despite these potential hazards, the coastal population is burgeoning. In fact, development in the coastal zone is proceeding at rates that more than double inland construction. Hundreds of thousands of beachfront structures (exquisite single-family houses, high-rise condominiums, and elegant hotels) have been built within a few hundred feet of an eroding ocean shore. Beachfront property is some of the most valuable real estate in the country, often exceeding $10,000 per linear foot of shoreline along the U.S. mid-Atlantic coast.

The present dilemma and developing disaster have resulted from the tremendous investment in coastal property at a time when most sandy beaches nationwide are eroding. Best estimates are that 90 percent of the U.S. sandy beaches are presently experiencing beach erosion. Accelerated sea-level rise will increase erosion rates and associated problems (Leatherman, 1988a).

Public attention is yet to be critically focused on the beach erosion problem. The recent (1988) drought and heat wave have brought about a dramatic awakening and interest of citizens in the greenhouse effect and climate change. Hopefully, a coastal disaster along an urbanized beach will not be necessary to promote public awareness of the sea-level rise phenomenon and its attendant impacts.

Sea level is the underlying cause of shore position, which translates to beach erosion when water levels are rising. While weather is subject to large-scale variations and hence climate change trends are difficult to measure, rising sea levels are relatively easy to discern and can be thought of as the dipstick of climate change, reflecting the integration of many earth surface processes.

The principal approach today of protecting coastal property and maintaining recreational beaches is nourishment through the introduction of new sand. Engineering structures, such as groins and seawalls, have often been shown to cause detrimental effects on adjacent beaches. Also, their construction and maintenance costs are quite high. Therefore, coastal communities have come to rely upon a "soft" engineering solution -- beach nourishment, since it is environmentally sound, aesthetically pleasing, and up-to-this-time, economically feasible. However, the projected accelerated sea level rise will cause more rapid rates of beach loss and could make even this alternative too costly for many resort areas along the United States coastline.

This paper summarizes the result of a report by Leatherman (1988b) for the Environmental Protection Agency regarding nationwide estimates for future beach fill projects. Of the approximately 7,000 miles of sandy shorelines in the U.S., 1,920 miles of beaches were evaluated in this study. These areas are considered to be the principal recreational beaches in the country. The contrasting option of retreat is also discussed for comparative purposes.

Beach Nourishment

The volume of sand and its cost to nourish all the major recreational oceanic beaches in the U.S., given various sea-level rise scenarios, was estimated. It is clear that coastal residents would prefer not to move back and abandon the coast and will attempt to stabilize the shore if at all possible. One approach is to place enough sand on the beach to maintain stable (non-retreating) conditions with rising sea levels. The quantity of sand required "to hold the line" is evaluated under various sea-level rise scenarios (rise/year combinations) at 20-year intervals from year 2000 to 2100.

Sand volume determinations were based on the direct approach of "raising the beach/oceanshore profile" as described elsewhere (Leatherman, 1988b). This approach overcomes objections to the Bruun Rule formulation regarding on/offshore sand transport. Also, other methodologies require considerable more data (e.g., Trend Analysis necessitates knowledge of historical shoreline change and sediment budget models involve site-specific information on transport rates; Leatherman, 1986).

Sand costs are estimated for the range of alternatives (e.g., various SLR scenarios evaluated at particular years when a certain sea level has been achieved). Values are based on current rates per cubic yard of material. A sand cost function was developed from past offshore dredging experience and applied to each coastal sector. While backbarrier lagoons and bays, mainland sites, and inlet sand shoals may be utilized for beach nourishment, the preferred borrow site areas are generally located offshore.

Nationwide estimates of sand quantities required with accelerated sea-level rise are arrayed in Table 1. It is apparent that tremendous quantities of good quality sand will be necessary to maintain the nation's major recreational beaches. Almost all of this sand must be derived from the offshore, but to date only enough sand has been identified to accommodate the two lowest scenarios over the long term. Even in these cases, the offshore sand is not evenly scattered along the U.S. coastline so that some areas will run out of local (the least expensive) sand in a few decades.

The costs of sand fill are based on current expense of offshore dredging and pumping onshore of locally-derived material (Table 2). Obviously, the costs will increase with inflation, but more importantly the expense could be greatly underestimated if sand must be acquired from a considerable distance of the beach requiring nourishment.

This study represents the first estimation of sand requirements necessary to stabilize the major U.S. recreational beaches in response to accelerated sea-level rise. The cost to stabilize the coast through the "soft" engineering approach of sand filling ranges from approximately $2.3 billion to $5.9 billion for Scenarios I to VI by the year 2020 on a nationwide basis (Table 2). Considering the enormous value of coastal property (e.g., Ocean City, Maryland is valued at over $2.6 billion alone), it is safe to assume that the densely developed areas will be nourished and maintained. However, the costs tend to increase in an exponential fashion due to the increasing rate of sea-level rise through time. What is unclear is at what point moderate-density areas will be forced by economic considerations to choose another approach (e.g., retreat from the eroding beach).

Retreat

Holding back the sea as water levels rise will almost always be technically feasible, but it may not be environmentally or economically sound. While it can be argued that stabilization measures can be deferred until made necessary by the accelerated rise, the fact is that most of the sandy beaches are presently experiencing erosion and decisions made today can have far-reaching impacts. Also, a planned decision to retreat would require a lead time of several decades.

Table 1. Nationwide Estimates of Sand Quantities Required with Sea Level Rise (million yd^3)

Year	I	II	III	IV	V	VI
	Scenarios					
2000	145.634	166.770	187.645	208.417	229.727	250.470
2020	404.697	531.097	654.255	777.742	900.743	1041.429
2040	749.914	1067.874	1394.713	1850.035	2272.343	2658.815
2060	1155.129	1925.232	2667.664	3390.477	4315.144	5428.242
2080	1772.567	2751.612	4314.381	6021.119	7469.329	9251.228
2100	2424.337	4345.477	6767.643	9070.906	11356.659	13655.708

Table 2. Nationwide Estimates of Cost of Sand Fill Required with Sea-Level Rise ($ millions)

Year	I	II	III	IV	V	VI
	Scenarios					
2000	837	958	1,073	1,192	1,310	1,428
2020	2,333	3,032	3,722	4,418	5,112	5,911
2040	4,277	6,073	7,896	10,956	13,497	15,873
2060	6,564	11,419	15,949	20,457	26,510	33,885
2080	10,524	15,874	26,528	37,525	47,672	59,502
2100	14,512	26,745	42,765	58,002	71,151	88,379

Retreat can occur as either a gradual process or through catastrophic abandonment. The Texas Open Beaches Act has the effect of promoting the former case as houses must be removed from the beach with shoreline recession. Perhaps the best approach is to anticipate a certain rate of beach loss and utilize building setbacks to prohibit construction too close to the waters edge. For example, North Carolina requires a setback of 30 and 60 times the annual erosion rate for single family houses and multistory buildings, respectively. This type of land use planning for erosion is now being adopted by other states (e.g., Maine, New Jersey, and Florida). This policy has the effect of promoting gradual retreat from an eroding shore, but does not preclude the eventual confrontation between buildings and storm surf.

The second approach is to rely upon storm destruction to be the ultimate arbitrator of housing location. This can result in catastrophic abandonment or emergency reconstruction and subsequent rebuilding in nearly the same location. Both responses have been exercised historically along the U.S. coasts. The rebuilding of the city of Galveston after the devastating hurricane of 1900 is the most prominent example of the latter. Also, Miami Beach was essentially swept clean by the 1935 hurricane. More recently, the 1962 northeaster (Ash Wednesday Storm) caused massive destruction along the U.S. mid-Atlantic coast, and emergency procedures were invoked to justify wholesale sand pumping and dune building operations. In recent time, abandonment per se has not been the method of choice, but historically small villages and even entire towns have been lost because of incessant beach erosion capped by episodic storm damage (e.g., Diamond City, N.C. in 1900 and Broadwater on Hog Island, VA by 1938 hurricane).

Discussion and Conclusions

Greenhouse-induced global warming and the attendant accelerated rise in sea level will have profound impacts on coastal areas. Development of a rational decision-making framework requires an understanding of both the complex physical coastal processes and the economics of the area. Without political or emotional considerations, economics will be the final arbitration in deciding whether to retreat or attempt to hold back the sea (National Research Council, 1987). Each approach is technically feasible, but the appropriate response will involve critical planning and timing of activities (Vellinga and Leatherman, 1989).

In the final analysis, there are three general responses to accelerated sea-level rise: retreat from the shore, armor the coast, or nourish the beach. The proper shore protection response is site-specific on a community or coastal sector basis due to large differences in environmental and socioeconomic factors. The abandonment alternative is not realistic for urbanized beaches. For less developed areas along eroding shorelines, planning decisions are less clear cut. Therefore, the costs and benefits of stabilization vs. retreat must be carefully considered as the cost in either case is likely to be quite high.

Acknowledgements

This research was supported by the Environmental Protection Agency as part of their report to the U.S. Congress.

References

Leatherman, S.P., 1986, Shoreline response to sea level rise: Ocean City, Maryland, Proceedings of Icelandic Conference on Coasts and Rivers, Reykjavik, Iceland, P. 267-276.

Leatherman, S.P., 1988a, Effects of sea level on beaches and coastal wetlands, Proceedings of the First North American Conference on Preparing for Climate Change, The Climate Institute, Washington, D.C., p. 140-146

Leatherman, S.P., 1988b, National assessment of beach nourishment requirements associated with accelerated sea-level rise, U.S. Environmental Protection Agency report, Washington, D.C., 74 pp.

National Research Council, 1987, Responding to Changes in Sea Level: Engineering Implications, National Academy of Sciences Press, Washington, D.C., 148 pp.

Vellinga, P. and Leatherman S.P., 1989, Sea level rise, consequences and policies, Climate Change, in press.

"The Threat to Federal Coastal Protection
Goals from Global Warming
And Accelerated Sea Level Rise"

by Lynne T. Edgerton, Senior Staff Attorney[1]
The Natural Resources Defense Council, Inc.
December 8, 1988

Because other panels have focused on policy considerations in limiting global warming, I will focus primarily on coastal protection considerations in adapting to accelerated sea level rise from global warming.[2] I emphasize, however, that policies designed to limit global warming are essential to achieving the nation's coastal protection goals. Furthermore, curbing greenhouse gas emissions may be more politically feasible and less costly to achieve than adapting along our coasts to accelerated sea level rise.

Existing national coastal protection goals provide a sound framework for assessing the significance of accelerated sea level rise and guiding our response to it. The achievement of these goals is seriously threatened by accelerated sea level rise.

I. WHAT ARE THE NATION'S EXISTING FEDERAL COASTAL PROTECTION AND GOALS?

Over the past three decades, Congress has committed the nation to an extensive set of environmental and coastal protection goals. Over 40 federal statutes now govern the management of our natural resources, their protection from pollution and the restoration of their chemical, biological and ecological integrity. Many of these statutes deal directly with coastal resources.[3]

The nation's three principal coastal protection goals are natural resource preservation; minimization of loss of property and life

---

1. This paper was prepared for presentation at the Climate Institute's Second Annual North American Conference on Preparing for Climate Change, held December 6-8, 1988, in Washington, D.C., and is subject to revision prior to publication in the Proceedings of the Conference.

2. There are important steps which state governments can take. NRDC will release a comprehensive report on federal and state policy responses to accelerated sea level rise in early 1989.

3. These statutes reflect the national awareness that the coastal environment has become increasingly degraded as increasing numbers of Americans have moved to settle along the coasts. Between 1950 and 1980, the coastal population swelled by over 30 million people. By 1990, it is expected that 75% of the nation's population will live within 50 miles of a coast.

from improper coastal development; and restoring the quality of our heavily polluted coastal waters.

The Coastal Zone Management Act, (CZMA), is the major piece of federal coastal law. The Coastal Zone Management Act, which was first enacted in 1972, establishes the national goal of wise management of coastal resources, and specifically seeks to preserve and protect natural resources, as well as to restore or enhance the resources of the nation's coastal zone for this and succeeding generations, where possible. 16 U.S.C.A. §§1451-2. Two coastal preservation and protection goals found in the CZMA are especially important in light of sea level rise from global warming: first, the goal of protection of natural resources, including wetlands, floodplains, estuaries, beaches, dunes, barrier islands, coral reefs, and the fish and wildlife and their habitat within the coastal zone; and second, the goal of management of coastal development to minimize the loss of life and property caused by improper development in flood-prone, storm surge, geological hazard, and erosion-prone areas and in areas of subsidence and saltwater intrusion; and by the destruction of natural protective features such as beaches, dunes, wetlands and barrier islands. 16 U.S.C.A. §1452. The twin goals of natural resource protection and wise coastal land use are reinforced in other statutes, such as the Clean Water Act,[4] which establishes a wetlands protection program under the Section 404 permit program, and the Coastal Barriers Resources Act, which prohibits federal subsidies for development of undeveloped coastal barriers.[5] These twin goals are threatened by accelerated sea level rise.

The Clean Water Act and the Safe Drinking Water Act establish water quality goals for both surface and groundwater resources. 33 U.S.C.A. §1251 <u>et seq</u>.; 42 U.S.C.A. §300f <u>et seq</u>. Accelerated sea level rise would aggravate pollution of rivers, estuaries, and groundwater, thus threatening water quality goals. Coastal sewage treatment plants would back-up and overflow; coastal aquifers would be salinized; and hazardous waste treatment facilities located in the coastal zone would be flooded or subjected to severe storm damage. These effects would jeopardize two and one-half decades of efforts to control water pollution.[6]

---

4. 33 U.S.C.A. §1251 <u>et seq</u>.

5. 16 U.S.C.A. §3501 <u>et seq</u>.

6. Many federal statutes seek to improve coastal water quality, eg. the Clean Water Act, 33 U.S.C.A. §§1251-1371; the Marine Protection, Research and Sanctuaries Act, 33 U.S.C.A. § <u>et seq</u>.; the Resource Conservation and Recovery Act, 42 U.S.C.A §6901 <u>et seq</u>.; the Comprehensive Environmental Response, Compensation and Liability Act of 1980, 42 U.S.C.A. §9601 <u>et seq</u>. <u>See</u> also the National Environmental Policy Act, 42 U.S.C.A. §4321 <u>et seq</u>.

II. WHAT ARE THE CONSEQUENCES OF ACCELERATED SEA LEVEL RISE?

According to scientists, even if greenhouse gas emissions are leveled within 10 years, (which is an enormously optimistic assumption), "sea level [from global warming] would probably rise 40 to 80 cm by 2060." (15.6" to 31.2").[7] Along the U.S. Eastern seaboard, the sea level would rise about 6 additional inches over the next 50-60 years, based upon other factors.[8] (In the past 100 years, relative sea level rise along the east coast has been about 1 foot.)

For policy development purposes, coastal planners should pick a number describing their best estimate of the amount of sea rise which will occur by 2050 - which is a reasonable planning horizon for major governmental investment. For this purpose, I have assumed that relative sea levels will rise by approximately one meter (3 feet) by 2050 along the U.S. Atlantic and Gulf Coasts, and will rise approximately 130 cm to 205 cm (49"-79") along the Louisiana coast.[9]

The consequences of a one meter sea level rise have been documented elsewhere.[10] There would be major losses of valuable natural resources such as beaches, dunes, estuaries and wetlands;

---

7. J. Hoffman, J. Wells, & J. Titus, "Future Global Warming and Sea Level Rise," Iceland Coastal and River Symposium (ed. G. Sigbjarnarson 1985). Accord, U.S. Environmental Protection Agency, "The Potential Effects of Global Change on the United States" (Draft) at 9-15 (October, 1988). _See also_ National Research Council, _Responding to Changes in Sea Level_ (1987).

8. Hoffman _et_. _al_., Projecting Future Sea level Rise at vi (EPA 1983). _See_ Gornitz, V. & Lebedeff, S., "Global Sea-level Change During the Past Century" in _Sea Level Change and Coastal Evolution_ (Nummedal, Pilkey and Howard eds.) (1987)

9. Saving Louisiana's Coastal Wetlands: The Need for a Long-Term Plan of Action at 36, EPA-230-02-87-026 (1987). Relative sea level rise in Louisiana from sources other than global warming is projected to be from 90 cm to 125 cm by 2050. The Louisiana coast appears to be most vulnerable to a rise in sea level. The Mississippi coastal wetlands are currently converting to open water at a rate of about fifty square miles per year due to the interaction between human activities such as building levees and navigation channels, and subsidence due in part to sediment deprivation. Current efforts to halt the loss and to restore the wetlands may be futile due to future sea level rise.

10. _See_ U.S. Environmental Protection Agency, "The Potential Effects of Global Climate Change on the United States (Draft) (October 1988); Barth & Titus, _Greenhouse Effect and Sea Level Rise_ (1984).

serious aggravation of existing water pollution problems, including salinization of coastal drinking water supplies; grave threats to already weakened coastal infrastructure such as roads, mass transit, wastewater facilities, solid waste facilities, and hazardous facilities, and of public works such as seaports, coastal structures and nuclear energy plants, and potentially enormous property damage and loss of human life from storms and storm surge. These findings necessitate a thorough evaluation of whether existing coastal protection measures suffice to achieve national environmental and coastal protection goals.

III. ASSESSMENT OF CURRENT FEDERAL POLICIES AND PLANNING

Very little planning is currently being done at the federal or state level to respond to sea level rise. For example, although past sea level rise and future sea level rise from global warming have "reasonably foreseeable environmental impacts" for many proposed federal actions along the coast, no federal agency has expressly discussed sea level rise from global warming in any environmental impact statement.[11] Indeed, the U.S Corps of Engineers has not evaluated sea level rise from global warming in designing major coastal projects.[12]

Most of the nation's coastal states have developed coastal zone management programs designed to manage and protect their important coastal regions. However, none of the coastal states, with the exception of Maine, has yet developed or implemented specific strategies to minimize the damage to their shorelines from climate-change induced sea level rise. Seven have initiated studies.

IV. WHAT POLICIES CAN BE ADOPTED TO ASSIST THE FEDERAL GOVERNMENT TO ACHIEVE THE NATION'S GOALS?

---

11. See Letter of A. Alan Hill, Chairman, Council on Environmental Quality to Senator Albert Gore (March 3, 1987). On November 9, 1988, the Natural Resources Defense Council, Inc. (NRDC) submitted a written statement to CEQ for the record urging CEQ to immediately publish a notice of proposed rule making announcing the amendment of the Council's regulations to require all federal agencies to fully examine the causes of and impacts from climate disruption, including accelerated sea level rise, for all major federal actions significantly affecting the quality of the human environment. Letter of David Wirth, senior staff attorney, NRDC, Inc. to A. Alan Hill, Chairman, CEQ (November 9, 1988).

12. Letter of March 21, 1986 from Earl Eiker, Acting Chief Hydraulic and Hydrology Division, Director of Civil Works to Lynne Edgerton, NRDC.

Two overriding considerations should guide federal policy in this area. First, one of the nation's highest priorities should be the development and adoption of effective measures to slow and limit global warming. Non-nuclear energy use options could cut future projected greenhouse emissions significantly, and thus avert some of the worst effects of a greenhouse-induced climate cataclysm.[13] By far the largest energy use saving would be achieved by implementation of least-cost energy planning techniques which would let energy conservation and efficiency improvements compete on an equal basis with other energy supplies.[14] For example, NRDC has demonstrated that improvements in the design of automobiles, houses, appliances and mass transit systems could plausibly deliver at least twice as much oil and oil substitutes as the National Arctic Wildlife Refuge plus all undeveloped undersea lands under federal control.[15]

Second, <u>new</u> building - whether public or private - should be sited landward (inland) of areas likely to be inundated, eroded or damaged from sea level rise. The nation should adopt policies to redirect growth out of areas which could be affected by 2050. Likewise, new building should be discouraged if its development would jeopardize the continued preservation of coastal natural resources in 2050, or aggravate existing coastal pollution. It is particularly important that this principle be pursued now for structures which have long use expectancies such as sewage treatment plants, large industrial facilities and hazardous waste facilities. This redirection of growth will reduce the adverse effects of accelerated sea level rise from global warming and minimize the need for extensive shore stabilization with its attendant costs and adverse environmental effects.

At the federal level, a nationally coordinated response to sea level rise is needed to assure that federal public investment decisions, federal public land management and other federal policies affecting the coasts 1) encourage the protection and preservation of natural resources, 2) minimize the potential property and human losses from the improper siting of development

---

13. A certain amount of warming is inevitable, however, due to past emissions of greenhouse gases. Consequently, a certain amount of sea level rise is inevitable. Furthermore, future emissions of greenhouse gases, even if drastically curtailed or stabilized, will continue to contribute indefinitely to a warming.

14. R. Cavanagh, D. Goldstein & R. Watson, "One Last Chance for National Energy Policy (Natural Resources Defense Council, July 1988).

15. R. Watson, "Oil and Conservation Resources Fact Sheet: A Least Cost Planning Perspective" (Natural Resources Defense Council, July 1988).

in coastal areas vulnerable to sea level rise, and 3) do not contribute to further water pollution by 2050.

To this end, federal funding of major new facilities and infrastructure located in areas vulnerable to a 3 foot sea level rise by 2050 should be eliminated.

New federal coastal installations, including defense installations, should be located inland of such areas to the maximum extent possible.

The Corps of Engineers should take the prediction of a one meter sea level rise by 2050 into account in planning their construction and stabilization projects. The Corps should specifically address a one meter rise by 2050 in its preparation of environmental impact statements on proposed projects, and identify whether the proposed project will affect natural resource preservation goals, encourage coastal growth in vulnerable areas, or potentially aggravate water pollution.

The Federal Flood Insurance Program should simultaneously be revised to include future sea level rise from global warming in its calculations of the coastal V-zones, 10-year and 100-year floodplain. Discontinuing the practice of issuing flood insurance policies for new development in the high hazard coastal V zones should be considered. The potential financial liability of the program due to increased insurance claims should be carefully assessed.

The Coastal Zone Management Act should include areas likely to be flooded by a 3 foot rise by 2050 in the definition of the coastal zone, and should specifically encourage state coastal zone management plans to begin the process of redirecting new growth outside of vulnerable areas. Additional financial incentives should be made available to assist the states with their efforts. These steps would be consistent with the Congressional mandate in CZMA to preserve natural resources and to minimize losses of property, lives, and natural features.

Consideration should be given to extending the scope of the Section 404 general permit program under the Clean Water Act to coastal areas which would potentially become wetlands if the sea rises one meter by 2050. Such an extension might secure buffer zones adjacent to existing wetlands in order to permit wetlands to migrate upland as the sea rises.

The Council on Environmental Quality should take steps to amend its regulations under the National Environmental Policy Act to clarify that all federal agencies are required fully to examine the causes of and impacts from climate disruption, including accelerated sea level rise, for all major federal actions significantly affecting the quality of the human environment.[16]

A federal program to help coastal communities assess the effect of sea level rise on their drinking supplies, including aquifers, should be considered. Funding could be provided to assist the communities to redesign their water systems, where necessary.

V. CONCLUSION

Although there is scientific uncertainty concerning the precise amount and timing of sea level rise from global warming, there is also a high probability that significant global warming and sea level rise will occur over the next 50-60 years, even assuming an aggressive global effort to curb greenhouse gas emissions is implemented in the next ten years. Consequently, federal agencies should begin immediately the process of adapting their coastal programs to assure that they will continue to further existing coastal protection and preservation goals into the 21st century. Such revisions should be designed to further natural resource protection, minimize economic and human loss, and encourage long-term stable economic growth along our coasts.

---

16. <u>See</u> Letter of David Wirth, NRDC, to A. Alan Hill, Chairman, CEQ (November 9, 1988).

IMPACTS OF GLOBAL CLIMATE CHANGE ON METROPOLITAN INFRASTRUCTURE

Ted R. Miller, Ph.D.
The Urban Institute, Washington, DC

Case studies suggest that a doubling in CO2 levels primarily will require urban infrastructure investments to ensure an adequate water supply and to prevent sea level rise from inundating coastal communities. Impacts are likely to vary substantially by region. Cleveland, and presumably other Great Lakes cities, seem likely winners, the garden spots of tomorrow. An anticipated one meter rise in sea level probably will require diking and pumping or raising the land surface in many urban coastal areas, including more than half of the 20 largest metropolitan areas. The cost in Greater Miami alone could exceed $600 million over the next 100 years. Northeastern cities might have to spend billions on new water sources. Inland cities primarily should be concerned about water supply and electric power, and the possibility of increased subsidence problems.

Climate Change Could Have Major Impacts on Fixed Capital Stock

Although actual practice varies, the nominal replacement cycle for most infrastructure is 30 to 50 years. Some water supply investments have 100-year lives. Thus, infrastructure investments lock communities into capital stock for long time periods and are likely to make them vulnerable to rapid climate change. Sea level rise, temperature change, and changes in precipitation patterns, for example, all could alter the water supply and demand balance between planned replacement cycles. The nature and pattern of precipitation could affect drainage requirements, as well as highway design and maintenance.

In coastal communities, sea level rise could stimulate extensive construction of oceangates. Oceangates are coastal defense structures -- dikes, levees, jetties, canals, locks, pumping systems. They prevent inundation by the sea, slow oceanfront erosion, control storm surges, reduce salt water intrusion into aquifers, and slow salt water advance up rivers. Currently, oceangates are such a minor category of infrastructure that they are not inventoried or included in needs assessments.

The heat wave of 1988 vividly illustrated some of the potential impacts of global climate change on infrastructure.

- Hundred-degree weather expanded railroad tracks, forcing Amtrack to cut speeds from 125 to 80 miles per hour between Washington and Philadelphia (Bruske, 7/17/88) and allegedly causing a train wreck that injured 160 people on a Chicago-Seattle run ("Warped Rails", 8/7/88).

- In the suburbs of Washington, DC, steel expansion joints bubbled along a 13-mile stretch of Interstate 66 (Lewis, 8/18/88).

- Thirty miles below New Orleans, U.S. Army Corps of Engineers contractor worked around the clock for two weeks to build a 560-foot wide, 30-foot high silt wall across the bottom 40 percent of the Mississippi River channel (Sossaman, 6/28/88, 7/15/88). This $2 million wall slowed an advancing wedge of salt water that threatened the water supply in New Orleans and nearby parishes. It was designed to wash away when spring snow melt demands the full capacity of the channel.

- In Manhattan, heat exacerbated the effects of long-standing leaks in 160 miles of steam pipes, causing the asphalt to soften. As vehicles kneaded the soft asphalt, thousands of bumps formed on city streets, requiring extensive repairs (Hirsch, 1988).

This Paper Uses Case Studies to Examine Probable Urban Impacts

This paper discusses the probable impacts of coming climate change on U.S. urban infrastructure. It focuses on case studies of the impacts on water supply in New York City and a variety of infrastructure in Miami and Cleveland.

The case studies were based on a review of existing infrastructure studies in the three cities, discussions about likely impacts with local infrastructure experts, analyses by these experts, and our own estimates of probable impacts.

We estimated the temperature and precipitation impacts of global climate change by applying the percentage changes by season indicated in the GISS and GFDL models of the impact of an effective doubling of atmospheric carbon dioxide to historical climate data from 1950 to 1980. Impacts on the number of heating and cooling degree days were computed as variations in average daily temperatures below and above 65 degrees F.

Because the study was based on only three cities, it could not cover the full range of potential problem situations. The analyses were preliminary. They revealed which infrastructure responses to global climate change might be expensive, but were not engineering analyses of the most cost-effective responses. The potential for reducing impacts through technological change also was not assessed.

Miami Will Experience Severe, Costly Problems

The Miami case study examined the probable impacts of climate change and sea level rise on Dade County's water supply, water control and drainage systems, building foundations, roads, bridges, airports, solid waste disposal sites, and sewage transport and treatment systems. The impacts on operating costs for pumping and air conditioning and on the need for beach nourishment were not analyzed.

Oceangates. Greater Miami is bounded by water on all sides in the rainy season. An extensive network of canals and levees has been built to control ocean and fresh water flooding and recharge the aquifer that lies beneath the area. One of the world's most porous aquifers lies less than 5 feet below the surface in one third of Miami's developed area. Because the aquifer extends under the ocean, the typical urban response to a rising sea -- diking the water at the surface and pumping out the seepage from ditches behind the dikes -- appears unworkable. Unless the dike extended down more than 150 feet, rising sea water pressure would cause the sea to rush into the aquifer below the surface and push fresh water up almost to the surface.

A preliminary appraisal suggests an oceangate strategy that includes raising the land in low-lying areas, upgrading levees and dikes with pumped outflows, selective retreat, and increasing the fresh water head roughly in proportion to sea level rise to prevent salt water infiltration of the aquifer.

Canals, Levees, and Drainage. The one-time capital costs of upgrading existing canals in response to a 3.3-foot rise in sea level would be about $12 million dollars because fill is virtually free. Almost $50 million in new control structures, including extensive pumping capacity, might be required on the canals used to maintain the fresh water head. Large-scale pumping also could involve substantial operating costs. Extrapolation from the estimates in Weggel et al. (1989) suggests an additional $48 million will be devoted to raising levees and dikes. Storm sewers and drainage would need upgrading, requiring investment of several hundred million dollars above normal replacement costs. Miami's airport also would need roughly $30 million in drainage improvements.

Foundations of Houses and Streets. Because houses are built on concrete slabs, most buildings in newer areas already are built on raised lots to meet Dade County's flood control ordinance, and the foundations of larger buildings often already extend into the water table, a local engineer felt that foundations generally should remain stable if the fresh water head were raised one meter. Conversely, the water table could infiltrate the base of about a third of Dade County streets, which would have to be raised or risk collapse. If sea level rose gradually enough that the streets and related sewer mains could be raised during scheduled reconstruction, a local official estimated the added public cost might approximate $250 million. Building owners would incur substantial costs to improve drainage, raise yards, raise lots at reconstruction, and pump sewage to mains. In addition, many houses would suffer aesthetically because they would be 3 feet below street level.

Bridges. A 3.3-foot rise in sea level would require raising most bridges to ensure adequate underclearances; to prevent erosion beneath abutments, lifting of steel and box culverts, and pavement failure in bridge approaches; and to reduce vulnerability to storm surges during hurricanes. An order-of-magnitude cost saving would be possible if the work were done during reconstruction, which typically occurs on roughly a 50-year cycle. Retrofit would be much more expensive.

Water Supply. Rainfall could decrease by 10 percent to 15 percent in the dry winter months, although aggregate rainfall might increase in response to increased hurricane intensity and frequency. Cooling degree days might increase about 50 percent, thus increasing evaporation from surface water storage and usage of water-consumptive commercial air conditioning. Because the scenarios suggest winds would decrease, evapo-transpiration loss from the aquifer and surface storage areas could decrease.

Increased hurricane activity would be a two-edged sword, largely resolving the potential water shortage while inflicting billions of dollars in wind and flood damage. Even under this alternative, some wellfields might need to be moved further inland. Absent increased hurricane activity, climate change probably would exacerbate future water shortages and lead Miami to start large-scale water desalination at three times current water prices.

Electricity. Linder and Inglis (1988) estimated that the added electric demand for air conditioning operation could require raising electric generating capacity by 20 to 30 percent in Miami and other parts of the Southeast. Improvements in operating efficiency would reduce the required increase.

Other Temperature Effects. Miami's sanitary sewer pipes would face increasing hydrogen sulfide corrosion if temperature rose. Little impact is likely on sludge treatment and solid waste disposal operations, although landfill subsidence might increase. Rising temperatures also should have minimal effects on local streets.

Summary. As Table 1 shows, global climate change could require $.6 billion, and perhaps substantially more, in capital investment in Greater Miami over the next 100 years.

Cleveland Could Emerge a Winner

The Cleveland case study examined impacts of climate change on snow and ice control costs, road construction and maintenance, heating and cooling costs and equipment needs, water supply, and storm and waste water transport. A preliminary analysis also was included of the impacts of a 2 to 5.5 foot drop in the level of Lake Erie.

Snow and Ice Control. Global climate change could cause annual snowfall in Cleveland to drop from 50 to roughly 8 inches, reducing annual snow and ice control costs by about $4.5 million. Further savings estimated at $700,000 per year were projected to result from decreased frost damage to roads and bridges.

TABLE 1
Probable Infrastructure Needs and Investment in Miami in Response to a Doubling of CO_2 (Millions of 1987 Dollars)

Raising Canals/Levees	\$60
Canal Control Structures	\$50
Pumping	not estimated
Raising Streets	\$250 added to reconstruction cost
Raising Yards	not estimated
Pumped Sewer Connections	not estimated
Raising Lots at Reconstruction	not estimated
Drainage	\$200-300
Airport	\$30
Raising Bridges	not estimated; retrofitting much more costly than raising bridges during reconstruction
Sewer Pipe Corrosion	minimal
Water Supply	depends on hurricanes
Electric Generating Capacity	20%-30% increase

Heating and Cooling Costs. Heating costs for public buildings could drop an estimated \$2.3 million dollars per year. Conversely, public air conditioning costs could rise by \$6.6 to \$9.3 million per year.

If temperatures rose to the levels predicted in the GFDL and GISS scenarios, Cleveland might spend about \$65 to \$80 million to add air conditioning to older schools and large non-office spaces like gyms and repair garages. Much of this expenditure would occur as buildings were replaced or refurbished and might have occurred even without climate change. The costs of adding air conditioning to buses as they are replaced should be minimal.

Road Construction and Maintenance. The rise in winter temperatures associated with a doubling in CO_2 probably would allow Cleveland to use thinner pavement. Savings of about 3 percent in road resurfacing cost and 1 percent in reconstruction cost might result. The net savings could average about \$200,000 per year, 1.3 percent of the city's 1988 capital budget. A decrease in roadway maintenance costs of about 7 percent, or roughly \$500,000 per year, also seems likely.

Water Supply and Navigation. A climate-induced drop in the level of Lake Erie probably would not adversely affect Cleveland's water supply, according to local experts. Dropping lake levels could require some dredging, especially in the Cuyahoga River. Keith et al. (1989) estimate the dredging costs at \$11 to \$29 million over the next century, depending on how severely lake levels drop. While gradually falling Great Lakes levels might not impose major costs, Changnon (1989) concluded from a study of Chicago's historical response to changing lake levels that widely oscillating lake levels could.

Sewer Capacity. Because Cleveland's rainy season is unlikely to get wetter, climate change appears unlikely to create a need for upgrading the city's combined storm and waste water collection system.

Summary. Overall, the impacts of global climate change on Cleveland probably will be modest and generally positive. The city's climate should become more hospitable. Cleveland's location on Lake Erie should insulate the metropolitan area from water supply problems, meaning it could become a more attractive location for water-intensive industry if water supplies in other areas become less reliable. Other Great Lakes cities also could become more desirable places to live. Resulting in-migration could bring growth-related infrastructure costs and a rise in property values.

As Table 2 shows, the net direct impact of a doubling in CO_2 on Cleveland's annual infrastructure costs could be negligible, although expenditures probably would shift between categories. In addition to the costs shown in Table 2 a one-time capital expenditure of $68 to $80 million to add air conditioners in public buildings could be required. Many private residences probably also would add air conditioners.

TABLE 2
Estimated Impacts of a Doubling of CO_2
on Cleveland's Annual Infrastructure Costs
(Millions of 1987 Dollars)

Cost Category	Annual Operating Costs
Heating	-2.3
Air Conditioning	+6.6 to 9.3
Snow & Ice Control	-4.5
Frost Damage to Roads	-0.7
Road Maintenance	-0.5
Road Reconstruction	-0.2
Mass Transit	summer increase offsets winter savings
River Dredging	+0.5
Water Supply	negligible
Storm Water System	negligible
Total	-1.6 to +1.1

If temperature rises several degrees, like Cleveland, most Northern cities probably could anticipate savings in snow and ice control, heating, and roadway construction and maintenance costs. These savings might approximately offset the increase in air conditioning costs. Communities south of Cleveland probably would experience modest budget increases. Especially in the

Midwest, however, only other Great Lakes cities are likely to share Cleveland's freedom from water supply problems.

New York City

New York City already has begun hedging against the possibility of sea level rise by raising an outlet on a drainage structure for a water tunnel it is building. Upgrading dikes and levees in response to a 3.3 foot rise in sea level could cost $85 million over a 100 year period (interpolation of Weggel et al., 1989). In addition, the city might have to protect underground infrastructure from sea water infiltration.

All sanitary and much storm sewage is pumped in New York (Schwarz and Dillard, 1988). The roughly 450 outfalls use gravity flow into tidal areas. With rising sea level, these outfalls would have to be inspected more frequently and pumps might run more, but the system capacity and design seem unlikely to need revision. Higher sea level, nevertheless, could increase sewer backups, ponding, and basement flooding when high tides coincide with high runoffs.

The effects of temperature change should be similar in New York City and Cleveland. Increased air conditioning use could raise peak electric demand by 10 to 20 percent in the New York metropolitan area (Linder and Gibbs, 1987) .

The most pressing and largest climate change problem facing the city probably is the impact on water supply adequacy.

Current Water Supply. The New York metropolitan area draws water from the Hudson and Delaware River Basins and from underground aquifers that serve much of New Jersey and Long Island. The water supply network is in deficit. The Mayor's Task Force (1987) preliminarily recommended remedying New York City's deficit through better demand management and reactivation of a water intake at Chelsea, a $223 to $391 million investment that would yield 100 to 200 million gallons of water daily.

Demand Impacts. The climate change scenarios suggested a rise in average temperatures in the New York area by 7 to 11.5 degrees F during most of the year, resulting in about a 20 percent increase in cooling degree days. In response, demand for water used in cooling large buildings and lawn watering during the summer could raise annual water needs by about 5 percent.

Supply Impacts. Higher temperatures could increase evaporation and evapo-transpiration, decreasing the ability to store water efficiently in surface impoundments. A loss of 10 to 24 percent of reservoir supply could result.

A 3.3-foot rise in sea level could place the proposed $300 million Chelsea intake below the salt line during the peak summer demand period in mild drought years, reducing supply another 13 percent. With a larger margin of error, it might prevent use of the existing Poughkeepsie intake during severe droughts, further reducing supply. In addition, subsurface infiltration would reduce the supply available from the Long Island aquifer.

Implications. In aggregate, doubled CO_2 could produce a shortfall equal to 28 to 42 percent of planned supply in the Hudson River Basin. Among the feasible alternatives evaluated by the Mayor's Task Force in 1987, the least expensive way to add that much capacity would be a $3 billion project to skim Hudson River flood waters into additional reservoirs.

National Policy Implications

The uncertain, yet potentially imminent impact of global climate change already has increased the riskiness of infrastructure investment. Application of design standards and extrapolation from historical data might not still provide reasonable assurance that water and power supply, dam strength and capacity, bridge underclearances, or storm sewer capacity will be adequate for the 35-, 50-, and 100-year design lives of these facilities. The National Flood Insurance Program's historically based maps identifying the 100-year floodplain and 500-year floodway might no longer provide a reliable basis for local building and zoning ordinances designed to minimize flood losses to life and property. Migration caused by climate change could radically alter the population projections underlying capacity decisions about highway and wastewater treatment systems.

Especially in coastal areas, the possibility of accelerating global climate change soon may require careful decisions about how and when to adapt the infrastructure. A strong emphasis on life-cycle costing and the courage to make expensive upgrades during reconstruction in anticipation of future changes could provide large cost savings.

Water Supply. Water supply is of particular concern because decades are required to plan and complete projects, which then might last 100 years. We already are building dams, reservoirs, and water intakes that global climate change might make obsolete or inadequate, and more are in the planning stage. Elsewhere communities may well be allowing development of land needed for reservoirs to meet the water shortages expected to result from climate change.

National policy intervention might be the best way to prevent expensive water supply mistakes and adapt in an orderly fashion. A sensible goal over the next ten years would be to study the probable impacts of a doubling of carbon dioxide and of two sea level rise scenarios (perhaps .8 and 3.3 feet) on the water supply of every major metropolitan area.

Floodplain Development. The Federal Insurance Administration's ambitious flood-plain mapping efforts might beneficially consider global climate change. It could be more cost-effective for coastal floodplain studies to assume a rise in sea level of perhaps 6 to 9 inches by 2040, requiring flood-proofing of new buildings on that basis. Prospective action might be easier and less expensive than trying to protect structures that already are built. In addition, pricing the costs of flood protection into construction could reduce economically unjustifiable overbuilding in future floodplains and ocean beds, avoiding future demands for public subsidy of flood protection.

Infrastructure Standards. The voluntary organizations that promulgate engineering design standards might be well advised to educate their committees about global climate change. Growing uncertainty about future temperature, precipitation, and sea levels might dictate a reassessment of existing standards and safety factors for ventilation, drainage, flood protection, facility siting, expansion capability, resistance to corrosion, etc. Conversely, prompt detection of lasting changes could allow adjustment of geographically based standards -- for example, on roadbed depth and home insulation levels -- and provide significant savings.

Infrastructure Needs Assessments and Coordination of Oceangate Investment. Congress has mandated regular reports on many types of infrastructure needs. On a one-time or routine basis, these reports could examine the sensitivity of the needs estimates to climate change. The Federal Highway Administration might save money by encouraging states to raise some bridges on the Federal-Aid Highway System prospectively during reconstruction rather than risking sea level rise. EPA might examine the likely impacts of climate change on its assessments of storm sewer needs and related water quality issues. The Corps of Engineers might examine the potential implications of climate change for dam safety. And the Office of Management and Budget might examine potential climate change impacts in the capital investment supplement to the annual Federal budget.

Designation of a lead agency for oceangates also might be helpful. Increasingly, information might be needed on oceangate technology, the nature and condition of the existing oceangate inventory, and methods of financing. Development of a small unit with expertise in the area could aid information flow and provide a base for quickly launching emergency coastal defense efforts.

Recommendations for Further Research

It could cost billions of dollars to meet water supply needs associated with global climate change. Research on technological change that might reduce these costs seems a priority. The possibilities include:

- Examining whether the engineering options and cost outlook for desalinization have changed since the Department of Interior's research program ended in the 1970s, and

- Encouraging use of drought-resistant grasses and garden plants suited to a lengthening growing season,
- Redesigning cooling towers to cut evaporation losses,
- Examining the feasibility, environmental implications, and cost-effectiveness of installing barriers to slow or prevent salt water infiltration into coastal rivers and aquifers during droughts.

More examination also is needed of metropolitan impacts generally. No one knows the probable impacts of global climate change on urban subsidence problems, for example, in Phoenix. Impacts on inland and West Coast cities have not been examined, nor have the potential for and consequences of salt water infiltration into pipes in older coastal communities.

Another issue arises from EPA's increasing use of variable-flow permits that make effluent discharge rates into a river a function of the river's flow rate. Where global climate change increases the frequency and intensity of low flow periods, treatment capacity may need to increase.

Research on urban oceangates needs to intensify. Cost-benefit analyses should compare preventive and remedial strategies. How substantially, for example, are the need for and costs of oceangates likely to be reduced by reassessing the floodplain maps every decade and incorporating best estimates of sea level rise over a 40- to 50-year time horizon into them, thus ensuring lots will be raised or flood-proofed during construction? What mapping costs and changes in construction costs would result?

Research is needed on the probable impacts of climate change on domestic and international migration flows and the infrastructure demands these flows produce. Some areas might become sufficiently dry or hot that people would move away, while others might flourish. Infrastructure investment in new water supply, for example, might be unnecessary in areas that would lose population, but extra capacity might be needed in areas where population would grow. Similarly, as climate change shifts the best growing areas for specific crops, new farm-to-market transportation networks might need to be developed. Rights-of-way for these systems might best be set aside now, before land prices rise in response to climate change.

ACKNOWLEDGEMENTS AND DISCLAIMER

The work described herein was supported by EPA Cooperative Agreement CR-814883-01-0 with The Urban Institute. J. Christopher Walker performed the New York City case study and, with G. Thomas Kingsley, the Cleveland case study. William A. Hyman performed most of the Miami case study. The opinions expressed and any errors are the responsibility of the author, who served as Principal Investigator.

BIBLIOGRAPHY

Bruske, Ed. "104 (Phew!) Degrees Hottest in 52 Years," The Washington Post 111:225, July 17, 1988, pp. A1, A6.

Changnon, Stanley A., Steven Leffler, and Robin Shealy. "Impacts of Extremes in Lake Michigan Levels along Illinois Shorelines: Low Levels," Appendix Report in EPA Global Climate Change Report to Congress, forthcoming, 1989.

Hirsch, J., "As Streets Melt, Cars Are Flummoxed by Hummocks," The New York Times 137 (47599). August 16, 1988, pp. B1, B5.

Keith, Virgil F., Carlos De Avila, and Richard M. Willis. "Effect of Climatic Change on Shipping Within Lake Superior and Lake Erie," Appendix Report in EPA Global Climate Change Report to Congress, forthcoming, 1989.

Lewis, Nancy. "Two More Heat Records Fall as Summer of 1988 Boils On," The Washington Post 111:257, August 18, 1988, pp. A1, A10, A11.

Linder, Kenneth P. Michael J. Gibbs, and Mark R. Inglis. Potential Impacts of Global Climate Change on Electric Utilities, ICF Incorporated 824-CON-AEP-86, December 1987.

Linder, Kenneth P., and Mark R. Inglis. "The Potential Impacts of Climate Change on Electric Utilities: Regional and National Estimates," Appendix Report in EPA Global Climate Change Report to Congress, forthcoming, 1989.

Schust, Sunny M. "Continuing Drought Slows Transportation," The AASTHO Journal Weekly Transportation Report, 88:25, June 24, 1988, pp. 6-7.

Schwarz, Harry E., and Lee Dillard. "Chapter III-D Urban Water," in AAAS Global Climate Change Report, Washington DC, 1988.

Sossaman, Bruce A. News Release, US Army Corps of Engineers, New Orleans District, June 28, 1988.

-----. News Release, US Army Corps of Engineers, New Orleans District, July 15, 1988.

"Warped Rails Checked in Amtrak Wreck," The Washington Post 111:246, August 7, 1988, p. A5.

Weggel, J. Richard, Scott Brown, Juan C. Escajadillo, Patrick Breen, and Edward L. Doheny. "The Cost of Defending Developed Shorelines along Sheltered Waters of the United States from a Two Meter Rise in Mean Sea Level," Appendix Report in EPA Global Climate Change Report to Congress, forthcoming, 1989.

Emergency Preparedness to Address Climate Change

by Sherry D. Oaks
Research Associate
Department of Geography and Institute of Behavioral Science
Environment and Behavior Program
University of Colorado

Emergency preparedness to address climate change needs to be considered within the broader context of hazard reduction. Hazard reduction strategies for a number of different hazards, including climate related ones, have been developed. Hazard reduction measures, including emergency preparedness, can be undertaken to mitigate the effects of climate change on vulnerable communities. In determining appropriate strategies and measures, interrelated aspects of climate change and climate change induced hazards should be recognized.

Not all projected climate change will be gradual enough to allow corresponding incremental adaptation. Projections for climate change induced effects include large scale changes such as drought. Other expected effects include increased variability and severity of extreme events like hurricanes, storm surge, and floods. Regardless of the uncertainly associated with these projections, climate change induced effects will not be unlike the climatic and meteorologically caused hazards urban areas already face. Losses associated with these events throughout the world currently are estimated at billions of dollars annually. Therefore, it is prudent for vulnerable communities to consider a number of hazard reduction strategies. These strategies will be necessary both for adjusting to present climate regimes and to those forecast in climate change projections.

In developing hazard reduction strategies, it is important to consider the interconnections between atmospheric processes, and between atmospheric and solid earth processes. More than one type of event can occur simultaneously or within short time horizons. Often, natural hazards result from a combination of meteorological and geophysical processes. Landslides occurring during earthquakes are often a function of the degree of ground shaking, the geomorphology of the landscape, the moisture content of the soil, the ground water table depth, and a number of other interrelated physical environmental factors. The hazards are also, of course, a function of human interaction with the environment. For example, in 1985, a landslide set off by a tropical storm killed 129 people in a densely populated hillside region of the Mamayes Barrio of Ponce, Puerto Rico (National Research Council 1987).

Urban areas are, and will be, vulnerable to coastal erosion, floods, storm surge, lake level and sea level rise, salt water intrusion, hurricanes, tornadoes, high wind, severe cold and snow, snow avalanche, landslides, debris and mud flow, tsunamis, drought, wildfire, and a number of other natural events. Cities will also be susceptible to a combination of technological and natural processes that will combine to produce air and water pollution, toxic waste poisoning, acid rain, and other large scale hazards. The combined effects of these natural and technological hazards will be detrimental to health and welfare of human populations and ecosystems. Effects will extend to the resource base which the urban areas depend for survival, to lifelines that provide cities with resources, and to social, economic, and political systems. In short, damage from climate change induced hazards will affect overall environmental and economic sustainability.

Due to the expected pervasiveness of projected climate change most, if not all, urban areas will be vulnerable to climate change induced hazards. The question is how to adequately address these expected changes in terms of emergency preparedness, or other hazard reduction measures, given the present scientific uncertainly concerning climate change scenarios.

Climate change will affect all natural and human systems making them more vulnerable to variability and severity of extreme events. Even more gradual changes will contribute to stress in these systems over long periods of time. Therefore, emergency preparedness for possible climate induced changes should be addressed in both an event oriented and ongoing hazard context. And, preparedness measures must be sustained in an ongoing, long term hazard reduction strategy.

Hazard reduction strategies that address location dependent and time dependent variables are not unlike the strategies currently needed by urban areas presently susceptible to a variety of hazards. Several factors, applicable to both short term and long term planning horizons, include the following:

1. The multi-hazard aspects of urban areas should be recognized. For example, urban areas may be susceptible to drought, floods, wind storms, or any number of meteorological hazards. In addition, the interconnectedness of meteorological and geophysical hazards must be accounted for in preparedness planning. For example, sea level or lake level rise in seismically vulnerable areas may increase the threat of flooding or inundation during earthquakes. In addition, structural attempts to mitigate the effects of climate change in these urban areas create other structural hazards. Dikes designed to hold back sea or lake level rise will be built on soils most susceptible to ground shaking from earthquakes and will be vulnerable to structural failure from ground shaking and ground failure. Since many coastal cities also lie in active seismogenic zones this is a severe problem. Multi-hazard planning frameworks, along with other hazard reduction strategies, should help sustain emergency preparedness in an ongoing hazard reduction context, especially if health and economic benefits can be shown.

2. Assessments of vulnerability must be based on a number of environmental, engineering, and societal factors (Oaks 1989). Current assessments of vulnerability which include probabilistic estimates of magnitude, location, or frequency of the natural event, or the number and type of vulnerable buildings and estimates of people potentially exposed to hazards are valid. In addition, vulnerability should be based on a whole set of interrelated social, economic, cultural, and political factors such as the debt burden of a given urban area, per capita wealth, social and cultural institutions, and political structure. These economic, social, cultural, and political factors need to be analyzed carefully (Kotlyakov et al. 1988). Knowledge about societal constraints and enabling mechanisms can be utilized in determining and shaping possible public and private strategies for hazard reduction.

In addition to assessing vulnerability within this interactive environmental and societal framework, it is also necessary to assess resiliency to climate change induced hazards. For example, an urban area may be susceptible to a variety of hazards, but other factors may account for resiliency to those hazards. Part of that resiliency might be embodied in hazard reduction measures. For example, hazardous building abatement programs or structural retrofitting may have been implemented. Strict land-use planning measures may be imposed. The economic ability to develop and sustain infrastructure may be possible. Emergency preparedness planning may be sustained.

Since prediction of hypothesized climate change induced hazards in not certain, emergency preparedness is one way to build resiliency into urban systems. Emergency preparedness in the 1990's and beyond could be restructured to include the expected effects of climate change induced hazards while still focusing on presently known and expected hazardous events. In this way, emergency preparedness could contribute to hazard reduction goals, regardless of whether hazards are climate change induced or not.

There are other powerful mechanisms for effecting hazard reduction strategies. One potentially effective strategy centers around the use of existing institutions for risk communication and for emergency preparedness (Oaks 1988). These in-place institutions often provide the most cost effective and socially acceptable means of accomplishing hazard reduction. Using socially and culturally based institutions that embody traditional, accepted value systems may be one of the most effective ways to work for increased hazard reduction, especially cross-culturally (Bolin and Bolton (1986). This is primarily because these institutions, and the functions they perform, are considered reliable by discrete social groups. In addition, some of the institutions already may have hazard reduction as part of their everyday activities, although it may not be defined as such. These activities need to be identified, enhanced if necessary, and emphasized.

In addition to using existing institutions and in-place programs or activities, it is also important to integrate hazard reduction into development and redevelopment planning (Mitchell 1988). Again, the cost effective benefits are immense compared to post-event responses and remedies. Economic incentives can be integrated effectively into economic development and redevelopment planning as long as attention is paid to socially, culturally, economically, and politically relevant factors (Lewis 1988). Within this context, it is important to consider economic sustainability of hazard reduction efforts. If preparedness planning can be integrated within social disaster planning (i.e. unemployment, ill health) there may be some shared protective benefits and potentially lower costs.

In working toward these preparedness measures within general hazard reduction strategies, it will be increasing important to build new constituencies for hazard reduction (Stanley Foundation 1988). Those who work to reduce negative effects of hazards, including climate change induced hazards, must be brought together in the planning and implementation phases of the hazard reduction process. Although climate change is global in nature, the effects of climate change will be felt regionally, nationally, and sub-nationally. While international and national leadership will be increasingly important in forging global and regional consensus on mitigating and adapting to climate change, implementation of climate change induced hazard reduction strategies will occur at the local level. Vulnerability assessments which take the totality of environmental and societal interaction into account will be vital to hazard reduction strategies in different urban areas as decisions are made in anticipation of and response to climate change.

If emergency preparedness can be successfully integrated into ongoing preparedness strategies for hazard reduction, and if hazard reduction can be fashioned to be environmentally and economically sustainable, then preparations for the multitude of economic, social, and environmental effects of climate change, climate change induced hazards, and other hazards can be integrated into a larger societal framework where there is a greater chance of successful implementation.

References

Bolin, R. P. and Bolton, P. 1986. Race, Religion, and Ethnicity in Disaster Recovery. Program on Environment and Behavior Monograph 42. Boulder, Colorado: Institute of Behavioral Science.

Kotlyakov, V.M., Mather, J.R., Sdasyuk, G.V., and White, G.F. 1988. "Global change: geographic approaches." Proceedings National Academy of Sciences, USA 85:5986-5991.

Lewis, J. 1988. "An open letter in response to Confronting Natural Disasters: An International Decade for Natural Disaster Reduction." Natural Hazards Observer 12(4):4.

Mitchell, J.K. 1988. "Report on reports - Confronting Natural Disasters: An International Decade for Natural Disaster Reduction." Environment 30(2):25-29.

National Research Council, Advisory Committee on the International Decade for Natural Hazard Reduction. 1987. Confronting Natural Disasters: An International Decade for Natural Hazard Reduction. Washington, D.C.: National Academy Press.

Oaks, S.D. 1988. "Decade for Natural Disaster Reduction." EOS Transactions, American Geophysical Union 69(31):753-757.

__________. 1989. "An interactive environmental/societal process model for the assessment of vulnerability to hazards." Paper to be presented at the Association of American Geographers Annual meeting. Baltimore, Maryland.

Stanley Foundation. 1988. Science and Technology Development. Report of the 19th United Nations Issues Conference. Muscatine, Iowa: Stanley Foundation.

Planning for Climatic Variability and Uncertainty

Mark W. Mugler
Eugene Z. Stakhiv
Hanna J. Cortner
Michael Rubino[1]

The role of the planner is to provide choices for the future. In the face of climate change, the water resources planner must respond to the variability of today's weather while accomodating the uncertainties of tomorrow's climate.

It is said that water is never in the right place at the right time in the right amount. The purpose of water resources infrastructure is to mitigate the effects of hydrologic variability, such as by storing water to regulate flows and by diverting water to where it is needed by off-stream or out-of-basin uses.

The climate of the future will be different than that of the present. Although we have general agreement among the global circulation models on changes in temperature, and less reliable projections on changes in evapotranspiration, we have little information on the temporal and spatial distributions of precipitation that the climate of tomorrow will bring.

We can be fairly confident that greater heat will drive numerous hydologic effects. These effects, in turn, drive changes in natural and human water uses, including forests, agriculture, water habitat and urban uses.

There will be more rapid declines in soil moisture and streamflows between precipitation events. In some regions, increased evapotranspiration will not be offset by increased and more frequent precipitation, resulting in increased water demand, reduced water availability, and reduced water quality. There will be more moisture in the atmosphere, perhaps resulting in greater and more frequent flood events in some regions.

We cannot project reliably for a given region how its vulnerability to drought, water availability constraints, flooding, and water quality degradation will be affected by climate change. Instead, we must rely on existing vulnerabilities to provide insight into the vulnerabilities of tomorrow.

---

[1] Respectively, Vice President and Director of Water Resources Programs, Apogee Research, Inc.; Senior Policy Analyst and Director, Risk Management Program, U.S. Army Engineer Institute for Water Resources; Associate Director for Information Transfer, Water Resources Research Center, University of Arizona; and Principal, Bluewaters, Inc.

For instance, we know that the arid regions of the Nation where withdrawals and consumptive use (such as for power, domestic use and irrigation) are stretching supplies are most vulnerable to drought, because drought has a proportionately greater impact on streamflows in arid regions, and because higher temperatures in the future will cause a more rapid decline in flows and soil moisture between rainfalls. Natural and human in-stream uses, such as habitat, navigation, recreation and hydoelectric power, are vulnerable to reduced flows or levels. Areas with water quality degraded by heat, wastewater, non-point sources, toxic materials, or irrigation salts are most vulnerable to reduced availability of water for dilution, but would also benefit from increased precipitation. Urbanized or urbanizing flood-prone regions are most vulnerable to increases in the magnitude or frequency of major floods.

Climate change has the potential to exacerbate conflicts among water uses: offstream uses versus instream uses that require minimum flows or volumes; in reservoirs, flood control (low pool) storage versus recreation (constant pool) versus water supply and instream uses (high pool); and offstream irrigation versus municipal, industrial, power and other offstream uses.

Our existing water management systems and institutions have the tools to respond to the weather variability of today. These tools include the following:

- o surface water development;
- o system optimization and regionalization to obtain greater yield and reliability;
- o water allocation and exchange through market mechanisms;
- o adoption of conservation technologies;
- o drought contingency planning; and
- o water quality protection.

Many of these tools, such as reservoir construction, have a long history of use. Others, such as system optimization, water markets, and water quality protection, have become prominent management tools only recently.

In addressing weather variability, water resources planners have traditionally assumed that climate is stationary, that is, hydrologic phenomena vary about a fixed mean, and variability may be known with a high degree of confidence. Under conditions of uncertainty associated with climate change, more sophisticated planning tools are required, and the menu of water management response and adaptation tools must be applied in a more flexible manner.

We are uncertain not only about climate and hydrologic conditions, but also about the response setting: demographic, economic, and technological conditions; values and preferences; management institutions. We know that society and its institutions will evolve, in part in response to the imperatives of coping with climate change. It would be a mistake to devise responses to climate change by superimposing the climatic conditions of tomorrow on the social setting of today.

The scientists among us urgently recommend reducing uncertainty so that we may begin to respond before the costs of responding have risen too high or the best options have passed us by. This is well and good, but in the meantime water resources management decisions are being made every day.

This returns us to the role of the planner to provide choices for the future, often in the face of uncertainty. The water resources planning community has developed and is developing technical tools for accomodating uncertainty.

One analytic tool is sensitivity analysis. Sensitivity analysis measures the sensitivity of a projection to errors in variables on which the projection is based. Sensitivity analysis is endorsed by the Principles and Guidelines, the "bible" of Federal water resources planning.

A second analytic tool is risk cost analysis. Essentially, risk cost analysis examines the costs of being wrong to various degrees on the high side and the low side. By assigning probability distributions to independent variables and expressing risk preferences, the planner can evaluate decision alternatives, such as alternative water plans.

Water projects and systems can be designed for robustness and resilience, robustness being the ability to perform under a range of conditions, and resilience being the ability to recover from performance "failure." Furthermore, it has been found that many existing systems were designed for extreme conditions using conservative design criteria, with the incidental result that they contain enough "buffering" to provide robustness and resilience under the proper operating rules. Designing projects that are fail-safe under extreme conditions costly, however, and "safe-fail" projects that perform under a variety of conditions may be satisfactory.

Where commitments of resources are reversible, a short time horizon can be adopted in water resources planning. Smaller, less capital-intensive projects that are staged and have short design life would result.

In conclusion, we do not always have to wait to respond to climate change until uncertainties have been eliminated. The water resources community can begin now to adjust to climate change in circumstances where there is little uncertainty, where uncertainty does not significantly affect the outcome of a decision, where the consequencies of being wrong are acceptable, or where the decision in question provides immediate benefits that justify the decision.

FACTORING CLIMATE CHANGE INTO CORPORATE PLANNING: Introduction to Panel Discussion

Roger Strelow
Vice President
General Electric

Our speakers today will address primarily the issue of corporate planning to anticipate and adjust to the economic changes (e.g., reduced crop yield in certain areas) that will result from the climate change, and from the immediate consequences thereof (e.g., sea level rise), which will occur in the decades ahead. These changes will take place to some considerable extent despite public and private efforts to ameliorate them. My brief, introductory remarks, in contrast, will cover an important related challenge for corporate planning--planning with respect to the public policy and technological steps that will be taken to limit the extent and severity of future climate change.

Clearly, corporate America--and the corporate World--must do both: plan for some significant amount of climate change and plan for dealing with government laws and incentives, and with new technologies, aimed at constraining climate change. My focus is on the second type of planning.

I will survey quickly:

- what is in place,
- what is being considered, and
- what is needed to achieve sensible results.

Laws in Place--Laws already in place include an international agreement on ozone layer protection (the "Montreal Protocol"), legislation and EPA regulations to carry out that agreement, and a broad "Global Climate Protection Act." This 1987 law directs EPA to propose to Congress a coordinated national policy on global climate change, calls upon the Secretary of State to pursue within the UN the designation of an International Year of Global Climate Protection, and asks the President to accord climate protection high priority on his agenda for US-Soviet relations.

Laws Being Considered--The new Congress, of course, will start in January 1989 with a clean slate. At a minimum, however, we can expect the climate protection bills introduced in the last Congress but not acted upon will be reintroduced, perhaps in modified form. And, undoubtedly, there will be some altogether new initiatives.

S. 2663, which retiring Senator Robert Stafford of Vermont introduced last year, would set up a regulatory framework to eliminate CFCs by the year 2000, reduce CO_2 emissions by 35% by the year 2010, initiate controls on methane emissions, and establish a national goal of generating all the nation's power from "non-polluting" sources by 2050. It would require a further tightening of automotive fuel efficiency standards. Senator Tim Wirth's S. 2667 would reduce CO_2 by 20% by the year 2000 and would authorize significant funding for faster development of energy technologies that do not generate CO_2 or other greenhouse gases.

What Is Needed--Ambitious regulatory mandates ("reduce emissions of X by Y% by Z date") are almost sure to be part of the ultimate legal/policy framework for trying to ameliorate global warming. But the frustrating experience to date with such mandates should caution policymakers to look carefully at alternatives as well. In fact, interestingly, Senator Wirth has been one of the leaders in a very recent effort, co-chaired by Senator Heinz of Pennsylvania, to evaluate possibilities for using economic incentives and disincentives to stimulate technology development and behavioral changes needed in a wide variety of environmental programs.

Bold, fresh thinking will be needed from politicians and from agency administrators. But it will certainly be needed as well from industry and its planners. Corporate leaders have an opportunity to help shape a more constructive, less confrontational debate in the climate arena than has been the case in the past with more localized environmental problems.

All involved in the climate policy formulation process must be intelligently aware of the truly global dimensions not only of the problems but of the potential solutions as well. As we saw with the CFC issue, it is critical for the U.S. Government to exercise both the patience and the firmness to get a sufficiently worldwide "buy-in" to solutions that only make sense when implemented around the planet.

At the same time, I am cognizant of the shrewd observation recently made by Climate Institute President John Topping in his October 1988 article for BNA's International Environment Reporter:

> Although long-term policies to constrain growth of greenhouse emissions will require technological and institutional breakthroughs to benefit the Third World, near-term constraints . . . must be concentrated on the industrialized nations, the source of most current emissions. Actions by the industrialized

> countries can have a tangible effect on current emissions. Moreover, they will set an example that may make a more credible case for Third World countries also acting to restrain growth of greenhouse gas emissions. (p. 560)

What about Technology--Much of the emphasis of any new legislation will be on development and deployment of energy production technologies that produce far lower greenhouse gas emissions than current fossil fuel combustion technologies do. Among other things, many people who had virtually written off the nuclear option are rethinking this position in light of climate concerns--especially if a new era of nuclear were accompanied by greater assurances against catastrophic risks to citizens living near the facilities.

But there is another whole area of technology which should not be overlooked despite its initial "science fiction" impression. These are technologies that could be deployed to literally combat or remove some of the climate-altering chemicals that will still be released into the atmosphere for many decades to come despite the best of efforts to minimize emissions.

The August 16, 1988 New York Times contained a fascinating article entitled "Scientists Dream Up Bold Remedies for Ailing Atmosphere." It mentioned, for example, the concept of "atmospheric processing," introduced by Dr. Stix at Princeton. Giant infrared lasers atop mountains would break apart CFCs in the lower atmosphere. He has calculated that a possible array of such lasers could destroy a million tons of CFCs each year--the amount that now annually flows into the atmosphere. Major potential stumbling blocks have yet to be answered, but the concept is tempting.

An alternative weapon against ozone depletion would be some method to "export" more ozone into the stratosphere. Scientists suggest producing of ozone on earth, then transporting it to the stratosphere by rocket, aircraft or balloons. Or, solar-powered ozone generators could be placed in high-altitude balloons.

Other technological dreams include possible means for increasing the reflectivity of the earth's atmosphere so that more sunlight is reflected back into space--or simply blocking some sunlight altogether. For example, some have suggested that a series of earth-orbiting satellites with an aggregate area equivalent to two percent of the earth's surface could block enough sunlight to offset a doubling of current CO_2 emissions.

Still other scientists contemplate how one might remove greenhouse gases from the atmosphere by massive biological action on earth. Conserving and planting forests, which absorb CO_2, is a theoretical possibility. More radical options include stimulating the massive growth of tiny ocean organisms that soak up CO_2.

While any one of these schemes may seem massively impractical or undesirable now, the scientific daring which they reflect should certainly be encouraged. Such bold approaches may well be part of the ultimate answer to the threat of significant global warming.

And getting back to industry planning . . . industry certainly has a role in developing and even formulating such hi-tech possibilities, just as it has had, for example, in the nation's space program.

With this broader perspective in mind, let's turn now to our four presentations on incorporating the inevitable fact of some significant climate change into corporate planning.

IMPLICATIONS OF CLIMATE CHANGE FOR THE INSURANCE INDUSTRY

D.G. Friedman
Travelers Insurance Company
Hartford, Connecticut

PURPOSE

For contingency planning purposes of insurance, emergency planning and hazard mitigation activities, it is prudent to attempt to ascertain the potentiality for sizeable changes in the production and severity of storm-caused natural disasters due to a Greenhouse-induced warming trend in the near term (immediately upcoming years) in addition to that in the longer term (upcoming decades). The future production and magnitude of weather-caused natural disasters in the United States would be very sensitive to a significant Greenhouse- induced warming that could occur near the end of the next half century. There is also a current need to ascertain if some sensitivity could even occur during the transitional climate phase, in the next decade or so, when lesser amounts of warming associated with the upward trending in global atmospheric temperature would take place.

An attempt has been made to quantify these potential near-term effects, utilizing presently available information that is consistent with the current state of knowledge about natural disasters caused by storms; regional storm characteristics (frequency, severity, location); and likely effects of a global atmospheric warming on these storms and their catastrophe-producing potential. Unfortunately, detailed implications regarding possible Greenhouse influences on regional storm characteristics (frequency, severity, location) based on global climate model simulations will not be available in the near future. Consequently, any inferences that presently can be made must be based indirectly on less pertinent and less credible information from other sources.

APPROACH

The approach was to determine the catastrophe production of various storm types in the present climatic regime and then to estimate changes attributable to a Greenhouse warming trend. Reference 1 provides details of the analysis. The representation for the present climatic period was based on an analysis of 800 weather events that occurred over the past 40 years which caused large amounts of insured damages and, as a result, were coded as weather-caused catastrophes by the insurance industry. The damage-producing potential of a recurrence in 1989 of each of these events to currently insured properties was simulated. Each recurrence would have caused at least $5 million in losses. Cause of loss was used as a criterion for grouping the catastrophes into one of three storm type categories: winter storms, hurricanes, severe local storms (thunderstorm-spawned perils of tornadoes, hail, and "straight-line" winds). Damage to "fixed properties" such as buildings, caused by inland and coastal flooding including storm surge is not a part of the catastrophe totals because this peril is primarily covered by National Flood Insurance, a federal program. Storm regions were defined.

Catastrophe production characteristics of each storm type during the present climatic regime were determined and summarized in Table 1. The number and magnitude of these events were grouped by day and month of occurrence so that the seasonal variation in catastrophe production could be determined for each storm type. An objective is to examine possible effects of a Greenhouse-induced warming trend in the near term which is defined to be a period of climatic transition extending from the present to ten years after the turn of the century (1990-2010). This interval would represent the changeover period between present climate conditions and those of a much warmer atmospheric environment (2010-2050) if the full effects of a Greenhouse warming are realized.

Table 1. Estimated total 1989 damages by storm type, resulting from a repeat of each of the catastrophe-coded weather events that have occurred in the past forty years and which would cause at least $5 million in damage to presently insured properties in the United States.

Present Climatic Regime

Storm type	Number of Catastrophes	Percentage	Total Damage Production (billions of dollars)	Percentage	Average Annual Damage Production
Hurricane	51	6.4	27.2	40.2	$ 680,000,000
Severe local storm	649	81.2	31.4	46.4	780,000,000
Winter storm	99	12.4	$ 9.1	13.4	230,000,000
Total	799	100.0	$67.7	100.0	$1,690,000,000

Regional temperature changes that would probably occur sometime during the transitional climate period have been used: hurricane region, +0.5°C (+0.9°F) in sea surface temperature; severe local storm region, +1.0°C (+1.8°F) northern sections and +0.5°C (+0.9°F) in southern sections in air temperature; winter storm, +1.0°C (+1.8°F) in air temperature. These increases are consistent with output of global climate model simulations reported in the scientific literature. Timing of the year of occurrence of the increases would depend on how the Greenhouse effect interacts with other sources of global atmospheric temperature variation. The effects of these assumed regional increases in average temperature upon the catastrophe production of each storm type were estimated by converting the temperature changes into a measure of the implied expansion or contraction of the seasonal distributions of the number and severity of catastrophes produced by each of the storm types.

Differences in the size and shape of these distributions between present climate and transitional climate conditions were used to denote the probable effects of a Greenhouse warming on catastrophe production. In the period of climatic transition, seasonal changes in the atmospheric circulation are assumed to be close to those of the present climate except that these changes would occur at a somewhat earlier date in the spring and a slightly later date in the fall. In effect, the characteristics of storm activity in the transition period would be similar to that in the present climatic period with a slight shift in time of occurrence. As a result, intervals of maximum activity could be shortened or extended depending upon storm type and season of occurrence.

HURRICANES

Seasonal changes in sea surface temperatures (SST), which were utilized as a climate change index for the hurricane region, would be much larger than the anticipated increase due to a Greenhouse warming during the period of climatic transition. Exhibit 1 illustrates seasonal variation under present climatic conditions. For reference, it is noted that tropical cyclones usually do not form when the sea surface temperature is less than about 80^{0} Fahrenheit, even though atmospheric conditions are favorable for storm development. Areas with sea surface temperatures of 80^{0}F or above are represented by double stipled areas in Exhibit 1. Exhibit 2 represents a typical seasonal sea surface temperature curve for specific locations in the hurricane region. The horizontal line on the chart denotes the approximate minimum sea surface temperature for tropical cyclone development. The intersection of this line with the sea surface temperature curve roughly defines the beginning and ending times of the tropical cyclone season (late May and late November). Refer to Exhibit 3.

It is assumed that during the transitional climatic period, the effect of a Greenhouse-induced upward trend in surface air temperature over the hurricane region will result in an eventual increase in sea surface temperature over the same area after an appropriate time lag of at least a year (Exhibit 4). The upward shift in the seasonal sea surface temperature curve, when an increase of +0.5^{0}C (0.9^{0}F) is added to the seasonal curve representing the present climatic period, is shown in Exhibit 5. This increase of +0.5^{0}C, in effect, would lengthen the tropical cyclone season as illustrated by the new location of the intersection of the modified sea surface temperature curve with the horizontal line representing the threshold temperature for tropical cyclone development. The overall length of the hurricane season would be increased by about 20 days.

The seasonal distribution of landfall dates of the 240 hurricanes is shown in Exhibit 6. There is a correspondence, although not necessarily a direct cause-effect relationship, between the seasonal variation in sea surface temperature and the number of hurricanes that cross the coastline of the United States (landfalling hurricanes). Because it is assumed that seasonal changes in the atmospheric circulation patterns will also shift slightly, the frequency distribution of the number of landfalling hurricanes, in effect, would be split into two segments at the time of highest sea surface temperatures. The early summer segment would move the equivalent of about 10 days toward earlier occurrence times. The late summer segment would shift about 10 days toward later occurrence dates. The result of these shifts would be an addition to the landfall frequency distribution of a period of about 20 days, during the time of highest sea surface temperatures, when more landfalling storms could occur. These occurrences would be over and above the ones experienced during the present climatic regime. This additional increment to the landfall frequency distribution, which is depicted by the crosshatched area in

Exhibit 7, represents an increase of about 33 percent in the annual number of landfalling hurricanes between the present and sometime during the period of climatic transition.

A plot of the number of times in the past 117 years that the estimated maximum winds of hurricanes at landfall attained given values suggests that there is a correspondence between the seasonal sea surface temperature and the occurrence of hurricanes of severe intensity (Exhibit 8). The higher the temperature, the greater the likelihood that the storm will be of stronger intensity. Similarly, a comparison between the percentage of September hurricanes (when the sea surface temperatures are near maximum) with that of non-September hurricanes also suggests a correspondence, although not necessarily a cause-effect relationship, between the seasonal sea surface temperature and the probable landfall location of an incoming hurricane. September storms are more likely than non-September storms to have a landfall on the Atlantic coastline and also to be more severe (Exhibit 9). This combination of storm intensity and landfall location is of importance to insurers because of its influence in the determination of whether a coded catastrophe will be produced and, if it does occur, what its resultant magnitude is likely to be.

Magnitudes of the coded catastrophes caused by hurricanes are closely related to both the physical characteristics of the storm and its path relative to coastal and inland concentrations of insured properties. Exhibit 10 graphically illustrates this interaction of the high wind speed pattern of a passing hurricane with the spatial distribution of insured properties. Coastline exposures, where the highest winds occur, are especially susceptible to damage (Exhibit 11). Computer simulation techniques can be used to approximate the damage-producing potential of a hurricane to a spatial array of properties (Exhibit 12). This approach was used to establish a benchmark measure of the catastrophe production by hurricanes of the present climatic regime using the entire hurricane climatology of 240 landfalling storms (1871-1988).

These present-day damage potentials which are based on a simulated recurrence of each of the 240 events, have been accumulated by date of landfall. The resulting frequency distribution of aggregate hurricane-caused loss potentials by date of occurrence is shown in Exhibit 13. A comparison with the seasonal sea surface temperature curve, which is also plotted on the chart, indicates that the largest amount of damage potential occurs during that portion of the hurricane season when sea surface temperatures are at a maximum. A 20-day extension of the hurricane season during the period of highest sea surface temperatures creates a combination of increased number and severity of hurricanes with landfalls that are at or near coastal concentrations of insured properties. This situation increases both the likelihood and probable magnitude of insurance industry coded catastrophes caused by hurricanes. It could increase the annual average loss expectancy due to hurricane-caused coded catastrophes by 40 percent over those expectancies in the current climatic regime.

SEVERE LOCAL STORMS AND WINTER STORMS

A Greenhouse-induced temperature increase could cause a time shift in seasonal activity of severe local storms which might reduce their damage producing potential by elimination of those catastrophes that occur during a two-week period centered at the midpoint of the cold season.

This shift in the timing of the storm patterns could result in an annual reduction in severe local storm damage production of about $5 million. However, the extension of severe local storm activity by an additional two-week period in midsummer because of the slight seasonal shift could contribute an annual increase of approximately $30 million per year. The net change in severe local storm catastrophe production between the present climate and transitional climatic regime would be an additional $25 million annually.

Contraction of the winter storm season could lead to an elimination of about two weeks of maximum storm activity near midseason in January. The average annual reduction in catastrophe damage production due to winter storms would be about $20 million per year.

OVERALL INCREASE IN CATASTROPHE PRODUCTION

Table 2 lists the resultant increases in damage production.

Table 2. Estimated change in the annual damage-producing potential of insurance industry coded catastrophes caused by the insured perils of wind and hail between the present climatic regime and an undefined time during the transitional climate period (1990-2010) when a projected warming due to the Greenhouse effect might be attained. Damage potential is expressed in 1989 dollars.

Storm type	Annual Damage based on Present Climatic Conditions	Change in Annual Damage Potential	Annual Damage based on Transitional Climate Conditions
Winter storm	$ 230,000,000	-$ 20,000,000	$ 210,000,000
Hurricane	680,000,000	+ 270,000,000	950,000,000
Severe local storm	780,000,000	+ 25,000,000	805,000,000
Total	$1,690,000,000	+$275,000,000	$1,965,000,000

The increase in the annual damage production of nearly $300 million represents a 30 percent increase over the insurance industry's current annual catastrophe loss based on an average over the past 3 years. Because of the large variability in the magnitude of individual catastrophes, the estimate of the average annual loss varies widely from one small sample to another. Over the past six years, the average annual loss is $1.6 billion. An estimate of a long-term average based on 40 years of experience is $1.7 billion as shown in Table 1. Using this estimate of annual damage for the present climatic regime results in about a 20 percent increase for the period of climatic transition.

CHANGE IN DAMAGE EXPECTANCIES DURING THE TRANSITIONAL CLIMATIC PERIOD

Table 3 lists the probable timing of increasing damage potentials during the transition period if the projected levels of regional warming are attained.

Table 3. Approximate timing within the transitional climatic period when various changes in damage potential relative to that of the present climatic regime might take place if assumed levels of Greenhouse-induced global atmospheric warming are realized. Increases are expressed in 1989 dollars.

Time Interval Year	Increase in Annual Damage Potential of Weather-Caused Catastrophes Over That of the Present Climatic Regime	Approximate Percentage Increase Over Present Climatic Regime
1995-1999	+$150,000,000	10%
2000-2004	+ 300,000,000	20%
2005-2010	+ 500,000,000	30%

CONCLUSION

This attempt to obtain "order-of-magnitude" estimates of possible near-term changes in weather-caused catastrophe production, which are caused by a Greenhouse warming trend, is based upon the meagre amount of pertinent information that is currently available. Results are subject to major modifications as more credible data is developed.

REFERENCE

1. Friedman, D.G., 1989, Implications of Climate Change for the Insurance Industry, Travelers Insurance Company, Hartford, Connecticut, 44 pages.

Exhibit 1. AVERAGE MONTHLY SEA SURFACE TEMPERATURE (DEGREES FAHRENHEIT)

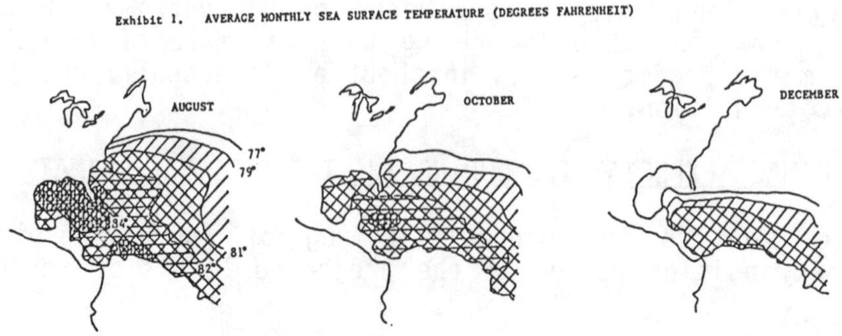

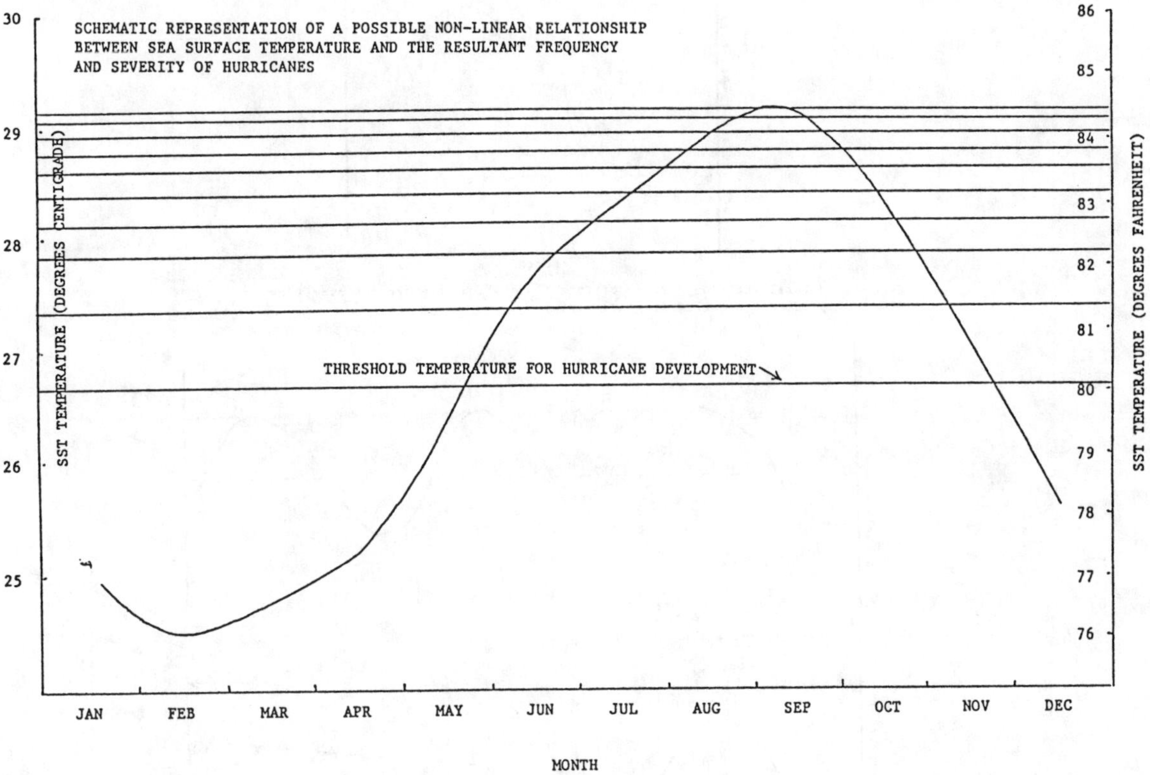

Exhibit 2. Seasonal changes in average sea surface temperature off southeast Florida.

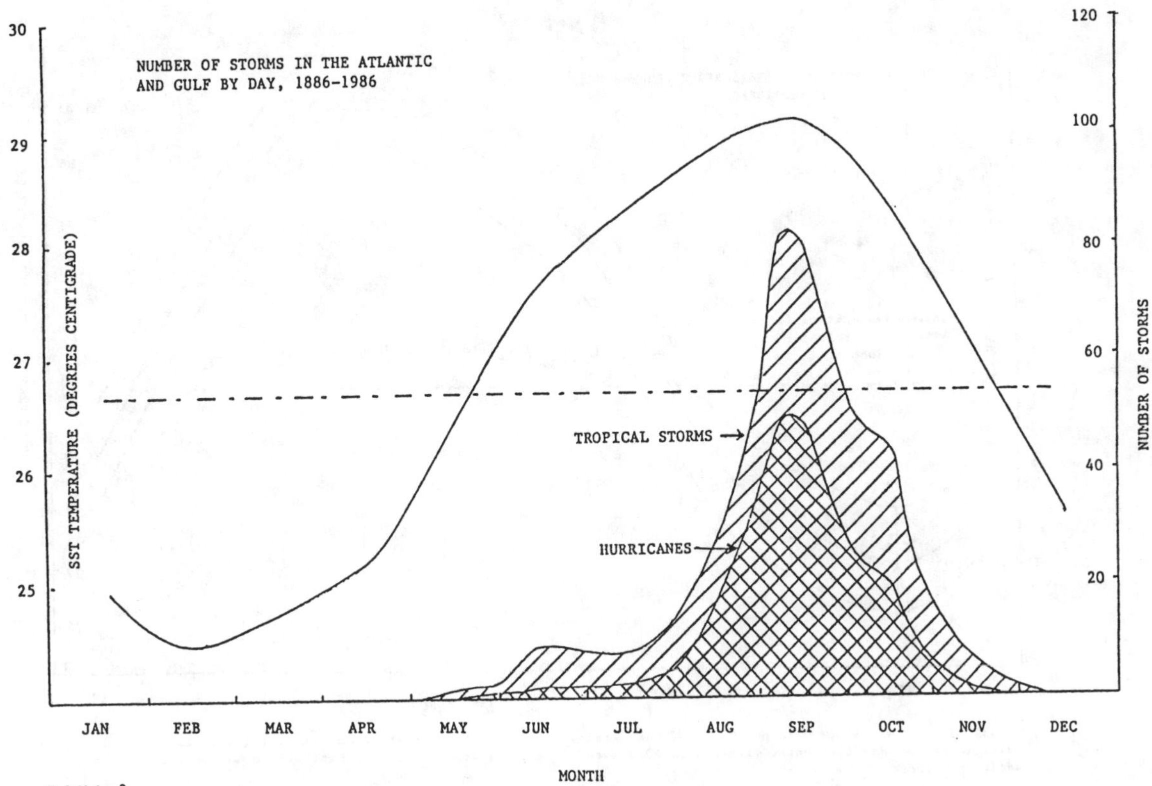

Exhibit 3.

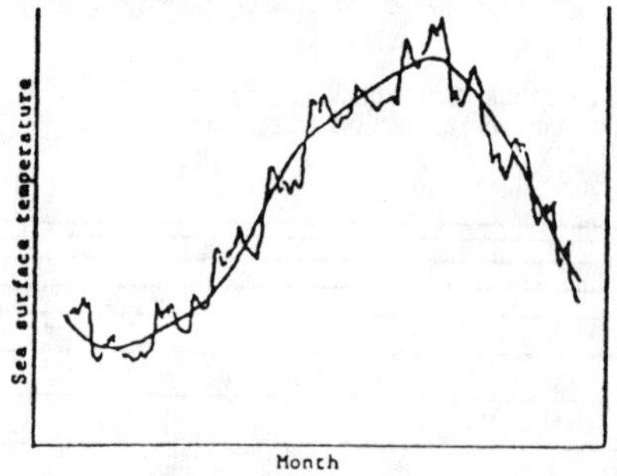

4a. Natural year-to-year temperature fluctuations average out over time to produce a smoothed seasonal variation.

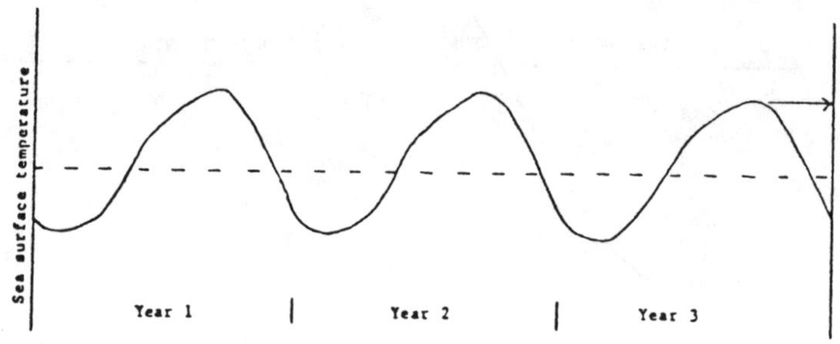

4b. Seasonal variation with no temperature trend

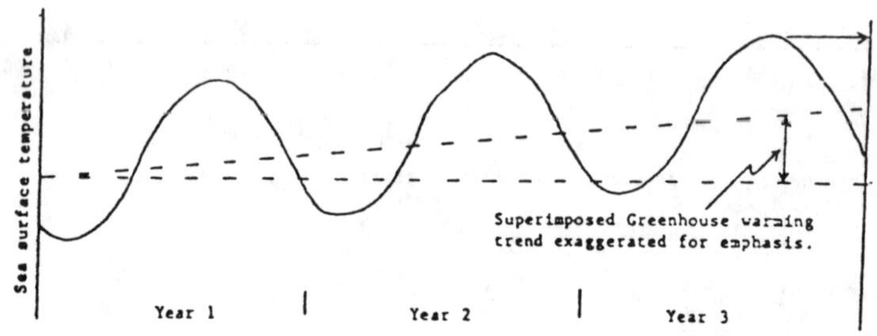

4c. Seasonal variation with assumed Greenhouse warming trend

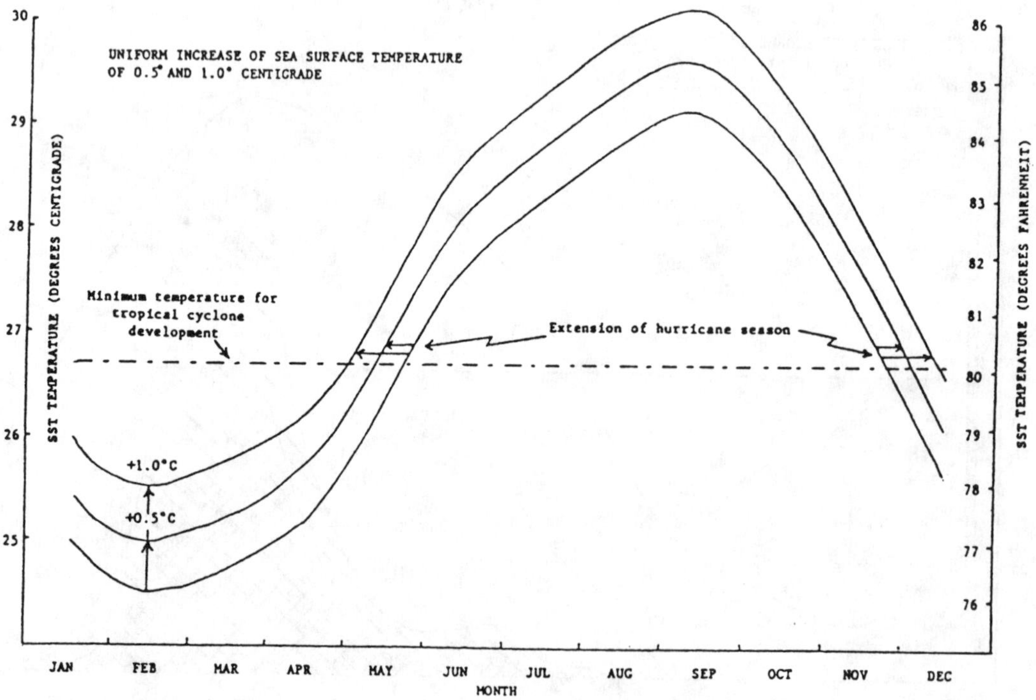

Exhibit 5. Assumed shapes and positions of the seasonal sea surface temperature curves during the transitional climatic period when Greenhouse warmings of +0.5°C and +1.0°C are added. Refer to Exhibits 1b and 1c for addition process.

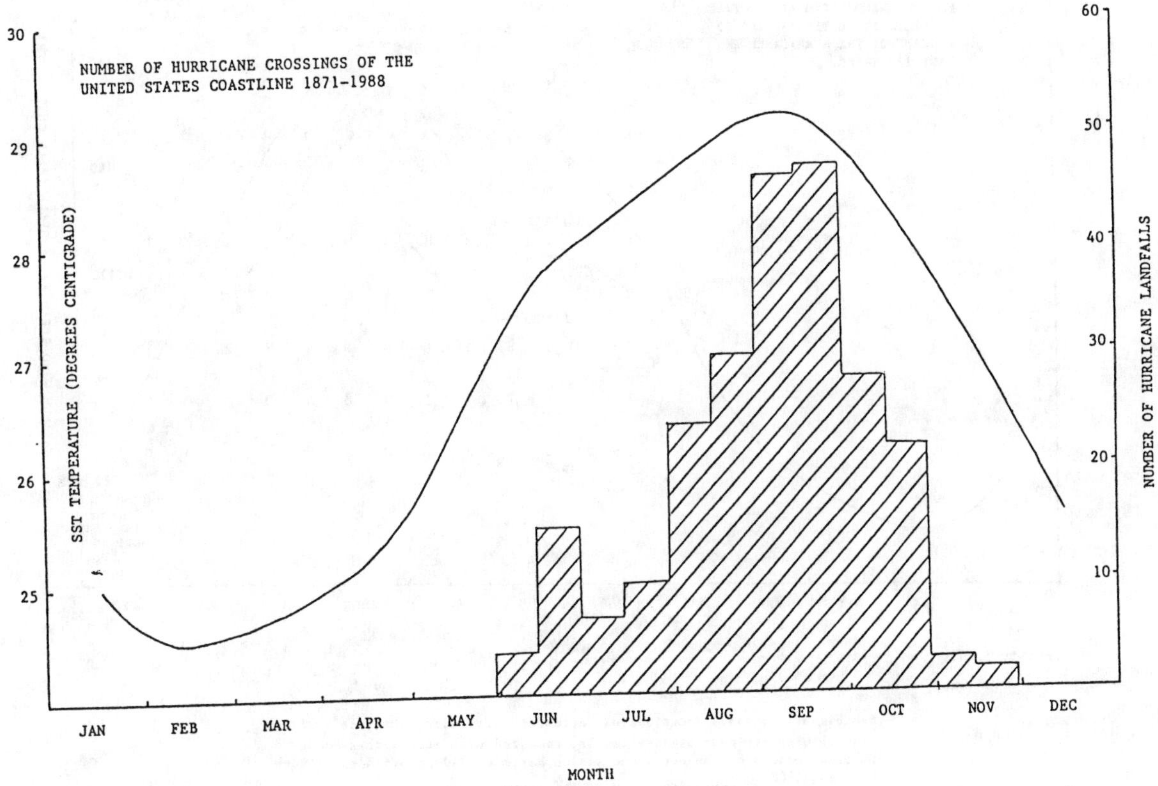

Exhibit 6.

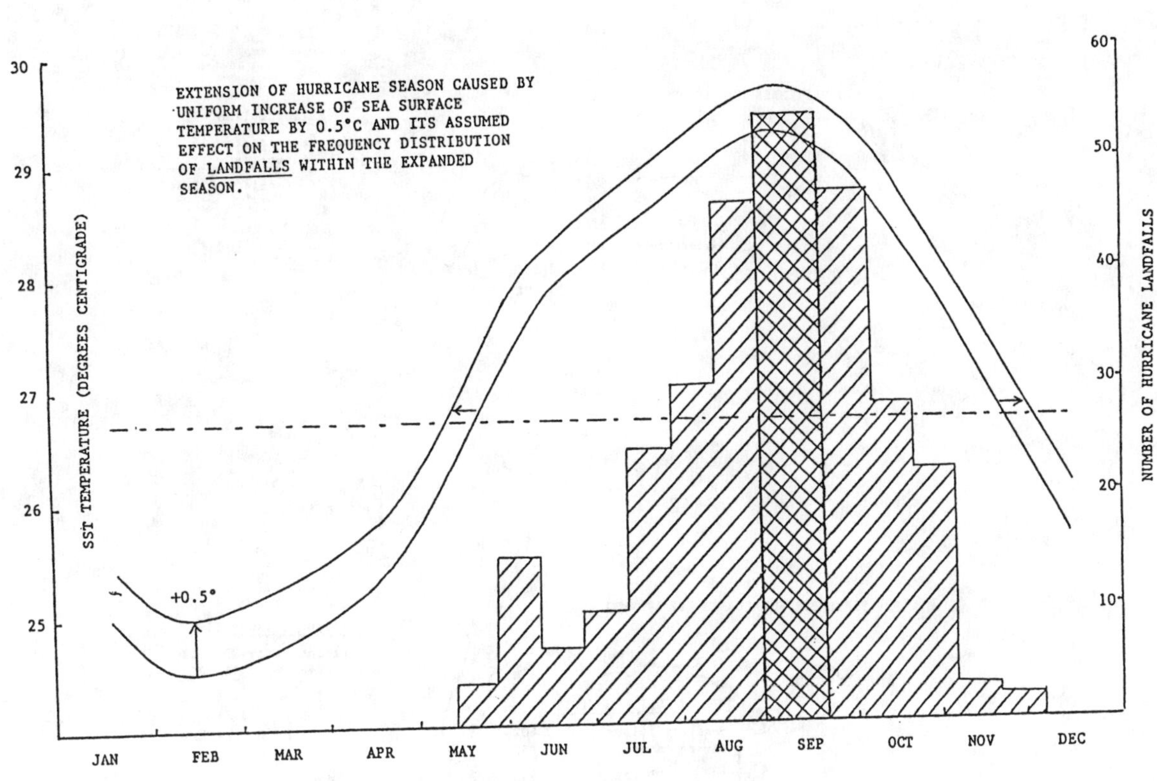

Exhibit 7.

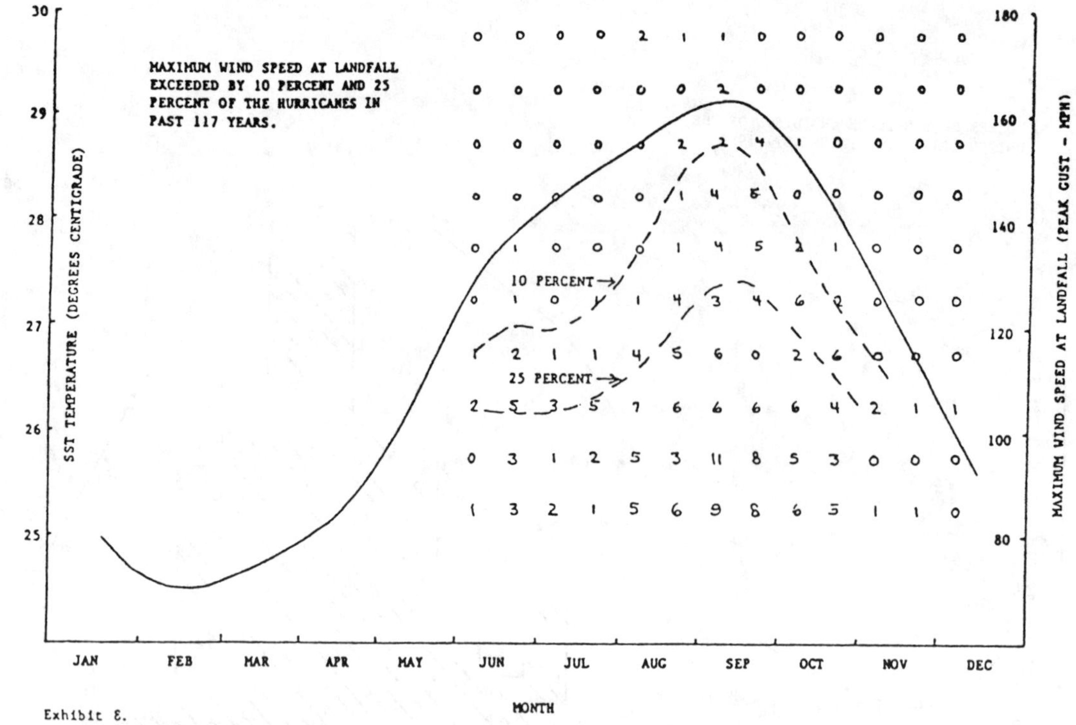

Exhibit 8.

Exhibit 9. Landfall locations of September hurricanes, when sea surface temperatures are near highest levels, compared with storms that occurred in other months of the hurricane season based on 240 occurrences between 1871 and 1988:

Storm Intensity	Non-September hurricanes		September hurricanes	
Moderate	120	81%	60	65%
Severe	28	19%	32	35%
Total	148	100%	92	100%

NON-SEPTEMBER HURRICANES (148 STORMS)

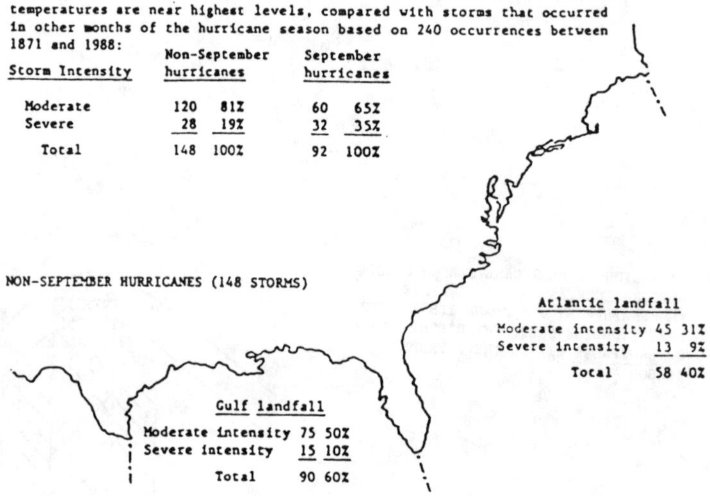

SEPTEMBER HURRICANES (92 STORMS)

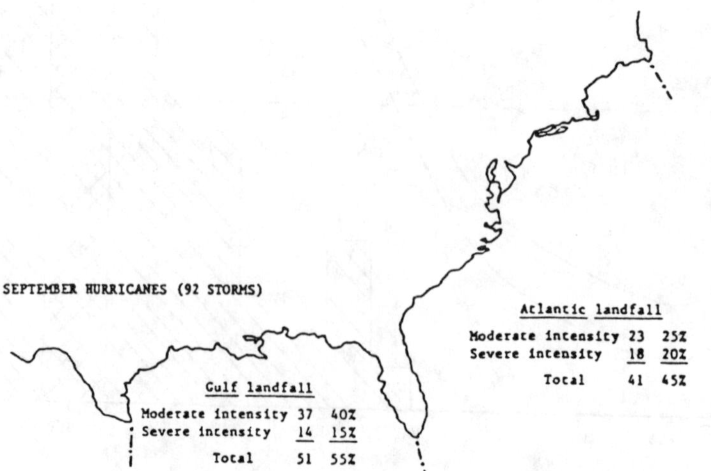

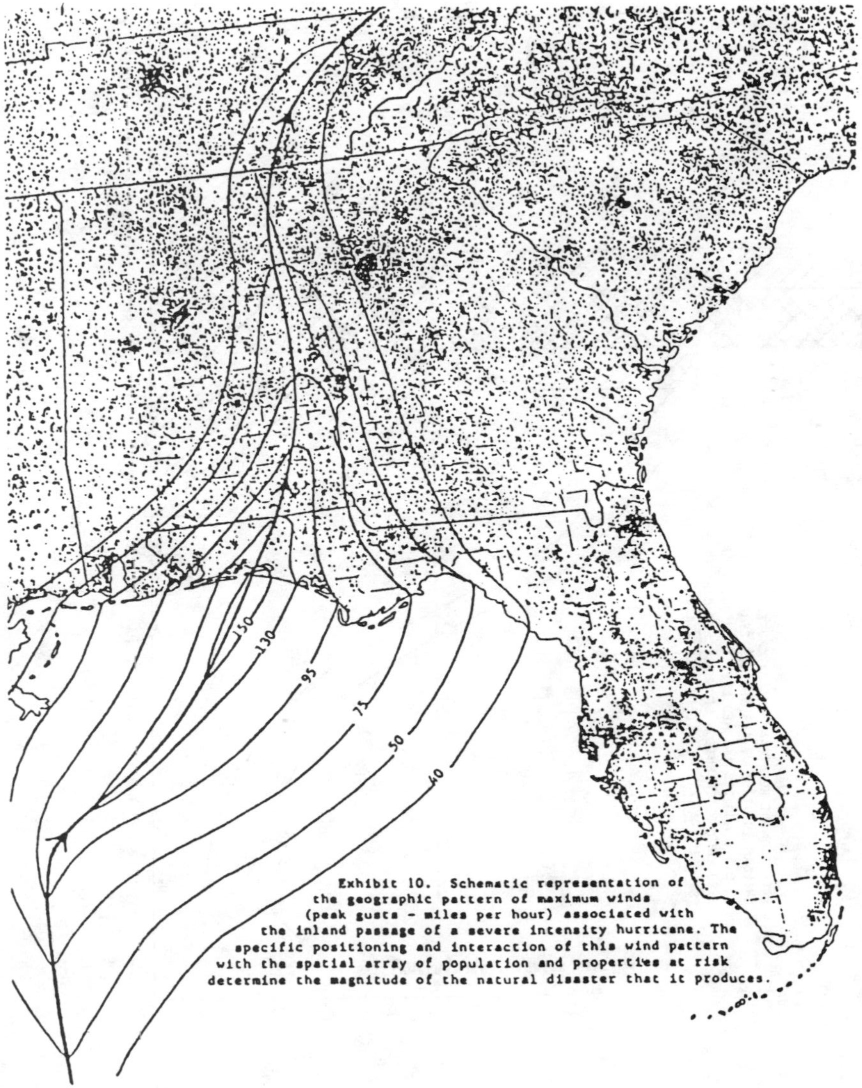

Exhibit 10. Schematic representation of the geographic pattern of maximum winds (peak gusts - miles per hour) associated with the inland passage of a severe intensity hurricane. The specific positioning and interaction of this wind pattern with the spatial array of population and properties at risk determine the magnitude of the natural disaster that it produces.

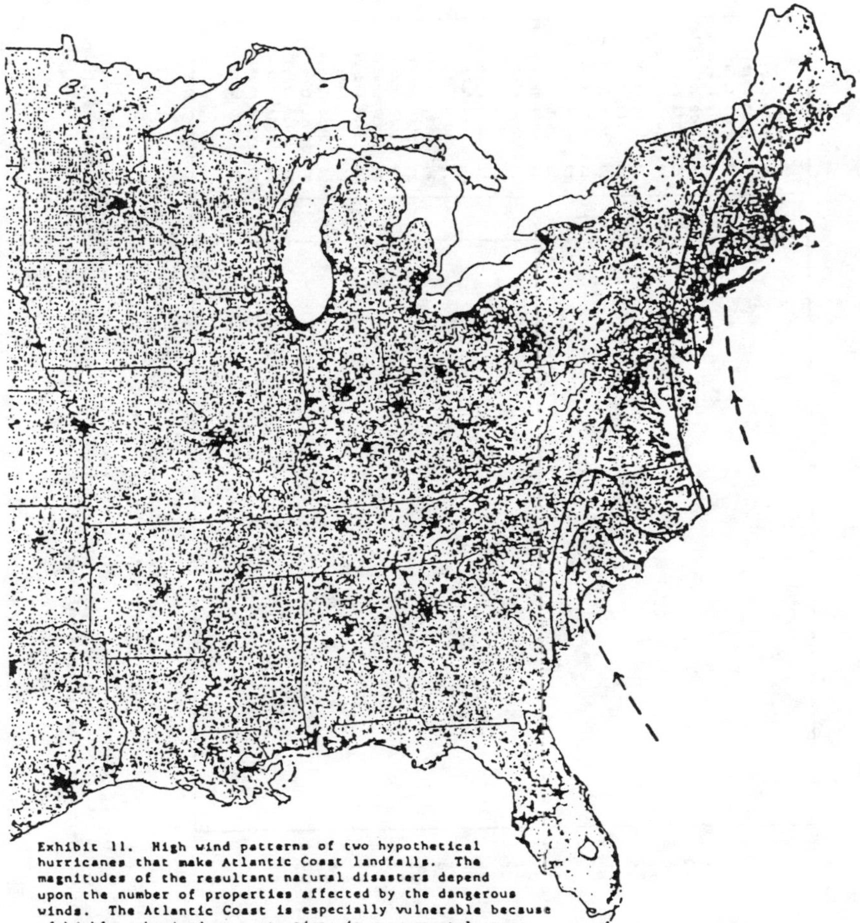

Exhibit 11. High wind patterns of two hypothetical hurricanes that make Atlantic Coast landfalls. The magnitudes of the resultant natural disasters depend upon the number of properties affected by the dangerous winds. The Atlantic Coast is especially vulnerable because of highly urbanized concentrations in many coastal areas. Gulf areas generally have fewer exposures.

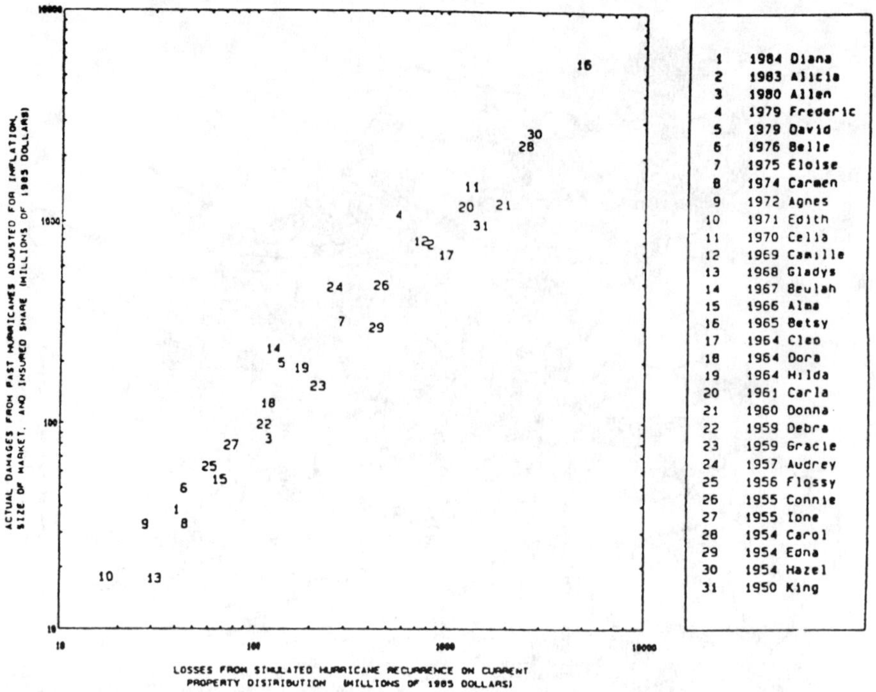

Exhibit 12. Example of the relationship between actual and simulated insured damages in hurricane-caused catastrophes that were coded by the insurance industry between 1950 and 1985. These results are plotted on a double logarithmic chart.

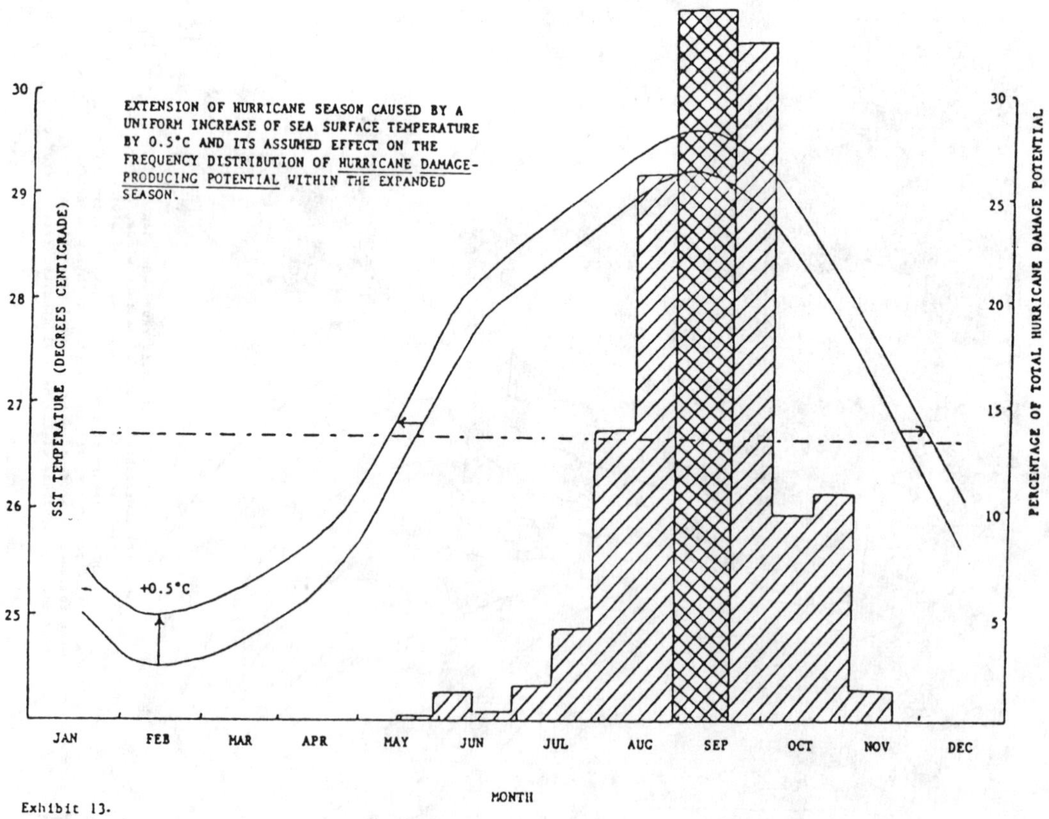

Exhibit 13.

REGIONAL AND NATIONAL EFFECTS OF CLIMATE CHANGE ON DEMANDS FOR ELECTRICITY

by

Kenneth P. Linder
ICF Incorporated

Global warming in response to increased atmospheric concentrations of "greenhouse gases" could affect electric utilities through several different pathways. One of the most critical impact pathways is likely to be the demand for electricity because of the importance of weather-sensitive demands in determining (1) the amount of generating capacity a utility must build and maintain to meet its customers' demands reliably and (2) the most efficient ways of utilizing the utility's resources. These determinations are important to the utility industry because of its capital-intensivity and its long planning horizon; uncertainty in future demands associated with potential changes in climate may pose substantial economic risks to the industry.

A study was conducted to estimate the potential impacts of greenhouse-gas induced temperature changes on U.S. regional and national demands for electricity, and the implications of these changes in demands for utility planning and operations. The study is an extension of an earlier analysis of two case study utility systems performed by ICF Incorporated. In the current study, climate change impacts are estimated for aggregations of utility systems in the Great Lakes, the Southeast, the Southern Great Plains, California, and the United States as a whole. Impacts are measured in terms of changes in utility peak demands and annual energy requirements, generating capacity requirements, electricity generation and fuel use, and capital and operating costs. The results of the study are reported in the U.S. Environmental Protection Agency's forthcoming Report to Congress, "The Potential Effects of Global Climate Change on the United States."

The steps in the analysis can be summarized as follows. First, two alternative climate change scenarios were developed for 47 state and sub-state utility regions. The climate change scenarios are based upon the results of two "transient experiments" developed by the Goddard Institute of Space Sciences (GISS) using a general circulation model of the Earth's atmosphere. The experiments differ primarily in the assumed atmospheric concentrations of greenhouse gases and the degree of volcanic activity. They are designated as "GISS A" and "GISS B." For purposes of this analysis, only changes in monthly mean temperatures from GISS A and GISS B were used to characterize future climate changes. Climate change scenarios were developed using both GISS A and GISS B for 2010, and using only GISS A for 2055.

Second, the weather-sensitivities of demands for electricity were estimated based upon findings of the case studies and analyses of historic temperature and demand data performed by individual utilities. Estimated parameters relating changes in temperature to changes in peak demand and annual energy requirements were developed from these sources. It was assumed that these relationships based upon analysis of historic data would continue into the future and would be valid over the range of temperature changes estimated in the GISS transient experiments (3.4-5.0 degrees C average annual increases in GISS A by the 2050s). Because of several uncertainties associated with the estimation of the weather-sensitivity parameters, an alternative set of values also was used in the analysis. The alternative increased the estimated weather-sensitivities by 50%.

Third, the results of these steps were combined to estimate changes in demands for electricity in 2010 and 2055 under the alternative climate change and weather-sensitivity assumptions. Estimated increases in peak demand on a national basis ranged from 2-6% in 2010, with substantial variation around these results on a regional basis. Changes in national annual energy requirements in 2010 were more modest, ranging from 1-2%. In 2055, peak demands nationally were estimated to increase by 13-20% and annual energy requirements by 4-6%.

Then, these results and a set of utility planning assumptions were used as inputs to a model which simulates utility resource planning and operations. The model was used to evaluate the implications of the changes in demand for generating capacity requirements, generation and fuel use, and electricity production costs. Climate change impacts were estimated by comparing (1) the planning model outputs assuming climate change occurs with (2) "base case" outputs assuming no climate change occurs. In addition to uncertainty being represented by alternative climate change scenarios and alternative weather-sensitivity parameter estimates, the rate of growth in base case electricity demands was varied, corresponding to alternative GNP growth rate assumptions.

Some of the results of the analysis are summarized in Figure S-1 and Figure S-2. Figure S-1 presents the estimated national impacts of changes in electricity demand induced by long-term temperature change on new generating capacity requirements (in gigawatts) and on electricity generation (in billions of kilowatt-hours). In 2010, new capacity requirements induced by climate change increase by 9-19%, or about 25 to 55 GW. The majority of the capacity increase is for peaking capacity as opposed to baseload capacity. This is a significant finding, representing an average increase of up to 1 GW per state.[1] The investment associated with these capacity increases is several billion dollars (in constant 1986 $). By 2055, the change in new capacity requirements increases in percentage terms and represents several hundred gigawatts. Under high GNP growth and high weather-sensitivity assumptions, the estimated increase due to climate change is almost 400 GW.

Annual generation increases are not quite as large in percentage terms, but nonetheless account for several hundred billion kilowatt-hours by 2055. Annual fuel and O&M costs associated with these increases in generation are hundreds of millions of dollars in 2010 and billions of dollars by 2055.

Figure S-2 presents some of the regional results. The figure indicates new capacity requirements in 2055 under (higher growth) base case conditions and under the higher weather-sensitivity climate change assumptions. The impacts range from an increase of about 20 GW in California to over 100 GW in the Southeast. As a percentage of base case demand, the results are particularly significant for the Southeast and the Southern Great Plains. In fact, the potential climate change impacts estimated for these two regions are similar in

---

[1] To provide some perspective, the generating capacity of a large nuclear or coal-fired baseload powerplant typically ranges from 0.6 to 1.0 GW. Natural gas or oil-fired peaking plants are much smaller.

FIGURE S-1
POTENTIAL IMPACTS OF CLIMATE CHANGE ON ELECTRIC UTILITIES UNITED STATES

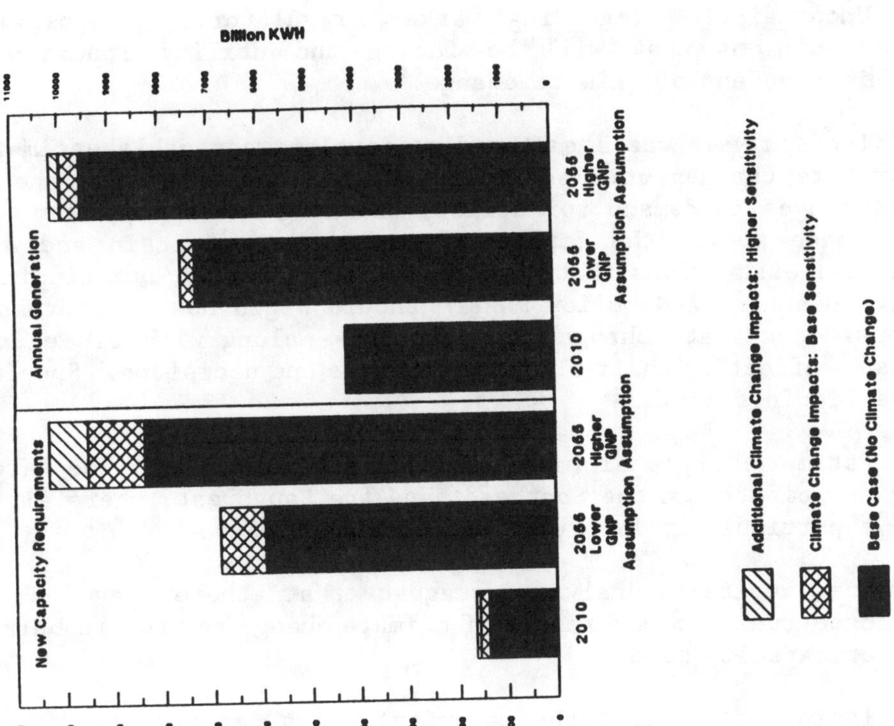

FIGURE S-2
POTENTIAL IMPACTS OF CLIMATE CHANGE ON ELECTRIC GENERATION CAPACITY REQUIREMENTS (2055)

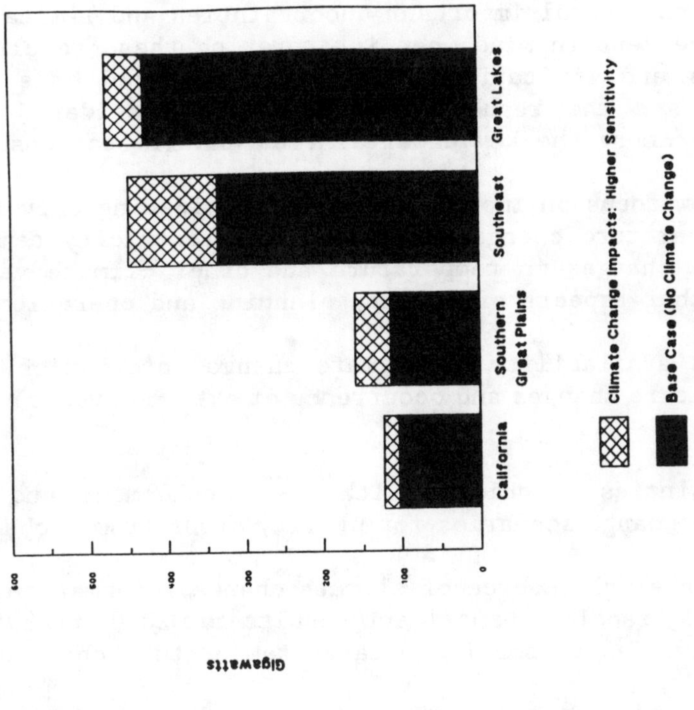

magnitude to the range of base case demand growth uncertainty represented by the alternative GNP assumptions.

There are several important uncertainties and limitations to the analysis that should be kept in mind when interpreting these results. Because of these uncertainties and limitations, the cases analyzed have been referred to as "scenarios," and the results should not be considered as "projections" or "forecasts." Among the key uncertainties and limitations are:

- A narrow focus on impact pathways, considering only the potential effects of temperature change on changes in electricity demand. Greenhouse-gas induced changes in temperature and other climate variables could affect many other aspects of utility planning and operations.

- Limited availability of climate change information (e.g., variations in temperature changes and occurrence of extreme events) important for utility planning.

- Uncertainties associated with the development and characterization of climate change scenarios for utility areas from global "grid square" data.

- Use of a single source of climate change information, although the use of the GISS transient experiment results for 2010 and 2055 indicates relative sensitivities to small and large temperature changes.

- Uncertainties regarding the concept, methods, and assumptions in developing and applying the estimates of parameters representing the weather-sensitivity of demand.

- Uncertainties regarding market, regulatory, technological, and other conditions that will be facing the utility industry in the future independent of climate change.

The analysis was structured to develop reasonable estimates of potential temperature changes on electricity demand, and to assess the implications of these changes in demand for utility planning and operations. Recognizing the points made above, the impact estimates are uncertain and are sensitive to particular assumptions that have been made. However, our findings indicate that utility planners and policy makers should begin now to address more fully and to consider climate change as a factor -- along with other uncertainties and issues -- affecting their planning analyses and decisions. Specific implications of the findings include:

- Estimated regional impacts differ substantially. The largest changes are anticipated in the Southeast and the Southwest, where air conditioning is a particularly important use of electricity.

- The regional analyses suggest that there may be important, new opportunities as a result of climate change for future bulk power exchanges or capacity sales.

- To the extent that the majority of new capacity requirements induced by climate change is for peaking capacity, this implies a new technological and market focus on these types of generating plants.

- A greater focus on short lead-time peaking capacity at the margin affords utilities more planning flexibility to respond to short-term changes in climate or other uncertain conditions.

- Increased electricity demands could increase the difficulty of achieving energy conservation goals which represent one policy option to contribute to atmospheric stabilization.

- Because of the nature and patterns of weather-sensitive demands, climate change could result in different overall fuel mixes for electricity generation than would be expected under base case conditions.

- The impacts of uncertain climate conditions in the long-term potentially poses significant planning and economic risks to the utility industry.

CLIMATE CHANGE: THE IMPLICATIONS FOR SECURITIES UNDERWRITING

Alfred Medioli
Vice President and Manager
Southeast, Regional Ratings
Moody's Investors Service

The topic of this conference is much broader in scope and longer term in its many implications than most people in the financial community are used to thinking about in their daily professional capacities. Global warming is not just another environmental problem that we typically deal with as the subject of a capital expenditure requiring public or private debt financing--the localized need, for example, to improve the quality of effluent release by one city's aging sewage treatment plant, or another city's need to construct a solid waste incinerator because its longstanding landfill is both reaching capacity and polluting the nearby aquifer. Here, today, I think, we might consider such normally quite important projects as no more than short term fixes on a scale relative to a phenomenon that threatens the very existence of our modern, technological way of life--a phenomenon that appears to tell us that the last several hundred years of our development, if not in retrospect wrong headed, have at least led us to a point that we might begin to develop away from very quickly. While I think that our investment bankers and credit analysts and other financial and underwriting professionals have a role in shaping this new development forward, so very much do our scientists and political leaders and philosophers.

Now, given the widely differing backgrounds and professions of people attending this conference, I should explain a little about my particular viewpoint in the financial or underwriting community that I've mentioned several times now. I am a Vice President and Manager in the Public Finance Department of Moody's Investors Service, which is a subsidiary of the Dun & Bradstreet Corporation. At Moody's, very simply, we rate the creditworthiness or risk of corporate and municipal bonds. You are no doubt basically familiar with the rating scale. Briefly, the Aaa designation means highest quality or lowest risk, while Aa or A rating denotes somewhat lesser though still above average credit quality. As the rating scale moves down through the B range it indicates increasingly speculative, below average quality, and then finally a C-rated bond is that which is in imminent danger of default. A bond rating is an essential guide to the purchaser of a debt instrument, given the proportionate relationship between risk and reward, and is thus also a key determinant of the interest rate and cost to the issuer; beyond this, the rating process itself is a critical element of full and adequate disclosure to the potential investor, particularly for municipal bonds where there is no SEC purview. As a rating agency, we are important participants in the process of capital formation. In Moody's Public Finance Department, where we rate municipal bonds, we have ratings on about 50% of the roughly 41,000 general purpose governmental entities in the United States that have issued debt.[1] In order to assign and maintain these ratings, our 120 analysts regularly assess the individual municipality's economy and tax base; its financial structure and performance; the quality and external limitations upon its administrative controls; and, since we are rating debt risk and not quality of life, the amount and structure of the municipality's current debt and future borrowing needs. Given this regular and standardized analysis applied to a very broad array of governmental entities, we at Moody's are keenly aware of the current trends and developments in the matters of finance, capital borrowing, and economic growth affecting local government across the United States.

Now, having reviewed, however briefly, our specific role and point of view in the underwriting community, it is easy to identify a handful of likely "implications" that global warming holds for us and specifically as municipal credit analysts. One is localized economic dislocation as, say, specific industrial activities are altered or stopped or subjected to capital intensive overhauls. This has already occurred to some extent with fluorocarbon propellants in aerosols, although substitute propellants appear to have mitigated any real economic dislocation. Coal mining is a possible and much more serious candidate here if you believe, as some do, that the greenhouse effect can be fought by switching back to nuclear power.

> [As an aside, there appears to be room for much debate here, given the still unsolved and serious problem of nuclear waste, and the possibility for more advanced emission controls intrinsic to fossil fuel generating stations as large point sources--as opposed to the far more complicated problem of auto exhaust.]

A second implication is massive capital outlay across a sector so essential that it is not localized in effect. The need to shift to mass transit to sharply reduce auto emissions, for

---

1 In approximate numbers, 60,000 such entities are able to issue debt.

example, would affect every municipality, and would certainly threaten the economic viability or at least some while others would be in a more advantageous situation. A third implication is the degree to which warming trends, irreversible for the near term, cause coastal flooding that involves significant broad range, if predictable, economic dislocation. A fourth implication is a combination of the first two--localized and sector wide dislocations as all the old rules change, involving the large scale redevelopment of our present patterns of settlement and economic activity across the United States and indeed much of the world.

These are not in any particular chronology or order of importance, and in fact they are so easy to identify only because they are very simple and vague. It is more useful to note that these "implications" are not results of global warming, as much as results of a serious, large scale, and concerted effort to avert global warming, an effort that causes us to change now fundamental ways of doing things. It is safe to say that such an effort would involve some degree of economic dislocation (as opposed to long term economic collapse, if nothing is done); more capital spending and borrowing by both public and private sectors; and increased financial pressures and administrative challenges resulting from economic change and increased capital programs, again both public and private. But how much change and pressure and dislocation, and how much capital borrowing? I think that the debate on the greenhouse effect has just warmed up (if you'll excuse the pun), and frankly it is impossible to say.

The point here is that long term climate change by itself--however probable and dangerous it may be--is irrelevant given the typical 20 year forward horizon for much capital financing. The specific effect upon the creditworthiness of a particular municipality's bonds, for example, is an analytic factor just as immeasurable, unpredictable, and inappropriate as is the possibility of nuclear war or earthquake. To be precise, Moody's A rating on the general obligation debt of Cheyenne, Wyoming, is a measure of bond holder risk reflecting, among other things, the effect of Wyoming's dominant but depressed energy sector, balanced by the city's state capital functions and the location of Warren Air Force Base, which are stabilizing economic factors. The fact that Warren is an MX missile site, such that Cheyenne is certainly a high priority target for Soviet ICBMs, is irrelevant as a credit factor; obviously, such a missile launch would affect not only Cheyenne but everyone in the United States. I do not mean to be flippant, but in this scenario of nuclear war, I expect that the markets would be closed, for good.

Similarly, Moody's does not penalize California credits because of the potential for earthquakes. While an earthquake in a given geographic area may be statistically predictable in a geologic time horizon, the all important particulars--when, where, how much damage--remain very much unpredictable. Moreover, we all point to the San Andreas fault as if earthquakes were limited to that region by some supernatural law. They are not, as the residents of New York's Westchester County discovered early one weekend several years ago, and indeed as Quebec discovered over this past Thanksgiving weekend. We also forget that the continent's greatest earthquakes occurred not in San Francisco but in Charleston, South Carolina, and New Madrid, north of Memphis on the Mississippi, in the 19th century.[2]

Now, why all of this rumbling on about earthquakes? There are two useful parallels to the current topic. The first is that just as earthquakes are specifically unpredictable if more certain over geologic time, the specific attributes of the greenhouse effect--the shape, speed, and detailed causes of this phenomenon--are as hazy as the overall trend seems definite. As I mentioned earlier, the debate, the period of discovery, of developing and weighing options, has only begun. Again, the specific causes and hence means of combatting climate change are by no means certain. A recent front page article in The Wall Street Journal highlighted the scientific debate that methane was perhaps as great a problem as carbon dioxide. If animal and vegetable metabolism are indeed as much or more of an issue than fossil fuel combustion alone, then obviously the shape of our response changes enormously.[3]

There is a second way in which the example of earthquakes and other geographically likely or statistically predictable natural disasters--such as Gulf Coast hurricanes--illuminate the current topic, and this is the extent to which they *have* become credit factors. You usually cannot build a house in a flood plain, unless perhaps on stilts and with insurance. Similarly, parts of California have strict building codes to limit earthquake damage, and most coastal areas in the United States also limit storm damage in the same way. Specialized insurance further protects the exposure of individuals and businesses, and is a legal requirement in some situations. The earthquake or the hurricane has thus become a credit factor insofar as it has been translated into an economic cost, a cost of living or doing business in these areas, influencing capital spending and borrowing, requiring administrative and financial management. (It is as much an economic cost as is central heating in New England, though the urge to stay warm in the winter requires no regulation.) In short, the need to counter the effect of these

---

[2] New Madrid earthquake, 1811-12, estimated to have been more than 8.0 Richter; Charleston earthquake, 1886.

[3] During the Climate Institute sessions, this point was perhaps most vividly illustrated by the TVA's Dr. Barbara Miller, whose presentation described the vastly different scenarios suggested by available climate models.

natural phenomena in geographically susceptible areas has been regulated into the market. To be just as succinct, the need to avert long term climate change will also become a definable credit factor, and a definable subject of capital financing, when it too has entered the market, which I would define as a combination of formal environmental regulation and rational economic response.

I hesitate to enter more deeply the issue of a combined legislative and free market response, first because it is a very broad topic for a political economist, and I think one very broad topic is enough for today. Second is because while we all accept the need for state intervention through environmental legislation, we also know that such laws by themselves are only so effective when economic realities are harsh. The water pollution legislation in the United States over the past two decades has, I think, been very successful; we have seen sewage treatment extended to most towns and cities, and now relatively sparsely settled rural areas are building treatment systems. Moreover, I think that any underwriter, credit analyst, consulting engineer, or municipal official who has worked on a sewer system financing accepts that treatment standards--and hence, capital costs--will be ratcheting upward for the foreseeable future. As a society, we have accepted this cost. But even so, the country's large cities, with the largest and most complicated sewage treatment problems, have only recently begun to complete their initially mandated projects. These projects simply cost too much to be implemented more quickly, and for many years such cities simply failed to meet federal standards. Similarly, New York and Los Angeles missed their federally mandated 1987 air quality deadlines and can only do so by shutting out a large volume of their daily auto traffic, or other economically difficult actions.[4] At the same time, we have seen private industries respond to antipollution legislation--and indeed antipollution itself in terms of equipment and consulting has become big business--while auto manufacturers continue to fight fuel efficiency standards. Whatever your viewpoint as to what mix of government intervention and free market response is most appropriate, I think it is clear that national government must take the lead, as it did in this country two decades ago. Already this summer, Senators Wirth of Colorado and Stafford of Vermont introduced measures in Congress that addressed global warming, and which are expected to set the groundwork for more legislative development in subsequent sessions. Wirth's proposed bill included a 20% reduction in carbon dioxide emissions over 12 years and a nearly half billion dollar program to develop non-fossil energy sources, while Stafford proposed banning some

---

[4] See "Los Angeles Weighs Change to Cut Smog," The New York Times, 12/19/88, p. A14.

chemicals and legislating more efficient automobiles and HVAC systems in new buildings by 1991. Once definite parameters similar to these are set and enforced through rules and charges, municipalities and corporations both can identify courses of action. I think it is at this point--implementation--that the financial or underwriting community becomes most directly and effectively involved.

I would like to close with a mild warning to the participants of this conference. Global warming, insofar as it is a much publicized, internationally visible environmental concern for the future based on projections of recent trends, is nominally reminiscent of the Club of Rome's widely known forecasts about "finite resources" some 20 years ago. Since then, of course, we have experienced the Green Revolution and the rather dramatic example of the price mechanism at work in the matter of oil, which now is relatively more plentiful and cheaper than it was when the Club of Rome's worries were announced. Let us all be more constructive than the Club of Rome, let us not cry wolf, and above all let us not lose sight of the ability of the market to respond once the necessary governmental discipline has been set. That discipline will probably encompass improved energy efficiency; a shift to fuels lower in carbon dioxide emissions; limitations upon specific climate-altering industrial activities; measures to conserve forestation; the encouragement of renewable energy sources; and chemical bans such as those upon fluorocarbons. But whatever shape this disciplinary framework takes, it is an absolutely necessary first step. The implications of global warming, or long term climate change for the securities underwriting community, finally, rest in this arena. Project finance, whether corporate or municipal, is only tactical, in the sense that it is a means to a larger, strategic goal. We are very much only at the beginning of formulating that strategy.

IMPLICATIONS OF CLIMATE CHANGE
FOR THE ENVIRONMENTAL ENGINEERING AND CONSTRUCTION INDUSTRY

Joseph F. Silvey

A. INTRODUCTION

As a representative of the environmental engineering community, I first want to define the term "environmental engineer", which appears in the title of my paper. An environmental engineer is first of all a futurist who is capable of foreseeing the future and accurately describing it for his clients. His clients are usually anxious to anticipate future regulatory policies so that they can undertake their long term projects with limited financial exposure and great certainty of success. Having accurately predicted the future, the environmental engineer must then become part negotiator/part arbitrator to serve his client in working out compromises with regulators and lawyers opposing his client's project. Throughout all of this the environmental engineer must retain the instincts of a riverboat gambler, knowing when to take a position and when to fold his hand, understanding at all times that his fortunes are very much dependent upon the significant uncertainties of the technical discipline in which he works.

Having established the framework in which I and my colleagues function every day, I intend to spend a little time advising you how we are beginning to address the problems likely to be posed by the issue of climatic change. All the aforementioned skills of the environmental engineer and his counterparts in the construction community will be necessary to address this new issue.

B. THE INDUSTRY'S CONTRIBUTION TO CLIMATIC CHANGE

The industrial clients served by the environmental engineering and construction community are rarely able to approach an environmental problem without having some responsibility for the problem in the first place. This is true again in connection with the climatic change issue. Climatic change, to the extent it is occurring, is directly related to a buildup of carbon dioxide, methane and nitrous oxides and other gases in the atmosphere. Although we humans are responsible for carbon dioxide emissions into the atmosphere with every breath that we expel from our lungs, and although our numbers are increasing in astounding fashion, we simply don't have as much of an impact on our climate as does the combustion of fossil fuels which accounts for approximately half the carbon dioxide in the atmosphere. Coal fired electric generating plants by themselves produce five to ten percent of the carbon dioxide levels, with other industrial operations contributing their proportional share. If a global warming trend is moving upon us, as scientific data suggests, then our industrialized society is clearly contributing to this trend in a very significant way.

As a resident of California, I feel obliged to admit that the human race also contributes directly to the build up of carbon dioxide through our fascination with motor vehicles. Although estimates of worldwide emissions from sources in any category must be viewed with some caution, it appears that the motor vehicle has as big an impact on climate change trends as does the coal fired power plant. If we are going to address the causes of global warming fairly, then the automobile and its use will be deserving of intense scrutiny that should worry the most laid back of California's freeway drivers.

C. NUCLEAR POWER SCENARIO

As some journalists have begun to note, the threat of climate change has begun to win some converts back to the side of nuclear power. As the old saying goes, "the devil you know is better than the devil you don't". In many ways we have come to understand the risks and impacts caused by nuclear power facilities. An intelligent society should recognize that properly controlled and regulated nuclear energy production must be a part of any electricity generating matrix that exists either today or in a future time when the global warming issue could be a concern.

Although the environmental engineering and construction community anticipates that the nuclear power industry will begin to make some type of comeback before the turn of the century, no one really can tell what that comeback will be. The most ardent proponents of nuclear power have constructed scenarios under which large numbers of nuclear generating stations would be built in order to replace today's aging coal fired power plants. Those scenarios are as unlikely as is a total non-nuclear future. However, as concerns about climate change continue to develop, the work community I represent expects to again be asked to address the environmental, engineering, and construction issues associated with nuclear power.

D. LIKELY SOCIETAL/INDUSTRIAL CHANGES PROVOKED BY CLIMATIC CHANGE

Many parts of this country are now undertaking major highway construction and expansion programs. However, if climatic change is occurring, society will soon be aggressively working to discourage automobile use and keep the highways as empty as possible. Mass transportation, work at home job situations, computerized buying and massed delivery of goods, electric automobiles, and the like will all become part of our future.

Try as we might, we are not going to be able to wish coal fired electric generating stations out of existence. Given the abundance of coal throughout the world, it's hard to imagine that such facilities will not continue to come into existence in many locales, even if climate change does happen. However, I expect that many of these facilities will use control devices and advanced combustion systems designed to limit the amount of carbon dioxide that escapes to the atmosphere. While these technologies don't exist today, some in the environmental engineering community are already beginning to plan for their development.

Although coal fired and nuclear power are likely to play important roles in our future, so also will conservation and new forms of electricity production. The oil crisis of ten to fifteen years ago fostered interest in these last two items, but the drop in the price of oil and the costs of some of these new technologies halted much of their development. Cogeneration has made its mark in some parts of the country, and we do have wind systems and solar thermal systems producing power. But, in reality alternative energy generation is still an infant industry. It is an infant industry that also has its share of environmental problems. Nothing in life is free. Wind systems have noise and visual impact problems that can make their use uncomfortable to many. Cogeneration systems make very efficient use of the forms of energy that are produced, but they still burn fossil fuel, and still contribute to some of the standard atmospheric pollution concerns. Even apparently passive systems such as the solar energy devices impact the environment by using significant quantities of land and requiring exotic materials which must be mined before use. Many problems will have to be solved before greater use can be made of these technologies.

E. WHAT MUST THE ENVIRONMENTAL ENGINEERING AND CONSTRUCTION COMMUNITY DO TO PREPARE FOR A FUTURE IMPACTED BY CLIMATIC CHANGE?

Our society is so regulated, it's natural to first talk of the regulatory changes that will be needed to enable climate change to be properly addressed. There are no regulations in effect today to address this problem. New regulations on pollutant emission levels will probably appear quickly, once society decides to address climate change issues, as will regulations on the use/siting of certain types of energy production and industrial facilities. Climate change issues will make their way into environmental impact statements, and a new segment of the environmental engineering community will evolve, made up of scientists and engineers with special expertise in atmospheric chemistry. The environmental engineering and construction community will have to be a part of this and help shape this regulatory development, and obviously must be knowledgeable about what does occur so that its clients can be properly served.

We are dealing with many unknowns when we deal with the problem of climate change. That's why the environmental engineering community must become an active participant in the research and development efforts needed to better understand the problems we may face and the solutions that are available to us. R&D monies need to be spent on atmospheric models, emission control technologies, improved combustion systems, energy efficient applications of electricity, new energy production technologies, improved mass transportation concepts, and the like. The environmental engineering community will be the recipient of funds from governmental entities and private industry seeking support in these research and development areas, and the environmental engineering community also should be prepared to provide some financial and resource support on its own to the effort.

Finally, as a total society we will all have to reevaluate the way in which we live if we are to avoid the most serious consequences of climate change. Transportation patterns, energy consumption, work/home selections, even entertainment choices will have to be viewed from new perspectives as the environment we live in adjusts to the changing chemistry of our atmosphere. The environmental engineering and construction community is part of society as a whole and should be able to help society modify its behavior by the manner in which it develops new projects for society and by the manner in which it presents these new projects to society.

F. AVOIDING MISTAKES OF THE PAST

I have been in the environmental business since I left the United States Navy almost two decades ago. I've seen several environmental problems arise and be dealt with by the regulatory agencies and private industries of the world. If we gave ourselves a report card for our performance on these past problems we probably wouldn't be too satisfied with our grades. We would give ourselves an "A" for effort and a perfect attendance mark, but in solving problems and keeping schedules and making solutions be cost effective we'd get a "C" at best. If climate change is indeed upon us, we owe ourselves a better performance on this problem than we have exhibited in the past.

If we have learned anything from the past, we should have learned about the dangers of overregulating an environmental problem. I say this not only as an individual who earns his living doing profitable work for industrial clients, but also as an individual who likes to go home in the evening and tell his children that his job can make our world be a better place in which to live. More often than I'd like to remember I've been asked to participate in environmental studies that were needed only to satisfy a bureaucratic regulatory requirement and not to make anything better or prevent anything from getting worse. We can't afford to treat climate change this way. The resources we need to apply to this issue are too precious to waste on mindless regulations.

Industry must be a willing participant in the solution to any climate change problem that we must face. Unfortunately, overregulation is often the result of industry intransigence when it comes to addressing an environmental issue. Then, overregulation leads to further intransigence and so on, in an endless spiral that only seems to end when a new environmental issue catches the public's attention. If today's industry does not take a more progressive view of the climate change issue than it has of other environmental issues, then the future we face is difficult. There are potential long term and perhaps irreversible aspects to the global warming situation that will require concerted action. A spiral of overregulation and intransigence will prevent that concerted action at the time when it is needed the most.

Until recently American society has been of the belief that anything could be fixed by pushing enough money towards the problem, and for a long time we seemed to have enough money to push at any problem. That is no longer the case. As we approach the environmental problems of the future we are going to have to do so while recognizing that the financial resources we can bring to bear on these problems are finite. We have to do other things with our financial resources including educating our children, caring for the sick, taking care of the homeless, and assuring our national defense. We cannot approach climatic change issues with a budget that knows no bounds. Fiscal controls must be a fact of our lives in everything we do, even in solving environmental problems.

Finally, we all must recognize that society is likely to change as the climate change issue evolves. We must be prepared to change ourselves. We can't adopt the "not in my backyard" or "NIMBY" philosophy and hope that someone else is going to pay the price necessary to make climate change disappear. The problem will impact all of us and all of our lifestyles. The environmental engineering and consulting community, as part of society, will have an opportunity to influence society's perspectives towards this new problem in an educated and professional way. This professional community cannot refuse to accept such responsibility.

G. CONCLUSION

In closing I'd like to note that my comments do not carry the scientific significance of many of my colleagues who are presenting papers at this conference. Moreover, I cannot report any significant research and development activities on the part of the environmental engineering area. The clients that we serve and the companies in my industry are only beginning to realize that the climate change issue may have an impact on their future. I hope through this paper to catch some of the attention of my colleagues so that their efforts to address global warming issues will be applied more quickly than otherwise might be the case. I hope also to provide some assurances to those of you having interest in this conference that the environmental engineering and construction community will be able to as effectively apply its resources to this new problem as it has to problems of the past.

Potential Coastal Effects of Climate Change in the Caribbean

by F. J. Gable and D. G. Aubrey
Woods Hole Oceanographic Institution

Introduction to Climate Change

Changes in climate are the norm when one studies the history of the earth. Examples of these changes include the glacial epochs and the more recent and largest contemporary climatic variation of the El Niño/Southern Oscillation. Many paleoclimatic records illustrate the results of natural variability. The present and arguably the future climate will be influenced by two components, those that are natural and those that are man-induced. Together, these forcings will cause future climate to evolve in uncertain ways (Schneider, 1987). It is the human activated contemporary sources/effects of climate change that the policy community must attenuate or adapt to. The cause of future global warming is presumed due in part to anthropogenic buildup of carbon dioxide and other radiatively-active trace gases, such as methane, nitrous oxide, tropospheric ozone, chlorofluorocarbons and water vapor in the atmosphere (Jaeger, 1988A). Carbon dioxide content of the atmosphere has increased by nearly 25 percent since the industrial revolution of the 1800's (Schneider, 1987), and more than 9 percent over the past 30 years (Rowland, 1988). Deforestation and increases in fossil fuel burning have contributed greatly to these increases (Woodwell, 1986).

Although the earth was about as warm in the 1930's and 1940's as it is today, the earlier warming was concentrated at high northern latitudes, while the recent warming is more global (and not in keeping with computer model predictions). The most recent global warming has been most discernible in the southern hemisphere and parts of the tropics (Parker and Folland, 1988). The warming since 1947 however, is much more significant in middle and low latitudes when year-to-year variability of real world global temperature trends, both land and marine are considered (Jones et al., 1988).

Future temperature trends resulting from trace gas and carbon dioxide buildup are unlikely to be uniformly distributed over the globe (Jones et al., 1988). The global warming in 1987, approximately as warm as 1981, the warmest year in the history of instrumental records, is attributed to a large increase in temperature (more than 0.4° Centigrade) at low latitudes (23.6°N - 23.6°S), which includes all of the Caribbean (Hansen and Lebedeff, 1988). Global warming can be compared with the last ice age when the average global temperature was 5°-6° colder than today.

Difficulties in Estimating Regional Coastal Effects of Sea-Level Changes

Global relative sea-level changes during the past century indicate a rise of abut 1.2 - 1.5 mm/yr (Gornitz and Lebedeff, 1987), part of which is due to sea levels and part to land levels. The global warming process generates higher relative sea levels by two major processes: thermal expansion of the upper ocean layers and an increase in meltwater from continental and alpine ice sheets (Aubrey, 1985). Historical relative sea-level data for the Caribbean (see Figure 1) illustrates a considerable variability in the magnitude and direction of sea-surface movements. One explanation is that the land level beneath the tide gauges is rising in some places and sinking in others (Aubrey et al., 1988). Therefore, the data indicate local patterns of tectonic earth movements have a considerable imprint upon the relative sea-level trends from historical records.

Several distressed coastal areas have been identified within the Caribbean region. In the Leeward Islands of the Lesser Antilles on the island of Nevis, field methods indicate that this island has been affected by relative sea level (Giese, 1988). Similarly, recent field investigations indicate that Tortola (British Virgin Islands) is also experiencing pronounced erosion in selected areas, in part, due to relative sea-level rise. Recent reconnaissance of Antigua reveals similar findings. On the Windward Islands there are similar reported examples (Gable, 1987). A relative rise in sea level does not in itself cause beach erosion; however, rising sea level (relative to the land) may be the single most important element fostering the erosion process. While erosion problems are often identified on a qualitative basis, surveys indicate a world-wide retrogression of sandy shorelines (Bird, 1987; Jaeger, 1988B), there is not enough quantitative evidence on a regional basis to quantify the rates and causes of erosion in the Caribbean.

Aubrey et al. (1988) have calculated average annual change of relative land level for Caribbean Islands from tide gauges. Most of the tide-gauge data depict subsiding land level (increasing relative water level) at a rate of 0.25 cm per year. Based on tide gauge station records, relative sea level throughout the wider Caribbean may rise on average between 5 cm and 25 cm by 2025 (UNEP, 1988). In general, an areal average relative sea-level rise on the order of approximately 10 cm over the next 40 years is anticipated (UNEP, 1988). Hoffman et al. (1986) have estimated that the *global* sea-level rise to be generated by the anticipated climatic changes would range from greater than 10 cm to less than 21 cm by 2025; superimposed on this global signature is a regional tectonic trend.

Sea-level series obtained from tide gauge station measurements are relative; that is, the series depicts changes in sea level relative to the adjacent land at that location. Several researchers have attempted to separate eustatic, oceanographic, and tectonic components of relative sea-level change derived from tide-gauge records. Tide-gauge data, however, are insufficient to permit complete separation of eustatic and tectonic effects on relative sea levels (Aubrey and Emery, 1986). Therefore, the projections of global sea-level rise do not separate quantitatively the effects of localized tectonic activity. Consequently, quantitative studies on a local scale are necessary in order to take into account geological and oceanographic factors such as tectonic movement and its influence on relative sea-level change, beach structure, offshore bathymetry, wave regime, and storm frequency and magnitude. Storm frequency and magnitude is particularly important because most erosion will take place during storm events, following an increment of short-term sea-level rise (Komar and Enfield, 1987).

Caribbean Region Recommendations

Several recommendations can be made for closely monitoring the change in sea level in the Caribbean region. First, establish a regional network for acquiring and archiving scientific data in a manner that allows for easy access. Included within the network should be additional monitoring stations for relative sea level and related meteorological measurements. The existing float-in-well tide gauge systems and paper punch tape recording devices are subject to inaccuracy, unreliability and time analysis problems which make the value of definitive sea-level observations obtained at any one location questionable and uncertain (Bossler et al., 1986). Moreover, in order to determine credible averaged long-term trends over a wide region or area a large enough ensemble of measurement series is necessary (Bossler et al., 1986). For a total systems approach, a Next Generation Water Levels Measurement System (NGWLMS) is being developed. This concept, employing near real-time data collection, and transmission will allow for greatly improved quality control procedures. The single measurement of a NGWLMS will be good to the 1± cm level, a tenfold improvement over the existing measurement type. The choice of which device should be used in the Caribbean depends on local expertise, finances, and environmental factors.

Second, other terrestrial measurements using space and satellite techniques should be employed to detect small changes in relative position. Very Long Baseline Interferometry (VLBI) and Satellite Laser Ranging (SLR) are two methods. Another more readily transportable and economic measuring program is based on the Global Positioning System (Done, 1988). While at present these techniques are in the research, development and evaluation period, they will likely become "on-line" in the Caribbean in the next decade. All of these systems should be used in concert with tide-gauge data to separate land-level from sea-level changes, for improved prediction capability.

Monitoring land-use changes and establishing land-use criteria particularly at the land-sea interface are keys to maintaining and preserving ecosystems that may be vulnerable to climate changes. Similarly, it is necessary to be able to identify the time-scales and magnitude of the large scale oceanographic and geological effects and to discern these from the wholly local signals, especially regarding coastal planning, land use controls and direct engineering solutions (Bossler et al., 1986).

The need for investigation of climate impacts on a regional scale is profound. For at least two important variables, precipitation and surface air temperature, there is little agreement between model simulations of present climate and real world regional climate(s) (Grotch, 1988). The research community needs improved understanding of the contributors of regional causes of relative sea-level change including both man-induced and natural long-term changes in atmospheric pressure, ocean currents, temperature, and wind patterns as well as land subsidence, emergence and tectonic movements.

Investigations to improve understanding of the processes that shape the shoreline (for instance, varying freshwater and sediment input from rivers and changes in storm climates) are needed. Local and regional governments and planning agencies that deal with coastal conservation and management, should give immediate recognition to the inevitability of climate change and its ancillary effects, which may include increasing relative sea-level rise. For countries having sandy shorelines, quantitative studies are necessary to ascertain the local effects of the likely scale of erosion due to relative sea-level rise, since the primary economy for many areas — tourism — is at risk.

Tourism, a major economic activity in most of the region, may be negatively impacted by man-induced global change. These changes may foster grave repurcussions on the expectation of economic development and growth (Blommestein and Singh, 1987). Caribbean tourism depends primarily on sandy beaches. Other factors such as shoreline alterations, pollution of coastal waters, and ecosystem/habitat destruction may exacerbate the anticipated climatic change effects upon wider Caribbean coastlines.

In closing, good reliable long-term data and selected *in situ* quantitative analyses would allow the research community to better predict the potential coastal effects of climate change on Caribbean shorelines. The uncertainty of contemporary climate conditions warrants the need for more study.

Acknowledgements

Research supported by the Richard King Mellon Foundation and the Coastal Research Center of the Woods Hole Oceanographic Institution. WHOI Contribution No. 6964.

SELECTED REFERENCES

Aubrey, D.G., 1985. Recent Sea Levels from Tide Gauges: Problems and prognosis. *In: Glaciers, Ice Sheets and Sea Level: Effect of a CO_2-Induced Climatic Change.* DOE|ER|60235-1, U.S. Department of Energy, Washington, DC, pp. 73-91.

Aubrey, D.G., and K.O. Emery, 1986. Relative Sea Levels of Japan from Tide-Gauge Records. *Geological Society of America Bulletin,* Vol. 97, pp. 194-205.

Aubrey, D.G., K.O. Emery and E. Uchupi, 1988. Changing Coastal Levels of South America and the Caribbean Region from Tide-Gauge Records. *Tectonophysics*, Vol. 154, pp. 269-284.

Bird, E.C.F., 1987. The Modern Prevalence of Beach Erosion. *Marine Pollution Bulletin*, Vol. 18, No. 4, pp. 151-157.

Blommestein, Erik, and Naresh Singh, 1987. The Impact of the Climatic Changes on Tourism. Paper presented at Annual General Meeting of the Caribbean Conservation Association, Tortola, British Virgin Islands, Sept. 9-12, 1987, 5 p.

Bossler, John D., John G. Hayes, Thomas Pyle, John M. Diamante, 1986. Global Sea Level Measurement Program. *In: Proceedings of Oceans '86*, pp. 1365-1371.

Done, Peter, 1988. Caribbean Tectonic Observations. *The Geographical Journal*, Vol. 154, No. 1, pp. 49-55.

Gable, F., 1987. Changing Climate and Caribbean Coastlines. *Oceanus*, Vol. 30, No. 4, pp. 53-56.

Giese, Graham S., 1988. Physical Condition of the Nevis Shoreline. Field Report to the Nevis Historical and Conservation Society, Coastal Research Center, Woods Hole Oceanographic Institution, (unpublished).

Gornitz, V., and S. Lebedeff, 1987. Global Sea Level Changes During the Past Century. *In: The Society of Economic Paleontologists and Mineralogists, Special Publication No. 41, Sea Level Change and Coastal Evolution*, pp. 3-16.

Grotch, Stanley L., 1988. Regional Intercomparisons of General Circulation Model Predictions and Historical Climate Data. U.S. Department of Energy, Technical Report 041, 291 p.

Hansen, James, and Sergej Lebedeff, 1988. Global Surface Air Temperatures: Update Through 1987. *Geophysical Research Letters*, Vol. 15, No. 4, pp. 323-326.

Hoffman, J.S., J.B. Wells and J.G. Titus, 1986. Future Global Warming and Sea Level Rise. *In:* Sigbjarnarson, G., (ed.), *Iceland Coastal and River Symposium, Proceedings*, pp. 245-266.

Jaeger, Jill, 1988A. Developing Policies for Responding to Climatic Change. WCIP-1, WMO/TD-No. 225, *World Climate Impact Studies Program, World Meteorological Organization,* 53 p.

Jaeger, Jill, 1988B. Anticipating Climatic Change: Priorities for Action. *Environment*, Vol. 30, No. 7, pp. 12-15 and 30-33.

Jones, P.D., T.M.L. Wigley, C.K. Folland and D.E. Parker, 1988. Spatial Patterns in Recent Worldwide Temperature Trends. *Climate Monitor*, Vol. 16, No. 5, Annual 1987, pp. 175-185.

Komar, Paul D., and David B. Enfield, 1987. Short-Term Sea-Level Changes and Coastal Erosion. In: *Society of Economic Paleontologists and Mineralogists Special Publication No. 41.*, pp. 17-27.

Parker, D.E., and C.K. Folland, 1988. The Nature of Climatic Variability. *The Meteorological Magazine,* Vol. 117, No. 1392, pp. 201-210.

Rowland, F. Sherwood, 1988. Chlorofluorocarbons, Stratospheric Ozone, and the Antarctic Ozone Hole. *Environmental Conservation*, Vol. 15, No. 2, pp. 101-115.

Schneider, Stephen H., 1987. Climate Modeling. *Scientific American*, Vol. 256, No. 5, pp. 72-80.

Woodwell, George M., 1986. Forests and Climate: Surprises in Store. *Oceanus*, Vol. 29, No. 4, pp. 71-75.

United Nations Environment Programme (UNEP), 1988. Draft Report on Implications of Climatic Changes in the Wider Caribbean Region. UNEP Caribbean Regional Coordinating Unit, Kingston, Jamaica, 183 p. and Annexes.

Figure 1. Geography of the Caribbean.

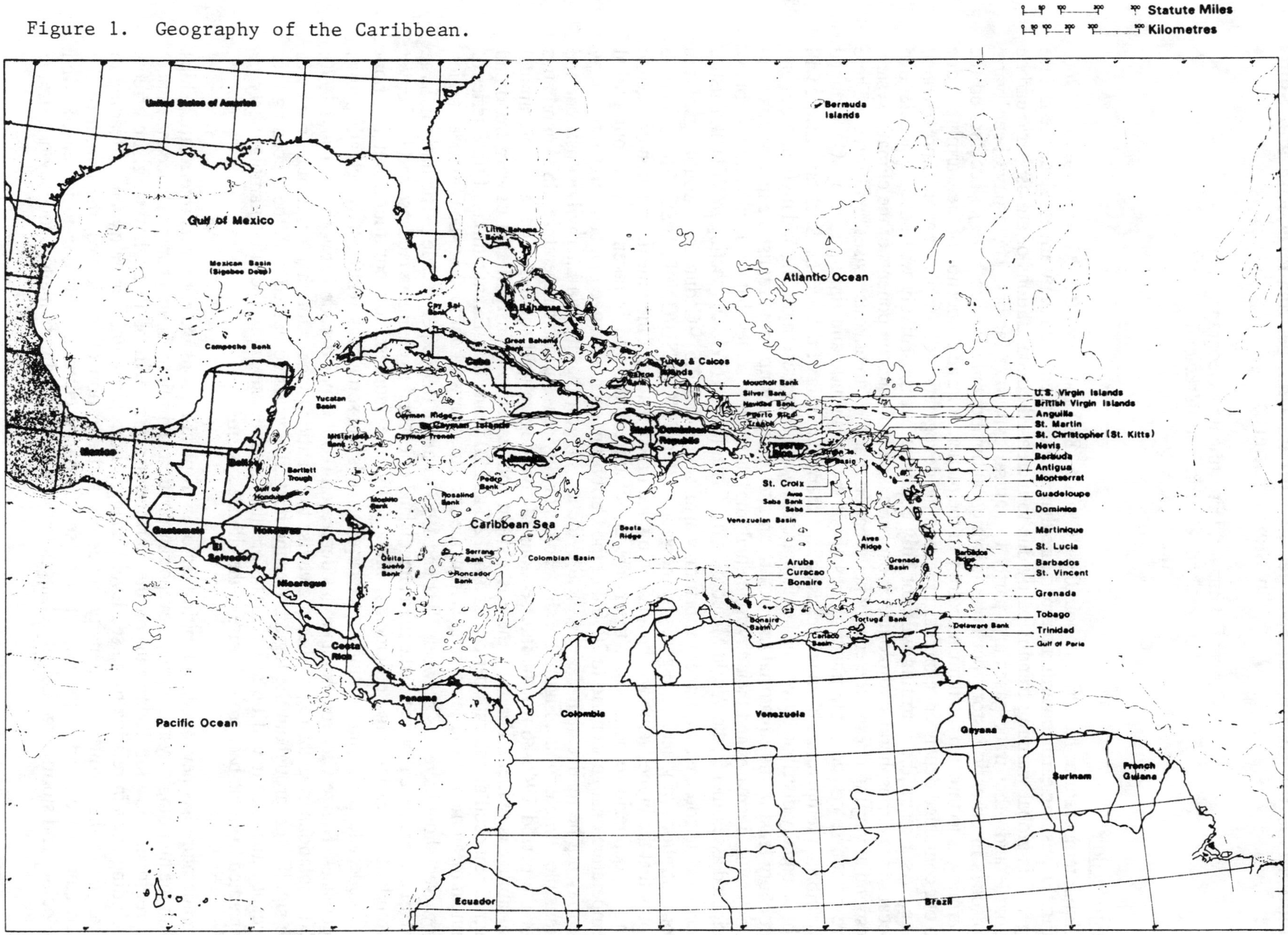

Implications for Caribbean Societies of Climate Change, Sea-level Rise and Shifts in Storm Patterns.

Orman E. Granger, Department of Geography
University of California at Berkeley
Berkeley, CA. 94720

Introduction

Global temperatures have risen about 0.6 °C since 1880 and about 0.3 °C since the mid-1960s (Fig. 1), pushing earth towards a temperature that may compete with the warmest since the last ice age. This global warming trend is partly due to increased concentrations of atmospheric carbon dioxide and other trace gases and partly to natural causative factors. There have been warm episodes in climatic history but none in which anthropogenic factors have so accelerated the warming as to threaten the human ecology of the planet. A large proportion of the warming is due to our continued dependence on fossil fuel and the devegetation of the planet at an unprecedented rate. As a result,the warming will not only continue but most likely will become accelerated because the atmospheric, hydrologic, cryospheric, and biologic components of the climatic system respond at varying rates to the energy perturbation that the "greenhouse gases" will force. . There are of course considerable uncertainties about the magnitude and timing of the CO_2 induced equilibrium temperature change because of the thermal inertia of the oceans. Such a warming will not be equally distributed over the hemisphere but will be greatest in middle and high latitudes and probably least in the equatorial regions. In an attempt to set up some realistic expectations as to temperature increases and their environmental impacts on time scales that are significant for policy-oriented assessments, the World Meteorological Organization (WMO) meeting in Villach, Austria accepted a forecast of an equilibrium warming of 1.5-4.5 °C due to a doubling of the concentrations of greenhouse gases by the year 2030 (WMO/ICSU/UNEP, 1985). There is some doubt that the higher value will be attained by that time because of the time distribution of oceanic and cryospheric responses but the range is not unrealistic and is taken here as our global temperature change scenario to 2030. For the Caribbean, such a change in mean temperature may or may not be realized but the empirical evidence suggest that it is an internally consistent and plausible climate future. Figure 1b. shows plots of July sea- surface temperature (SST) anomalies for the period 1946-86 for the intersections of five-degree grid squares. The large interannual variability is evident. For July, there is a slight and statistically insignificant upward trend and significant oscillations with periods of 5-6, 12.5 and 20 years. that. Examinations of time series of annual mean monthly air temperatures from the terrestrial network throughout the Caribbean region however, show increases between 1931 and 1958 of about 0.9°C per decade particularly in large urban centers probably due to the urban heat island effect and to changing land-use and land cover attendent on expanding settlements. Since 1965 the increase has continued but at about 0.3°C/decade. Some rural stations in the larger islands and some coastal cities have actually shown decreases. Belize City for example has decreased 0.66°C per decade between 1951 and 1970. Extrapolations from the regional data give rise to a cautious suggestion that temperature rises will be significant throughout the region over the next fifty years compared with the preceding four decades and that it will be a decade before the 'greenhouse signals' could unambiguously be separated from the background noise of the climate system.

Temperature has never been a limiting factor in tropical regions except for its interrelationships with the hydrologic system through evapotranspiration and soil moisture change. Hence the impact of regional and global warming on the Caribbean region will be through synergistic effects including rises in relative mean sea level ; changes in tropical atmosphere dynamics leading to changes in the frequency, intensity, trajectory and duration of hurricanes, the length of the hurricane season and to the impact of these on coastal environments due to increased wind velocities and storm-surge height; changes in the magnitude, duration and frequency of extreme precipitation, runoff and floods and the frequency, intensity and rate of onset of droughts. For the Caribbean region as elsewhere, the vulnerability of human societies to disasters precipitated by these extreme events depends upon the economic, political and social conditions of those societies and to the historical antecedents of those conditions. Moreover, social organizations and human behavior contribute to the devastation. In what follows we shall examine the impact of changing climate on the Caribbean region and assess the implications for Caribbean societies. We begin with rises in relative mean sea level.

Changes in Relative Mean Sea Level

Changes in sea level in the Caribbean region are difficult to assess because of the paucity of long-term gauge data and the extent of both eustatic and tectonic processes there. Despite large regional variations in the estimates of projected rise in sea level, (Gornitz *et al*,1982; Klige, 1982) the consensus values appear to be 61 to 183 cm (2 to 6 feet) globally during the next century with a rise of 30 cm (1 ft.) during the next 40 years. The WMO/ICSU/UNEP, Villach scenario envisages a global sea-level rise of 20-140 cm due to a doubling of the concentrations of greenhouse gases. Gable (1987) estimates that during the next 40 years, the rise in relative mean sea level for part of the Caribbean region will be 15-20 cm greater than the global average because of natural subsidence and subsidence due to petroleum and natural gas extraction, groundwater pumping and sediment compaction. By a series of linear extrapolations based on fitted linear relations between global temperature and sea level changes, Gornitz *et al* (1982) after due cautionary notes, arrived at between 16 and 30 cm/^{o}C to give changes in sea level of between 25 and 165 cm (0.82ft and 5.4ft). Revelle (1983) estimates that at least 30 cm of that will be due to expansion of the water above the thermocline over the next century for a 4.5^{o}C warming. Around 7.0 mm lowering of sea level will accrue to increased atmospheric vapor content with most of the added evaporation occurring over the tropical oceans. For the Caribbean region, relative sea level has been rising at about 36 mm/year in some areas over the past three decades for which tide-gauge data are available. However, there are many regions where the changes have been negative. Because of the high spatial variability, a regional value for sea level rise in the Caribbean region as a whole must be taken with caution, but 30 cm over the next 40 years (Gornitz, 1982) is not an unreasonable estimate. According to the 'Bruun Rule' (Bruun, 1962), which tends to underestimate, such a rise translates into shoreline retreat of around 30-50 meters (98 - 164 ft) and a significant loss of valuable real estate and opportunities for employment and investment. With that degree of transgression, tourism areas such as Grand Anse in Grenada, Vigie Beach in St Lucia, tourist areas in western Barbados from Bridgetown to Speightstown and the coastal boundaries of the Vale, and in Antigua, Nevis-St Kitts and Guadeloupe will be adversely affected because most of the beaches especially on the low islands are not much more than 100 ft wide at high tide. In other areas coastal inundation will affect centers of large urban populations such as Georgetown, Guyana. For island nations that have predicated their recent economic diversification on tourism and associated industries, and for whom negative balance of payments and external debt servicing overburden their limited GDP, the impact could be devastating. For these nations many of the costly technological options such as beach replenishment and engineering works open to the more affulent nations are foreclosed leaving retreat and abandonment as the only viable options. Of course there are always transnational conglomerates ready to provide capital in exchange for a loss of economic self-determination and political and national integrity.

Changes in atmospheric dynamics: Hurricanes, storm surges, floods and droughts.

Hurricanes, devastate some part of the Caribbean region each year during the northern hemisphere summer. Between 1871 and 1980, the average duration of a north Atlantic hurricane was about 8 days with a modal duration of 6 days (Neumann, *et al*, 1981). During that period, the average occurrence was 4.9 per year. In the period 1910-1930 north Atlantic hurricanes averaged 3.5 per year but between 1944 and 1980 they averaged 6.0 per year. Since 1960, frequency has decreased but intensity and magnitudes have increased significantly with hurricanes David, Frederic, Allen and Gilbert being among the severest this century. There is evidence to suggest that the temporal variability of hurricanes is related to fluctuations in sea surface temperature (SST), sea-level pressure (SLP) and the 500mb and 700mb surfaces such that increases in hurricane activity are associated with decreased SLP and increased SST especially off the West Africa coast. For the Eastern Caribbean specifically, the number of hurricanes that have tracked through between 1871 and 1980 are given in Tables 1 and 2. There have been years in which as many as 5 storms (Fig. 2) traversed the region,with August and September being the preferred month of occurrence with 47 each (Table1). It is clear that the number of occurrences for any given island or group of islands is by no means negligible and every year one hurricane on the average has passed through the Eastern Caribbean devastating at least one island or group of island. Cumulative discrete probability distribution and recurrence intervals of Eastern Caribbean hurricanes during 1900 and 1980 (Fig.3) show that the threat of hurricane disaster is greater in the Eastern Caribbean than on the central Texas coast. The maximum sustained winds and the large magnitude and high intensity rainfall in hurricanes are crucial for risk assessments particularly as they pertain to destruction of structures including buildings, forests, agricultural crops and beach frontages. A Category 2 hurricane is to be expected every year on average; Category 4 have a high probability of occurring once every five years while Category 5 hurricanes are expected to recur at least every 15 years on the average. In the Eastern Caribbean there were at least 3 hurricanes of Category 5, 6 of Category 4 and 6 of Category 3 in the period 1951-1980 for an average of 2 hurricanes per year capable of causing death and destruction. Notable Category 5 to devastate the Eastern Caribbean since 1955 include Donna (1960), Inez (1966) and David (1979). During that same period six Category 4 traversed the region.

Hurricanes are an integral part of the general circulation of the troposphere in lower latitudes and serve to transport accumulated energy and momentum together with precipitable water mass and latent heat polewards. They derive kenetic energy from the latent heat of condensation. Some of the heaviest rain in the world have occurred in hurricanes that halt in their tracks. Surface pressure differentials, wind velocity , momentum transfer, evaporation and condensation rates in hurricanes are all functions of surface temperature and the thermal structure of the lower and mid troposphere. Consequently if the warming over the Caribbean Sea and north Atlantic Ocean that have been projected in earlier sections materialize, atmospheric thermodynamics and hydrodynamics and the principles of dynamical meteorology dictate that hurricane climatology including size, frequency, intensity and duration, length of season, and intensity of attendent rainfall will change. Precise projections are risky at best because of oceanic dynamics and the response of the mid-latitude troposphere to increased tropical heating, eg., the response of the Trade-wind Inversion and ocean currents driven by an intensified Trade Wind belt, or the influence of a warmer Africa and eastern Atlantic on Easterly Waves in the tropical troposphere all of which are important to hurricane formation and persistence. From all the available evidence it is logical to expect that hurricane intensity though not necessarily frequency will increase so that the probability of Category 5s will become significantly higher and hence incidences of wind devastation and disaster will increase markedly. Emanuel (1987), using a GCM, estimates a 40-50% increase in the damage potential of mean August and September hurricanes due to a doubling of carbon dioxide concentration. The model estimated increased kinetic energy and surface pressure depressions to as low as 800mb. Because evaporation from the warmer ocean surface is sure to increase, precipitable water mass will be greater and hence rainfall amounts and intensity will increase as would the incidences of severe flooding in the interfluves and coastal plains and soil erosion on steep devegetated slopes and on slopes where poor agronomic practices prevail. Next to heavy rains in terms of potential for inundations of coastal areas and interfluves is the storm surge which results from a number of complex, dynamically related factors including wind stress, convergence and vorticity and the inverse barometer effect. Its height is affected by the angle of incidence of the hurricane track to the coastline, the convergence of wind-driven currents at sea, the shape of the coastline and the continental shelf and the mean elevation of the ocean surface. The superelevated sea levels and large waves scour coastlines, destroy piers, jetties and other docking facilities and are responsible for the majority of drownings during hurricanes. If therefore, mean sea level rises in the next 40 years at the same time that there is an increase in the kinetic energy and sustained winds in the more frequent Category 5 hurricanes that will affect the Caribbean region, coastal and estuarine inundations will become more prevalent leading to the destruction of aquatic habitats, wetlands, coastal highways particularly in the mountainous islands, tourist resorts and fishing facilities. In addition, the essentially monocultural agricultural system and tourism which form the backbone of many island-nation economies will be severely impacted. In the Eastern Caribbean, bananas are highly vulnerable but resilient unlike coffee, citrus and other tree-crops that are the mainstay of the agrosystem. The disaster impacts of 'greenhouse induced' climate changes on these increasingly vulnerable but decreasingly resilient societies in the next thirty years will lead to both structural and processal changes in those societies.

Implications for Societies of the Region

Contemporary events are best understood as the outcome and continuation of the historical past and any given social or cultural realization is governed by the society's previous experience. Social impact scenarios must therefore be envisioned as modifications of the social matrix already in place and for the Caribbean that matrix is interestingly complex. The region has the highest concentration of mini-states in the world because of the patterns of conquest and colonization by European powers that introduced a wide array of languages and religions, forced the development of different customs among African, Indian, Chinese, indegenous Amerindians and various Semitic and Hispanic peoples. The Plantation and Hacienda systems together with slavery and later indentured immigration determined the nature and structure of present-day societies including the persistent poverty of their prolitariat. These have set the stage for class structures predicated on race and color, for land ownership and tenure systems that have proven inimical to the viability of agriculture, and for the political and economic woes that have beset the region in the past four decades. In the last two decades, rates of economic growth have been small or negative. For the region as a whole, per capita real income was no higher in 1986 than in 1970 and in some nations as much as 60% lower. The regional economies have been unable to diversify because of protectionism in the north, internal misdirection of savings and resources, the narrow productive base of the region's economy, diseconomies of scale and dependence on foreign capital, technology and skills. In addition there is the 'double whammy' of the disintegration of the agricultural subsistence system and difficulties with traditional export crops and the rapid rise of the Trans-national Corporate agro-processing industries based on imported inputs. Unemployment has been increasing (15-50 % of the existing labor force) and the distribution of wealth, income and access to productive resources are becoming worse. Moreover, there is the accelerated foreign control of domestic resources despite nationalization programs in some countries, increasing corruption, nepotism and clientilism, and the continued flight of skilled workers. Diversification, as a whole failed in its primary objectives: it has done little for capital formation and investment and it did not result in the acquisition and dissemination of technological expertise that was supposed to have formed the basis of economic self determination, all it did was to provide low paying jobs to local workers while exporting large proportions of the profits. Economic diversification into tourism did even less while creating water resources problems and social divisiveness by raising the materialist expectations of the citizenry beyond the capacities of their own economies thus giving rise to frustration. Internal and external monetary imbalances have increased the indebtedness of the region (about US$6 Billion) to non-commercial bilateral and multilateral aid donors and lending institutions. These contemporary socio-economic conditions increase the region's extremely high vulnerability to the impacts of disasters catalysed by extreme climatic events due to the greenhouse effect.

In the small island-nations of the Caribbean the resource base of the societies is limited so that internally generated aid cannot be counted on. Their smallness ensures that the whole country rather than communities within it will be devastated and yet must bear the brunt of rehabilitation and reconstruction or depend on external aid whose attached requirements themselves contribute to modifications in government, educational and monetary institutions. In addition, family and kinship structures are destabilized and unemployment forces both intra- and inter-regional migration. With the contemporary frequency of disastrous hurricane the problem is formidable, but with increased frequency the society will still be digging out when the next one hits thus undermining confidence in eventual restitution of the social equilibrium. Furthermore, post disaster efforts to effect social and economic changes have resulted in decisions that turned out to be detrimental in the long run to social stability such as restructuring agriculture to emphasize food production programs rather than a one-crop export system, a more equitable distribution of land resources or granting concessions to foreign corporations in the hope of attracting investment and employment opportunities in resort-type development including casinos, or in the hope of attracting the capital that will revitalize the economy in the area of housing, agriculture, forestry, small industries and public and private-sector financing. In the best of times, Caribbean countries are unable to cope with the present frequency and magnitude of geophysical disasters therefore increasing that frequency will put unimaginable strain on those countries that can only be relieved by massive external aid unless a regional Disaster Relief Agency be made responsible for the collection and disbursment of aid similar perhaps to a mini-UNDRO. It must be noted however, that every aid package exacts its own price: no country can hope to survive on foreign largesse while maintaining its sovreignty and national integrity. What is more, aid can be addictive leading to a loss of self determination and control. It discourages planning by politicians, administrators, and research workers - roads get repaired, medical facilities are set up and staffed, for a time at least, by foreign personnel, schools are built, electricity networks extended, fertilizers and seeds are distributed and preferential treatments in markets given, - all politically attractive to government officials but adversely affect the development of local expertise, but also does away with the need to plan education systems, and release time for civil servants for upgrading of technical and managerial skills in the context of long-term goals.

Conclusion

The impact of 'greenhouse warming' on the Caribbean region will be evident in changes in the oceanographic and near-shore environments including coral reefs, mangrove swamp ecology, and other aquatic environments; in the atmospheric environment through changes in hurricane climatology including changes in the frequency, intensity, trajectory of storms and the destructiveness of storm surges and coastal flooding; in the spatial and temporal distribution of floods and droughts and marine transgressions into the interfluve due to sea-level rise that will destroy valuable agricultural lands on coastal plains and contaminate freshwater aquifers that are now important sources of potable water. Disasters emanating from these environmental changes will catalyse revolutionary changes in the societies of the region.

References

Bruun, P., 1962; Sea level rise as a cause of shore erosion. Amer. Soc. Civil Engr. *Proc., J. Waterways and Harbors Div., 88, 117-30.*

Emanuel, K. A., 1987; The dependence of hurricane intensity on climate. *Nature*, 326, 483-85.

Gables, F., 1988; Changing Climate. *Oceanus,* 30(4), 53-56

Gornitz, V., S. Lebedeff and J. Hansen, 1982; Global sea level trend in the past century. *Science*, 215, 1611-14.

Neumann, C. J., G. W. Cry, E. L. Caso and B. R. Javinen, 1981; Tropical cyclones of the north Atlantic ocean, 1871-1980. *U. S. Dept. of Commerce, NOAA-National Climatic Center, Ashville, N. C., 174pp.*

WMO/ICSU/UNEP, 1985; Report of the Int. Conference on the assessment of the role of carbon dioxide and other greenhouse gases in climate variations and associated impacts. Villach, Austria, 9-15 October, 1985. *WMO Ref. No. 661. 1-4; 1986.*

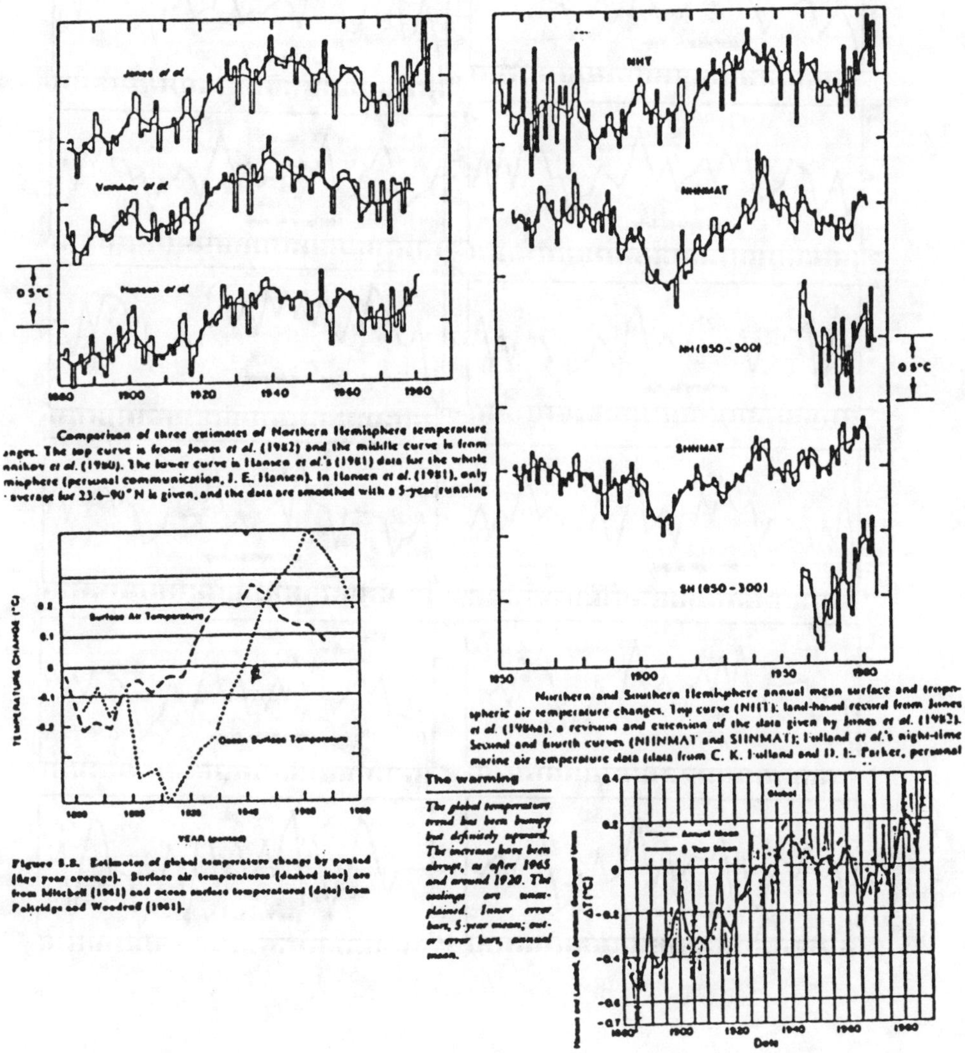

Figure 1: Estimates of temperature changes from various sources. See sub-captions

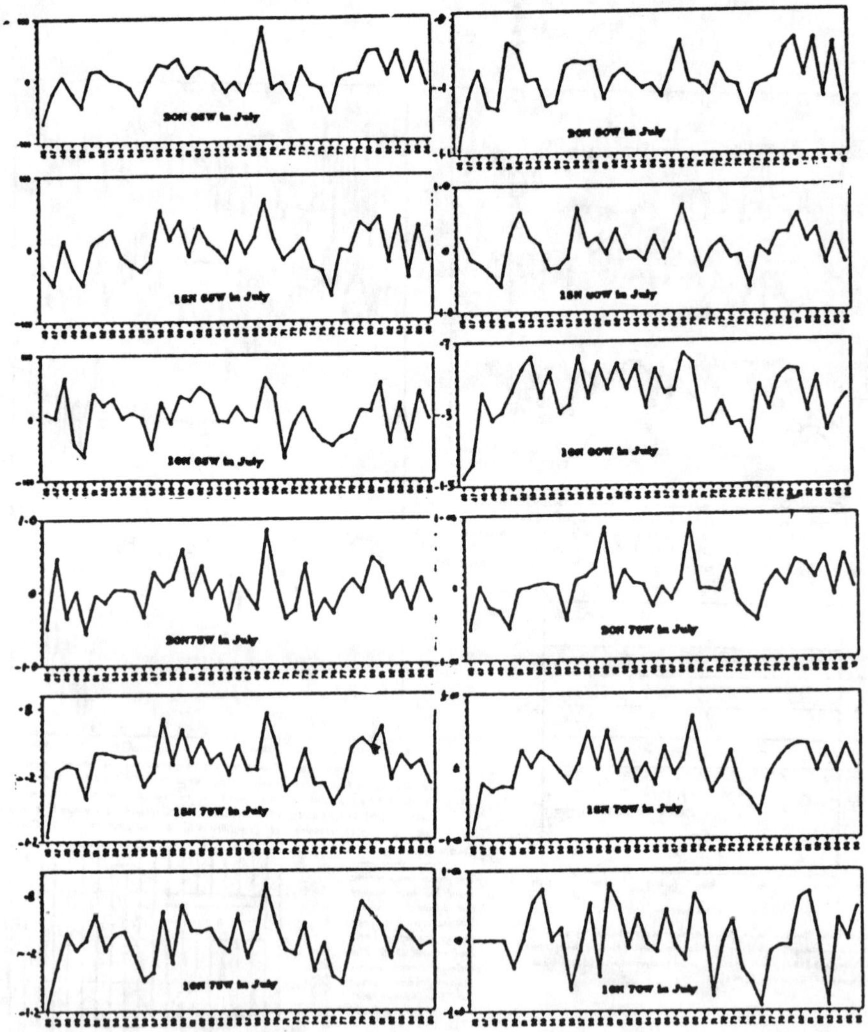

Figure 1b: Sea surface temperature (SST) anomalies for the five-degree grid positions shown in (C), 1946-1986.

TABLE 1

HURRICANES IN THE EASTERN CARIBBEAN BY MONTHS, 1871-1980

YEAR		Mar.	Apr.	May.	Jun.	Jul.	Aug.	Sep.	Oct.	Nov.	Dec.	Total
1871-1900	0	0	1	0	1	14	17	7	1	0	41
1901-1930	1	0	0	0	2	11	11	1	0	0	26
1931-1960	0	0	0	1	1	15	10	5	0	1	33
1961-1980	0	0	0	0	2	7	9	0	1	0	19
TOTALS	1	0	1	1	6	47	47	13	2	1	119

TABLE 2

COMPARISON OF 20-YEAR PERIODS OF HURRICANE OCCURRENCES IN THE EASTERN CARIBBEAN, 1880-1980.

PERIOD	MEAN	STD. DEV.	VARIANCE	CV(%)
1881-1900	1.3	1.05	1.11	123
1901-1920	0.9	1.14	1.29	79
1921-1940	0.7	0.83	0.69	90
1941-1960	1.3	1.01	1.01	129
1961-1980	0.9	0.74	0.55	128

TABLE 3.

RECURRENCE (Years)	CENTRAL PRESSURE (mb)	MAXIMUM SUSTAINED WIND (m/sec.)	MAXIMUM SUSTAINED WIND (mph.)
5	948	63.6	142.5
10	936	68.1	152.5
15	930	70.6	158.1
20	927	71.8	160.8
30	923	73.5	164.8
40	920	74.8	167.6
50	918	75.7	169.6
100	910	79.4	177.9

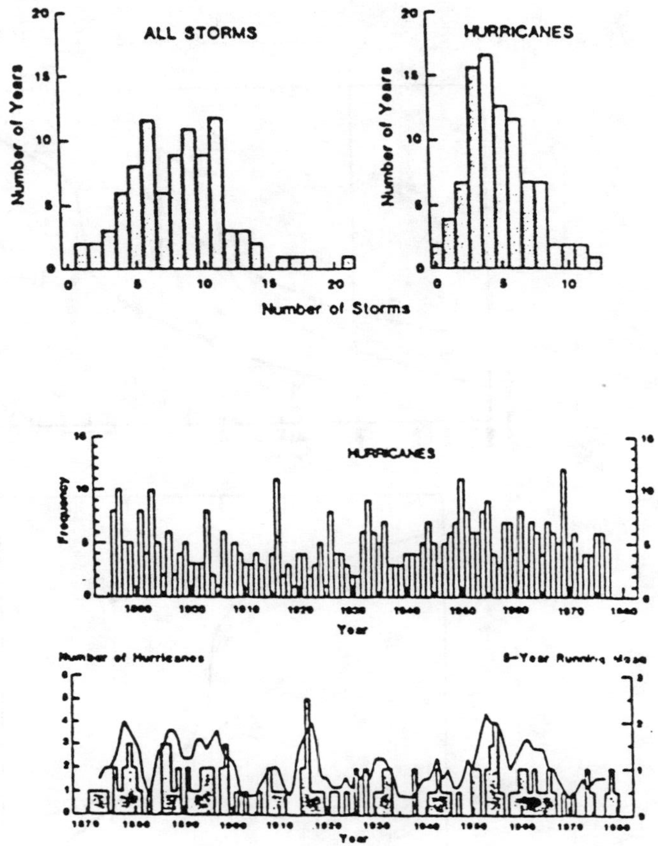

Figure 2: Hurricaneclimatology for hurricanes that have affected the Eastern Caribbean, 1880-1986.

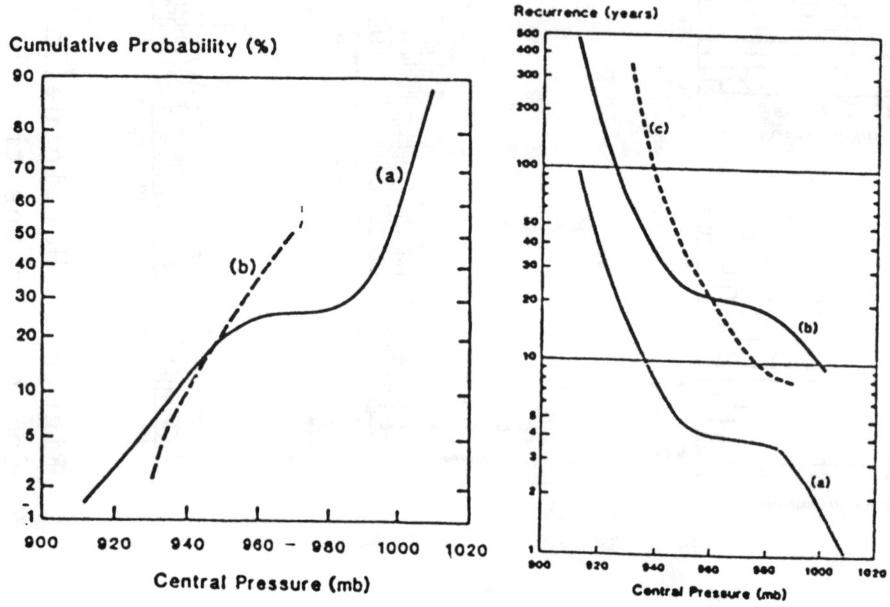

Figure 3: Probability assessments and recurrence intervals for hurricanes that passed through the Eastern Caribbean, 1880-1986.

IMPLICATIONS OF CLIMATIC CHANGES IN THE WIDER CARIBBEAN REGION

Including the Caribbean Sea, the Gulf of Mexico, the Bahamas, and the north coast of South America

A REPORT BY THE TASK TEAM OF EXPERTS
UNITED NATIONS ENVIRONMENT PROGRAMME/
INTERGOVERNMENTAL OCEANOGRAPHIC COMMISSION

George A. Maul
Chairman and Editor

IMPLICATIONS OF CLIMATIC CHANGES IN THE WIDER CARIBBEAN REGION: AN OVERVIEW

by George A. Maul[1]
Task Team Chairman

INTRODUCTION

The history of modern civilization is inexorably related to Earth's climate. Climatic changes have occurred that have influenced our literature, raised and toppled empires, altered our view of God, modified economies, forced mass migrations of both humans and animals, caused hunger and starvation; the list is nearly endless (*e.g.* Bryson and Murray, 1977). Yet as the quincentennial of the European discovery of the Caribbean approaches, little more is known about climate in the region than what the early explorers told their sponsors. Assessing the impact of climate *change* then is a particularly challenging problem.

This study emphasizes the marine and coastal environment and addresses the implications of climatic change in the Wider Caribbean Region. As was the perspective of the early explorers, it is limited to a sailor's view of the coastline, reefs, passages, harbors, deltas, estuaries, and deep waters of the Caribbean Sea, the northeast coast of South America (excluding Brazil), the Gulf of Mexico, and the Florida-Bahamas region of the western North Atlantic Ocean (Figure 1). Nevertheless, such a perspective presents a formidable challenge involving meteorology, oceanography, geology, economics, sociology, medicine, law, and ecology.

To address such challenges, the United Nations Environment Programme (UNEP) was founded in 1972, and within two years UNEP established its Regional Seas Programme. An action plan for the Caribbean Environment Programme was adopted in 1981, and five years later the Regional Coordinating Unit (RCU), in consultation with the Intergovernmental Oceanographic Commission (IOC) and its Subcommission for the Caribbean and Adjacent Regions (IOCARIBE), began addressing marine environmental

[1] National Oceanic and Atmospheric Administration, Miami, FL 33149 USA

MAP OF THE WIDER CARIBBEAN REGION

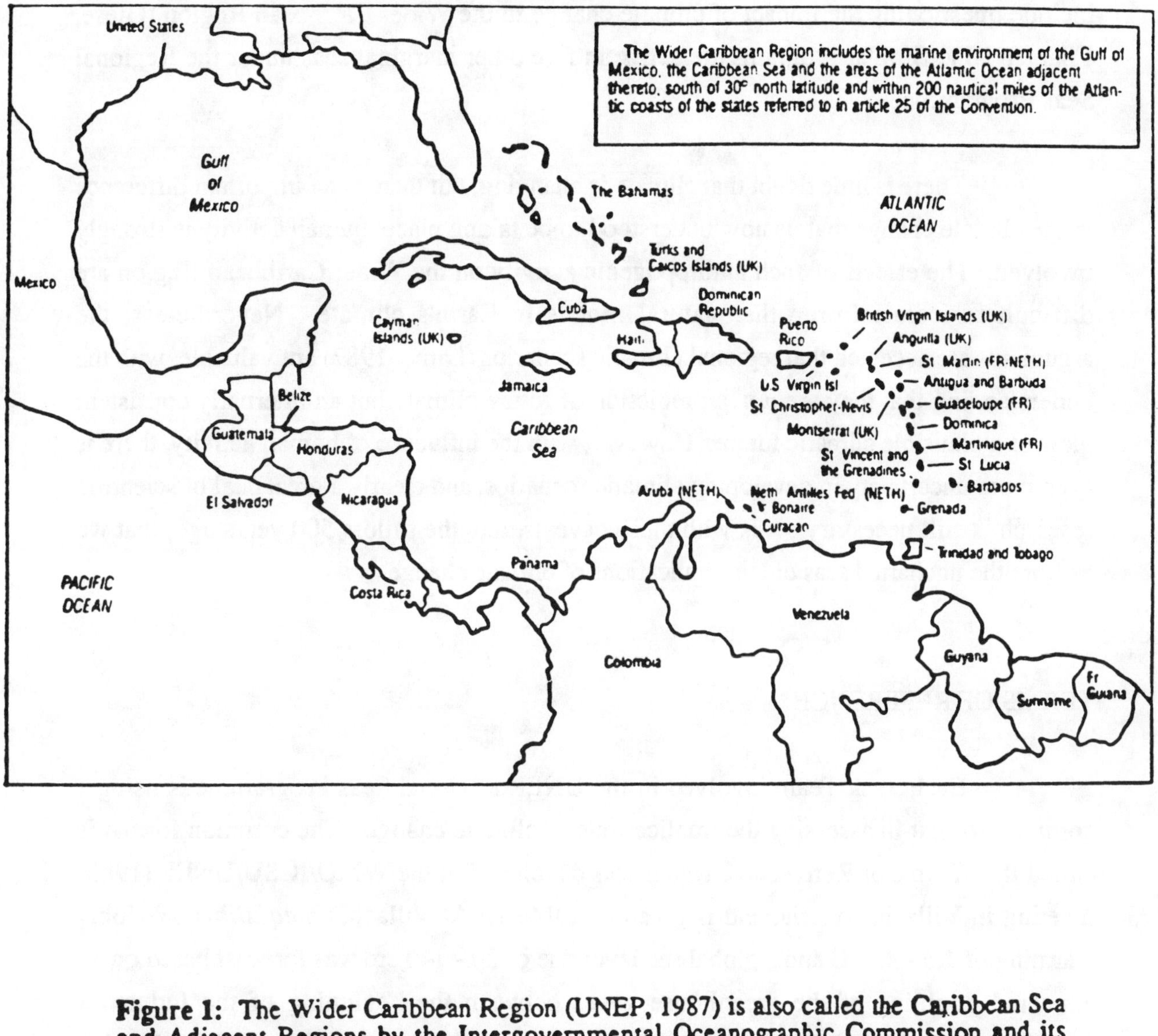

Figure 1: The Wider Caribbean Region (UNEP, 1987) is also called the Caribbean Sea and Adjacent Regions by the Intergovernmental Oceanographic Commission and its Subcommission IOCARIBE. Early European oceanographers mistakenly called it the American Mediterranean, but the author prefers the term IntraAmerican Sea to emphasize its unique geography, climate, and culture.

issues from the RCU's new offices in Kingston, Jamaica. In concert with the recommendations of the World Meteorological Organization (WMO) / International Council of Scientific Unions (ICSU) / UNEP sponsored meeting in Villach, Austria (WMO/ICSU/UNEP, 1985), the RCU extended its marine environmental interests to include questioning the impact of climate change in the Wider Caribbean Region (Giese, Chapter 1.1). Similar programs are active in five other marginal seas under the Regional Seas Programme.

There is little doubt that climate is changing, but there is an important difference in the climate change that is now understood to be taking place: human activity is strongly involved. The effects of such anthropogenic activity on the Wider Caribbean Region are difficult to isolate from other natural moods in Earth's climate. Nevertheless, the arguments are accepted that regional climatic scenarios (Lamb, 1987) are valuable, with the understanding that they are *not* a prediction of future climate but an internally consistent view of a plausible climatic future. However, with the influence of human activity, there is even more uncertainty in developing climatic scenarios, and clearly a great deal of scientific research is still necessary. It is with these caveats, like the sailors 500 years ago, that we explore the uncharted seas of the implications of climate change.

TERMS OF REFERENCE

Each Task Team involved in the UNEP Regional Seas Programme is using a common format in assessing the implications of climate change. The common format is called the "Terms of Reference", which was developed at the WMO/ICSU/UNEP (1985) meeting in Villach, Austria, and is given in Table 1. At Villach, an *equilibrium* global warming of 1.5 - 4.5 °C and a global sea level rise of 20 - 140 cm was forecast based on an expected doubling of the greenhouse gases between the beginning of the Industrial Revolution and the year 2030. The determinations to be made by each Task Team involve a common scenario, and although not specified in the Terms of Reference, the Task Team for the Wider Caribbean Regions chose to separately question the validity of the scenario as it applies to the local marine and coastal environment.

Many climate change scenarios have been made, as a quick reading of the references in Lamb (1987) and others will show. For this UNEP study, a rise in temperature of 1.5 °C and a rise in sea level of 20 cm by the year 2025

TABLE 1

TERMS OF REFERENCE

Under the overall guidance and supervision of the UNEP Oceans and Coastal Areas / Programme Activity Centre (OCA/PAC), the Caribbean Area Region / Regional Coordinating Unit (CAR/RCU) in cooperation with IOC, will establish a Task Team on the implications of expected climatic changes in the Wider Caribbean Region:

(a) to examine the possible effects of the sea level changes on the coastal ecosystems (deltas, estuaries, wetlands, coastal plains, coral reefs, mangroves, etc.);

(b) to examine the possible effects of temperature elevations on the terrestrial and aquatic ecosystems, including the possible effects on economically important species;

(c) to examine the possible effects of climatic, physiographic, and ecological changes on the socio-economic structures and activities;

(d) to determine areas or systems which appear to be most vulnerable to the above changes; and

(e) to prepare a comprehensive, well documented report reflecting the points above. The report should also include several detailed case studies, which would constitute the substantive annexes of the report. The region intended to be covered by the report would include the coastal and marine environment of the States participating in the Caribbean Environment Programme.

∞ ∞ ∞ ∞ ∞ ∞ ∞ ∞

(WMO/ICSU/UNEP, 1985) is the baseline scenario. The lower values *vis a vis* the WMO/ICSU/UNEP scenario are chosen because by 2025 it is not expected that climatic equilibrium to CO_2 doubling will have been reached due to the thermal inertia of the oceans. Deliberations by the Task Team emphasized the point that 1.5 °C and 20 cm are a *global* change scenario, which is interpreted to mean a global average change, that may or may not be realistic for the Wider Caribbean Region. While the Task Team used these baseline values in addressing the implications, two reports (Chapters 2.1 and 2.7) asked "Does the historical record support such values for the Wider Caribbean Region?"; the answer is yes and no.

The next issue in interpreting the Terms of Reference is defining how the Task Team understands the terms "climate" and "climate change". Climate is understood to be limited to the time averaged meteorological and oceanographic conditions of the marine and coastal environment in the Wider Caribbean Region; although agriculture and forestry are briefly discussed, they are only considered in their relation to the ocean. Climate change is understood to mean the decadal-scale scenario of a 1.5 °C temperature rise and a 20 cm sea level rise; short time scales, such as seasonal and high frequency aspects of interannual change, are not considered, although certain aspects of El Niño - Southern Oscillation (ENSO) events are explored.

<u>Temperature Rise of 1.5 °C by 2025</u>

Climatologists use a variety of means to describe Earth's past climate. In the U.S. National Academy of Science report "Understanding Climatic Change" (NAS, 1975), a wealth of information is given on the subject. To illustrate what is thought to be known about temperature change, Figure 2 shows the temperature record for the last 1,000 years (upper panel), and for the last 100 years (lower panel). Although the records are from different sources, they show that Earth's temperatures have varied significantly since the European discoveries of North America, and one can see that when the Caribbean Sea was discovered, Europe was in a cold period known as the Little Ice Age. At present, Earth's climate is warmer than it has been in the last 1,000 years, but by no means is it as warm as the longer records (last 1,000,000 years) show. So while an interpretation of the last 100 years shows increasing global temperatures on average (Hansen and Lebedeff,1988; lower panel in Figure 2), the decades of 1940-1970 had declining temperatures. Translating the global records such as shown in Figure 2 to the regional level was one important challenge for the Task Team.

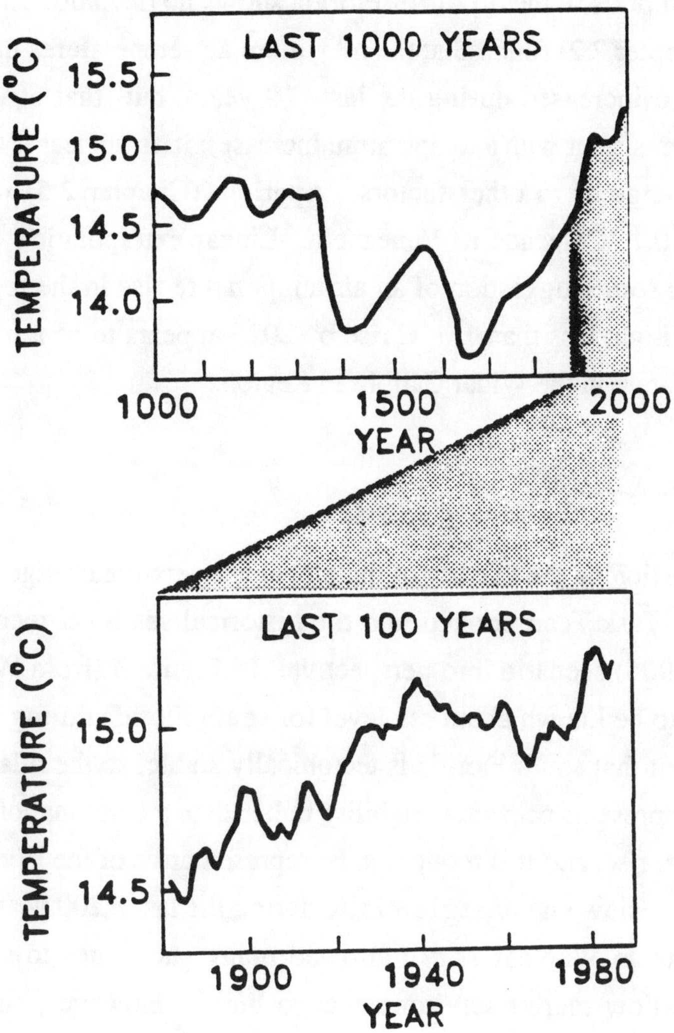

Figure 2: Temperature change for the last 1,000 years based on records in eastern Europe (upper panel; from NAS, 1975), and a global estimate (lower panel; from Hansen and Lebedeff, 1988). Temperature scale for the last 1,000 years is only approximate and is chosen to be in agreement with the record for the last 100 years which is based on direct observations.

Does the historical record support the WMO/ICSU/UNEP (1985) scenario of a 1.5 °C temperature rise in the Wider Caribbean Region by the year 2025? Data to assess a rise of sea surface temperature was considered scarce, so Hanson and Maul (Chapter 2.1) decided to analyze air temperature at Key West, Florida. The 136 year record gives evidence that a warming has occurred between 1890-1950, but the last 30 years or so have been relatively steady at +0.3 °C above the longterm mean; a similar analysis of air temperature from ship reports in the Straits of Florida shows no deviation from constancy of the mean. Gray (Chapter 2.2) found that the maximum air temperatures in Jamaica and in Trinidad and Tobago increased during the last 10 years, but that evaporation had decreased (which is inconsistent with a temperature increase); it is unclear if these changes are due to climatic change or to other factors. Aparicio (Chapter 2.5) reports an air temperature trend of +0.1 °C/decade in Venezuela. Linear extrapolation of these case studies leads one to see some suggestion of an air temperature rise in the region, but that 1.5 °C seems to be too high; less than 1.0 °C rise by 2025 appears to be a more plausible picture of future temperature in the Wider Caribbean Region.

Sea Level Rise of 20 cm by 2025

As with questioning the validity of the global temperature change for the Wider Caribbean Region, the Task Team looked into the historical sea level record to put the WMO/ICSU/UNEP (1985) scenario into perspective. In Figure 3 (from Wanless *et al.*, 1988) what is thought to be known about sea level for south Florida during the last 7,000 years is shown. It is felt that south Florida is tectonically stable, so the relative sea level rise show in Figure 3 represents oceanic variability rather than a combination of both land subsidence and sea level rise, and it is thought to be representative of the Wider Caribbean Region. The remarkably slow rate of sea level rise during the last 3,200 years (only about 0.04 cm/year according to Wanless *et al.*) allowed many shorelines to stabilize or to expand, and many shallow marine environments to build. However, since the early 1930's, sea level records from many sites around Florida show much faster rates of sea level rise, very similar to the rate during the period 3,200 - 5,500 years ago, when there was a rapid retreat of the shoreline. The data in Figure 3 gave the Task Team a benchmark against which to judge the WMO/ICSU/UNEP global scenario of 50 cm rise per 100 years.

Does the modern historical record of the Wider Caribbean Region support the WMO/ICSU/UNEP (1985) scenario of a 20 cm sea level rise by the year 2025? To answer

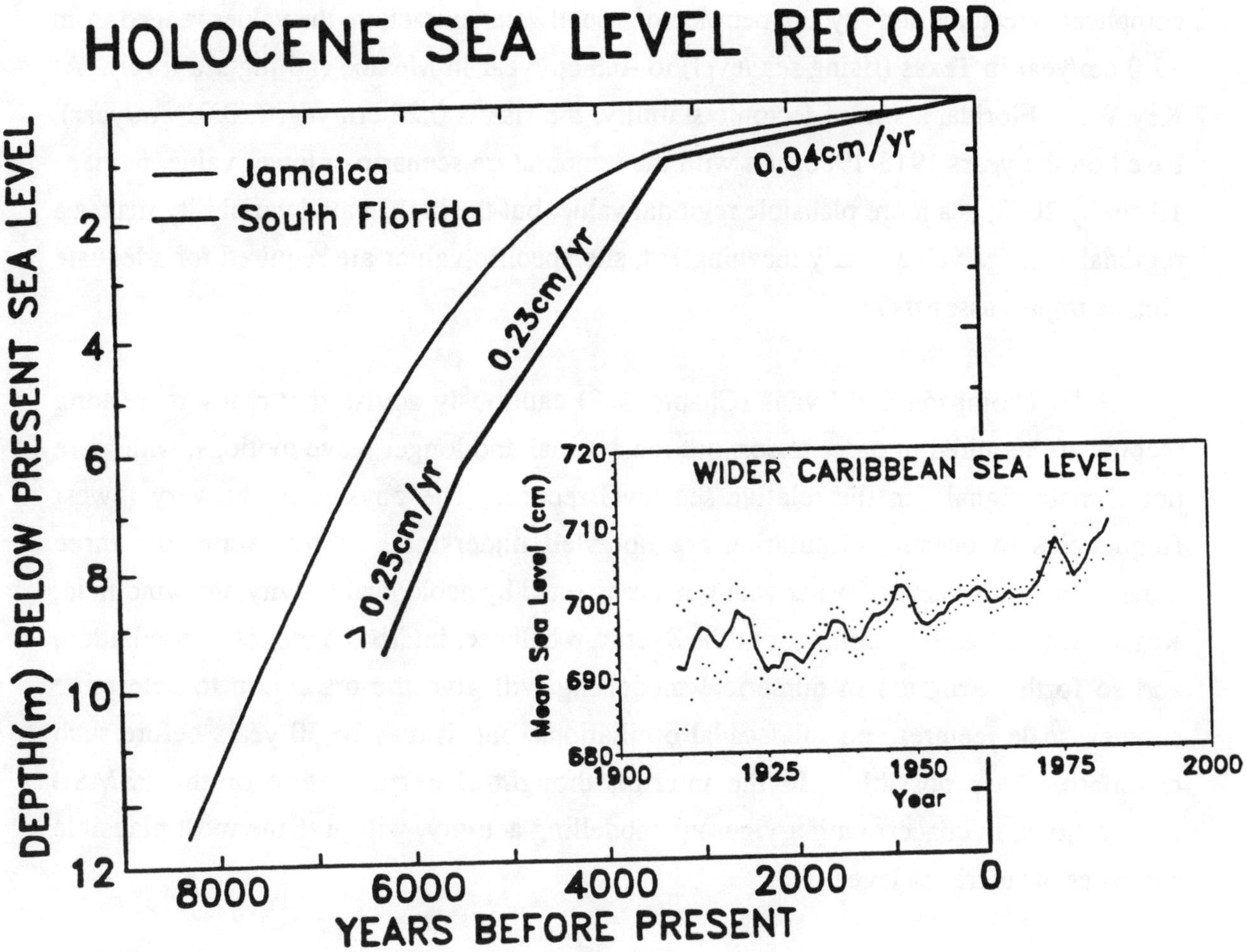

Figure 3: Sea level change based on geological records taken in south Florida by Wanless *et al.*, 1988, and from Jamaica by Digerfeldt and Hendry, 1987. The relative rise of 18 cm per 100 years shown in the inset is based on direct observations at all PSMSL (Pugh, *et al.*, 1987) RLR sea level stations in the region; dots show range of standard error of the mean. The shoreline retreat associated with sea level rise occurs in storm-driven steps and is often-times not immediately seen as beach erosion or lowland inundation.

this question, the highest quality (revised local reference) data on file with the Permanent Service for Mean Sea Level (PSMSL) were studied by Maul and Hanson (Chapter 2.7) for the wider Caribbean region, and from Venezuelan records by Aparicio (Chapter 2.5) for the coast of Venezuela. For the longest records, Maul and Hanson found that sea level is rising on average at about 0.36 cm/year (± 0.25 cm/year) over the last 30 years, but due to complicated tectonic activity and petroleum/groundwater extraction, the values ranged from +1.0 cm/year in Texas (rising sea level) to -0.3 cm/year in Mexico (falling sea level). At Key West, Florida, a site of tectonic stability, the rise is 0.22 cm/year (±0.01 cm/year), based on the years 1913-1986. As with the temperature scenario, a lower value, perhaps 10 cm by 2025, is a more plausible regional value, but the high spatial variability makes a regional average value nearly meaningless; site specific values are required for adequate climate impact assessment.

Thompson and Evans (Chapter 2.8) cautiously advise that many more long records are required in order to sort out the decadal and longer wave motions, which are not climate signals, in the relative sea level record. The physics of the very lowest frequencies in oceanic circulation are not well understood. Circulation, the three dimensional movement of water with time, is affected by geological activity, the wind field, Rossby waves, periodic behavior of the Sverdrup balance, interbasin modes of oscillation, and so forth. Progress in numerical modelling will give the resolution to determine submesoscale features and subdecadal oscillations, but it may be 10 years before such calculations are possible. In the interim, thoughtful extrapolation of the PSMSL observations, in concert with a vigorous modelling activity, will give the most plausible estimates of future sea level.

EFFECTS OF SEA LEVEL CHANGES ON COASTAL ECOSYSTEMS

The first task in the Terms of Reference (*q.v.* Table 1) is to examine the possible effects of the sea level changes on the coastal ecosystems. Gable (Chapter 4.1) gives an overview of the ecosystems in the Wider Caribbean Region, and *Oceanus* (1987/88) is an issue devoted to Caribbean marine science. The variable of interest in this section is relative sea level (RSL) rise, that is the net effect of tectonic uplift or subsidence plus expansion or contraction of the water column. During the Holocene (last 10,000 years) in Jamaica for example (Hendry, Chapter 2.4), sea level rise is less than the 0.5 cm/year implied by the assumptions at the International Conference in Villach, and in the last 3000 years RSL rise has been almost nil. All other things being equal, 0.5 cm/year is expected to place an unusual stress on coastal ecosystems.

Deltas

In the Wider Caribbean Region there are four major river deltas: the Mississippi (USA), the Rio Grande (Mexico/USA), the Magdalena (Colombia) and the Orinoco (Venezuela). Deltas are particularly vulnerable to erosion enhanced by sea level rise because the sediments are unconsolidated muds subject to subsidence and compaction. One might expect, according to the Bruun Rule, a shoreline retreat up to several meters horizontally for each centimeter RSL rise; this translates into thousands of hectacres of lost land. The problem is exacerbated by potential increased storm activity (Gray, Chapter 2.2; Shapiro, Chapter 2.3), since most shoreline erosion occurs during storms, and by subsidence such as in the case of Louisiana. However, Mercado (Chapter 2.6) argues that the 20 cm RSL rise scenario will be of no practical consequence on storm surge modelling, as far as it might introduce increased surge heights. Delta benthic systems, particularly seagrass beds, would be most affected or destroyed by the expected RSL rise. In contradistinction, RSL near the Orinoco Delta may be falling (Chapters 2.5 and 2.7), but more measurements are needed to document this preliminary result.

Estuaries

The RSL effect on estuaries, as with deltas and many other geomorphological features, must be considered on a case-by-case basis in order to make meaningful impact assessments in the Wider Caribbean (Vicente, Singh, and Botello; Chapter 4.1). Because of local uplifting, the following areas are expected to have reduced increase in RSL due to climate change: east coasts of the Cayman Islands; north coast of Jamaica; southeast coast of Cuba; north coast of Bahia, the southwest coast of Haiti; Barbados; north coast of the Dominican Republic; and the southwestern Gulf of Mexico. In addition to the subsidence experienced in the deltas, other Caribbean areas experiencing downwarping include: the Maracaibo region of Venezuela; the entire northern Gulf of Mexico from Texas to Florida; the estuary of Port au Prince; and the western Gulf of Honduras. Coastal lagoons, salinas, and estuaries, depending on their location, could all suffer from saline intrusion, but lagoons should be able to support their usual nurseries; salinas on the other hand (Vicente, *et al.*, Chapter 4.1), could be flooded over continuously, and lose their economic value.

Wetlands

The ability of a wetland to sustain vertical growth is a balance between sedimentation and RSL rise. In the tectonically complicated Wider Caribbean Region, no single definitive statement is possible, but in the last 5,000 years, many Caribbean wetlands have been able to keep pace with rising sea level. In areas where marked subsidence is occurring, particularly if there is canalization of organic silts and clays away from the wetland into the marine environment (Hendry, Chapter 2.4), wetlands will be submerged and lost to productivity. Where wetlands are bounded by steep-sided basins, as is the case in many of the Caribbean islands, it is unlikely that they will be replaced as sea level rises; on gentler island and continental floodplains the problem may be less severe. Loss of wetland economies such as shellfish industries is expected to occur occur with the 20 cm RSL rise scenario.

Coastal Plains

The primary effect on coastal plains will be increased flooding during storms (either from raised sea level and/or from heavy rainfall). Unfortunately many storm surge models differ markedly in their predictions (the variability in their predictions being several orders of magnitude larger than the 20 cm scenario; Mercado, Chapter 2.6). Shore migration (both erosion and accretion) will vary depending on the substrate, and sandy beaches will be more affected than rocky coasts. No single rule can be applied for the Caribbean as a whole, but modelling on a local scale is required to account for differences such as tectonic displacement, beach structure, offshore bottom topography, and storm frequency and magnitude. The concentration of population in the poorly drained low-lying coastal plains is a source of concern in many Caribbean countries. Special attention should be paid to areas where subsidence is evident, as it will exacerbate the flooding problem. Port au Prince, Haiti, Puerto Cortes, Honduras, and the Galveston centered area of the U.S. Texas-Louisiana coast, are coastal plains areas that are particularly vulnerable to flooding from sea level rise and storms.

Coral Reefs

The second largest coral reef system in the world dominates the offshore area of the western Caribbean Sea, and all but the northern Gulf coast have extensive reef systems. Growth of individual coral organisms is estimated between 1-20 cm/year (Vicente *et al.*,

Chapter 4.1), and reef growth rates as a whole are known to be up to 1.5 cm/year (Hendry, Chapter 2.4). Not all reefs accumulate at these rates, but if they did, they could keep pace with the rise in RSL of 20 cm by 2025 if other factors do not alter growth conditions. Environmental stress on the reefs from other variables (storms, sedimentation, disease, rainfall, radiation, turbidity, overfishing, mass mortality in algal grazers, *etc.*) is causing some to be unable to keep pace with rising RSL, resulting in alteration of the nearshore hydrodynamics. The issue is further complicated by consideration of the type of reef, coastal geomorphology, reef depth, and ecological state of the reef in question. Accurate predictions on the effect of RSL rise may be possible in reefs that have already been physically and biologically monitored, such as in Panama, Jamaica, and Puerto Rico.

Mangroves

Mangrove forests are a unique feature in coastal lowlands of the tropics and subtropics; their prop roots stabilize the soil, dampen wave energy, provide habitat shelter for organisms, and form the base of the marine food web. RSL of 20 cm by 2025 will affect different mangrove species differently. The black mangrove (*A. germinans*) has a root system that is different from the other three common Caribbean species, and it should be particularly monitored. There is some evidence (S. Snedaker, U. Miami, personal communication) that red mangrove replace black mangrove when mean water elevations increase. In general however, if the mangroves are not juxtaposed to a steep coastline, they are expected to be able to tolerate the anticipated RSL change (Vicente *et al.*, Chapter 4.1), provided other environmental stress factors don't change the natural balance.

Seagrass Beds

Seagrasses are a bottom feature throughout the Wider Caribbean Region that are of importance in stabilizing the bottom sediments, serve as nurseries for juveniles, and like mangrove roots provide surfaces upon which many organisms attach. A 20 cm RSL rise *per se* is not expected to seriously affect the six common species, but if there are other changes such as in the quality of light, influence of herbivores, substrate, wave energy, or bottom slope, the beds may be impacted.

Fisheries

The impact of sea level rise on fisheries is not expected to be great unless turbidity increases due to erosion from higher water. Turbidity increase could have a negative impact on fisheries particularly during the early life history stage (W. Richards, NOAA/NMFS, personal communication). Estuarine dependent species in areas such as Mississippi, the Everglades, Guyana and the Orinoco Delta, may be particularly vulnerable to sea level rise especially if salinity changes are involved.

EFFECTS OF TEMPERATURE ELEVATIONS ON ECOSYSTEMS

As discussed above, there is considerable question if trace gas induced temperature elevation can be seen in the records at Key West, Venezuela, Jamaica, or in Trinidad and Tobago. Temperature change however is only one aspect of the meteorology that will effect terrestrial and aquatic ecosystems. Hanson and Maul (Chapter 2.1) found no evidence for changes in precipitation at Key West during the last 101 years; similarly, Aparicio (Chapter 2.5) found none along the southern Caribbean. In the Wider Caribbean Region, Gray (Chapter 2.2) found decreased rainfall in the last 20 years, which he associates with decreased hurricane activity (a major contributor to the total). An increase of 1.5 °C in sea surface temperature could increase the number of hurricanes by as much as 40% (Shapiro, Chapter 2.3), and the maximum wind speed by 8%; Shapiro is quick to point out that there is considerable uncertainty in these figures (40% increase means on average +1.6 ± 1.2 hurricanes/year). Many other factors are important in hurricane analysis, and it may be that the storm formation location and track are more important than changes in strength or frequency.

In the sense that Lamb (1987) develops climate change scenarios as *plausible* future events, Gray argues for the following likely effects (*cf.* Aparicio, Chapter 2.5): rainfall will continue to decrease, air temperatures will continue to increase, surface wind speed will continue to increase, and evaporation will increase. Caution must be exercised in applying these changes as anything other than persistence forecasting. It is unknown for example if the lowered number of large hurricanes over the last two decades is really a long term trend or part of some cycle as yet not understood. Hurricanes are an important contributor of rainfall; is the decrease in precipitation merely a reflection of fewer large storms? Increased temperature may affect the drag of wind on water, but Mercado (Chapter 2.6) sees no clear indication of a significant change in storm surges or waves associated with elevated temperature. With these thoughts in mind, the second item in the Terms of Reference (*q.v.* Table 1), effects of temperature elevation on the ecosystem, including on economically important species is considered.

Agricultural Resources

Saline intrusion is expected to have more impact on agriculture in the coastal plains than elevated temperature, particularly on rice production along the Guyana coast. Soil erosion will probably increase, but poor management practices are probably more contributive than temperature elevation and saline intrusion. Finally, Vicente *et al.* (Chapter 4.1) argue that it is unlikely that inland and hilly forests will be affected much by increased temperature, although Gray (Chapter 2.2) warns of increased erosion due to increased winds and decreased precipitation. However, warmer temperatures could be a significant factor in forest fires, particularly if precipitation decreases. Human settlements are unlikely to be affected significantly by 1.5 °C weather changes, except where RSL is important.

Coastal Systems

In the tropics, marine organisms live closer to their maximum thermal tolerance than those in more temperate climates. Although the 1.5 °C temperature rise scenario would raise the summertime mean temperature to 30.5 °C over much of the Wider Caribbean Region, most migratory organisms are expected to be able to tolerate such a change. Some corals will be affected (*viz.* the 1983 and 1987 bleaching events), but it is expected that other environmental stresses will be more important. Littoral-supralittoral organisms such as mangroves are adapted to withstand high temperature, and unless the 1.5 °C increase affects the reproductive cycle, the temperature elevation will likely cause unmeasurable results. Similarly, only seagrass beds that are located in thermal stress situations already (*i.e.*, in shallow lagoons or near power plant effluents), are expected to become negatively affected by the projected WMO/ICSU/UNEP (1985) temperature rise.

Fisheries

The blue, clear waters of the Caribbean are relatively poor in nutrients, and most of the fisheries are concentrated on Campeche Bank and along the northern coast of the Gulf of Mexico, at the Mosquito Bank off Honduras and Nicaragua in the Caribbean Sea, and in the Gulf of Paria and the coastal waters of the Atlantic Ocean off Suriname, Guyana, and French Guiana. It is not expected that a modest increase in temperature will significantly affect the fisheries except in some shallow lagoons where hypersalinity may affect productivity particularly if juveniles have a critical dependence on salinity or temperature. Increased alongshore winds however could lead to increased coastal upwelling along some continental coasts (Aparicio, Chapter 2.5) or other circulation changes (Thompson and Evans, Chapter 2.8), and thus to increased productivity (*cf.* Gable, Chapter 4.1).

Tropical fish eggs hatch very quickly (12-48 hours) and development is associated with temperature. Just as "cold snaps" can be devastating, so can "hot snaps", particularly during early juvenile stages. Extrema in temperatures usually get averaged out in climate analysis, but with increased temperature, the likelihood of "hot snaps" increases; the 1987 Caribbean coral bleaching event was attributed to "hot snaps" by some researchers (W. Richards, NOAA/NMFS, personal communication). The complexities of the ecosystem could be greatly effected by slight temperature changes. It is unknown for example why fish stocks decline or increase by orders of magnitude except due to early life history events - either directly by the physical world or indirectly through complex chains in the ecosystem dynamics. Temperature effects on tropical fisheries remains an important and unanswered question, although there is some evidence of fish migration associated with increased coastal temperatures.

POSSIBLE SOCIO-ECONOMIC CHANGES

This section concerns the third item in Table 1, examining the possible effects of climatic physiographic and ecological changes on socio-economic structures and activities. Climate change will have socio-economic impacts on both the micro-economic or localized level, and on the macro-economic or economy-wide level. The smaller or the more coastal-oriented an economy is, the greater will be the impact of sea level rise (Alm, Blommestein, and Broadus; Chapter 3.1). The Wider Caribbean Region, with its many small island-based economies such as fishing and/or tourism, is most vulnerable to the physical changes associated with changing climate. Some climatic changes will benefit certain sectors of an economy (raising RSL may benefit the construction industry), while being detrimental to others (beach erosion may cause a loss in tourism). A climate change induced benefit to the construction industry reflects a *transfer* of benefits and costs rather than the *creation* of new benefits and costs. The net sum of costs and benefits must be assessed on an individual basis because it is the true cost due to climate change only that is of interest.

Agriculture and Forestry

Islands usually have small, coastal aquifers, and sea-level rise will impact water quality in aquifers that have hydrological continuity with the sea. Loss of agricultural land in low-lying coastal plains will be a minor, but perceptible impact, particularly in those areas where saline intrusion affects the water supply, such as on the leeward side of small mountainous islands; continental areas are not expected to be seriously affected. Differing permeability in aquifers can cause great variability in the effect of rising RSL. Relocation of wells, construction of weirs, water storage schemes, and barging of water are all possible socio-economic responses. In regards to forestry, as noted in an earlier section, the expected climate impact is small compared to proper management impacts on the industries and people involved.

Fisheries and Coastal Zones

Most fishing in the coastal zone in the Wider Caribbean Region is artisanal except for a few larger industries such as the menhaden fishery in the Gulf of Mexico. The WMO/ICSU/UNEP (1985) scenario of 1.5 °C and 20 cm increases by 2025 AD are not expected to create any significant changes in the fisheries, although to the artisanal fisherman, a displacement in traditional fishing sites may be perceived as being important (Alm *et al.*, Chapter 3.1). There does remain an unanswered question of the effect on fisheries of extreme temperature events. Aquaculture in the region is considered undeveloped at the present. The critical issue of shoreline migration, which is the most important impact on the coastal zone, is discussed in the following sections.

Tourism

The single most important industry in the Wider Caribbean Region is tourism, especially in Florida, The Bahamas, Jamaica, and the Lesser Antilles. Of all the possible climate change impacts that affect tourism, none can be so clearly demonstrated as being important as is beach erosion (*q.v.* Hendry, Chapter 2.4). Shoreline migration will create new areas of economic benefit as new beaches are built, but the protection, replenishment, and stabilization of existing beaches, at least until major existing tourist investments are amortized, will be an increasing socio-economic impact. It is difficult to estimate the impact of climate induced sea level rise, in addition to the erosion associated with the relentless interaction of the sea on the coast, that is not associated with climate change; in addition, certain sand mining practices (such as in Trinidad and Tobago) already are considered important. Indirect socio-economic effects on tourism due to increasing pollution, coral reef mortality, and storm damage are also involved.

Settlements and Structures

Up to a certain point, structures will be worth building to protect settlements and facilities. Navigation and port facilities normally have to be reconstructed and maintained, so the socio-economic impact of a 20 cm sea level rise is not considered serious (Alm *et al.*, Chapter 3.1). Some nearshore roads, seawalls, and bridges will have to be increasingly repaired, and if the RSL rise is augmented by increased storm activity, the impacts will be serious, particularly in countries with marginal economies. As with agriculture in low lying lands that depend on well water, many municipal water supplies and drainage and sewerage systems, will have to be modified; areas of particular concern in this regard are coastal cities of Guyana and Belize (Vicente *et al.*, Chapter 4.1). The most damaging socio-economic aspect is climate change coupled with population growth and migration to coastal cities. Oftentimes the population growth is into areas most likely to be impacted by water level changes, and in periods of extreme weather events, serious public health impacts are probable in addition to physical danger (Hardin, 1971).

Public Health

Both temperature and sea level rises are expected to have an effect on human health; temperature because many diseases and acute effects are associated with elevated temperatures, and water levels, because water is a principal agent for many diseases and organisms that carry disease (de Sylva, Chapter 3.2). If higher temperatures are coupled with higher humidity as Gray expects (Chapter 2.2), heat related health stress and mortality will increase. Human health changes are related to a wide variety of considerations including: mortality and morbidity related to weather and climate; extreme weather events; airborne materials; seasonal diseases caused by microorganisms; parasitic diseases; nutrition; water quality and abundance; and changes in the marine environment including population shifts in dangerous fish such as sharks, and toxic organisms (de Sylva, Chapter 3.2). Socio-economic effects relate not only to increased spread of tropical diseases and their associated shift in costs and benefits to the health industry, but also to potential losses in other industries due to health related absenteeism. It is anticipated that there will be a transfer of costs and benefits associated with climatic change to public health in the Wider

Caribbean Region, but that the health-care delivery systems will be capable of keeping pace with the climate related aspects to the year 2025; whether or not the systems are capable of coping with other social changes is uncertain.

MOST VULNERABLE AREAS OR SYSTEMS

In order to determine areas or systems which appear to be most vulnerable to changes in sea level and temperature (*q.v.* Terms of Reference), and their impact on ecological and socio-economic structures and activities, three broad topics are addressed: Physical Processes; Ecological Aspects; and Socio-Economic Issues. Much of the material in the following sections is drawn from notes and recommendations made during the deliberations of these three working groups of the Task Team. For a more detailed listing of the recommendations and deliberations, see Annex 5.2, which was substantively written by B.A. Miller.

It was clear to the Task Team that detailed information on the wide variety of areas or systems which are most vulnerable to climate change in the Wider Caribbean Region could not be prepared without additional substantial effort and support. In order to ultimately provide such detail, the consensus was the following:

1. Strengthen existing institutions rather than creating new ones.
2. Improve communication and information exchange particularly through the use of electronic media and "personal computer" technology.
3. Reduce uncertainties in the regional impact of the global 1.5°C/20cm scenario by data generation, case studies, and modelling, obtaining probability estimates on sea-level rise and other climate change.
4. Continue the interdisciplinary interaction of the Task Team in order to provide *quantitative* information to member states.

The latter point of quantifying results, based on the best physical or economic models, is considered the penultimate goal of this Regional Seas Programme.

The overwhelming consensus of the Task Team was that "case studies" were the most cost effective method of quantifying the effects of climatic change. Several sites that are deemed to be at high risk from sea level change and/or temperature rise should be selected for an extensive and intensive study. Site selection must be based on the applicability of the results to as many other similar sites as possible; it therefore must be representative of a class of problems rather than an individual problem. The following short list suggests five important case study themes: Coastal Erosion; Groundwater/Petroleum Extraction; Tropical Storm Frequency/Intensity; Wind Stress and Ocean Transport; Riverine Discharge Quantity/Quality. As discussed below, case studies must address three interrelated subjects: physical processes, ecological aspects, and socio-economic issues.

Physical Processes

Climate change involves much more than RSL rise and temperature increase; precipitation, evaporation, humidity, wind velocity, hurricanes, cloudiness, insolation, ocean currents, waves, mixing, riverine input, *etc.,* are all important variables. In order to strengthen quantitative information transfer to states, regional climate models nested in coupled ocean-atmosphere global circulation models are needed, along with a vigorous, stable, longterm *in situ* verification program, coupled with an active multidisciplinary research effort which should include examination of the historical, geological, and archeological records in order to supplement direct measurements. Understanding future shoreline migration is arguably the first priority in the region based on current information, but if precipitation changes (for example) are markedly underestimated, the impact on agriculture and coastal ecosystems could be far more important. To this end, participation in efforts such as the World Climate Programme, with significant international visibility by scientists from the Wider Caribbean Region, is absolutely necessary to improve the physical basis upon which quantitative information is provided to ecologists, sociologists, economists, politicians, and managers.

Ecological Aspects

Identification of the most vulnerable ecosystems requires more of a microscale approach than the mesoscale thinking required of the physical processes discussed above. Preparation of a regional map with a classification scheme showing areas and ecosystems most vulnerable to climate change is a massive, but necessary undertaking. Seagrass beds, coral reefs, mangroves (particularly the black mangrove), and coastal lagoons are probably the most critical habitats to be mapped. Associated with the critical habitats are species that utilize them as feeding and/or nursery grounds. Of vital concern to these critical habitats are climate related impacts from the sewerage and toxic wastes of nearby human population centers and agricultural regions. Conversely, impacts of saline intrusion on local fresh water supplies and inundation particularly during storms of seaside population centers, are critical concerns to local residents; high population density islands such as Barbados, and cities with rapidly rising RSL such as Galveston, Port au Prince, Puerto Cortes, and Cartagena, are particularly vulnerable.

Socio-Economic Issues

Before effective socio-economic responses to climatic changes can be initiated, there is a need to reduce significantly the degree of uncertainty about the likelihood, extent, and direction of such changes. The most vulnerable "system" in the socio-economic and health sectors is the credibility of those making impact assessments. Governments and institutions will revert to procrastination as the most viable response to weak forecasting, rather than to improving information development and dissemination, risk spreading and diversification, or to reducing levels of fixed commitments. Some states, Costa Rica for example, have already established new building set-back laws for construction along the coasts; others, Florida for example, have locally opted for massive beach replenishment programs. Most small island states, which numerically constitute a substantial fraction of governing units, do not have the financial resources nor the technical expertise to develop appropriate socio-economic responses to climate change. Probably the greatest single socio-economic scenario that individual governments must prepare for is a significant international migration of populations from highly vulnerable locales to areas where safety and the quality of life is deemed to be better. To prepare for such future change, a catalog of institutional responses needs to be developed along with specification of conditions under which those responses should be implemented.

Synthesis

H.L. Mencken once said "For every complex problem there is a solution that is simple, neat, and wrong." With this caveat in mind, an attempt to synthesize much of the implications of climatic changes in the Wider Caribbean Region is given in Table 2. Three subjective levels of vulnerability to rises in sea level and temperature are chosen and assigned to the ecosystems and socio-economic topics outlined in the Terms of Reference (*q.v.* Table 1). While on a site-specific scale many of the estimated impact levels will be different, on a regional scale vulnerability to most climatic changes *per se* is judged low to moderate. However, due to other pressures on the marine environment, and to human

TABLE 2

Implications of Climatic Changes in the Wider Caribbean Region*

(L) Low Impact (M) Moderate Impact (H) High Impact

Terms of Reference	(a) RSL 20 cm	(b) SST 1.5 °C
Ecosystems	Level of Vulnerability	
Deltas	H	L
Estuaries	M	M
Wetlands	M	M
Coastal Plains	M	L
Coral Reefs	L	M
Lagoons	M	M
Mangroves	M	L
Seagrass Beds	M	M
Fisheries	L	M
Agriculture	L	L
Forests	L	M
Rivers	L	M
Coastal Lakes	L	L
Beaches	H	L
Socio-Economic	Level of Vulnerability	
Coastal Zones	L	M
Tourism	M	L
Settlements & Structures	M	L
Public Health	L	M
Tropical Storms	L	H
Human Migration	L	M
Cultural Heritage	M	L

*These levels of vulnerability reflect only the WMO/ICSU/UNEP (1985) climate scenario detailed in the Terms of Reference, and must be considered as issues that exacerbate other problems such as population pressure, pollution, subsidence, coastal erosion, construction, warfare, etc.

efforts to deal with the effects of these pressures, the vulnerability of society to climatic changes increases. In many cases the future impacts on society of non-climatic factors may far exceed those due to climatic changes. It is important therefore for policy considerations, to view this synthesis in a proper context, which is, climatic changes will exacerbate environmental changes already ongoing and documented in other studies.

MODELS OF FUTURE CLIMATE

In Chapter 1.2, Wigley and Santer give a very detailed discussion of the possible future climate of the Wider Caribbean Region. They compare the results of four numerical models that predict future surface air temperature change and precipitation change for each of the four seasons. Each model calculates the effect on temperature and rainfall of doubling all the greenhouse gases, expressed as doubled CO_2. All four numerical models are *global* models, but Wigley and Santer only report the *regional* results of interest herein.

Figures 4 and 5, from Wigley and Santer (personal communication) show the model results for temperature and precipitation respectively. Results from atmospheric General Circulation Models (GCMs) for future climate on a regional scale must be interpreted very cautiously because of the limitations in numerically simulating such a complex problem as climate. Cautiously then, the range of annual average modeled temperature change (Figure 4) and annual average modeled precipitation change (Figure 5), are discussed below.

The annual average temperature change caused by effective CO_2 doubling shows a fairly consistent result in each GCM: an increase of 2°C to 4°C is calculated. Details of increased annual averaged temperature change are different from GCM to GCM, but in general the continental boundaries of the Wider Caribbean Region are modeled to have higher annual average temperatures than the islands. Annual average precipitation changes due to effective CO_2 doubling also shows significant variability between GCMs, but each model shows that the zero contour (in millimeters per day), which runs through the center of the Caribbean Sea, is the dominate feature of the calculation. Precipitation in the Wider Caribbean Region is strongly influenced by tropical storms however, which are not in these GCMs.

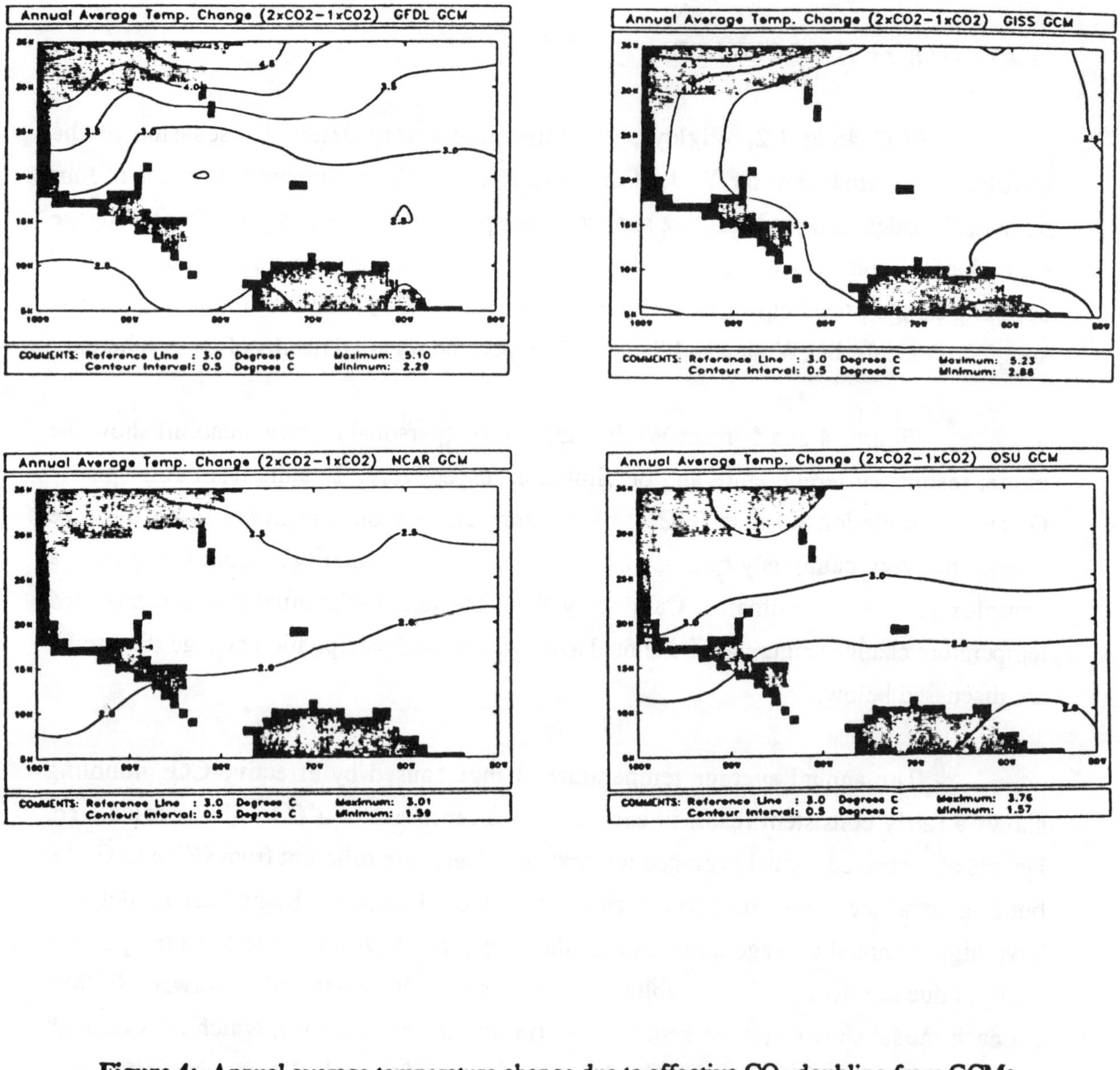

Figure 4: Annual average temperature change due to effective CO_2 doubling from GCMs of the Geophysical Fluid Dynamics Laboratory (GFDL), Goddard Institute for Space Studies (GISS), Oregon State University (OSU), and the National Center for Atmospheric Research (NCAR). Maps courtesy of Wigley and Santer (*cf.* Chapter 1.2).

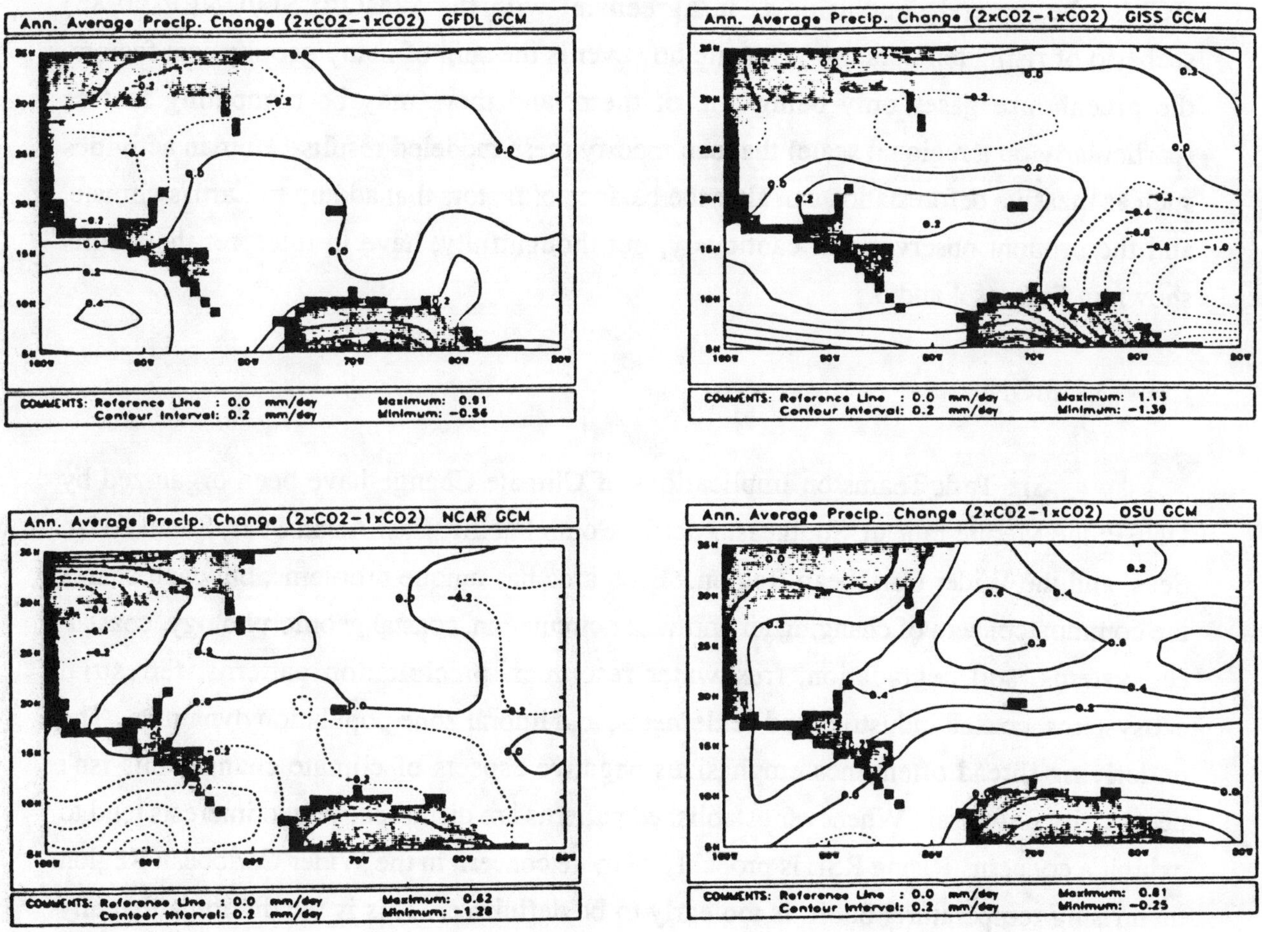

Figure 5: Annual average precipitation change due to effective CO_2 doubling from GCMs of the Geophysical Fluid Dynamics Laboratory (GFDL), Goddard Institute for Space Studies (GISS), Oregon State University (OSU), and the National Center for Atmospheric Research (NCAR). Maps courtesy of Wigley and Santer (*cf.* Chapter 1.2).

Although the results shown in Figures 4 and 5 vary considerably between GCMs, the general conclusion is in agreement with the WMO/ICSO/UNEP (1985) scenario of rising temperature. Climate however is the sum of many geophysical factors, the greenhouse gases only being one of them, and there may be competing factors (particularly on a regional scale) that can modify these modeled results. Human activities such as massive deforestation can alter the balance of factors that add up to Earth's climate, and the prudent observer will cautiously, but thoughtfully, have to interpret the results shown in Figures 4 and 5.

CONCLUSION

Six Task Teams on Implications of Climate Change have been organized by UNEP: the Mediterranean, Southeast Pacific, South Pacific, East Asian Seas, South Asian Seas, and the Wider Caribbean Region. Each area has unique problems, but each shares the common concern of changing air and water circulation, coastal geomorphology, coastal ecosystems, soil degradation, freshwater resources, precipitation patterns, terrestrial ecosystems, coastal industries and settlements, and littoral zone population dynamics. The underlying thread oftentimes emphasizes negative aspects of climate change; this isn't necessarily universal. Whenever established patterns are disturbed, vested interests tend to exhibit a concern. Rising RSL is probably of more concern in the Wider Caribbean Region than rising temperature, but it is too early to be definitive. This is the dilemma that any forecaster must confront.

Of primary concern to forecasting is the availability of adequate data. The Caribbean sea level network, which was in such good repair for earlier regional programs such as BOMEX in 1969, is now in substantial disarray. From a climate perspective, a sea level observing network must be reestablished and include marine meteorological data, geodetic leveling data, sea water chemistry data, and ancillary site-specific information. Because of the many short records of sea level and weather, and the difficulty of making conclusions based on them, a concurrent program of geological, archeological, and historical data analysis is considered a cost-effective means of strengthening those conclusions. There must also be rapid and free exchange of the observations, a basin-wide commitment to common problems, a responsibility to calibrate and intercompare measurements, and adequate sustained funding. Establishing and *maintaining* a modern sea-level/weather observing network is absolutely necessary to document and ultimately forecast climate change impacts. Of particular importance in such an observing system is the ability to record extrema in sea level and temperature of both the water and air; it is in the extreme events that climate change impact may be most noticeable.

Finally, we choose not to argue for wholly negative impacts. There is a realistic expectation that certain positive benefits may accrue; the local response to global change is simply not predictable at this time. What may be perceived as negative to one sector of society in the Wider Caribbean Region may be beneficial to another. Two examples: a change in precipitation associated with a temperature rise may allow the introduction of different crops but perhaps at the sacrifice of others; an increase in the alongshore component of the wind could increase coastal upwelling and be a benefit to fisheries, yet it may be a cause for concern to agronomists dealing with aerial erosion. A truly challenging and interesting problem will be to identify and explore the legal and institutional implications under the diverse systems and governments which characterize a region that has been influenced by so many European and native cultures.

REFERENCES

Bryson, R.A., and T.J. Murray, 1977. *Climates of Hunger*. University of Wisconsin Press, Madison, WI, 171 pgs.

Digerfeldt G., and M.D. Hendry, 1987: An 8,000 year Holocene sea-level rfecord from Jamaica: Implications for the interpretation of the Caribbean reef and coastal history. *Coral Reefs, 5*: 165-169.

Hansen, J., and S. Lebedeff, 1988: Global Surface Air Temperatures: Update Through 1987. *Geophys. Res. Lett., 15*(4), 323-326.

Hardin, G., 1971: Nobody Ever Dies of Overpopulation. *Science, 171* (3971), pg. 527.

Lamb, P.J., 1987: On the development of regional climatic scenarios for policy-oriented climatic-impact assessment. *Bull. Amer. Meteorol. Soc.*, *68*(9), 1116-1123.

NAS, 1975: Understanding Climatic Change, A Program for Action. U.S. National Academy of Sciences, Washington, DC, 239 pgs.

Oceanus, 1987/88: Vol. 30(4). Caribbean Marine Science. Woods Hole Oceanographic Institution, Woods Hole, MA 02543.

Pugh, D.T., N.E. Spencer, and P.L. Woodworth, 1987: Data Holdings of the Permanent Service for Mean Sea Level. Bidston Observatory, Birkenhead, Merseyside L43 7RA, UK, 156 pp.

United Nations Environment Programme (UNEP), 1987: *Action Plan for the Caribbean Environment Programme*. UNEP Regional Coordinating Unit, 14-20 Port Royal Street, Kingston, Jamaica, 24 pgs.

Wanless, H.R., J.J. Dravis, L.P. Tedesco, and V. Rossinsky, Jr., 1988. *Field Guide to the Carbonate Sediments of Caicos Platform with an introductory comparison with South Florida*. Amer. Geophys. Union, International Geological Congress Field Guide Series (in Press).

WMO/ICSU/UNEP, 1985: Report of the International Conference on the Assessment of the role of Carbon Dioxide and of other Greenhouse Gases in Climate Variations and Associated Impacts, held in Villach, Austria, 9-15 October, 1985. WMO Ref. No. 661, pp: 1-4, 1986.

The Arctic Sea Ice Record from Satellites--Is There Evidence of a Polar Warming?

(And What Would the Likely Consequences of a Warming Be?)

Claire L. Parkinson
Oceans and Ice Branch/Code 671
NASA Goddard Space Flight Center
Greenbelt, MD 20771

Abstract

Through satellite passive-microwave technology, it is possible to monitor global sea ice distributions on a routine basis. Satellite imagery of the Arctic region is available for most of the period since December 1972 and reveals a strong seasonal cycle with considerable interannual variability in the sea ice cover but no consistent long term trends. The lack of evidence of an Arctic warming is counter to first-order expectations from numerical modeling predictions that greenhouse-induced low-level atmospheric warming should be greatest in the polar regions. However, the short length of the satellite record limits the conclusions that can be drawn from it regarding long-term trends. In the event of a significant warming in the Arctic, sea ice should retreat, leading to more absorption of solar radiation at the sea surface and greater heat exchanges between ocean and atmosphere, thereby presumably intensifying the climate warming. Some consequences would be favorable, such as improved conditions for Arctic shipping operations, but the climate consequences could be very unfavorable and of uncertain magnitude.

1. Introduction

In the mid-1970s there was a marked retreat of the Southern Ocean sea ice cover, with the annual average ice extent decreasing by about 18% from 1973 to 1977 (Kukla and Gavin, 1981). This was briefly heralded as possibly the first geophysical evidence of a CO_2 warming of the atmosphere. However, it soon became apparent that this interpretation was premature, as the ice cover rebounded in the late 1970s and early 1980s (Chiu, 1983; Zwally et al., 1983). In fact the decreases in the mid-1970s, although fairly smooth and prominent for five years, were part of an oscillation in the ice cover which included comparable increases in both the preceding (Kukla and Gavin, 1981) and succeeding (Zwally et al., 1983) years.

The above example is illustrative of both the expectations regarding sea ice as a possible early indicator of climate change and the pitfalls which can be encountered when attempting to extract climate implications from a very short data record. Proceeding from that example, I will attempt in the ensuing sections to answer in turn four questions, beginning with: (1) Why should a global warming (or cooling) be reflected in the sea ice cover? and (2) Why might we be able to see changes in the sea ice cover before we see changes in other climate variables? The answer to question 2 hinges on the fact that we now have the technological capability of monitoring the sea ice cover routinely from satellite observations, leading to the third question: (3) What does the satellite record indicate regarding the Arctic ice cover and the existence or nonexistence of long-term trends in it? Finally, I will briefly address the several issues involved in: (4) How does the Arctic sea ice record from satellites relate to the larger climate record, what are the likely implications for the ice cover of a global warming, and what further consequences could be expected in the event of a significant retreat of the Arctic sea ice?

2. Why Global Temperature Changes Should Be Reflected In The Sea Ice Cover

The answer to the first question is relatively trivial. Sea ice forms because the water temperature reaches its freezing point or below, causing the water to freeze. Sea ice melts due to the action of solar radiation and various energy fluxes from the atmosphere and oceans. Should local atmospheric and ocean temperatures increase, less sea ice will form and there should be more energy available for ice melt, leading to a retreat of the ice cover. If sea ice occurred in only small, isolated locations, its extent might correlate very poorly with global temperatures because isolated local temperatures might well be out of phase with global temperatures. However, three-dimensional climate modeling studies consistently indicate that greenhouse warming should be particularly strong in high latitudes (e.g., see the comparisons in Schlesinger and Mitchell, 1987); and sea ice is not restricted to small, isolated locations but instead extends over a large portion of the polar oceans. The global sea ice extent at any given time almost always exceeds the combined area of the United States (9.5 x 10^6 km^2) and Canada (10 x 10^6 km^2) and often exceeds the area of the North American continent as a whole (24.4 x 10^6 km^2). In view of this large areal extent, and the expectation of an enhanced greenhouse effect in the polar regions, it would be unlikely that a persistent major global warming would not eventually be reflected in a decreased areal extent of global sea ice.

3. Observing The Global Sea Ice Cover

Reports of global temperature trends tend to be based on data from selected meteorological stations and ship reports (e.g., Jones et al., 1986, and Hansen and Lebedeff, 1988) rather than on a uniform global record. This creates several complications, such as the possibility that warming trends at some of the stations could be reflecting local urbanization and the urban heat island effect rather than larger-scale changes. Temperature not being a variable which can (so far) be observed globally, it becomes desirable to seek other variables as potential early indicators of climate change for which global depictions are possible.

Thirty years ago, sea ice could scarcely have been considered a potential early indicator of climate change. The remoteness of the polar regions and the harshness of the polar conditions made sea ice one of the most difficult of climate variables to observe in the large-scale, so that it would have been extremely difficult to identify with assurance interannual changes in the ice cover other than on a local basis. That situation has changed dramatically, however, with the availability today of satellite passive-microwave imagery.

Satellite passive-microwave data record the amount of microwave radiation received by a satellite instrument from the surface and intervening atmosphere. They are particularly useful for sea ice studies for the following reasons: sea ice emits far more microwave radiation than does open water, allowing the sea ice cover to be clearly delineated; non-precipitating clouds tend not to absorb or emit much microwave radiation, so that a cloud cover tends not to obscure the surface information received by the satellite; microwave radiation is emitted and can be recorded independently of sunlight, so that satellite passive-microwave coverage is obtainable throughout the day/night and summer/winter cycles; satellite passive-microwave imagery is obtained with a spatial resolution on the order of 30 km, which is fine enough to depict many interannual variations in the sea ice cover and yet coarse enough to allow routine global coverage; the imagery is obtainable with an excellent temporal resolution of about 1 day. As a result, the global sea ice cover can be routinely monitored from satellite observations, so that when large-scale changes in the sea ice cover occur these can be quickly known and analyzed. Unfortunately this has not been the case for a long enough period to establish a true climate data base, but at least a beginning has been made.

Satellite passive-microwave data are available for the polar regions for most of the period since December 1972, when the Electrically Scanning Microwave Radiometer (ESMR) was launched on NASA's Nimbus 5 satellite. The ESMR transmitted good quality data for most of the four-year period from its launch until the end of 1976. In October 1978, NASA's Nimbus 7 satellite was launched, carrying a Scanning Multichannel Microwave Radiometer (SMMR) that transmitted good quality data on an every-other-day basis for most of the period from October 1978 through August 1987. The Special Sensor Microwave Imager (SSMI) was launched on a satellite of the Defense Meteorological Satellite Program (DMSP) in June 1987 and continues to transmit passive-microwave data. Descriptions of the conversions of the ESMR and SMMR radiometric data to sea ice information can be found in Parkinson et al. (1987) and Cavalieri et al. (1984), respectively. The algorithm being used for the SSMI data is based on the Cavalieri et al. algorithm for the SMMR data, with only minor modifications. It is these passive-microwave instruments that are allowing the routine observation of global sea ice distributions, thereby opening the possibility that identifiable large-scale year-to-year changes in the sea ice cover might be observable before changes in many other climate variables, especially if these changes are occurring at the same time.

4. The Arctic Sea Ice Record From Satellites

The satellite passive-microwave record of the Northern Hemisphere sea ice cover reveals the expected dominating seasonal cycle, with the ice extent typically ranging from a minimum of 8×10^6 km^2 in September, when the ice covers much of the Arctic Ocean plus large portions of the Canadian Archipelago and the northwestern Greenland Sea, to a maximum of 15×10^6 km^2 in March, when the ice extends outward into many of the peripheral seas and bays, covering almost all of the Arctic Ocean, Canadian Archipelago, Hudson Bay, Baffin Bay, and Kara Sea, plus large portions of the Greenland Sea, northern Barents Sea, Sea of Okhotsk, northern Bering Sea, and Davis Strait (Parkinson et al., 1987).

Compilation of the available satellite passive-microwave data from the ESMR and SMMR instruments into time series of monthly areal sea ice extents reveals noticeable interannual variability in the Northern Hemisphere ice cover but no consistent long-term trend (Figure 1). When the data are further averaged to yearly averages and linear least-squares fits are calculated, the fitted line for the ESMR years has an upward slope of 75×10^3 km^2/yr (signifying a percentage change of 0.6% per year) and the line for the SMMR years has a milder downward slope of 23×10^3 km^2/yr. Not only is neither slope large, but the data points for both lines are scattered (Parkinson and Cavalieri, 1989).

The lack of a convincing long-term trend in the Northern Hemisphere satellite sea ice record is further apparent when the data are examined regionally. The Northern Hemisphere sea ice area was divided into eight regions by Parkinson et al. (1987), and this division was used by Parkinson and Cavalieri (1989) to examine regional sea ice trends separately over the ESMR and SMMR years. None of the eight regions was found to have a persistent trend, although the lines of least-squares best fit had negative slopes for the Sea of Okhotsk, Greenland Sea, and Kara and Barents Seas and positive slopes for the Bering Sea, Hudson Bay, Baffin Bay/Davis Strait, and the Arctic Ocean. The Canadian Archipelago had an ice cover which on a yearly averaged basis remained essentially constant over the ESMR/SMMR time period. The ice extent in the region with the strongest upward trend over the SMMR years, Baffin Bay/Davis Strait, peaked

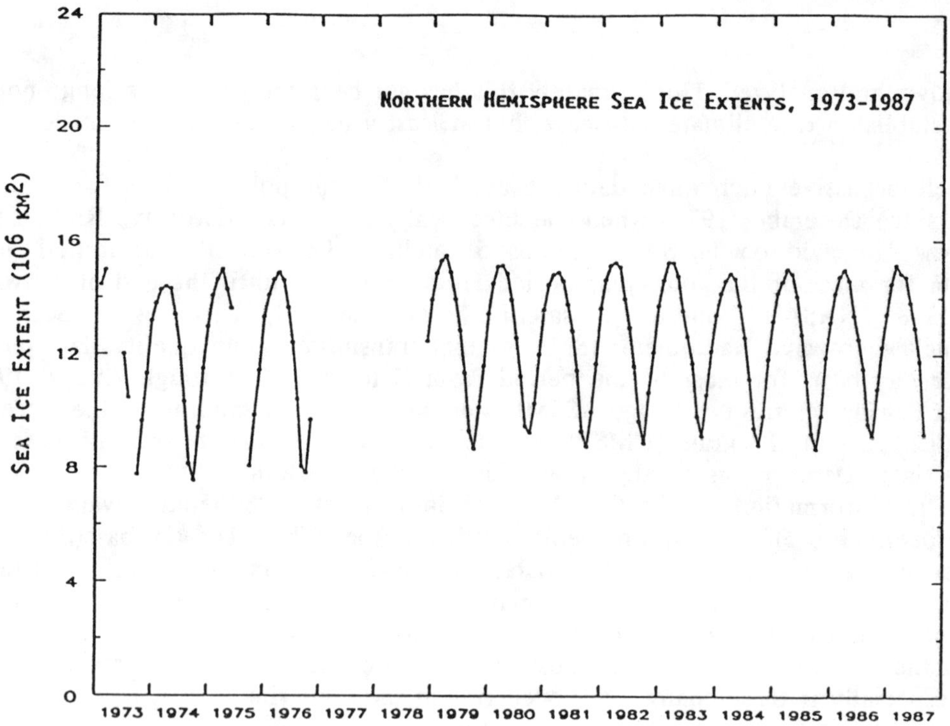

Figure 1. Time series of monthly averaged Arctic sea ice extents from the data of the Electrically Scanning Microwave Radiometer (ESMR) for January 1973 through October 1976 and the Scanning Multichannel Microwave Radiometer (SMMR) for November 1978 through August 1987. The extents plotted are for sea ice with areal concentrations of at least 15%. ESMR data are unavailable for March-May and August of 1973 and for June-August of 1975. [Modified from Parkinson and Cavalieri (1989).]

in 1983; and the ice extent in the region with the strongest downward trend over the SMMR years, the Kara and Barents Seas, showed significant increases after 1984. Both these facts further confirm the lack of a convincing, continuing trend, either upward or downward, in the ice extents of the Northern Hemisphere over the period of the past decade and a half. The reader is referred to Parkinson and Cavalieri (1989) for further information on Arctic ice extents as revealed in the combined ESMR/SMMR data set and to Parkinson et al. (1987) for much greater detail and analysis of the interannual variability over the shorter four-year ESMR record.

5. Discussion

Major general circulation models, including those developed at the National Center for Atmospheric Research, the Goddard Institute for Space Studies, and the Geophysical Fluid Dynamics Laboratory, consistently indicate that the low-level atmospheric warming from greenhouse-gas increases should be greatest in the polar and subpolar regions [see Schlesinger and Mitchell (1987) for a summary and comparison of results from several of the major model simulations]. Sea ice models in turn indicate that significant polar warming should lead to a noticeable retreat in the Arctic sea ice cover (e.g., Parkinson and Kellogg, 1979). Hence it would be expected that a greenhouse-induced warming would be accompanied by a retreat of the Arctic ice cover. The fact that no significant Arctic ice retreat is seen in the satellite data would present a difficulty for greenhouse-warming scenarios if that lack of a retreat were to continue for several decades. However, the lack of evidence of a retreat at this point could well be the result of having too short a data record.

Should global warming occur and the sea ice cover retreat as anticipated, this would have several further ramifications. To first order, a reduced sea ice cover should produce a positive feedback on local atmospheric temperatures, because the removal of sea ice would (1) result in more solar radiation being absorbed at the ocean's surface, as the high-albedo ice would be replaced by very low-albedo water, and (2) remove an effective insulator between the ocean and atmosphere. It could also have significant consequences on ocean circulations, especially bottom water circulations, as a large portion of the world's bottom water is believed to be generated in the polar regions, in the vicinity of the sea ice edge (Killworth, 1983). On a more immediate practical level, shipping operations and oil exploration in the polar regions would be easier with a reduced sea ice cover, and submarine acoustics would no longer suffer as much ice-caused interference.

To conclude: We now have the technological capability, through satellite passive-microwave imagery, of monitoring global sea ice distributions on a routine basis, so that changes in the sea ice cover can be seen more readily than changes in most other climate variables. The satellite passive-microwave record to date shows strong seasonal, regional, and interannual variability in Arctic sea ice, but it covers a period of only about 15 years, which is far too short to allow firm conclusions regarding long-term climate trends. These data do not show any persistent increases or decreases in the Arctic sea ice extent over the 1973-1988 period of the data record. However, should a major warming occur, the Arctic sea ice will almost certainly retreat, leading to further consequences of unknown magnitude.

References

Cavalieri, D. J., P. Gloersen, and W. J. Campbell, 1984: Determination of sea ice parameters with the Nimbus 7 SMMR. *Journal of Geophysical Research*, 89, 5355-5369.

Chiu, L. S., 1983: Variation of Antarctic sea ice: An update. *Monthly Weather Review*, 111, 578-580.

Hansen, J., and S. Lebedeff, 1988: Global surface air temperatures: Update through 1987. *Geophysical Research Letters*, 15, 323-326.

Jones, P. D., T. M. L. Wigley, and P. B. Wright, 1986: Global temperature variations between 1861 and 1984. *Nature*, 322, 430-434.

Killworth, P. D., 1983: Deep convection in the world ocean. *Reviews of Geophysics and Space Physics*, 21, 1-26.

Kukla, G., and J. Gavin, 1981: Summer ice and carbon dioxide. *Science*, 214, 497-503.

Parkinson, C. L., and D. J. Cavalieri, 1989: Arctic sea ice 1973-1987: Seasonal, regional, and interannual variability. *Journal of Geophysical Research*, submitted.

Parkinson, C. L., and W. W. Kellogg, 1979: Arctic sea ice decay simulated for a CO_2-induced temperature rise. *Climatic Change*, 2, 149-162.

Parkinson, C. L., J. C. Comiso, H. J. Zwally, D. J. Cavalieri, P. Gloersen, and W. J. Campbell, 1987: *Arctic Sea Ice, 1973-1976: Satellite Passive-Microwave Observations*, NASA SP-489, National Aeronautics and Space Administration, Washington, D.C., 296 pp.

Schlesinger, M. E., and J. F. B. Mitchell, 1987: Climate model simulations of the equilibrium climatic response to increased carbon dioxide. *Reviews of Geophysics*, 25, 760-798.

Zwally, H. J., C. L. Parkinson, and J. C. Comiso, 1983: Variability of Antarctic sea ice and changes in carbon dioxide. *Science*, 220, 1005-1012.

AN OVERVIEW OF POTENTIAL EFFECTS OF RAPID WARMING ON THE CANADIAN ARCTIC

Dr. Stephen C. Lonergan

Introduction

The general circulation models used to forecast global temperature and precipitation patterns that will accompany a doubling of carbon dioxide levels in the atmosphere have found that the greatest changes will occur in high latitude zones and, more specifically, in the Canadian Arctic. These findings generated an intense concern to assess what impacts - physical, biological, economic and social - might accompany such changes and stimulated a considerable effort on the part of the government to consolidate existing and past research on changing climate and to promote future impact assessment studies. The purpose of this paper is to discuss the relevant aspects of some of these studies and to report on the author's present work on economic and social impacts in the Arctic that might result from climate warming. The paper is divided into three sections corresponding to physical, biological and socio-economic impacts, although most of the discussion will center on the last section. There has been a dearth of research on both the biological and socio-economic components; many of the assessments are highly speculative, therefore, and are relevant only insofar as they use some of the more extensive results from research on the physical impacts of climate change as explicit assumptions to forecast social and economic effects.

It is important to note at the outset that it is very much a misnomer to speak of the Canadian Arctic as a single homogeneous region. There exist wide differences in climate, topography, soils, hydrology and vegetation, and, certainly, in social and economic activity. Detailed impact studies, by their nature, are location specific, and no attempt will be made here to generalize to the Canadian Arctic in its entirety. Most of the region's population and economic activity are, however, centered in the Mackenzie River Valley and the Mackenzie Delta, and much of the discussion will conentrate on these regions. In addition, most of the material in the sections on physical and biological impacts was based on a workshop sponsored by the Canada Climate Program and reported in French, et al. (1986).

Physical Impacts

Snow and Ice

The increased temperature and precipitation expected in the Arctic as a result of climate warming will have significant impacts on the presence of snow and ice. The most direct relationship will be to freshwater ice, and lakes in northern Canada should experience a shorter ice season of up to 20 days in the fall and

15 days in the spring (Barry, 1986). Previous work has shown that a fall season temperature increase of one degrees Celsius would delay freezing from 3 - 10 days (Palecki and Barry, 1986).

Some of the impacts on ice, however, might be mitigated by increased snowfall that is forecasted for the Arctic. Overall, snowfall is expected to increase although the snow season will be shorter due to higher temperatures. Scenarios developed with the Goddard Institute model (GISS) predict snowfall increases of 22% or greater at 70 degrees north, but snow cover duration will be shortened by 30 - 50 days as one moves south to 54 degrees (Barry, 1986). These snowier winters will augment glacier accumulation, but warmer temperatures will tend to overcompensate for this effect and net ablation should occur. Typical warming scenarios predict mass losses on the order of 38 cubic kilometers per year, which equals the negative balance which occurred in 1961-62 (Koerner, 1985).

Because of the complex dynamics of sea ice, little is known about the potential effect of climate warming on this system, but it may have important consequences for marine travel and Canadian security. Bilello (1961) predicted a 34% reduction in winter ice in the Beaufort Sea corresponding to a warming of 10 degrees Celsius. The GISS warming scenario implies a 20 day advance in ice breakup in Baffin Bay, but the dynamics of sea ice growth and decay and their relationship to climate is the least understood of any of the physical parametres.

Permafrost

The Canadian Arctic is underlain almost entirely by continuous permafrost and, in the southwest, by discontinuous permafrost. Changes in permafrost could have significant implications for vegetation, major development projects and local communities and is worth considerable attention. The simplified view is that as temperature increases, permafrost at the southern boundary disappears. The permafrost boundary has not been static in the past, having moved almost 200 miles northward in the Mackenzie Valley since 1850 (McKay, 1975). Certainly as temperature increases, the continuous permafrost zone would thin and become warmer and the melting of ground ice would lead to thermokarst activity. Probably the greatest changes would occur at the margin; where soil temperatures are near freeing. At this point the permafrost is very much affected by vegetative cover, the surface organic layer and snow cover. Additionally, as permafrost warms, the load bearing strength decreases dramatically.

So what will be the effect of rapid warming on permafrost? Certainly, large amounts of permafrost would eventually disappear. Even small increases in temperature would affect the active layer (which sits on top of the permafrost and ranges from one-half to three meters in depth) and decrease the load bearing strength. But since large amounts of energy are required to melt ice-rich permafrost, any degradation would be very slow (Smith, 1986).

The Biological System

The effects of rapid warming on Arctic vegetation has not been studied to date and all speculations are based on our knowledge of the spatial distribution of flora during the last global warming 8000 years BP. During that time the boreal forest extended 250 kilometers north of its present location. Warmer temperatures would imply a longer growing season and many plants could "re-extend" their ranges northward (Edlund, 1986). An increase in available moisture as well that is being predicted by the GCMs would result in an expansion of wetland communities. Any drastic change in ecosystems could also bring in new species, although this is unlikely unless conditions become much drier.

A similar situation exists with respect to animal life. The Canadian Arctic has a very low diversity of species. It contains only 15 of the world's 32 mammals, 70 of the world's 8600 birds and 50 of 23,000 fishes. This implies a simple and quite vulnerable food chain that may be very sensitive to changes caused by climate warming. Although the warmer water temperatures and increased nutrient flows could increase the number and species of fishes, mammals such as caribou and muskox could be threatened. Warmer, wet winters have caused problems in the past to these animals, and lemmings, the so-called "hamburgers of the north" may also be threatened. While migratory birds may expand their nesting ranges, some animals would find their migration patterns disrupted by the increase in open water (Harington, 1986).

The expected changes in the biological system are very speculative and the time horizon very uncertain. Many of the expected changes are simply extrapolations of past occurances, either long periods of warmer temperatures, as existed in the Holecene some 6000 years ago, or examples from more recent, extreme events such as the warm winter of 1973 that resulted in many caribou deaths. Far more research on the impacts of climate change on vegetation and wildlife is needed.

Social and Economic Effects

Similar to the situation discussed above, there has been a dearth of study on the social and economic effects of rapid warming in the Arctic. This is being rectified somewhat, as the author, in collaboration with Dr. M-K. Woo of McMaster University is undertaking a major study of the social and economic effects of climate warming in the Mackenzie River Valley and the Mackenzie Delta. Much of what is reported below represents preliminary results of this study.

Forecasted temperature and precipitation scenarios under a doubling of carbon dioxide in the Mackenzie region are illustrated in Figures 1 and 2. Temperature under all three of the GCM's listed is expected to be higher than the past 30 years during all seasons, with the GFDL model showing the greatest overall increase.

freezing average temperature, as opposed to the negative six or seven that has existed during the past three decades. This could have significant effects on the permafrost and ice and snow melting, as discussed above. The annual average temperature could be as much as eight degrees or more higher than at present.

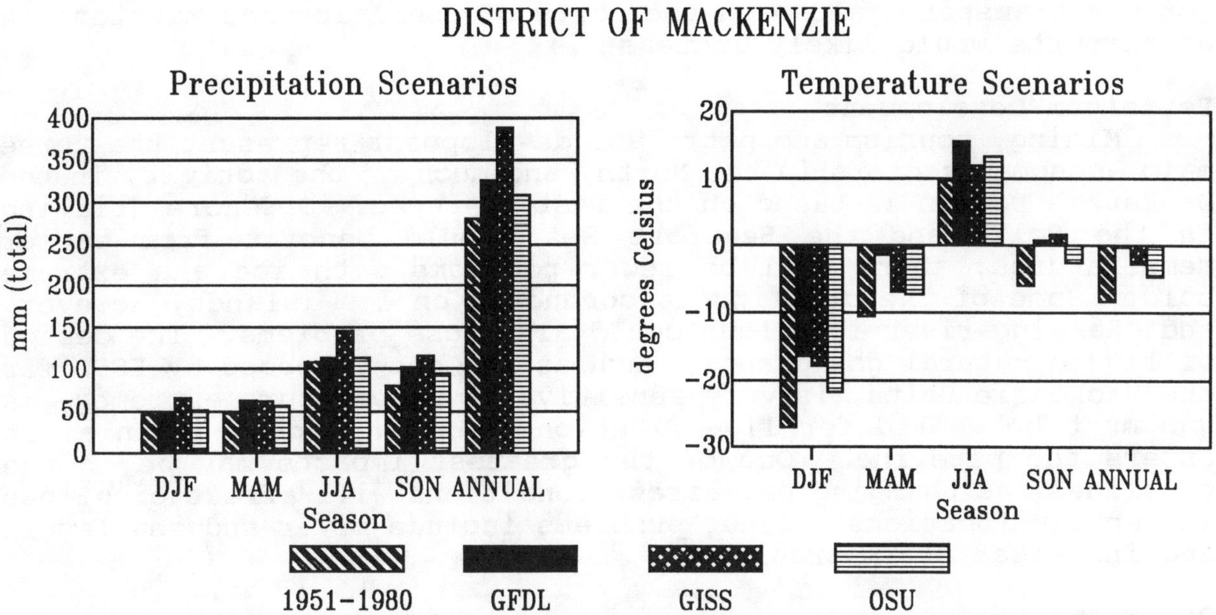

Forecasted precipitation shows a similar pattern, but with the GISS model predicting the greatest increase (over 35% more than at present). Much of this would come in the summer or early fall, but a greater number of storms would be expected throughout the year. More important than simply an increase in annual average temperature or precipitation, however, are the frequency and magnitude of extreme events; the droughts, floods and severe storms that appear as variations around the average. Preliminary results show that these events are expected to become more frequent and of greater magnitude. It is felt that these extreme events will have a greater economic and social impact than the simple averages, or at least will require greater planning and policy consideration. Listed below are some of the impacts that are expected to occur.

Transportation

Warmer temperatures should benefit marine transport as an extended shipping season of up to six to eight weeks is likely and reduced ice flows will be present. Rougher seas could pose a problem, particularly since much of the increase in precipitation is expected for the summer months and more icebergs from ice calving may counteract some of the beneficial impacts. A similar situation would occur for barge traffic in the Mackenzie River, as the season would be extended due to later freeze-up and earlier break-up. The season now runs four months, with a crucial six week period during break-up in the spring where no barge or truck shipments can be made.

The winter roads, which extend down the Mackenzie and throughout the Delta, will experience problems with greater snowfall and a shorter season. With barge traffic non-existent in the winter, these roads are a major source of commodity shipments in the North, and climate warming will pose a hinderance to their operations. Air shipments could be affected as well with more storms likely, but to date weather has not posed a major problem for air transport, although the costs of operation and maintenance at airports would likely increase.

Petroleum Development

Mining, tourism and petroleum development represent the three main economic sectors in the North, and much of the activity in the Mackenzie region is based on the last of these. Offshore drilling in the Delta and the Beaufort Sea should benefit from warmer temperatures; there will be fewer problems with ice and extreme cold. Some of the drilling is conducted on ice islands, however, and changing river ice flows could also pose problems. The design of liquid natural gas plants, such as the one proposed by ESSO for the Mackenzie Delta, is very sensitive to temperature extremes, as gas must be cooled for liquification and then cooled again as it enters the pipeline. One of the greatest impacts may be on the pipelines, as changing permafrost conditions will affect pipelines in certain locations. Other problems include river and sea levels and increased storm problems.

Other Impacts

The Arctic represents a very minor economy relative to the rest of Canada, with far less than one percent of the population and of national GDP. Nevertheless, the region is an important cultural and natural resource, and the impacts on this region are important for the entire country. Improved capabilities in mining, petroleum and agriculture may generate some economic benefit for native groups and residents, as will the increased tourism that can be expected from warmer temperatures. With increased activity, however, might be negative impacts on the indigenous society as well, with construction "booms" as exploration and development activity picks up (and corresponding "bust" periods), the advent of more tourists and changing lifestyles as animal and plant species change.

The impacts of rapid warming on the Arctic, like other regions, will be extensive and varied. With the expectation that this region will experience the greatest increases in temperature, there is reason to be concerned over the potential impacts. This paper has provided some insight into a few of these expected impacts and noted that substantive work on biological and socio-economic impacts in particular has just begun. We have only confronted the tip of the iceberg, so to speak!

(Complete citations for the references listed above can be found in: French, H.M. (ed.), 1986. Climate Change Impacts in the Canadian Arctic, Proceedings of a Canadian Climate Program Workshop, Atmospheric Environment Service, Downsview, Ontario.)

Impacts of Climate Change on California Water Resources

Joseph B. Knox and Robert W. Buddemeier

Lawrence Livermore National Laboratory
P.O. Box 808
Livermore, CA 94550

California Water Resources Management

The California water system is highly engineered and managed. Although the state appears to possess abundant water resources, demand is separated from supply in both time and space. Most of the precipitation falls during winter in the mountains of northern and central California, while population and irrigated agriculture are concentrated along the coast and in the valleys of central and southern California – often in semi-arid to arid environments with high summer irrigation demands. This has resulted in a series of projects to capture, store and transport large volumes of water from catchment to end users.

This water system is economically vital; the system and its components have become the cornerstone for state-wide municipal, industrial and agricultural development. California has a rapidly growing population, a "GNP" larger than that of many nations, and an agricultural economy that accounts for over 80% of the state's net water use and produces farm income on the order of $15 billion annually.

Consideration of the function of the state's water system and its potential vulnerability to climate change is complicated by three important features. First, the system is neither hydrologically nor administratively unified, even though we discuss it as such for convenience—in reality, it consists of a collage of overlapping agencies and jurisdictions representing all levels of government. Second, the major facilities and agencies are multipurpose—in addition to supplying water users, they have responsibilities involving flood control, hydropower, recreation, environmental preservation/protection, fisheries, navigation, etc. Finally, we note that as a consequence of both its importance and its complexity, the water system design and operation is politically sensitive—designed more by compromise and conflict than by consensus, and often more influenced by politics than by hydrologic and engineering considerations.

Present policies for water system management are based on existing water law and contractual arrangements, and on the previously credible assumption that the water supply of the future can be predicted from the past weather/climate observations. Because of the high economic and political stakes, divided authority, and the complex network of historically developed water law and policy, management policies tend to be very conservative and oriented to present and past conditions. Present policies will be influential far into the future, however, because of the long lead times required for legal and political change, and for authorization and construction of major engineering projects. It is the thesis of this paper that very probable and reliably predictable aspects of climate change (the greenhouse effect) could seriously affect California's water distribution system. We address in this paper some of the policy, management, and research needs that will have to be met if the state's water system is to anticipate, prepare for, and operate resiliently in, a changing environment.

Climate Change and Water Resource Impacts

When the effects of all greenhouse gases (CO_2, CH_4, CFCs, etc.) are combined, it looks like we are headed for a combined effect on our planet's climate system equivalent to a doubled atmospheric CO_2 concentration by as early as 2030. There is consensus that unless there are unknown or unpredictable compensating factors (e.g., increased vulcanism or unanticipated climatic feedback mechanisms), this implies an increase in global mean temperature in the range of 1.5 to 4.5°C (with greater warming at high latitudes than in the tropics) occurring around the middle of the next century. Elements of the 3D–"fingerprint" of response of the climate system will begin to be perceptible in the near future. A further manifestation of this warming is projected to be a global sea level rise of 0.5 to 1.5 m that is predicted to occur by the year 2100. At a more detailed level, there are substantial uncertainties in the rate, timing, and pattern of climatic shifts, the nature of greenhouse climate conditions in specific regions (e.g., precipitation changes, storminess, variability), responses of oceanic circulation patterns to the climatic forcing, and human responses. Although resolution of these uncertainties and more reliable regional predictions are most highly desirable, we can rather confidently predict a number of direct impacts on California water resources as a result of the probable warming. The discussion that follows assumes a warmer world and explores its implications.

Gleick (1987) has shown that warmer winters result in a higher snowline and a shorter snow season in the mountains. This shifts a larger fraction of the total runoff into the winter months at the expense of the spring and summer runoff; this change in the relative runoff patterns is insensitive to details of whether annual precipitation increases, decreases, or remains the same. California rivers will thus experience reduced natural flows during most of the year unless the shift in runoff is artifically compensated by release of additional captured water. The reduction in volume flow during the spring and summer will decrease the amount of water available for extraction from the river system at precisely the time of year when water demand is at its peak. In addition, reduction in volume flow presumably translates into lower linear flow rates and therefore longer residence times; this may be detrimental to water quality because of reduced flushing and dilution of contaminants and salts. Higher peak winter flows increase flood probability and also require capture of a larger fraction of the winter runoff to augment flows later; the increased collection requirement is probably in excess of present and foreseeable on-stream reservoir capacity, and the need for earlier runoff retention may seriously conflict with prudent flood control strategy.

Warmer temperatures result in increased evaporation; this affects both water supply and demand. Reduced soil moisture levels during the growing season (Gleick, 1987) increase demand for both agricultural and landscape irrigation. Less certain but possible are impacts on natural groundwater recharge and local (e.g., coast range) runoff collection if winter and spring temperatures rise enough so that evapotranspiration produces significantly greater soil dryout between storms than is presently the case. Higher temperatures increase evaporative losses from surface water; not only will this further reduce supply, but water quality is likely to be reduced by evaporative concentration of dissolved and suspended solids. Water quality effects may also be exacerbated by temperature-induced changes in solubilities and reaction rates, as well as the reduced flow rates noted above. In addition to the terrestrial effects of temperature rise, additional water system impacts will result from the sea level rise due to warmer temperatures. Encroaching ocean water threatens the quality and therefore the usable quantity of fresh water resources because of increased salt water intrusion into coastal aquifers and low-gradient rivers and estuaries. Elevated water levels also increase the vulnerability to flooding or storm damage of water facilities (levees, pipelines or canals, treatment plants, etc.) in low-lying areas.

Should the "Future Climate" Determine Present Water Policy?

It is our contention that the reliably predictable impacts of climate change on the California water system are at once so highly probable and so potentially detrimental if they do occur that all levels of government should factor them into water policy and planning at the earliest opportunity. We have avoided speculative or uncertain predictions, and have focused on the effects of the temperature rise that are considered very likely to occur over the next several decades; although the details of timing or magnitude may be in question, there is little doubt about the general nature of the expected changes.

The California water system is extremely vulnerable to the combined effects of these changes. The most striking and potentially critical example is the Sacramento-San Joaquin Delta. The Delta channels, formed by unstable levees more than a century old, serve as a presently irreplaceable transport facility to move water from the northern mountains to the Central Valley, southern California, and the San Francisco Bay area. Salt-water intrusion into the lower Delta is presently a problem, as is maintenance of the levee system. If California experiences reduced summer runoff, there will be reductions in both the quantity and quality of water available to meet the increased summer demands occasioned by rising temperatures and even greater problems of salt water intrusion because of both summer flow reductions and the effects of rising sea level. Although these threats to the function of the Delta as a transport facility are serious in their own right, climate change also threatens its physical survival. Increases in winter runoff increase the probability of flooding; storm damage and rising sea level amplifies this threat to the integrity of the levees. Loss or serious impairment of the Delta's water transport facility would have enormous economic and social consequences for the state of California and indeed for the nation.

The significance of the future threat to the water system is enhanced by two compounding variables. First of these is rising population: the state's population is expected to double by the time the "doubled CO_2" condition is in effect, and the rates of growth are highest in the areas farthest removed from the sources of water. In addition to increasing demands for new residents, another important consideration relates to the relative rates of change in climate and in water system facilities and operations. The present State Water

Project is already over 25 years old and only half complete according to its original standards and planning assumptions; on both the political and the engineering level, systems of this magnitude take decades to design, authorize and implement. If we accept that major climate change is highly probable on a time scale of less than half a century, it is clear that we cannot be prepared for that future unless active planning starts almost immediately.

Policy and Information Needs

We believe that the first step in developing a pro-active climate-oriented water policy is the decision to have a policy. The state and federal governments should adopt as a water policy guideline the finding that given the present state water resource system, the probable climate changes may be expected to reduce both the availability and the manageability of California water resources in the future, and that all water planning and development activities shall be required to be consistent with this finding. As an implementation step, the various levels of government should establish a coordinated administrative framework to develop, evaluate and implement policies to minimize the negative impacts of changing climate on California's water resources. To be effective, this will have to be not simply another agency or task force, but an innovative approach to bridging interagency barriers by combining political and economic guidance with scientific research and assessment on a continuing basis.

Among the strategies that should be assessed or implemented on a priority basis we can identify:

1. Assessment of system capabilities for dealing with larger and more frequent extreme events (floods, protracted periods of high or low runoff, etc.);
2. Exploration of the potential for technical, legal or economic measures to reduce both total water use and peak demands in the agricultural and municipal-industrial sectors;
3. Increase in system capacities to divert or extract and store peak winter runoff for subsequent use.

Although effective water use and development policies can and should be developed on the basis of our present knowledge of climate change, we must concurrently continue to improve our understanding of both the nature of climate change and of the types and interactions of its specific impacts.

One of the necessary features of future work is the continuation and intensification of present efforts to improve the General Circulation Models (GCM) now in use, especially with respect to focusing down to local and regional predictions—for example, the scale of the GCM grids is presently much larger than the size of California watersheds. This work will involve the progressive incorporation of air-sea interactions, finer grid sizes, more realistic topographic effects and other improvements, and will yield gradual improvements in regional climate scenarios over time.

We can, however, achieve substantial improvements in our understanding of many of the impacts of climate change without waiting for a perfect computer model or a consensus from several GCM's being continually but independently improved. Gleick (1987) has demonstrated how robust predictions (e.g., of rising temperatures) can be combined with basic physical principles and simple but well-proven models to yield information that is vitally important to planners. It is our opinion that Gleick's initial results are, in fact, conservative with respect to estimates of the magnitude of the effects on water resources, and that more detailed calculations involving additional parameters will lead to more pessimistic conclusions than reported in that study. A high priority should be placed on expanding this type of approach to address such issues as water quality impacts, groundwater recharge changes, future water supply variability, and the probable frequency, intensity and manageability of extreme events.

We believe that the policy community, the public, and indeed perhaps the technical community have been unnecessarily distracted by controversy about the limitations of the surface-based climate data base, and its possible bias from the so-called urban heat island effect. They may also be confused by the charge that the observations from the real climate system and the assessments from theory do not agree. If this were indeed the case, it would be imprudent to modify present water planning. However, recent studies of the global warming of the free troposphere and the stratospheric response with cooler temperatures averaged over a decade are impressive in that this evidence is obtained from layers far removed from any urban heat island effect or pollution island effect of cities (Angell, 1988). This evidence is consistent with theory.

In regard to the conflicts of observation and theory, we note that the theory is incomplete and the system is more complex than represented by the best of models. Hence, inconsistencies are to be expected; the analyst is challenged to detect the expected greenhouse signal incrementally as the fingerprint of response emerges from the noise. The analysis of large-scale climate variables suggests strongly that change is in progress and is in the direction of a warmer world.

California has traditionally been one of the nation's leaders in environmental and resource management. Climate change presents a particularly credible and serious threat to the state's water system – one of its most critical and most vulnerable resources. We anticipate that California will be in the vanguard of efforts to understand, anticipate and respond to the problems of a changing climate.

References

Angell, J. (1988) "Variations and Trends in Tropospheric and Stratospheric Global Temperatures, 1958–1987," presented at the Climate Trends Workshop, Sept. 7–9, 1988. National Climate Program Office, NOAA.

Gleick, P. H. (1987) "Regional Hydrologic Consequences of Increases in Atmospheric CO2 and Other Trace Gases." *Climatic Change, 10,* 137-161.

THE IMPACTS OF CLIMATE CHANGE ON THE SALINITY OF SAN FRANCISCO BAY

by
Philip B. Williams
Philip Williams and Associates
Pier 35, The Embarcadero
San Francisco, CA 94133

Introduction

San Francisco Bay is the largest estuary on the U.S. Pacific Coast. Its estuarine ecosystem is dependent on the amount and timing of freshwater inflows and the resulting salinity distribution throughout the Bay. The upper part of the estuary, the Sacramento-San Joaquin Delta, acts as a conduit for water supply transfers from the northern part of California to the arid San Joaquin Valley and Southern California. Consequently, salinity management in the estuary is vitally important to protecting the estuarine ecosystem and the quality of water exported for agriculture and urban water users.

Projected climate change due to the greenhouse effect will greatly alter the salinity distribution in the estuary. There are two primary impacts:

- The timing and amount of freshwater inflow will change.
- Sea-level rise will alter the tidal characteristics of the Bay, with dramatic changes taking place if levees protecting low-lying areas are allowed to fail.

Methodology

Three steps were required for the analysis of these changes:

- Define the future morphometry (shape) of San Francisco Bay with sea-level rise.
- Determine the tidal exchange characteristics for San Francisco Bay for the future morphometry.
- Determine the salinity response of future tidal characteristics to future Delta outflow scenarios.

Because water released from reservoirs for salinity repulsion (carriage water) affects future Delta outflows, new carriage water flows for future tidal characteristics were developed. With these new carriage water requirements, the monthly Delta outflows for three different doubled carbon dioxide climate change scenarios were modelled.

Two separate doubled carbon dioxide scenarios of the morphometry of the estuary were used to investigate the range of possible salinity responses:

- 1-meter sea-level rise, no levee breaks.
- 1-meter sea-level rise with all levees failed.

The changes in estuarine morphometry with an increase in sea level have a significant effect on tidal circulation, which in turn is a major determinant of salinity distribution.

Fischer's one-dimensional finite element tidal hydrodynamic model (Fischer 1970) was used to simulate the tidal exchange characteristics at various points in the estuary. This input was then used in a simple mixing model (Denton & Hunt 1986) to simulate the response of salinity in the estuary to different Delta outflows for future conditions.

The mixing model was run for steady-state (constant freshwater outflow) conditions for the two sea-level rise scenarios, as well as a scenario with levee failure at existing sea-level. The steady-state condition is a theoretical calculation of salinity that would seldom actually occur because of the large seasonal change in runoff, but is useful for systematically analyzing sea-level rise effects on carriage water requirements.

Using the new carriage water requirement in a reservoir operation simulation, Sheer (1988) simulated the monthly Delta outflow for three climate change scenarios for the hydrologic base period 1951 to 1980. The three scenarios were developed from the following General Circulation Models (GCM's):

- o Geophysical Fluid Dynamics Laboratory (GFDL)
- o Goddard Institute of Space Science (GISS)
- o Oregon State University (OSU)

All of these scenarios have greater annual freshwater inflow to the estuary than occurs at present. In addition, all have a significant seasonal shift in runoff from the spring to the winter.

The mixing model was run again using the simulated monthly Delta outflows as input. The output included average monthly and average annual salinities in different bays within the estuary under the different scenarios.

Results

Failure of the perimeter levees would have a significant effect on the physical character of San Francisco Bay. It could cause a tripling in area from approximately 1,100 to 3,500 square kilometers and a doubling in volume from 7 to 14 cubic kilometers (see Figure 1). Most of this increase would occur from failure of Delta island levees. These levees are the most fragile and poorly maintained, and are susceptible to failure even at the existing sea level. The exact amount of future sea-level rise is less important hydrodynamically than whether or not the Delta levees are preserved. If all levees are preserved, the area of the Bay could increase 30% and its volume 15%.

The tidal transport or average tidal velocity increases dramatically with increasing volume. If levees are maintained, the deeper water tends to make the tidal channel hydraulically more efficient, and this tends to compensate for the increased energy losses due to the larger velocities. The net result is that tidal ranges do not appear to change dramatically from the existing condition.

If all levees fail, the large increase in tidal prism (the volume of water between high and low tides) in the Delta increases the velocities and energy losses in downstream constrictions in Carquinez Strait and Chipps Island and greatly reduce the tidal range. This reduction in tidal range largely compensates for the increased tidal volume. It is important to note that this analysis assumes rigid boundary channels. In fact, there would be a considerable scouring in response to the increased tidal velocities in the lower part of the Delta, causing deepening and widening of channels. These changes could substantially affect the tidal hydrodynamics and allow for an increased salinity in the future.

The steady-state salinity analysis for the two levee failure scenarios indicates that approximately double the Delta outflow is required to maintain the same salinity as present-day conditions. Consequently, the carriage water requirement to maintain low-salinity water for export would be approximately doubled.

The mixing model simulation results are illustrated in Figure 2. In all scenarios, the average spring and summer salinities are higher, due to the combined effects of sea-level rise and reduced runoff. Only in the winter is the greatly increased runoff of the GFDL scenario sufficient to compensate for the effect of sea-level rise. Sea-level rise by itself causes the average salinity to migrate about 15 km inland; however, the increased winter runoff partially compensates for this. The combined hydrologic and sea-level rise effect of climate change results in a net inland migration of the average annual salinity of up to about 10 km (6 miles). This 10-km shift in salinity is approximately the same magnitude as that which has already occurred due to upstream water development in the last century (Williams and Fishbain 1987).

Policy Implications

The question of the influence of the failure of the Delta island levees on salinity intrusion may be crucial in the justification for public expenditures to improve the existing levees. An estimate to improve these levees to protect against flooding from existing sea level is approximately $4 billion. The total valuation of property in the Delta is about $2 billion.

With accelerated sea-level rise causing major increases in salinity, the policy question becomes: whether the incremental value of maintaining levees in the Delta to prevent further increases of salinity, in addition to protecting farmland and infrastructure, justifies the cost.

Only in the last decade have the freshwater inflow requirements for maintaining the estuarine ecosystem been quantified. These requirements have not yet been incorporated into the water allocation and reservoir operating criteria of California's water system. With the major changes in estuarine hydrology, morphology, and hydrodynamics, it will become even more important to modify water resources management to incorporate protection of the estuarine ecosystem. This could have positive benefits, for example: increased reservoir releases for salinity repulsion may also have substantial ecological benefits for San Francisco Bay.

The vulnerability of the State Water Project and Central Valley Project to salinity intrusion into the Delta was the genesis of the controversial and now-abandoned Peripheral Canal plan. The high cost of large diversion schemes that bypass the Delta probably means that salinity will continue to be managed by releases from reservoirs. Sea-level rise and climatic change will require that these releases be increased. The larger releases will require reallocation of water contracts and revision of present water management practices. These revisions may lead to consideration of structural modifications to the physical system such as channel modifications or off-stream storage.

Because of the large economic, social, and environmental costs of these policy implications, the value of a greater degree of certainty in this kind of analysis is high. A major research effort needs to be initiated to analyze the estuarine morphology, sediment dynamics, and hydrodynamics.

<u>References</u>

Denton, Richard A. and James R. Hunt. "Currents in San Francisco Bay: Final Report." University of California, Berkeley, California, 1986.

Fischer, Hugo B. "A Method for Predicting Pollutant Transport in Tidal Waters." University of California, Berkeley, Water Resources Center, Contribution No. 132. 1970. 143 pp.

Sheer, D. 1988. "Reservoir management of freshwater runoff for the California Central Valley and Allianta." For EPA. In press.

Williams, Philip B. and Larry Fishbain. "Analysis of Changes in Suisun Bay Salinity Due to Existing and Future Water Development." Report 412-2. San Francisco, California, 1987.

Williams, Philip B. "The impacts of climate change on the salinity of San Francisco Bay." Report prepared for ERL Corvallis, U.S. EPA. 1988.

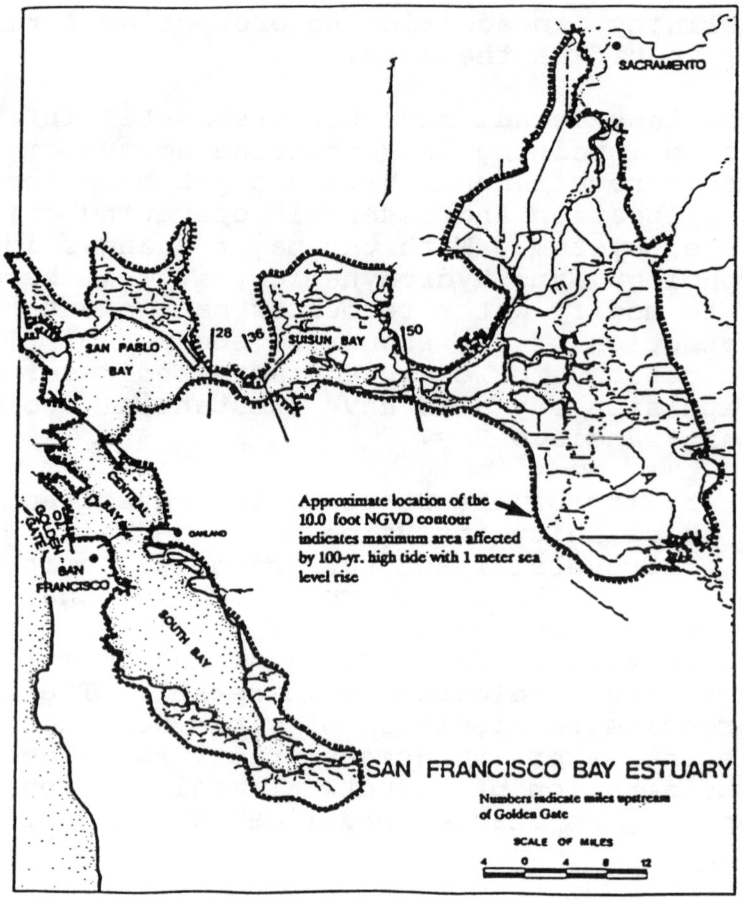

FIGURE 1

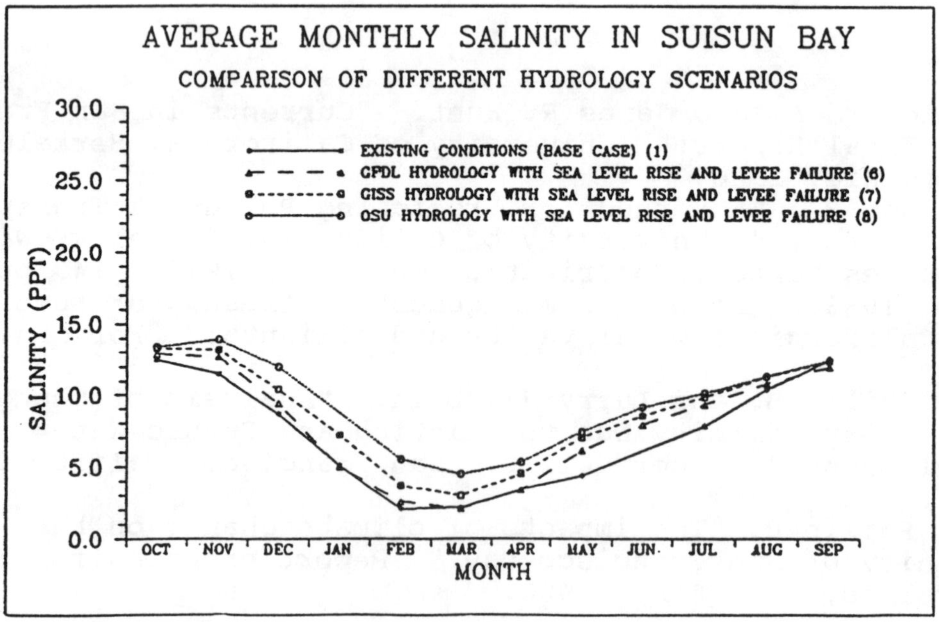

FIGURE 2

POTENTIAL CLIMATE CHANGE EFFECTS ON IRRIGATED AGRICULTURE IN CALIFORNIA

by
Daniel J. Dudek
ENVIRONMENTAL DEFENSE FUND
New York, New York

Introduction to Study Area

This study assesses the impacts of climate change on California's agricultural and water resource systems. Previous studies (Dudek, 1987) have identified irrigated arid regions as potentially vulnerable to climate change. Since California agriculture annually produces about 10% of total cash farm receipts in the United States, it was selected for a more detailed case study. In 1986, California's farm income was first in the Nation at $14.5 billion, followed by Iowa and Texas with $9.1 and $8.5 billion respectively. Grapes, cotton, hay, lettuce, almonds, tomatoes, strawberries, oranges, broccoli, walnuts, sugar beets, peaches, and potatoes represent, in descending order, the major irrigated crops ranked in terms of gross receipts. As a sector, California agriculture produces 3-4% of total state income.

The State of California is also an important example of a highly developed irrigated system with jurisdiction over 1,188 dams and reservoirs with a gross storage capacity of 19.7 million acre-feet. The federal government operates an additional 125 dams and reservoirs with 22.9 million acre-feet of capacity. Various estimates can be made of the economic value of this installed capacity, but a very rough indicator of the magnitude of the investment in this system is $15-$20 billion using an average cost of $500 per acre-foot of installed water supply delivery capacity.

Methodology

The methodology used in this study (Dudek, 1988) explicitly links predictions of climate changes from general circulation models (GCMs) of the atmosphere with an agricultural productivity model. These productivity impacts are introduced into the California Agriculture and Resources Model (CARM) which determines the economic and market implications of such changes. CARM measures aggregate economic values in terms of changes in values received by both producers and consumers, i.e. producers' and consumers' surplus (CPS). The climate changes assessed include temperature, evapotranspiration, precipitation, and cloudiness. The implications of such changes for water resource supplies were separately evaluated for the Sacramento and San Joaquin Valley Basins by a team of hydrologists (Lettenmaier, 1988; Sheer and Randall, 1988).

To assess climate change driven productivity impacts, an existing agro-ecological zone model (De Wit, 1965; Doorenbos and Kassam, 1979) was adapted. For several general crop groups, this method estimated that climate changes, in general, would reduce yields with the greatest impacts in interior southern regions on cool season crops such as sugarbeets. Overall, yield reductions under the GISS 2xCO_2 scenario were greater than under projections from the GFDL GCM owing to the greater temperature increases for California produced by the GISS model. Reductions in crop productivity from climate changes ranged from modest 4% declines to more serious 40% impacts for the crop groups and regions selected. Including CO_2 enrichment effects significantly alters the overall result. Net productivity changes range from a 41% increase to a 3% decline. This radical change reflects the differential ability of crops to utilize the increased CO_2.

Results

The economic system exerts compensating influences which tend to modulate the ultimate influence of direct physical productivity changes. Market incentives act to reallocate production activities and resources to maximize returns to society under the environmental conditions specified in the scenarios. Scenarios for the economic model were constructed from the climate changes predicted by alternative GCMs, by the specific set of productivity impacts analyzed, by the availability of water resources, and assumed social response to such changes. Results from CARM indicate that even after market responses function, statewide average yields would be significantly reduced for all crop groups as a result of climate changes. Vegetables would be least severely impacted with average yields reduced from 6-15%. Fruit and nut crops would be hardest hit with average declines from 23-33%. After CO_2 enrichment is factored in, vegetables and cotton yields show positive increases over the base, but the remaining crops still have yield reductions ranging from 1-11%.

Regional changes in the production of agricultural commodities and in the use of resources by agriculture dominate the results. Increased temperatures in California would produce more rain and less snow during the winter. In addition, the snowpack would melt earlier in the spring reducing total effective reservoir storage and available supplies. The hydrologic study team has estimated that for a thirty year simulation of hydrologic flows under alternative future climates, SWP deliveries were reduced by 25-28% on average. Federal Central Valley Project (CVP) service area deliveries are not similarly affected due to differences in the seniority of water rights.

Figure 1 portrays the spatial pattern of resource use changes that might result from climate changes depicted by the GISS and GFDL GCMs. The values depicted are an index constructed from the ratio of scenario results to the 1985 base period. Overall, the Imperial Valley is the region most severely affected experiencing nearly uniform reductions in crop acreage, ground and surface water use. In acreage terms, both the Delta and South Coast regions are consistent gainers. The results display the north to south and coast to inland pattern of impacts.

The most dramatic changes occur in groundwater use in the Northern San Joaquin Valley between the scenarios based upon climate change impacts only and those which include CO_2 effects. These shifts are driven both by crop acreage increases in general and by alfalfa hay acreage differences in particular. Alfalfa is only exceeded by rice in its water requirements. For those regions relatively less disadvantaged by climate changes, in both productivity and water supply terms (northern and coastal regions), resource changes are beneficial. For example, in each of these regions groundwater is more expensive than surface water and so groundwater pumping is reduced. These reductions are accomplished through a combination of cropping pattern and irrigation efficiency changes.

Overall, climate change effects reduced the net economic well-being (CPS) produced from agricultural operations between 14-17% (see Figure 2). Statewide crop acreages were reduced between 3.5 and 6% depending upon scenario from a base level of slightly more than 9 million acres. Regionally, acreage declines were greatest in the Imperial Valley. Although no assessment was made of future groundwater stocks or pumping lifts, pumping declined roughly 20% statewide as a result of reduced crop profitability. Overall surface water use declined roughly 16% as a result of both supply and demand changes. Regional water use changes were most dramatic throughout the San Joaquin and Imperial Valleys. Factoring in CO_2 effects produced significantly different results. Total crop acreage was broadly similar to the 1985 base with net economic well-being showing a slight 1% decline to a 2% improvement.

The last phase of this study evaluated social responses to climate change and their potential to reduce the impacts of such changes. This analysis focused on the introduction of irrigation water markets and improved on-farm irrigation management. All publicly provided surface water supplies were offered for sale as long as transfer charges were profitably covered. For all scenarios, including the base, water marketing produced net economic benefits. Overall, the results after water marketing are broadly similar; irrigated acreage changes little, groundwater pumping is increased and surface water use declines. The bottom section of Figure 2 describes the changes after water marketing in reference to the corresponding scenarios without marketing. For the climate change scenarios, crop acreage is slightly reduced (less than an additional 1%), but consumers' and producers'

Figure 1. Regional Resource Changes

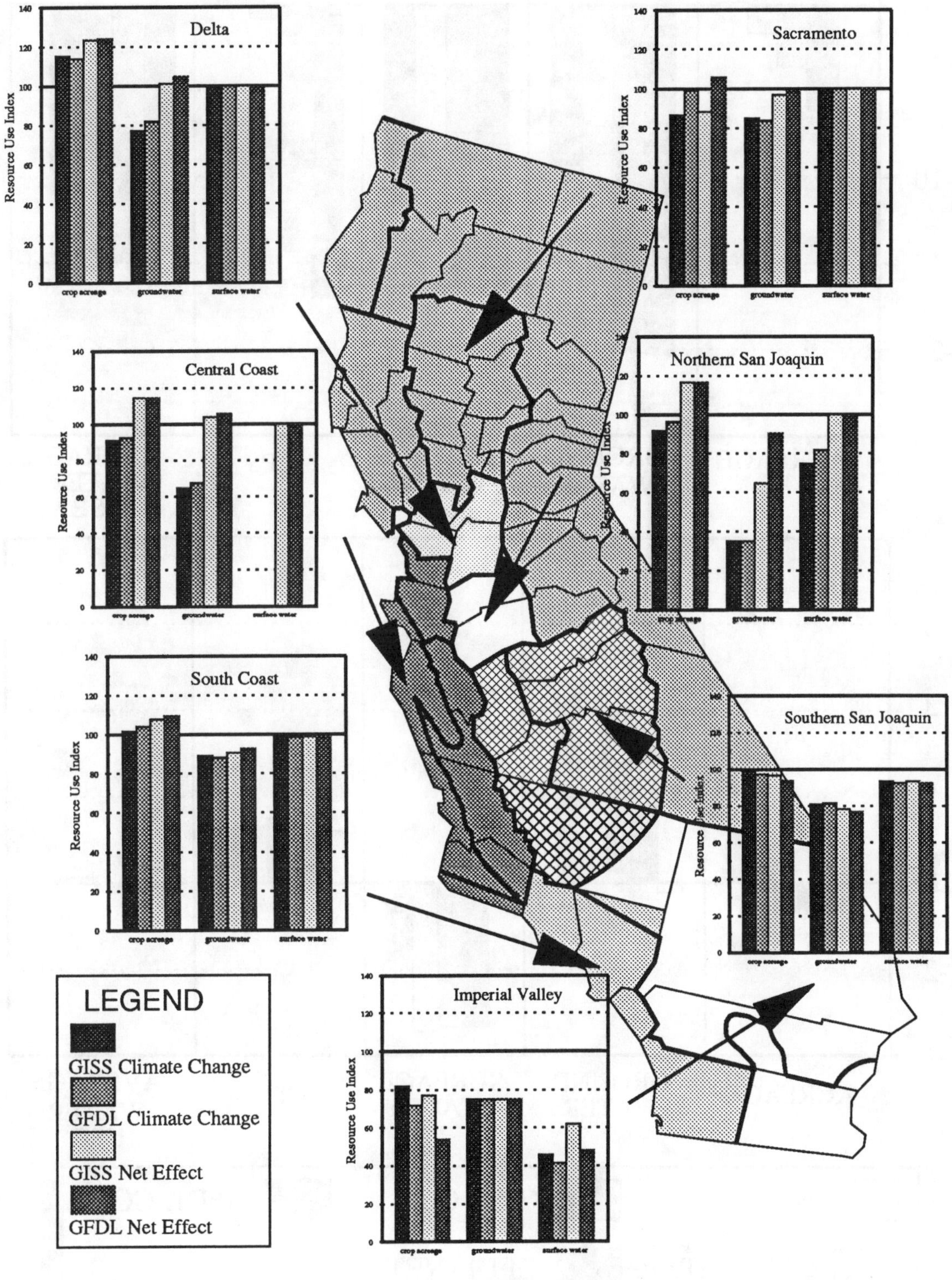

Figure 2. Overall Results

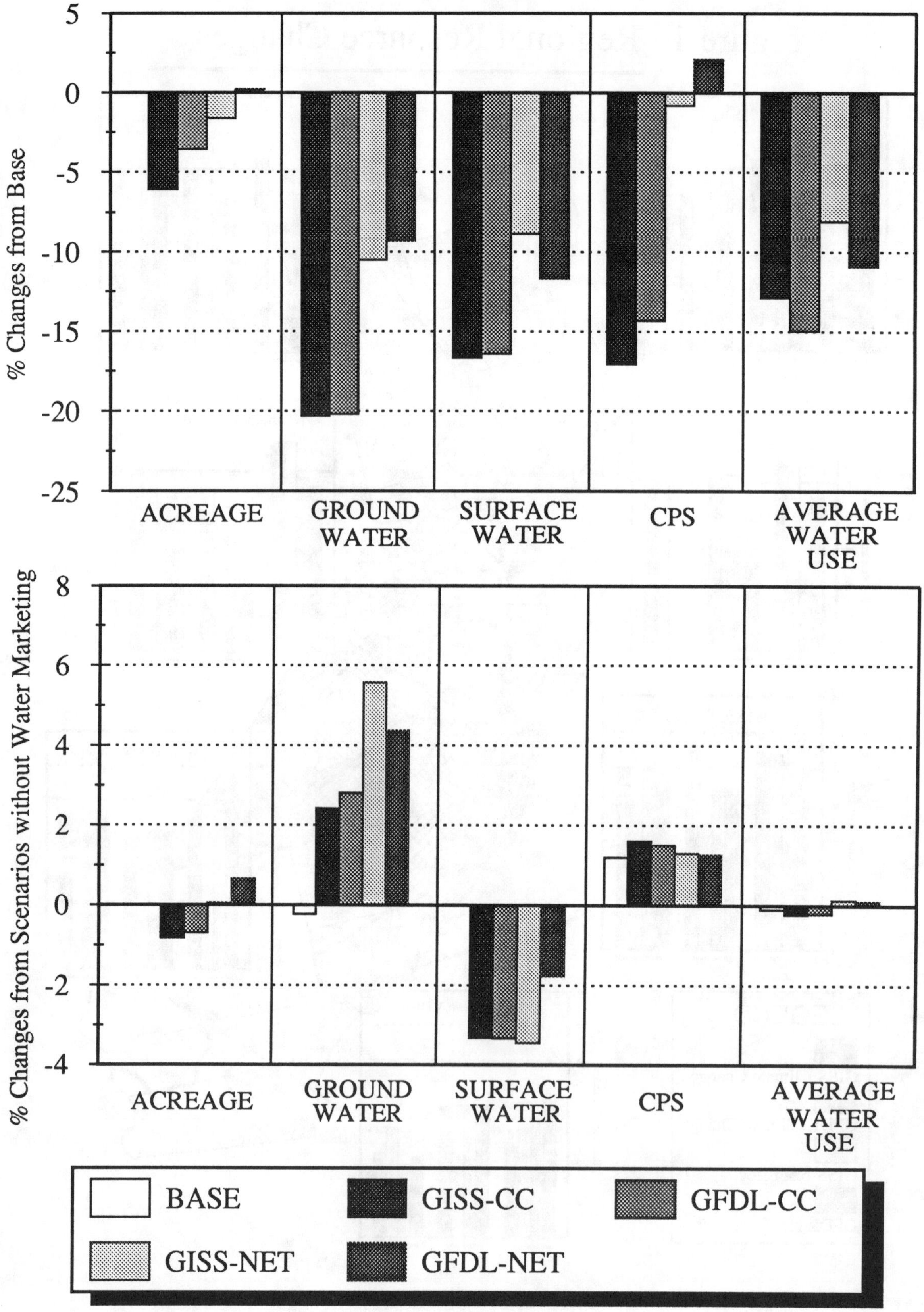

(CPS) surplus increases more than compensate. Total water use remains approximately the same but shifts occur between groundwater and surface sources.

Conclusions

In summary, this case study has demonstrated the importance of including all related climate change impacts within a single analytical framework. In particular, it is not possible to translate physical crop productivity or water supply changes directly into impacts without accounting for the effect of market forces. Commodity markets operate to induce shifts in crop locations producing average yields close to the minimum biologic impact. Introducing markets for water resources similarly offsets supply reductions by improving the efficiency of use of what is available.

Market forces play a crucial role in creating the flexibility to respond to climate changes and in mitigating their ultimate impact. However, in the absence of radical changes in agricultural production technologies or environmental management institutions, nonmarket effects such as nonpoint source pollution will be exacerbated as agriculture relocates in response to climate changes. Some problems, such as drainage and salinity, may improve marginally as acreage and average water use are reduced. However, groundwater overdraft problems are likely to be exacerbated in the San Joaquin Valley although it is not known whether future energy prices or groundwater levels would support these withdrawal rates. Nonetheless, water marketing remains an important example of policies that can be adopted now to mitigate the potential impacts associated with climate change. Further, as illustrated by this analysis, it is a policy which can relieve existing demand pressure on surface water supplies.

Climate changes and increased competition for water likely will have negative impacts on existing aquatic ecosystems. Increased temperatures and altered flow regimens in managed and free-flowing river systems may change species composition from cold to warm water varieties. Altered precipitation patterns would reinvigorate interest in large scale public works including expansion of both the federal Central Valley Project and the State Water Project. Increased reservoir capacity would also affect fishery resources and the mix of recreational opportunities. Intensification of cropping and water use could negatively impact migratory waterfowl in critical Pacific flyway habitats. In short, many of the most severe consequences of climate change for California's environment remain to be studied.

References

De Wit, C.T., "Photosynthesis of Leaf Canopies", Agricultural Research Report 663, Pudoc, Wageningen, 57 pp., 1965.

Doorenbos, J. and A.H. Kassam, "Yield Response to Water", FAO Irrigation and Drainage Paper No. 33, FAO, Rome, 1979.

Dudek, Daniel J., "Assessing the Implications of Changes in Carbon Dioxide Concentrations and Climate for Agriculture in the United States", paper presented at the First North American Conference on Preparing for Climate Change: A Cooperative Approach, Washington, D.C., 28-29 October 1987, 26 pp.

Dudek, Daniel J., "Climate Change Impacts Upon Agriculture and Resources: A Case Study of California", completion report submitted to U.S. Environmental Protection Agency, Office of Policy, Planning and Evaluation, June 1988, 44 pp.

Lettenmaier, Dennis, "Hydrologic Implications of Climate Change in Sacramento and San Joaquin Valleys", University of Washington, Seattle, draft U.S. Environmental Protection Agency report, May 1988.

Sheer, Daniel P. and Dean Randall, "Methods for Evaluating the Potential Impacts of Global Climate Change on Water Related Activities: Case Studies from Atlanta, Georgia and the State of California", Water Resources Management, Inc., Columbia, MD, May 1988.

CLIMATE CHANGE AND WATER RESOURCES MANAGEMENT: ASSESSING CAPACITY FOR INSTITUTIONAL ADAPTATION IN THE SOUTHEAST

by

Mark Meo
Robert E. Deyle
Lani L. Malysa
Laura A. Wilson

Science and Public Policy Program
University of Oklahoma
Norman, Oklahoma 73019

INTRODUCTION

Awareness about the continued loading of radiatively active trace gases into the atmosphere has resulted in heightened public concern about the potential environmental and socioeconomic impacts of global climate change. Because of the relatively rapid warming predicted to result from the "greenhouse effect," natural and social systems may be at risk from the increasing frequency of extreme climatic events such as drought, storms, or shifts in patterns of precipitation and temperature. As a result of the uncertainty that surrounds current forecasts of climate change, deliberative bodies have urged that climate impact assessments be undertaken to characterize the range of effects and impacts that could arise and which would necessitate public and private responses.

As part of the research sponsored by the EPA for its report to Congress on the potential effects of climate change on the United States, the Science and Public Policy Program undertook a study of the policy implications of potential climate change impacts on three specific regions in the southeastern U.S. In a manner consistent with the overall EPA research strategy, each case study followed a hydrological emphasis and was based upon independent biophysical impact assessments using general circulation model projections of global climate altered by a doubled concentration of carbon dioxide.

The overall aim of the policy study was to assess the institutional implications of potential impacts within each of the three study regions, weigh the vulnerabilities and strengths of individual public sector organizations and their capacity to respond to projected impacts, and develop a set of recommended actions for water management institutions to enhance their ability to adapt to different climate change impacts if and when they arise. See Meo, et al. (1988) for a complete discussion.

CASE STUDIES

The three case studies examined include the Tennessee Valley Authority reservoir system, the Apalachicola River estuary in Northwest Florida, and South Florida from Lake Okeechobee through the Everglades National Park. Of the three cases, two had biophysical impact assessments performed in conjunction with the policy analysis: Miller and Brock (1988) for the TVA reservoir system and Livingston (1988) for the Apalachicola River, estuary, and bay. The South Florida study was conducted without an independent biophysical impact assessment.

A major institution in the Southeast, the TVA is a federal agency that is unique in its purpose, organization, and role in the Tennessee Valley as well as the United States. As the largest electric power utility in the nation, it has commanded close attention from analysts for many years. However, the absence of comparable organizations has limited the application of policy research findings. Yet, its technological leadership in power production and multipurpose water resource management has provided many valuable lessons in energy and environmental policy.

The federal system of governance is well represented in the Apalachicola case study with agency involvement in water resources management at the local, regional, state, and national levels. The acknowledged importance of the oyster fishery to the local and regional economies has motivated development of an intricate collaborative network of public institutions in Northwest Florida directed toward preserving the environmental quality of the river, estuary, and bay areas.

The dominant institutions in the South Florida case study are the South Florida Water Management District (WMD), a substate agency, and the U.S. Army Corps of Engineers. In sharp contrast to the Apalachicola case study, the South Florida hydrologic basin lies entirely within the boundaries of the Water Management District and the region is highly populated with a growing economy. South Florida has also experienced significant environmental alteration and environmental abuse in the past.

METHODOLOGY AND ANALYTICAL FRAMEWORK

Results from the independent climate impact assessments were complemented with published literature and structured interviews with key analysts and decisionmakers at each study site. In the absence of a formal impact assessment for the South Florida case, interviewees were asked to respond to hypothetical climate change scenarios and discuss aspects of climatic variability and natural resource sensitivity.

In order to gauge the vulnerability of case study regions to climate change and assess capacity for adaptation, a set of qualitative scoring criteria was used to rank specific attributes of key organizations. Measured attributes include financial resources, technical expertise, statutory authority and political power, geographic jurisdiction, regulatory enforcement authority, public accountability, responsiveness, use of scientific and technical information, degree of organizational flexibility and coordination, and awareness of potential climate change impacts. Each set of attributes was ranked for climate change-induced conditions of water shortage, water surplus, water quality, and sea level rise where appropriate.

ASSESSMENT OF CAPACITY FOR INSTITUTIONAL ADAPTATION

TVA

Institutional capacity examined in the TVA case study is divided between power and nonpower operations. As discussed in the case study (Meo, et al. 1988), the TVA can be described as two separate organizations operating under one board of directors. Power operations are underwritten by a separate financing authority and paid for by TVA utility ratepayers. Achieving economic competitiveness with other utilities plays a key role in TVA strategic planning. The recent corporate restructuring was undertaken at the behest of the new Board Chair in order to stabilize utility rates and enhance TVA's ability to attract new customers.

Power operations score very well in financial resources under conditions of water surplus and scarcity as well as for water quality. Expertise scores are high for both power and nonpower operations, except for nonpower activities in times of water shortages.

Of the political/social/legal factors, TVA scores less well with respect to public accountability, an issue that has been controversial for TVA in recent years. For water quality, TVA scores poorly in these attributes. This is a result both of the omission of water quality as an explicit goal in the TVA Act as well as inadequate attention given to the issue by Valley states. Accordingly, TVA scores poorly in institutional responsiveness to water quality issues while attaining high rankings in proactive planning for both water surplus (its original mission) and water shortage (as a result of its leadership role in recent droughts).

In the acquisition and utilization of technical information, TVA scores highest for water surplus and poorest for water quality. In the last four years, TVA has demonstrated its competence in the acquisition and use of information for water shortages, especially during the 1988 drought.

Reflective of its mandate, TVA has utilized historic climate variability in its planning and investment criteria related to water surplus. Nonpower operations rank more highly than power operations on this attribute for water quality conditions, but both power and nonpower activities score poorly under water shortage conditions.

Apalachicola

The overview of institutional capacity to adapt to climate change for the Apalachicola River and estuary focuses on the Florida agencies with primary jurisdiction in the Apalachicola area, the Mobile District of the U.S. Army Corps of Engineers, the Apalachicola Estuarine Research Reserve, and the local Franklin County government. Pertinent climate change impacts include sea level rise and water shortage due to decreased flow in the Apalachicola-Chattahoochie-Flint (ACF) River basin. Relevant Florida state agencies include the Department of Environmental Regulation, Department of Community Affairs, Department of Natural Resources, Game and Freshwater Fish Commission, and the Northwest Florida Water Management District.

Historically, the Florida agencies have been well equipped to contend with the possible impacts of greenhouse gas-induced sea level rise and water shortage. The agencies have a history of responding proactively to environmental issues by: (1) protecting the ecosystems of the Apalachicola area through a combination of land acquisition programs and a national estuarine research reserve designation; and (2) initiating the Interagency Management Committee to investigate sea level rise issues as they relate to Florida and to build up sea level rise technical expertise.

Both issues also speak to other strengths of the Florida agencies to adapt to climate change impacts; specifically, their in-house expertise and their excellent internal and external coordination. Much expertise and coordination, relating to the Apalachicola area, exists within each agency. Also, the Florida agencies are involved in external collaborative arrangements with states and other federal and local agencies. The 1983 ACF Memorandum of Agreement between the states of Florida, Alabama, and Georgia, and the Corps of Engineers, and the on-going "308" ACF basin-wide study are examples.

The Mobile District of the Corps of Engineers has a well-developed institutional capacity to adapt to climate change impacts related to sea level rise. Since the Corps is primarily a reactive rather than proactive agency it undertakes water resources planning and management actions in response to congressional directives or initiatives from state, county, or local governments under authorized programs such as the beach renourishment program. The Corps' greatest adaptive capacities are their in-house technical expertise and experience relating to coastal flood protection and erosion protection structural measures and their financial resources. The Corps co-funds with the individual states, structural projects germane to its mission. Much of this coastal protection work would not be done if the individual states had sole responsibility for financing or design.

A combination of institutional attributes of the Apalachicola National Estuarine Research Reserve suggest that its capacity to adapt to climate change impacts related to sea level rise and water shortage is also growing. The research reserve, while not having any regulatory authority, has the institutional benefit of excellent geographic jurisdiction and a proactive mission towards research and public education.

The local Franklin County government ranks low in its institutional capacity to adapt to climate change impacts related to both sea level rise and water shortage. While it possesses excellent geographic jurisdiction for sea level rise impacts and less so for water shortages, it lacks adequate financial resources and in-house expertise. More importantly, it lacks adequate external coordination mechanisms to allow it to work closely with other agencies in Florida. This is largely because Franklin County government and the local townspeople of the Apalachicola Bay area are critical and wary of the increased regulatory and enforcement presence of Florida agencies in the Apalachicola area.

South Florida

The overview of institutional capacity to adapt to climate change for South Florida focuses on the South Florida Water Management District, the Jacksonville District of the U.S. Army Corps of Engineers, and county governments in the region. The assessment of state agencies included under the Apalachicola Estuary and Bay case study is applicable to South Florida for the most part.

Awareness of the potential impacts of climate change is a major constraint across all of the institutions for all four impacts. None of the institutions is ranked as well developed for this attribute. Only the South Florida Water Management District is ranked as having a relatively high level of awareness for all four climate change impacts. Institutional capacity appears adequate for potential water shortage, water surplus, and sea level rise impacts, although the capacity is primarily that of the Water Management District in the case of water surplus and shortage and the Corps of Engineers for sea level rise. Institutional adaptive capacity is least assured for potential water quality impacts of climate change. This reflects the current fragmentation of responsibility and authority for water quality management in the state. While recent trends suggest an increasing role for the South Florida Water Management District, the district lacks sufficient regulatory authority at present.

The South Florida Water Management District is exceptionally well equipped to contend with the possible impacts of greenhouse gas-induced climate change, with the exception of sea level rise where its statutory jurisdiction is limited. The direct effects of sea level rise in coastal areas on property, infrastructure, and public services due to inundation and increased coastal storm impacts are outside the district's domain except where those impacts are associated with the primary drainage system and water supply. Within this limited span of sea level rise impacts, the district has well developed adaptive capacity.

The district's ad valorem tax base, broad statutory powers, and extensive expertise provide an excellent foundation for responding to climate change. The inter-organizational and intra-organizational boundary spanning role of the district's Office of Resource Assistance, the use of matrix task forces for planning and problem solving, and the organization-wide commitment to proactive consensus building will serve the district well in contending with the potential impacts of climate change.

The only areas where the district's capacity is less than optimal are in water quality impacts and water surplus. In the case of water quality, the district is just beginning to develop a significantly increased role as a result of the 1987 State Surface Water Improvement and Management Act. Thus its expertise, data acquisition and utilization, and planning are not as fully developed as for example, water shortage. The district's capabilities also are constrained by its limited regulatory power over water quality. Direct regulatory control of water pollution sources is exercised primarily by the state Department of Environmental Regulation. Regulatory control of land uses that may affect water quality through nonpoint sources is primarily the domain of county and local governments.

The Water Management District's institutional capacity to adapt to water surplus impacts of climate change are limited only by the lack of a comprehensive analysis of the capacity of the flood control system to account for major changes in land use since design and construction of the system in the 1960s.

Because a comprehensive analysis of the system is planned over the next several years, this is only a short-term constraint.

Awareness of potential climate change impacts within the district is relatively high compared to many other state and regional water management institutions. Key individuals within the organization have a sense of the broad implications of climate change for water resources management in South Florida and information gaps that must be filled to raise climate change to a higher level of priority in the district's planning and decisionmaking process.

The Corps' greatest adaptive capacities are in those areas of water resources management where it traditionally has been involved in South Florida: flood control and coastal erosion control. Its involvement in water quality and water supply management in South Florida has been recent and tentative. The Corps' awareness of potential impacts of climate change also appears to be recent and not significantly diffused through the organization at either the district or national levels.

FINDINGS

The case study results suggest that different levels of government may have particular roles in climate change adaptive strategies that are best suited to their resources, powers, and jurisdictional authorities. Coordination among public agencies at different levels of government is also essential. A proposed scheme of appropriate local, state, and federal roles follows.

Local Government

- Coordinate or centralize (e.g., at the county level) the management of climate-sensitive natural resources that are currently managed by subcounty governments and special districts (water supply and drainage districts).

- Integrate climate change adaptive strategies into the planning process with specific attention to land and water use, economic development, capital (infrastructure) development, and natural hazard (e.g., hurricane, flooding) mitigation planning in cooperation with substate regional, state, and federal agencies with complementary powers and jurisdiction (e.g., Florida water management districts, state emergency management agencies, TVA, regional planning councils).

- Revise regulatory mechanisms to effect the objectives of strategic planning for climate change adaptation, especially in areas where local government jurisdiction is primary such as land use control.

- Participate in the development and operation of a state-level information acquisition and utilization network with substate regional, state, multi-state regional, and federal organizations for climate change information pertinent to local government roles in climate change adaptation.

State Government

- Develop or enhance basin-wide water resources management and planning capabilities within the state.

- Develop coordinating mechanisms with other states and appropriate federal agencies where drainage basins extend beyond state boundaries or into major federal land holdings.

- Develop the capacity to achieve state-wide land and water management, objectives through regional and local planning and regulation that is consistent with state policies.

- Promote conjunctive management of the quality and quantity of surface and ground water resources.

- Coordinate state-wide strategic planning for climate-sensitive resources (e.g., hurricane mitigation, coastal erosion mitigation, flood protection, drought management) that involves substate regional and local agencies.

- Develop and maintain an information acquisition and utilization network, including the necessary technical expertise, with substate regional, state, multi-state regional, and federal organizations for climate change information pertinent to state and substate government roles in climate change adaptation.

- Sponsor and fund research on climate-sensitive state resources needed to supply information for the state-level climate change information system (e.g., fully characterize state ground water resources, determine the climate sensitivity of major natural resources-based economic sectors, vulnerability of coastal areas to accelerated sea level rise).

Federal Government

- Participate with the states to develop a climate change information acquisition and utilization network, including the necessary technical expertise.

- Sponsor and fund research and demonstration projects to enhance the ability to predict and detect climate change impacts on a regional level and to enhance development of planning and management strategies by federal, state, and substate governments for climate change adaptation, including structural mitigation alternatives.

- Strengthen coordination mechanisms among federal agencies and between federal agencies and the research community for detecting climate change impacts and for conducting and sponsoring climate change research.

- Encourage and support the development of water resource management coordinating mechanisms among states and appropriate federal agencies where drainage basins extend beyond state boundaries or into major federal land holdings.

IMPLICATIONS FOR THE SOUTHEAST

The case study research findings, in conjunction with the above mentioned roles for institutional adaptation, suggest several implications for the larger Southeastern region. Although any findings based upon three case studies are subject to extensive qualifications, some generalizations appear consistent with the evidence. First, environmental planning and management agencies would be well advised to examine closely Florida's system of water management districts and adopt similar guidelines and procedures where appropriate. The district approach, based upon hydrologic regions and ad valorem taxing powers, provides a socially responsive and environmentally sound basis for managing water resources. The major limitations to the WMD approach are encountered in areas such as Apalachicola where much of the watershed lies outside the regional WMD's jurisdiction. An exception to this condition is characterized by the TVA, which presently favors increased use of basin wide management strategies for water quality enhancement.

A second implication for the Southeast concerns local government capacity. Apalachicola's primary economic resource as well as its surrounding environment are likely to be severely damaged by potential climate change impacts. With the demise of the oyster industry from heat stress and saline intrusion, the town of Apalachicola will enter into yet another phase in its historical pattern of changing livelihoods. The distinction in this case is that anticipatory plans and programs can be developed ahead of time, rather than remain unprepared. In this regard, the state agencies have demonstrated a keen level of program planning and coordination. For other locales similar to Apalachicola, Florida is well-equipped to undertake actions which could minimize the negative impacts associated with climate change. For the Southeast in general, local governments will be subject to similar threats. A prudent strategy for southeastern states to pursue would be to develop and strengthen state-local government coordination and planning capabilities. Included in this list would be measures for public involvement, acquisition and use of technical information by qualified professionals, and a clear mandate to address issues germane to climate change.

The Apalachicola case study also highlights the importance of interstate coordination and cooperation for developing coherent responses to climate change. In regions where drainage basins encompass more than one state coordination among and between water resource supply and demand activities is likely to become increasingly important for adaptation.

Finally, the case study results reinforce a leading role for the federal government. Federal agencies such as the National Oceanic and Atmospheric Administration and the U.S. Army Corps of Engineers perform vital and important functions in the case study regions. The TVA performs similar roles in its seven-state region. Federal agencies can provide a firm basis for research, sponsor demonstrations, and foster public awareness. They are also important as a credible source of detecting the onset of climate change impacts. As more detailed studies of potential climate change impacts are initiated, federal agencies, such as the EPA, can be expected to play a larger role in assisting states and regions conduct research, plan alternative courses of action, and undertake structural or nonstructural responses.

LITERATURE CITED

Livingston, Robert J. 1988. "Projected Changes in Estuarine Conditions Based on Models of Long-Term Atmospheric Alteration." Center for Aquatic Research and Resource Management, Florida State University, Tallahassee, FL.

Meo, Mark, Steve Ballard, Robert E. Deyle, Thomas E. James, Jr., Lani L. Malysa, and Laura A. Wilson. 1988. "Policy Implications of Global Climate Change Impacts Upon the Tennessee Valley Authority Reservoir System, Apalachicola River, Estuary, and Bay, and South Florida." Science and Public Policy Program, University of Oklahoma, Norman, OK.

Miller, Barbara A. and W. Gary Brock. 1988. "Potential Impacts of Climate Change on the Tennessee Valley Authority Reservoir System." Engineering Laboratory, Tennessee Valley Authority, Norris, TN. Report No. WR 28-1-680-101.

GLOBAL CLIMATE CHANGE - IMPLICATIONS FOR THE TENNESSEE VALLEY AUTHORITY RESERVOIR SYSTEM

B. A. MILLER* and W. G. BROCK*

ABSTRACT: Increased global temperatures resulting from rising concentrations of greenhouse gases are likely to be accompanied by changes in regional hydrologic cycles. The response of the Tennessee Valley Authority (TVA) reservoir system to two climate scenarios is illustrative of the significance of these changes. The wetter, warmer scenario increased water availability for power production, recreation, minimum flow requirements and water supplies, but significantly increased flood potential. The drier, warmer scenario decreased flooding, but reduced water availability for power production, recreation and water quality. Both scenarios would impact the ability of the TVA system to meet present project purposes and would necessitate a reevaluation of current reservoir operating philosophy.

INTRODUCTION

Scientists and policymakers are increasingly concerned about global climate changes resulting from increased concentrations of atmospheric carbon dioxide (CO_2) and other greenhouse gases. A doubling of CO_2 concentrations (or its radiative equivalent from all greenhouse gases) is expected to raise global mean surface temperatures by 3 to 8°F (1.5 to 4.5°C) by the mid-21st century (National Academy of Science, 1987). Increased global temperatures will likely be accompanied by changes in the magnitude and distribution of key climatic variables, including precipitation, evapotranspiration, runoff, etc. The occurrence of extreme hydrologic events such as floods and droughts is also likely to be modified. These combined changes would significantly alter regional hydrologic cycles and the ability of water resource management systems to meet competing water use demands.

As part of a national assessment of the potential effects of climate change in the U.S. (Smith and Tirpack, 1988), the Environmental Protection Agency (EPA) contracted with TVA to evaluate the impact of two climate scenarios on the TVA system. TVA was selected for a case study because it represents a large, comprehensively managed, multipurpose water resource system. The objective of the study was to evaluate the sensitivity of the TVA reservoir system to extreme climate changes and to identify the implications of these changes. Significant results of this project are summarized in this paper, while a more in-depth analysis is provided by Miller and Brock (1988).

THE TVA RESERVOIR SYSTEM

TVA, a multipurpose Federal Agency, operates a range of economic development and water resource programs in the Tennessee River Basin within a 41,000 square mile area in seven southeastern states. The TVA reservoir system, which includes 42 major dams and reservoirs, is comprehensively

*Tennessee Valley Authority, Engineering Laboratory, Norris, TN 37828

managed as an integrated system. Primary operating objectives of the reservoir system include navigation, flood control and hydropower generation, although recreation, water supplies, water quality and environmental management are also important considerations.

All major dams and reservoirs in the TVA system serve multiple functions; however, major projects can be categorized based on their primary purpose--power production, mainstem multipurpose reservoirs, and tributary multipurpose reservoirs. The 12 single-purpose power projects have small reservoirs with minimal flood storage capacity. The 9 mainstem multipurpose reservoirs on the Tennessee River were built primarily to support navigation and generate hydroelectric power. Water level fluctuations on the mainstem reservoirs are minimal, varying only 5 to 6 feet from low winter flood control levels to higher summer full pool levels.

The 21 tributary multipurpose reservoirs were built to provide flood control, power generation, and adequate flows for mainstem navigation requirements. These reservoirs are relatively deep and provide key flood storage capacity for the system. Tributary reservoir operations, which are dominated by flood control concerns, closely parallel the regional annual streamflow cycle. Typically, tributary reservoirs are lowered to flood control levels by January 1 to provide storage for heavy winter-spring flows, rapidly filled by spring rains from mid-March until full-pool levels are reached towards June, then lowered through the summer and fall months when streamflows are low and power demands high. Water levels in the tributary reservoirs fluctuate up to 60 feet between low winter flood levels and full pool summer levels.

Scheduling daily and seasonal water releases from TVA reservoirs is a complex process which accounts for: the quantity of water in storage; travel time through the system; unregulated local inflows; and daily weather, streamflow, and power demand variations. The daily rate and quantity of water released from each dam is determined to maximize benefits while meeting seasonal operating goals. Operating philosophy and reservoir guide curves are based on over 50 years of operating and flood control experience. As illustrated in Figure 1, significant climatic changes would not only necessitate a reevaluation of current reservoir operations, but would also impact the primary functions of the TVA reservoir system.

METHODOLOGY

Potential climatic change impacts to the TVA reservoir system were assessed using the Weekly Scheduling model (WSM), a reservoir operations and planning model (Shane, 1984). Based on a linear programming algorithm, WSM simulates weekly variations in water level, discharge and power production for 42 reservoirs operated within the system. The model objective is to minimize deviations from historical normal operating levels, subject to reservoir release and level constraints imposed to meet navigation, flood control, power generation, water supply, water quality and recreational requirements.

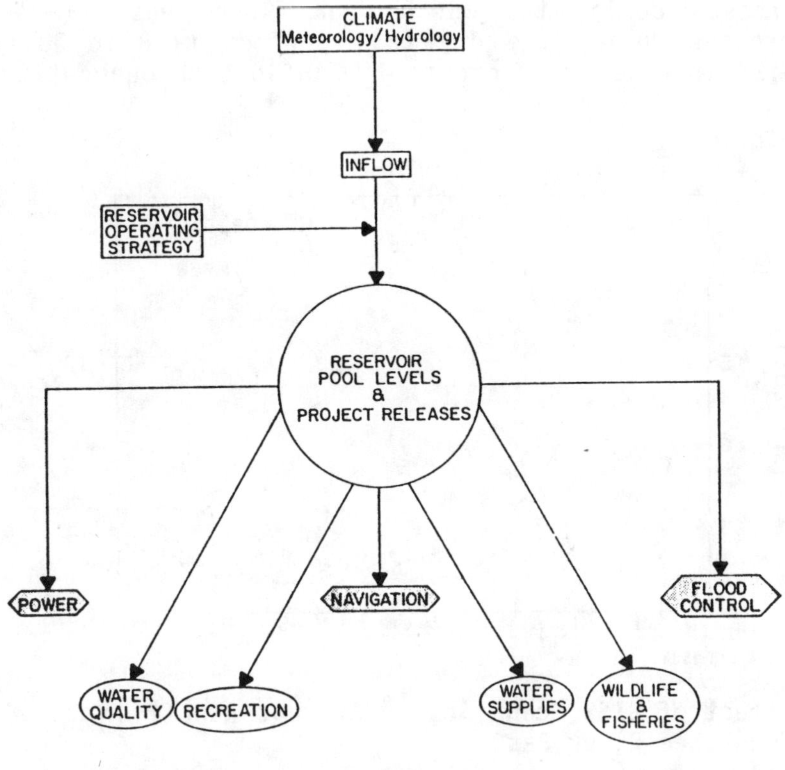

Figure 1. Primary Functions of the TVA Reservoir System.

A primary input into WSM is historical local flow (unregulated overland and tributary flow) into each reservoir in the TVA system. Assuming changes in local flow are directly proportional to runoff, monthly ratios ($2*CO_2$/CONTROL) for surface runoff provided by EPA were used to adjust weekly historical local inflows to each project for the 30-year study period (1951-1980). The monthly ratios were assumed to apply equally to each week in the month. These runoff ratios, generated by NASA's Goddard Institute for Space Studies (GISS) global circulation model, represent the endpoint effects of a doubling in CO_2 concentrations from 1958 levels of 315 mg/L to 630 mg/L.

The first scenario, GISS, predicts a warmer and wetter climate for the Tennessee Valley. East of Chattanooga, average annual temperatures would increase by 7.1°F, with monthly variations from 3.7°F in August to 11.2°F in March. In this eastern basin, which contains all the large storage tributary reservoirs, average annual runoff increases by 31 percent (see Figure 2). Monthly variations range from +73 percent in March to -28 percent in November, thereby exaggerating peak flows during the traditional flood season and further reducing flows during a dry period of year.

Other global circulation models have predicted a drier climate for the Tennessee Valley. Consequently, EPA recommended utilizing the inverse of the GISS runoff values for the second scenario. This GISS Inverse scenario predicts a warmer and significantly drier climate for the Tennessee Valley. In the eastern basin, average annual runoff is decreased by 31 percent, with substantial decreases during the March peak flood season (-75 percent) and only modest increases during the dry fall periods (+18 to 28 percent). The net effect of GISS Inverse is to create low inflows throughout the year in the Tennessee Valley.

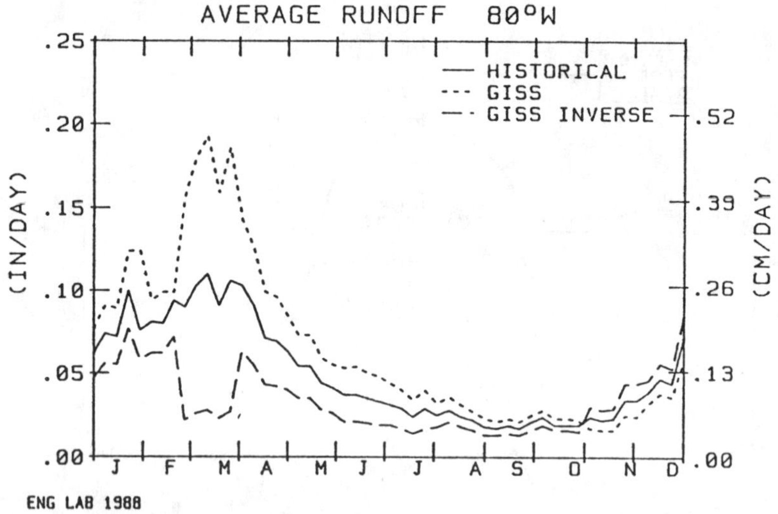

Figure 2. Effect of GISS Scenarios on Average Runoff in the Eastern Tennessee River Basin.

The project methodology provides a good overview of the integrated response of the TVA reservoir system to changes in historical local inflows. Important project limitations, however, include: existing guide curves and current operating policy were used in the WSM, runoff ratios provided by the GCM were used directly, current water use and power demand patterns were assumed in the analysis, and changes in the spatial and temporal distribution of runoff and precipitation events were not considered. In future studies, hydrologic modeling and more sophisticated impact analyses should be performed.

RESERVOIR SYSTEM RESPONSE AND IMPLICATIONS

GISS Scenario. The increased runoff predicted by the wetter GISS scenario would result in higher reservoir levels throughout the year at all major projects in TVA. Most significantly, at the large storage reservoirs in the eastern Tennessee River Basin (see Figure 3): full pool summer levels are maintained for an extended period of time, normal operating levels are exceeded in the fall due to increased storage earlier in the year, and normal maximum levels are exceeded in wet years during the traditional flood season.

Exceedence of normal maximum reservoir levels would likely result in the major adverse impact of the GISS scenario--spillage from dams and increased flood potential in the Tennessee Valley. Flood prone areas on the Tennessee River such as the city of Chattanooga, Tennessee, and low lying agricultural lands would be particularly vulnerable to significant flood damage. Safety issues at dams and nuclear power plants, resulting from higher sustained flows and possible changes in the Probable Maximum Flood (PMF), would need to be reevaluated.

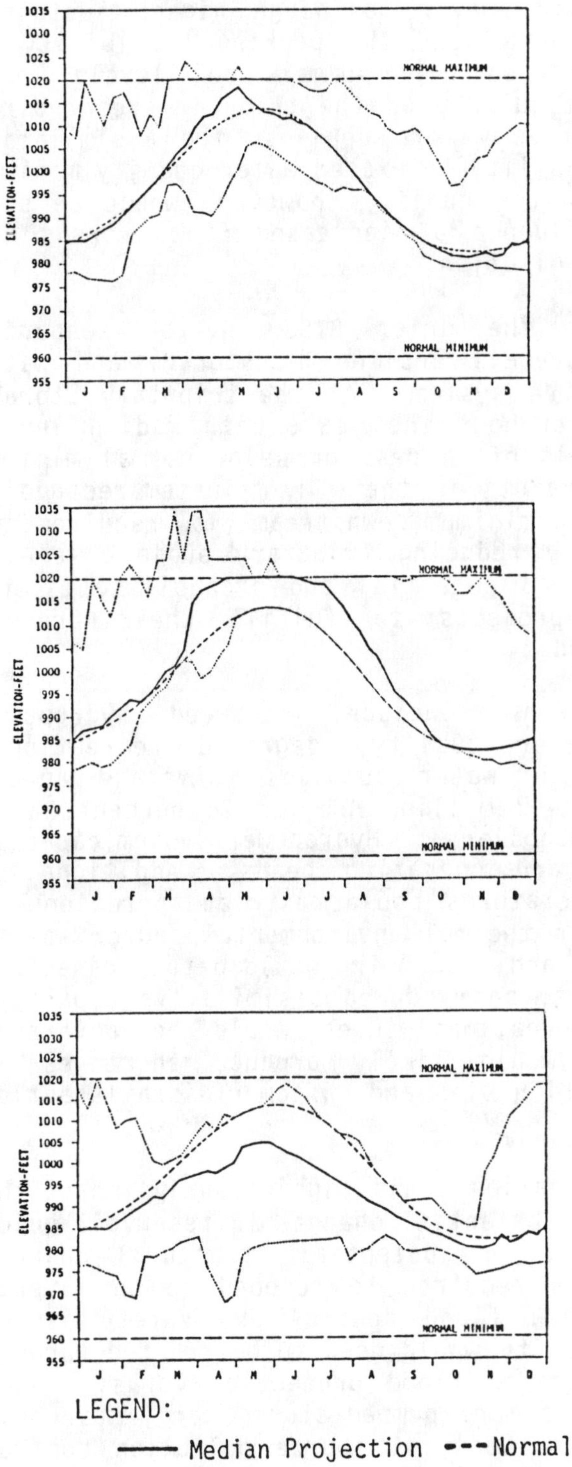

Base Case

GISS Scenario

GISS Inverse Scenario

LEGEND:

—— Median Projection - - - Normal Operations ···· Upper & Lower Envelope

ENG LAB 1988

Figure 3. Probabilistic Pool Level Forecasts for Norris Reservoir.

Primary benefits of the GISS scenario include increased hydroelectric power production, enhanced recreation, and improved water availability. Average annual system generation increased by 3.2 million MWh, or 16 percent, valued at $54 million (1988 dollars). However, although higher sustained flows increase based load hydropower production, the peaking flexibility of these projects would be limited. Extended full summer pool levels would enhance recreational opportunities, particularly on the tributary reservoirs. Increased water availability would improve water supplies, the assimilative capacity of lakes and streams, and the ability to exceed water quality minimum flow requirements. The net effect on water quality, however, would be site specific depending on the relative influence of increased flows versus the potential for increased nonpoint source pollution.

GISS Inverse Scenario. Under the drier GISS Inverse scenario, decreased runoff would result in an overall decline in storage and water availability at major projects in the TVA system. At the tributary storage reservoirs, lake levels are lowered throughout the year, with median levels reduced up to 30 feet and minimum levels often near or below normal minimum pool levels in dry years. Due to constraints in the WSM, mainstem reservoirs are filled to normal operating levels and minimum downstream flow requirements are met, but at the expense of severely reducing tributary project levels. Although this scenario would reduce flood potential and probability of dam failure, the ability of many TVA projects to fulfill their present multipurpose functions could be threatened.

Adverse impacts of GISS Inverse include: reduced hydropower generation and capacity, impaired water quality, degraded recreational opportunities, and decreased storage for water supplies. Average annual system power generation is decreased by 4.7 million MWh, or 24 percent, at a replacement value of $87.2 million (1988 dollars). Hydropower system capacity is also reduced due to lower flows and operating heads. Additionally, decreased flows and elevated water temperatures could restrict operations at several fossil and nuclear plants due to thermal environmental and/or safety constraints. Water quality in TVA lakes and reservoirs would deteriorate from reduced DO levels, increased temperatures, and reduced assimilative capacity. Aquatic biota, fish, wildlife and recreational uses would be adversely impacted. Reduced water availability would likely produce shortages in groundwater supplies and operational difficulties and customer dissatisfaction in reservoir water supplies.

Implications. Both climatic scenarios could significantly impact the operation of the TVA reservoir system. Substantial changes in reservoir guide curves and operating philosophy, as well as potential structural changes and/or additions to the system, would be required to respond to an altered climate. Under the wetter GISS scenario, flood control and safety issues would predominate. Additional flood capacity would need to be created through operational changes, dam modifications and/or flood protection works. Added turbine capacity could be justified, while more pumped-storage projects could be needed to satisfy peak power demands. Nonpoint source pollution control programs would need to be expanded, while enhanced recreational opportunities would encourage economic development.

Under the drier GISS scenario drought related issues would increase in significance. Difficulty in satisfying project multipurposes would necessitate a reordering of TVA priorities. Alternative sources of energy would need to replace lost hydropower potential, while fossil and nuclear plants may not be able to meet environmental/safety limits under a warmer climate. Industrial and municipal treatment facilities could be subject to more stringent waste standards. Increased power costs, decreased recreation revenues, and increased industrial restrictions could have significant adverse impacts on the Tennessee Valley economy.

ADDRESSING THE GLOBAL CLIMATE CHANGE ISSUE

The consequences of climate change reported in this study, which represent the effects of a doubling in CO_2 concentrations, may not be felt for 50 to 100 years. The key climatic change issue, therefore, becomes what to do in the intervening years when the direction, magnitude and rate of regional climatic and hydrologic changes are uncertain. In the face of these challenges, it is still cost effective and wiser to prepare for climate change than to ignore its possibility. Water resource agencies and/or power utilities, such as TVA, can respond to climate change on an immediate basis by developing strategies for education, assessment, adaptation, and mitigation.

Education of pertinent staff about climate change issues and potential implications is imperative. Impact assessments should identify the sensitivity of major agency programs to changes in climatic variables. The impacts of seasonal shifts, the occurrence of more extreme events, increased air and water temperatures, and changes in the magnitude and distribution of key hydrologic variables such as rainfall and runoff should be considered.

Immediate adaptation strategies should be based on the incorporation of climate change uncertainty into long range planning. Data collection and analysis procedures should target and/or monitor climate change trends. In reservoir systems, flexible guide curves, based on a changing climate rather than a past historical record, need to be developed. Hydrologic analyses such as flood forecasting, estimation of the Probable Maximum Precipitation (PMP) and Flood (PMF), and dam and nuclear plant safety must also account for climate change uncertainty. In power systems, a range of temperature and hydrologic scenarios should be used to forecast long-range power supply and demand. The economics of nuclear power plants, which emit no CO_2, should be reevaluated in light of potential CO_2 emissions controls on fossil fuel plants. Conservation programs should be encouraged in the production and consumption of energy.

Mitigation strategies to control CO_2 emissions should focus on improved conservation and efficiency, alternative energy production and technologies, and emissions control. Conservation, or a reduction in energy requirements, can be encouraged through load management to reduce peak loads and through consumer conservation. Increased use of nuclear, hydroelectric and pumped-storage projects would lessen the dependence on fossil fuels. Reduced fuel consumption can also be achieved by improving the efficiency of fossil fuel energy production, energy transmission and distribution, and consumer products. Among combustion materials, coal generates the most CO_2 per Btu of energy produced. Substitution of coal with natural gas or petroleum would decrease CO_2 emissions, as would the development of alternate technologies such as atmospheric fluidized-bed combustion (AFBC), fuel cells, wind or solar power, etc. Flue gas treatment (scrubbing) technologies would reduce CO_2 air emissions, but would create a land disposal problem. Some of these mitigation strategies could be implemented in the near term, while others require additional research. Long term research needs include the development of new materials to withstand higher boiler temperatures, superconducting materials, and uses of CO_2 removed in flue gas scrubbing.

Adaptation and mitigation strategies should be geared towards increasing the efficiency and productivity of water resources and power systems. Until GCM's or other climatic models improve their simulation of regional climatic impacts, these strategies should make sense regardless of the magnitude and direction of changes in local hydrologic and climatic variables. The key to addressing the global climate change issue is to increase our understanding of climate sensitive activities and to improve our flexibility to deal with future climatic changes.

REFERENCES

Miller, B. A., and W. G. Brock, 1988, "Sensitivity of the Tennessee Valley Authority Reservoir System to Global Climate Change," Report No. WR28-1-680-101, TVA Engineering Laboratory, Norris, Tennessee.

National Academy of Sciences, 1987, "Current Issues in Atmospheric Change: Summary and Conclusions of a Workshop, October 30-31, 1986," National Academy Press.

Shane, R. M., "Weekly Scheduling Model for the TVA Reservoir System," WR28-1-500-126, TVA Engineering Laboratory, Norris, Tennessee, 1984.

Smith, J. B., and D. A. Tirpack, 1988, "The Potential Effects of Global Climate Change on the United States," Draft Report to Congress, Executive Summary, EPA Office of Policy, Planning and Evaluation.

THE IMPACT OF CLIMATE CHANGE ON WATER QUALITY IN THE SOUTHERN U.S.A.: STREAM WATER TEMPERATURE

E. Cooter, Oklahoma Climatological Survey
and
W. Cooter, Oklahoma Conservation Commision

Temperature scenarios generated by three Global Climate Models (GCMs) have been utilized to construct a range of possible temperature related water quality scenarios throughout the south and southeast. GCMs include: the Goddard Institute of Space Studies (GISS; Hanson et al., 1986), the Geophysical Fluid Dynamics Laboratory model (GFDL; Manabe and Wetherald, 1980) and the Oregon State University model (OSU, Schlesinger and Zhao, 1988). Temperature changes from a historical base at each model grid point were computed as described in Johnson et al. (1988) and resultant increases range from 3 to 10 degrees C throughout the study area. GCM modeled changes in precipitation are far less certain than those of temperature. Some of the issues associated with making impact estimates involving processes that are highly sensitive to precipitation delivery characteristics such as intensity and duration period are also discussed in Johnson et al. (1988).

Equation 1 (Edinger and Geyer, 1965) was used to compute daily average July equilibrium water temperatures at GCM grid points as a function of ambient air temperature, solar radiation, water vapor, wind, cloud cover and vegetative cover. Figure 1 contains contoured grid-point estimates of computed diurnal average July water temperatures using GCM base and climate change scenario conditions and Equation 1. Water temperature changes under the three GCM scenarios range from a decline from base conditions of 2 degrees C under the GFDL scenario to an increase of 14 degrees C over base conditions under the OSU scenario.

The next step is to relate computed water temperatures to water quality. A common measure of water quality is the level of dissolved oxygen (D.O.). For this example, D.O. is estimated by a generalized Streeter-Phelps waste load model given in Equation 2 (O'Connor, 1984). This model estimates the impacts of a municipal wastewater discharge on a stream with certain characteristics. The assumptions associated with its application at selected water temperatures are found in Table 1. Figure 2 summarizes the results for four water temperature levels while assuming constant flow. A commonly encountered water quality standard of 5 ppm of dissolved oxygen is indicated by the dashed line. In all but the 25°C water temperature case, D.O. is estimated to fall below this standard. Figure 1 suggests that water temperatures would exceed 25°C throughout the south and southeast under all three GCM scenarios. It can be expected, then, that some level of waste water treatment could be required in stream and discharge situations such as that described by Table 1.

A collorary to Figures 1 and 2 might be, "Can present water treatment technology respond to waste load changes resulting from GCM climate change scenarios?" Figure 3 contains D.O. curves for the example stream. It illustrates that even at mean daily temperatures as high as 40°C, current advanced waste water treatment is capable of maintaining stream water at a D.O. close to the 5ppm standard as long as stream flow does not also decline. The implications are that municipal discharges to stream systems with characteristics such as those of Table 1 may require increased levels of treatment in the future under even the most favorable climate scenario examined.

An alternative to expensive advanced treatment facilities is to modify the ambient stream flow temperature above the wastewater discharge point. Although not practical on broad streams and waterways, riparian vegetation may be a useful tool on smaller streams and tributaries. The effect of vegetative shading on equilibrium water temperature is presented in Figure 4 at nearby grid-point locations for the three model scenarios and base (historical) conditions. Under base conditions, each 25% increase in vegetative cover results in a 2.7 degree water temperature decline. Response is most rapid under GISS scenario conditions at approximately 6.5°C for every 25% increase in vegetative cover. Response is most lethargic under GFDL scenario conditions at 1.42° for every 25% increase in vegetative cover. At 25% shading OSU and GFDL scenario water temperatures are approximately equivalent. At 50% shading GISS and GFDL water temperature are approximately equivalent. At greater than 75% shade GISS climate change conditions could approximate those of current base conditions. In most cases, the use of riparian vegetation at the sample location would not completely ameliorate the effects of climate change on stream water temperature and subsequent D.O. conditions, but it could serve to build added flexibility into the stream/waste water system, better enabling such a system to accommodate unknown but potentially costly alternative futures. If stream flows were to decline, maintenance of appropriate riparian vegetation could also act as wind breaks and humidity blocks to decrease surface evaporative losses. If stream flows were to increase at the same time water temperatures increase, riparian vegetation could act as an effective means of bank stabilization as well as filtering turbid side streams and tributaries before they discharge into larger waterways and lakes.

SUMMARY

Given grid point GCM estimates of July mean daily temperatures, resultant climate change induced water temperatures have been computed. The implications of these changes for meeting EPA dissolved oxygen standards have been explored for a specific example stream defined in Table 1. It has been found that without increased levels of treatment, the D.O. of the receiving sample stream would likely fall below current water quality standards at temperatures above 25°C. However, it is further noted that with current advanced treatment technology, the example stream D.O. could be maintained within standards at diurnal average temperatures in the neighborhood of 40°C as long as flow is maintained. The use of riparian vegetative cover may be a viable method of building increased system flexibility which could respond to such changes if suitable heat tolerant species can be identified and successfully established along the banks of small to moderate size streams and tributaries.

REFERENCES

Edinger, J and J. Geyer. Heat Exchange in the Environment, John Hopkins University, 1965.

Hanson, J., A. Lacis, D. Rind, G. Russel, I. Fung, P. Ashcraft, S. Lebedeff, R. Ruedy and P. Stone. "The Greenhouse Effects of Changes in Stratospheric Ozone and Global Climate." In: Effects of Changes in Stratospheric Ozone and Global Climate, Volume I: Overview, USEPA/UNEP, Washington, D.C., 1986. pp199-218.

Johnson, H., E. Cooter and R. Sladewski, "Impacts of Climate Change on the Transport of Agricultural Chemicals Across the USA Great Plains and Central Prairie," U.S. Environmental Protection Agency, Washington D.C., 1989. 63pp (in press).

Manabe, S. and R.T. Wetherald. "On the Distribution of Climate Change Resulting from an Increase in CO_2 Content of the Atmosphere." Journal of the Atmospheric Sciences, 37: 99-118, 1980.

O'Connor, D.J. Waste Load Allocation Seminar Notes, USEPA, 1984.

Schlesinger, M.E. and Zong-ci Zhao. "Seasonal Climatic Changes Induced by Doubled CO_2 as Simulated by the OSU Atmspheric GCM/Mixed-Layer Ocean Model." Oregon State University, Climatic Research Institute, Corvallis, OR, 1988. 51pp.

EQUATION 1

$$E = \frac{-0.05\,E}{K} + \frac{H_r - 1801}{K} + \frac{K - 15.7}{K\,(.26 + B)}\left(e_a - C(B) + 0.26\,T_a\right)$$

where: E = equilibrium temperature (^{o}F)

H = net incoming short and longwave radiation (btu/ft^2/day)

T_a = ambient air temperature (^{o}F)

e_a = water vapor pressure of ambient air at air temperature (mmHg)

B = proportionality constant (mmHg/^{o}F)

C(B) = value dependent on B (mmHg)

K = thermal exchange coefficient (btu/ft^2/day/^{o}F)

EQUATION 2

$$D = D_0 * e^{-K_a (x/u)} + CBOD + NBOD$$

where: D = oxygen deficit (mg/l)
D_o = initial oxygen deficit (mg/l)
K = reaeration rate (day^{-1})
x = distance from source (km)
u = streamflow (km/day)
CBOD = carbonaceous oxygen demand (mg/l)
NBOD = nitrogenous oxygen demand (mg/l)

Table 1. Waste load allocation characteristics for a sample stream in the Southern U.S.A.

	Effluent	Headwater
Dissolved Oxygen (mg/l)	5.0	7.0
Flow (m^3/day)	1.07×10^6	2.14×10^6
BOD5 (mg/l)	5.0	1.0
Nitrate (mg/l)	2.0	0.1

Stream length = 32.26 km
Velocity = 52.8 km/day
Stream D.O. standard = 5 mg/l
D.O. saturation = 6.6 mg/l

BOD decay rate = 1.252 day^{-1}
Reaeration rate = 5.943 day^{-1}
Nitrification rate = 0.985 day^{-1}
BOD settling rate = 0.0 day^{-1}

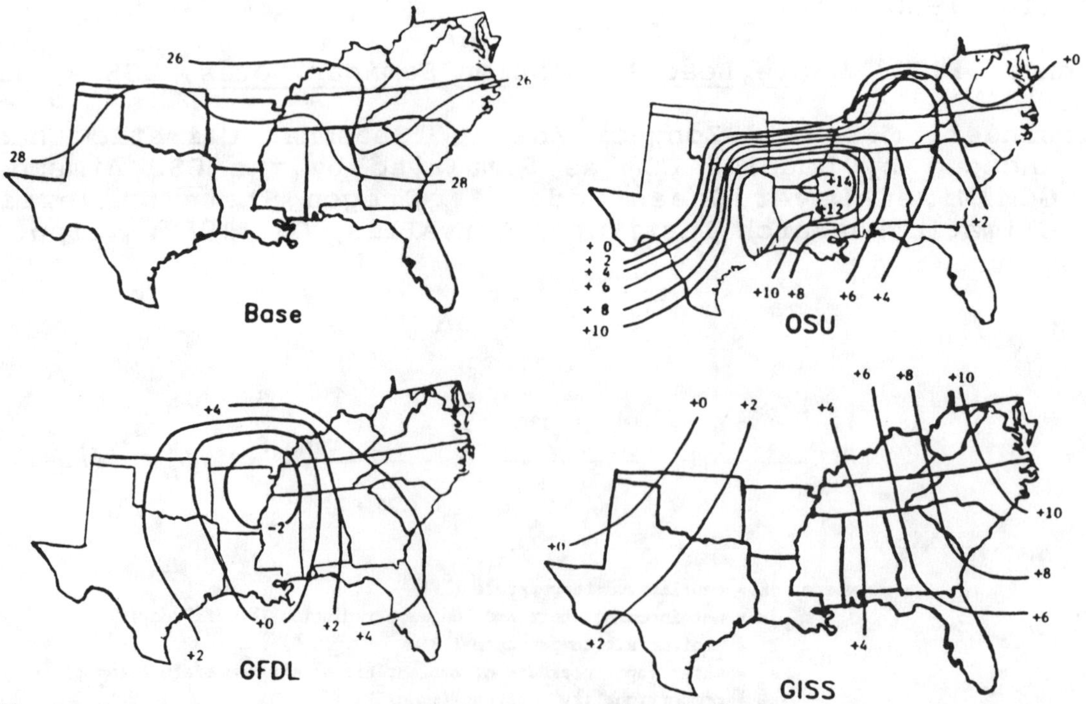

Figure 1. Historical base water temperatures and changes from that base associated with three climate change scenarios.

Figure 2. The impact of water temperature on dissolved oxygen (ppm).

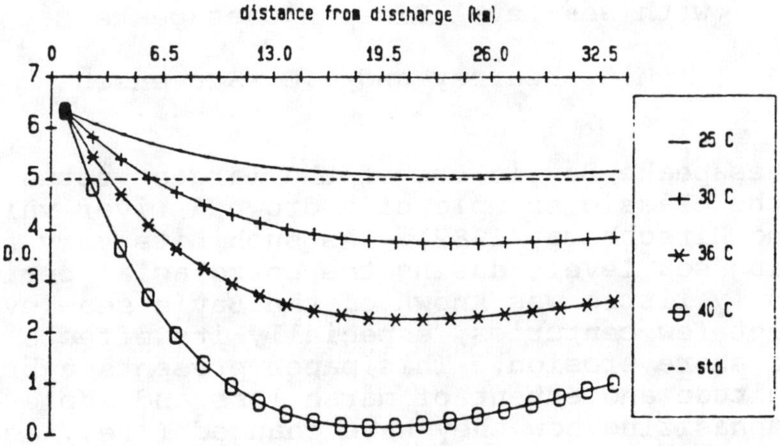

Figure 3. The effect of treatment and flow on dissolved oxygen under elevated water temperature conditions (ppm).

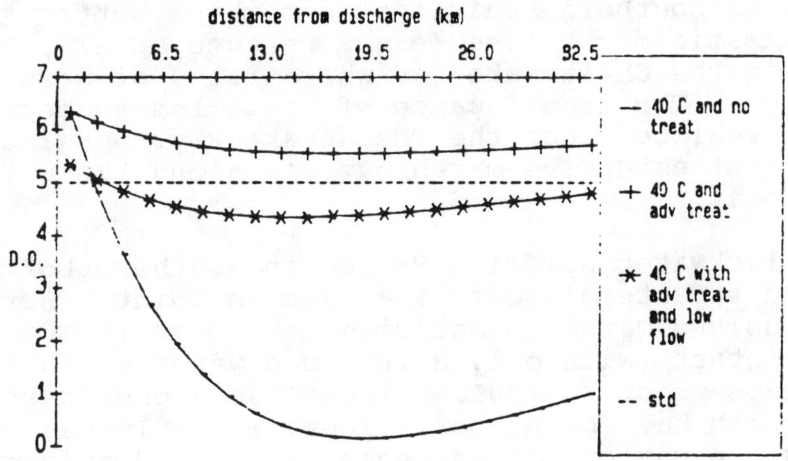

Figure 4. Response of water temperature to vegetative shading (GISS = 35°lat x 80°long; GFDL = 33°lat x 82°long; OSU = 34°lat x 80°long; Base = Columbia, SC = 34°lat x 81°long).

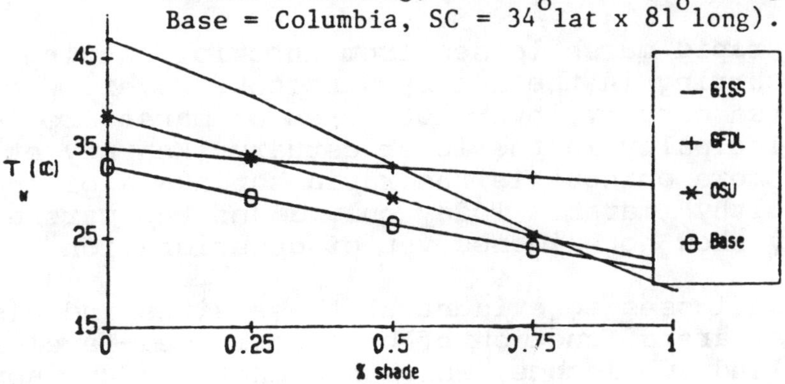

Marsh Loss and Shore Erosion with Sea-Level Rise in Chesapeake Bay

M.S. Kearney and J.C. Stevenson

The Chesapeake Bay is one of the largest estuaries in the world and the classic example of a drowned river valley estuary (Schubel and Hirschberg, 1982). As such, its very existence is due to rising sea levels during the postglacial period. However, until recently little was known of the Bay's sea-level history over the last few centuries, especially its effects on coastal marshes and shore erosion. This paper presents a brief synopsis of the magnitude and extent of marsh loss and shore erosion in the Bay, emphasizing how they have changed (i.e., accelerated) as a function of sea-level rise (and land subsidence), and ultimately look at what these processes tell us about sea level itself in the recent past.

Coastal Marshes

Although southern Louisiana is perhaps better known as an area of dramatic marsh loss (e.g., DeLaune et al., 1983), there are sites in the Chesapeake Bay where rates of marsh loss are just as high. The significance of these losses becomes apparent when it is realized that the Chesapeake Bay contains some of the single largest estuarine marsh systems along the U.S. middle Atlantic Coast.

The Blackwater Wildlife Refuge in southern Dorchester County on Maryland's Eastern Shore is a case in point. Here, a large submerged upland marsh, comprising once over 10,000 acres, was basically intact (with only a few open water areas) according to accurate maps prior to 1900. Since 1938 alone, a total of 5700 acres of marsh have been lost. These marsh losses have occurred through the formation, coalescence, and rapid enlargement of interior ponds (Stevenson et al., 1985). At present, the only extant areas of marsh in the center of the refuge consist of remnant vegetation adjacent to major tidal creeks and interior ponds, where higher inorganic sediment inputs are able to sustain marsh growth.

Equally rapid marsh losses from interior ponding processes have been occurring in the nearby Nanticoke River. In this once marsh-dominated estuary, over 100 acres of marsh are lost annually, principally in the lower estuary (Kearney et al., 1988). Even more ominous is the rapid shrinkage of the remaining areas of "healthy" marsh. Today only 3% of the marshes in the lower estuary show no evidence yet of deterioration.

The marsh losses so evident at these sites and elsewhere in Chesapeake Bay are a function of both world sea-level rise and rapid local land subsidence, which has enhanced the apparent submergence rate. Stevenson et al. (1985) have shown that certain areas in the Blackwater marshes have incurred an 8 cm accretionary deficit versus sea level since 1940. Similar accretionary deficits also characterize many of the Nanticoke marshes. Moreover, the highly organic nature of the marsh sediments in these systems inherently predisposes them to high rates of loss. As progressive "drowning" occurs from the cumulative accretionary deficits and the surface peat mat fragments as the plants die, wave activity readily erodes the underlying unconsolidated organic ooze. Interior ponds thus are able to enlarge dramatically in a few years.

Shore Erosion

If marsh loss has been proceeding at a catastrophic pace in some areas of Chesapeake Bay, principally as a result of rising sea level, similar trends appear to characterize rates of shore erosion.

As a generalization, shore erosion (largely by cliff retreat) is the principal mechanism of shoreline retreat on the Bay's western shore; on the low-lying Eastern Shore it is augmented by simple submergence. Present rates of shore erosion in Chesapeake-Bay are nevertheless highly variable, as might be expected along such a complex coastline. In the middle Bay stem, up to 50% of the shoreline is eroding at rates of between 2 to 8 ft/year. More significantly, fully 20% of the shoreline is retreating at rates of greater than 8 ft/year (Conkwright, 1975) in several areas. These latter rates of shoreline retreat are especially significant when they are compared to the 2 ft/year of retreat that typifies most of Fenwick Island on the nearby open coast (Leatherman, personal communication).

Yet, the comparatively high rates of shore erosion in the Chesapeake Bay appear to be only a recent phenomenon. Existing historical maps and charts of the Bay limit realistic assessments of shoreline trends before essentially the middle 19th century. Changes in the acreage of Bay islands, many of them settled since the mid-17th century, partially offset this lack of information on earlier shore erosion rates. These data (Kearney and Stevenson, in preparation) suggest that rates of land loss of Bay islands were relatively slow prior to the early-middle 19th century. Since then, rates of land loss have greatly accelerated. For example, Sharp's Island on Maryland's Eastern Shore, consisting of over 400 acres in 1848, had been reduced to 53 acres by 1910, and to only 6 acres by 1946 (Singewald and Slaughter, 1949). The island presently no longer exists.

Perhaps even more telling has been the human adjustment to the rapid erosion (and submergence) of these islands in this century. Erosion was apparently so rapid in the first decades of the century that, although new churches, schools, and businesses were often still being built on the island as late 1910-1915, most had been abandoned by 1930s. This abandonment was undoubtedly accelerated by two major hurricanes that passed over the area in 1933 (Stevenson et al., 1988). In many instances where shore erosion had not outright destroyed the islands, progressive submergence or conversion of uplands to marsh had made continued habitation untenable (Kearney and Stevenson, in preparation).

The acceleration in the rate of local sea-level rise argues persuasively for sea level as the driving force underlying these processes. When it is realized that rates of regional subsidence in the Bay area among the highest of any locality along the U.S. Atlantic Coast, it is easy to foresee the dire consequences for Bay marshes and shorelines that would result from any acceleration in the global sea-level trend. Even at present rates of loss, most Bay marshes may be in imminent danger of disappearance by the end of the coming century. Of course, all the complex ecological functions that marshes perform in maintaining the Chesapeake Bay as viable biotic entity would necessarily cease. In addition, as the marshes erode, the sediments previously stored in them will be transported into the estuary, creating less than ideal habitats for traditional estuarine species. We would be facing a much altered Bay, apart from the pollution problems that already beset it.

References

Conkwright, R.D. 1975. Historical shorelines and erosion rate atlases: Maryland Geol. Survey, 4 vols.

DeLaune, R.D., R.H. Baumann, and J.G. Gosselink. 1983. Relationships among vertical accretion, coastal submergence, and erosion in a Louisiana gulf coast marsh. J. Sed. Petrol. 53: 147-157.

Kearney, M.S., R.E. Grace, and J.C. Stevenson. 1988. Marsh loss in Nanticoke estuary, Chesapeake Bay. Geogr. Rev. 78: 205-220.

Schubel, J.R., and D.J. Hirshberg. 1982. The Chiang Jiang (Yangtze) Estuary: Establishing its place in the community of estuaries. In: Kennedy, V.S. (ed.), Estuarine Comparisons. New York, Academic Press, pp. 649-666.

Singewald, J., and T.H. Slaughter. 1949. Shore erosion measurement of Tidewater Maryland: Maryland Dept. of Geology, Mines and Mineral Resources Bull. 6, 118 p.

Stevenson, J.C., M.S. Kearney, and E.C. Pendleton. 1985. Sedimentation and erosion in a Chesapeake Bay brackish marsh system. Mar. Geol. 67: 213-235.

Stevenson, J.C., L.G. Ward, and M.S. Kearney. 1988. Sediment transport and trapping in marsh systems: implications of tidal flux studies. Mar. Geol. 80: 37-59.

POTENTIAL EFFECTS OF CLIMATE CHANGE ON CHESAPEAKE BAY ANIMALS AND FISHERIES

Victor S. Kennedy
University of Maryland
Center for Environmental and Estuarine Studies
Horn Point Environmental Laboratories
Cambridge, Maryland 21613

Chesapeake Bay lies in the Atlantic Temperate Region, a transition zone between the cold-water Boreal Region north of Cape Cod and the sub-tropical Atlantic Warm-Water Region south of Southern Florida (1). This Temperate Region is divided by Cape Hatteras into the northern Virginian Province, which includes Chesapeake Bay, and the southern Carolinian Province which is more intimately warmed by the Gulf Stream (2). Vagrant boreal marine organisms may swim or be carried southward past or into the marine mouth of the Bay. As vernal warming occurs, mobile vagrants retreat northward along the coast and less mobile forms perish. In contrast, sub-tropical marine vagrants move or are carried northward to or past the Bay and respond to autumnal cooling by retreating or perishing. In this transition zone, there are more resident species of animals with tropical affinities than with boreal affinities, but the fauna of the Temperate Region is impoverished in comparison with the Warm-Water Region (1). These biogeographic realities need to be considered in any attempt to predict the effects of climate change on Bay biota.

Now, although it is clear that climate will change over ensuing decades (3), there are uncertainties as to the extent of climate warming and the magnitude and direction of changes in rainfall and wind patterns. Prediction of effects of climate change on an extensive shallow estuary such as Chesapeake Bay is constrained by such uncertainties. The amount of temperature increase of Bay waters and the time scale involved are unknown. Concomitant changes in rainfall may affect salinity concentrations, water circulation patterns (also affected by wind), and other physical and biological factors important to the maintenance of species that use the Bay as a spawning area, nursery, feeding ground, or residence. In addition, sea level rise will create new (aquatic) habitat and destroy old (terrestrial) habitat, but the extent of change is unclear. With these uncertainties in mind, I will turn first to examine possible effects of climate warming, then follow with an examination of the effects of changes in rainfall and sea level.

Predictions of the effects of higher temperatures are hampered because our knowledge of the tolerances of even a few Bay species to increased temperature is modest (these are mainly commercial species or those thought to be important to energy flow and nutrient flux in the system). We know little or nothing about the thermal tolerances of most of the Bay's other inhabitants, many of which may be more "influential" than we think. We are just beginning to realize that the Bay is influenced perhaps more by microscopic than macroscopic organisms, yet the study of this micro-realm is barely underway. Thus, given the uncertainties of the extent of climate change and our limited biological knowledge of the structure, function, and thermal responses of the Bay ecosystem, predictions of the effects of climate warming on the Bay's aquatic animal life will be speculative.

Some general predictions that are least speculative are those based on known physical and biological principles. For example, warmer conditions will result in increased metabolism of cold-blooded animals, but because warmer water holds less oxygen, the associated demand for oxygen may exceed the supply. This is of special concern in Chesapeake Bay because of present low oxygen concentrations in summer; these hypoxic and anoxic conditions would worsen with increased thermal stratification and resultant vertical stability of the water column. Warmer temperatures may be lethal to species with northern affinities, but heat-tolerant species may also be at risk because their food supply may be less heat-tolerant than they are and may die out. And, even if higher temperatures are non-lethal, organisms may be stressed to the point that they fall prey to more robust predators or pathogens.

Other general predictions are more speculative because they are based on analogy. For example, one can look south at warmer estuaries of the southeast Atlantic or the Gulf of Mexico. These southern estuaries are as warm as Chesapeake Bay may become, and have many species in common with the Bay. However, species with northern affinities are rare or missing (e.g., soft clam, winter flounder; see below) and these species will also be lost from the Bay as it warms. Unfortunately, it is probably "easier" to eliminate a species than to replace it, especially with one that has a similar ecological role. That is, elimination from a range may occur over a few years. Contrarily, expansion of a species' range may take longer because successful expansion involves not just the provision of a physically tolerable habitat but also requires that suitable food be present and that the predators and pathogens be familiar. Importantly, species must be able to reach the new environment in order to colonize it, yet estuarine species may not tolerate the marine salinities of coastal waters that intervene between the home estuary and the target estuary, and many species (especially invertebrates) may not be suitably mobile. There is virtually no chance that such organisms could move along the seacoast into new habitat unassisted by human actions in anything less than many decades, if not centuries or millennia. Yet climate warming is predicted to occur over a few decades and some species may disappear from the Bay not long thereafter. It would take a long time for the Bay to come to resemble its southern estuarine counterparts (as they are now, for they too will change as climates warm), and in the interim it would have a lower diversity of resident estuarine species, with unpredictable consequences.

Some species found to the south of the Bay have a reproductive season that starts and ends earlier, or starts earlier and ends later, than in the Bay (e.g., see striped bass and oyster below), so a warmer Bay may result in a similar such season for them here. However, the success of a spawning season is measured by the survival and growth of offspring. If the food of the young is not available in the earlier or the expanded portion of the spawning season, the young may not survive. For Bay species, this might happen for those that would spawn earlier in spring or later in autumn than at present. At these times, light conditions may be sub-optimal for photosynthesis (light is often a limiting factor in temperate regions - see 4) so herbivorous larvae produced earlier or later than at present may find phytoplankton to be poor in quantity or quality. Thus, there may be no advantage to spawning under these light and food conditions, and energy and DNA would be wasted.

The general principles discussed above can be applied to four abundant and commercially important species in Chesapeake Bay that are reasonably representative of many Bay species in their response to higher temperature. The four species include: striped bass (ranges from Gulf of St. Lawrence to Florida and Gulf of Mexico; floating eggs and larvae appear upriver in spring; juveniles spend some years in the Bay before leaving as sub-adults; migratory adults return to the Bay to spawn); blue crab (ranges from about Cape Cod to the Gulf of Mexico; eggs hatch in high salinities at the Bay mouth; larvae and juveniles re-enter the Bay to grow and then reproduce as mobile adults; most adults remain within the Bay); oyster (ranges from Gulf of St. Lawrence to Gulf of Mexico; eggs and larvae present throughout the Bay in summer; juveniles and adults attach permanently to hard substrate); and soft clam (ranges from Labrador to Cape Hatteras; behaves as does the oyster except it lives in soft sediments).

Of the four species, I expect that the soft clam will be lost to the Bay if temperatures rise by 3-4°C. The species lives near the southern end of its range here and experiences high mortalities in warm summers (5). Its elimination will end an important fishery. (The winter flounder is a northern species of fish that may disappear from the Bay in like fashion). The remaining three species range south to the Gulf of Mexico, so they may not suffer increased mortality from higher temperature. However, the striped bass may spawn some weeks earlier (6), as it does in Georgia, whereas the oyster might expand its spawning period from the present two months to six months or more as in the Gulf (7), dependant upon suitable food to support gamete production and larval survival in the Bay (see above). The blue crab may be active year round, rather than becoming inactive as it does now over winter. If the soft clam disappears from the Bay, the blue crab will have lost a prey species (it preys upon young soft clams), but it is a generalist feeder and may find sufficient alternate prey.

Of concern in the case of striped bass is the extent and duration of temperatures above the adult's "upper avoidance temperature" of 25°C (6). This temperature is now exceeded in the upper Bay in early July, with the volume of unacceptably warm water increasing through August and dissipating in September. Coutant (6) has used two climate-change models (GISS, GFDL) to predict that water above 25°C may appear in the Bay by early June, with the entire Bay estimated to exceed 25°C from July through mid-September. Such conditions will force striped bass to seek cooler conditions. However, continuation or expansion of hypoxic conditions such as are now experienced in deeper parts of the Bay in summer may render these somewhat cooler waters unacceptable, while at the same time shallower waters with acceptable oxygen levels may remain too warm for striped bass. The species would then be faced with the "squeeze" that Coutant (6) suggests constrains its habitat space. Although adults and sub-adults (especially those in the lower Bay) may be able to leave for cooler coastal waters, resident juveniles may not.

It is not clear if the blue crab's habitat space would be "squeezed" like that of the striped bass. If it is, blue crabs may not be able to avoid sub-optimal conditions by leaving the Bay; males and juveniles appear to be largely resident. Because the species is sensitive to low oxygen conditions (8), the celebrated "crab wars" that find crabs crawling onto Bay shores in milling masses in summer may be their response to low dissolved oxygen in deeper water. Blue crabs are more tolerant of higher temperatures (8) than are striped bass and may find refuge in shallow waters if these have suitable oxygen concentrations, but this constriction of habitat space could lead to smaller population size and harvest.

Unlike the mobile striped bass and blue crabs, oysters and soft clams cannot move away from stressful conditions. Their swimming larvae will be no help in escaping sub-optimal conditions because larval mobility is limited to modest active vertical swimming and to passive entrainment in horizontal currents. However, the oyster is more temperature tolerant than the soft clam, as evidenced by its more southerly distribution, and should survive warming conditions.

Turning to other components of climate change, I will consider briefly the effects of changes in rainfall and in sea level. Increased spring rainfall and runoff in the watersheds of the Bay will have a deleterious effect because intensity of spring runoff has been linked to intensity of low dissolved oxygen conditions in the Bay (9). This is the result of the over-riding of denser saltier water by the less dense freshwater inflow from rivers. The resultant increased density gradient inhibits diffusive transport of oxygen from the atmosphere to deeper waters where biological oxygen demand is high, with hypoxia or anoxia ensuing (9). Immobile hypoxia-intolerant organisms will die and mobile forms will move to such suitable conditions as may exist. However, shallow habitats that are now spared hypoxic conditions may be subjected to lowered oxygen concentrations during all or most of the summer if climate conditions change as predicted.

Decreased rainfall and runoff will allow higher salinity water to penetrate further up the Bay, accompanied by coastal marine organisms that will increase species diversity in the lower Bay. However, some of these invaders will have negative effects. At present, oysters in Maryland's portion of the Bay are largely safe from predators found in higher salinity waters (20 ppt or more), such as in Virginia. These predators include starfish and species of snails that bore holes into oyster shell to kill and eat the oyster within (7). Also, most oysters living below 12-15 ppt salinity are safe from diseases such as 'MSX' and 'Dermo', which have taken a recent heavy toll of oyster populations in saltier water (7). If decreased runoff allows these predators and pathogens access to Maryland's oysters and to those oysters in lower salinity portions of Virginia's tributaries, the stock of oysters will decrease. Similarly, other Bay organisms, commercial and otherwise, may be at risk in the future from predators and disease presently held in check by low salinities.

As sea level rises due to thermal expansion of water and to melting of ice, marsh lands will be submerged and shallow waters will deepen. Marsh plants will be lost to flooding and, if light fails to penetrate the deepening waters to support photosynthesis, so will rooted aquatic plants such as seagrass. These plants and their assemblages provide food and shelter for a variety of estuarine organisms. Because humans have built to the edges of marshes, the landward march of the sea may not be balanced by an inland movement of marshes. More likely, marshes will disappear in all but lightly-populated regions. Their loss will affect their animal dependents. Rooted aquatic plants may be able to invade new shallows successfully, but this will take time because the substrate will be former marsh soils that may not become satisfactory to support rooted aquatic plants for years. Although seagrass meadows contain more diverse fish communities than do tidal marshes (10), the loss of marsh habitat will mean the loss of many animal species and the decline of habitat, such as the nesting and feeding grounds of waders, gulls, terns, and the waterfowl that support the hunting industry.

I conclude, then, that warming of Chesapeake Bay over the next half-century will result in loss of heat-sensitive species, slow influx of mobile southern species (mostly salinity tolerant and predominantly fish), and "habitat squeeze" for other species, of which the sessile forms will be at a disadvantage. The soft clam will probably disappear, but oysters, blue crabs, and striped bass should persist, probably in lesser abundance in nature. Their continued harvestability will require both strict pollution controls to restore higher oxygen concentrations to avoid "habitat squeeze" and strong encouragement of the farming of these species by aquaculturists to maintain their market share. Increased runoff will exacerbate low dissolved oxygen conditions whereas decreased runoff will expose up-estuary populations to increased predation and disease from invading high-salinity predators and pathogens. Sea-level rise will decrease marsh habitat, with perhaps limited change in area of submerged aquatic plants. For an unknown period, the estuarine portion of the Bay will be poorer in species than it is now, to the probable detriment of the system in general and commercial and sport harvests in particular. Recovery to a semblance of more southerly estuaries as they are now will not be rapid.

References and Acknowledgments

(1) Gosner, K.L. 1971. Guide to Identification of Marine and Estuarine Invertebrates. Wiley-Interscience, New York; (2) Hall, C.A., Jr. 1964. Ecology 45:226-234; (3) Schlesinger, M.E. and J.F.B. Mitchell. 1987. Rev. Geophys. 25:760-798; (4) Pennock, J.R. 1985. Est. Coastal Mar. Sci. 21:711-725; (5) Kennedy, V.S. and J.A. Mihursky. 1971. Chesapeake Sci. 12:193-204; 1972. Chesapeake Sci. 13:1-22; (6) C. Coutant, ORNL, Oak Ridge TN, personal communication; (7) Kennedy, V.S. and L.L. Breisch. 1981. Maryland's Oysters:Research and Management. MD Sea Grant Publication No. UM-SG-TS-81-04; (8) Millikin, M.R. and A.B. Williams. 1984. NOAA Tech. Rep. NMFS 1:1-39; (9) Officer, C.B. et al. 1984. Science 223:22-27; (10) Weinstein, M.P. 1985. Pages 285-310 in A. Yanez-Arencibia (Editor). Fish Community Ecology in Estuaries and Coastal Lagoons. UNAM Press, Mexico City, Mexico.

I acknowledge the critical comments of B. Baldwin, C. Coutant, H. Ducklow, S. Leatherman, and M. Roman on earlier drafts of this manuscript. Contribution number 1981 HPEL from the Center for Environmental and Estuarine Studies, University of Maryland.

PREPARING POLICYMAKERS TO ADDRESS THE PROBLEM OF CLIMATE CHANGE

Greg Watson
Executive Director
Massachusetts Office of Science and Technology

About two months ago representatives from a number of Massachusetts state agencies began meeting informally twice a month during lunch to discuss the possible impacts of global climate change and stratospheric ozone depletion on the citizens, environment and economy of the Commonwealth. Agencies represented included the offices of the Secretary of State and the Lieutenant Governor; the Executive Offices of Environmental Affairs, Economic Affairs, Energy Resources and Communities and Development; and the Departments of Environmental Quality Engineering, Coastal Zone Management, Public Utilities and Transportation.

We understood that strategies designed to mitigate the effects of climate instability will ultimately have to be resolved internationally as was done with the Montreal Protocol that addresses the problem of ozone depletion. However, scientists tell us that regardless of what we do, we are committed to a certain degree of climate change as a result of the carbon dioxide, nitrous oxide and other heat-trapping gases already deposited in the atmosphere. Federal, state and local governments will therefore ultimately have to deal with whatever national and regional problems accompany these changes. If this is in fact true, there are some nagging questions that must be addressed immediately.

Given the uncertainty that still exists with regard to the eventual outcome of humanity's inadvertent experiment on the atmosphere, we want to know what immediate actions can and should be undertaken that will best insure that human suffering and environmental degradation are minimized? What effective steps to this end can be implemented at the state and local levels? Who should take the reponsibility for proposing these recommendations?

It is obvious that if we wait for absolute verification of the various predictions made by scientists' computer models, we will have delayed too long to act. The reason I'm here is to see if I can get a little closer to discovering some answers to these extremely important questions.

Clarifying these issues will require the close cooperation of scientists, policymakers, the private sector and the general public. Scientists will be hard pressed to give their best assessments of when, where and how the climate is expected to change. Decision makers will be asked to take this information and draft legislation and/or regulations that will, in their estimation, best protect the health and safety of their constituents. Who bears the burden of proof here?

Hopefully these decisions will be made in a timely manner. When they are, the public (and that includes the press) will undoubtedly have many questions concerning them and assumptions that underlie them. Rightfully so. In these days of austere budgets, if some of the suggested proposals are approved and implemented, a reorganization of funding priorities will certainly be required. For example, are Bostonians willing to postpone projects to depress their central traffic artery and to build a third harbor tunnel in favor of projects designed to mitigate damage from a possible rise in sea level? Should resources be directed to support massive soil remineralization and tree planting programs in an attempt to stave off or at least minimize climate instability? Citizens would surely express their reaction to such proposals in the voting booth. That's why some elected officials may be reluctant to commit themselves to specific strategies until "all of the data is in and analyzed."

The private sector may be asked to voluntarily modify their production processes in lieu of regulations (as some computer companies have done by substituting aqueous solutions for CFCs to clean electronic circuits). In order for all this to occur in time to avoid human suffering and economic disruptions, an unprecedented public education effort will have to be mobilized--one that stresses the urgency of the situation enough to motivate action while avoiding unnecessary fear and panic.

Most of us will need to adjust both the time frame and the context that we're used to planning with. Our time frame will have to be expanded beyond the short term business and election cycle, and our geographical domain must literally encompass the whole Earth. In short, we'll have to adopt the concept of "Thinking Globally and Acting Locally" as our standard operating procedure.

THE RELATIONSHIP BETWEEN RELATIVE SEA-LEVEL RISE AND COASTAL UPLAND RETREAT IN NEW ENGLAND

GRAHAM S. GIESE AND DAVID G. AUBREY
WOODS HOLE OCEANOGRAPHIC INSTITUTION

INTRODUCTION

In 1986, the Massachusetts Coastal Zone Management Office (MCZM) identified a need for upland retreat studies after reviewing the Commonwealth's coastal policies in light of predictions that global climatic warming in the 21st Century would produce increased rates of sea-level rise. The MCZM review found that the state's wetlands were relatively well protected by statutes and by an extensive body of wetland regulations, and it found that studies of wetland response to relative sea-level rise were appearing regularly in the professional literature. In contrast, however, the review revealed a more critical situation in regard to coastal uplands. Restrictions on upland development were largely confined to wetland-boundary setback requirements and FEMA flood zone regulations. Despite large and increasing pressures for development of coastal upland areas, little information was available to help planners quantify the rates of upland loss that would accompany hypothetical future sea-level rises.

The term "upland" refers to terrain landward of wetland that has not been altered appreciably by coastal processes - tides, waves and winds, for example - while "wetland" refers to features such as dunes, beaches and marshes that have been formed or reworked by coastal processes. When relative sea-level rise produces upland retreat, the upland lost is usually accompanied by wetland gain; but, since wetland use is restricted, the area available for most societal uses, such as farming or commercial development, is reduced by the amount of upland lost. The present report summarizes the results of our recent studies of coastal upland retreat in response to relative sea-level rise. Our objectives are: (1) to examine the conditions and processes that control the rate of upland retreat in response to relative sea-level rise, and (2) to quantify the rate of upland retreat under those conditions and processes.

NEW ENGLAND RELATIVE SEA-LEVEL RISE SCENARIOS

During the past 40 years, relative sea-level in New England has been rising at a mean annual rate of between 2 and 3 mm per year (Aubrey and Emery, 1983). Approximately 2 mm per year can be attributed to land subsidence (Braatz and Aubrey, 1987). For present purposes, we assume the global rate of rise to be 1 mm per year, which is consistent with the recent results presented by Gornitz and Lebedeff (1987). We also assume the rate of relative sea-level rise in New England to be 3 mm per year. Should this rate continue unchanged into the future, by the year 2125 relative sea level will be 13.5 cm above its 1980 level.

Within recent years, however, a rapidly increasing body of data has supported the hypothesis that global climatic warming within the next century will produce increased rates of global sea-level rise. Hoffman et al. (1983), for example, have projected global sea-level rises between 1980 and 2025 ranging from 26.2 cm to 39.3 cm for "mid-range" estimates. Adding these values to the total New England subsidence expected during that period (9.0 cm) produces relative sea-level increase estimates of between 35.2 cm and 48.3 cm between 1980 and 2025. We argue neither for nor against the validity of these estimates; they are used solely as hypothetical scenarios.

UPLAND LOSSES DUE TO WAVE-DOMINATED EROSION

Coastal upland retreat takes two interrelated forms arising from distinct processes: the active wave-produced erosion of exposed shores, and the passive losses that result from gradual inundation of protected shores due to relative sea-level rise. We discuss first the effect of relative sea-level rise on the erosion of exposed shores in New England. The scouring action of Wisconsinan glaciation left northern New England relatively bare of overburden and its coast characterized by exposed crystalline bedrock. South of central Massachusetts, however, the coast is largely dominated by deposition features, such as thick glacial outwash and morainal deposits. Observations of the exposed rocky coast of New England show that it is remarkably resistant to wave action and has changed little during the past 50 years (Shepard and Wanless, 1971) despite a relative sea-level rise of some 15 cm. Based on those observations, we expect little change in the exposed rocky New England coast by 2025 regardless of whether relative sea level continues to rise at the present rate or at one of the accelerated rates suggested by Hoffman.

In contrast, the unconsolidated glacial deposits that form much of the exposed southern New England (and Long Island) coast have eroded significantly within historical times. Perhaps the best-studied example of this process is the retreat of the sandy marine bluffs of Outer Cape Cod. Studies by Marindin (1889), Zeigler et al. (1964a), and Leatherman et al. (1981) have produced long-term mean bluff erosion estimates of 0.9, 0.8 and 0.7 meters per year respectively. Since all three estimates fall well within one standard deviation of the scatter of the data from which they were derived, no special significance should be attached to the fact that the values progressively decrease over the time period. It is safe to say, however, that there is no evidence of an increase in the bluff erosion rate over the past century at this location.

A recent study of short-term bluff erosion (Giese and Aubrey, 1987a) has produced evidence that retreat of the Cape Cod marine scarp at any specific location and time depends to a large extent on the volume of the beach deposits lying in front of the bluff. Since these beach deposits are derived from materials eroded from the bluffs by wave action (Zeigler et al., 1964b), they provide a negative feedback control on bluff erosion. The beach deposits do not consist of sediment in transit in an offshore direction as would be expected if the "Bruun Rule" were to be applicable for calculation of shore retreat as a function of relative sea-level rise. Rather, the beach sediment is in transit along the shore, and therefore the rate at which it is removed from the bluff area depends upon the alongshore gradient of energy flux. Thus, while it would be reasonable to expect some increase in the rate of unconsolidated marine scarp erosion in New England to accompany an increase in the rate of relative sea-level rise, the increase in erosion is not likely to be as great as the concurrent increase in sea-level rise. That is to say, the linear relationship between shore retreat and sea-level rise expressed by the Bruun Rule would not be expected in this case. In fact, it may be that changing storm climate accompanying global climatic warming will affect unconsolidated marine scarp erosion as much as increasing sea-level rise.

PASSIVE UPLAND RETREAT DUE TO INUNDATION

As relative sea-level rises, it produces a concurrent rise in the associated tidal datum levels (e.g., mean spring high water, mean annual high water). Therefore, along protected shorelines the level of the boundary separating upland from wetland rises at the same rate as relative sea level, regardless of the local elevation of that boundary, which varies considerably from place-to-place due to varying exposure to waves and tides. We refer to upland loss resulting from increasing elevations of the upland/wetland boundary due to relative sea-level rise as "passive retreat" of coastal upland.

Measuring the passive retreat of coastal upland presents special problems. The linear retreat rate at any one point is simply the product of the inverse of the local upland slope and the rate of relative sea-level rise. But using this method to calculate the rate at which upland is lost by an entire community would require an extraordinarily large effort. As an alternative, we calculated the rate of passive retreat of Massachusetts coastal communities by use of "hypsometric curves", cumulative frequency diagrams that present the distribution of upland area of each community with respect to elevation.

There is a striking variation between communities in the shape of their hypsometric curves, reflecting variation in the geological processes that formed them. For example, communities in the glaciated sections of northern New England that have been largely stripped of unconsolidated overburden, such as Manchester, Massachusetts, have curves with steep slopes at low elevations, giving them a "convex-up" form (Figure 1), while communities on the glacial outwash plains of southern New England - for example Marion, Massachusetts - have flatter slopes at low elevations and an overall "concave-up" form (Figure 2).

Making use of these hypsometric data, we have calculated the upland areas that each community would lose given a specific change in relative sea level. By the the year 2025, given a continuation of the historical mean annual relative sea-level rise rate of 3 mm/yr, all Massachusetts communities combined will lose a total of 1,200 hectares (3,000 acres) of upland due to passive retreat. The loss per town varies greatly: rocky Manchester will lose only 0.1 % of its total upland area, while low-lying Marion will lose 1.4 %. Given Hoffman's "mid-range low" estimate of global sea-level rise, the total upland loss in Massachusetts would be about 3,000 hectares (7,500 acres), while his "mid-range high" estimate would result in a loss of about 4,000 hectares (10,000 acres). The major point to note here is that the rate of passive upland retreat increases at the the same rate as (is linear with) the rate of increase in relative sea level.

DISCUSSION AND SUMMARY

The results of this preliminary survey indicate that the rocky, glaciated coasts prevalent in northern New England (but also present at scattered locations farther south) are much less susceptible to the effects of rising sea level than are the sandy coasts formed of unconsolidated glacial deposits. Upland loss due to wave attack is minimal along rocky shores, and projected relative sea-level increases by the year 2025 should not produce much of an increase. Because glaciated coasts tend to be characterized by relatively small areas of upland at low elevations, passive retreat of upland due to relative sea-level rise also will be relatively small.

In contrast, communities along the sandy coasts of southern New England are presently losing upland at an appreciable rate. Most visible and widely discussed is the active wave-produced erosion of unconsolidated marine bluffs. In fact, however, most of these communities lose more upland as the result of passive inundation than from active erosion. It has been estimated, for example, that under present conditions Cape Cod loses about 3.5 hectares (9 acres) of upland per year from active erosion as opposed to about 9.6 hectares (24 acres) per year due to passive retreat (Giese and Aubrey, 1987b). The present results suggest that the differential between the rates of active and passive upland retreat in southern New England will increase with increasing rates of

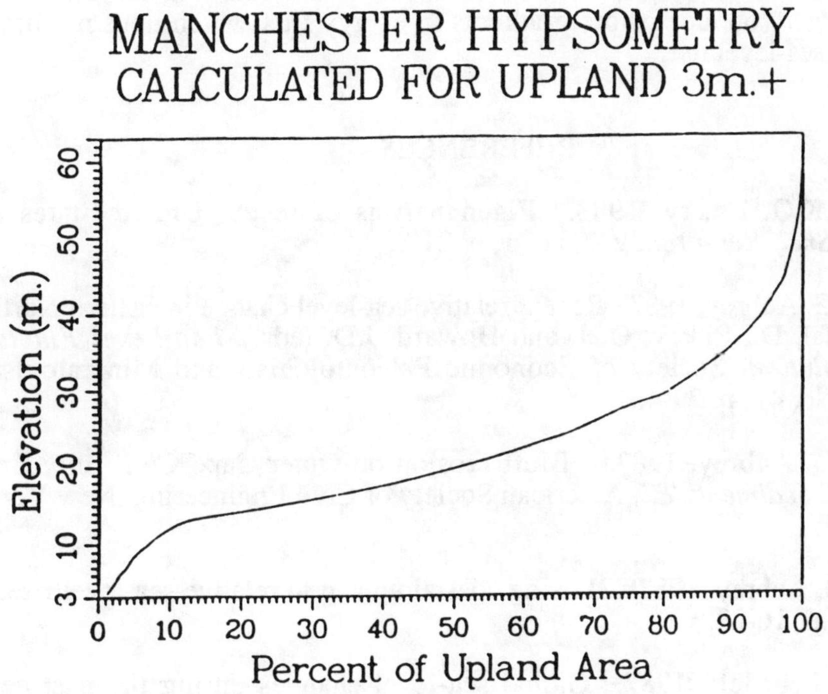

Figure 1

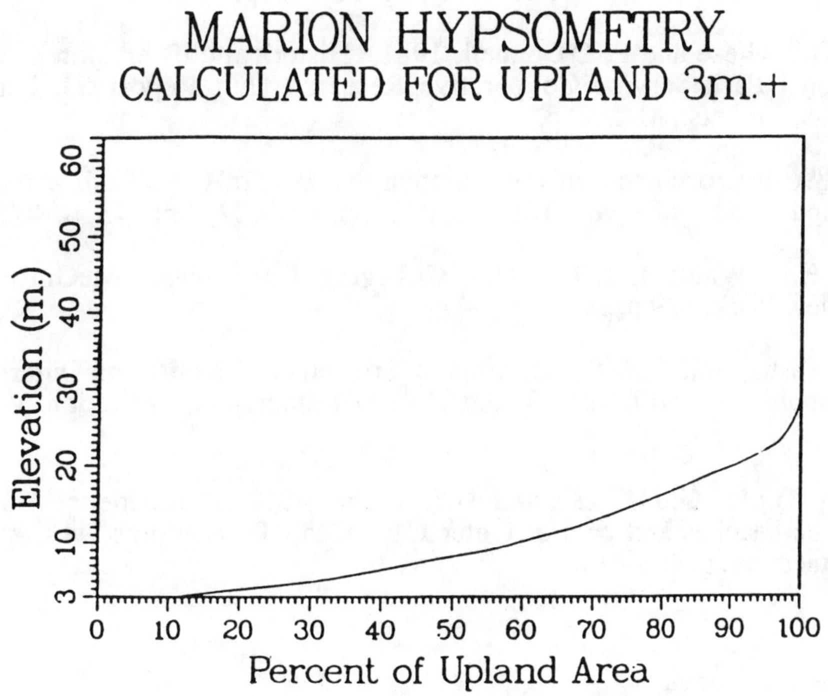

Figure 2

relative sea-level rise. This is because passive upland retreat increases at the same rate as sea-level rise, but active upland erosion is buffered by the beach sediments that erosion supplies; their removal depends largely on alongshore gradients in energy flux which does not have a simple direct relationship to sea-level rise.

REFERENCES

Aubrey, D.G. and K.O. Emery, 1983. Eigenanalysis of recent United States sea levels. *Continental Shelf Research*, v. 2, no. 1, p. 21-33.

Braatz, B.V. and D.G. Aubrey, 1987. Recent relative sea-level change in eastern North America. In: Nummedal, D., Pilkey, O.H. and Howard, J.D. (eds.), *Sea-Level Fluctuation and Coastal Evolution,* Society of Economic Paleontologists and Mineralogists, Special Publication No. 41, p. 29-46.

Giese, G.S. and D.G. Aubrey, 1987a. Bluff erosion on Outer Cape Cod. In: Krause, N.C. (ed.), *Coastal Sediments '87*, American Society of Civil Engineering, New York, p. 1871-1876.

Giese, G.S. and D.G. Aubrey, 1987b. Losing coastal upland to relative sea-level rise. *Oceanus*, v. 30, no. 3, p. 16-22.

Gornitz, V. and S. Lebedeff, 1987. Global sea-level changes during the past century. In: Nummendal, D., O.H. Pilkey and J.D. Howard (eds.), *Sea-level Fluctuation and Coastal Evolution*. Society of Economic Paleontologists and Mineralogists, Special Publication No. 41, Tulsa, Okalhoma, p. 3-16.

Hoffman, J.S., D. Keys and J.G. Titus, 1983. *Projecting Future Sea-Level Rise*. U.S. Environmental Protection Agency Report 230-09-007, 121p.

Leatherman, S.P., G.S. Giese and P. O'Donnell, 1981. Historical cliff erosion of Outer Cape Cod. National Park Service Cooperative Research Unit Report 53, University of Massachusetts, Amherst, MA, 50 p.

Marindin, H.L., 1889. Encroachment of the sea upon the coast of Cape Cod. Annual Report, U.S. Coast and Geodetic Survey, 1889, App. 12, p. 403-407, App. 13, p. 409-457.

Shepard, F.P. and H.R. Wanless, 1971. *Our Changing Coastlines*. McGraw-Hill Book Company, New York, 579 p.

Zeigler, J.M., H.J. Tasha, and G.S. Giese, 1964a. Erosion of the cliffs of Outer Cape Cod: tables and graphs. Report 64-21, Woods Hole Oceanographic Institution, Woods Hole, MA.

Zeigler, J.M., S.D. Tuttle, G.S. Giese and H.J. Tasha, 1964b. Residence time of sand composing the beaches and bars of Outer Cape Cod. Proceedings 9th Conference on Coastal Engineering, p. 403-416.

STRATEGIES TO RESPOND TO CLIMATE CHANGE AND SEA LEVEL RISE IN ATLANTIC CANADA

Peter K. Stokoe
School for Resource and Environmental Studies
Dalhousie University
Halifax, Canada

Introduction

Over the past four years, the Canadian Climate Program has sponsored a series of studies into the possible socio-economic impacts and policy implications of climate change for major economic sectors that would be most immediately affected in each region of Canada. This paper is based on such a study for the Atlantic Region of Canada, which focusses on five sectors of the marine economy: fisheries, marine transportation, energy development, coastal infrastructure, and tourism and recreation.

Small coastal communities are especially dependent on these economic sectors. Consequently, the socio-economic and policy implications of climate change are especially acute for these communities. There are over 1300 coastal communities with populations of less that 10,000 in the Atlantic Provinces of Canada (Poetschke, 1984). Together, these communities account for about a quarter of the Region's total population of just over two million.

In Phase 2 of this study (Stokoe et al., 1988), which carries on from Phase 1 (Stokoe, 1988), reported here last year (Stokoe, 1987), we have given special attention to small coastal communities, particularly eight communities which were selected as case studies. Our effort to produce a longitudinal study at this level of specificity revealed wide-ranging measures which could be taken to mitigate impacts and exploit opportunities from climate change. This paper draws some generalizations from these measures, suggesting some overall strategies for responding to climate change and sea level rise in Atlantic Canada.

Basic Sectors of Coastal Communities

Coastal communities in Atlantic Canada are usually situated on bays or estuaries. Supplemented by coastal infrastructure, these natural features provide harbours, which serve as bases for marine-related activities. While marine transportation and offshore oil and gas development are important to a few coastal communities, the major traditional economic activity for most has been the capture fisheries. Although access to a freshwater supply is generally important to such communities for the needs of the residents, it is especially important to those communities which also require fresh water to operate fish processing plants. In addition to these basic sectors, some of the larger communities have also become commercial and service centres for their surrounding areas. In recent years, these traditional sectors have been supplemented by tourism and now aquaculture.

Climate Impacts on Basic Sectors

The sectors of importance to most coastal communities are shown in the left column of Table 1. The second column lists some of the direct impacts of local climate change on each sector. Global climate change may also have indirect impacts on each sector, as indicated in the third column. Potential outcomes depend not only on impacts induced by climate change, but also on trends in other contributing factors which will be operative over the same time period, as suggested in the fourth column. The "indirect climate impacts" or the "other contributing factors" may augment or offset direct climate impacts; possible outcomes from summing all of these influences are suggested under "cumulative impacts" in the fifth column. Finally, the sixth column lists some potential policy responses to these cumulative impacts.

Coastal Infrastructure

For coastal infrastructure, flooding, as a direct effect of a rising sea level, is the dominant impact. Climate change may also increase needs for infrastructure development in areas that become more attractive for economic development generally. In response to potential risks to coastal infrastructure, committees of engineers and planners should be authorized to propose amendments to engineering standards and building codes, as well as to zoning and building permit processes, to ensure that sea level rise is duly taken into account in new construction. Mitigation strategies for coastal resources (e.g. wetlands and beaches) should also be considered.

Freshwater Supply

Climate change scenarios derived from general circulation models suggest that the most populated areas of Atlantic Canada may experience decreased average precipitation and increased evaporation, leading to reduced freshwater availability and a lower water table. A lower water table and higher sea level would both be conducive to greater saltwater intrusion into coastal estuaries and groundwater. Therefore, climate change may further reduce the availability of potable water (especially groundwater) in coastal communities, where supplies are already becoming degraded by contaminants (e.g. gasoline, pesticides, dry cleaning fluid and other toxic chemicals). The prospect of climate change increases the urgency of measures to conserve freshwater supplies and to protect them from contamination.

Capture Fisheries

With the oceanographic consequences of climate change still very uncertain, it will likely be some time before predictions can be made about increases or decreases in the abundances of fish species. From past observations, however, it appears likely that climate warming would cause a northward shift in the latitudinal ranges of many species. The ability of different

SECTOR	DIRECT CLIMATE IMPACTS	INDIRECT CLIMATE IMPACTS	OTHER CONTRIBUTING FACTORS	CUMULATIVE IMPACTS	POLICY RESPONSES
Coastal Infrastructure	flooding from sea level rise reduced ice stress	shifts in locations and types of infrastructure required	expected lifetimes of existing infrastructure	damage, and possible loss, of some existing infrastrucuture demands for replacement and new infrastructure	consideration of sea level rise, and changing requirements, in planning new infrastructure, and upgrading or replacing old infrastructure
Freshwater Supply	reduced supply lower water table saltwater intrusion	change in acidic, and other polluting, deposition?	groundwater contamination	reduced freshwater availability declining water quality	conservation planning water quality protection
Capture Fisheries	increase in abundance? greater interannual stock variability? northward shift in fish species distributions longer ice-free fishing seasons	change in world fish supply and demand?	capital intensification inshore vs. offshore competition changing markets	need to adapt to new fishing opportunities and areas increased fishing pressure?	more research on fish stock responses to climate change enhancement of abilities for gathering real-time information on stocks prudent stock management, given uncertainty
Aquaculture	faster growth of fish expansion of geographic range of suitable sites	increased mobilization and uptake of toxics and pathogens?	changes in nearshore environmental quality increasing uses of coastal and nearshore areas	increased opportunities, but possible increase of risks of contamination more conflicts between aquaculture and other coastal activities	protection of nearshore environmental quality anticipation and mitigation of potential use conflicts
Tourism/Recreation	increased summer tourism potential	potential losses of wetlands, wildlife and other coastal resources	general growth of tourism	increased tourism with greater environmental impacts on declining resource base	mitigation of tourism impacts consideration of climate change in conservation

Table 1 - Potential Policy Responses to Impacts of Climate Change in Atlantic Canada

components of the harvesting sector to respond to these (and other) changes will depend on their adaptability, mobility and access to real-time information on stock sizes and locations. With the substantial uncertainties remaining at this stage and for the foreseeable future, policy responses would most effectively be directed to improving capabilities for prediction and real-time information provision. Some scientists believe that variability in the abundance of some species (especially the pelagic species) may increase with changes, or greater variability, in climate; this should be considered in prudent management of these species.

Aquaculture

The direct effects of climate warming for aquaculture in Atlantic Canada should be very beneficial: warmer sea temperatures should allow faster growth of fish, and less risks of fish mortality during overwintering in coastal sites, as well as reduced energy costs in shore-based facilities. Warmer waters should also greatly expand the geographic range of suitable sites for aquaculture. Unfortunately, warmer waters may also be more conducive to outbreaks of algae and bacteria which render molluscs unfit for human consumption. If such outbreaks are great enough to make benthic areas anaerobic, mobilization and uptake of toxic heavy metals in sediments could also be greatly increased. Apart from these biological effects, the pressure for expansion of aquaculture into new areas could also increase conflicts between aquaculture and other uses of coastal sites. Therefore, achievement of the potential benefits of climate change for aquaculture could depend on protection or improvement of nearshore environmental quality, and anticipation of possible future uses of sites for aquaculture in coastal zone management.

Tourism and Recreation

Climate warming could also be beneficial for tourism in Atlantic Canada by lengthening the summer season, when most tourists visit. A warmer, drier climate should also be more favourable for summer outdoor recreation. These conditions should be especially favourable relative to those which are expected to prevail in other populated areas of North America, where climate warming may make summers less pleasant. Less favourable indirect impacts may occur because climate change and sea level rise may result in the loss of some natural and cultural resources which support tourism and recreation. Sea level rise may result in losses of coastal wetlands, and the wildlife they support. Public beaches and areas of natural and cultural interest may also be threatened. Lower water levels in lakes and streams would be less favourable for freshwater fish, especially the highly-valued migrating species, such as salmon. Lower water levels may also hamper canoeing, swimming and cottage activities. These unfavourable impacts would be exacerbated if they combine with other factors which are contributing to deterioration in resources, amenities and environmental quality.

Conclusions

Climate warming and sea level rise would be detrimental to coastal infrastructure and freshwater supplies, and would require major adaptations in the capture fisheries industry. Climate warming offers potential benefits to aquaculture and to tourism and recreation. The sectors considered here appear to exemplify a general principle that there are strong, even multiplicative, interactions between climate change impacts and other environmental factors in producing overall cumulative effects. General deterioration in the environment and natural resource base will tend both to exacerbate the costs of climate change (e.g. by imposing further losses of freshwater supplies and wetlands), and limit or preclude the potential benefits of climate change (e.g. by making potential sites for aquaculture or recreation no longer fit to support these activities). Therefore, the prospects of climate change increase the urgency and stringency with which we must address general problems in environmental and resource conservation.

References

Poetschke, T. 1984. Community Dependence on Fishing in the Atlantic Provinces. The Canadian Journal of Regional Science. Vol. 7(2): 211-226.

Stokoe, P. 1987. Adaptability to Climate Change: the Case of the Marine Economy of Atlantic Canada. In: Preparing for Climate Change (Proceedings of the First North American Conference on Preparing for Climate Change: A Cooperative Approach, October 27-29, 1987. Washington, D.C.)

Stokoe, P. 1988. Socio-Economic Assessment of the Physical and Ecological Impacts of Climate Change on the Marine Environment of the Atlantic Region of Canada - Phase 1. Climate Change Digest. CCD 88-07. Ottawa: Environment Canada.

Stokoe, P., M. LeBlanc, C. Lamson, M. Manzer, P. Manuel. 1988. Socio-Economic Assessment of the Physical and Ecological Impacts of Climate Change on the Marine Environment of the Atlantic Region of Canada, Phase 2: Implications for Small Coastal Communities. Draft Report submitted to the Canadian Climate Program.

Acknowledgements

The Canadian Department of Fisheries and Oceans, especially Mr. Tim Hsu and his colleagues, supported this work by the provision of information. Funding of this research is gratefully acknowledged from the Canadian Climate Program, administered by the Atmospheric Environment Service of Environment Canada.

PREPARING FOR CLIMATE CHANGE IN THE GREAT LAKES: INTRODUCTION TO PANEL DISCUSSION

Stewart J. Cohen
Canadian Climate Centre
Atmospheric Environment Service
Downsview, Ontario CANADA M3H 5T4

Presented at the Second North American Conference on Preparing for Climate Change: A Cooperative Approach, December 6-8, 1988, Washington, D.C.

The Great Lakes region is well known for its tremendous supply of fresh water, a resource that is exploited in many ways by governments, industry, and individuals in eight states in the United States and two provinces in Canada. Its status as an international basin is recognized through the efforts of federal, state and provincial agencies in both countries, and by the International Joint Commission (IJC), a bi-national organization created by the Boundary Waters Treaty of 1909. The IJC has an important role in international efforts to manage the water resources of this region.

Unlike many other regions in North America, the Great Lakes has received a lot of attention over the past few years regarding possible impacts of projected global warming. Since 1983, more than 20 studies have been published on a wide range of topics, including hydrology (e.g. lake levels), agriculture, hydroelectric power production, commercial shipping, recreation, fisheries and energy demand. Most of the Canadian studies are reviewed in Cohen and Allsopp (1988). Others include Meisner et al. (1987), Regier et al. (1987), Cohen (1988) and Marchand et al. (1988). Recent U.S. studies include Quinn (1987), Linder and Gibbs (1987), and Raoul and Goodwin (1987).

In the near future, there will be a number of significant additions to this literature, thanks in large measure to three major projects which are at or near completion:

a) a research project organized by the U.S. Environmental Protection Agency (EPA) entitled "The Potential Effect of Global Climate Change on the United States," which includes 15 studies of the Great Lakes;

b) the IJC Great Lakes Water Levels Study, which includes five study groups: 1) Hydraulics, Hydrology and Climate, 2) Coastal Zone Ecology, Resources, Uses and Management, 3) Socio-Economic and Environmental Assessment, 4) Public Participation and Communications, and 5) Cross System Impact Evaluation; and

c) the First U.S.-Canada Symposium on Impacts of Climate Change on the Great Lakes Basin, jointly organized by the National Oceanic and Atmospheric Administration (U.S.), EPA, and the Atmospheric Environment Service (Canada), which took place on September 27-29, 1988, in Chicago.

The panel that has been assembled for today's discussion consists of individuals who have participated in the organization and/or research activities associated with each of the above initiatives. Their presentations will provide us with an update on these activities, as well as an assessment of future needs.

As a result of these broad multi-disciplinary efforts, we are at a somewhat more advanced stage in understanding what the potential problems are. That does not mean we know what the solutions are. However, we can take advantage of what we have learned so that we can begin to assess the feasibility of possible response options. Hopefully, this will lay the groundwork for future management of this important international resource, so that all of us will be able to enjoy its benefits, regardless of what the future climate may be.

REFERENCES

Cohen, S.J. 1988. Great Lakes levels and climatic change: impacts, responses and futures. In Glantz, M.H. (ed.), Societal Responses to Regional Climatic Change: Forecasting by Analogy. Westview Press, Boulder, 143-167.

Cohen, S.J. and T.R. Allsopp. 1988. The potential impacts of a scenario of CO_2-induced climatic change on Ontario, Canada. Journal of Climate, 1, 7, 669-681.

Linder, K.P. and M.J. Gibbs. 1987. The potential impacts of climate change on electric utilities: project summary. In: Preparing for Climate Change, Proceedings of the First North American Conference on Preparing for Climate Change: A Co-Operative Approach, October 27-29, 1987, Washington. Government Institutes Inc., Rockville, 284-293.

Marchand, D., M. Sanderson, D. Howe and C. Alpaugh. 1988. Climatic change and Great Lakes levels: The impact on shipping. Climatic Change, 12, 107-133.

Meisner, J.D., J.L. Goodier, H.A. Regier, B.J. Shuter and W.J. Christie. 1987. An assessment of the effects of climate warming on Great Lakes Basin fishes. Journal of Great Lakes Research, 13, 340-352.

Quinn, F.H. 1987. Likely effects of climate changes on water levels in the Great Lakes. In: Preparing for Climate Change, Proceedings of the First North American Conference on Preparing for Climate Change: A Co-Operative Approach, October 27-29, 1987, Washington. Government Institutes Inc., Rockville, 481-487.

Raoul, J. and Z.M. Goodwin. 1987. Climatic changes--impacts on Great Lakes levels and navigation. In: Preparing for Climate Change, Proceedings of the First North American Conference on Preparing for Climate Change: A Co-Operative Approach, October 27-29, 1987, Washington. Government Institutes Inc., Rockville, 488-501.

Regier, H.A., J.A. Holmes and J.D. Meisner. 1987. Likely effects of climate change on fisheries and wetlands, with emphasis on the Great Lakes. In: Preparing for Climate Change, Proceedings of the First North American Conference on Preparing for Climate Change: A Co-Operative Approach, October 27-29, 1987, Washington. Government Institutes Inc., Rockville, 313-327.

Summary of the US-Canada Great Lakes Climate Change Symposium

W.C. Bolhofer
National Climate Program Office

The First United States-Canada Symposium on the Impacts of Climate Change on the Great Lakes Basin was held in Chicago, September 27, 28, and 29, 1988. More than 120 people attended. The Symposium brought together scientists, government officials, business and industry leaders, urban planners, agriculture and forestry experts, transportation and public utility managers to discuss ways to mitigate, or adapt to man-induced and natural climate changes expected to affect the Great Lakes by the first half of the next century. The Symposium was jointly organized by the National Climate Program Office (NOAA), the Canadian Climate Program Office (AES), and the Office of Policy Analysis (EPA). The Illinois State Water Survey in conjunction with the midwest regional climate center was the local host.

The Great Lakes represent a shared strategic fresh water resource between Canada and the United States that supports a highly developed and integrated infrastructure of urban environments. Urbanization and industrial activities in the area require energy which is primarily achieved through relatively inexpensive hydro-electric generation. Commercial activities require the movement of raw materials and finished products through regulated lake transportation systems. Finally, the lakes support various ecologies ranging from wetlands to deep fresh water species.

The projected increase in greenhouse gases such as CO_2 over the next five or more decades suggests, through regional interpretations of global climate modeling studies, that temperatures over the basin will be higher than current climatology. However, net basin runoff into the lakes will be less due to increased evaporative processes. Studies show the Great Lake levels dropping anywhere from 1/2 meter to two or more meters. The regions highly urbanized infrastructure is very dependent on relatively stable lake levels thus making the area very vulnerable to future climate induced lake level change.

The scope of the Symposium was broad allowing for a wide variety of recommendations. The Symposium attempted to:

(1) review and critique the state of knowledge of climate change and its implications for the Great Lakes Basin; (2) inform decision makers of potential impacts and management issues likely to develop over the next few decades because of climate change; (3) identify needs and sensitivities for future research assessment and policy review; (4) serve as a case study of regional climate impacts assessment; and, (5) begin U.S.-Canada coordination of an integrated climate impact study on the Great Lakes region. Some of these items were accomplished during the Symposium. A Symposium report is currently in press which more thoroughly reviews (than this presentation) the discussions, conclusions, and recommendations of the various panels.

The plenary sessions, through keynote speakers, identified and clarified for the diverse audience climate change and climatic impact topics. Four concurrent panels were organized: i) energy and transportation; ii) municipal water supply, lake pollution and lake levels; iii) recreation, conservation, and wetlands; and, iv) agriculture, urban land use, and forest management. The panels, each consisting of a chairperson, rapporteur, designated panelists, and other members of the Symposium, were charged with identifying key issues that would most likely emerge due to climate change over the basin. The panels provided recommendations to assist planners develop policy guidelines that would help ameliorate the effects of climate impact on the Great Lakes region.

The panels' recommendations fell into four broad issue areas. (1) There exists at this time considerable uncertainty about potential physical, socio-economic impact, and response to climatic change. This is because it is very difficult to predict the evolution of the urban environment 75 years into the future. (2) The panels agreed that any recommendation put forth is intrinsically tied to current global climate model studies. Such models have coarse grids making regional definition difficult. Also, the models are not coupled to ocean circulations which appear to play an important role in climate forcing. (3) There already exists a number of provincial, state, and federal coordinating and regulatory bodies for the Great Lakes System. Studies suggest that these bodies in their present configuration would probably be inadequate to address the effects of basin-wide climatic changes. (4) As mentioned earlier, the Great Lakes basin is a shared resource. As such, there exists the potential for disagreement between Canada and the United States primarily because the lakes are strategic and finite.

The final plenary session synthesized panel recommendations as a guide for the development of future planning meeting agendas as well as for immediate and ongoing studies. A major point was that global climate change models need to be modified to yield more specific outlooks of future climatic change in the Great Lakes basin region. This may require the development of regional climate models. Centralized climatic, ecological, and economic data bases are also required for integrated studies on physical and socio-economic impacts of climate with or without change. Such studies would include methodologies for coping or adapting to future climate change. The Symposium agreed that there is a potential for conflicts of interests between Canada and the United States over Great Lakes water resources. There may be need for modifications of the lakes' international water and environmental management regulations to accommodate possible future lake level declines. Finally, all agreed about the importance of keeping the regional community informed of progress on understanding climate change and subsequent climate impacts on the Great Lakes Basin.

The broad and challenging extent of the above recommendations for studies, assessments, research, and changes in various public and private activities led attendees to the concept of a plan that was seen to fit the needs of the Great Lakes Basin communities and the evolving international concerns over climatic change. The attendees strongly recommended two actions:

1. Develop a US-Canada integrated study of the Great Lakes Basin as a regional pilot project on how to respond to global climatic change.

2. Establish a joint planning group to plan and develop the pilot project. The recommended activity should be integrated with and built upon two major on-going basin efforts - the Regional Area Pollution Studies (RAPS) Program for water quality concerns, and the ongoing International Joint Commission (IJC) Lake Levels Reference Study. Both programs contain activities and elements that should be considered in the planning and development of the recommended global change pilot project.

Presently, both Canada and the United States are actively engaged in determining how best to implement the recommendations of the Symposium. A planning meeting at government level is tentatively scheduled for late Winter or early Spring 1989. Much of the initial planning activities for future working meetings on this important regional topic will be jointly undertaken by both the National Climate Program Office and the Canadian Climate Program Office as part of the Memorandum of Understanding between the United States and Canada for climate related projects.

An Overview of the EPA Studies of the Potential Impacts of Climate Change on the Great Lakes Region

by

Joel B. Smith
U.S. Environmental Protection Agency
Office of Policy Analysis

Residents of the Great Lakes region are well aware of how climate affects their natural resources. Twenty-five years ago, dry weather caused the levels of the Great Lakes to drop to record lows. Docks and harbors became unusable; shipping was impeded; and hydroelectric power production was cut back. In 1986, however, a series of cool and wet years raised levels to record highs. Apartment buildings on the Chicago shoreline were flooded and many shoreline structures were damaged. Landward resources have also proved vulnerable to climate fluctuations. For example, the 1988 drought devastate corn yields in such states as Ohio and Illinois. These were short term impacts. The greenhouse effect, which may raise average temperatures, could produce long-term changes in many of the region's resources.

Following a series of hearings on the greenhouse effect in 1986, Congress asked EPA to examine the health and environmental effects of climate change. EPA met the second request by examining the potential effects on a number of areas, including coastal resources, agriculture, forests, health, and water resources. We focused on four regions: the Southeast, the Great Plains, California, and the Great Lakes. The actual analyses were performed by more than fifty-five scientists using existing models and approaches. EPA specified the scenarios of climate change to be used.

This paper discusses the results of EPA studies on the potential effects of climate change on the Great Lakes region. These studies examined changes in lake levels, ice cover, shorelines, shipping, lake thermal structure, fisheries, agriculture, forests, and other systems.

Climate Change Scenarios

The researchers estimated the impacts of climate change based on regional climate change scenarios associated with a doubling of carbon dioxide levels over preindustrial levels. The climate change caused by a doubling of carbon dioxide would probably not occur until the latter half of the next century. These scenarios were based on three general circulation models (GCMs): 1) Goddard Institute for Space Studies (GISS); 2) Geophysical Fluid Dynamics Laboratory (GFDL); and 3) Oregon State University (OSU). The scenarios assume that average climate conditions change, but climate variability remains the same as in 1951-80. The average temperature and precipitation changes from the three models are displayed in Figures 1 and 2. All three models show temperature rising, although they disagree on the magnitude. The models all show annual precipitation rising, but disagree on the direction of seasonal changes. In addition to the doubled CO_2 scenarios, EPA developed transient scenarios of how climate may change between now and the middle of the next century. These were based on transient runs from the GISS model.

Lake Level Changes

Croley and Hartmann (1989) used the Great Lakes Environmental Research Lab's water supply and lake level model of the Great Lakes to estimate lake level changes. The analysis did not account for changes in withdrawals due to climate change. Results are displayed in Table 1.

Table 1

Reduction in Great Lakes Levels from 1951-80 Averages (meters)

Scenario	Superior	Michigan/ Huron	Erie	Ontario
		$2XCO_2$ Scenarios		
GISS	-0.4 to -0.5	-1.3 to -1.3	-1.0 to -1.2	N/A
GFDL	N/A	-2.5	-1.7 to -1.9	N/A
OSU	-0.4 to -0.5	-0.9 to 1.0	-0.6 to -0.8	N/A

Source: Croley and Hartmann

Under the doubled CO_2 scenarios, average lake levels were estimated to fall in all lakes and below historic lows in Lakes Michigan and Erie. The plan for regulating levels in Lake Ontario was estimated to fail under all three scenarios. The Lake Superior regulation plan would fail under the relatively hot and dry GFDL scenario. The results appear to be driven by higher temperatures, which would reduce the snowpack and increase evaporation. The results are very sensitive to estimates of evaporation, which in turn is sensitive to changes in wind speed and humidity (Cohen, 1987). Croley and Hartmann also found that enough heat would reside in the lakes to maintain water surface temperatures at a sufficiently high level through the year, so that the lakes would not thoroughly mix every year.

Shoreline Effects

Changnon (1989) studied the impacts of Croley and Hartmann's lake level estimates on the Illinois shoreline. He found that among the adjustments to lower lake levels would be dredging harbors, lowering docks, and extending water supply sources and stormwater outfalls. These adjustments would cost $270-540 million. The figures would be reduce by $50-110 million if normal replacement costs incorporate lower lake levels.

Ice Cover

Assel (1989) estimated changes in ice cover for Lakes Superior and Erie. The results for Central Lake Erie are displayed in Figure 3. Assel found that ice cover could be slightly reduced in 1981 to 2009 and could be significantly reduced in later years. While Central Lake Erie currently has 83 days of ice cover per year, it would have 6 to 19 days under doubled CO_2 conditions. Assel also found that ice cover in Lake Superior, which currently lasts about four months, could be reduced to 1 to 2 1/2 months a year.

Shipping

Lower lake levels would cause commercial ships on the lakes to lower their cargo loads. On the other hand, reduced ice cover would allow for a longer shipping season. Keith et al. (1989) examined how the estimated lake level and ice cover changes would affect commercial ships using several ports in Lakes Erie and Superior, assuming that the fleet and cargo do not change. If there is no dredging, which is unlikely, ships in Lake Erie ports would reduce their cargoes by 5 to 27%. With a longer shipping season, there could be more voyages, which could make up for the reduction in capacity. Keith et al. found that, for example, in the port of Buffalo, the longer season would allow for the transport of more cargo under the smaller lake level drops in the OSU and GISS scenarios. However, it would not be long enough to make up for the cargo reductions associated with the GFDL scenario and annual tonnage would be reduced. Dredging costs for Buffalo alone were estimated to be $11-31 million.

Water Quality

Two studies examined changes in thermal mixing of Lakes Michigan and Erie. McCormick (1989) found that southern Lake Michigan could be stratified two months longer than it is now. Blumberg and DiToro (1989) found that the Central Basin of Lake Erie could be stratified two to four months longer than currently. Results from both studies, however are very sensitive to changes in winds and storm frequency.

Blumberg and DiToro also found that higher temperatures would increase the growth of algae and other aquatic species. The enhanced biological activity would combine with the longer stratification to lower dissolved oxygen levels, especially in the lower depths of the Central Basin. The area of the lower level of the Central Basin that is estimated to be anoxic (having no dissolved oxygen and no life) is shown in Figure 4. The specific results are based on the high pollutant loadings from the early 1970s. Even with lower pollutant loadings, we would expect dissolved oxygen levels to be reduced under global warming.

Fisheries

Using the results of the McCormick and the Blumberg and DiToro studies, Magnuson et al. examined potential response of Great Lakes fishes to higher temperatures. They found that fish generally would benefit from higher temperatures. The production of phytoplankton and zooplankton would increase several fold and the thermal habitat for fishes during non-summer months would expand. For example, lake trout habitat could more than double. Assuming that sufficient prey are available, higher temperatures would also increase body growth. If prey do not increase, body growth could decrease.

Magnuson et al. identified some potential sources of stress for Great Lakes fishes. Hotter temperatures could reduce summer habitats, especially in shallow areas such as in parts of Lake Erie or in streams. There also could be intense pressure on the forage base in the summer. Furthermore, warmer temperatures could make the lakes more suitable for introduction of new species, which could lead to the loss of important stocks.

Forests

Botkin et al. estimated that a warmer climate would bring about major changes in forest composition throughout the region. In southern Michigan, dry sites, which now have oak and sugar maple could be converted to oak savannas or even prairies. Wet sites could suffer a reduction in density. Minnesota, which is currently a mixture of northern hardwood and southern boreal species, would become all northern hardwood (see Figure 5). Such a shift in species could have major implications for the forest industry. Perhaps most interesting, Botkin et al. found that under the transient scenarios, these changes in forest composition could become evident in three to six decades.

The rate of climate change may be faster than the ability of forests to migrate. Zabinski and Davis (1989) studied climate change and migration for four species currently found in the Great Lakes region: sugar maple, yellow birch, hemlock, and beech. They found that these species would migrate out of the southern Great Lakes region under the doubled CO_2 scenarios. They also found that the southern boundaries of the forest would shift northward by around 500 kilometers (due to climate change), while the forests would only migrate northward by about 100 kilometers. Thus, the inhabited range of these species could shrink considerably.

Agriculture

Ritchie et al. used crop yield models to estimate changes in yields of corn and soybeans in the regions. Results from selected sites, assuming no CO_2 fertilization effects, are shown in Table 2.

Table 2
Effects of Climate Change on Corn and Soybean Yields in Great Lakes States
(% change from base)

Site	CORN Dryland GISS-GFDL	CORN Irrigated GISS-GFDL	SOYBEANS Dryland GISS-GFDL	SOYBEANS Irrigated GISS-GFDL
Duluth, MN	+49 - -30	+86 - +36	+109 - -6	+153 - +96
Green Bay, WI	-7 - -60	-3 - -44	-3 - -65	+3 - -26
Flint, MI	-17 - -48	-14 - -38	-6 - -51	+6 - -11
Buffalo, NY	-26 - -47	-18 - -38	-21 - -53	+6 - -6
Fort Wayne, IN	-11 - -51	-15 - -48	-2 - -58	0 - -19
Cleveland, OH	-26 - -50	-19 - -43	-16 - -59	-1 - -14
Pittsburgh, PA	-22 - -55	-19 - -45	-13 - -59	0 - -13

Source: Ritchie et al.

Yields increase in northern areas such as Duluth because higher temperatures significantly increase the growing season. In southern areas, however, yields would be lower because higher temperatures reduce soil moisture and increase heat stress. Ritchie et al. also examined the combined effects of climate change and higher CO_2 levels. (The CO_2 effects were probably overestimated. See Rosenzweig in these proceedings.) They found that the combined effects could increase yields if there is sufficient rainfall. If rainfall does not increase significantly, as in the GFDL scenario, results are mixed, with yields falling in some areas.

The response of farmers to climate change may depend in large part on how the relative productivity of their crops changes compared to crops elsewhere. Adams et al. (1989) used estimates of yield changes of grain crops across the United States to estimate regional shifts in production. Generally, yields were relatively higher in northern areas than in southern areas. Adams et al. estimated that across the country, there would be a northward (and westward) shift in agricultural acreage. In the Great Lakes region, there would be little change in acreage in the southern states, but a slight increase in acreage in the northern states. This increase would be limited by glaciated soils in the area, which would constrict expansion of agriculture.

POLICY IMPLICATIONS

Climate change raises many issues that regional natural resource managers will have to confront in the future. International, federal, regional, and state institutions, such as the International Joint Commission, the U.S. Army Corps of Engineers, the EPA, the U.S. Department of Interior, the U.S. Department of Agriculture, and the Great Lakes Fisheries Commission will need to address whether to adapt to climate change as it occurs or to minimize impacts by anticipating change. Among the issues they will need to address are:

- Lake Regulations. Lower lake levels would necessitate changing lake regulation plans to balance the needs of users of the lakes and users of the St. Lawrence River and Seaway.

- Diversions. Pressure for diversion of water out of the basin will probably increase with climate change.

- Water Quality. Lower lake levels, reduced dissolved oxygen levels, and the potential for increased dredging activity may make it more difficult to meet water quality standards.

- Fisheries. Although there may be positive impacts on fisheries, many problems could occur due to stress situations and the potential for introduction of new species. More intensive fisheries management may be needed.

- Shoreline Management. Lower lake levels presents opportunities and challenges. Will exposed areas be developed or will they be kept undeveloped to serve as recreation areas and buffers against fluctuating lake levels?

- Management of Forests and Agriculture Shifts. Forests may die back in southern areas. Will more southern species be replanted or the lands be put to other uses? Northern areas may see shifts in forest composition and some expansion of agriculture. How will these potential conflicting uses of land be managed. How can the pristine environment in northern areas be maintained?

These are among the challenges to be faced in coming decades by regional resource managers.

Bibliography

The following studies were prepared in support of the EPA Report to Congress on the Potential Impacts of Global Warming on the United States. U.S. Environmental Protection Agency, Office of Policy, Planning and Evaluation and Office of Research and Development. Washington, D.C. (Draft). All are in draft and will be published by EPA.

Adams, R.M., J.D. Glyer, and B.A. McCarl, 1989. The economic effects of climate change on U.S. agriculture: a preliminary assessment. Oregon State University.

Assel, R.A., 1989. Impact of global warming on Great Lakes ice cycles. Great Lakes Environment Research Lab.

Blumberg, A.F. and D.M. DiToro, 1989. The effects of climate warming on Lake Erie water quality. HydroQual, Inc.

Botkin, D.B., R. A. Nisbet, and T.E. Reynales, 1989. Effects of climate change on forests of the Great Lakes states. University of California, Santa Barbara.

Changnon, S.A., S. Leffler, and R. Shealy, 1989. Impacts of extremes in Lake Michigan levels along Illinois shorelines: low levels. University of Illinois.

Croley, T. F. and H.C. Hartmann, 1989. Effects of climate changes on the Laurentian Great Lakes levels. Great Lakes Environment Research Lab.

Keith, V.F., C. DeAvila, and R.M. Willis, 1989. Effect of climatic change on shipping within Lake Superior and Lake Erie. Engineering Computer Optecnomics, Inc.

Magnuson, J.J., H.A. Regier, B.J. Shuter, D.K. Hill, J.A. Holmes, and J.D. Meisner, 1989. Potential responses of Great Lakes fishes and their habitat to global climate warming. University of Wisconsin and Toronto.

McCormick, M.J., 1989. Potential climatic changes to the Lake Michigan thermal structure. Great Lakes Environment Research Lab.

Ray, D.K., K.N. Lindland, and W.J. Brah, 1989. Effects of global warming on the Great Lakes: the implications for policies and institutions. The Center for the Great Lakes.

Ritchie, J.T., B.C. Baer, and T.Y. Chou, 1989. Effect of global climate change on agriculture: Great Lakes region. Michigan State University.

Zabinski, K. and M. Davis, 1989. Hard times ahead for Great Lakes forests: a climate threshold model predicts responses to CO_2-induced climate change. University of Minnesota.

Other literature:

Cohen, S.J. 1987. Sensitivity of water resources in the Great Lakes region to changes in temperature, precipitation, humidity, and wind speed. In: The Influence of Climate Change and Climatic Variability on the Hydrologic Regime and Water REsources (Proceedings of the Vancouver Symposium, August 1987) IAHS Publ. no. 168, 1987.

Figure 1

TEMPERATURE SCENARIOS FOR THE GREAT LAKES REGION
2xCO2 LESS 1xCO2

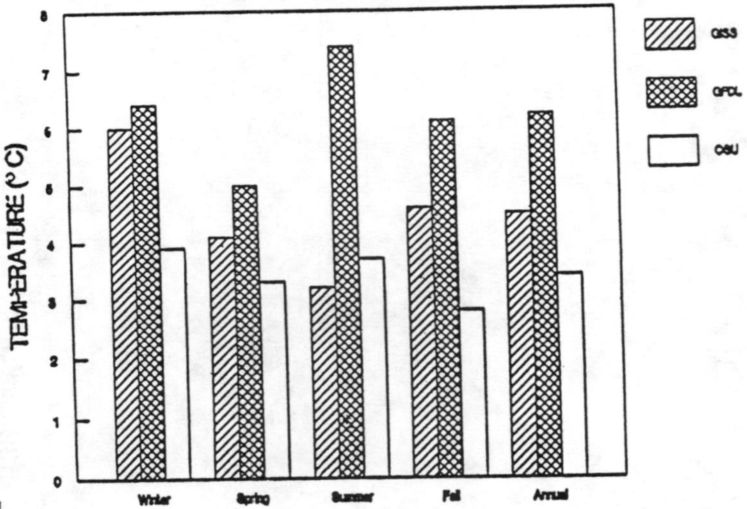

Figure 2

PRECIPITATION SCENARIOS FOR THE GREAT LAKES REGION
2xCO2 LESS 1xCO2

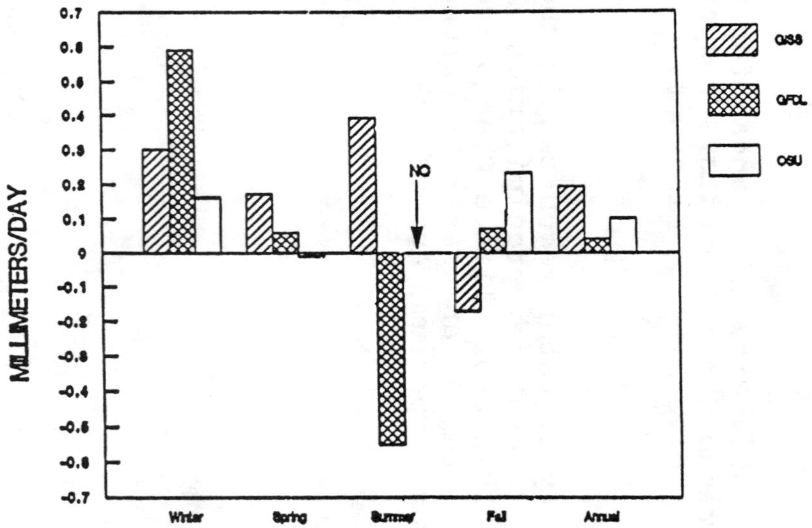

Figure 3

ANNUAL NUMBER OF DAYS OF ICE COVER IN CENTRAL LAKE ERIE

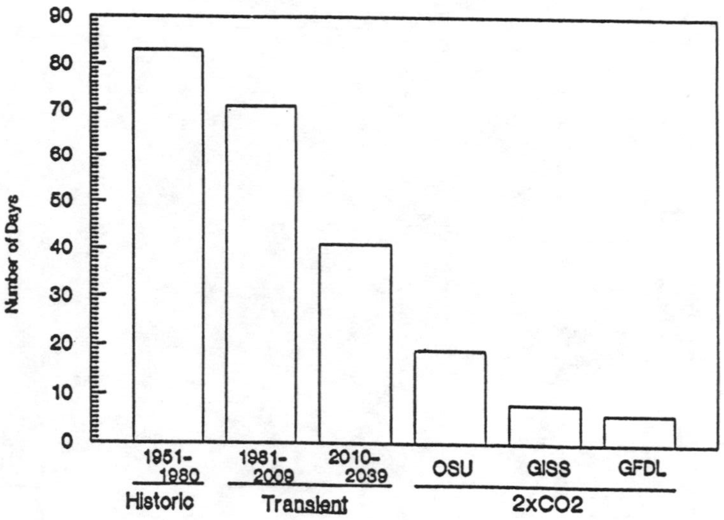

Figure 4

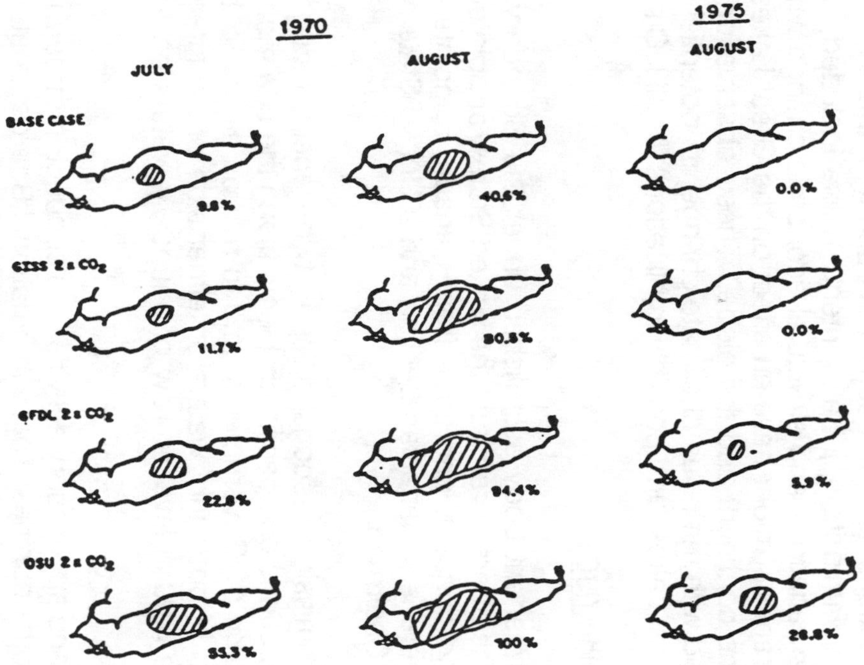

Area of central basin of Lake Erie that becomes anoxic.

Source: Blumberg and DiToro

Figure 5

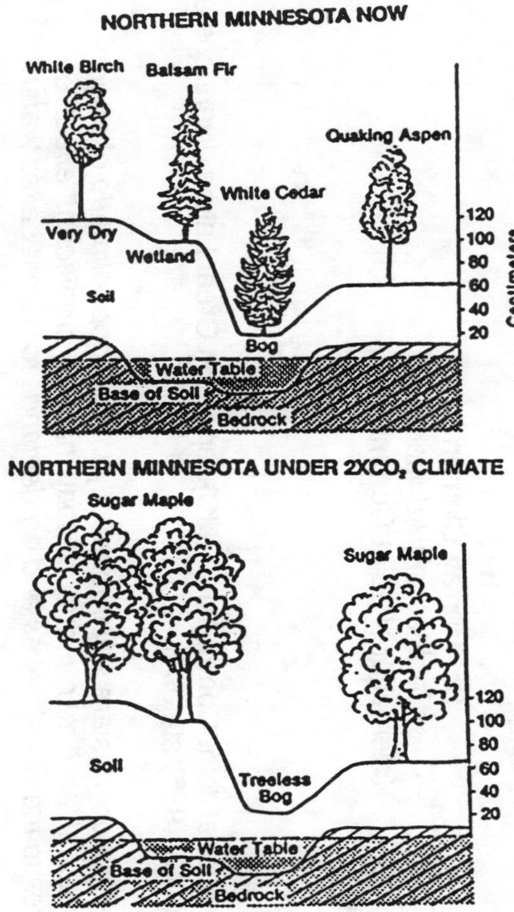

Effects on northern Minnesota forests (Virginia, Minnesota).

Source: Botkin et al.

CLIMATE CHANGE AND THE IJC

by
Dr. Murray Clamen
Engineering Adviser
International Joint Commission

Introduction

A 1985 International Joint Commission (IJC) report on Great Lakes diversions and consumptive uses (IJC, 1985) states that:

> *"While its workings are still imperfectly understood, climatic change, both natural and man-induced, may have a significant effect on supplies to the Great Lakes and on demands for Great Lakes water."*
>
> *"The economic, social and environmental implications of climate change could be severe and largely unavoidable, so that anticipation and preparedness for adaptation are required."*

The report was prepared in response to a 1977 reference from the Governments of Canada and the United States. Part Two of that report went beyond providing immediate, practical recommendations to the questions posed in the reference, to consider a wider range of issues within its spirit and intent. These included, amongst other issues, longer-term climatic variations and structural economic change which, the Commission noted in the same report, "might have an effect on the Great Lakes region as well as on other regions of the continent and the globe." These observations signal that the Commission is concerned with the issue of climate change, its potential impacts on such a major region as the Great Lakes, and possible implications for all IJC activities.

The International Joint Commission (IJC)

The Commission is a permanent body established under the Boundary Waters Treaty of 1909. It consists of six members, three from the United States, appointed by the President with Senate advice and consent, and three from Canada, appointed by the Governor General in Council. There is a Canadian Section and a United States Section of the Commission, each with a Chairman.

The Treaty gives the Commission responsibilities in two general categories: to investigate and make recommendations on specific problems along the common frontier which are referred from time to time by Governments; and to rule on all applications for the use, obstruction, or diversion of boundary waters on either side of the international boundary which would affect the natural level or flow of boundary waters on the other side.

Investigative, surveillance and supervisory activities are conducted through Boards made up of qualified experts in both countries. These international Boards created by the Commission operate as a single unit under joint chairmen, one from each country, and provide technical advice and assistance on a case-by-case basis.

Regulation of Lakes Superior and Ontario

The outflows of Lakes Superior and Ontario have been modified by man, within limits, by altering the sequence and magnitude of releases. Since long-term weather forecasting techniques are not sufficiently advanced, the Commission and its Boards rely on historical hydrological data to devise regulation rules and indices to estimate probable water supplies. The regulation of Lakes Superior and Ontario requires the application of prescribed rules to manage the variable water supplies so as to meet the conditions set forth in the Commission's Orders of Approval.

The purpose of these rules is to provide levels and flows that result in generally beneficial conditions without unacceptable adverse effects on any one interest. The difficulties of devising regulation rules that provide beneficial conditions for this purpose are compounded by unprecedented water supplies and severe storms. Undoubtedly there will be further complications if, as is predicted by climate researchers, climate change takes us outside the boundaries experienced in the last centuries.

Lake Superior Regulation- Lake Superior outflows have been regulated within limits under Commission Orders since 1921. The Orders establish the basic objectives for and limits to regulation and are implemented by the IJC-appointed International Lake Superior Board of Control (IJC, 1979).

Based upon regulation experience since the 1920's and various studies, the Commission amended its Order in 1979 in an effort to provide systemic regulation, taking into account conditions both upstream and downstream of the St. Marys River control works.(The original objective focused exclusively on Lake Superior and attempted to maintain its levels within a narrow range.) Under normal conditions, the Board calculates outflows using the procedure set out in the Commission-approved regulation plan which is designed to meet the objectives set forth in the Orders of Approval. This procedure, known as "Plan 1977", works very well as long as water supplies to the lakes are within normal historic ranges.

The Order provides that in the event of extreme supplies the Commission will indicate the appropriate outflows from Lake Superior. This provision was used for the first time during the 1985/86 high water period. However, should climate change produce frequent deviations from normal conditions the regulation plan likely will "fail" more often. The Commission will have to determine whether it wishes to continue to apply this emergency provision essentially in an ad-hoc way (as it did in 1985/86) or adopt a better regulation framework. More than likely the regulation plan would be modified - the key question being how and when. A Commission study in progress pursuant to a 1986 reference from Governments is expected to provide timely advice on this matter.

Lake Ontario Regulation- The Commission approved the construction of the St. Lawrence Power Project by way of a 1952 Order of Approval, amended in 1956 (IJC, 1956). The Commission established the International St. Lawrence River Board of Control to ensure compliance with the provisions of the Order.

Since regulation began in 1960 there have been three plans used to regulate the lake. All the plans were designed to satisfy the criteria and other requirements in the

Orders, and tested using the historic water supply period of 1860-1954 which contains periods of extreme low and high water supplies. In the mid-sixties and early seventies the lakes received, respectively, water less than and greater than that received during the plan design period. During these periods the regulation plan in effect (Plan 1958-D) failed to satisfy the stipulated criteria and objectives. On each occasion the Board, with IJC approval, took action under one of the criteria in the Order (Criterion k) to deviate from the plan. Under the climate change scenarios now being studied, projections are that Plan 1958-D will "fail" more frequently. As might be the case for Lake Superior,if Plan 1958-D is increasingly unable to cope with the changed supplies resulting from climate change, the Commission will have to determine when and how it should change the regulation regime.

The Great Lakes Fluctuating Water Levels Reference

Beginning in 1985 and continuing through 1986, water levels on all the Great Lakes, except Lake Ontario, established record highs in response to continued above average precipitation over the Great Lakes basin. Considerable shoreline flooding, erosion and associated coastal zone problems increased pressure on governments for action. Amongst other responses, the Governments of the United States and Canada on August 1, 1986 agreed to request the IJC "to examine and report upon methods of alleviating the adverse consequences of fluctuating water levels", in terms of both the high level crisis occurring at that time and the long-term consequences of levels, both high and low (IJC, 1986).

To date the Commission has responded to the reference in a number of ways (Clamen, 1988). The Commission considers that this reference calls upon it to undertake a broader Great Lakes water level study than it has ever been asked to carry out, encompassing as it does major issues not considered in previous IJC studies on the Great Lakes. An important new and critical issue is the uncertainty of Great Lakes climate in the future.

Within the reference study team, one task of the Hydraulics, Hydrology and Climate Group is a hydroclimatic study of the Great Lakes basin to assess the impacts of a doubling of atmospheric CO_2 on Great Lakes water supplies. To evaluate the future hydrologic impacts of the various climate change scenarios (three have been assumed to date), supplies were routed and the resulting levels and flows compared to the levels and flows developed from a set of base condition supplies. Preliminary modelling indicates that we are looking at a possible reduction in lake levels in the <u>range</u> of up to three feet under the two times CO_2 scenario. This varies by lake and depends on whether we are referring to averages, maximums, minimums or ranges of levels. For example, the current estimates suggest a decrease in Lake Erie mean levels of between two to six feet, and an expansion in the range of Erie levels by one to two feet. Previous studies (for example, Sanderson & Wong, 1987) show similar results.

Some of the scenarios produced such low supplies to the lakes that Plan 1977 (Lake Superior) and Plan 1958-D (Lake Ontario) failed. To be able to continue hydrologic evaluations, natural outlet conditions for Lakes Superior and Ontario had to be assumed. In reality however, one assumes that there will be changes made to these regulation plans before they "fail". For one thing, decreased lake levels will affect such interests as commercial shipping, recreational boating, hydropower production, municipal water supply agencies and others. Should climate change itself bring about an increased frequency of low levels, or indirectly through an increase in consumptive use demands for water in the basin, such interests will undoubtedly complain to governments (and the IJC) for relief. One objective of the review of these plans is to define how and when they might be modified as a result of changing water supply conditions in the basin.

Modifications to the regulation plans may be satisfactory as far as the regulated lakes are concerned, but what about the unregulated lakes Michigan, Huron, St. Clair and Erie? From a structural viewpoint, outlet control works or underwater dykes or weirs in the Connecting Channels could be constructed to reduce outflows and raise levels upstream. However,environmental and other impacts in the downstream lakes and rivers certainly would have to be considered. Issues such as these continue to challenge those involved in this latest study.

The Great Lakes Water Quality Agreement

In addition to Great Lakes water level investigations, the IJC has studied Great Lakes water pollution several times over the years at the request of the two governments. Using recommendations from IJC and other studies,Canada and the United States developed and signed the first Great Lakes Water Quality Agreement in 1972. It was renewed in 1978 and amended by a Protocol signed in November 1987. The Agreement expresses the commitment of each country to restore and maintain the chemical, physical and biological integrity of the Great Lakes Basin Ecosystem, and includes a number of objectives and guidelines to achieve these goals. It reaffirms the rights and obligations of Canada and the United States under the Boundary Waters Treaty.

The Agreement has become a major focus of IJC activity. The IJC monitors and assesses progress under the Agreement and advises Governments on matters related to the quality of the waters of the Great Lakes system. The Agreement also calls upon the IJC to assist the Governments with joint programs, and provides for two binational Boards - the Great Lakes Water Quality Board and the Great Lakes Science Advisory Board - to advise the Commission.

The Commission therefore is very interested in predictions by some that water quality problems on the lakes will become worse due to the effects of climate change. Water quality changes that have been suggested include variations in water temperature and thermal stratification, increased occurrence of algae blooms, and degraded water quality due to lower water levels. It has also been suggested that Great Lakes Areas of Concern (areas where Agreement objectives currently are not being met), could specifically be affected by the impacts of climate change. Whatever the impacts may be, the issue of climate change is adding emphasis to the need for a better understanding of the Great Lakes basin ecosystem.

Other Activities

There are presently twenty-three active Boards of the Commission through which the IJC conducts its various investigative, surveillance and supervisory activities.Those whose responsibilities encompass the Great Lakes region currently are the focus of considerable Commission attention. However others across the boundary whose activities could be affected by global climate change, clearly will need to be aware of any significant shifts in water supplies.

For example the IJC has apportionment responsibilities on the Souris River and for the waters of the St. Mary and Milk Rivers. There are IJC Orders of Approval relating to Kootenay and Osoyoos Lakes on the west coast, and the St. Croix River on the east coast, as well as water quality surveillance responsibilities on the Red, Rainy and St. Croix Rivers that could be affected by climate change.

Summary

The impacts of global climate change on the IJC could be considerable. Potentially all regions of both nations along the common boundary might be affected, thereby involving virtually all IJC activities. Although the implications of climate change have not yet been developed and quantified well enough for immediate application by the Commission,studies in progress on the Great Lakes region should help clarify this issue.

Literature Cited

Clamen, M., 1988. Fluctuating Great Lakes Water Levels:Expectations and the IJC. Paper presented at the 24th annual conference of the American Water Resources Association, Milwaukee,Wisconsin.

International Joint Commission, 1956. Orders of Approval as amended for Regulation of Lake Ontario. Ottawa, Ontario and Washington,D.C.

International Joint Commission, 1979. Orders of Approval as amended for Regulation of Lake Superior. Ottawa,Ontario and Washington,D.C.

International Joint Commission, 1985. Great Lakes Diversions and Consumptive Uses. Ottawa,Ontario and Washington,D.C.

International Joint Commission, 1986. Reference request from the Governments of Canada and the United States on the adverse consequences of fluctuating water levels in the Great Lakes - St. Lawrence River Basin. Ottawa, Ontario and Washington, D.C.

Sanderson, M. & Wong, L., 1987. Climate Change and Great Lakes Water Levels.Proceedings of the Vancouver Symposium on the Influence of Climate Change and Climatic Variability on the Hydrologic Regime and Water Resources. Vancouver, B.C.

INCENTIVES FOR CFC SUBSTITUTES: LESSONS FOR OTHER GREENHOUSE GASES

ALAN S. MILLER, ESQ.

As the risks of global climate change have received more attention and definition from the scientific community, policymakers have begun to ask what can be done to limit the buildup of greenhouse gases. The more expensive and difficult this task is expected to be, the greater the resistance that might be expected. These issues will be a central focus of the inter-governmental panel of policy responses chaired by the United States under United Nations auspices.

The recent international agreement to reduce global emissions of chlorofluorocarbons ("CFCs") provides a useful precedent for examination of the impact of government policy on the development and introduction of environmentally acceptable substitutes. CFCs are widely used around the world for a wide variety of economically important activities, including refrigeration and air conditioning, production of soft and rigid foams for furniture and insulating materials, as a solvent for cleaning electronic components, and as an aerosol propellant. Use of CFCs has grown at least as fast as GNP in most parts of the world except for the mid-1970s to early 1980s, when aerosol uses were regulated by the U.S. and some other nations. In the U.S. alone, the value of products made with CFCs exceeds well over $100 billion.

The successful negotiation of the Montreal Protocol in September 1987 limiting CFC production followed shortly after some promising corporate announcements about the availability of substitutes. In the year following the Protocol, the announcement of potential alternatives has occurred almost daily. In Janaury, AT&T announced a competitively-priced solvent substitute for CFC 113 suitable for many electronics applications. In March, DuPont announced a corporate goal of a full phase-out of CFCs by the end of the century. In April, Pennwalt Corporation announced it would produce a chemical substitute in commercial quantity by late 1990, the food packaging industry announced they would eliminate use of CFCs by the end of 1988. A May 1988 article in the New Scientist noted promsing claims by some English companies, making clear that the substantial international competition to develop substitutes. Summarizing the situation, an industry official commented that "I think you're going to see a lot of diversity in terms of R&D and new technology. It's a real race by companies to find the most cost-effective way out of the situation."[1] In September 1988, DuPont announced it will build a new plant to produce a CFC substitute as part of a larger plan to phase out its existing production at a cost of more than $1 billion over the next decade.

Industry observers remain deservedly skeptical about some of these early announcements which may reflect marketing claims that ultimately do not pan out. Nevertheless, this "can-do," competitive atmosphere is exactly what may be crucial to bring about the technologies necessary to reduce emissions of CO2 and other greenhouse gases. However, analysts of the Protocol differ about a critical question--did the development of substitutes precede and make possible the Protocol, or was the introduction of substitutes announced in anticipation of government regulation? If the former, the precedential value of the Protocol

[1] R. Montasterksy, "Decline of the CFC Empire," Science News, vol. 133, April 9, 1988, pp. 234-236.

is obviously limited since similar progress toward development of substitutes would be a precondition to control of other gases. For example, Steve Rayner of Oak Ridge National Laboratory argues that the Montreal Protocol would not have been possible without three factors: discovery of the Antarctic ozone hole, "the belief of major manufacturers like DuPont that they were on the verge of producing a substitute," and the US ability to threaten other countries with trade sanctions.[2]

There is no doubt considerable truth to the view that announcements about potential substitutes facilitated agreement on the Protocol. However, to some extent this begs the question; why did companies develop and go public with their hope for substitutes that compete with their existing products? DuPont, for example, already controls a substantial percentage of the CFC market in the U.S. and worldwide and could hardly hope to improve its market position. The decision becomes rational, however, if regulation was considered likely. In that case, an early announcement of potential alternatives would (and did) give the Company a jump on its competitors, and as a non-trivial bonus, substantial public relations benefits.

A more important point emerges from an examination of the prior history of corporate announcements concerning the availa-bility of CFC substitutes. Many of the substitutes touted as so promising today were identified years ago. DuPont, for example, summarized results of six years of research costing in excess of $15 million in a June 1980 newsletter--not coincidentally shortly after EPA publicly stated its intent to propose further regulations as a fllowup to the earlier aerosol ban. The Company concluded that "Seven to ten years may be necessary to reach commercial production for most alternatives, assuming that all technical and toxico-logical programs yield favorable results."[3] This estimate was remarkably similar to DuPont's September 1986 announcement that commercial production of substitutes was possible in about the same time period.

The striking fact is that between June 1980 and September 1986, the chemical companies and CFC users vigorously asserted that alternatives to CFCs had been explored and found wanting. As late as February 1986, a paper prepared by the DuPont Company and reviewed by an industry panel[4] for presentation at an EPA workshop concluded as follows:

---

[2]S. Rayner, "Opening Remarks at Symposium on Global Environmental Change and the Public," First International Conference on Risk Communication, KFA Julich, Federal Republic of Germany, October 20, 1988.

[3]DuPont Corporation, "Fluorocarbon/Ozone Update," June 1980.

[4]The committe that reviewed the paper included representatives of the two major US CFC producers, DuPont and Allied, as well as two lawyers listed as counsel to the industry trade association opposing regulation of CFCs.

> "[Industry's] efforts began in 1975 to develop <u>acceptable</u> replacements compounds [for CFCs] should their use be required.
>
> At DuPont, this effort lasted several years and cost over $15 million. Unfortunately, <u>all promising coumpounds identified have one or more limitations</u> . . .Consequently, we conclude that fully satisfactory fluorocarbon alternatives will not become available in the forseeable future. The status of DuPont's program was formally reported to EPA in 1978 and 1980 . . .The list of candidates became progressively smaller and smaller as the studies progressed. Consequently, we are updating the results of this alternatives program and its objectives, parameters, and status in this final report."[5]

Lest there be any doubt about plans for further research, a preface to the report states, "No further studies, beyond those described herein, are planned by DuPont. We are unaware of any information that conflicts with the results that we describe in this report."

Another example of industry thinking is evident from statements in 1984 concerning recycling technology, now considered one of the most promising means of reducing CFC emissions. An industry review of this possibility concluded that costs would be "intolerable" and could not even cite any ongoing studies of this technology.[6]

The CFC industry rejected even the need to make the effort as protection against the possibility that the ozone depletion hypothesis was borne out, pointing to modeling estimates that no depletion would occur at then current emission rates and atmospheric measurements that had not found measurable depletion.[7]

I want to emphasize that my point is <u>not</u> that industry had alternatives to CFCs all along but simply withheld them. To the contrary, there are still many barriers to the development and dissemination of fully acceptable chemical and process alternatives to CFCs. Indeed, those who argue that the Montreal Protocol was possible because finding substitutes for CFCs is "easy" arguably know very little about the technical issues still to be addressed. New chemicals are themselves subject to stringent environmental and safety requirements and must prove themselves compatible with equipment developed for the performance

---

[5] Alliance for Responsible CFC Policy, <u>A Search for Alternatives to the Current Commercial Chlorofluorcarbons</u>, February 24, 1986.

[6] J. Gordon Logue, Chairman, Steering Committee, Alliance for Responsible CFC Policy, letter to Fitzhugh Green, Associate Administrator, EPA, March 28, 1984.

[7] See, e.g., Florocarbon Program Panel, Chemical Manufacturers Association, "Comments on the U.S. Position Paper on the UNEP Convention for the Protection of the Ozone Layer," April 1984.

characteristics of CFCs.[8] (As one example, the Department of Energy is currently evaluating the potential energy penalty associated with CFC substitutes as an issue in setting efficiency standards for refrigerators.) The important difference between pre- and post-Montreal is that industry is commited to making the necessary effort to find alternatives, with the competitive incentive brought forth by a recognition that those companies who develop the best alternatives will capture a multi-billion dollar world market.

Thus, the industrialized world has commited to replacing CFCs without clear solutions simply because it is necessary to protect the environment. One can look for other more crass explanations--for example, CFC producers may have felt they risked potential future legal liability following evidence of the Antarctic ozone hole--but the importance of the forcing effect of government policy remains. This lesson should not be lost in thinking about the global warming problem.

In some respects, the obstacles to replacing CFCs are arguably greater than with respect to CO2. The substitutes so far identi-fied for CFCs are in almost all cases inferior products--they cost more, don't perform as well, or suffer from toxicity or flamma-bility problems. In contrast, there are today many opportunities to reduce energy consumption in both industrialized and developing countries that are cost-effective and environmentally superior to fossil fuel consumption.[9] Moreover, while use of CFCs has until recently been growing faster than the use of energy, governments will most likely seek to eliminate almost all use of CFCs in only a decade while much more gradually reducing the fossil fuel use.

No existing economic activity or service is beyond replacement if time, brains, and money are put to the task. We hope adequate alternatives to CFCs will be available no later than the decade allowed by the Protocol; they are not available today. Similarly, time will be needed to develop substitutes for the other greenhouse gases, but we err in looking only at current options and costs in advance of any serious commitment to identify alternatives. A serious effort requires a substantial program, equivalent to DuPont's billion dollar commitment to CFC alternatives.

How can the government provide the necessary incentives most effectively? The evolution of EPA's CFC regulations raises some important issues. EPA has already issued regulations providing for production quotas, but the Agency has asked for comment on the possibility of adopting taxes on CFCs as an additional measure. For several reasons, taxes are a highly desirable supplement to regulation.

---

[8]Ironically, the CFCs were themselves developed as a safe substitute refrigerant for ammonia. M. Jones, "In Search of the Safe CFCs," New Scientist, May 26, 1988, pp. 56-60. Some serious attention is being given to the possibility of going back to ammonia taking advantage of greatly improved technology.

[9]See, e.g., J. Goldemberg et al, Energy for a Sustainable World (Washington, D.C.: World Resources Institute, 1987).

First, fees preserve the "polluter pays" principle. EPA estimated that production quotas will provide CFC producers with additional revenues between $2 and $6 billion by the year 2000. Some firms contest this figure, but whatever the truth it will still be a lot of money directed to the companies responsible for delaying regulation. They should not be given a financial advantage relative to their competitors.

Second, taxes can be justified as "user fees" consistent with free market theory--consumers should be alerted to the environ-mental cost of goods and services. The revenues can be applied directly to finding alternatives, providing a double benefit.

A third benefit is that fees can be graduated to reflect degrees of risk, as opposed to the sense of moral judgment attached to most regulation, or at least perceived by the regulated. Gasoline taxes would promote more efficient cars without most of the burdens of regulation most objectionable to the automobile industry.

Finally, fees communicate a clear price signal, whereas production quotas have produced considerable uncertainty about the future price and availability of CFCs. Some firms for whom CFCs are a relatively low input cost may find it cheaper to pay three or four times current prices rather than change their current production. Other firms may not invest in alternatives out of concern that the price of CFCs will not rise enough to justify the risks.[10]

---

[10]Potential new entrants must be willing to take on the existing CFC producers despite the advantages provided by profits due to the production quotas, superior knowledge of the characteristics of equipmented designed to use CFCs, and long established business relationships with the CFC using industries.

HALON SUBSTITUTIONS

T. CARL JEWELL
HALON RESEARCH INSTITUTE

SLIDE - HOW HALONS ARE USED
EMISSION REDUCTION PROGRAMS
DEVELOPMENT OF ALTERNATIVES

For the next few minutes I will briefly review halon usage and estimated emissions, the progress being made to reduce unnecessary emissions, and the prospects for the development of alternative fire suppression agents.

SLIDE - NO PRESENTLY AVAILABLE SUBSTITUTES

Halons provide rapid fire extinguishment, are safe for use when people are present, are electrically nonconductive, do not cause secondary damage to equipment or materials and require no clean up after use. They are used to provide fire and explosion protection for many critical applications for which we have no presently available substitutes.

SLIDE - UNNECESSARY EMISSIONS MUST BE REDUCED

Although halons are used in a wide variety of applications critical to our present needs for fire protection, emissions for causes other than fire extinguishment are unacceptably high.

Based upon an industry survey of total usage and causes of emissions in the US, the following slides show how halons are used and the causes of emissions.

SLIDE - 1211 USAGE PATTERN

For Halon 1211 approximately four percent of usage was to extinguish fires. Eighty four percent was banked or stored in containers for standby use. Nine percent was used in training, primarily for training of fire fighters. Three percent of usage were emissions from causes such as equipment servicing, transfer losses, research and development, inadvertent or unwanted discharges, equipment testing, and similar causes.

SLIDE 1301 USAGE PATTERN

Halon 1301 usage resulted in the following:

Fire suppression accounted for seven percent of usage. Approximately 71% was banked or stored in containers in systems for standby use. Nine percent was used for discharge testing to prove operational capability of newly installed systems. Six percent of usage was the result of unwanted, false or accidental discharges of systems. Eight percent of usage was related to servicing, transfer losses, equipment testing, research and development and similar causes.

SLIDE - CONTROLLABLE HALON EMISSIONS
TOTAL WEIGHT BASIS

The total controllable emissions on a per pound basis are shown on this slide.

Six percent of the total pounds of halon emissions is Halon 1211 emitted for miscellaneous causes. Twenty three percent of the total pounds emitted is Halon 1211 used for training purposes.

Halon 1301 emissions accounted for 71% of the total pounds of halon emissions. Twenty five percent was for miscellaneous causes, twenty eight percent for discharge testing of systems and eighteen percent for unwanted or accidental discharges.

SLIDE - CONTROLLABLE EMISSIONS ON AN O.D.P. BASIS

This slide provides the clearest picture of where we can obtain the greatest benefit by controlling or reducing emissions for causes other than fire suppression.

On an ozone depletion basis Halon 1211 emissions only account for eleven percent of the total O.D.P.

Halon 1301 discharge testing accounts for thirty six percent of the total ozone depletion potential of 1301 emissions. These emissions can be substantially reduced or totally eliminated.

Thirty one percent of the controllable emissions are from fire suppression and miscellaneous causes. Twenty two percent are caused by accidental or unwanted discharge of systems..

The previous slides show halon usage and emission causes. I would like to review what the industry has been doing to reduce unnecessary emissions.

SLIDE - CLOSED RECOVERY SYSTEMS
IMPROVED TRAINING

Through a cooperative effort of the FEMA and NAFED trade associations, the manufacturers of portables and the distributors who service this equipment are expanding their use of closed recovery systems. These closed recovery systems recapture halon expended due to testing required during manufacturing or losses of halon due to subsequent servicing operations.

Individual manufacturers, as well as trade associations, are implementing improvements to the training provided personnel which will reduce emissions during manufacturing and field servicing operations.

SLIDE - ALTERNATIVE TEST AGENTS AND
ROOM PRESSURIZATION TESTING

A program to test Halon 121 as an alternative test gas for discharge testing Halon 1301 systems was implemented earlier this year. Another test gas, sulphur hexafluoride, or SF6, has been tested for use in a program sponsored by the US Navy.

Additional testing will be required to determine the acceptability of either or both products as a substitute test agent. It is expected this program will be completed within 4 to 6 months.

A room pressurization (door fan) test program was implemented this fall. The program was sponsored by HRI, Northern Telecom and Environment Canada. It is being managed by the NFPA Research Foundation.

The objective of both of these programs is to provide alternative means of testing systems so that discharging Halon 1301 to prove system function and integrity will no longer be required.

I would expect that by the end of 1989 most of the usage of Halon 1301 for testing of systems can be eliminated. This will result in reducing current emissions by over 1/3 of the total O.D.P presently being emitted to the atmosphere.

SLIDE - IMPROVED TRANSFER EQUIPMENT

Individual manufacturers are developing improved transfer pumps and equipment that will achieve a higher percentage of halon recovered from a container. This unrecovered "heel" is presently discharged into the atmosphere when halon must be removed from a container.

As the use of improved equipment, as well as improved procedures for handling and transfer of halon is implemented, significant reduction in unnecessary emissions due to transfer and servicing operations will be achieved.

SLIDE - 1211 USE REDUCED FOR TRAINING

The military has drastically reduced the use of 1211 during training. This has been achieved by reducing the amount used per training exercise, as well as extending the frequency of training intervals. Civilian usage for training and demonstration is also being reduced.

Because training use accounts for over half of Halon 1211 emissions, it is estimated that these changes will reduce 1211 emissions by a substantial percentage within the next year.

SLIDE - RECOVERY AND RECYCLING

Recovery and recycling of halon from existing systems and equipment taken out of service and halon that has become contaminated with moisture or other contaminants must be implemented to prevent future emissions. Although some work is being done at present in a program recently started by the Air Force, as well as programs implemented by some Distributor servicing organizations to recycle and reuse halons, considerable work remains to be done.

Implementation of a comprehensive program to recover and recycle existing halon, or provide for environmentally safe disposal will require solving several problems. The costs associated with collecting, handling, transfer and transportation exceeds the value of the product at the present time. An additional problem is making the owner of a few portables or a system aware that halon should be salvaged rather than discharged into the atmosphere.

Although these will be difficult problems to solve, I am confident they can be. When these problems are resolved, potential future emissions from the existing bank of halons can be avoided.

SLIDE - STATISTICS

The Halon Research Institute in cooperation with producers, users and industry organizations is undertaking a comprehensive program to gather production, usage and emissions data on a worldwide basis. It is intended that this will provide better information to the scientific community for use as a base line for monitoring atmospheric changes. Once the data is obtained for past production and emissions, it will be continued and updated annually. This statistical data base will also provide a basis to monitor progress in emission reductions.

SLIDE - DEVELOPMENT OF ALTERNATIVES

Most if not all of the present producers of halons are individually committing substantial resources to develop alternative fire suppression agents. Although no specific materials have been publicly identified at this time, it is my understanding that there are several potentially promising candidate chemicals being investigated.

In addition to the work being done by individual firms, an international conference on Halon, The Ozone Layer and Research on Alternative Chemicals was held November 15 through 17 at Tyndall Air Force Base. The conference was jointly sponsored by the US Environmental Protection Agency and the US Air Force.

The conference was attended by representatives from the UK, Sweden, Finland, France, the Peoples Republic of China and Canada, as well as the US. The majority of attendees were scientists from private industry, governmental agencies and universities. The group represented specific expertise in chemistry, atmospheric science, toxicology and various engineering disciplines.

There was a high level of confidence expressed by the conference participants that alternative chemicals can be developed to replace the existing halons. The major unknown factors involve the questions of toxicity and implementation of commercial production of new chemicals.

There will be a follow up meeting of this group in early January to address specific questions raised at the Tyndall conference and to explore programs for cooperative research to speed up development of alternative fire suppression agents.

Although alternatives to the present halons are considered to be a problem that can be solved, it is recognized that it will take time to implement environmentally acceptable substitutes.

From a pragmatic standpoint, a realistic time frame would be 7 to 10 years. However, I have the utmost confidence that with the expertise that can be marshalled to solve the problem, it can be done within this time frame or possibly earlier.

SLIDE PROJECTOR OFF

In summary:

There are no presently available substitute chemicals which can be used to replace halons used for fire and explosion suppression.

Substantial resources have been committed with prospects for increased cooperative efforts devoted to developing environmentally acceptable alternatives for these agents.

In the interim, considerable progress has been made to reduce unnecessary emissions and provide means to effectively manage the existing bank of these materials to minimize future emissions. Additionally, when acceptable alternatives do become available it is expected that existing materials can be disposed of in an environmentally safe manner to avoid release of banked materials to the atmosphere at some future time.

There is a saying that a problem well defined is half solved. The problem has been defined. Significant progress has been made to not only reduce current and future emissions, but to develop environmentally acceptable alternatives for our future use.

UNCERTAINTIES IN ENERGY MODELS

JAE EDMONDS
Battelle, Pacific Northwest Laboratory
Washington DC

Energy and Greenhouse Gas Emissions: The "greenhouse" issue exists because the scale of global human activities has expanded to the point that the concurrent worldwide release to the atmosphere of CO_2 and other radiatively important gases, including CH_4, CO, N_2O, NO_x, and the chlorofluorocarbons (CFC's), is changing the global atmospheric concentration of these gases. Man-made emissions of these gases take place against a natural background of emissions and absorptions which range from non-existent, for the CFC's, to enormous, for CO_2. CO_2 is generally considered to be the most important of these gases, accounting for about half of current and anticipated radiative forcing. Fossil fuel use is the most important present and anticipated source of CO_2 emissions, though land-use change (deforestation) is another important source of net release.

Historic CO_2 Emissions: Since 1860 global annual emissions of fossil fuel CO_2 have increased from 0.1 to more than 5.0 PgC/yr.[1] During the period 1945 through 1979 the rate of CO_2 emissions from fossil fuel use grew at 4.5%/yr. Emissions declined after 1979 until 1983. Emissions have risen subsequently. The US, USSR and PRC account for half of the world's fossil fuel CO_2 emissions. US fossil fuel CO_2 emissions accounted for more than 40% of global emissions in 1950. This share has steadily declined to less than 25% today. US CO_2 emissions peaked in 1973 and have not regained that level.

CO_2 Release by Fuel Type: Carbon content varies by fuel. Of fossil fuels, natural gas is lowest (13.7 TgC/EJ)[2]; coal is highest (23.8 TgC/EJ); and oil falls between the two (19.2 TgC/EJ). The mining of oil shales in carbonate rock formations would add an additional stream of CO_2 to the atmosphere. The transformation of primary fossil fuel energy, as for example from coal to electricity or from coal to synoil or syngas, releases carbon in the conversion process. Energy technologies such as hydroelectric power, nuclear power, solar energy, and conservation emit no CO_2 to the atmosphere. Traditional biomass fuels, such as crop residues and dung, release CO_2 to the atmosphere, but are in a balanced cycle of absorption and respiration whose time frame is short. The use of other biomass fuels such as firewood may provide either a net annual source or sink for carbon depending upon whether the underlying biomass stock is growing or being exhausted. There are approximately 560 PgC of terrestrial biomass.

The Size of the Fossil Fuel Resource Base: The fossil fuel resource base provides no constraint on future atmospheric CO_2 release. The present atmospheric stock of carbon is approximately 740 PgC (1988). The estimated resource of fossil fuels is huge by comparison. While the carbon content of conventional oil and natural gas resources is only slightly more than half as large as the current atmospheric stock of carbon, coal resources are an order of magnitude larger. The carbon content of unconventional oil resources is 55 times larger than the current atmospheric stock of carbon. The pool of carbon available for combustion might be constrained to 4000 PgC by considering only those resources recoverable under present technologies. Even this severely constrained resource definition provides no physical constraint on climate change from fossil fuel use. Approximately 80% of the coal resource base is thought to be in three countries: the US, USSR, and PRC. There are approximately 800 PgC in the form of coal, recoverable with today's technologies, within the jurisdictional boundaries of the world's other countries.

CO_2 Emissions Forecasts: The extrapolation of time trends in which CO_2 emissions from fossil fuels, growing at 4.5%/yr, led early researchers to expect atmospheric CO_2 concentrations to reach 600 ppmv[3] shortly after the turn of the next century. The rate of growth of fossil fuel emissions is now expected to grow at less than 1%/yr. Recent analysis of uncertainty in future fossil fuel CO_2 emissions pushed the expected (median) date, by which 600 ppmv of CO_2 would be reached, to late in the 21st century and possibly not until the 22nd century. (I note there that other emissions sources, in particular land-use changes were not included in that analysis.)

Uncertainty in Future Fossil Fuel CO_2 Emissions: Uncertainty regarding the future rate of fossil fuel use remains great. Forecasts of emissions for the year 2075 vary by almost two orders of magnitude, with associate atmospheric concentrations varying by a factor of 3 to 4. The uncertainty in CO_2 emissions grows out of underlying uncertainties surrounding major human activities such as economic growth, the rate of energy efficiency improvement, and the type and rate of economic development experienced by the developing world.

The Timing of Radiative Forcing When All Greenhouse Gases and Emissions Sources Are Considered: We would not be here today if the conclusion of our story was simply that we expected an atmospheric concentration of 600 ppmv CO_2 sometime near the year 2100. We are here because recent research indicates that the combined accumulation of CO_2 and other greenhouse gases may yield radiative effects equivalent to that of a doubled concentration of CO_2 early in the 21st century. Such studies relied on time trend extrapolations for non-CO_2 greenhouse gases. Uncertainty surrounding the time when the buildup of greenhouse gases would be equivalent to that of 600 ppmv CO_2 is high. We define a "doubling window" as the range of years within which radiative forcing by all greenhouse gases yields a radiative forcing equivalent to that of 600 ppmv CO_2 under "business as usual" conditions. While no formal analysis has been conducted to approximate a "doubling" window, the timing might be placed within the period 2030 to 2075. Key uncertainties affecting estimates of the timing of the "doubling" event include: international policies precipitated by concerns other than the greenhouse, including those affecting chlorofluorocarbon use and land-use changes; uncertainties in source/sink relationships of greenhouse gases; and uncertainties with regard to economic growth, the rate of energy efficiency improvement, and the type and rate of economic development experienced by the developing world.

Analysis of Emissions Control Options: The rate of climate change could be altered through alternative energy, land-use, agriculture, and CFC practices. The analysis of options to control emissions of greenhouse gases is at an early state of development. To date studies have focused on options to alter CFC and energy production and use. CFC policies directed toward the control of stratospheric ozone may have a substantial effect on the rate and timing of climate change.

Five different classes of options to alter the rate of fossil fuel CO_2 emissions exist:

1. Pre-combustion carbon removal,
2. Fuel substitution,
3. Energy conservation,
4. Changing the composition of final demand,
5. Post-combustion carbon recovery.

Fuel Substitution, Energy Conservation and Changing Final Demand: Initial studies of options to control fossil fuel CO_2 emissions lead to two conclusions: 1. Taxes proportional to the carbon content of fuels as high as 100% of the cost of coal provided only marginal changes in the rate of fossil fuel CO_2 emissions, and 2. Emissions reduction strategies implemented by the US alone were never effective. "Bottom up" studies, which concentrate on strategies to accelerate the rate of improvement in end-use energy efficiency have obtained more optimistic results. Such studies assert that technologies exist which simultaneously improve energy efficiency and reduce CO_2 emissions, both at a net economic advantage to society. To accelerate introduction of such technologies, policies such as CAFE standards, energy taxes, and/or building code changes, are recommended by such studies. These studies have generally been microeconomic in nature and have not assessed feedback consequences of policies within a market equilibrium setting. Key issues remaining to be explored include the effect of accelerated investments on capital markets, economic growth, competitiveness, and energy prices.

Post-Combustion Removal--CO_2 Scrubbing: CO_2 scrubbers are also technically feasible, although costs are high for central power stations (electric power generation costs would double), and prohibitive for small scale energy use. Alkanolamine scrubbing has been used to co-generate power and CO_2 for commercial sales at a natural gas fired plant in Lubbock, TX. Injection of CO_2 into salt domes or directly into deep oceans are the two major repositories currently considered. Insufficient information exists to evaluate these strategies' full costs, efficacy and environmental risks. Tertiary oil recovery does not provide a repository of a scale comparable to global emissions.

Post-Combustion Removal--Biomass Plantations: Biomass plantations have also been suggested as a mechanism for CO_2 removal. The scale of the effort required to remove an additional 5 PgC/yr from the atmosphere is approximately equal to doubling the net annual yield of all the world's closed forests or planting new fast growing forests over an area equivalent to the total of global forest clearing to date. The cost of such a scheme is immense but needs to be compared with the costs of other approaches to dealing with atmospheric CO_2 or of coping with the attendant changes in climate.

Pre-Combustion Removal: It is possible to use coal as a feedstock for hydrogen, removing the carbon for return to the mine. The feasibility, economics and environmental consequences of such technologies have not been addressed.

Conclusions:

1. Global greenhouse gas emissions are likely to continue to grow in the future.

2. It appears likely that an accumulation of greenhouse gases equivalent to a 600 ppmv concentration of CO_2, will occur in the middle of the next century though the exact timing remains uncertain.

3. Altering energy, land-use, agriculture, and CFC practices could change the rate of emissions.

4. The costs and benefits of altering the rate of greenhouse emissions are not well understood.

---

1 gC = grams of carbon; 1 PgC = 1 petagram of carbon = 10^{15}gC = 1 gigatonne C = 1 billion metric tonnes of carbon.

2 TgC = 10^{12}gC; EJ = exajoule = x10^{18}J = 0.948 Quads.

3 ppmv = parts per million volume in the atmosphere.

Energy Strategies to Restrict Emissions Growth:

GOVERNMENT STRATEGIES TO LIMIT BUILDUP OF GREENHOUSE GASES

by

David J. Bardin
Arent, Fox, Kintner, Plotkin & Kahn
Washington, D.C. 20036-5339
December 1988

Abstract

This paper weighs government strategies to limit build up of greenhouse gases in the context of energy strategies to restrict emissions growth. It considers the relationship between strategies dealing with energy and non-energy gases. Government strategies may concern taxation, regulation, governmental lending and investment, guidance of government-owned and private companies, and government funding of research and development. Greenhouse gas emissions strategies could be integrated with strategies concerning such factors as clean air, flood control and shore protection, environmental diversity, employment in certain sectors, economic rationalization, development programs and geopolitical policies. Failure to integrate these strategies diminishes or cancels out their effectiveness.

Present Strategies

What are the existing government strategies to limit buildup of greenhouse gases? In the energy sector, although there are a number of recent proposals, virtually no such strategies have yet been adopted in principle, much less been set "in place" for implementation.

CFCs. In the non-energy sector, the Montreal Protocol and the proposed EPA implementing regulations would phase down (but not out) the production and use of CFCs. Although the initial motive for concern about CFCs was their effect on thinning the stratospheric ozone shield against harmful ultraviolet radiation. An additional benefit, however, is now recognized to be reducing the emissions of a "new" greenhouse gas family, since CFCs unchecked are thought to be a major component (perhaps as much as a quarter) of the excess greenhouse warming effect that is now building up.1/ In this context, we now have serious proposals to amend the Montreal Protocol to eliminate CFCs altogether. Ironically, these strategies upset a related energy conservation strategy

1/ Historically, CFCs were not a significant part of the century-long global warming trend.

since CFCs are widely and efficiently used in insulation, including energy-consuming appliances such as refrigerators.

Tropospheric ozone (smog). Under the Clean Air Act (CAA) the EPA has adopted a regulatory program for limiting the build up of ground level, and near ground level ozone. Adoption of the program was based on health considerations. Global warming considerations did not play a part in adoption or design of the program, and they never seem to be mentioned by EPA in defence of the program or in assessing benefits to be secured. Compliance is supposed to be secured by state regulatory actions, with federal actions as a fall back. The program set a deadline of December 1987 for compliance with a 0.12 ppm ambient standard. Most metropolitan areas are not in compliance, to varying degrees. EPA has suggested a variety of national and local measures, including energy-related measures such as switching some vehicles from gasoline to other fuels, as means to approach compliance. Some of the measures would tend to reduce ozone in rural areas which are in compliance with the CAA as well as in non-compliance, largely urban areas. Measures generally involve cut backs in reactive hydrocarbon emissions. Some argue for NOx cut backs as well. Carbon monoxide is also a precursor, rarely mentioned. EPA Administrator Thomas sometimes calls ozone the agency's toughest regulatory issue. Global warming and a thinned stratospheric ozone layer which lets more ultraviolet get through will each make tropospheric ozone build up a worse problem to solve.

EPA regulation of sulfur dioxide emissions. Some proposed regulations for small boilers will, if adopted, tend to shift boilers to natural gas, thereby incidentally reducing carbon dioxide emissions. Other sulfur emission reduction programs were designed to maximize coal use.

Energy Conservation. The U.S. government has several energy conservation programs, notably including the vehicle fuel efficiency standards applicable to each manufacturer's total car output each year (CAFE standards) and appliance efficiency standards. Congress mandated the latter over President Reagan's veto. Many state public utility regulatory agencies encourage utility conservation activities, including subsidization of high-efficiency appliances and investments in conservation equipment. The Department of Transportation cancelled a scheduled tightening of the CAFE standards for this year. The Reagan Administration has favored abandonment of the CAFE standards. Congress has neither agreed to abandon existing standards nor has it forced the Administration to accept tougher standards. In addition, the last Congress repealed in substantial measure the nationally-prescribed maximum 55 miles per hour speed limit, which was defended in part on energy savings grounds.

Renewable Energy and Nuclear Power. Tax credits once offered for renewable energy facilities (e.g., residential solar) have all expired with the following three exceptions of credits which are due to expire at the end of 1989: business solar (10% credit)· geothermal (10%); and ocean thermal (15%). Proposals to close down working nuclear power plants before the end of their useful life have been repeatedly rejected (most recently by the Massachusetts referendum vote). On the other hand, some state authorities have discouraged the completion of nuclear power plants under construction and, in the one case, have sought to dismantle a plant already complete but not yet started up (Long Island Lighting Company's Shoreham Plant, for which a remarkable potlatch has been proposed by Governor Cuomo but not enacted by the Legislature). Although public opinion polls indicate national majority support for nuclear power, local politics seem to vary. A tax of 1 mill per kwh of nuclear power is building up a fund for spent fuel disposal and certain decommissioning activities.

Taxation and User Fees. There are no existing user fees aimed directly or indirectly at restraining greenhouse emissions or likely to have that effect.

Research and Development Budgets. Congress has approved, overcoming initial Reagan Administration resistance, a five-year, $2.5 billion Clean Coal Technologies program for the Department of Energy. Some of the approved projects would increase the fuel efficiency of using coal. R&D budgets for renewable energy and other forms of conservation are more modest. The R&D budget for lower-carbon forms of fossil fuel (e.g. natural gas), particularly for geoscience R&D, is virtually non-existent.

National Environmental Policy Act. So far as I can determine, no NEPA-compliance environmental impact statement or assessment (or equivalent state-law EIS) has yet treated global warming as a truly significant, much less decisive, consideration. Moreover, major state and federal initiatives are under way without an EIS that seriously weighs climate impacts. For example, the State of California seems committed to having most of the future vehicles its citizens use burn a synthetic fuel, methanol, without assessing climatic effects that change might have as compared to the alternatives.

General Approach to Future Strategies

Some suggest that the measures which should be taken now as to global warming are those which we should undertake "anyhow," for independent reasons. When it comes to government strategies, alas, there are significant disagreements as to what should be done "anyhow" -- as the above paragraph on energy conservation indicates. So we cannot use a neat formula to obviate the need for substantive debate. Anyone who believes that accelerated global warming is likely to cause severe and unacceptable costs to human societies and the planetary environment in which we subsist should welcome an evaluation of options to ameliorate the trend, including an assessment of their costs. In selecting strategies, of course, there are good reasons for seeking multiple benefits.

Government strategies should be founded on a general approach that takes account of the basic knowledge and uncertainties. Governmental response to the global warming issue should be based upon --

recognition of the uncertainties;

global perspectives, geographical and industrial;

rational priorities, integrating multiple goals;

timely and patient identification of measures; and

cautious implementation.

One must carefully consider when to push a panic button, and when to take a second look; when to charge ahead and when to delay. Above all, one should build into any strategy a high measure of <u>flexibility</u> -- for the industry sectors affected as well as for the strategic planners and policy makers.

A. <u>UNCERTAINTY:</u> Recognize the element of uncertainty, debate out the scientific issues, educate and be educated, but do not expect thereby to achieve certainty or near-certainty.

1. Reach a first-round of policy decisions on the basis of present knowledge.
2. Encourage decades of continuing scientific and policy debate, and be prepared to adjust policies adopted in light of new knowledge.
2. Fund a reasonable level of effort to improve atmospheric and climatic knowledge and understanding.

B. <u>GLOBAL PERSPECTIVE:</u> Start from a global perspective, geographically, chemically, and industrially.

1. Geographically, emissions originating in any country may influence its climate and that of others. To the extent practicable, a realistic and effective policy should probably have significant international backing -- whether adopted in a multi-national context or sequentially, by one nation after another.

In any event, policy-makers need to understand the international context of proposed measures.
2. The energy sectors should not be singled out. Their emissions should be examined together with other pertinent emissions before rational policy decisions are reached. The energy sectors may best be called upon to perform their appropriate share once they are assured that every other sector has also been examined, impartially.

C. PRIORITIES: Set rational priorities based on factors such as cost-effectiveness, minimal disruption, congruence with other policy goals. Allow all affected parties maximum flexibility in selecting means to achieve the policy goals.

D. TIMELY AND PATIENT IDENTIFICATION AND CAUTIOUS APPLICATION: Rapidly identify global warming impacts so as to enable policy-makers to work out appropriate responses. They should avoid hasty decisions of vast environmental or energy import until they have made a painstaking analysis of the global warming implications. Then they can sort out which existing programs need early adjustment and which new program deserves early adoption, rejection or modification.

Specific Issues for Additional Governmental Strategies

Chemical Emissions of Non-Energy Sectors. Both environmental logic and avoidance of arbitrary concentration on the energy sector dictate continued progress on phasing out CFCs altogether.

Anticipation of New Chemicals. The CFC experience also teaches the importance of assessing potential greenhouse significance of new man-made chemicals before their significant introduction into the atmosphere. An effective guidance strategy needs to be devised which will identify and allow timely action without bogging down industrial progress. The development of such a strategy could enlist scientific resources of NASA and the U.S. Air Force. The Office of Technology Assessment should be asked to come up with a plan.

Carbon Emission User Charge. If one decided to restrain the rise in carbon dioxide emissions, the most efficient means would be to levy a charge on emissions, proportional (or roughly proportional) to the carbon in such emissions. Solar and other renewable energy, as well as nuclear, would not bear such a fee (either because no carbon is emitted or, in the case of biomass, because carbon recycling would prevent any net emission). An increase in the federal excise on gasoline, which has been proposed by Ways and Means Committee Chairman Rostenkowski for budget balancing purposes, could be one component of such a broader program (but would be second best if substituted). A carbon emission user charge would fall more heavily on emissions from coal burning than on those from oil burning or gas burning. Considerations of fairness would support allocating part of the revenues from such a carbon user charge to assure the pensions and health care of miners and dependents of miners who had toiled underground and risked their lives to help light the streets, homes and workplaces of America. Other portions of the user charge revenue could be allocated to deficit reduction. The charge would be designed to stabilize (rather than eliminate) carbon emissions.

Regulation of Carbon Emissions. Command-and-control regulation must prove less efficient than a user charge which lets market forces drive particular carbon emitter decisions. If the latter is mandated, however, the regulations should leave fuel users the maximum flexibility to choose methods they find cost effective. For example, a "compromise" version of the "acid rain" legislation discussed this past summer would have mandated use of wet scrubber technology to reduce sulfur dioxide emissions. That rigid approach would have resulted in derating coal-burning power plants and would have led to an increase in carbon dioxide emissions as those plants were run harder in many hours of the year to overcome the derating.

Renewables and Energy Efficiency. Research and development efforts and the appropriateness of subsidization could be reconsidered, especially if carbon emission user charges were rejected. The CAFE standards issue could be revisited, particularly if American vehicle fuel prices remain well below those of major industrial competitors.

More Efficient and Less Polluting Vehicles and Stationary Fuel-Using Facilities. Government strategy could prod the development of improved vehicles, including a personal vehicle for the 21st Century which will take advantage of progress in strength of materials and electronics to improve fuel efficiency radically and also to cut nitrogen oxide and reactive hydrocarbon emissions. The effort may have to combine capabilities of several companies and industry sectors, so that a limited antitrust immunity may be in order. In such an effort, consideration should seriously be given to development of an advanced natural gas vehicle technology which exploits the 130 octane character of natural gas. If advanced natural gas vehicles prove appropriate, they would be a boon to developing countries which, in case after case, have discovered very large resources of natural gas which their economies often cannot absorb. (For example, space-heating uses are not relevant in many climates.) In terms of thermal input, for example, the emissions from direct combustion of various automotive fuels may be summarized as follows:2/

Fuel:	Ratio Relative to Natural Gas:
Natural Gas	1
Propane	1.21
Gasoline	1.4
Diesel	1.46
Methanol from Natural Gas	1.42
Methanol from Coal	1.92-2.56

The relationships for stationary fuel-burning sources are comparable, however, the output efficiencies may be markedly different.3/ Efficiency gains could sometimes be accelerated

Nuclear Power. Revival of a nuclear option could slow down the growth of carbon dioxide emissions and of tropospheric ozone build up. Several electric utilities have demonstrated the ability to build nuclear plants on time, on budget and in accordance with all safety requirements, even under present design, licensing and construction practices. These practices could be improved and new development could design a standard nuclear unit suitable to future utility needs and able to operate safely with a minimum of human intervention.

Methane Emissions from the Energy Sector. So far as I can determine, the growth in methane concentrations in the atmosphere is virtually unrelated to energy industry methane (which constitutes the bulk of natural gas). That being the case, and given the significant advantages that greater methane use could bring as well as the selfish interest of the energy sector to minimize loss of methane to the atmosphere, no special regulation of the energy sector seems appropriate on this score.

2/ Taken from MacDonald, Greenhouse Consequences of Using Methanol as a Vehicle Fuel, Table 5, presented to South Coast Air Quality Management District, November 29, 1988.

3/ For example, natural gas used in a combined cycle plant to generate electricity has significantly more of its input fuel energy converted to output electricity than does either gas or coal used in a conventional steam boiler.

Addendum on Global Warming and "Greenhouse" Effect

The greenhouse effect is both widely understood as established reality and seen as a relatively novel notion about inexorable change. Conventional science holds that the blanket of atmospheric gases moderates the earth's climate by absorbing and limiting the escape of low-frequency, infra-red radiation after allowing high frequency solar radiation to penetrate and become converted to infra-red. There is also widespread understanding that slight increases in certain of the "greenhouse" gases could tend to enhance this "greenhouse" or blanketing effect[1/] of the atmosphere and that atmospheric concentrations of some of these gases, including carbon dioxide (CO_2), methane (CH_4), tropospheric ozone (O_3) at less than 10 kilometers, chlorofluorocarbons (CFCs) and nitrous oxide (N_2O), have been on the rise for several decades.[2/]

Moreover, the earth's surface appears to have warmed by over 1° F. during the last century, a warming that proponents of "greenhouse" theory ascribe to the changes in atmospheric gases. (Some of these changes -- e.g. more CO_2 and CFCs -- have accompanied industrialization; but the reasons for other changes -- e.g. more CH_4 -- are not understood.) Skeptics still question the theory, pointing out for instance, the occurrence of a cooling mini-trend (approximately from 1940 to 1965) in the midst of the longer-term warming trend; but these skeptics have not come up with an alternative theory to explain the longer-term trend.

Proponents of the "greenhouse" change theory predict a continued warming trend, at a far faster rate than the earth has previously experienced throughout geological time -- unless deliberate measures are adopted to slow down the trend. Some proponents foresee an accelerating warming trend that may continue for centuries, until the shifts in earth's

---

1/ The term "greenhouse" is misleading, since the architectural model provides a controlled climate inside. In the atmospheric model there is no implication of controlled climate whatsoever. Climate, influenced by growing accumulation of "greenhouse" gases remains wild and chaotic, perhaps even more so because of the perturbations caused by change. Only the direction of change may be predictable.

2/ By and large, these "greenhouse" gases appear to disperse throughout the atmosphere, rather than concentrate at one level. It may take up to a few years, however, for gases generated in the northern hemisphere to disperse fully to the southern hemisphere. The survival period of "greenhouse" gases varies from a few days to well over a century.

orbit around the sun would lead to the anticipated onset of the next ice age, in 10,000 to 12,000 years.

Skeptics argue that with climatic chemistry and physics so complex, so poorly understood, and so prone to feedback relationships, at least some of which may cancel each other out, one should not foreclose the possibility that global warming will slow down much sooner, rather than accelerate further, even without deliberate human efforts to restrain build up of energy-related greenhouse gases. Proponents of societal actions argue that mankind cannot afford the luxury of waiting much longer to find out, because the potential effects of rapid warming (including effects on drought, rainfall, severe storms and how fast ocean levels will continue to rise) are dramatically unpleasant. It is reasonable to assume that such debates, and the uncertainties they may engender, will continue. Indeed, as scientific understanding grows and deepens, it seems quite likely that new uncertainties will arise and new investigations will be proposed.

State Initiatives. In a nation as large and varied as the United States, uniformity is often neither possible nor desirable. Yet state-by-state consideration of some issues (such as highway speed limits will be practicable, if those who are deeply concerned about global warming will press an agenda for action).

MORE EFFICIENT TECHNOLOGIES AND FUEL SWITCHING: THE NEAR-TERM PREVENTION STRATEGY

William Fulkerson
A.M. "Bud" Perry
David B. Reister
Oak Ridge National Laboratory

ABSTRACT

Worldwide CO_2 emissions continue to increase year by year. The rate of increase has been significantly lower in the 70's and 80's than it was before the oil embargo--~1.5% per year over the past decade (1977-1987) versus ~4.6% per year during 1953-1973. Nevertheless, it is still increasing, and much of the increase comes in the developing countries. To stabilize CO_2 emissions at something like the present 5-6 GtC per year would not be an easy task, yet it would still not prevent the CO_2 concentration from increasing. To reduce CO_2 emissions appreciably, as called for by the summary statement of the Toronto Conference, would be even more difficult and would require deep cuts in emissions by industrial nations to offset some further increase in emissions by developing nations. To reduce CO_2 emissions without inhibiting economic growth will require both very large gains in energy efficiency and massive switching from fossil to non-fossil energy sources. In the near term, only efficiency improvements, together with some shifting to natural gas, appears ready technologically to make a significant impact on reducing CO_2 emissions. Nevertheless, sustainable control of CO_2 emissions will require much better nonfossil sources.

By *prevention* we mean actions to reduce the increase of greenhouse gases in the atmosphere. This is in contrast to *adaptation* which is the strategy of managing to cope with the greenhouse effect whatever it turns out to be. The focus of this paper is on controlling CO_2 emissions by management of energy technologies. Although CO_2 is the principal greenhouse gas being affected by anthropogenic activities, others must also be managed--namely methane, CFCs and N_2O.

Reducing CO_2 emissions will be difficult to accomplish at best for several reasons:

First, worldwide demand for primary energy continues to increase, particularly that of developing nations as their economies grow, and the economic growth is fossil-fueled. This is illustrated in Fig. 1 where CO_2 emission rates for the world and various nation groups are plotted and indexed by their 1973 values. In fact, if one extrapolates the CO_2 emission rates of the past decade for these same groups of nations, one finds that for a number of different growth rate scenarios, the CO_2 emissions of developing nations will exceed those of the OECD nations by about the year 2005, as shown in Fig. 2.

Second, the best non-fossil sources are not yet very good competitors to fossil fuels, despite the fact that the world technical communities have been working for decades on various non-fossil sources. None can be deployed at sufficient scale and at competitive costs to displace a substantial fraction of fossil fuels. Nuclear power is the nearest, but its large scale deployment is in doubt because of issues of waste management, safety and proliferation. Renewables are either too expensive, such as photovoltaics, and solar thermal electric, or too limited, such as hydropower and biomass, although biomass may prove to be a very important future source of liquid fuels for transportation. Fusion is still decades away from prototypical demonstration.

So, while we learn to improve non-fossil sources, prospects for reducing CO_2 emissions rest primarily on using fossil fuels more efficiently, and also perhaps on some use of natural gas to replace coal where practical.

Fortunately, improving the efficiency of energy use and conversion should be attractive for several other reasons:

(1) The technical opportunities for cost effective efficiency improvements appear very substantial all across the energy system, so efficiency improvement can be a low cost strategy for increasing energy services with economic growth.

(2) More efficient use of energy which is economic can improve international competitiveness for nations which achieve it.

(3) It can relieve stress on oil and gas markets.

(4) It can reduce environmental stresses generally in addition to the greenhouse effect.

Thus, to the extent it is economic, improving the efficiency of energy end use and conversion should be a desirable strategy for all nations regardless of the greenhouse effect. This is one of the important messages developed in the remarkable book, *Energy for a Sustainable World*, by Jose Goldemberg, Thomas Johansson, Amulya Reddy and Bob Williams (Goldemberg, et al., 1988).

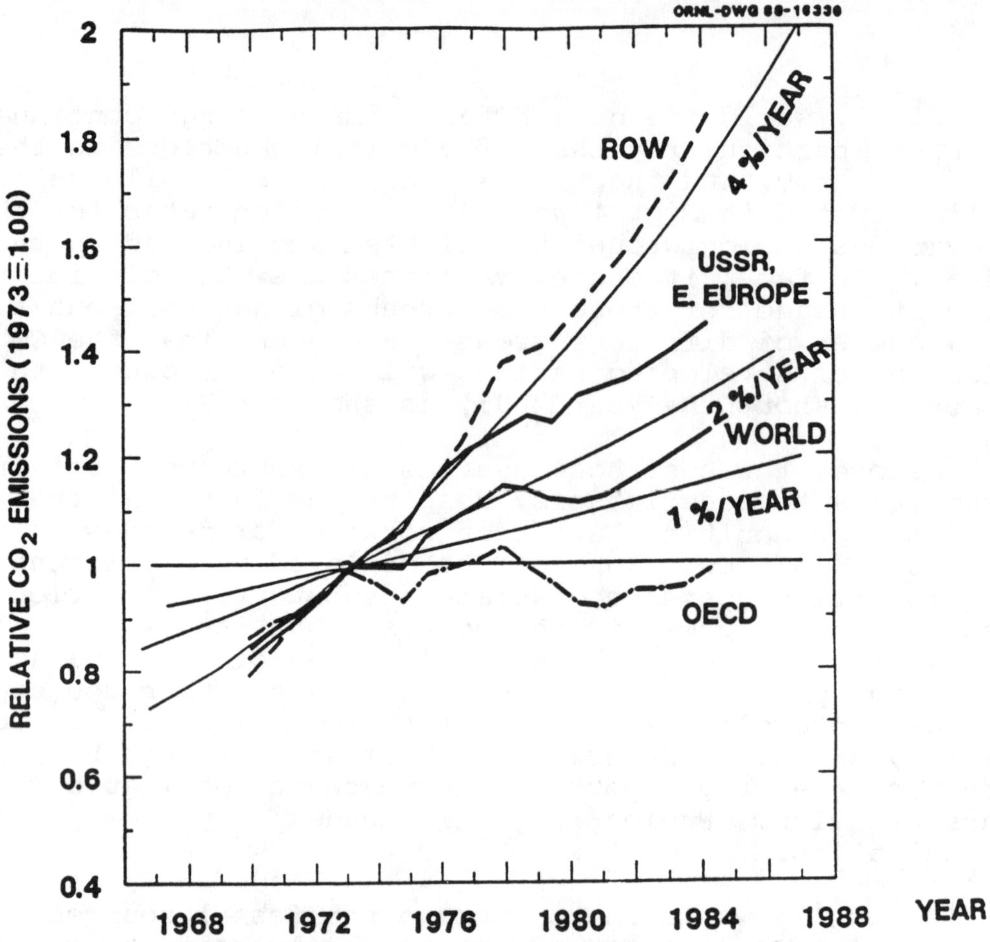

Fig. 1. CO_2 emissions from fossil fuel combustion indexed to unity for 1973 for the world and various nation groups: The Organization of Economic Cooperation and Development (OECD) nations, the USSR and Eastern European nations, and the Rest Of The World (ROW) (mainly the developing nations, including China).

CO_2 EMISSIONS FROM FOSSIL FUELS, 1980 - 2040,
UNDER ALTERNATIVE ASSUMPTIONS FOR GROWTH RATES

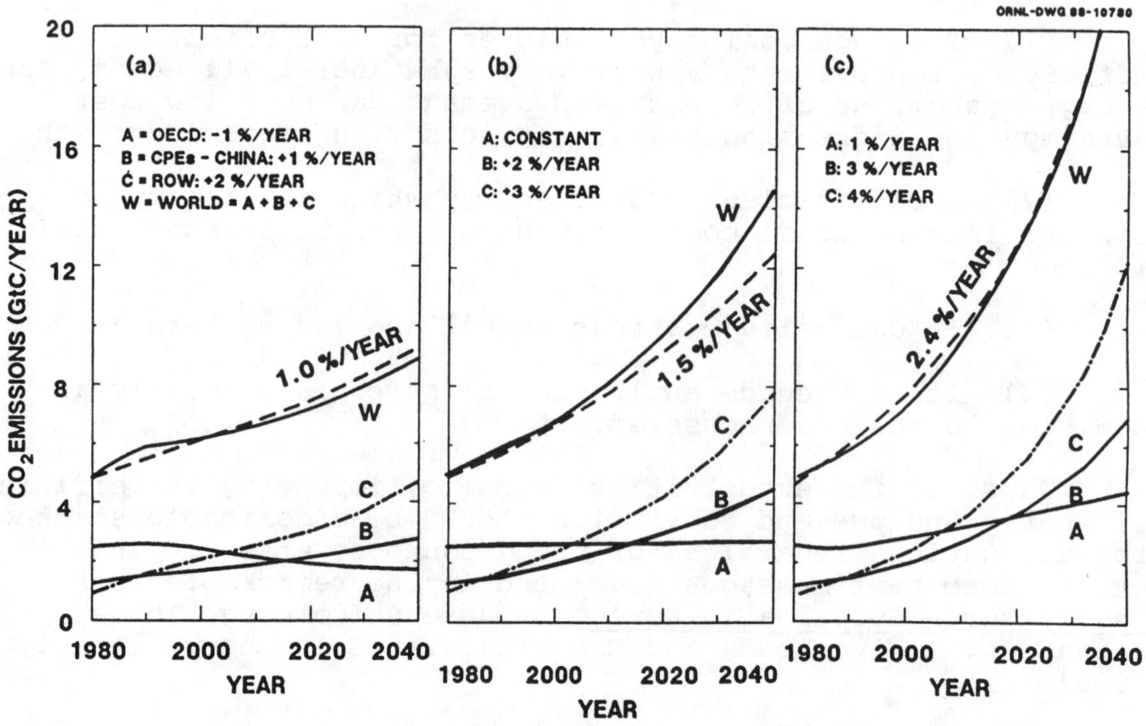

Fig. 2. Projected CO_2 emissions for various nation groups assuming various exponential growth rates for the emissions. These rates represent a range around recent trends (see the notes to the figure).

NOTES FOR FIG. 2

CO_2 emissions trends, 1977-87 (GtC per year and growth rate)[a)]

		Oil	Gas	Coal	Total	Assumed Annual Emissions in 1990
A	1977	1.405	0.416	0.689	2.510	
	1987	1.209	0.419	0.856	2.484	2.5
	1987/77	0.860	1.007	1.242	0.990	
	Av, %/year	-1.50	0.07	2.17	-0.10 (OECD av ~ 0%)	
B	1977	0.384	0.192	0.645	1.221	
	1987	0.419	0.353	0.713	1.485	1.6
	1987/77	1.091	1.839	1.105	1.216	
	Av, %/year	0.87	6.1	1.00	1.96 (CPEs - China ~ 2%)	
C	1977	0.386	0.057	0.496	0.939	
	1987	0.514	0.115	0.818	1.447	1.6
	1987/77	1.332	2.018	1.649	1.541	
	Av, %/year	2.86	7.0	5.00	4.32 (ROW av ~ 4%)	
W	1977	2.175	0.666	1.830	4.671	
	1987	2.142	0.887	2.387	5.416	5.7
	1987/77	0.985	1.332	1.304	1.159	
	Av, %/year	-0.15	2.87	2.66	1.48 (1977-87 world av. %)	

a) A = Organization for Economic Cooperation and Development
B = Centrally Planned Economies (excluding China)
C = Rest of the world (including China)
W = The world

Sources: Fuel use taken from BP Statistical Review of World Energy, British Petroleum Company, June, 1988; CO_2 emission coefficients (17.4g carbon/MJ for oil, 13.7 gC/MJ for gas, 23.9 gC/MJ for coal) are from Marland and Rotty, 1983.

One figure from this book illustrates the potential for efficiency improvement worldwide. Fig. 3 shows a comparison for 2020 of the high-efficiency calculations by Goldemberg, et al., with the high and low projections by IIASA, 1981, and the World Energy Conference (WEC, 1983). Comparison to actual energy use in 1980 is also shown. Goldemberg and his colleagues assumed the worldwide application of the best available or soon to be available technology. So, their calculation represents technical feasibility, and it shows the enormous potential for doing things more efficiently while still accommodating population growth and economic growth equivalent to that represented by other scenarios. Furthermore, Goldemberg, et al., argue that energy services in developing nations could be raised to those equivalent to the average amenities enjoyed in Western Europe (in the 1970's) with only a very small increase in per capita primary energy consumption. The big difference between the scenarios lies in the energy used by the industrialized nations with the calculations of Goldemberg, et al., showing a possible decrease by a factor of two and IIASA and WEC scenarios showing increases by about the same factor or more.

Of course, the CO_2 emission rates for the scenarios are quite different. We have simulated the emission characteristics for two cases using the Edmonds and Reilly (ER) model (Edmonds and Reilly, 1986) which is a general equilibrium model of the world energy system (Fig. 4). One case, the base case, is similar to the IIASA or WEC low scenarios, and the other case is similar to Goldemberg's, et al., calculations. Use of the model allows us to extrapolate the results beyond 2020, but we are skeptical that efficiency can continue to improve as the lower extrapolation would require. It would mean an average decrease in E/GNP for the world economy between 2000 and 2050 by a factor of three. Thus, these curves should simply be taken as a rough indication of the potential for reducing CO_2 emission rates by improved efficiency of energy use.

Prevention by the United States

We have examined similar scenarios for the United States. We looked at two scenarios for the years 2020 and 2040 and compared CO_2 emission rates to those in 1987 (Fig. 5). For each year the taller bar is the ER base case using median coefficients[1]. The shorter bar for 2020 is a very high

---

[1] i.e., median values of all of the several dozen variable parameters in the model, such as price and income elasticities, technical change coefficients, supply parameters, etc.

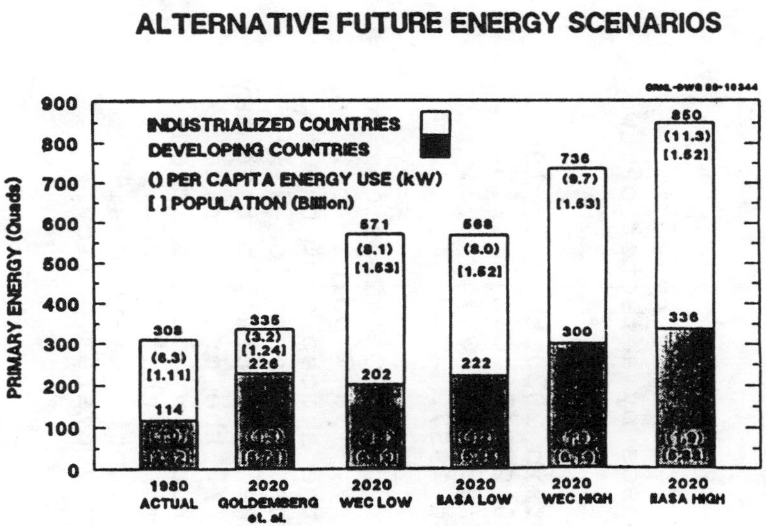

Source: J. Goldemberg, T.B. Johansson, A.K.N. Reddy and R.H. Williams, Energy for a Sustainable World, World Resources Institute, Washington, D.C. 1987.

Fig. 3. Alternative scenarios of global energy use by industrialized and developing nations for 2020 after Goldemberg et al., 1987. Note that world energy use in 1987 was 352 quads of primary energy including biomass.

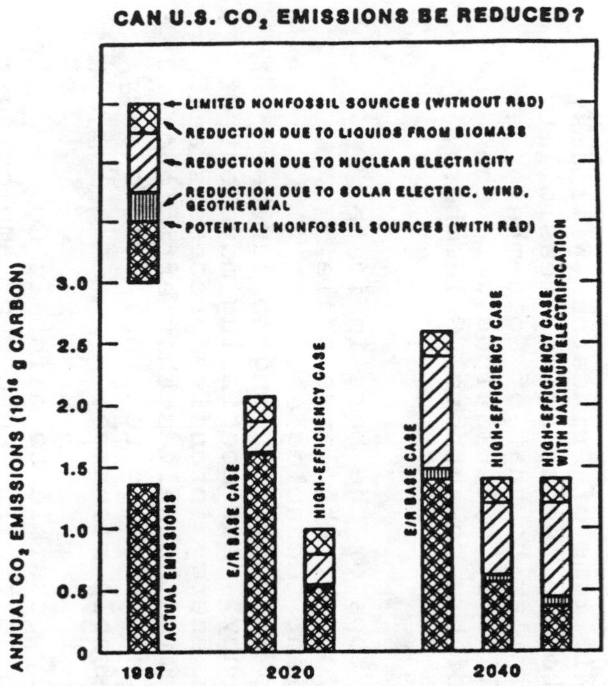

Fig. 5. Estimated potential of efficiency increase and nonfossil energy sources for reducing U. S. CO_2 emissions from fossil fuel combustion.

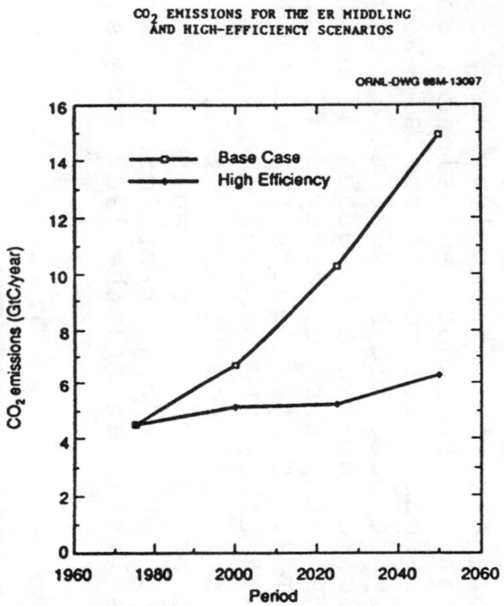

Fig. 4. CO_2 emissions calculated using the Edmonds and Reilly Model for the ER base case (or middling scenario) and a high-efficiency scenario.

efficiency case designed by Bob Williams of Princeton (Williams, 1987). His calculation is a bottom-up technological feasibility analysis for the United States similar to that of Goldemberg, et al., for the world. In fact, Williams' calculations were the basis for some of the estimates of his co-authors, Goldemberg, Reddy and Johansson.

The differences between the tops of the bars indicate potential for reducing CO_2 emissions by the large scale implementation of the most efficient technologies. We extrapolated Williams' results to 2040 by assuming no further technical improvements in efficiency,[2] but continuing movement of the industrial sector toward less energy intensive processes and activities. Our purpose is to remind ourselves that efficiency improvements are likely to yield diminishing returns. Efficiency improvement is in a sense a depletable resource.

We conclude it is technically possible to maintain CO_2 emissions about constant over the next half century by employing more efficient technology very aggressively. However, it is possible to reduce emissions a good bit further by substituting non-fossil for fossil fuels. With successful R&D we estimate it may be possible by 2020 to supply 10 quads of biomass derived liquid fuels and 16 quads of nuclear power, based on an advanced LWR with passive safety features and a fully passively safe MHTGR. Sixteen quads (thermal) is supplied by about two and one half times the present nuclear power capacity.[3] By 2040, non-fossil sources, particularly nuclear, can grow sufficiently to bring emissions down to 1987 levels even for the ER base case, and for the higher efficiency case, emissions might be reduced to as little as 1/3 the 1987 levels.[4] This assumes R&D can make

---

2 Note: Williams' scenario called for a 4-fold decrease in E/GNP from 1980 to 2020 (GNP/capita increases by a factor of two and energy/capita decreases by 1/2).

3 Capacity factor is assumed to be higher in 2020 than in 1987. <u>Capacity</u> is only up by about a factor (230 GWE/91 GWE) = 2.5. <u>Generation</u> is up by a factor ($1500*10^9$ kwh/$455*10^9$ kwh) = 3.3. Capacity factor in 2020 ~74% versus 57% in 1987.

4 Of course, it may not be necessary to reduce CO_2 emissions to zero. As John Firor (Firor, 1988 and Perry, 1984) points out for any given future concentration of CO_2 in the atmosphere, there will exist an emission rate below which the concentration will not increase further. That rate may change with time, but it may be substantial, e.g., 2-3 GTC/year about 1/2 the present rate. For example, the permitted use of fossil fuel may be sufficient to power the world's transportation system, if that system becomes very efficient. Finding out more about this "fossil" ration is obviously very important to energy system policy.

non-fossil sources acceptable with regard to cost and other attributes. The final bar represents using the full non-fossil electrical potential in 2040 to replace natural gas in buildings and provide some electricity for powering vehicles, in addition to the other electrical loads called for in the high-efficiency case.

Both of these cases, the base case ER projection and the Williams high-efficiency scenario, represent significant decreases in E/GNP from today's value. For the ER base case, E/GNP is decreased by 36% from the 1987 value by 2020, while it is 68% less for the Williams case, as shown in Fig. 6.

A very important conclusion appears obvious from this analysis. Prevention of further large increases in the CO_2 emissions rate will require either continuing dramatic increases in the efficiency of energy use or better non-fossil sources or both. Reducing emission rates substantially below the present value, however, will surely require both high efficiency, and better non-fossil sources. We may want to reduce U.S. emissions below current levels to give developing nations more maneuvering room with fossil fuels. Furthermore, sustained control of CO_2 emissions will require better non-fossil sources.

R&D Opportunities--Efficiency Improvement and Fuel Switching

Although technology already exists for substantially improving the efficiency of energy use, R&D can improve the potential even further and make it more economical and otherwise attractive.

The DOE Office of Conservation (DOE, 1988) has recently estimated the potential impact of more efficient technologies being developed under its support and by others such as GRI and EPRI. Overall, the deployment of these technologies could reduce primary energy use by about 20-34% in the year 2010 from the DOE base case projection.

In Fig. 7 we list some specific R&D opportunities which could, in our judgment, significantly improve efficiency and hence reduce CO_2 emissions. These apply to all three end use sectors and electricity.

In transportation, improved engines and more nearly continuously variable drive trains have great potential. Likely advances include low heat rejection turbo charged engines, two stroke designs of the Otto cycle, and the direct injection stratified charge engine. Advanced materials and better emissions control are required. Much improved aircraft efficiency seems assured through better materials, unique aerodynamic designs and ultra high bypass engines.

U. S. ENERGY CONSUMPTION PER UNIT OF GNP
(Relative to 1970 ≡1.00)
HISTORICAL (1960 - 1987) AND PROJECTIONS

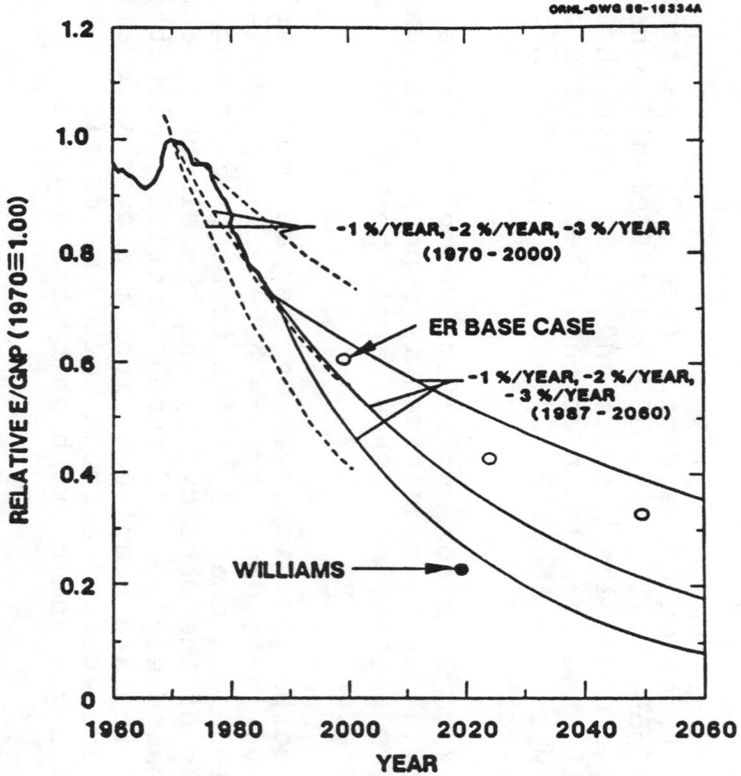

Fig. 6. U.S. primary energy use per unit of GNP normalized to unity for 1970 and showing the Edmonds and Reilly base case scenario and the calculations of Williams for 2020.

Fig. 7

R&D OPPORTUNITIES FOR REDUCING CO_2 EMISSIONS BY IMPROVING EFFICIENCY OF ENERGY END USE

TRANSPORTATION

- Advanced Engines
- Continuously Variable Transmission
- Improved Aircraft Efficiency

BUILDINGS

- Heat Pumps
- Smart Systems - Sensors and Controls
- Improved Envelopes
- Existing Building Retrofit

INDUSTRIAL

- Sensors and Controls
- Heat Management and Cogeneration

ELECTRICITY

- Power Electronics

In the building sector, both electric and gas fired heat pumps that are 50% more efficient than current models are likely, assuming CFC substitutes are found. Smart systems based on better sensors and controls should allow buildings to be optimally controlled and optimally integrated from HVAC systems and appliances to the utility grid. Better envelope design and materials can make all envelope components more efficient and more easily factory-manufactured. Innovative measures to retrofit existing buildings may be the largest near term efficiency potential.

In the industrial sector, use of better sensors coupled to smart controls should continue to have a major impact on many processes, not only on their energy efficiency but also on the efficient use of other production factors. Better management of heat flows, e.g., by using the pinch point method (Linnhoff and Vredeveld, 1984) and innovative cogeneration technologies promise significant energy savings.

Finally, power electronics is a cross-cutting area of research which should improve efficiency of electricity use, e.g., in the optimum control of motors. Advances in microelectronics, sensors and smart process control and in materials have much to do with our optimism about success of many of these areas of R&D. For most of the end use, R&D opportunities listed in Fig. 7, efficiency improvements of 20-50% over best available technology today seem likely.

R&D can also improve the potential for fuel switching, although natural gas is a resource limited in magnitude and availability.

Several examples of fuel switching are analyzed in Fig. 8. These include: (1) replacement of an advanced electric heat pump

FIG. 8 - FUEL SWITCHING EXAMPLES

SUBSTITUTIONS	REDUCTION IN CO_2 EMISSIONS (GTC/QUAD OF GAS)	REDUCTION COMPARED TO BURNING ONE QUAD OF COAL
A) BUILDINGS		
GAS HEAT PUMP* FOR ELECTRIC HEAT PUMP*	0.013	52%
B) TRANSPORTATION		
CNG FOR GASOLINE	0.008	31%
METHANOL FOR GASOLINE	0.0006	2%
C) ELECTRICITY GENERATION		
GAS TURBINE (47%) FOR COAL (37%)	0.015	59%

*ASSUMING EQUAL HEATING AND COOLING LOADS AND THE FOLLOWING COEFFICIENTS OF PERFORMANCE (COP): GAS ABSORPTION HEAT PUMP: COP = 2.5 FOR HEATING, 1.5 FOR COOLING; ELECTRIC HEAT PUMP: COP = 3.0 FOR HEATING, 4.0 FOR COOLING. CO_2 EMISSIONS FROM ELECTRICITY GENERATION WRE COMPUTED FOR THE 1987 U.S. MIX OF ELECTRICITY SOURCES; REGIONAL DIFFERENCES, E.G. MUCH LARGER PERCENTAGE CONTRIBUTIONS FROM HYDRO OR NUCLEAR STATIONS, COULD ALTER THE EFFECT OF SUBSTITUTIONS IN PARTICULAR REGIONS.

with a gas absorption heat pump; (2) using CNG or methanol derived from natural gas to replace gasoline in automobiles; and (3) replacing a pulverized coal power plant with an aeroderivative gas combustion turbine, the so-called intercooled steam injected, gas turbine or ISTIG described by Williams & Larson (Williams & Larson, 1988). These examples are chosen to represent a range of uses of natural gas to reduce CO_2 emissions. The results show that the largest reductions in CO_2 emission can be obtained by ISTIG and by absorption heat pumps. Natural gas converted to methanol for vehicles may perhaps be useful to reduce ozone, CO or particulate emissions in cities but not to reduce CO_2. CNG is a better choice.

Adoption of high-efficiency

It is one thing to develop the technologies for much more efficient use of energy or for fuel switching. It is quite another to achieve sufficient deployment to make a difference. We know pitifully little about the dynamics of the commercial penetration of new technology. We do not fully understand and are inadequately equipped to deal with various market barriers and imperfections.

Barriers include the difficulties of obtaining credible information about the performance of new products and expert assistance for their adoption. In other words, substantial transaction costs may bar the use of new, more efficient products. The effects of some of these transaction costs can be overcome by innovative institutional arrangements. Utilities are getting more and more involved in least cost planning. The costs of helping consumers invest in improvements to reduce demand are compared to the costs of additional generating capacity and the least expensive alternatives are picked to meet rising energy demands. ORNL experience with evaluating gas and electric utility residential programs is that cost effective efficiency improvements can yield 15-30% energy savings. These evaluations involve only a few utilities located mostly in the northern parts of the country, so the generality of our observations is quite limited.

Perhaps the most important barrier is the propensity of consumers to select least first cost alternatives rather than least life cycle cost ones. The uncertainty about future fuel prices compounds the problem of calculating future savings and payback. Additionally, life cycle costs may only weakly depend on efficiency so that fuel saving is a weak force in selection of technologies. For all of these causes, efficiency improvement in practice will lag what may be economically optimum from the societal viewpoint.

The final factor influencing the adoption of higher efficiency technology is the fuel price feedback. The more universally more efficient technology is applied, the lower fuel prices will become. The adjustment to lower prices will inhibit continued decrease in demand until a new equilibrium is established, lower to be sure, but with considerable takeback. Pressure to continue reducing demand can come by additional technological improvements and/or by government actions such as those proposed recently at ACEEE in the book, Energy Efficiency: A New Agenda (Chandler, Geller & Ledbetter, 1988) and in a variety of bills introduced during the last session of Congress (U.S. Congress, 1988).

In Summary

1. Prevention of the greenhouse effect by reducing CO_2 emissions will be difficult to achieve.

2. In the near to mid-term (the next two decades) efficiency improvements and some fuel switching are likely to be the principal available strategies.

3. Non-fossil sources are just not ready to be deployed at reasonable costs and at large scale but will be essential in the long term for sustainable control of CO_2 emissions. Efficiency improvements and fuel switching can provide the time to do the R&D needed to improve and develop better non-fossil sources.

4. The technological opportunities for efficiency improvements and fuel switching are very large and can be substantially increased by R&D.

5. Achieving high efficiency of energy use near that which is technically feasible will be difficult because of various market barriers and imperfections and because achieving the necessary level of efficiency to control CO_2 emissions adequately may not be economic since efficiency will drive fuel prices down. Government policies will be required.

6. CO_2 emissions from developing nations will exceed those by industrialized nations early in the first decade of the next century if current trends continue. What developing nations do about cost effective, more efficient use of energy is critical to the prevention strategy.

REFERENCES:

British Petroleum Company, 1988, "BP Statistical Review of World Energy," June, 1988.

Chandler, William U., Howard S. Geller, and Marc Ledbetter, 1988. Energy Efficiency: A New Agenda, American Council for an Energy-Efficient Economy, July, 1988.

DOE, 1987. Five Year Research Plan 1987-1991 National Photovoltaics Program, U.S. Department of Energy, Photovoltaic Energy Technology Division, May 1987.

Edmonds, J., and J. Reilly, 1986. The IEA/ORAU Long-Term Global Energy--CO_2 Model. Personal Computer Version A84 PC, ORNL/CDIC-16, CMP-002/PC, Oak Ridge National Laboratory, Oak Ridge, Tenn.

Firor, J., 1988. "Public Policy and the Airborne Fraction," Climatic Change 12, 103-105.

Goldemberg, J., T. B. Johansson, A. K. N. Reddy, and R. H. Williams, 1988. Energy for a Sustainable World, Wiley Eastern, Ltd., New Delhi; also Energy for a Sustainable World, 1987, World Resources Institute, Washington, D.C.; and Energy forDevelopment, 1987, World Resources Institute, Washington, D.C.

IIASA (International Institute of Applied Systems Analysis), 1981. Energy in a Finite World: A Global Systems Analysis, Report by the Energy Systems Program Group, Wolf Hafele, Project Leader, Ballinger, Cambridge.

Linnhoff, B. and D. R. Vredeveld, 1984, "Pinch Technology Has Come of Age," Chem. Eng. Prog., July, p. 33.

Marland, G. and R. M. Rotty, 1983, "Carbon Dioxide Emissions from Fossil Fuels: A Procedure for Estimation and Results for 1950-1981," Department of Energy Report DOE/NBB-0036 (TR 003), June, 1983.

Perry, A. M., 1984. Atmospheric Retention of Anthropogenic CO_2: Scenario Dependence of the Airborne Fraction, EPRI Report EA-3466, Electric Power Research Institute.

U.S. Congress, 1988. Proposed CO_2 Legislation includes: S 2614: National Global Change Research Act of 1988 by Senator Ernest Hollings, Democrat, South Carolina designed to increase the coordination of the federal agencies working on global change; S 2666: Global Environmental Protection Act of 1988 calls for cuts in CO_2 by the year 2000 and is authored by Senator Robert

REFERENCES (Continued)

Stafford, Republican, Vermont; S 2667: National Energy Policy Act of 1988 by Senator Tim Wirth, Democrat, Colorado and 15 other senators is a comprehensive piece of legislation aimed at reducing the global greenhouse effect by efficiency improvement, inherently safe nuclear power, and other means. The companion measure in the House is HR 5380 introduced by Representative Les AuCoin, Democrat, Oregon; HR 5460: Global Warming Prevention Act of 1988 introduced by Representative Claudine Schneider, Republican, Rhode Island and 31 colleagues has some similarities to the Wirth bill but without nuclear R&D.

Williams. R. H., 1987, "A Low Energy Future for the United States," <u>Energy 12</u>, 929-44.

Williams, R. H. and E. D. Larson, 1988. "Aircraft Derivative Turbines for Stationary Power", PO/CEES Report #226, The Center for Energy and Environmental Studies, Princeton University, Princeton, New Jersey.

World Energy Conference, 1983. <u>Energy 2000-2020: World Prospects and Regional Stresses</u>, ed. J. R. Frisch; Graham and Trotman.

CARBON EMISSIONS TRENDS IN CANADIAN TRANSPORTATION

By Philip S. Jessup

Introduction

Transportation is the largest source of Canada's carbon emissions. In 1985, Canadian cars, trucks, and other vehicles produced 34.5 megatonnes of carbon, about 30 percent of Canada's total emissions of 118 megatonnes. Industry was next, with 33.3 megatonnes. Electricity generation, by comparison, produced 25.4 megatonnes of carbon. About 80 percent of transport's carbon emissions come from personal and commercial light-duty vehicles, with the balance from heavy-duty vehicles like trucks and buses, with minor contributions from marine, rail, and air transport.

If Canada should embark in the future on an effort to reduce carbon emissions, policymakers would do well to look at transportation as a sector where significant reductions might be achieved quickly. The main reason is that the rapid turnover of vehicles and their high energy use presents significant opportunities for reducing per capita consumption of petroleum products for energy, assuming consumers replace their old cars with more fuel efficient new cars.

Another reason is that significant improvements in the efficiency of automobiles produced by North American manufacturers have already occurred over the past 15 years, the result of consumer reaction to rising gasoline prices and manufacturer response to both changing consumer preferences and federal fuel efficiency standards in both the U.S. and Canada. Technological advances should permit further advances in new car efficiency over the next fifteen years.

This paper asks three questions:

- What are the prospects for carbon emission from light-duty vehicles towards 2005, should petroleum prices remain stable and the federal government shy away from new fuel efficiency standards?
- What are the prospects for achieving significant carbon emissions reductions from light-duty vehicles by 2005?
- What policy options should government explore to achieve such reductions?

The information presented here comes from Statistics Canada's Socio-Economic Resource Framework (SERF) database. In addition to providing historical data for the period 1961-1981, the database simulated two scenarios.

The first "business as usual" scenario projects gasoline consumption—a proxy for carbon emissions—by personal and commercial light-duty vehicles up to the year 2005, assuming that current fuel efficiency ratings of new cars and light-duty trucks remain constant at 27.5 miles per gallon after 1988.

The second "fuel efficiency" scenario projects gasoline consumption up to the year 2005, but assumes the efficiency of new vehicles—after remaining constant from 1988-1992 at 27.5 miles per gallon—improves after 1993 at an average annual rate of 5.8 percent. This is the rate that Canada's voluntary Corporate Average Fleet Economy (CAFE) standards asked manufacturers to meet during the 1980s.

Description of SERF

The Socio-Economic Resource Framework (SERF) database was developed by Statistics Canada in 1980 to describe and simulate physical attributes of the Canadian economy, such as population, households, and consumer purchases.

This report is based on information and simulations made by SERF's "automobile transport calculator," whose data comes from four sources: provincial annual new car registrations; annual sales of new autos reported by manufacturers; auto ownership information from Statistics Canada's annual "households facilities survey;" and Statistics Canada's annual "fuel consumption survey" begun in 1979. The primary strengths of the auto database are the physical information it contains about automobiles being bought and driven in Canada from year to year and its ability to link each year's auto stock and driving trends to observable population and labour force trends.

A word about terminology. The car stock discussed here is divided into two categories: compact and standard sizes. Compact size includes compacts, sub-compacts and minicars. Standard size includes large and intermediate cars, as well as light trucks and minivans. Compact and standard stocks include vehicles used for personal and commercial use.

Historical Trends, 1961-81

New vehicle purchases and driving habits in recent decades have been strongly related to two post-World War trends: the baby boom and growth of the two-income household. In Canada, since 1970 the proportion of the driving age (16 or older) population has been increasing, and the female labour force has been climbing faster than the male labour force. The result was been a more rapid increase in the labour force in the 1970s than the 1960s and the growth in multiple-car households.

These trends have helped promote a steady increase in the number of autos in Canada. Indeed, Canada's personal auto stock has been climbing faster than households over the 1961-1981 period. Since 1976 there have been more personal autos in Canada than households. In addition, the total distance Canadians drive their cars, both for personal and commercial uses, has steadily increased, while the average mileage each per person aged 16 drives their car has also risen over the period.

These socio-economic trends produced steadily rising gasoline consumption in the 1960s. The price effects of the 1973 and 1981 OPEC oil shocks, coupled with new Canadian road taxes, and U.S. and Canadian Corporate Average Fleet Economy (CAFE) standards caused gasoline consumption to level off in the late 1970s and begin declining in the early 1980s, as more fuel efficient cars went on the road.

Two Scenarios, 1981-2005

The SERF model ran two simulations for the period 1981-2005. The first "business as usual" scenario assumes that the weighted average fuel efficiency of new compact and standard cars sold each year would meet Canada's voluntary CAFE standards for 1981-87. (Experience shows they did.) From 1988-2002, the fuel efficiency of new cars was held constant at 29 mpg for compacts and 21 mpg for standard size cars, an average of about 27.5 mpg. Therefore, the first scenario assumes no new standards or dramatic price shifts would occur. (The National Energy Board of Canada projects relatively flat gasoline prices out to the 2005).

This "business as usual" scenario projects gasoline consumption will rise to 892 petajoules in 2005, up 23 percent from 727 petajoules in 1988.

The second "fuel efficiency" scenario factors a renewed market for fuel efficient autos in the mid-1990s into the baseline case. Beginning in 1993, each year's new auto stock is assumed to improve in average fleet economy about 5.8% per annum, the rate of improvement Canada's voluntary CAFE standards that called for in the early 1980s. This scenario assumes that the design and technologies necessary for such improvements are or will be commercially feasible soon and that manufacturers will feel compelled, either by market incentives from new fuel taxes or by new standards, to produce such vehicles.

In this second scenario, therefore, manufacturers would produce vehicles with the following average fuel efficiency ratings.

	1995	**2000**	**2005**
Compacts	35 mpg	47	64
Standards	26	35	47

The "fuel efficiency" scenario projects there will be a significant reduction in gasoline fuel consumption by the year 2005—537 petajoules of gasoline, a 26 percent reduction from 1988, or a 49 percent reduction from gasoline consumption in 2005 projected by the baseline scenario.

Such fuel efficiency ratings are achievable. A number of automobiles with such fuel economy ratings are already available commercially, mostly in Japan and Europe. They are generally sub-compacts or minicars, and have engine displacements lower than two litres. Most of the Japanese micro-minicars were introduced there in response to consumer demand for second cars that were inexpensive and easy to operate in traffic-clogged Japanese cities.

More advanced prototype cars designed mostly by European manufacturers with maximum fuel economy as a primary goal also show much promise. They include Volvo's LCP project, for instance, which designed a car to appeal to consumers that could be mass produced, with a city fuel economy of 63 mpg and a highway fuel economy of 81 mpg. Despite the LCP's high use of magnesium to lower weight, Volvo engineers projected a "break even point" of 20,000 units per year and prices comparable to today's.

It's likely, however, that major new initiatives by government, in the form of fuel taxes and/or standards will be necessary to shape a future strong market for fuel efficient automobiles.

Policy Options

Economists who have studied the "elasticity" relationship between the price of gasoline and gasoline fuel consumption say that consumers respond to increased fuel prices by buying more fuel efficient cars. One study of 25 countries suggested that long-term fuel price elasticities for auto fuel economy are about -0.33, that is, a 10 percent increase in fuel price was associated with a 3 percent improvement in fuel economy. Another suggested that long-term elasticities were higher, about 0.5 percent. A 10 percent increase in fuel price leads to a 5 percent improvement in fuel economy of new cars.

Studies of Canadian gasoline price elasticities show a wide range of results. Studies respectively showed long-term gasoline demand price elasticities from -0.5 to 1.04 in a recent study of Quebec gasoline consumption for the period 1962-1982 at Université de Laval. The Laval study showed that a 10 percent increase in gasoline price brought a corresponding 10 percent improvement in fuel economy of new automobiles.

If the range of long-term from -0.33 to 1.04 were incorporated into the "fuel efficiency" scenario in the form an annual fuel tax rise beginning in 1993, an annual fuel tax increase ranging from 5 to 17 percent would be required to drive the scenario, assuming constant petroleum prices. The price of gasoline in Canada would range in 2005, as a result, from about $1.01 per litre or $3.85 per U.S. gallon—about twice today's price—using to about $4.15 per litre or $15.74 gallon, or about 8 times today's price.

Apart from the wide range of observed results they report, these elasticity studies have one main weakness: they do not account very well for the influence of national fuel economy standards on new car stocks. In other words, the cars which manufacturers offered for sale beginning in 1978 not only sought to meet changing consumer preferences, but to satisfy federally mandated standards in the U.S. Although Canadian voluntary standards followed several years later, the U.S. standards no doubt had a spillover effect in Canada after 1978, due to the integration of the North American auto industry under the free trade agreement known as the U.S.-Canada Auto Pact. It is during this same period that marked correlations between price fluctuations and improvements in new car fuel efficiency in the U.S. and Canada were observed and reported by the various studies.

Another uncertainty relates to the impact of increasing fuel efficiency on the cost of gasoline as a part of the cost of owning and operating a car. The contribution of fuel purchases to the cost of driving diminishes as fuel economy increases. Particularly beyond 35 mpg increases in the price of gasoline by orders of magnitude do not significantly affect the proportion of cost attributed to fuel, therefore, the gasoline price incentive to buy more fuel efficient autos weakens as a result.

Despite these uncertainties, there seems little doubt that consumers would purchase more fuel efficient autos if the price of gasoline were to rise significantly, either because of fuel shortages or deliberate government policies. How certainly fuel tax policies would actually work or how politically palatable they would be to a public generally opposed to tax hikes is very uncertain, however.

Another policy approach would be for the federal government to mandate new fuel economy standards for automobiles, light trucks, and vans, to take effect in the early 1990s. A strong argument for such standards is they worked in the past when gasoline price rose, and they would probably work in the future, especially if combined with new fuel taxes.

Such standards have flaws, however. When CAFE standards mandated by the U.S. went into effect in 1978, for instance, they discriminated against American manufacturers, who customarily carried a full line of cars, and in favour of Japanese manufacturers, who already specialized in small, more fuel efficient cars. In addition, after 1979 manufacturers were not allowed to include domestic sponsored imports in their domestic fleet average. The intention of the stringent American law, which proved burdensome to American manufacturers, was to force domestic manufacturers to develop the capacity to manufacture fuel efficient cars at home, so jobs wouldn't be shipped overseas.

The Canadian CAFE standards went into effect in 1980, pretty much following the lead of the U.S. by adopting the same fuel economy goals. Canada's CAFE program is more flexible than the American CAFE program. Apart from its voluntary nature, it allows Canadian manufacturers to calculate corporate average fleet economies irrespective of the country of origin of the automobile, and there is no gas guzzler tax, as there is under the American system. However, unlike the U.S. Canada has not rolled back its 1986 and 1987 standards to 26 mpg from 27.5 mpg.

If government should turn to new fuel economy standards to achieve carbon reductions in the 1990s, they should be carefully designed to minimize the problems which occurred, especially in the U.S., under CAFE. Cars are now manufactured with parts from all over the world, and the distinction between domestic and domestic sponsored imports is no longer as strong as it once was. Manufacturers should have more flexibility to meet standards with a variety of models regardless of country of manufacture, and the standards should be designed to avoid discriminating in favour of companies that specialize in small cars, for instance, with fuel efficiency goals based on interior volume.

Even better, standards and fuel taxes could be designed to work together. For instance, gradual increases in annual fuel taxes intended to increase demand for fuel efficient cars could be levied, but if they didn't achieve the desired goals that had been set, mandatory standards based on the goals of the fuel tax program would take effect, with tax penalties on car models to follow. Or vice versa: voluntary standards could be set by government for manufacturers, with fuel taxes taking effect in years that manufacturers were unable to meet the goals.

ENERGY EFFICIENCY: A NEW AGENDA

William U. Chandler
Howard S. Geller
Marc R. Ledbetter

American Council for an Energy-Efficient Economy
1001 Connecticut Avenue, N.W.
Washington, D.C. 20036 USA

Three major priorities in the U.S. and elsewhere--environmental quality, economic competitiveness, and energy security--provide a new and urgent rationale for saving energy. None of these ends can be fully attained without also achieving an energy-efficient economy.

Environmental protection demands energy efficiency. Energy use is the principal factor driving climatic change. At least three-fourths of the carbon which remains in the atmosphere each year originates in fossil fuel combustion. Fortunately, energy efficiency improvements can greatly slow the rate of warming over the next 100 years.{1}

Economic competitiveness also demands greater energy efficiency. Cutting energy costs in manufacturing would permit companies to produce goods more cheaply, making them, by definition, more competitive. Nevertheless, U.S. industry has not cut its energy intensity as quickly or as much as its trade competitors, especially Japan, West Germany, and the United Kingdom.{2}

The large U.S. trade deficit demands energy efficiency. Oil imports now account for about one-quarter of the U.S. merchandise trade deficit. If government projections are correct, the annual U.S. oil import bill of about $40 billion could increase to $100 billion per year by 2000.{3} Efficiency improvements can greatly cut this serious drain on our economy.

REGAINING MOMENTUM

The United States has made impressive achievements in energy efficiency since the 1973 oil price shock. The nation has reduced energy intensity in every major energy-consuming sector. Industry has made the largest gains, cutting energy requirements per unit of output by 30 percent between 1973 and 1984.{4} Households cut energy use by about one-fifth over the same period, while owners and operators of commercial buildings cut energy intensity by more than one-tenth. And the fuel economy of automobiles improved by 35 percent between 1973 and 1985.{5}

The country, however, is losing momentum in energy efficiency. The United States reduced energy use per dollar of output by an average of 2.7 percent per year between 1976-86. But the energy intensity of the entire U.S economy did not fall in 1987, and energy use rose over 7 percent between the first quarter of 1987 and the first quarter of 1988.{6}

There are two main reasons for the slowdown in efficiency improvements. First, energy prices in general and the price of oil derivatives in particular have fallen in recent years. Corrected for inflation, the average price of gasoline in 1987 was nearly half that in 1980. Natural gas and electricity rates have also fallen. The price impetus for consumers to conserve has thus been diminished.

Second, the federal government substantially reduced its support for conservation during the 1980s. Automobile fuel economy standards were rolled back; conservation research funding and federal support for state and local conservation programs were dramatically reduced; and implementation of national building and appliance efficiency standards was resisted.

U.S. leadership has failed to fill the policy vacuum formed by unstable energy prices. Today's energy market does not reflect high security, economic, and environmental costs, and government has not moved to correct this deficiency. Filling the void requires a reasonable goal for energy efficiency and a complementary set of policies for achieving the goal.

POLICY RECOMMENDATIONS

Energy efficiency offers U.S. leaders a practical strategy for ameliorating the environmental, economic, security risks related to energy waste. No new energy supply option competes with efficiency in terms of cost, effectiveness, or environmental acceptability. For these reasons, we suggest that the United States set an overall goal of reducing its energy intensity--the rate of energy used per dollar of economic output--by at least 2.5 percent per year well into the next century. This will mean holding energy use constant or achieving a small decrease in total energy use by 2000 if the economy grows at 2 to 2.5 percent per year as now anticipated. To provide policymakers with a means of converting energy-efficiency's potential into reality, we propose 21 specific policies designed to accomplish the policy priorities set out. Specific policy proposals follow.

These policy priorities have been selected for four reasons. First, they address problems that the marketplace cannot completely solve. Second, they represent the largest of many opportunities to save energy and accomplish other objectives cost effectively. Third, they ameliorate critical problems that energy use causes for the global environment, the economy, and national security. And fourth, they are compatible with efforts to reduce the federal deficit.

HOW CAN ENERGY EFFICIENCY PROTECT THE ENVIRONMENT?

Because unchecked fuel use could irrevocably alter the global environment, we urge the United States to sponsor the first and most important of a series of protocols to control greenhouse gas emissions: An agreement among nations to reduce the energy intensity of the world economy. A reasonable and effective goal is to reduce global energy use per unit of economic output by roughly two percent per year for at least the next three decades. This could be achieved if most countries reach the 1985 efficiency levels of Denmark, Switzerland, or Japan by the year 2025.{7}

In conjunction with this agreement, we propose a major cooperative effort between the United States, the Soviet Union, and OECD countries to share research and information on technologies for improving energy efficiency. The nations should sponsor exchanges of energy experts from research institutes, universities and industry, focussing on technologies for saving energy, means of fostering energy-efficiency policy, and application of conservation to ameliorate environmental problems.

We also propose that urgent research be undertaken to develop technologies that simultaneously save energy, reduce greenhouse warming, and cut the use of ozone-depleting chlorofluorocarbons (CFCs). Roughly 40 percent of CFCs produced in the U.S. are used energy-related applications such as refrigeration equipment and foam insulation. Presently available CFC substitutes are less efficient; i.e., they would lead to an increase in energy use.{8} This research is needed to avoid solving one environmental problem (stratospheric ozone depletion) while exacerbating others (through increased energy use).

WHAT SHOULD BE DONE TO CUT OIL IMPORTS?

Cars and light trucks account for one-third of U.S. petroleum consumption. Even though the fuel economy of cars has improved since the 1973 oil crisis, fuel economy can be cost-effectively doubled with existing technology.{9}

A complementary set of policies should be established to double automobile and light truck fuel economy. Economic incentives are needed to pull the market; performance standards are needed to push it. Specifically, we propose a new transportation fuels tax to more nearly reflect the real, long-run cost of consuming oil. We suggest increasing federal gasoline and diesel fuel taxes by 10 cents per gallon yearly for at least three years. At the same time, a portion of the revenue collected as a result of this tax increase should be rebated to low-income households.

Second, we propose new fuel economy standards that will raise the mileage of new cars and light trucks to 45 and 35 miles-per-gallon, respectively, by the end of the century. And finally, we propose an expanded gas-guzzler tax along with rebates provided to buyers of highly efficient cars produced in North America. The latter will encourage the production of innovative vehicles that achieve a high level of fuel economy and meet other criteria, such as safety and low emissions.

HOW CAN THE UNITED STATES REDUCE THE COSTS OF UTILITY SERVICES?

Utilities accounted for nearly 15 percent of total investment in new plant and equipment in recent years. Moderating growth in electricity and natural gas demand would lower this level of investment, thereby reducing the risk of over-building expensive plant capacity and lowering the cost of capital. Likewise, cutting the cost of heat, light, refrigeration, and other energy services saves consumers money, and helps the United States become more competitive.

To reduce investment by utilities and lower the cost of energy services, state utility commissions should redesign financial incentives provided to the utilities. Specifically, state utility commissions should offer financial incentives to public utilities based on their achievements in providing utility services at the least cost. This will encourage utilities to invest in cost- effective sources of generating capacity and end-use efficiency. The federal government can assist utilities and states by providing analytical support and information.

We also recommend that efficiency investments be allowed to compete with supply options when a utility obtains new resources through bidding procedures, that long-term interstate power sales be made consistent with least-cost utility plans, and that federal support be provided for experiments and evaluation to improve efficiency programs.

HOW CAN GOVERNMENT SUPPORT INDUSTRY'S EFFORTS TO BECOME MORE COMPETITIVE?

Despite impressive gains in the past decade, American industry is less energy efficient than most of its major competitors. And although the U.S. is pre-eminent in basic science, it lags in applying scientific breakthroughs and commercializing innovative processes and technologies. Improving the energy efficiency of American products and processes will make our nation more competitive and open new markets for American goods.

First, we recommend a renewed federal government commitment to cooperate and share the cost with industry of researching, developing, and demonstrating energy-saving products and processes. This cooperative effort can be conducted by expanding the U.S. Department of Energy's conservation research and development program. During the past seven years, the budget for this program was cut more than 50 percent. This occurred despite many successful R&D projects which will eventually yield billions of dollars of savings.{10}

Second, we propose that the federal government together with private industry create research centers in energy-intensive industrial processes. These centers would investigate common industrial processes such as metal casting or chemicals separation--key elements of our industrial infrastructure. The centers would not strictly focus on energy efficiency but would look towards innovations with multiple benefits, thereby increasing the likelihood of commercial application.

HOW CAN BUILDINGS BE MADE MORE EFFICIENT AND AFFORDABLE?

We propose a variety of policies to make buildings more efficient. First, we recommend efficiency standards for lighting products. This priority derives from the fact that lighting consumes one-fifth of all U.S. electricity, and cost-effective savings are readily available. Minimum efficiency standards on incandescent and fluorescent lamps could save consumers about $2 billion per year and reduce peak electric demand by 5,000 megawatts by the year 2000.{11}

Second, we recommend increased funding for low-income housing weatherization programs. We also recognize the difficulties this program has encountered and suggest ways of increasing its effectiveness. For example, research and technical assistance has helped some states to increase the energy savings achieved without spending more money per household.{12} Also, we urge greater reliance on non-profit energy service companies for delivering weatherization services.

Third, we propose that the federal government lead the way in energy efficiency by setting ambitious goals for energy conservation in its own buildings. The goals would be supported by special incentives, guidelines, and technical assistance.

Finally, we propose expanding federal technical assistance for a variety of state and local conservation efforts. One area is assistance to utilities as well as state and local governments adopting building rating and information programs. Another area is technical and managerial support for state and local governments adopting energy-conserving building codes.

HOW SHOULD THE UNITED STATES FOSTER EFFICIENCY IN THE THIRD WORLD?

The United States has a major stake in the economic development of poor countries. Growth in developing countries means increased trade and jobs for the United States. But energy waste will stymie that growth, raise the danger of a new oil crisis, and contribute to greenhouse warming. The United States would thus benefit by speeding the transfer of technology and skills for saving energy to developing countries.

We propose exchange programs and assistance to develop lasting capabilities in energy efficiency among developing country scientists, policymakers, and industrial leaders. We also recommend changes in the energy planning and lending policies of the World Bank and the other multilateral development institutions. Specifically, we urge U.S. representatives to advocate use of least-cost planning principles, to provide financing and technical assistance for conservation, and to support policies within LDCs that will foster greater efficiency.

In summary, the United States possesses the technical ingenuity and institutions necessary to create an energy-efficient nation. But doing so requires political determination. Energy efficiency improvements will continue to slow in the absence of leadership and coordinated national policies. By adopting a broad set of policies in support of energy efficiency, the next President and Congress can help create a prosperous, secure, and environmentally sound nation. Moreover, the United States would do its part in a worldwide energy-efficiency campaign, a vital endeavor for limiting climate change and maintaining economic growth.

REFERENCES

1. I.M. Mintzer, A Matter of Degrees: The Potential for Controlling the Greenhouse Effect (Washington, D.C.: World Resources Institute, April 1987); William U. Chandler, Energy Productivity: Key to Environmental Protection and Economic Progress (Washington, D.C.: Worldwatch Institute, 1985).

2. International Energy Agency, Energy Conservation in IEA Countries (Paris: Organization for Economic Cooperation and Development, 1987).

3. U.S. Department of Energy, "Patterns of U.S. Energy Demand," DOE/PE-0076, Washington, DC, August 1987.

4. Energy Information Administration, Energy Conservation Indicators 1986, DOE/EIA-0441(86) (Washington, D.C.: U.S. Department of Energy, Feb. 1988).

5. EIA, Energy Conservation Indicators 1986.

6. Energy Information Administration, Monthly Energy Review (Washington, D.C.: U.S. Department of Energy, June 1988).

7. IEA, Energy Conservation in the IEA Countries. Also, World Resources Institute and the International Institute of Environment and Development, World Resources 1987 (New York: Basic Books, 1987).

8. S.K. Fischer, F.A. Creswick, and J. Dieckmann, "Energy-Use Impact of Chlorofluorocarbon Restrictions in Refrigeration and Buildings Applications", draft, Oak Ridge National Laboratory, Oak Ridge, TN, Nov. 1987.

9. See, generally, Office of Technology Assessment, Increased Automobile Fuel Efficiency and Synthetic Fuels (Washington, D.C.: U.S. Government Printing Office, 1982), and C. Gray and F. von Hippel, "The Fuel Economy of Light Vehicles", Scientific American, May 1981.

10. Geller, et al., "The Role of Federal Research and Development in Advancing Energy Efficiency".

11. H. Geller, American Council for an Energy-Efficient Economy, Washington, D.C., working paper.

12. M. Ternes, et al., "Field Test Evaluation of Conservation Retrofits of Low- Income Single Family Buildings in Wisconsin", ORNL/CON-228/P1, Oak Ridge National Laboratory, Oak Ridge, TN, Oct. 1987.

STATUS OF PHOTOVOLTAIC SOLAR TECHNOLOGY

Dan E. Arvizu

Sandia National Laboratories
Albuquerque, New Mexico 87185

INTRODUCTION

Historical Perspective

Although the photovoltaic effect was first observed more than 150 years ago and the first laboratory device was demonstrated more than 30 years ago, it was the energy crisis of the early 1970s that spawned a new era in U.S. photovoltaic research. At that time, national attention focused on exploring all energy options as a matter of national energy security. As a consequence of this national attention, the U.S. Department of Energy began its Photovoltaic Research Program in the mid-1970s.

At first, all the research attention was focused on improving the solar-to-electric power conversion of photovoltaic devices, also known as solar cells. Later, the emphasis of the DOE Photovoltaic Program was directed at demonstrating viability at the system level and achieving lower cost. This emphasis led to a number of system demonstrations and a concerted program to reduce the cost of raw and processed silicon material. To a large extent these two facets of the DOE's program were successful (1). Development of new large scale silane production plants and the fluidized bed reactor demonstrations did reduce silicon costs significantly. Likewise, the systems demonstrations produced evidence that photovoltaics was a reliable electric power technology (2).

To address processed silicon cost, the DOE program also began exploring alternate materials and silicon growth options. This part of the program later evolved into the present day program whose emphasis is primarily on basic materials and device research. The DOE Program today supports research on amorphous silicon and other polycrystalline thin-film materials, crystalline silicon approaches including ribbons and sheets, concentrator technology, and a core program in systems development and technology transfer. Despite declining budgets, DOE sponsored research is directed in a manner consistent with providing support for U.S. industry in order to accelerate the commercialization of photovoltaics for utility markets (3).

This work was supported by the U.S. Department of Energy, under contract DE-AC04-76DP00789.

Present Energy Climate

The present photovoltaic markets are in consumer products and remote power systems. In these markets the value of electricity is very high and photovoltaics today offers the best, most reliable source of energy. These worldwide markets presently represent a 25 megawatt business that is growing steadily every year. However, until photovoltaics enters the residential and utility power markets it will not have a significant impact as an electric power technology.

Today there is no utility requirement for new generating capacity since utilities either have more generating capacity than required or they are able to buy inexpensive electricity from other utilities who have excess energy to sell. Although energy prices remain relatively low, industry planners are concerned that the current supply and demand situation will change abruptly. Their concerns arise because, as the demand for electricity continues to grow, the U.S. still has a significant dependence on foreign oil.

Utility power forecasters are currently predicting that significant new generation capacity will be required in the mid-1990s. Best estimates are that 55,000 megawatts or more of new electrical generation capacity will be required by 1995 (4). Since much of this new generating capacity is not presently under construction, the choices for what technology will be available is limited to those approaches with short deployment periods. Utility planners make it clear that whenever the time does come for the utilities to add new generating capacity, the technology that makes the best business sense for their needs will be used. For example, a modular power plant that can grow as requirements grow will be preferable to a large power plant that requires a large initial capital investment and years of development time.

Finally, new sources of energy must contend with a growing national concern over safety and environmental impacts. In the past, insufficient attention has been given to long-term harmful effects of energy supply options on our environment. However, the scientific community, in response to a growing public concern, is now beginning to evaluate these impacts and their long-term costs, for appropriately assessing our future energy supply options and their real costs (5).

New Opportunity

A tremendous opportunity for photovoltaic technology appears on the horizon if one considers the following points. 1) A large utility need for new generating capacity appears to be eminent, 2) photovoltaic technology has been successfully demonstrated in utility scale projects, 3) target costs required for photovoltaic systems' to enter the utility markets are within sight, 4) the environmental impacts and associated costs for fossil and nuclear options are substantial, and 5) research results continue to promise improved cost-effectiveness of photovoltaic technology. It should also be pointed out that photovoltaic technology offers modular power and an excellent match for providing power when many utilities have their peak demands. During these peaking periods the value of electricity is sometimes double that of electricity for off-peak conditions and for many utilities this value is near 12 ¢/kWh. Utility estimates are that 70 gigawatts of new peaking power will be required by the year 2000.

TECHNOLOGY ASSESSMENT

Because of the lead time required to build a manufacturing plant and to put the plant in full-scale operation, it is necessary to have a technology identified today in order to be ready for the mid-1990 opportunities. This situation provides crystalline silicon based photovoltaic technology a distinct advantage over other photovoltaic materials that are currently part of the broad based U.S. Photovoltaic Research Program (6). For this reason, this paper will concentrate on the prospects of crystalline silicon technology, recognizing that as other photovoltaic options mature, it is possible that the most cost-effective photovoltaic technology may well include one of the many thin-film materials presently under evaluation.

Crystalline silicon photovoltaic technology has matured to the point where it is now possible to make reasonable economic assessments of the technology potential. In particular, we now can rely on data from actual operating experience of installed systems. Furthermore, a correlation between advances in the laboratory and progress in improving the cost-effectiveness of commercial photovoltaic systems is emerging. See Figure 1. From this historical account of silicon cell efficiency, one can observe that there is a four to five year lag before laboratory cell achievements find their way into production. In addition, we can estimate the efficiency compromises necessary in a production versus laboratory environment.

Field Experience

Experience with utility scale crystalline silicon photovoltaic systems has been very favorable. Most design problems identified in first generation systems have been corrected and subsequent generation installations have performed admirably. A further point to make concerning systems experience is that systems architecture and designs have been developed and checked in the field for low cost and reliability.

Figure 2 graphically presents an assessment of the technology potential based on operating experience of some of the major photovoltaic system installations since 1980. Along the x-axis of Figure 2 nineteen systems are shown at the approximate time of their installation. The system cost, given in levelized energy cost, is given by the left y-axis while the system performance, efficiency, is given by the right y-axis. System costs are based on the total known cost of the project, with adjustments made for the research and development effort involved. The system performance is based on the system efficiency measured at various times after installation. Assumptions are described in reference (7).

Present Economics

In the past few years, a variety of economic analyses have been performed for photovoltaic systems. One such analysis, performed at Sandia focuses on the primary main cost component, the photovoltaic module. Since area and power related balance-of-system costs for fixed and passively tracking systems are, at present, approaching the necessary cost targets for cost-effectiveness in a utility market, they have been assumed constant at the target values in the analysis. Including further assumptions given in reference (6), Figure 3 presents system levelized energy cost parametrically as a function of module cost and performance (right hand and top axes), the cross-hatched area is representative of today's third generation photovoltaic technology.

For the bottom and left hand axes of Figure 3 an assumption that approximately one-half of the module cost can be attributed to the cells has been made. Present flat-plate technology has typical average production cell efficiencies in the 13 to 15% range. Likewise, cell costs in the 2 to 2.5 ¢/cm^2 range are possible under full capacity production. Since module fabrication costs are expected to be below \$100/m^2, the real challenge for flat-plate technology is to produce cells with improved efficiency, near 16%, at less than 1 ¢/cm^2. This efficiency improvement will reduce the levelized energy cost of flat-plate systems by a factor of two from today's systems, and thus become attractive in utility peak-load markets.

Research Status and Direction

In the past ten years, the U.S. photovoltaic program has made excellent progress in the research laboratory. Figure 4 shows this progress for all photovoltaic cell materials. At present our emphasis is on long-range, high-risk research. However, the outstanding progress in the laboratory needs only to be effectively transferred to industry in order for photovoltaics to contribute in the near-term electric power markets. Unfortunately, this transfer process will take significant government support if we expect photovoltaics to contribute to our electricity requirements before the year 2000. With U.S. photovoltaic budgets now at a ten year low (\$35M versus a high of \$150M) and Japanese and European governments now spending more per year than the U.S., it is imperative that we once again put renewable energy, including photovoltaics, on the national energy agenda, and address the task of commercialization.

SUMMARY

Photovoltaics has made significant progress and is now ready to contribute in U.S. utility markets. Performance is no longer the key question, instead, economies of scale and industrial competitiveness are the primary issues. These issues will need attention in order to accelerate the use of photovoltaics as a power source that will address some of the nation's and the world's most pressing energy needs.

REFERENCES

[1] "Eleven Years of Progress", Final Contract Reports of the Jet Propulsion Laboratory to the DOE, 1987.

[2] D. F. Menicucci & A. V. Poore, Today's Photovoltaic Systems: An Evaluation of Their Performance, U. S. DOE and Sandia Report, 1988.

[3] U.S. Department of Energy, Five Year Research Plan 1987-1991, National Photovoltaic Program, April 1987, DOE/Ch 10093-7.

[4] National Electric Reliability Council, Reliability Review, 1986.
[5] H. Hohmeyer, "Social Costs of Energy Consumption", Contract Report to the Commission of the European Communities, Springer-Verlag, New York, 1988.
[6] D. E. Arvizu, "Crystalline Silicon Photovoltaic Cell Technology: Meeting the Challenge for Utility Power", 20th IEEE Photovoltaic Specialists Conference, Las Vegas, NV, September 17-30, 1988.
[7] H. P. Post & G. T. Noel, A Comparative Evaluation of New Low-Cost Array Field Designs for Grid-Connected Photovoltaic Systems", Proceedings of the 19th IECEC, San Francisco, August 1984.

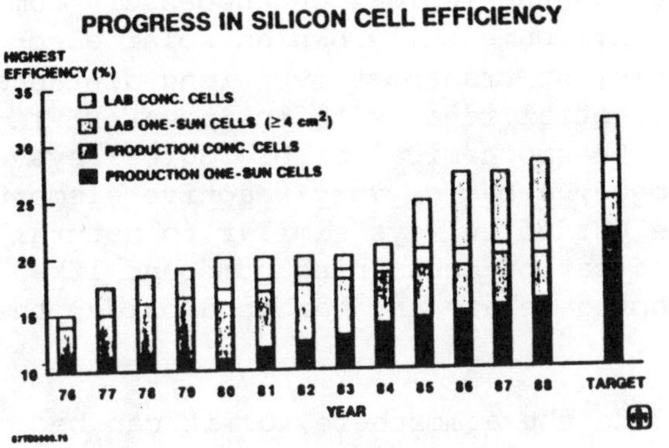

Figure 1. Silicon solar cell efficiency progress since 1976. Lab refers to research demonstrations.

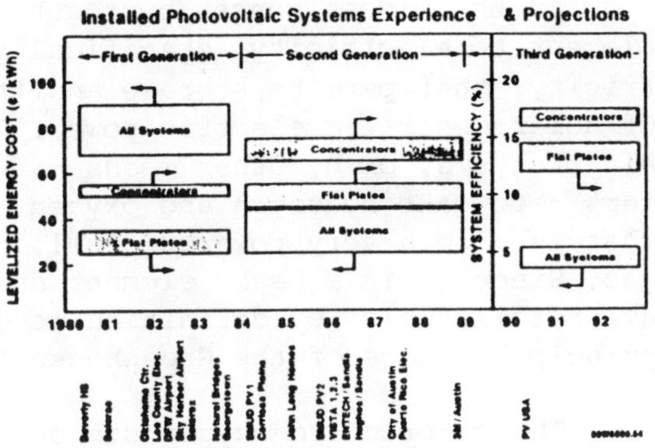

Figure 2. Photovoltaic systems experience. Cost and efficiency are estimated from Sandia data. See reference (6) for assumptions on calculations converting to levelized energy cost.

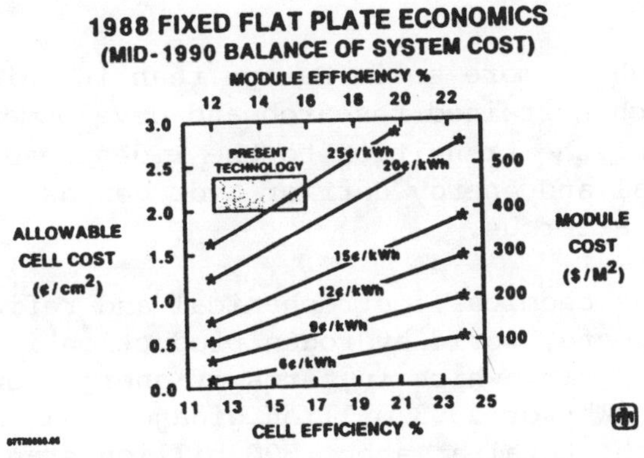

Figure 3. Levelized energy cost derived parametrically for module and cell, efficiency and cost for fixed flat plates. See ref. (6) for assumptions.

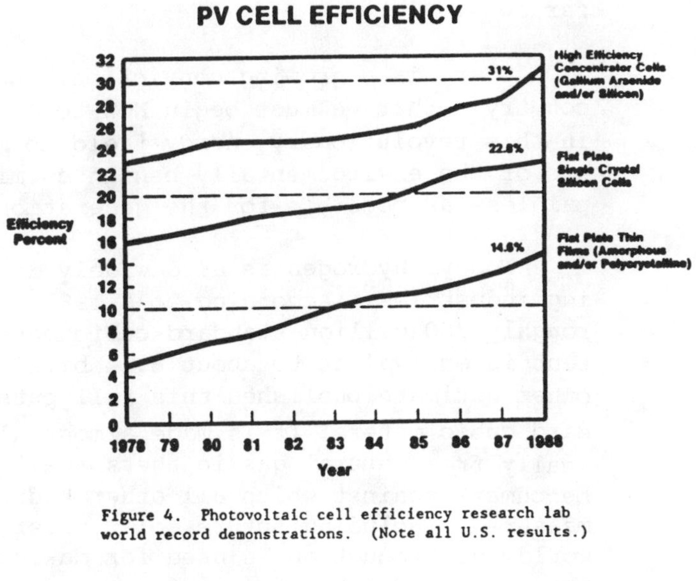

Figure 4. Photovoltaic cell efficiency research lab world record demonstrations. (Note all U.S. results.)

Potential Energy Uses and Greenhouse Implications of Hydrogen

Peter Hoffmann
The Hydrogen Letter
Hyattsville, MD 20781

Hydrogen is regarded as the ultimate long-term option when it comes to environmentally benign chemical fuels - the master fuel at the end of the line when everything else has been depleted, found unworkable or environmentally objectionable. Its introduction, more and more scientists and energy planners believe, is both unavoidable and desirable, and its widespread introduction will not only revolutionize the way we handle our energy affairs, but it would also profoundly affect our views on national, and global environmental concerns such as the Greenhouse Effect.

In the widest sense, hydrogen is seen as the logical and necessary complement to electricity, specifically, to environmentally benign solar electricity, that permits storage and the efficient transport over long distances of solar-generated electric power. The essential idea is to split water via electrolysis, or by other means - such as thermochemical or biological systems - to make hydrogen and oxygen. Hydrogen, a basic, very reactive element that is also a very powerful fuel, can be burned in ways similar to natural gas. Since it is a basic element and not a carbon-containing compound like all fossil fuels, by definition it does not generate any carbon dioxide, the principal menace of the Greenhouse Effect.

The co-produced oxygen can be vented to the atmosphere, or it can be used productively to restore and recover polluted bodies of water, for instance. Once hydrogen burns with the air's oxygen, it recombines into water vapor, condensing out into clouds and rain and returns to the natural weather cycle - an elegant, environmentally benign, regenerative energy system. However, other than in space flight and rocketry, hydrogen does not play a role in the national and the international energy arena today, and it is not likely to do so until after the turn of the century, but that's not really that far away.

There is a growing conviction, though - more so in Europe than in this country - that we must begin NOW to launch sustained research and development in this revolutionary energy field to make the transition to the widespread use of the environmentally benign chemical and energy carrier hydrogen as painless as possible for the generations after us.

Today, hydrogen is used widely in the chemical, petrochemical and refining industries. According to a 1987 estimate, world hydrogen production is roughly 350 billion standard cubic meters/year, which in terms of energy content is equivalent to about 8.14 billion kWhr or 29.3 million gigajoules; another estimate published this fall puts the total at about 500 billion standard cubic meters. It is made almost always from fossil fuels - most economically from natural gas in the so-called steam-reforming process, an economic benchmark against which all other hydrogen production processes have to be measured, including future solar water-splitting processes. About 90% of worldwide production is used for making fertilizers and in processing petroleum and other fossil fuels.

It is also widely used in the refining industry, where it is needed to break down heavy fuel fractions, for instance. It is also used as a cooling gas for large generators in the utility industry, to harden fats and oils, and as a protective atmosphere in electronics manufacture to assure surface purity of electronic materials, for instance.

Hydrogen is not an "energy source", but a so-called "energy carrier", analogous to electricity. Neither electricity nor hydrogen exist "raw" in nature, free for the asking and immediately usable. They must be "manufactured", as it were -- electricity in power plants and generators, hydrogen in chemical plants or by electrolysis, for instance, a process in which a current is run through water: hydrogen bubbles up at the negatively charged cathode, and oxygen - the stuff we breathe - at the positively charged anode. If the process is reversed, hydrogen and oxygen combine, producing energy and water: if you burn it, you get heat, plus water. If you combine it in a fuel cell, you get electricity and water. Thus, hydrogen produced from water and electricity -- photovoltaics or perhaps advanced, safer nuclear techniques, still much discussed in some quarters -- really complements electricity, opening prospects of genuinely integrated utilities that can provide electricty and chemical fuels, and even this basic chemical, as needed.

The need to look at hydrogen as possible ultimate replacement for fossil fuels prompted the West German government to appoint a special 12-member commission two years ago to look into hydrogen as a future energy option. West Germany is regarded by the international hydrogen community as perhaps the nation furthest advanced in moving ahead with hydrogen R&D, and the country is spending considerably more than $100 million a year on exploring all aspects of a future hydrogen economy, including areas that are not strictly hydrogen-related such as advanced combustion and energy efficiency techniques and the like.

The commission concluded this spring that a solar hydrogen economy is a feasible option for the next century, but it stopped short of recommending drastic political measures at this time. The commission said solar hydrogen is a "conceivable" option world-wide -- assuming that the tentative information about the need to limit the increase of gases in the atmosphere due to the burning of fossil fuels turns out to be correct.

That cautious conclusion came before last summer's drought, and before the testimony last June by NASA meteorologist James Hansen in Congress who said he was 99% convinced that the Greenhouse Effect is for real. Had the German commission come out with its report after Hansen's testimony and after the North American drought, the language might have been a little stronger. But nevertheless, the commission recommended that as a prudent, cautious measure the country should spend, in round numbers, one billion Deutsch Marks (about $650 million dollars) between now and 1992 - the year Europe is supposed to come together as a real European Community - to explore the hydrogen option.

PRODUCTION R&D: This is one of the main areas of concern - how to economically produce hydrogen. At the 7th World Hydrogen Energy Conference held in Moscow in late September 1988, almost half of the papers were devoted to production. Research is going on in most industrial countries as well as developing countries like India, Iraq, Brazil, and even Sri Lanka. This field is broken broken down into electrochemical methods, which includes electrolysis of water - the technically most certain method now - and thermal production methods, including hydrogen from fossil fuel resources (A number of researchers believe fossil fuels offer a relatively rapid back-door approach to making hydrogen, including ways to capture carbon dioxide before it is released into the atmosphere, thus utilizing existing resources and at the same time avoiding adding to the Greenhouse Effect).

The most interesting news there was the report of a multi-year, (Can.) $4.2 million feasibility study that will explore the idea of shipping electricity across the Atlantic in form of hydrogen -- cryogenic liquid hydrogen or in some other form yet to be determined -- from the province of Quebec to Europe, most likely to Germany, with Hamburg apparently the preferred choice so far as the European terminal.

Partners in this study are the province of Quebec, with its vast reserves of hydropower, and the European Community where the idea was hatched by researchers at the EC's Joint Research Center in Ispra, Italy. The idea would be to use cheap Quebec hydro-electricity to electrolyze water in a large 100 MW electrolyzer, liquefy the hydrogen and ship it in cryogenic tankers, similar to those used in liquid natural gas transport to Europe. The hydrogen would be used to make electricity, operate a fuel cell for natural gas enrichment, but also used directly to experimentally power some Hamburg city buses, for example, and perhaps for testing of future supersonic and hypersonic aircraft.

Also, last year in Germany, the Munich-based utility "Bayernwerke" announced it will build a DM 62 million (about $27 million) pilot 500 kW solar-hydrogen prototype plant in Bavaria in an economically underdeveloped area close to the Czechoslovak border. Half of the funding comes from the government. Partners in the "Solar-Wasserstoff-Bayern GmbH" (roughly, Solar Hydrogen Bavaria Inc.) are carmaker BMW, aerospace maker Messerschmitt-Boelkow-Blohm, and electrical equipment maker Siemens. Hydrogen will be used to fire boilers and operate fuel cells, catalytic heaters and internal combustion engines. The entire facility is viewed as a sort of permanent testing facility where future hydrogen technologies will be tried out as well. Planning and design is going on, and construction apparently will start next year.

Earlier, in 1986, West Germany and Saudi Arabia signed an agreement for a similar joint 350 kW solar hydrogen production facility that will be built near Riyadh. A smaller facility will be built for testing and training purposes in Stuttgart, and construction apparently is well under way. The operating partners are the German aerospace agency DFVLR, the University of Stuttgart and the German Research and Technology Ministry, and Saudi Arabia's King Abdulaziz City of Science. Other Arab nations are beginning to show interest as well, apparently: DFVLR officials say Algeria, Morocco as well as India are interested in cooperating with the agency in similar projects.

But electrolysis, the technologically surest method of water-splitting is not the only way of making hydrogen that is being studied. Other production R&D subjects include:

*** Thermochemistry: Thermochemical concepts are still being studied in Japan, mostly, the United States and Europe. Basically, the idea is to split water in two or three separate steps involving relatively high temperatures derived from solar or advanced nuclear sources plus very reactive chemicals such as bromine or iodide. The idea is to get away from electricity with its inherent conversion inefficiencies and use heat directly. Thermochemistry was pronounced dead in the late seventies and early eighties, mostly because of the problems of shifting large volumes of highly toxic chemicals around, but the concept seems to be making a comeback.

*** Photobiology: Oak Ridge National Laboratory is investigating photobiological hydrogen production, via genetically engineered micro-organisms that break up water into hydrogen and oxygen. Similar work is going on at the University of Florida, and also in Japan and England.

*** Various hydrogen-related synthetic fuels production research programs are supported by the Chicago-based Gas Research Institute, including the use of advanced metal and semiconductor catalysts.

*** Advanced fossil-based hydrogen production: Texas A&M's Center for Electrochemical Systems and Hydrogen Research, the only one of the handful of American hydrogen research institutions that got seed money from the National Science Foundation some years ago to investigate hydrogen, is researching hydrogen production from coal, from natural gas, but also the use of electrocatalysts to split water with sunlight, and other approaches.

TRANSPORTATION: The most intriguing recent development was a report this spring that the Soviets had converted a commercial tri-jet airliner, the TU-154, similar to the Boeing 727, to fly partially on liquid hydrogen -- the right engine had been modified. The initial 21-minute flight took place in April, and Soviet press accounts said the whole venture had been a nine-year, high-priority project. The United States did the same thing thirty years ago when the U.S. Air Force converted a twin-jet B-57 bomber, operating one engine on liquid hydrogen. The main difference was that the Soviet jet used hydrogen in the entire flight cycle -- take-off, cruise, landing -- while the B-57 switched to hydrogen only after it was safely aloft.

The most important aspect of the Soviet hydrogen aviation work, though, is that the Soviets have designed a new generation commercial jet which is supposed to make its maiden flight this month, the TU-204. Inititally, this plane will use conventional kerosene, but eventually it will be outfitted with engines now under development that will use cryogenic fuel -- presumably liquid natural gas at the outset because the Soviets have plenty of that right now, but later, once economics and technology are right, liquid hydrogen.

In the U.S., the National Aerospace Plane (NASP) program conducted by DARPA and the Air Force represents the leading edge of hydrogen R&D. This concept calls for the use of liquid hydrogen or something called "slush" hydrogen to power a combination aircraft and space vehicle that would take off from a conventional runway and be propelled all the way into orbit and return to a normal runway in a single stage, without the need for multi-stage throw-away boosters, and tanks. Commercial spin-offs of that concept are being studied now in Britain, West Germany, Japan and maybe also in the Soviet Union. They envision high-speed hypersonic transport aircraft that would not fly into space but that would operate at altitudes of 100,000 feet or more.

Road transport: At present, two West German carmakers are actively pursuing hydrogen programs, Daimler-Benz and BMW. However, at the Moscow hydrogen conference, Soviet researchers presented papers on hydrogen-powered cars, vans, busses and fork lifts, including plans for hydrogen taxis in Kharkov in the Ukraine. One liquid hydrogen-powered mini-van was parked outside the conference building for several hours on the last day of the conference. Also, hydrogen-powered busses and vans have been successfully operated in Riverside, Cal. A Denver-based consultant is working on a hydrogen-powered underground mining vehicle under contract with the U.S. Bureau of Mines and a manufacturer of mining vehicles.

In addition, companies in Belgium and the United States are working on fuel cell-powered busses. There was also a report in Moscow on a Soviet-Hungarian joint project for a fuel cell-powered bus, and tests on new, compact, car-sized fuel cells are scheduled to start in Canada next spring.

This spring Daimler-Benz concluded a three-and-half year fleet test involving five station wagons and five vans in West Berlin. All were powered by gaseous hydrogen, stored on board in hydride storage tanks. The vehicles worked well, there were apparently no accidents or technical problems due to the use of hydrogen, but their range was limited to about 100 miles. The cars were heavier and more sluggish than their gasoline-powered equivalents mostly due to the heavy hydride storage systems with attendant more wear and tear on components. Also, the cars have to carry additional water needed for injection into the cylinders during combustion to prevent backfire, something that happens with hydrogen under certain conditions.

Right now, the company's hydrogen plans seem to be a bit on hold. At the Moscow conference, Daimler-Benz presented a paper discussing various options for a hydrogen-powered bus, and that may be the direction the company is looking; there have been some preliminary talks with interested partners in Switzerland and Sweden. Large fleet vehicles - busses, vans, trucks and the like - seem to offer the best chance for hydrogen operation because of the ability to set up a central fueling operation and because of fewer space and weight restraints posed by hydrogen.

Unlike Mercedes, which so far has opted for gaseous hydrogen and hydride on-board storage, BMW is going straight for the space fuel, liquid hydrogen. BMW is considering a major fleet test involving perhaps 100 vehicles in the early 1990s, with special liquid hydrogen "gas" stations to be built at selected key cities throughout Germany, but much will depend on whether the German government is willing to help finance such a program.

Also, Canada is quietly looking at the idea of powering railroads by liquid hydrogen, perhaps with large, yet-to-be-developed fuel cells. The basic idea is to "electrify" the rail system in the sense that once more cheap Canadian hydro electricity would be the prime source, but converted to storable, transportable liquid hydrogen. Savings would come by not having to install overhead power lines for the locomotives, or large electric generator stations every few hundred miles to provide the current.

In uses other than transportation, at least two groups, Germany's Fraunhofer Institute of Solar Energy Systems and Chronar Corp. in the United States are developing self-contained hydrogen energy systems for homes and buildings. Essentially they would tap solar energy via photovoltaic panels and make hydrogen to store the energy and to produce warm water, heat or cool the house, and operate catalytic oven burners in the kitchen.

As to the cost of hydrogen as fuel, there are no real data, because nobody is using hdyrogen as fuel for terrestrial applications. To give at least some idea, NASA at present is paying $1.05 per pound of liquid hydrogen for its Shuttle but again, this is liquid hydrogen produced from fossil natural gas - not exactly the wave of the future. NASA has paid as much as $1.40 per pound, but 15 years ago, NASA was paying as little as 35 cents per pound.

For water electrolysis, gaseous hydrogen can be produced at a cost of roughly $2 for the amount of energy equal to the energy in a gallon of gasoline. This number is based on current rock-bottom electricity prices for industrial users of about 3 cents per kilowatt/hour. Liquefaction would add another 25% to 100%. Nevertheless, at least one German researcher at DFVLR has concluded that for acceptable vehicle ranges of 250 miles or so, liquid hydrogen is more efficient because of initial lower vehicle weight and consequently lower fuel consumption.

Two Princeton University researchers, Joan Ogden and Robert Williams, say photovoltaically-produced solar hydrogen could be cost-competetive by the year 2000, especially if environmental benefits were factored in. Ogden and Williams, who will present their full analysis in a book to be published next year by the World Resources Institute, believe such hydrogen could be an environmentally benign and economically acceptable alternative to coal-based synthetic fuels. (Their analysis has met some crticism as being overly optimistic, however).

They base their conclusions on the "rapid advances in amorphous silicon solar cell technology" achieved in recent years. They say if industry projections about the progress in cell efficiency and their cost prove to be correct, hydrogen could be produced via PV-powered electrolysis in sunny areas such as the southwestern United States at costs of $9-13 per Gigajoule, or roughly $1.56 to $2.12 in terms of energy equivalent to a gallon of gasoline. That may sound high, but that's what most European's pay today anyway, and even more.

PV hydrogen production would require relatively small areas, they claim. Producing hydrogen equivalent to total U.S. oil use would require a collector field of some 24,000 square miles (64,000 square kilometers), equal to 0.5% of the total U.S. land area - more land than would be needed to make fuel from strip-mined coal, but much less than from biomass. Put another way, Ogden and Williams say a circular tract of land with a diameter of 240 miles in, say, New Mexico, would be sufficient to make enough photovoltaic hydrogen equivalent in energy to the entire fossil fuel-derived energy - oil, natural gas and coal - consumed in 1986 in the United States. At the same time, they say, even a relatively small fraction of vehicles powered by hydrogen could make a difference in the severity of air pollution in the post-2000 city of Phoenix, for instance: "We estimate that converting 25% of fleet cars (2% of total cars) in Phoenix from gasoline to hydrogen would reduce air pollution by about 5%," they write. If 50% of the cars switched to hydrogen by 2015, "emissions could be maintained at about half the 1988 levels, despite a projected threefold increase in the number of vehicle miles travelled," they say.

Would a future widespread use of hydrogen and a consequent increase of water vapor in the atmosphere also contribute to the greenhouse effect? According to EPA there is essentially no problem with water vapor - steam - buildup in the troposphere. Apparently, no studies have been conducted but EPA experts say that water content in the troposphere is basically a self-regulating mechanism: excess water vapor condenses out as rain.

The wide-spread use of hydrogen as a fuel for future generations of very-high flying high-speed commercial transports at altitudes of 30- 40 kilometers - is a different matter. The stratosphere is essentially very dry, and the water vapor production of a large fleet of high-flying hydrogen-powered advanced supersonic and hypersonic jet aircraft at those altitudes could conceivably lead to the formation of clouds where there are none today, which in turn could lead to a situation similar to the one over Antarctica with its ozone hole. However, none of this is certain, and EPA says it may initiate a study of the problem. Also, there were hints at the Moscow hydrogen conference that similar concerns are surfacing in Western Europe: the representative of one major chemical and fuel manufacturer indicated that a similar study may be awarded to a German meteorological institute before industry commits itself to begin planning for a fuel infrastructure for such types of aircraft.

OCEAN THERMAL POWER EFFECT ON GREENHOUSE GASES

by J. HILBERT ANDERSON

In the ocean thermal power process electrical power is generated from the thermal energy available in the ocean. An ocean thermal power plant generates power by transferring heat from a warm body thru a heat engine to a cold heat sink. In this case, the source of heat is the warm surface water in the tropical ocean, and the cold heat sink is the cold water deep in the ocean.

The cycle diagram on Fig. 1 shows the power cycle (ref 4) developed by Sea Solar Power, Inc. for generating power from the temperature differences in the ocean. Warm water at the surface is pumped thru a heat exchanger which transfers the heat from the warm water to boil a working fluid, in this case, the ordinary household refrigerant designated R-22. The fluid boils into a vapor at high pressure, and flows upward thru a turbine driving a generator that generates electricity, which is transmitted from the floating plant to a shore based station.

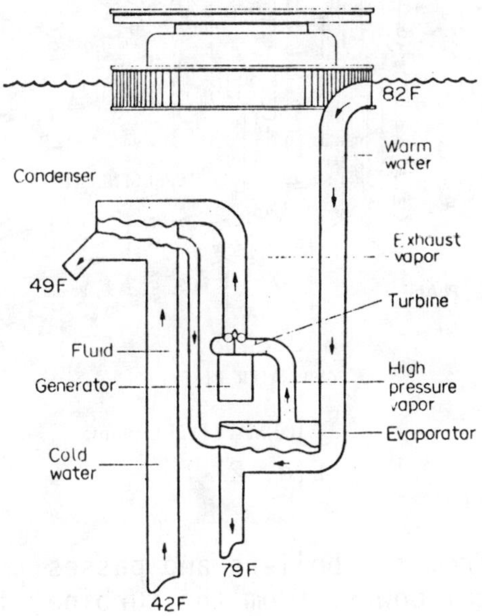

Ocean thermal energy conversion—closed power cycle.

Fig. 1

After passing thru the turbine, the vapor pressure is reduced. The vapor then flows to condensers that are cooled by cold water pumped from the ocean depths. The heat of condensation from the vapor passes to the cold water, and the vapor is thereby condensed to a liquid in the same way that your breath condenses on a cold window pane. The liquid then flows by gravity down to the boiler to be boiled again. Thus a continuous cycle of power generation is sustained. In this cycle there are no discharges of working fluid to the atmosphere, since the R-22 merely circulates thru the system much as it does thru your kitchen refrigerator, except in reverse.

Fig. 2 shows a conceptual sketch of a 100 megawatt Sea Solar Power plant (ref 4). This plant works in the same way as shown on the diagram. Warm water is drawn thru a screen from the warm mixed layer at the surface of the ocean. This water is pumped down to the boilers at the bottom of the power plant, and from there discharged into the ocean.

Ocean thermal power—the coming energy revolution

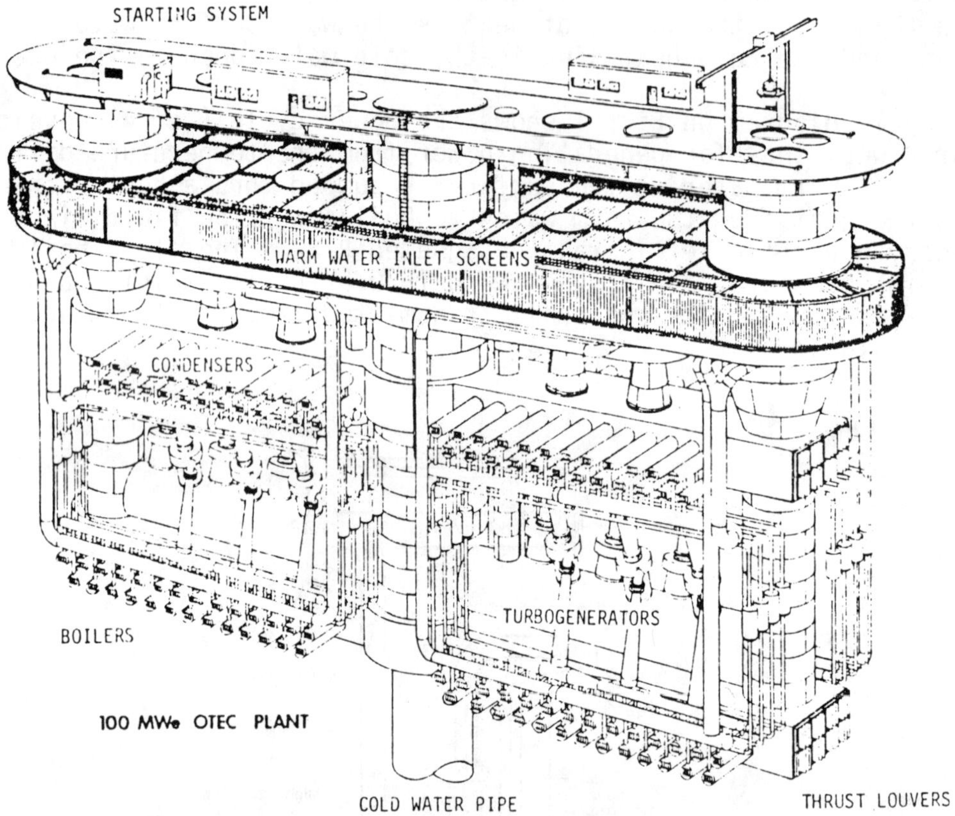

100 MWe OTEC plant.

Fig. 2

The R-22 vapor rises from the boilers and passes thru the turbogenerators where it generates electrical power. From the turbine, it exhausts upward to the condensers where it is condensed into a liquid, and flows by gravity down thru the small pipes to the boilers at the bottom of the plant.

The cold water is pumped thru a pipe approximately 8.5 meters in diameter and 900 meters long suspended from the ocean platform. The water is pumped thru the condensers into the ocean at a level approximately 35 meters below the surface.

A typical 100 megawatt power plant weighs approximately 25,000 metric tons, including the cold water pipe and the working fluid. The plant is approximately 110 meters long and the bottom of the plant is approximately 60 meters below the surface.

Since there are no gases discharged from the plant itself, the only possible effect on the atmosphere is that of discharging dissolved gases from the water, or absorption of gases from the atmosphere.

It is difficult to estimate whether CO2 will actually be discharged from the water coming out of a Sea Solar Power plant. The cold water is discharged at a level higher than the warm water, but both discharges are into the upper mixed layer in the ocean. Therefore, we can assume that the net effect is that of mixing warm and cold water. There are two possible effects.

The first effect is that which will occur if you assume that both the warm and cold water streams are saturated solutions of CO2.

The second effect is that of photosynthesis. Since the cold water contains a large percentage of nutrients including both nitrogen and phosphorus compounds, then it has been well established, both in nature and by experiment (ref 5), that increased photosynthesis does occur, and marine life is greatly increased.

The solubility of CO2 in sea water is much more sensitive to the pH value than to the temperature. If we use Takahaski's values (ref 3) for the CO2 dissolved in the cold water, and assume that the warm water is saturated, as per the data from the Office of Saline Water (ref 1), then by mixing the two streams in the upper mixed layer we should theoretically discharge .00179 grams of carbon in CO2 per kilogram of water passing thru the plant. The total amount of water passing thru the plant is 136 cu. meters per second of cold water, and 248 cu. meters per second of warm water. This total amounts to 12.4 billion metric tons per year. The theoretical amount of carbon released would then amount to 22,200 metric tons per year.

Next, let us estimate the amount of carbon that would be absorbed into marine life by photosynthesis.

Roels (ref 3) estimates that one kilogram of fish would be produced for each 300 metric tons of cold water brought to the surface. On the basis of 4.4 billion metric tons of cold water per year pumped thru the power plant the weight of fish produced would be 14,700 metric tons per year. If we conservatively assume that an equal tonage of plankton is produced to supply food for the fish, and assume that the plankton are 44% carbon, typical of cellulose, then photosynthesis would absorb 6500 metric tons per year.

Based on these conservative estimates the maximum net discharge of carbon into the atmosphere from a 100 megawatt power plant would be 22,200 minus 6500 equals 15,700 metric tons per year.

This would appear to be the maximum possible release of carbon to the atmosphere from a 100 megawatt power plant. The probability is that the discharge would be considerably less than this, both because the water is likely not to be fully saturated with CO2, and because the plankton consumption by fish is likely to be much more than that estimated here. Therefore, CO2 coming out of the solution should be less, and carbon absorbed by photosynthesis should be higher than I have estimated here.

George Woodwell (ref 2) reports that the net increase in carbon in the atmosphere is approximately 2.3 billion metric tons per year. If we calculate the amount of carbon released by coal fired power plants, then we can calculate how many coal fired power plants must be replaced by ocean thermal plants in order to eliminate this increase in atmospheric carbon.

Assuming 30% thermal efficiency, a 100 megawatt coal fired power plant will discharge approximately 250,000 metric tons of carbon per year into the atmosphere. If we assume that we will replace sufficient coal fired power plants to eliminate completely the net increase in carbon in the atmosphere, the number of 100 megawatt power plants that would need to be replaced by ocean thermal plants would be 2.3 billion divided by 250,000 equals 9,000 power plants. 9,000 100 megawatt power plants would generate 900,000 megawatts of electrical power. This is roughly equivalent to one and one half times the total electrical power generated in the United States. It is quite conceivable that this many power plants could be replaced by ocean thermal plants in a period of 25 years.

Another approach toward reducing carbon dioxide in the atmosphere is to substitute electric driven heat pumps for wood stoves used for heating homes. Ponte (ref 8) reports that 33,000,000 acres of forest land are being denuded each year. Most of these forests are being destroyed for use in heating rather than for use as lumber. Therefore, we can calculate how many ocean thermal power plants would be required to replace the wood burned from 33,000,000 acres or 13,300,000 hectares of forest land, which is now being destroyed.

Tropical forest land should absorb at least 5.6 metric tons of carbon per hectare per year (ref 2) by photosynthesis. On this basis the total area of forests destroyed each year would correspond to 74.5 million metric tons per year of carbon that would otherwise be absorbed from the atmosphere. If we assume that electric heat pumps could heat a house with a coefficient of performance of 2.0, and also assume that wood is burned for house heating at 60% efficiency, then each kilowatt hour would replace 286 grams of carbon. A 100 megawatt plant would then eliminate 250,000 metric tons per year that would be discharged into the atmosphere from domestic heating. On this basis it would require 297 power plants to replace the heat produced by 13,300,000 hectares of forest land burned for domestic heating. Note that these calculations are based on the assumption that burning the forests produces heat equivalent to 5.6 metric tons of carbon per hectare year. Actually, when a forest is destroyed by burning, much more than 5.6 tons of carbon per hectare is discharged to the atmosphere. Therefore, it would require many more than 297 ocean thermal power plants to cancel out the carbon discharged by burning 13,300,000 hectares per year.

If this land were allowed to continue to absorb carbon from the atmosphere, the total reduction in carbon from the atmosphere would be 149 million metric tons per year produced by the 297 ocean thermal power plants.

If we subtract 149 million tons from the 2.3 billion tons net increase in the atmosphere, then this would leave 2.15 billion tons yet to be accounted for

The number of power plants to replace 2.15 billion metric tons produced by coal fired power plants would be 8540 plants. Adding 8540 plus 297 equals 8837 ocean thermal plants required if plants are used for domestic heating equivalent to the forests being destroyed and the replaced coal fired power plants. This is somewhat less than the figure calculated above, based only on replacing coal fired power plants to eliminate the increase in carbon in the atmosphere.

Although replacement of coal fired power plants is by far the most important effect producible by ocean thermal plants toward reducing carbon in the atmosphere, the effect on food supply for the world should also be considered.

The following table shows the approximate world food production in million tons per year (ref 9). The table also shows food production in the United States (ref 6) in tons per hectare for various of the more important foods used for human consumption.

WORLD FOOD PRODUCTION

Item	World Million Tons	U.S. Tons per Ha.
Meat	144	
Eggs	30	
Fish	75	
Dairy products	483	
Potatoes	260	31.2
Other roots	300	
Other vegetables	350	
Fruit	280	
Corn	450	6.69
Rice	410	7.89
Wheat	460	2.69
Other cereals	340	
Total	3582	

If we use the figure of one kilogram of fish per 300 metric tons of cold water required for ocean thermal power plants, then 8837 plants would produce 129,600,000 metric tons of fish per year. This is one and three quarter times the total present world production of fish. This would obviously have an enormous impact on the world's food supply, and would probably accentuate the present trend toward substituting fish for meat in our diet. If this results in a continued decrease in the production of meat from grazing animals, then this would also reduce the methane production by grazing animals (ref 7), which is now considered to be an important factor in increasing the methane content of the atmosphere.

In addition to generating a huge additional food supply, each 100 megawatt ocean thermal plant should produce approximately 18,000 cubic meters of fresh water per day. This occurs primarily because it is desirable to remove some of the oxygen from the water entering the heat exchangers so that barnacles will not grow and attach themselves to the interior passages in the plant.

At this rate of fresh water production, the total production from 8837 plants would be 60 billion cubic meters per year. This would irrigate 3.3 million hectares of land at the same rate of water usage as in the Imperial Valley in California (ref 10). This acreage is equivalent to approximately 17 times the cultivated area in the Imperial Valley. Since the Imperial Valley is one of the most productive food and grain areas in the United States, it is obvious that this could have an appreciable impact on the food supply.

If we assume that irrigated food producing land would absorb three metric tons of carbon per hectare per year, then the carbon absorbed by this irrigated land would amount to 9.8 million metric tons per year.

If this irrigated land were used for producing potatoes, the yearly tonnage would amount to 102 million tons per year or almost half the present world's production.

From the foregoing crude estimates, it becomes clear that ocean thermal power plants can provide by far the most important solution for reducing the increase in carbon dioxide in the atmosphere. It can provide many benefits in addition to this. There is no question that Ocean Thermal Power can and will become the most important factor in providing means for controlling our change in climate.

Ocean Thermal Power is no longer a scientist's dream. It will soon become a practical reality. We already have commitments for installing four 100 megawatt power plants in the tropics, and are negotiating for furnishing some 50 more power plants. We estimate that one shipyard alone could produce four of these power plants per week. Therefore, it is easy to see that we could indeed solve the problem of carbon increase in the atmosphere with a very reasonable time frame such as 20 or 25 years. The important point is that we need to start this program with all possible speed.

References

1. Office of Saline Water, Saline Water Chart, p. 12,581.
2. George Woodwell, "The Carbon Dioxide Question", Scientific American, Jan. 1978.
3. Takahashi, Dept. of Energy report DOE/EA-0147.
4. J. H. Anderson, "Ocean Thermal Power, The Coming Energy Revolution", Solar and Wind Technology, Jan. 1985, The Pergamon Press.
5. Othmer and Roels, "Power, Fresh Water and Food from Cold, Deep Sea Water", Science 12 Oct. 1973.
6. Agricultural Statistics, U.S. Dept. of Agriculture, 1985.
7. Jerry Bishop, Methane as Culprit in Greenhouse Effect, Wall Street Journal, 24 Oct. 1988.
8. Lowell Ponte, "What's Wrong with our Weather", Reader's Digest, Nov. 1988.
9. "Farmed Animals", Torstar Books.
10. Private Correspondence from R.F. Carter, General Mgr., Imperial Irrigation District.

The Potential for Geothermal Energy

J. W. Tester
Massachusetts Institute of Technology
Cambridge, Massachusetts

D. W. Brown
R. M. Potter
Los Alamos National Laboratory
Los Alamos, New Mexico

Motivation and Scope

Energy use and its effects on the environment are among the key issues that the United States and the other countries on this planet must confront and begin resolving during the remaining decade of this century. This report considers several aspects of future national energy production as they relate to local and world-wide energy supply and environmental degradation. One critical problem is our large dependence on foreign oil -- currently almost 40% of our total energy budget. Given our ever-diminishing domestic oil and gas reserves, our dependence will undoubtedly increase even with continuing conservation measures. To maintain proper economic stability and alleviate a significant portion of our balance-of-payments problem, the US must become more energy independent. Because oil and gas discoveries within the US have not been able to maintain sufficient reserves to supply our current and future needs, other energy sources are needed to achieve this independence. This shortfall of conventional oil and gas supplies in the US could be regarded as a harbinger of future instability that requires a reassessment of national energy policy. Synthetic fuels generated from oil shale and tar sands, gasified and liquified coal, and other synthetic fuel sources may provide some relief, but the economics and environmental consequences of their utilization may prove to be sufficiently unacceptable as to limit their widespread development in the US.

In addition to the supply issue, continuing studies have shown, with various degrees of confidence, the almost always negative environmental effects of twentieth century methods of energy production and its application to needs such as transportation. Of primary concern are the deleterious effects resulting from the production of vast amounts of CO_2 from the combustion of fossil fuels, and smaller amounts of other gases such as SO_2 and NO_x that are liberated in the combustion process (Brown et al., 1988; Lovins et al., 1981; Shepard, 1988; and Mathews, 1987). These gases play major roles in the processes that cause global warming (the Greenhouse Effect) and pH changes in regional precipitation (Acid Rain). How energy is produced and consumed worldwide is directly related to a most important climatological index, the earth's average temperature. Just as in biological systems, this average temperature may determine, to a large degree, the "health and vitality" of our entire planet. Because of the tremendous uncertainty in predicting the future changes in this climatological index and the resulting consequences, the natural tendency is to minimize change in attempting to preserve the present world state.

It appears that the only practical solution to this dilemma is to significantly reduce the use of fossil fuels in electrical generation, direct thermal applications, and transportation systems. Active conservation programs may be able to achieve additional gains, but the real solution seems to lie with the application of alternative energy generation technologies (Chandler et al., 1988). Many studies have proposed significantly increasing the use of nuclear energy and non-fossil energy options as rapidly as possible. Alternative options mentioned typically include solar, fusion, biomass, and wind turbines in the energy generation field, and the use of direct or indirect electrical methods for transportation needs. Conspicuously absent (at least to the authors of this paper) is the inclusion of geothermal energy as a potential fossil fuel replacement for a major portion of our future energy needs.

The role of geothermal energy in our national energy policy

Geothermal energy, as exemplified by the extraction of high temperature water or steam from geologic formations, is providing an economic and trouble-free source of electric energy for many regions of the earth. A hallmark of these naturally-occurring geothermal systems is the presence of a system of connected open fractures through which water can circulate and be heated by conduction from the surrounding rock. On a worldwide scale, these natural geothermal systems are confined to a relatively few regions of the earth where they are present in sufficiently high grade to supply base-load energy needs; for example, in Iceland, New Zealand, Northern Italy, and in California in the US (Armstead, 1983, and DiPippo, 1980). Even with their distribution limited to regions where these geologic anomalies occur, the magnitude of the US resource base is still substantial at some 44,000 quads of energy, which is the equivalent of 7200 billion bbl of oil (National Research Council [NRC], 1987). Worldwide, the geothermal resource base is estimated to be about 130,000 quads (Armstead and Tester, 1987). In contrast to other alternative energy sources that have yet to be developed commercially, geothermal energy is currently providing over 2200 MWe of baseload electricity in the US and over 5000 MWe worldwide. In addition, in many areas geothermal energy is used for direct heat supply. In the short term, natural systems will continue to be exploited as they provide an economically and environmentally acceptable alternative to fossil fuels and nuclear.

However, of much greater ultimate importance in the long term will be the extraction of larger amounts of geothermal energy from accessible regions of hot rock beneath the earth's surface which do not contain sufficient natural porosity or permeability. As depicted in Fig. 1, this can be accomplished by creating artificially fractured reservoirs which emulate natural geothermal systems. The primary technique for engineering these so-called Hot Dry Rock (HDR) geothermal reservoirs utilizes fluid pressure to open and propagate fractures from an inclined well, creating artificial permeability within a fracture network as illustrated in Fig. 1. This hydraulically-stimulated region is then connected to a second well to complete the underground system. Heat is extracted by circulating water from the surface, down one well, through the fractured rock network, and up the second well. The heated water then passes through an appropriately-designed power plant on the surface where, for instance, electricity or process steam is generated. The cooled fluid is then reinjected to complete a closed-loop cycle. Thus, effluents from HDR systems are practically non-existent. In effect, we would be mining heat in an analogous fashion to the way other resources are obtained (Smith et al., 1975, and Armstead and Tester, 1987).

Because hot dry rock systems do not require the presence of contained hot fluids and high permeability, the natural HDR resource is much larger and more widely distributed than natural geothermal systems. Numerous estimates place the accessible HDR resource base somewhere between 10,000,000 and 13,000,000 quads in the US and over 100,000,000 quads worldwide (for example see White and Williams, 1975; Muffler, 1979; HDRAP, 1977; Rowley, 1982; NRC, 1987; and Armstead and Tester, 1987). In this case, accessibility is defined as being reachable by conventional drilling technology to depths ranging from 6 to 10 km. Figure 2 provides estimates of the geothermal temperature gradient distribution across the US and clearly shows that the moderate grade (30° to 45°C/km) HDR resource is well distributed across the US. Recent studies by Kron and Heiken (1980) estimate the high grade US HDR resource base with gradients greater than 45°C/km to be in excess of 650,000 quads or about 116 trillion bbl of oil. Thus on almost any basis, the amount of potentially useable thermal energy in the HDR resource is vast -- literally orders of magnitude larger than the sum total of all possible fossil and fissionable resources (see Fig. 3). However, these estimates are clearly upper limits as the key question remains as to just how much of this vast resource base can be recovered within existing technical and economic constraints. As a point of comparison, usually less than half of the oil in an underground reservoir is actually produced. At this stage of its development, it is not known what fraction of the HDR resource base is recoverable. However, even if only 0.1% or 0.01% of the accessible HDR resource base is ultimately extracted, the impact on the US energy supply would be far-reaching. In fact, it is this great potential for HDR that has encouraged many to advocate its inclusion as a major component of the US energy development program.

A major research effort to develop techniques for creating such reservoirs in essentially impermeable hot granitic rocks has been carried on for the past 16 years by the Los Alamos National Laboratory, under the sponsorship of the US Department of Energy (USDOE). The Federal Republic of Germany and the Government of Japan have also supported a portion of this work under a collaborative agreement with the US. In addition to the US effort, the British have been involved in the development of HDR technology for about 10 years at a site in Cornwall (Batchelor, 1984). The US program has already successfully met several crucial technical goals that are required before large-scale commercialization of HDR can occur. These are:

- <u>Accessibility to the resource</u> - Successful adaptation of conventional oil and gas drilling methods and capabilities to the harsh environment encountered at depth in hard crystalline rock where useful temperatures (about 150° to 300°C) are found. Figure 4 shows trends in drilling costs for HDR wells in comparison to oil and gas wells to similar depths.
- <u>Creation of usefully-large fracture systems</u> - Again, successful adaptation of oil industry methods of hydraulic stimulation has resulted in the creation of permanently opened and connected, artificial and natural fracture systems, all contained within very large volumes of hot rock.
- <u>Efficient extraction of thermal energy</u> - Numerous tests have shown that the heat extraction process in these fracture systems is both predictable and suitable for commercial use.

- Determination of the physical structure of the created fracture system - Measurement of the acoustic emissions accompanying the creation of these fracture systems has provided estimates of the location, size and orientation of the fractured and interconnected region; that is, the HDR reservoir.

At present, most of the HDR R & D field efforts are concentrated at two locations: at the Fenton Hill site in northern New Mexico in the United States and at the Rosemanowes site in Cornwall in the southwestern region of the United Kingdom. However, useful work on a smaller scale is also being done in Japan, France, and the USSR, but it is too early to forecast their effects on this emerging technology. In the United Kingdom, following the results of testing at Rosemanowes, consideration is now being given to a deeper, hotter commercial-scale heat extraction test at some suitable location within the Carmenellis granite of Cornwall.

Realistically, however, the future of HDR technology development in the US, at least in the decade ahead, seems to lie solely with the fate of developments at Fenton Hill. Perhaps, too much emphasis has been placed on achieving a commercially-viable proof of concept in this first attempt in the field. With competition from other energy projects, and a decreasing federal budget for advanced energy research within the USDOE, funding for the development of the potentially large HDR resource has decreased to a point where it is below a critical threshold level. Over the past 16 years, the pioneering field efforts at Fenton Hill have provided the basis for developing the necessary stimulation and extraction technologies. These achievements in the field, and remaining technical and economic barriers, need to be evaluated with a long-term perspective rather than a shortsighted, near-term view. The potential impact of HDR as a major non-fossil component of the US energy supply in the next century needs to be carefully compared to other resources of comparable magnitude, such as fusion and solar. New HDR development sites, in addition to Fenton Hill, should be considered for funding as a part of the USDOE program; for example at other high-grade sites such as the Geysers/Clear Lake region in California or Roosevelt Hot Springs in Utah for electric power generation; and at other lower-grade HDR sites in the eastern U.S. for integrated combined-cycle cogeneration and/or direct-heat applications.

In the US alone, electricity production from hydrothermal natural geothermal sources, located mostly in California and other regions of the far west, is predicted to increase from over 2200 MWe in 1988 to possibly more than 4200 MWe by the year 2000 (Kruger, 1987). This projected growth in hydrothermal resource utilization will create the industrial infrastructure involving field developers, producers and end-users needed for large-scale commercialization of HDR.

With increased Federal support, it is likely that one or more commercial-scale HDR developments will be in full-scale operation during this century. Given today's economic climate, the required level of federal funding for these follow-on HDR projects at new sites will approach 100% of the total development costs, with significant private-sector participation. However, within the next decade or two, a growing number of successful Federally-supported HDR projects would catalyze privately-funded developments in much the same way as fission reactors were developed in the post-World War II era with substantial federal subsidy. Under these assumptions, HDR, could assume an ever-increasing role in supplying the country's energy needs.

HDR energy, with its unique characteristics, could be a major replacement for fossil-fuel fired energy supply systems, providing the following benefits:

Minimal Environmental Impact

No fluids or gases are released to the environment, since HDR systems operate in a closed-loop circulation mode (not unlike an automobile engine's cooling system). Thus, HDR energy would not contribute to increased levels of atmospheric CO_2. Natural hydrothermal systems on the other hand, depending on the specific characteristics of the resource and the surface plant design may emit some CO_2. However as shown in Fig. 5, the ranges of CO_2 emission per unit of energy for hydrothermal systems are significantly lower than the amounts generated from fossil fuel based technology. Waste heat rejected from geothermal electric power plants is so widely dispersed that its effect on any local environmental condition would be minimal. The drilling of access wells and laying of pipelines temporarily disrupts the land surface, but obviously on a very much smaller scale than that associated with the strip mining of coal, for instance.

Available Everywhere in Continuous Supply

HDR energy is available everywhere on the earth's surface, with the resource temperature inexorably increasing with depth (see Fig. 2). The HDR resource base is defined by the thermal energy above a certain minimum temperature stored in rock beneath the surface to depths accessible using conventional drilling technology. Therefore, in developing the HDR resource, the primary parameter determining the local grade of the resource is the average temperature gradient, or the drilling depth required to reach a suitable temperature for the specified application.

Readily Usable as Produced

HDR production is in the form of hot water under pressure, which can be used directly for local space or process heating needs, or for lower-temperature heating in electric power plants. Various end-use options are shown in Figs. 6 and 7. HDR can be used as a single supply of energy to a power plant, or used in conjunction with coal, oil or natural gas in a hybrid plant to improve overall efficiency. In the first case, fossil fuel consumption and CO_2 generation rates per unit of electricity produced are entirely eliminated and in the second case, they are significantly reduced.

Each Reservoir Represents a Large Thermal Resource

Field experience has shown that fractured HDR reservoirs can be created economically, and that their size can be almost arbitrarily large, depending primarily on the volume of injected water. For example, by increasing reservoir size, one could provide for multiple cogeneration needs by producing both power and process heat, with only a very modest increase in reservoir development costs.

Large Increases in Domestic Energy Production from HDR are Possible

With the vigorous application of existing technology, HDR geothermal energy could provide a significant portion of our domestic energy requirements early in the next century. Within less than 20 years, our country could be well on the way to energy independence, rather than slipping into an ever increasing dependence on foreign sources of petroleum.

Future Technological Advances Will Increase the Available HDR Resource

As has occurred during the past 10 years, advances in drilling technology and supporting services will allow for the development of ever deeper -- and hotter -- HDR reservoirs with minimal cost increases.

Applicable to Pumped Storage Use of Off-Peak Electric Power

Because of their elastic nature, HDR reservoirs can be utilized for off-peak pumped storage (see Fig. 7e), with the potential for increasing the effective size of our domestic electric generating capacity. As a major advance over other pumped storage concepts, the stored fluid is returned as heated water which can be used in hybrid power plants to produce more power than was used in the storage process. In this sense, HDR pumped storage is essentially an "electric power multiplier".

HDR geothermal energy can play an early and substantial role in national policies directed towards an overall reduction in fossil-fuel use. As an abundant, but low-grade source of heat, one probable use will be in the preheating of working fluids used in the electric generating process. However, of much greater importance will be its potential for supplanting major quantities of natural gas and heating oil now used for both space heating and lower-temperature industrial processes. This in turn would preserve our more limited quantities of high-grade fuels (of high thermodynamic potential) for use in electric power generating systems operating at high conversion efficiency (40-50%). In this mode, HDR could more than double the lifetime of our domestic reserves of petroleum and natural gas.

If an HDR system were to be designed to meet only direct-use space heating needs, reservoir temperatures of 100°C to 85°C would be sufficient. In the eastern US, this temperature level can normally be reached by drilling to about 3 to 4 km. A program of HDR direct-heat utilization from these depths could be developed on a national scale within a reasonably short time, by utilizing the excess capacity of our lethargic drilling industry (only 917 operating drilling rigs as of 9/26/88) with only a minimum of reorientation. New home subdivisions and/or new industrial complexes could provide a ready space-heating market to drive the development of the HDR resource, particularly with favorable tax incentives.

In addition to these immediate direct-heat uses, HDR also can play a substantial future role in the generation of electricity. Almost surely, advances in hard rock drilling technology during the next decade will permit the drilling of HDR boreholes to depths of 10 km (equivalent to present oil and gas well maximum depths). Such capability ensures that geothermal temperatures in excess of 200°C will be accessible almost everywhere in the US. Rock temperatures above 200°C are attractive for HDR-augmented electric power generation.

The ubiquitous distribution of the HDR resource is ideal for more diffuse, less concentrated electric power and thermal energy production. The resulting dispersion of energy production, if HDR were introduced on a national scale, could be of great practical and sociological importance, helping to ameliorate past environmental and social problems resulting from the use of mega power plants, often sited near population centers. HDR, with its potential for providing both the electrical and thermal power needs locally for smaller communities and their associated industries, might be a strong force in revitalizing the nation. Of key importance to all of these issues are the questions of (1) the roles of both the Federal Government and the private sector in the further development of the HDR concept and (2) the level of funding from public and private sources allocated to this development.

We believe that the best overall national energy policy would call for Federal funding levels for the various alternate energy programs that are consistent, first with the potential magnitude and time scale of that alternative's impact on US energy needs and second, with the physical needs of the country. Adherence to such a policy in the case of HDR would imply a continuing strong role for the Federal Government during the next phase of development. During this phase, a Federally-funded program, with industry involvement in the construction and testing of several HDR demonstration systems at geologically different sites, would provide the private sector with sufficient hard data for HDR commercialization on a scale needed to significantly influence national environmental and economic problems.

Although it is a generally accepted premise that no single energy source or system can adequately meet either the nation's or the world's energy needs now or in the future, we feel that the HDR resource, with its tremendous range of thermal quality and relatively straightforward implementation technology, can be drawn upon in increasingly sophisticated ways to provide, both directly and in concert with other fossil and non-fossil thermal sources, a major portion of the world's energy supply for many thousands of years.

Further detail and documentation to substantiate our claims can be obtained from other publications dealing with HDR technology, for example see Smith, Aamodt, Potter and Brown (1975); Murphy, et al. (1982); Tester et al. (1979); Armstead and Tester (1987); and Cummings and Morris (1980); and Tester, Brown, and Potter (1989).

Acknowledgement

This study was performed under the auspices of the US Department of Energy, in cooperation with the Energy Laboratory of the Massachusetts Institute of Technology.

References

Armstead, H.C.H., Geothermal Energy, E.F. Spon, London, 2nd ed, 404 p (1983).

Armstead, H.C.H., and J.W. Tester, Heat Mining, E.F. Spon, London, 460 p (1987).

Batchelor, A.S., "Hot Dry Rock Geothermal Exploitation in the United Kingdom," Modern Geology, 9, 1-41 (1984).

Brown, L.R. et al., State of the World 1988, W.W. Norton & Co., New York (1988).

Chandler, W.U., H.S.Geller, and M.R.Ledbetler, "Energy Efficiency: A New Agenda", Amercian Council for an Energy-Efficient Economy, Washington, D.C., 76p (1988).

Cummings, R.G. and G. E. Morris, "Economic Modeling of Electricity Production from Hot Dry Rock Geothermal Reservoirs: Methodology and Analysis," EPRI report EPRI-EA-630, Palo Alto, CA (1979).

DiPippo, R., in A Sourcebook on the Production of Electricity from Geothermal Energy, US Govt. Printing Office, Washington, DC (1980).

Hot Dry Rock Assessment Panel (HDRAP), "Hot Dry Rock Geothermal Energy-Status of Exploration and Assessment," Report #1, June, 1977, Energy Research and Development Administration, ERDA-77-74, Washington, D.C. (1977).

Kron, A. and G. Heiken, "Geothermal gradient map of the conterminous US," Los Alamos National Laboratory, Los Alamos, NM, (LA-8476-MAP)(1980).

Kruger,P., "1987 EPRI Survey of Geothermal Electric Utilities," Proc. 11th Annual EPRI Geothermal Workshop, Berkeley, CA (1987).

Lovins, A.B., L. Hunter Lovins, F. Krause, and W. Bach, Least-Cost Energy-Solving the CO_2 Problem, Brick House Publishing Co., Andover, MA (1981).

Mathews, J.T., "Global Climate Change: Toward a Greenhouse Policy," Issues in Science and Technology, p57 (Spring, 1987).

Muffler, L.J.P. (ed.) "Assessment of Geothermal Resources of the United States - 1978," U.S. Geological Survey Circular 790 (1979).

Murphy, H.D., R. Drake, et al., "Economics of a 75 MWe Hot Dry Rock Geothermal Power Station Based on the Design of the Phase II Reservoir at Fenton Hill". Los Alamos National Laboratory report LA-UR-9241-MS, Los Alamos, NM (1982).

National Research Council (NRC), "Geothermal Energy Technology: Issues, R&D Needs, and Cooperative Arrangements," National Academy Press, Washington, D.C. (1987).

Rowley, J.C., "Worldwide Geothermal Resources," Chapter 2 in Handbook of Geothermal Energy (Edwards,L.M., Chilingar, G.V., et al. (eds.)) Gulf Publishing, Houston, TX (1982).

Shepard, M., "The Politics of Climate," EPRI Journal, p5 (June, 1988).

Smith, M.C., R.L. Aamodt, R.M. Potter, and D.W. Brown, "Man-made Geothermal Reservoirs," Proc. 2nd US Symposium on Geothermal Energy, San Francisco, CA, USA, 1781-1787 (1975).

Tester, J.W., G.E. Morris, et al., "Electricity from Hot Dry Rock Geothermal Energy: Technical and Economic Issues," Los Alamos National Laboratory report LA-UR-7603-MS, Los Alamos, NM (1979).

Tester, J.W., "Energy Conversion and Economic Issues for Geothermal Energy," in Handbook of Geothermal Energy, (L.M. Edwards, et al.,eds.) Gulf Publishing Co., Houston, TX (1982).

Tester, J.W., Brown, D.W., and Potter, R.M., "Hot Dry Rock Geothermal Energy--A New Energy Agenda for the 21st Century" Los Alamos National Laboratory report, Los Alamos, NM (1989).

White, D.F. and D.L. Williams, (eds.) "Assessment of Geothermal Resources of the United States - 1975," US Geological Survey Circular 726 (1975).

List of Figures

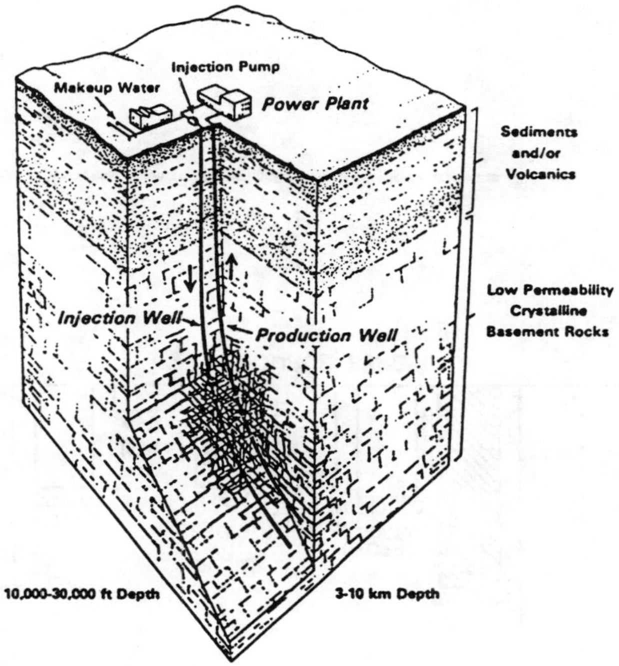

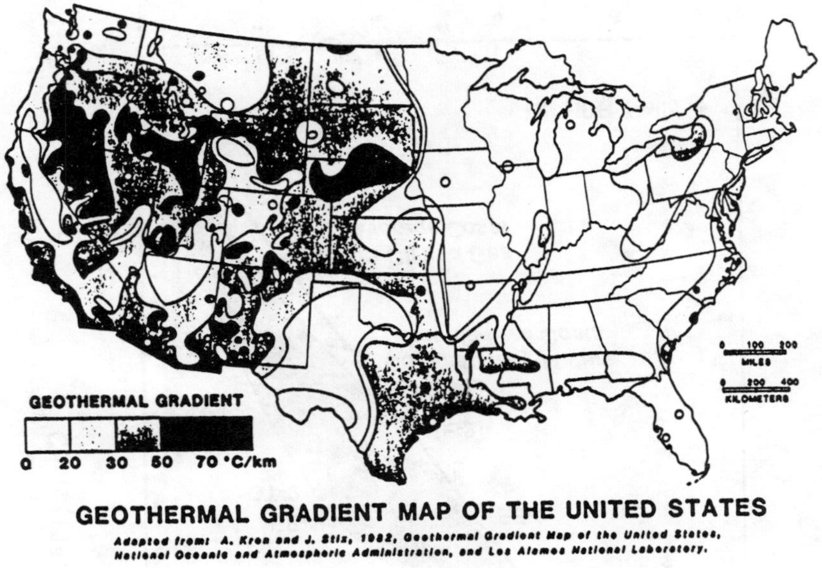

Fig. 1 Hot dry rock (HDR) geothermal system concept for low permeability formations.

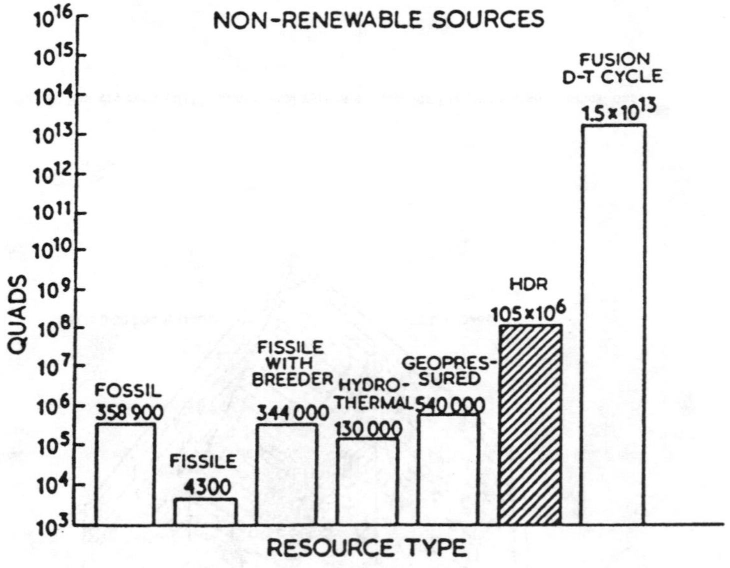

Fig. 3 Worldwide resource base estimates for non-renewable resources.

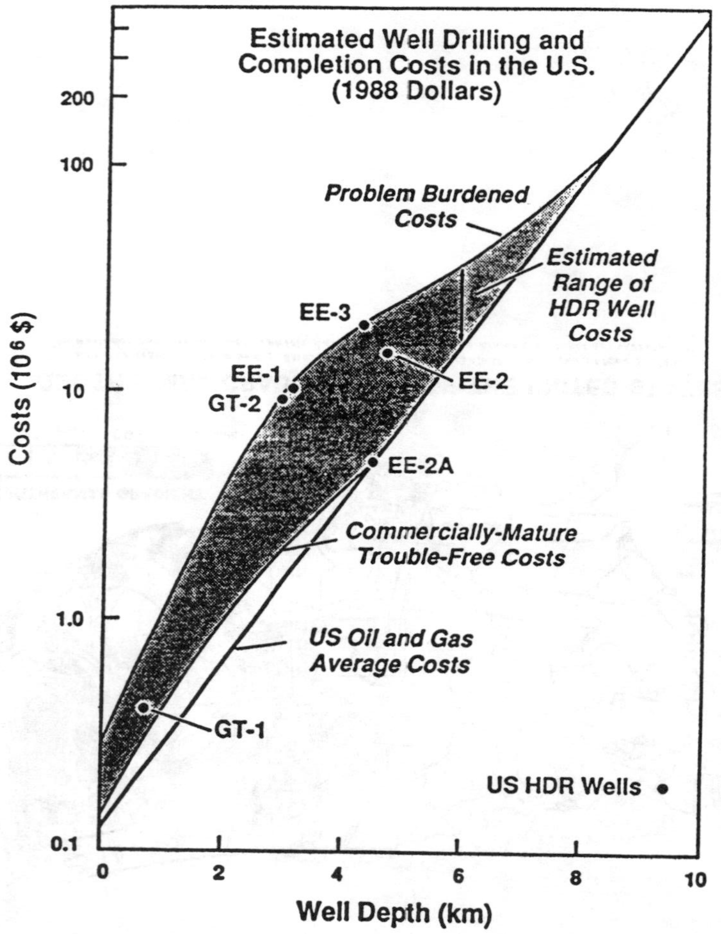

Fig. 4 HDR well drilling costs in comparison to oil and gas wells to similar depth.

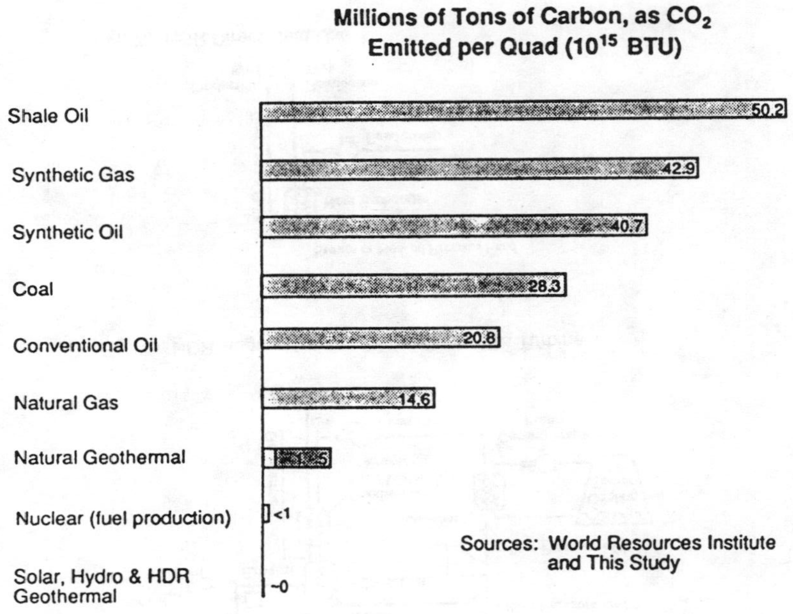

Fig. 5 Carbon dioxide emitted per quad of thermal energy produced for different fuel sources.

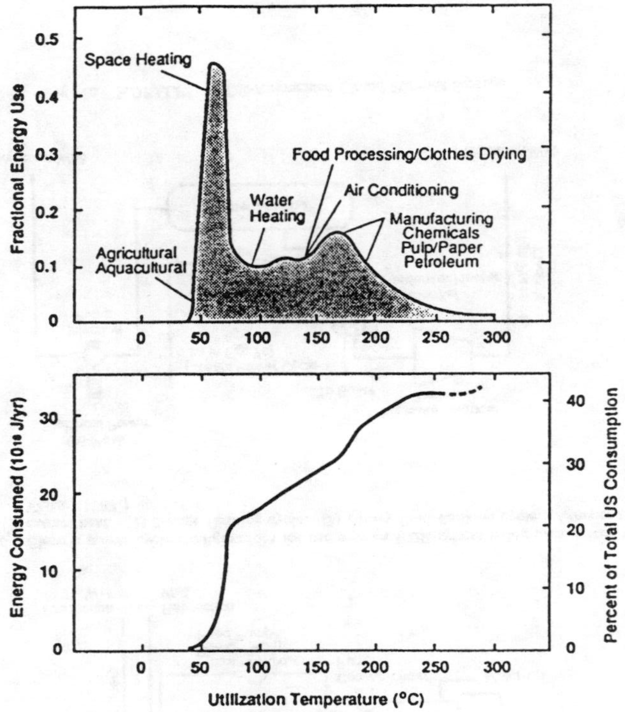

Fig. 6 Distribution of energy use in the US as a function of the temperature of utilization.

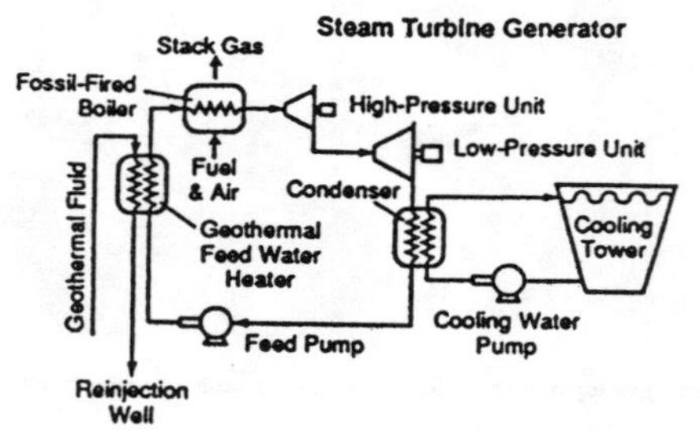

Fig. 7a Hybrid Steam Power Cycle with a Fossil-Fired Boiler and HDR Preheater

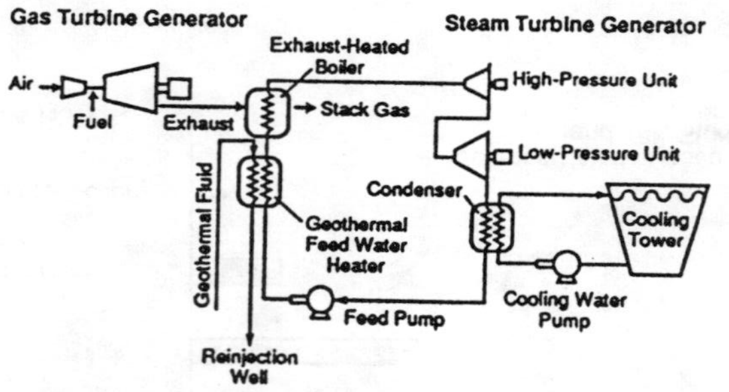

Fig. 7b HDR-Augmented Combined-Cycle Gas Turbine

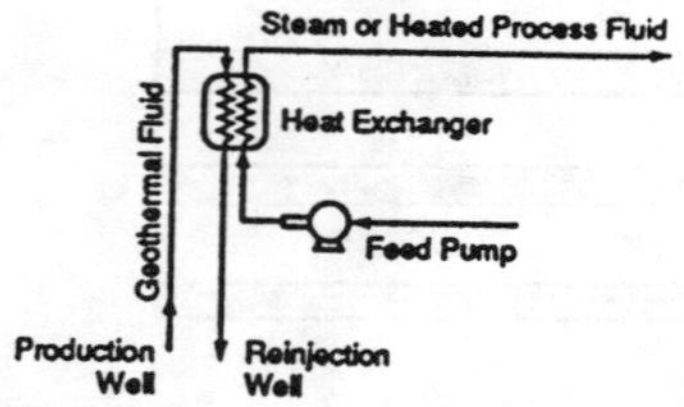

Fig. 7c HDR Direct Heat Use

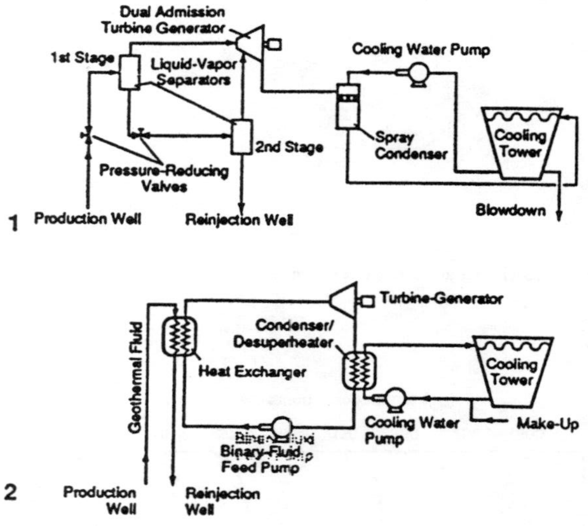

Fig. 7d Electric power cycle configurations for use with an HDR system using pressurized water to extract heat. (1) 2-stage flashing cycle. (2) Binary-fluid Rankine cycle. (Armstead and Tester, 1987.)

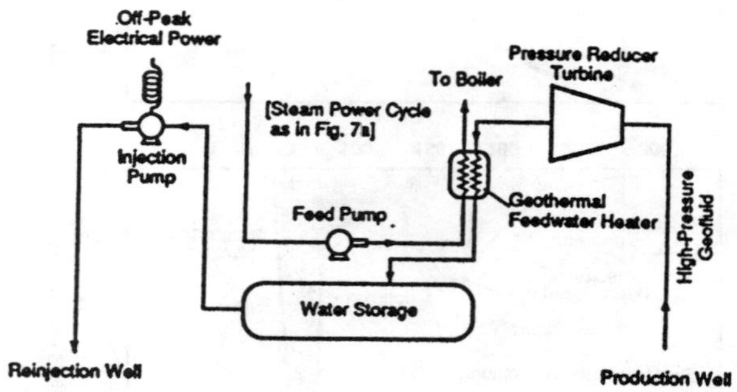

Fig. 7e HDRALPS: HDR-Augmented Liquid Pumped Storage

DIRECTIONS IN ADVANCED REACTOR TECHNOLOGY

Michael W. Golay
Professor of Nuclear Engineering
Massachusetts Institute of Technology
Cambridge Massachusetts 02139
6 December 1988

The purpose of this paper is to examine the directions being pursued in nuclear reactor technology development worldwide, in the context of possible future global warming, the 'greenhouse' effect. Current reactor development programs are categorized according to whether they are primarily concerned with enhanced economic performance or safety, although both factors must be addressed satisfactorily in any concept.

The two energy technologies which show substantial promise as substitutes for use of fossil fuels are solar and nuclear power. Nuclear power uses either the fission or fusion reaction. However, only fission power has been developed to the point of practicality.

Considering the various energy demand sectors, neither solar nor nuclear power serve them all well. Solar power has to this point been useful mainly for distributed, low-temperature heating. Civilian fission power has been developed mainly as a source of electricity, however, it could also be exploited as a source of concentrated, high temperature (T > 1400 F) heat. It does not appear to be easily adaptable as a source of distributed, low-temperature heat, or for non-electrical powering of transportation needs. The latter two needs could be served by substitution of hydrogen for fossil fuels. Nuclear power could be used for bulk generation of hydrogen. This could be done by splitting water molecules, either electrically or thermally. However, at the moment no reactor development program is underway focussed upon hydrogen production. The remainder of this paper is concerned with worldwide reactor development efforts, all of which are focussed upon electricity production. The major emphases in development of different reactor concepts are summarized in Table 1.

The greenhouse effect is a global problem. Thus, it is logical to consider the prospects for future nuclear power development on a global basis. There are four major categories of countries in terms of nuclear power development. The major (but not exclusive) emphasis of development in each category is as follows:

- o Passive safety (Federal Republic of Germany, Sweden, United States)
- o Economic performance (Federal Republic of Germany, France, Japan, United Kingdom, United States)

- o Having unresolved emphasis between safety and economics (The Soviet Union)

- o Lacking significant domestic nuclear technology development programs (all other countries).

Depending upon national priorities and values the emphases of reactor development programs may differ greatly from one country to another. In most cases the priorities of these programs do not reflect specific concerns with alleviation of the greenhouse effect. Rather, they are focussed upon nearer-term issues of public acceptance (in the case of passive safety) and economic benefits which can be captured over less than 20 years. The three major reactor types differ according to the fluids used for heat removal. The categories are light water reactors (LWRs, cooled by H_2O), gas-cooled reactors (GCRs, cooled by helium) and liquid metal-cooled reactors (LMRs, cooled by sodium).

Passive Safety: The United States is pursuing different major nuclear power development programs focussed alternatively upon safety and economics. These parallel emphases reflect the divisions of opinion in our society regarding the net value of nuclear power. The Federal Republic of Germany also has a dual emphasis in its programs. Sweden has an exclusive emphasis upon passive safety as the theme of its future nuclear power program. It has also adopted a phased moratorium upon future use of nuclear power. The United States has had an effective moratorium on new plants since 1974. Concerns among the public over the level of safety achieved with the current generation of nuclear power stations have motivated reactor designers to improve their new concepts greatly. Passively safe concepts involving all three classes of reactors have been invented and are being developed.

Protection of the public, or attainment of nuclear safety, requires that the radioactive material, which is created within the reactor fuel during the fission process, be isolated from the biosphere. The primary means of doing this are the following:

- o Maintain the integrity of the reactor fuel

- o Capture escaped radioactive materials within the reactor coolant system

- o Capture escaped radioactive materials within a containment building.

The main threat to reactor fuel integrity arises from overheating, when the heat of radioactive decay within the fuel is not removed adequately.

The modular HTGR: The modular HTGR concepts emphasize maintenance of fuel integrity by using high-temperature resistant fuel. The fuel is expected to remain intact even when the available means of cooling are highly impaired, but still able to utilize natural convection in the atmosphere or radiative heat transfer to the reactor's surroundings.

LWRs and LMRs: Passively safe versions of LWRs and LMRs rely less upon thermal toughness of the fuel than upon providing highly - reliable natural convection of the reactor coolant in order to cool the fuel.

An important aspect of passive reactor cooling is use of lower power levels than is typical in the economic-performance oriented concepts. This imposes capital cost disadvantages upon the passively safe concepts which must be overcome if they are to be employed. For example, to save money all of the passively safe reactor concepts proposed to-date in the United States would be built without a containment building. The justification provided is that a containment would be unnecessary since the probability of fuel-damage would be very low. This argument will be examined very carefully in light of the experience of the Three Mile Island and Chernobyl reactor accidents. Both reactors sustained substantial core damage, but the former - which had a containment - caused no public injuries, while the latter - which had no containment - caused many injuries.

The main proposal for achieving competitive economic performance with the passively safe concepts is to employ factory-fabrication of a large portion of the power plant. The modules produced would be shipped to the station site for integration into a complete power plant. These modules would then be operated remotely from a central control station. It is argued that improvements in work quality and construction scheduling would be sufficient to make the modular passively safe concepts economically attractive.

Semi-passive Safety: In a variation upon the emphasis upon passive safety two semi-passively safe LWR concepts are being pursued in the United States. A passively safe concept provides the reactor cooling function without the intervention of an 'active' system (e.g., a human action or an action by a device such as a pump or valve which must change its state). A semi-passive system involves a small number of 'active' components, but in a way which will be very reliable (e.g., by requiring very small energetic stimuli in order to activate the system). A BWR (the ASBWR) and a PWR (the AP-600) of 600 MWe each are designed to provide assured core cooling, but using typically two valve realignments in order to do so. By doing this some of the economic penalties of the purely passive designs are avoided. It is expected, but not yet proven, that the marginal detrimental effects upon safety of this design approach are small compared to the passively safe alternatives.

Economic Performance: In most countries worldwide interest remains focussed upon economic performance-oriented designs. These mainly concern LWRs and LMRs.

LWRs: With LWRs concerns for evolutionary improvements in economics and safety are evident. Efforts for future improvements are focussed upon aspects of economic performance which were considered to be of lower concern when the current generation of plants was designed. Among the more important are improvements in plant operational availability and shortening the plant construction duration.

LMRs: With LMRs outside the United States the universal concern remains that of designing highly efficient breeder reactors. In most countries this concern is driven by a lack of nuclear fuel resources. The United States is the only country with a LMR program which is not focussed upon breeding as the primary objective. In all of the LMR countries listed in Table 1 the programs have been in place for many years, and are at the stage of gaining experience through operation of small prototype reactors. The largest such reactor is the 1200 MWe Superphenix in France. Over time scales greater than a century the ability of nuclear fission to contribute to alleviation of the greenhouse effect will depend greatly upon efficient use of nuclear fuel resources. This will require use of breeder reactors. From this perspective the focus of the United States LMR program upon passive safety is seen to be a preoccupation with near-term issues, mainly that of public acceptance of nuclear technology. With this focus the benefits of LMR technology for alleviating the greenhouse effect are not being advanced.

The Soviet Union: The Soviet Union is the only Communist country with a reactor development program. Until the Chernobyl accident in 1986 the Soviet Union had been focussed upon economic performance with its reactors. Since then their development priorities have appeared to be ambiguous. They have ceased development of new Chernobyl-type RBMK reactor concepts. However, development of improved, economic performance-oriented PWR and LMR concepts continues. During 1988 they have also agreed to build a passively-safe XXX MWe HTGR, of West German design and manufacture. The details of this project are still being determined, but undertaking it may indicate a major change in the orientation of the Soviet development program. Since the Soviet Union is a major nuclear power station supplier to Eastern European and third world countries such a change could have a major effect upon future versions of nuclear power technology.

The Third World: The developing and underdeveloped nations of the world are consumers of nuclear power technology, but not to a large degree. These countries will have large roles in determining the future severity of the greenhouse effect. This is because they account for most of current deforestation activities, much of which provide biomass fuels; and they are largely reliant upon fossil fuels. If these countries continue this reliance and become affluent their emissions of greenhouse gases will dwarf current worldwide contributions. For example if the people of only China and India were to attain the United States standard of living their CO_2 emissions would exceed twice those of North America, the Common Market Countries and Japan combined. In order for nuclear power to be effective in serving the third world it must be developed in forms very safe to operate and adaptable to hydrogen production. Using electrical hydrogen generation the latter task is feasible, but the former will be very difficult. Developments in this direction are underway with passively-safe reactor concepts, but will still be inadequate for most of the third world. Satisfying their needs may eventually be the greatest challenge to nuclear power in stemming the greenhouse effect.

Table 1

WORLDWIDE PROGRAMS OF NUCLEAR POWER TECHNOLOGY DEVELOPMENT

Programs Emphasizing Passive Safety

Federal Republic of Germany
- 100 MWe modular HTGR (Siemens, Brown Boveri)

United States
- Modular HTGR (130 MWe, General Atomic)
- Modular LMR (130 MWe PRISM concept, General Electric)
- 750 MWe PIUS-BWR (Oak Ridge National Laboratory)
- 600 MWe LWRs (semi-passive safety)
  - ASBWR (BWR, General Electric)
  - AP-600 (PWR, Westinghouse)

Sweden
- 500 MWe PIUS-PWR (ASEA-Brown Boveri)

Programs Emphasizing Economic Performance

Japan
- 1250 MWe LWRs
  - ABWR (Tokyo Electric Power, General Electric, Toshiba, Hitachi)
  - APWR (Kansai Electric, Mitsubishi, Westinghouse)
- 714 MWt LMR (Monju LMFBR project)

France
- 1400 MWe PWR (N4 project, Framatome, Electricite de France)
- 1200-1450 MWe LMR (Superphenix 1 project and successor Superphenix 2, Novatome, Electricite de France)

United Kingdom
- 1000 - 1400 MWe PWR (Sizewell B, Hinkley Point C projects and successors)
- 1450 MWe LMR (successor to 254 MWe Dounreay LMFBR Project)

Federal Republic of Germany
- 500 MWe HTGR (successor to 300 MWe THTR project)
- 300 MWe LMR (SNR 300 LMFBR project)

United States
- LWR Requirements Document Project (Electric Power Research Institute)
- 1250 MWe ABWR (General Electric)
- 1250 MWe APWR (Westinghouse)

Table 1 (cont')

<u>Soviet Union</u>

- Emphasis upon passive safety
  - 100 MWe Modular HTGR
  - Chernobyl-type RBMK reactor series discontinued
- Emphasis upon economic performance
  - 950 MWe PWR
  - 550 MWe LMR (LMFBR type)

<u>Abbreviations</u>

LWR: light water reactor
BWR: boiling water reactor
PWR: pressurized water reactor
PIUS: Process Inherent Ultimately Safe (version of LWR)
HTGR: high temperature gas-cooled reactor
LMR: liquid metal-cooled reactor
LMFBR: liquid metal-cooled fast breeder reactor (version of LMR)
PRISM: Power Reactor Inherent Safe Modular (version of LMR)

GLOBAL ENERGY STRATEGIES AND CLIMATE CHANGE

William U. Chandler
Battelle, Pacific Northwest Laboratories

INTRODUCTION

The scale of human activities may now be capable of changing the earth itself. (Clark, 1987) The bulk of gases forming Earth's climate and sustaining life is only 36 kilometers deep, a dimension which makes it vulnerable to human pollution in ways scientists do not fully understand. Indeed, if Earth were an apple, the atmosphere would be no thicker than the peel. Evidence that anthropogenic emissions into this thin but dynamic atmosphere could induce global climatic change has led two major international scientific conferences to call on leaders to cut carbon emissions, which are intimately related to energy use, by at least 20 percent. (Toronto, 1988; Hamburg, 1988)

The U.S. National Research Council reported in 1983 that human-induced climate change is a likely prospect over the next century. (NRC, 1983) Current trends in greenhouse gas emissions could warm the earth's atmosphere by 1-5 degrees Kelvin (K) by about 2050. This warming could shift storm track patterns and significantly diminish soil moisture levels in major grain-producing areas, including those of China, the United States, and the Soviet Union. (Kellogg, 1987) (See Figure 1 [at the end of the paper].) Thermal expansion of the oceans and some melting of sea ice could raise the average sea-level, causing serious problems for low-lying nations such as Bangladesh. Worse, non-linear, or unexpected changes in the earth system might occur. (Broecker, 1987) Thus, the future may hold some major, even catastrophic surprises. The question is no longer whether humans will alter the earth's climate, but by how much. Energy economics has a central role to play in both assessing and planning a response.

Carbon dioxide--a major by-product of fossil fuel combustion--is the most important greenhouse gas. (See Table 1.) In the atmosphere, it absorbs infrared energy and emits it back toward the ground and thus warms the earth. Without carbon dioxide in the atmosphere, the earth would be several tens of degrees colder, and so there is little dispute that the greenhouse effect is real. The more important question relates to the speed with which human activities are changing the atmospheric concentration of carbon dioxide and other trace gases.

Since 1960 the global concentration of carbon dioxide gas has increased from 316 parts per million to about 345 ppm. (See Figure 2.) Most scientists agree that if this level of concentration passes 600 ppm, climate change will be inevitable. Current trends in energy use suggest that this level will be reached sometime in the next century.

Methane, which constitutes natural gas and which is also released in coal production, is responsible for one-quarter of the warming effect estimated already to have been caused by anthropogenic greenhouse gases since the industrial revolution. (Dickinson and Cicerone, 1986) (See Table 2.) Its concentration has doubled over the last 200 years, largely as a result of human activities. (See Figure 3.) Atmospheric methane continues to increase at an annual rate of about 1 percent. (Cicerone and Oremland, in press) If unabated, this trend will lead to a temperature forcing effect which could raise global average surface temperature by approximately 0.1 K by the year 2020. (Hansen, et al, 1988) The combined temperature forcing of methane and other greenhouse gases, especially carbon dioxide and the chlorofluorocarbons, could dramatically affect the global environment. (EPA, 1988)

TRENDS IN REGIONAL CARBON AND METHANE EMISSIONS

Fossil energy use produces about 5.5 billion tons of carbon (~20 billion tons of carbon dioxide) worldwide each year. (Rotty, 1987) This total is roughly four-fifths of anthropogenic carbon emissions, or three to ten times releases from deforestation, which are highly uncertain. (Crutzen, 1987) Developed countries account for the largest share of carbon from energy use, with the United States and the Soviet Union each producing about one-fifth of the world total. Though developing nations represent three-quarters of the world's population, they generate only one-quarter of global carbon emissions. (See Figure 4.) Developed nations produced the bulk of the anthropogenic carbon that has accumulated in the atmosphere and upper oceans.

Carbon emissions have been relatively flat in the OECD nations since the first oil shock, but in developing countries, the Soviet Union, and Eastern Europe, emissions have increased sharply. (See Figure 5.) Although growth in energy use and income was largely decoupled in the West during the last fifteen years (IEA, 1987), growth in energy use--and probably carbon emissions--will be necessary to raise incomes in poorer nations. Incomes in most developing nations remain very low, averaging less than one-fifteenth that of the developed nations. Energy consumption per capita also remains an order of magnitude below that of the rich nations. Growth in incomes and energy use are both necessary for improving living conditions in the Third World, and developing nations will be the chief source of growth in carbon emissions. (Goldemberg, et al, 1987)

The need for Third World growth was explicitly acknowledged in the Montreal Protocol to Protect the Ozone Layer, as developing countries were permitted increases in use of ozone-depleting chlorofluorocarbons. (Bierbaum, et al, 1987) That type of approach, however, needs to be reexamined both for protection of the ozone layer and for reducing carbon emissions. Energy demand growth in the developing countries could, in the absence of major policy initiatives, double global carbon emissions by the year 2020. (Edmonds and Reilly, 1983; Edmonds and Reilly, 1985a; Chandler, 1985) It will thus be necessary to include developing countries in any plan to control carbon emissions.

Anthropogenic sources account for two-thirds of the methane lost to the atmosphere each year, and energy activities are responsible for one-third of these human sources. (Cicerone and Oremland, in press; Chandler, Barns, and Edmonds, 1988) The largest source of methane from coal mining is China, the world's largest coal producer with about one-fifth of the world total. The United States produces a similar share of the world's coal, and the Soviet Union uses about 15 percent. Methane emissions from natural gas consumption have also been estimated, with the United States, the Soviet Union, and the Middle East accounting for most of the gas lost at the wellhead, in transmission, and by intentional venting and flaring. (Chandler, Barns, and Edmonds, 1988) (See Table 3.)

FUTURE ENERGY-RELATED GREENHOUSE GAS EMISSIONS

Analysis of future energy use requires application of a model--and even a hunch in one's own head about the future is a model of sorts. A valid model for analysis must be reproducible and based on transparent and valid assumptions and principles. The Edmonds-Reilly Energy-Economic Carbon Dioxide Model meets these criteria and has become a widely-used tool for climate policy research. (Edmonds and Reilly, 1983; Edmonds and Reilly, 1985b) The capability to project energy and waste-related methane emissions from coal mining, natural gas production and transmission, woodfuel use, and municipal waste landfilling has also recently been added. (See Figure 6.)

The strengths, weaknesses, and sensitivities of this model have been detailed elsewhere, and will not be repeated here. (Chandler, in press) The model serves best in screening for the useful "solutions" to emissions problems as they are defined. For example, Edmonds-Reilly can be used to discover whether a particular set of policy objectives--reducing emissions while maintaining incomes, for example--can be better served by energy taxes or population control measures, if at all. It has been applied recently to assess the practicality of meeting the Toronto Conference goals. (Chandler, 1989)

The model's wide and continued use stems from three attributes, which are prerequisites for valid economic modelling. First, it is based on sound economic principles. Economic activity, or GNP, drives energy demand. The world economy is simply represented, as is appropriate for modelling over the very long term--exogenous variables such as population growth are difficult to predict accurately, and assumptions for price and income elasticity of energy demand are controversial. (Gibbons and Chandler, 1981) Second, the model is transparent. The model's key assumptions and economic principles have been clearly articulated, and they are open to view and modification. Third, the model is reproducible. The modelling philosophy, principles, and technical detail have been well-documented. It is also readily obtainable, permitting a variety of users to conduct policy experiments in a common framework.

POLICY EXPERIMENTS IN EMISSIONS REDUCTIONS

A Base Case scenario was constructed for this analysis from mid-range assumptions about key future energy and economic variables. (Edmonds and Reilly, 1985) (See Table 4.) A number of specific policy options were then produced for comparison with the Base Case. Note that each scenario is independent--that is, the policies and their effects are not additive across the scenarios.

1. World Base Case: a scenario generated from "medium" values for all assumptions, with these values being selected by Edmonds and Reilly (Edmonds and Reilly, 1985). It assumes a 1 percent annual rate of energy efficiency improvement in all regions. (See Table 4.)

2. World Low-Cost Nuclear Energy: tests a case in which nuclear energy meets a cost target worldwide of $0.037 per kilowatt hour (at the busbar) by the year 2005.

3. OECD Energy-Efficiency Standards: assumes the OECD nations attain 2 percent per year energy productivity improvements (with 1 percent per year elsewhere). This scenario helps differentiate the contributions to ameliorating the CO_2 problem that Western countries can make relative to developing and centrally planned nations.

4. U.S., Soviet Union, and China Coal Ban: tests the ability of these three major coal consumers to address the carbon dioxide problem with a hypothetical three-nation ban on coal use and exports after the year 2000.

5. Developing Nations Efficiency Standards: a scenario of 2 percent per year energy productivity growth in the developing world, with a 1 percent rate in the OECD, Soviet Union, and Eastern European countries.

6. World Efficiency Standards: a case in which world-wide energy efficiency improvements--through technical change or by cooperative regulation, averaged 2 percent annually, permit the world as a whole to reach current Japanese or Danish levels of energy efficiency by the year 2025.

Global fossil-fuel carbon emissions in the Base Case scenario would by 2075 total over 19.1 gigatonnes per year (a 1.5 percent annual rate of increase), almost four times the level of the 1980s. If 55 percent of the carbon dioxide remained in the atmosphere, the atmospheric concentration in the year 2075 would be about 600 ppm, compared with 345 ppm currently. Global per capita income would--in the absence of climatic feedbacks--reach almost twice current U.S. levels.

The World Low-Cost Nuclear scenario assumes that cost reduction targets for nuclear electricity--busbar costs of less than $0.037 per kilowatt hour--are met. (EPRI, 1987) This effort alone, however, would not appear to hold carbon dioxide concentrations far below 600 ppm. This scenario does not contribute larger reductions because even with lower costs, relatively cheap coal-fired power would reduce the market penetration of nuclear power. (See Table 5 for selected scenario results.)

Non-fossil fuel energy sources are often touted as solutions to the carbon dioxide greenhouse problem. Advocates of hydroelectric power, nuclear power, and solar energy sometimes argue that the climatic benefits of using these energy systems outweigh any other negative aspects. But a scenario (not presented here) in which China, for example, completely abandoned new sources of hydroelectric power would have little effect on the greenhouse problem. The difference amounts to 100 megatonnes of carbon, or 1 ppm atmospheric carbon dioxide concentration by the year 2075.

A scenario of high efficiency for developed Western nations acting alone, the OECD Efficiency Standards scenario reduces carbon emissions by 2.7

gigatonnes compared to the Base Case, and holds the atmospheric concentration in the year 2075 to about 575 ppm. A much greater reduction--more than 8 gigatonnes (to 525 ppm)--results in the Developing Nation Efficiency Standards scenario. This case assumes a 1 percent energy efficiency improvement rate in the Western nations and a 2 percent rate in the rapidly growing economies of developing nations. Income levels in most regions would rise slightly in either scenario.

The combined efforts of the United States, the Soviet Union, and China to ban coal use both for domestic use and export after the year 2000 would reduce year 2075 Base Case carbon emissions by 9.6 gigatonnes. Atmospheric carbon dioxide concentration would thus be held to just over 510 ppm for the year 2075. Chinese incomes would fall almost $1,000 compared to the Base Case, and thus China would not reach current U.S. income levels by the year 2075.

A global ban on coal might yield an atmospheric CO_2 concentration of about 465 ppm of carbon dioxide. In that case, a country like China might lose almost $1,800 in per capita GDP, compared to the Base Case scenario. This reduction in per capita GDP would result from the economic burden of using higher-cost energy, in the absence of major efficiency improvements.

It is worth re-iterating that these scenarios are not predictions but are policy experiments. What is most important about them is certainly not their precision, but the direction and magnitude of the impacts they suggest.

MANAGING ENERGY DEMAND

The Toronto Conference on the Changing Atmosphere called for carbon emissions to be cut by 10 percent by the year 2005 specifically by applying existing energy-efficiency measures. (Toronto, 1988) One study, at least, suggests that this goal could be met for the United States. (Edmonds, et al, 1988) Achieving a global 2 percent rate of energy efficiency improvement would hold carbon emissions to 7 billion tons in the year 2020, compared to 10 billion in the Base Case.

However, rates of efficiency improvement much higher than the 2 percent case tested above would be necessary. The developed nations, for example, would have to achieve about a 3.5 percent rate to cut carbon emissions by 10 percent by the year 2005 (assuming GDP grew an average of 2.5 percent per year). (See Figure 7.)

Even a 2 percent global rate of energy-efficiency improvement would be high compared to overall world trends but plausible compared to gains in the European Community and China. China apparently has reduced energy intensity at a rate well above 3 percent per year for a decade now. (Bentjerodt, et al; Institute of Energy Economics, 1987; Lu, 1986) And in the United States, a dramatic decoupling of energy and GNP growth took place with a 2.5 percent annual rate of energy efficiency improvement between 1973-88. (See Figure 8.) This accomplishment--although it has lately been reversed--has probably done more to reduce the rate of carbon emissions growth than any other factor.

While this rate is technically achievable, it would require unprecedented reductions in energy intensity. If the 3.5 percent annual rate of reduction in energy intensity could be continued to the year 2020, this policy would alone reduce carbon emissions to roughly 4.5 billion tons. This scenario, though it assumes an optimistically low population of 7 billion in 2030, does allow an ambitious global economic growth rate. The Chinese, for example, would increase their current income levels from about $600 per capita in 1985 to $8,000 by the year 2020.

The only scenarios considered in this study which met the Toronto goal for the year 2005 of cutting carbon emissions by 20 percent utilized very hefty rates of energy-efficiency improvement and very low cost nuclear power. The efficiency rate, 3.5 percent per year, is probably technically plausible, but would require acceptance of the urgency of acting to reduce the risk of climatic change, dramatic policy action to combat it, and unprecedented international cooperation. Even so, it is probably implausible to assume, as was done for this scenario, that solar or nuclear power can be reduced in cost to the level necessary to replace fossil fuels in the time permitted. In addition, the other problems of nuclear power must be overlooked to place confidence in this scenario.

The overriding importance of energy efficiency is demonstrated in these scenarios. Moderate efficiency and very inexpensive nuclear power together fail to meet the Toronto targets. Very high efficiency alone, on the other hand, reduces emissions below current levels by the year 2005 and keeps them there through 2020. Adding very cheap nuclear or solar immediately (say, through subsidies) would slightly improve the year 2005 picture, but would not meet the overall Toronto target for that year. Neither would it meet the 2020 Toronto target, but in this case global emissions would be reduced from 5.3 to 4.7 billion tons per year. This level could be reduced by another 100 million tons by reducing global population from 7 to 6.5 billion persons. Note, however, that holding population levels to the higher figure will require major policy intervention.

COST-EFFECTIVENESS IN REDUCING GREENHOUSE EMISSIONS

Few national studies have been made of the costs and benefits of aggressive energy efficiency options. But coupling macroeconomic with microeconomic-engineering analysis suggests that a large potential for energy efficiency exists. (Hafemeister, et al, 1986 SERI, 1981; OTA, 1981) Many nations compare poorly in terms of energy (and carbon) per unit of economic output, regardless of considerations of stage of economic development. (Chandler, 1985; Chandler, 1986; Chandler, in press) (See Figure 9.) If these comparisons are valid the world as a whole could maintain a 2 percent annual rate of energy efficiency improvement for 35 years without exhausting the existing energy-intensity reduction potential. That is the time which would be required to bring the world average energy intensity down to the level of the most-efficiency nations today such as Japan.

Technical analysis tends to corroborate the macroeconomic comparisons. The steel industry worldwide uses an average of 26 gigajoules (GJ) per tonne of steel produced, far above the 19 GJ used by best-available virgin-ore (BAT-V) technology. (Chandler, 1985) Recycling, moreover, can cut this requirement to 10 GJ per tonne (BAT-R), but 100 percent recycling is not possible because of impurities. The thermodynamic minimum for reducing iron from ore is about 6 GJ per tonne. A two percent annual rate of improvement in the energy efficiency of steel-making worldwide would necessitate going beyond the thermodynamic limit by the year 2075, which is of course impossible. The two percent rate could, however, readily be maintained for the next four decades. Further improvements after about the year 2025 would require knowledge that we do not possess or some steel-substitute that was much less energy intensive. (See Figure 10.)

A few simple appliances--furnaces, water heaters, refrigerators, and air conditioners--account for the bulk of energy use in residences. In the commercial building sector, lighting and space conditioning represent the majority of energy applications in terms of energy use. Fortunately, the considerable progress has been made in appliance efficiency in recent years. (Geller, 1988) Large and cost-effective reductions in energy use in these appliances remain possible. [See Figure 11.]

Similarly, automobiles and light truck fuel economy could be cost-effectively doubled with existing technology. (Bleviss, 1988) This is significant because two-thirds of the fuel used in the U.S. transport sector, and light vehicles account for 20 percent of U.S. carbon emissions. U.S. vehicles alone account for 5 percent of global energy-related anthropogenic carbon releases. Indeed, Americans emit as much carbon using their cars as the average citizen of the world does from all activities.

The economic costs of carbon emissions reductions can be compared for nuclear electricity, natural gas, and a variety of energy-efficiency measures against the cost of coal-fired electricity. (See Figure 12.) In general, energy-efficiency measures have negative costs--that is, they cut carbon emissions and save consumers money at the same time. More-efficient refrigerators, for example, save electricity typically at a cost of $5 per GJ, while delivered electricity costs $10 per GJ (in primary energy). When society invests extra money in an efficient refrigerator, and when the electricity saved is generated with coal, it saves 23.9 kg of carbon <u>and</u> $5 for every GJ saved. The cost is thus -$0.21 per kg of carbon emissions reduction.

Cutting carbon emissions by substituting natural gas for coal in electric power generation would cost essentially the difference in the cost of the two fuels. Natural gas contains only about half as much carbon per unit of energy, but costs about $1 per GJ more than coal. Substituting natural gas would thus save carbon at a cost of $0.10 per kg. Electricity generated from natural gas rather than coal in conventional power plants would cost an additional $0.011 per kWh, about 10 percent of average delivered costs.

The delivered price of nuclear power in the United States is--at the margin--about $0.10 per kWh, or $0.01 per kWh more than coal-fired electricity. Substituting nuclear for coal would cut carbon emissions at about $0.04 per kg. If the EPRI cost-reduction target for nuclear power could be met, this cost would be reduced by one-third. Note, however, that this estimate does not include waste storage or decommissioning costs, which according to some estimates could double the cost of nuclear power.

If automobiles were required to achieve 50 mpg, carbon emissions would be reduced at about $0.60 per gallon of gasoline saved. Since gasoline contains 19.6 kg of carbon per GJ, the cost of carbon reduced would total -$0.03 per kg. That is, the effect on the consumer and on society would be a net savings exclusive of the benefits due to reduced risk of climatic change. The fuel economy of the world's automobile fleet could, according to some sources, easily be doubled. (Bleviss, 1988)

POLICY ISSUES

There remains the question of how to change energy-using behavior to economically-justifiable levels. What kinds of policies would be required to achieve a global rate of 2 percent energy efficiency improvement? Results using the Edmonds-Reilly model suggest that even 200 percent price increases--through a carbon tax or otherwise--would not alone suffice. Additional policies, perhaps in the form of end-use efficiency regulations, would still be required.

Incorporating external costs of energy use--that is, the environmental cost of climatic change--into the price of energy will be difficult. Society does not even pay the production cost of energy, mainly because half the world's energy is produced in non-market economies where prices rarely reflect opportunity costs. Such nations are invariably the least energy-efficient. Of course, centrally planned nations could always mandate energy efficiency improvements, but there is little evidence that this will occur. It is thus not clear how efficiency can be achieved in the absence of signals that convey to the consumer the economic value of energy. (Chandler, 1986)

It should be noted, however, that many nations already have goals to achieve high rates of energy efficiency. The European Community has an official goal of reducing the energy-intensity of its members by 20 percent by the year 1995. (EC, 1988) China, having achieved a 3.7 percent rate during the eighties, plans to maintain a 2.2 percent annual rate of energy-efficiency improvement for the rest of this century. (Institute for Energy Economics, 1987; Lu, 1986; Xu, 1988)

One strategy to control carbon dioxide emissions would be to wring first the maximum cost-effective levels of energy efficiency out of the world's economies and then turn to non-fossil renewable energy sources. This strategy is feasible, but it would be difficult to achieve without artificially restricting fossil fuel use to force market penetration of new supply technologies. Fossil fuels are likely to remain competitive for decades with renewable technologies, because the latter require large capital investments that will probably make them expensive even compared to scarce fossil fuel. Imposing the 200 percent tax on fossil fuel--that is, tripling 1975 prices--might accomplish much of the task. But effecting such a policy would require policy makers at a minimum to expand substantially their understanding and acceptance of human-induced climatic change as a major problem to be addressed.

Another strategy would be to pursue maximum cost-effective energy conservation investments through the year 2025 while simultaneously investing in basic research in the hope of finding brand new processes that would permit either energy efficiency improvements to continue at high rates, or low-cost, benign energy supplies to substitute for fossil fuels. The rationale for this approach is that fundamental breakthroughs in processes will be necessary, but that there are no reasons *a priori* to assume that new energy supply processes will provide a better solution than new processes that improve efficiency.

CONCLUSION

A truly international effort will be required to keep atmospheric carbon dioxide levels well below a doubling relative to pre-industrial levels. From a policy perspective, this finding is reason for concern; obtaining a consensus for effective and timely action on so profound and controversial an issue will be difficult.

Most significantly, results suggest that climatic change can be slowed with measures that are, for reasons unrelated to climate, in many nation's own self-interest. Energy efficiency scenarios both significantly reduce carbon emissions and increase per capita incomes. International cooperation for transfer of technology and knowledge for energy efficiency could provide global benefits. The choice to ameliorate climate change, therefore, need not be laden with unacceptable risks or costs. The decision, however, belongs to citizens and their leaders. Science cannot and should not make the determination whether carbon dioxide control policies are worthwhile. At best, science can advise society on the benefits and costs of such efforts.

Table 1: Anthropogenic Gases and the Greenhouse Effect

Constituent	Estimated Effect of Doubling on Temperature (Deg. Celsius)	Current Rate of Increase (% yr^{-1})	Residence Time in Atmosphere (years)
Carbon Dioxide	0.2-2.3	0.5	2[a]
Carbon Monoxide	0	1-2[b]	3
Tropospheric Ozone	0.9-1.1	(c)	na
Nitrous Oxide	0.3-1.1	0.2	150
CFC-11	0.1	5.0	75
CFC-12	0.02	5.0	110
Methane	0.3-1.4	1.0	7-11

a. This may be misleading because CO_2 is part of a natural cycle.
b. One source reports "no trend" in CO accumulation.
c. Measures of total column ozone indicate a depletion rate ranging from - 0.2 to - 5.0 percent per year, with most reduction coming in the stratosphere.

SOURCES: Jae Edmonds and Gregg Marland, "The Energy Connection to Global Climate Change: Gaseous Emissions," Institute for Energy Analysis, 1986, mimeo; NASA, "Present State of Knowledge of the Upper Atmosphere," 1986; and T.F. Malone and J.G. Roederer, eds., Global Change (London: Cambridge University Press and the International Council of Scientific Unions, 1985).

Table 2: Anthropogenic Greenhouse Warming by Gas, 1985

Gas	Thermal Trapping Compared to Pre-Industrial (Watts per square meter)
Carbon Dioxide	1.3
Methane	0.6
Tropospheric Ozone	0 to 0.2
CFC-12	0.12
CFC-11	0.06
Nitrous Oxide	.05
TOTAL	~2.2

SOURCE: Robert E. Dickinson and Ralph J. Cicerone, "Future Global Warming from Atmospheric Trace Gases," Nature, 9 January 1986.

Table 3: Estimated Current Annual Sources of Atmospheric Methane

Source	Total (Range) (million tons)	Anthropogenic Share (percent)	Total Share (percent)
Natural Wetlands	115 (100-200)	-	~20
Rice Paddies	110 (60-170)	~30	~20
Animals	80 (65-100)	~20	~15
Biomass Burning	55 (50-100)	~15	~10
Natural Gas	45 (25-50)[a]	~10	~8
Landfills	40 (30-70)	~10	~8
Termites	40 (10-100)	(b)	~8
Coal Mining	35 (25-45)	~10	~8
SUB-TOTAL	520 -	~95	~97
ALL SOURCES	540 (400-640)	~66	100

NOTES:

a) Recent research by Barns suggests that methane emissions from all natural gas production and transmission total only about 20 teragrams per year.

b) Termite decomposition of cleared forests could also be considered anthropogenic.

SOURCE: Modified from R.J. Cicerone and R.S. Oremland, "Biogeochemical Aspects of Atmospheric Methane," Global Biogeochemical Cycles, in press.

Table 4: Major Assumptions for Scenario Analysis

Population:

World total, year 2075 = 7.1 billion with aggressive family planning; 8.4 billion without.

Labor Productivity:

Developed countries = .017 growth per year; developing countries = .029 growth per year.

Income elasticity of energy demand:

Initial values of 1, 1.25, and 1.4 for developed western, developed eastern bloc, and developing countries, respectively; converging toward 1 by the year 2075.

Price elasticity of energy demand:

-.7 for all nations for all sectors.

Energy price-GNP Feedback elasticity:

-.15 and -.2 for developed and developing countries, respectively.

Fossil fuel resources:

Oil = 21,250 exajoules (EJ)
Natural gas = 13,950 EJ
Coal = 271,000 EJ.

Table 5: Year 2075 Scenario Results: International Carbon Dioxide Control Policy Options

Scenario	Global Carbon Emissions (C in gigatonnes)	Chinese Income Per Capita ($1975)	Atmospheric Concentration[1] (ppm)
0 1975 Actual	4.5	360	339
1 World Base Case	19.1	7236	601
2 Low-Cost Nuclear Energy	16.3	7398	576
3 OECD Efficiency Standards	16.4	7428	577
4 US-Soviet-China Coal Ban	9.5	6277	512
5 Developing Nation Eff. Stds.	10.8	7561	525
6 World Efficiency Standards	5.9	7739	473

1 Assumes an atmospheric carbon retention rate of .55.

See text for definitions.

SOURCE: Author

Figure 1

Changes in Precipitation as a Consequence of Anthropogenic Climatic Change

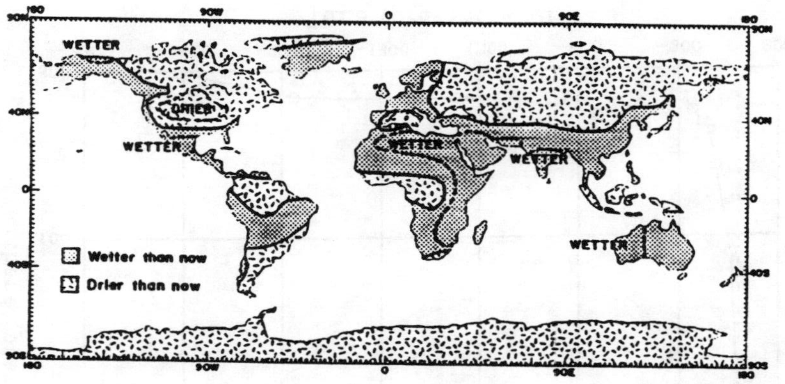

SOURCE: Kellogg, 1987

Figure 2

Trends in Atmospheric Carbon Dioxide Concentration

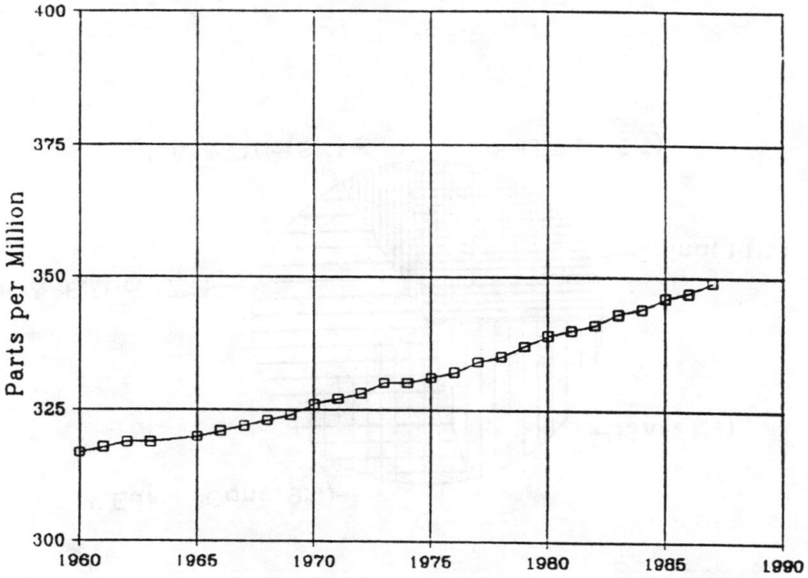

SOURCE: Scripps Institute of Oceanography

Figure 3

Trends in Atmospheric Methane Concentration

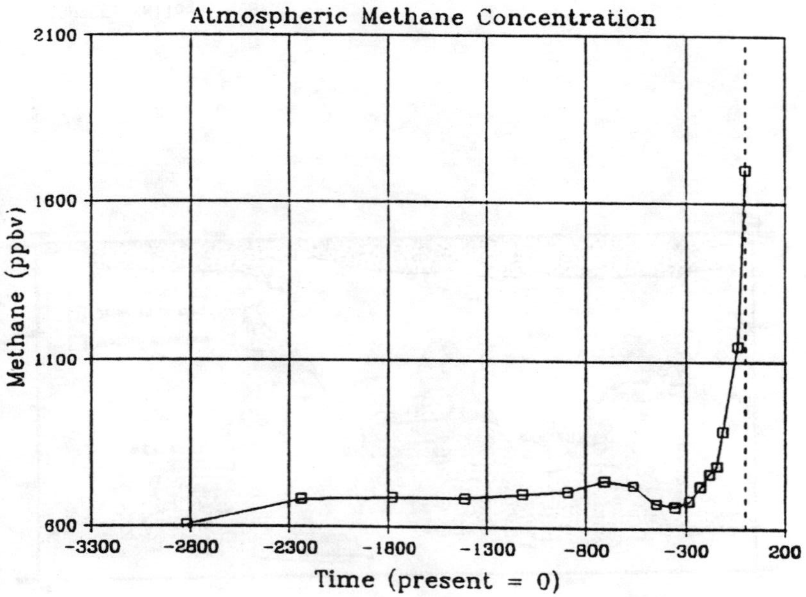

SOURCE: Rasmussen and Khalil, 1984

Figure 4

Carbon Emissions by Region, 1986

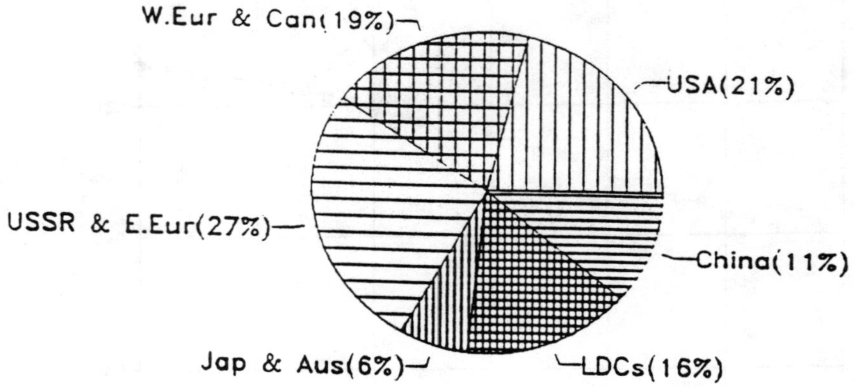

Figure 5

Trends in Carbon Emissions by Region

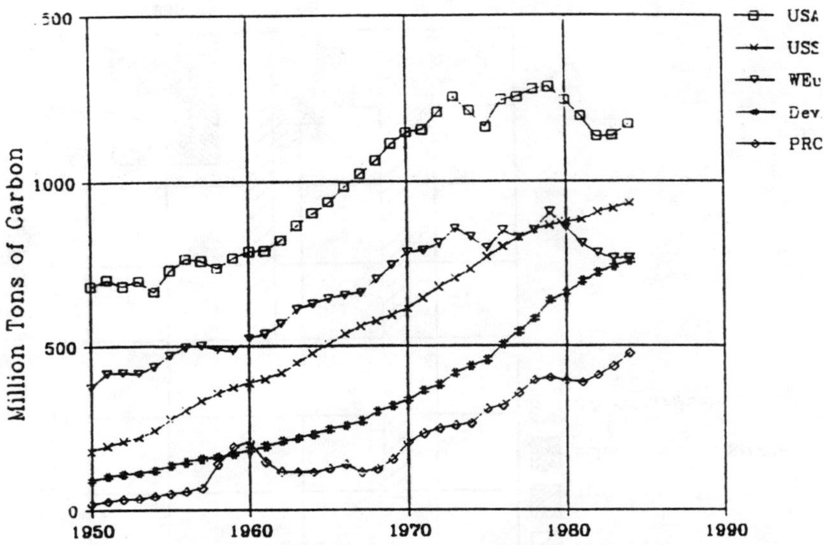

SOURCE: R. Rotty, UN data

Figure 6: Schematic of the Edmonds-Reilly Model

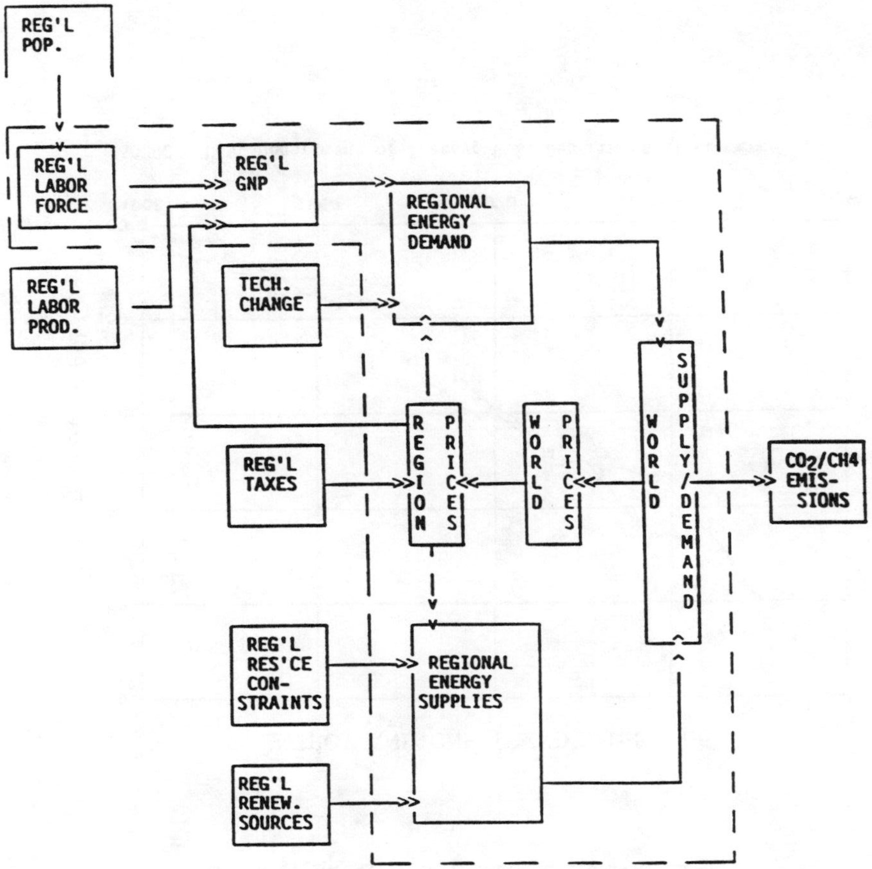

Figure 7

Options for Meeting the Toronto Goal

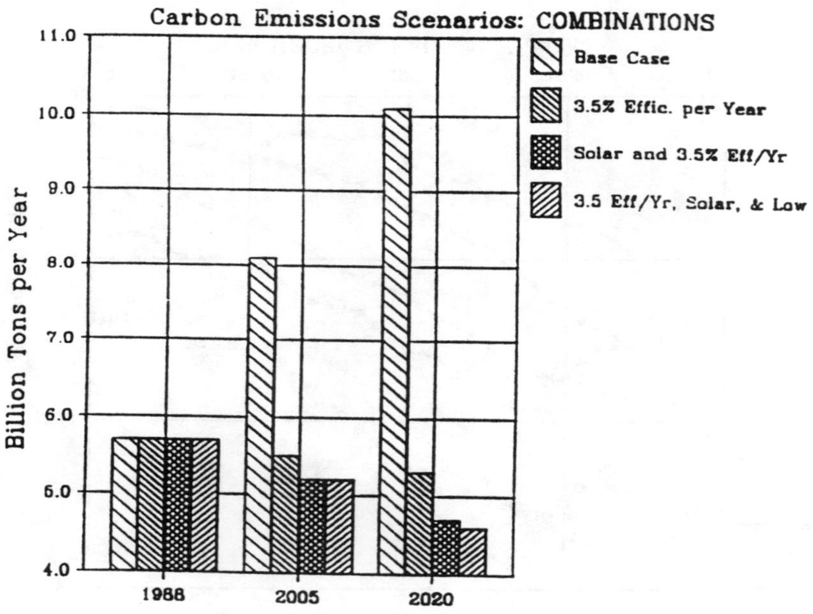

SOURCE: Author

Figure 8

Energy Intensity in the United States, 1950-88

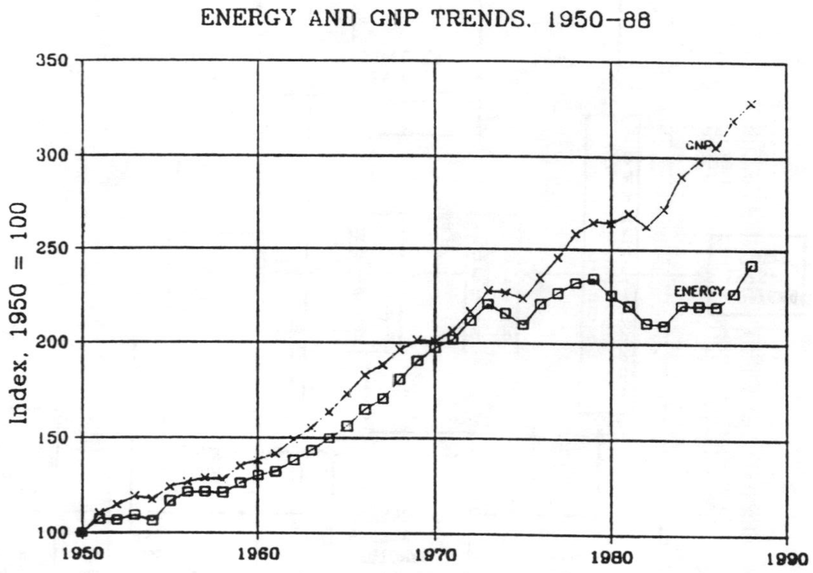

SOURCE: U.S. Department of Energy; U.S. Department of Commerce

Figure 9

The Energy and Carbon Intensity of Selected Nations

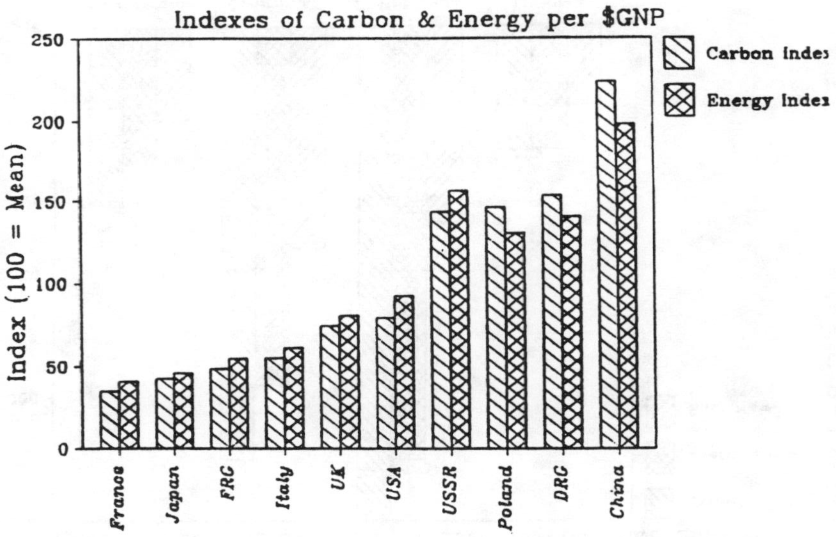

SOURCE: Author

Figure 10

Energy Intensity of Steel Production

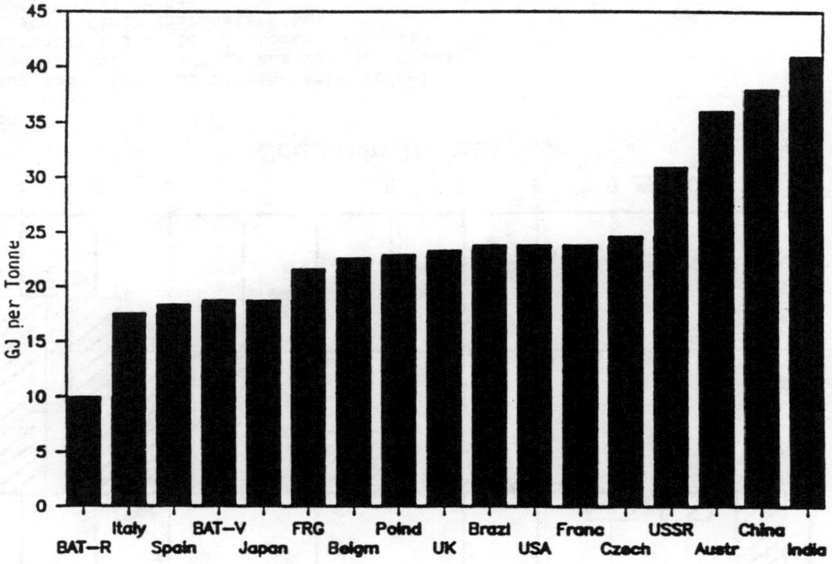

SOURCE: Author

Figure 11

Energy Efficiency Improvements in U.S. Appliances

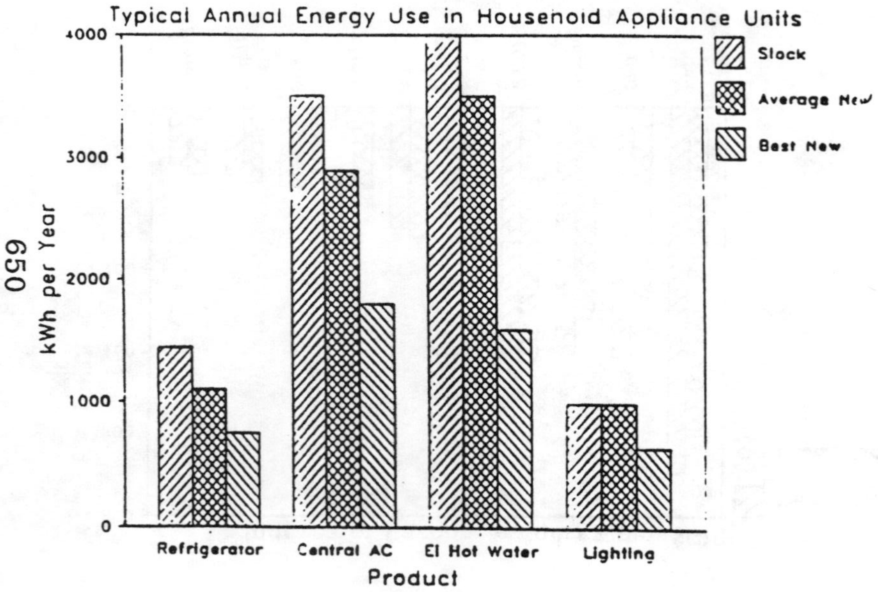

SOURCE: Geller, 1988

Figure 12

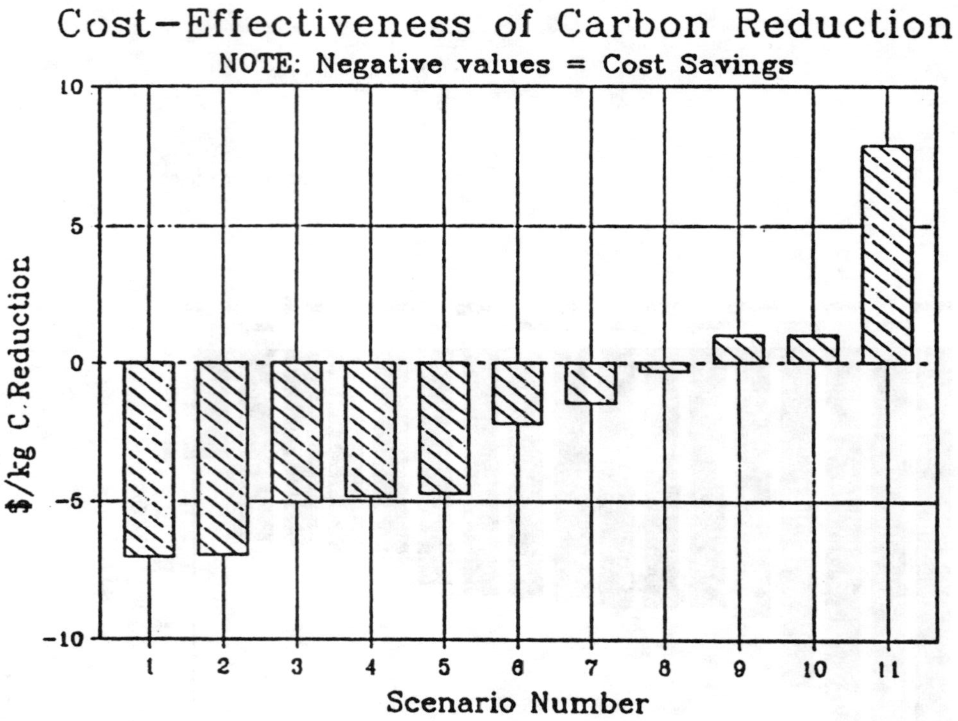

Option

1 Buy Most Efficient Refrigerator (Elec)
2 Install Hot Water Flow Restrictor (Elec)
3 Retrofit Industrial Evaporators (Gas)
4 Install Clock Thermostat (Gas)
5 Switch Coal to Gas
6 Incr. Attic Insul., R-11 to R-19 (Gas)
7 Buy 50 mpg Car
9 Install Wall Insulation, R-11 (Gas)
9 Caulk Worst House Shell Leaks (Gas)
10 Install Storm Windows (Gas)
11 Switch Coal to Nuclear

Assumes 16.7, 19.2 and 23.8 kg/GJ of gas, oil, and coal, respectively
Assumes $5.5, $8, and $5.50/GJ of gas, oil, and primary energy for delivered electricity, respectively

SOURCE: Author

REFERENCES

Roberto Bentjerodt, et al., China: The Energy Sector (Washington: The World Bank, 1985).

See R. Bierbaum, N. Sundt, and R. Friedman, "An Analysis of the Montreal Protocol," Staff paper, Office of Technology Assessment, U.S. Congress, Washington, D.C., December 1987.

D.L. Bleviss, The New Oil Crisis and Fuel Economy Technologies: Preparing the Light Transportation Industry for the 1990s (Greenwood Press, 1988).

W. Broecker, "Surprises in the Greenhouse?", Nature, 1987.

William U. Chandler, Energy Productivity: Key to Environmental Protection and Economic Progress (Washington, D.C.: Worldwatch Institute, 1985).

William U. Chandler, The Changing Role of the Market in National Economies (Washington: Worldwatch Institute, 1986).

William U. Chandler, "Assessing Carbon Emission Control Strategies: The Case of China," Climatic Change, in press.

William U. Chandler, "An Evaluation of International Measures for Controlling Carbon Dioxide Emissions," Draft, Battelle, Pacific Northwest Laboratories, 1989.

William U. Chandler, David W. Barns, and James A. Edmonds, "Atmospheric Methane Emissions: A Summary of Sources and Policy Issues," draft, Battelle, Pacific Northwest Laboratories, 1988.

R.J. Cicerone and R.S. Oremland, "Biogeochemical Aspects of Atmospheric Methane," Global Biogeochemical Cycles, in press.

W.C. Clark, editor, The Sustainable Development of the Biosphere (London: Oxford University Press, 1987).

Paul J. Crutzen, "Role of the Tropics in Atmospheric Chemistry," in R.E. Dickinson, editor, The Geophysiology of Amazonia (New York: John Wiley, 1987.

Robert E. Dickinson and Ralph J. Cicerone, "Future Global Warming from Atmospheric Trace Gases," Nature, 9 January 1986.

Jae Edmonds and John Reilly, "A Long-Term Global Energy-Economic Model of Carbon Dioxide Release from Fossil Fuel Use," Energy Economics, April 1983.

James A. Edmonds and John M. Reilly, "Future Global Energy and Carbon Dioxide Emissions," in John R. Trabalka, ed., Atmospheric Carbon Dioxide and the Global Carbon Cycle (Springfield, VA: National Technical Information Service, 1985). (Edmonds and Reilly, 1985)

Jae Edmonds and John Reilly, Global Energy (Cambridge: Cambridge University Press, 1985). (Edmonds and Reilly, 1985b)

Electric Power Research Institute (EPRI), "Technical Assessment Guide, 1: Electricity Supply 1986," Palo Alto, CA, 1986.

Commission of the European Communities (EC), "The Main Findings of the Commissions's Review of Member States' Energy Policies," Brussels, 1988.

Howard H. Geller, "Residential Equipment Efficiency: A State-of-the-Art Review," American Council for an Energy-Efficient Economy, Washington, D.C., May 1988.

John H. Gibbons and William U. Chandler, Energy: The Conservation Revolution (New York: Plenum Press, 1981).

J. Goldemberg, T.B. Johansson, A.K.N. Reddy, and R.H. Williams, Energy for a Sustainable World (Washington, D.C.: World Resources Institute, September 1987.

Jose Goldemberg, Thomas B. Johansson, Amulya K.N. Reddy, and Robert H. Williams, "Energy for Development," World Resources Institute, September 1987.

Hamburg Conference on Climate and Development, "Hamburg Manifesto," 10 November 1988.

The Institute of Energy Economics (Japan), "China's Energy Situation, Present and Future," NIRA Report, Vol. 4, No. 1, 1987.

International Energy Agency, Energy Conservation in IEA Countries (Paris: OECD, 1987).

David Hafemeister, Henry Kelly, and Barbara Levi, editors, Energy Sources: Conservation and Renewables (Washington, D.C.: American Institute of Physics, 1986).

William W. Kellogg, "Mankind's Impact on Climate: The Evolution of an Awareness," Climatic Change, 10 (1987) 113-136. See also, Zong-ci Zhao and William W. Kellogg, "Sensitivity of Soil Moisture to Doubling of Carbon Dioxide in Climate Model Experiments: Part II: The Asian Monsoon Region," Journal of Climatology and Applied Meteorology, in press.

T.F. Malone and J.G. Roederer, Global Change (Cambridge: Cambridge University Press and the International Council of Scientific Unions, 1985).

R.A. Rasmussen and M.A.K. Khalil, "Atmospheric Methane in the Recent and Ancient Atmospheres: Concentrations, Trends, and Interhemispheric Gradient, Journal of Geophysical Research, 89, D7, 1984.

Ralph Rotty, private communication, 1987.

Solar Energy Research Institute (SERI), Building a Sustainable Energy Society (Washington, D.C.: U.S. Government Printing Office, 1981).

U.S. Congress Office of Technology Assessment (OTA), Industrial Energy Use (Washington, U.S. Government Printing Office, 1981).

Lu Yingzhong, "Economic Growth and Energy Use in PRC," Working Paper, Institute for Energy Analysis, Oak Ridge Associated Universities, 1986.

Joel B. Smith and Dennis A. Tirpak, editors, The Potential Effects of Global Climate Change on the United States, draft report to Congress, U.S. Environmental Protection Agency, October 1988

Toronto Conference on the Changing Atmosphere, 27-30 June 1988, "Conference Statement," Environment Canada, 1988.

U.S. National Research Council (NRC), Changing Climate: Report of the Carbon Dioxide Assessment Committee (Washington: National Academy Press, 1983).

U.S. National Aeronautics and Space Administration (NASA), Present State of Knowledge of the Upper Atmosphere, An Assessment Report (Washington: NASA, 1986).

World Bank, World Development Report 1987 (New York: Oxford University Press, 1987).

Xu Shoubo, Remarks by Professor Xu Shoubo, Vice Director, Scientific Committee, Technical Economics Research Institute, Chinese Academy of Social Science, at "Workshop on Developing Country Energy Strategies: Implications for the Greenhouse Problem," U.S. Environmental Protection Agency, Washington, D.C., 28 April 1988.

WORLD RESOURCES INSTITUTE
A CENTER FOR POLICY RESEARCH

1709 New York Avenue, N.W., Washington, D.C. 20006, Telephone 202-638-6300

DEVELOPING COUNTRY ENERGY STRATEGIES
An Overview of the Climate Implications of Energy Strategies

Dr. Irving M. Mintzer, Senior Associate
Climate, Energy and Pollution Program

During the last five years, a strong consensus has emerged in the international scientific community that the continuing buildup of radiatively-active trace gases in the atmosphere will ultimately lead to a dangerous warming of Earth's surface and to large potential changes in regional climatic regimes.[1] Recent analyses of the historical record of surface temperature measurements suggest that the planet has already warmed by about 0.5 to 0.7 degrees C, an amount consistent with the level of warming predicted by models of the greenhouse effect.[2] The observed increase in temperature is believed due to the combined effects of increases in atmospheric concentrations of carbon dioxide (CO_2), nitrous oxide (N_2O), methane (CH_4), the chlorofluorocarbons (CFCs), and tropospheric ozone. Table 1 summarizes the measured changes in concentrations of greenhouse gases over the last century.

TABLE 1

KEY GREENHOUSE GASES

Present concentrations in air, rates of increase and pre-industrial era concentrations. ppm = parts per million, ppb = parts per billion, ppt = parts per trillion.

GAS	Concentration in air Pre-Industrial	1986	Present Rate of Increase (per year)
Carbon Dioxide (CO_2)	275 ppm	346 ppm	1.4 ppm (0.4%)
Methane (CH_4)	0.75 ppm	1.65 ppm	17 ppb (1.0%)
Fluorocarbon-12 (CCl_2F_2)	Zero	400 ppt	19 ppt (5.0%
Fluorocarbon-11 (CCl_3F)	Zero	230 ppt	11 ppt (5.0%)
Nitrous Oxide (N_2O)	280 ppb	305 ppb	0.6 ppb (0.2%)
Ozone, Tropospheric (O_3)	15 ppb?	35 ppb	0.3 ppb? (1.0%) Northern Hemisphere only
Other Fluorocarbons	Zero	see text	(5.0 to 15%)

ENERGY USE IN DEVELOPING COUNTRIES: The Real Energy Crisis

Low gasoline prices and the apparent recovery of the Western industrial economies have not changed the desperate situation faced by the half of the world's population that is starved both for food and for commercial energy resources. More than two-thirds of the human population lives in developing countries, most of which have per capita incomes that are 10 percent or less of those in the industrialized world. Food shortages are widespread and increasing; over 20 percent of the population of these countries is in imminent danger from hunger or malnutrition.

In most developing countries, per capita energy use is about one kilowatt per year. This is less than one-sixth of the average level in the industrialized world. Almost half of the energy use in developing countries comes from traditional solid biomass-derived fuels, including fuelwood, dung, and crop residues. Unfortunately, more than half of the developing world now lives in areas of fuelwood deficit.

The Environmental and Economic Costs of Conventional Energy Strategies

For the past several decades, energy planning in developing countries has emphasized the expansion of conventional energy supplies.[3] Geared to increasing the volumes of commercial fuels available to city-dwellers in the modern sector of these economies, the energy needs of the much-larger rural populations have gone largely unaddressed. Nonetheless, these programs have led to a doubling of per capita commercial energy use in less than twenty years. In the process, the cost of imported oil has become an ever-larger share of total export earnings. For low and middle-income developing countries, oil imports rose to 61 and 37 percent, respectively, of total earnings by 1981.[4] During the 1970s, investments to expand domestic energy supplies, measured as a fraction of total Gross Domestic Product, roughly doubled in these countries.

Despite the dismal results of many of these investment programs, the end is nowhere in sight. The trends continue unabated, with frightening prospects ahead. If these trends continue, per capita commercial energy demand in 2020 will be more than the four times the level of 1980 (2.3 vs. 0.55 kilowatt-year per capita. In aggregate, rising populations would thus increase commercial energy demand in developing countries from 2 terawatt-years per year to 15 terawatt-years per year. Meeting this demand would require impossible amounts of foreign exchange for these economies and would be likely to replicate the poisonous episodes of atmospheric pollution that have characterized development of the Western market economies.

Plants Won't Grow Fast Enough

In this type of high energy growth scenario, it will be impossible to fuel development from traditional renewable sources. Biomass production for energy could not be expanded fast enough to meet the need for combustible fuels without sacrificing food production as well as burning down all the available forest areas. Hydropower, the principal new source of electricity for many developing countries, has only limited availability and varies from region to region. Guilford[5] has estimated that although only a small fraction of the hydropower resource has been exploited commercially in developing countries, the total economic potential is less than one terawatt-year per year.

Many developing countries have abundant resources of solar energy which could potentially be exploited through direct and indirect conversion technologies. Wind resources, an indirect form of solar energy, are widely available but unevenly distributed and expensive to tap. Photovoltaic conversion of solar energy to electricity is totally feasible and widely used but much too expensive today to be an economical source of bulk power. Prices of photovoltaic and wind systems are likely to drop in the future but not as quickly as had once been hoped.

Sustainable Development: A Feasible Alternative

Despite this depressing outlook, there is reasonable cause for optimism. Goldemberg, et al,[6] point out that end-use oriented analysis of energy demand can suggest a way out of the dilemma that will sustain the prospects of sustainable development. Using this method, the need for energy can be separated from appetites for energy.

Experiences during the last fifteen years in the United States, Western Europe, Japan, the Soviet Union, the People's Republic of China, Korea, Taiwan, and many other countries have shown that GNP growth is not inexorably tied to growth in energy use. End-use oriented analysis can help to identify additional opportunities for squeezing more useful work and economic value out of each pound of coal or barrel of oil burned. By focusing on the question of what services energy is needed for, and emphasizing the importance of increasing the efficiency of energy use, it may be possible to identify development strategies that minimize the requirements for hard currency inputs, support integrated rural development, and encourage a smooth transition toward efficient industrialization of developing countries.

As an added benefit, the technologies required for this set of development strategies will minimize the long-term environmental damages from energy supply and use. In particular, they will minimize the contribution to future global warming from the industrialization of developing countries. In the process, they will reduce the risks of rapid climate change that would otherwise undermine the prospects for successful and sustainable development.

SLOWING THE BUILDUP

Numerous policy options exist for reducing the rate of future emissions growth for each of the principal greenhouse gases. This section highlights some of the options related to energy strategy.

CARBON DIOXIDE (CO_2). The three main approaches to slowing the rate of CO_2 build-up are: (1) improving the efficiency of energy supply and use; (2) shifting the fuel mix away from coal toward less CO_2-intensive fuels; and (3) reducing the rate of CO_2 emissions from biotic sources.

Two recent studies published by the World Resources Institute (WRI) demonstrate that an energy strategy based on increased energy efficiency can sustain economic growth in both developing and industrialized countries.[7] Many technological opportunities now exist for improving energy efficiency. On the demand side, these range from the introduction of more efficient light bulbs in commercial buildings to the construction of better-insulated buildings and the manufacture of more fuel-efficient vehicles. In each area, the best technology available today requires 50 percent or less energy than the typical devices currently used in the United States. In most cases, these newer devices cost more to buy but have lower total life-cycle costs.

The mix of energy supply options can also be shifted away from such carbon-intensive sources as coal. Direct combustion of coal releases 26.7 million tons of carbon per exajoule.[8] Delivering the same amount of energy from oil releases about 70 percent as much CO_2 while burning natural gas emits only about half as much CO_2 than burning coal directly does. If the energy is produced in nuclear reactors or from such renewable energy sources as hydropower, wind, or solar technologies, however, no CO_2 is released. To the extent that energy supply can be shifted to these less CO_2-intensive technologies, emissions of CO_2 can be reduced without reducing energy supply.

Biotic sources contribute between 20 and 40 percent as much CO_2 to the atmosphere each year as does the burning of fossil fuels. The principal means of reducing biotic emissions are limiting tropical deforestation and making biomass use more efficient. The destruction and burning of the Earth's tropical forests contribute to the build-up of atmospheric CO_2 both by reducing the stock of forests that act as a sink for CO_2 and by increasing emissions through decay and direct combustion.

Cooking with fuelwood over an open fire is the largest single end-use of energy outside of industry in many developing countries. The thermal efficiency of cooking can be improved from current averages of 10 percent or less to nearly 70 percent by substituting newer stove designs and switching to high-quality fluid fuels.[9] If cooking is more efficient, pressure on forests will decrease along with the drudgery of fuelwood collection. The widespread use of such stoves could also noticeably reduce global CO_2 emissions from biotic sources.

NITROUS OXIDE (N_2O). Nitrous oxide is released in both bacterial processes and the combustion of fossil fuels, especially coal and fuel oil. Discouraging growth in the use of coal and coal-derived synfuels will slow the rate of N_2O emissions from fossil fuel use. Not enough is currently known about the biotic processes leading to N_2O emissions to identify control policies for the future, but some recent research suggests that deforestation (especially clear-cutting) can increase local emissions by two orders of magnitude.[10] To the extent that policies can be implemented to limit these practices, future releases of N_2O may be further reduced.

METHANE (CH_4). The sources and sinks of CH_4 in the atmosphere are not completely understood today.[11] As a consequence, the ability of policy to slow the build-up of CH_4 is limited, but a few options remain. Losses from fossil-fuel extraction and the transport of natural gas may contribute as much as 15 percent of global emissions of CH_4.[12] About three percent of the natural gas mined in the United States is unaccounted for by gas companies. Providing incentives to improve pipeline maintenance practices and imposing strong penalties for leakage will reduce this source of emissions.

Another important option is to make biomass combustion more efficient. When wood or other biomass fuel is burned in inefficient cookstoves or open fires, some of the carbon in the fuel is released as CH_4 rather than as CO_2. Limiting tropical deforestation and introducing more efficient cookstoves could significantly reduce this source of CH_4. The magnitude of these effects cannot be estimated yet, but important research underway in Brazil promises to provide some valuable new data.

A third option for slowing the build-up of atmospheric CH_4 involves preserving the natural stock of free hydroxyl (OH) radicals in the atmosphere, the principal sink for CH_4. Carbon monoxide itself is not a greenhouse gas. However, CO combines readily with free hydroxyl radicals in the atmosphere, depleting the natural sink for CH_4. By removing OH radicals, CO emissions extend the atmospheric lifetime of CH_4 molecules. CO is produced primarily during inefficient combustion of biomass and during the incomplete combustion of hydrocarbon fuels. It is released in large quantities worldwide as a component of the exhaust gases produced by cars and light trucks. Policies that encourage the introduction of more efficient cookstoves will reduce the first family of sources. The introduction of standards and performance goals for auto and truck engines will reduce the second.

CHLOROFLUOROCARBONS (CFCs). CFCs are manufactured molecules used as aerosol propellants, blowing agents for plastic foams, refrigerants, and solvents. The most dangerous members of this family of chemicals are fully-halogenated compounds, such as CFC-11, CFC-12, and CFC-13 -- extremely stable compounds with lifetimes in the atmosphere of 75 to 150 years. On a per-molecule basis, they are approximately 10,000 times as efficient as greenhouse gases as CO_2. Four policy strategies can be used to slow the build-up of CFCs in the atmosphere. THe incentives for each of the following options may take the form of statutory limits on the production and use of these compounds or the imposition of taxes or fees.

The first strategy involves incentives (or penalties) to improve equipment maintenance and encourage the more efficient use of CFCs in applications (such as residential refrigeration) where their special properties make substitutes impractical today. The second option involves policies that encourage recapture and recycling of these compounds rather than release and replacement. The third option is to provide incentives for the introduction of alternative, safer formulations of CFCs in such uses as building insulation. Several formulations of the new, safer CFCs (e.g., CFC-22, CFC-142b, and CFC-152a) are now available and others (e.g., CFC-134a, CFC-123, and CFC-141b) are under development. All are less dangerous than such traditional formulations (such as CFC-11 and CFC-12) because they have much shorter atmospheric lifetimes or they contain no chlorine. To date, they have only been produced in laboratory quantities; full-scale production processes have not yet been developed and tested. As a result, these alternative compounds remain 1.5 to 5.0 times as expensive to produce and the more dangerous, traditional CFCs (Dupont, 1987).

The fourth options is to encourage the use of non-CFC substitutes. One application where non-CFC substitutes could replace CFC-based products with little loss of utility is in insulated food packaging. Use of waxed paper, aluminum foil, or cardboard cartons in food packaging in lieu of CFC-blown plastic foams is often less expansive, equally effective, and equally appealing to the consumer.

DEVELOPING COUNTRY ENERGY OPTIONS: The Key to Sustainable Development

The message emerging from the increasing literature of end-use oriented analysis is that the quantity of energy a nation needs is largely determined by what it is trying to do. Production of raw commodities for export, especially basic industrial materials, for example, is much more energy intensive and provides fewer jobs per dollar of invested capital than does production for domestic consumption to meet basic human needs.[13] But even within the realm of domestic consumption, not all needs are created equal. Many developing societies, like their industrial counterparts, are stratified societies. At the top of the heap in many developing countries is a small elite, living in the modern sector of their societies on a heavy diet of commercial fuels. This fuel consumption supports a way of life that would not be out of place on Park Avenue in New York or along the Champs Elysee in Paris. It maintains air conditioners, stereos, computers, microwave ovens, and big, gas-guzzling cars. It lives side-by-side, as in America, with squalor, poverty and deprivation.

Developing countries, like the industrialized countries of the West, face a fundamental choice about the shape of things to come. If we choose to build a world around these attractive indulgences for the few, the cost will be high and the pleasures will be brief. On the other hand, if we can find the political courage and the moral discipline to limit our appetites, there are ample resources to support strong economic growth, minimize environmental damages and bring sustainable development to the rural base that supports all our urban wealth.

Agriculture and Rural Development: Biomass and the Neglected Majority

The central challenge in such a strategy is the need for investment of energy and other resources to improve the living standards if the rural poor. Investment in agriculture will be necessary to support the growing world population, whose size is expected to double by the time we reach the doubled CO_2 environment. To meet these needs, the Food and Agriculture Organization of the United Nations has proposed doubling agricultural production by the year 2000. How can this be accomplished? It cannot occur by transporting the California miracle of motorized agriculture to the thin soils of the tropical world. Even if it worked from a technical standpoint, it would put half the population out of the only jobs they know.

Another strategy will be necessary. It will have to be based on the modernization of traditional agricultural practices, supplementing native varieties with high-yield, high-resilience cultivars. It will have to focus agricultural production on intensive use and preservation of the best lands to grow food and fiber, not extensive use of fragile and marginal lands. As has been demonstrated with alternative methods for increasing yields with rain-fed rice, modern methods don't necessarily mean dependance on machines instead of men or chemicals instead of animals.[14]

One cautionary note is essential. In any such large-scale enterprise, careful and systematic attention would have to be given to the displacement of existing uses of biomass as soil amendments. Substitutes, such as the sludge from biogas digesters, would have to be found and applied or soil productivity would plummet.

You Don't Have to Be Rich to Be Wasteful: The Crucial Role of Energy Efficiency

The success of any plan to meet energy needs in an affordable and sustainable fashion will hinge on efforts to improve energy efficiency. Misled by many into believing that waste is a sign of wealth, developing countries have emulated their more affluent cousins in the West. As even the richest have recently learned, it is faster, cheaper, and safer to save a kilowatt-hour than it is to generate the next one. Faced with shortages of capital as well as energy resources, in the future, developing countries will be absolutely compelled to drain every useful kilowatt-hour from the resources they consume.

Opportunities for energy efficiency improvements abound today in developing countries, as they do in the West. In India, for example, it takes about eight kilowatts of installed generating capacity to deliver one kilowatt of electricity, on average. This compares unfavorably with the approximately two kilowatts of installed capacity that is required in Japan or the United States to deliver the same amount of energy service. Theft, misuse, and lack of maintenance cripple the Indian power sector making the reliability of energy supply even more problematic than its cost.

In Brazil, the situation is different but not dissimilar. A major program of capacity expansion has been underway in the Brazilian power sector for decades. A recent study by the World Bank and the Compagnia Energetica da Sao Paulo (the world's largest electric utility) has demonstrated that a $2 billion investment in cost-effective improvements in electricity efficiency could offset the expected demand for 22 gigawatts of new generating capacity by the year 2000. The cost of the new capacity is forecast to approach $44 billion, if inflation can be kept under control.

The largest end-use of energy in developing countries is for cooking in households. The poor who cook with wood require almost ten times as much fuel as their colleagues who use fluid fuels.[15] For the urban poor who cannot gather wood but must buy it in local markets, the acquisition cost is painfully high. In parts of India, for example, the urban poor pay almost 20 percent of their income for cooking fuel, a burden as back-breaking as the hours spent walking and gathering fuelwood in the rural areas. By making high-quality energy carriers like kerosene, natural gas, or LPG available at affordable costs, great reductions in demand can be achieved.

High-Tech Efficiency and the Road to Modernization

In the modern sector of developing countries there are also great opportunities to improve efficiency. Developing countries need not be saddled with the outmoded industrial technology of their industrialized benefactors. If we in the North remain concerned about the effects of industrialization in these countries on our common environmental resources, we must be certain that they have access to the most efficient technology of production which we can transfer to them.

In the process of effecting these transfers, we must be careful that the technologies under consideration are well-suited to the physical, economic, and social environments of the receiving countries. Conditions are fundamentally different and these differences must be respected. Developing countries have ample supplies of hydropower and biomass, for example, available at many decentralized locations. They are less likely to have large unused supplies of fossil fuels or omnipresent electrical grids. Social settings also differ widely between industrial and developing countries and among developing societies. Capital and energy intensive technologies may be a great fit for Switzerland but be worse than useless in Swaziland.

Absolutely critical to the overall process of development will be the emergence of advanced, high-efficiency technologies for biomass utilization. Advanced gasification systems, combined with highly efficiency combustors and turbines, could fuel the industrialization of many developing countries rich with biomass resources. New aeroderivative gas turbines, for example, offer the promise of high conversion efficiency and low emissions. Their small modular sizes may be a perfect fit for the decentralized demands of rural electrification far from the urban power grid. The number of options and variations is as large as the diversity of cultures in the developing world. We just need to find the right fits.

MAKING THE TRANSITION:
A Challenge to Policy Makers

The legacy of today's industrial activities will be future climate changes, and the extent of those changes will be determined largely by policy choices made in the next few decades. Atmospheric concentrations of greenhouse gases increase every day. If national governments conclude that a warming commitment greater than 1.5 to 4.5 degrees C will cause unacceptable social risks, then remedial actions will be needed soon. The longer the delay before policy options and their implications are explored, judgments made, and choices implemented, the more extreme the policies imposed will have to be to keep temperature increases within some prudent upper bound.

The challenge facing policy makers and managers today is to identify the policy options that will limit greenhouse gas emissions without substantially slowing economic growth over the long term. THe task is complicated by the significant and persistent uncertainty in regional analyses of the impacts of climate change. Substantial methodological difficulties must be overcome to develop new analytic approaches and tools. These tools will need to have long time horizons if scientists are to evaluate adequately the policies that must be implemented over 50 or 100 years. Conventional methods of discounted cash flow analysis are clearly not adequate.

The task for policy makers will be neither simple nor quick and the choice is not between preventing or adapting to climate change. The challenge is to find those policies that, in the circumstances peculiar to each region and nation, will slow the rate of change and allow societies to adapt to the climatic changes that cannot be avoided. This task is politically difficult because the costs of preventing or adapting to climate change are in the present and the potential benefits are both uncertain and far off. But the risk of ignoring the challenge, however, seductive, is enormous.

Scientists worldwide agree that, if left uncontrolled, growth in the emissions of greenhouse gases will commit human societies, within the lifetimes of many of us alive today, to more radical and disruptive environmental changes than anything experienced during written human history. Coastlines will be inundated, farmlands will be lost, and forest yields will decline significantly. On the other hand, the benefit of meeting the challenge directly and with imagination is to sustain economic growth and preserve the health of natural ecosystems. The resulting innovations in energy use and industrial processes will reduce many forms of pollution and may just provide societies enough "breathing space" to allow a smooth transition to a new and productive climatic era.

REFERENCES

1. See, for example, U.S. National Academy of Sciences, 1983; World Meteorological Organization, 1985; Bolin, et al, 1986; and World Meteorological Organization, 1986.
2. Jones, et al., 1988.
3. Goldemberg, J., T.B. Johansson, A.K.N. Reddy, and Robert H. Williams. 1987a. Energy for Development. (World Resources Institute: Washington, DC).
4. Goldemberg, et al., 1987a.
5. Guilford, 1981.
6. Goldemberg, J., T.B. Johansson, A.K.N. Reddy, and R.H. Williams. 1987b. Energy for a Sustainable World. (World Resources Institute: Washington. DC).
7. Goldemberg, et al., 1987a, 1987b.
8. One exajoule equals 10^{18} joules or approximately one quad of energy.
9. Goldemberg, et al., 1987a; Baldwin, S. 1987. Improving Cookstove Performance: A Technical Reference and Design Manual. (Volunteers in Technical Assistance: Arlington, VA).
10. Bowden, W.B. and F.H. Bormann. 1986. "Transport and Loss of Nitrous Oxide and Soil Water After Forest Clear-Cutting" in Science, 233(4766):867-769.
11. World Meteorological Organization. 1986. Report of the International Conference on the Assessment of the Role of Carbon Dioxide and of Other Greenhouse Gases in Climate Variations and Associated Impacts, a Conference sponsored by the United Nations Environment Programme, the International Council of Scientific Unions, and the World Meteorological Organization. Report published by the World Climate Programme. (World Meteorological Organization: Geneva, Switzerland).
12. Natural gas is more than 98 percent methane by volume.
13. Goldemberg, et al., 1987a.
14. Ibid.
15. Ibid.

GREENHOUSE IMPLICATIONS OF ENERGY POLICIES OF MULTILATERAL DEVELOPMENT INSTITUTIONS

David A. Wirth
Senior Attorney
Natural Resources Defense Council

Through this exciting conference, the Climate Institute has brought together some of the most insightful thinkers from around the world to discuss the science of the "greenhouse" effect and policy responses for arresting and adapting to the impending climate modification. I do not want to repeat points made by previous speakers. However, I would like to provide some perspective on how we at the Natural Resources Defense Council (NRDC)--a private, nonprofit, environmental organization based in the United States with 95,000 members--are thinking about what is probably the most significant environmental issue of the next several decades.

The fundamentals of the greenhouse phenomenon are now well understood and the need for swift implementation of policy responses firmly established. Failure to respond to the threat of greenhouse warming would amount to an affirmative decision to wager the health and well-being of our own and future generations against overwhelming odds. There may still be some uncertainty concerning the rate and magnitude of warming anticipated. However, it is clear that the sooner necessary action is taken, the more effective it will be. Conversely, the longer implementation of a policy response is delayed, the greater the warming that will have accumulated "in the bank" and the more radical the measures that will be required to prevent further climatic upheaval. Moreover, once a crisis has been reached, it will be too late to act.

NRDC has recently launched an institution-wide "Atmosphere Protection Initiative" involving no fewer than 20 lawyers and scientists dedicated to crafting public policies to reduce or eliminate the serious threat of climate chaos. A major component of that effort is a study we have just begun at the request of the United States Department of Treasury. Our report will (1) examine the energy lending priorities of the World Bank and the regional development banks; and (2) make recommendations for structural and institutional reform of these multilateral lending institutions. Today I would like to describe the most recent conclusions of our work in progress--which, I must emphasize, are preliminary and subject to revision in light of future work. I would also like to extend an invitation to those present today and others with significant expertise or experience in this area to participate with us in this important endeavor.

EFFICIENCY AND CONSERVATION AS ALTERNATIVES TO CLIMATE-MODIFYING ENERGY INVESTMENTS IN THE THIRD WORLD

Encouraging sound energy policies in the developing world is crucial to solving the greenhouse warming problem. As economic development accelerates, Third World countries may account for the preponderance of greenhouse gas emissions as soon as the year 2000. Forest burning in Brazil, according to some sources, means that that country is already the third largest emitter of greenhouse gases, after the United States and the Soviet Union. Developing nations, with fewer of the extensive resources needed for successful adaptation to environmental disturbances at their disposal, also stand to suffer disproportionately from the effects of climate deterioration. There is, however, a potential equity issue lurking here. Developing countries have caused little of the problem, for which prosperous industrialized countries must bear the bulk of the blame.

The answer to this apparent dilemma is that the "greenhouse" problem is only one of several compelling reasons that require a swift reexamination of priorities for energy investments in Third World countries. For one, it is a practical impossibility to service the demand for energy services in the developing world through increased power generation capacity. Because of serious constraints on available capital, at best about half the projected energy needs in developing nations over the next 20 years can realistically be supplied by increased power generation.[1] Moreover, power generation projects in the Third World--as anywhere--often carry profound environmental risks. For instance, land degradation and local air pollution caused by mining and burning coal are all too often the high price paid for a new fossil fuel-fired power installation. Large dams often destroy forest and wetland ecosystems and displace and undermine the livelihoods of the poor and powerless.

At the same time, there is tremendous potential for supplying energy needs through conservation and improved end-use efficiency in many countries of the developing world.[2] It is a common fallacy that increased energy use is a necessary

---

1 <u>See</u>, <u>e.g.</u>, United States Agency for International Development, Power Shortages in Developing Countries: Magnitude, Impacts, Solutions, and the Role of the Private Sector (1988).

2 <u>See</u>, <u>e.g.</u>, World Bank, End-Use Electricity Conservation Options for Developing Countries (Energy Department Paper No. 32, 1986); Goldemberg, Johansson, Reddy & Williams, Energy for a Sustainable World (1987); Goldemberg, Johansson, Reddy & Williams, Energy for Development (1987).

consequence of economic growth. In fact, according to Jose Goldemberg of Brazil, by the year 2020 it is possible to achieve a universal standard of living equivalent to that of Western Europe without increasing global energy consumption from today's levels.

Most developing countries, however, are still highly inefficient users of energy. Macroeconomic policies, such as electricity price subsidies, in many developing countries actually discourage conservation measures and efficiency improvements. One firm in Brazil, where electricity prices are highly subsidized, manufactures an energy efficient air conditioner for the export market, but a cheap, inefficient model for domestic consumption.

Investments in efficiency and conservation improvements are extremely attractive from many points of view. They require little capital, pay for themselves rapidly, ultimately contribute to the productivity of a country's overall economy, and typically service energy needs more cheaply than creating new generating capacity. Developing countries could avoid $1.4 trillion in power supply expansion costs between now and the year 2008 through efficiency and conservation improvements. Efficiency is a particularly cost-effective option in many developing countries, where significant new energy infrastructure that is now being put in place can benefit from state-of-the-art advances available in the industrialized world only through expensive retrofit programs.

THE ROLE OF THE MULTILATERAL BANKS

Efficiency and conservation investments are a major opportunity--to date nearly totally untapped--for donors such as the United States and the World Bank to assist developing countries in making wise energy choices and avoiding mistakes already made in the developed world while reducing risks to the entire planet from greenhouse warming. Nonetheless, development assistance in the highly environmentally sensitive energy sector often exacerbates the threat of greenhouse warming by heavily emphasizing "traditional" sources of energy, such as massive fossil-fuel fired power installations.

The World Bank, for instance, controls an annual energy lending portfolio of nearly $3.5 billion. This leverage is magnified considerably by co-financing arrangements and the tremendous influence the Bank's priorities have on development agendas in the Third World. The Bank requires a "least cost" study to precede approval of a loan in the energy sector. These analyses, however, consider only supply-side strategies for providing energy.

The Bank has no requirements for the consideration of non-CO_2 demand reduction measures, such as efficiency or conservation improvements, as alternatives to proposed loans to support increased energy supplies. Indeed, only a minuscule portion of the World Bank's energy portfolio has ever been devoted to conservation, efficiency, and renewable energy sources. In fact, the Bank has an explicit policy prohibiting investments in "new" or "alternative" technologies in the energy sector. According to a World Bank official with the title "Principal Industry and Energy Specialist,"

> In the world of energy, as in other areas, [the Bank's] role is not to create innovative technical solutions, or to help countries to gamble on new processes, but to identify the best practices that have been fully proven in practice and will work in a developing country situation, and encourage their wider adoption where merited by circumstances.[3]

Even the structure of the World Bank's energy and industry department guarantees that alternative energy investments receive short shrift. The only technical expertise on conservation and end-use efficiency at the Bank resides in an internal "consulting" group, which provides advice only upon request and does not have the power to propose alternatives to investments intended to increase energy generating capacity. Although Bank staff have refused to quote us a figure, it is widely agreed that considerably fewer than 10 of the institution's approximately 3,000 professionals have been trained in end-use energy efficiency and conservation technology.

NRDC'S STUDY OF THE DEVELOPMENT BANKS' ENERGY SECTOR LENDING PRIORITIES

As I said, we have recently undertaken a study of the environmental consequences--with a particular emphasis on implications for the world's climate--of the energy sector priorities choices made by the multilateral development banks (MDBs) of which the United States is a member. These include the World Bank, the Inter-American Development Bank, the African Development Bank, and the Asian Development Bank. Our primary emphasis at this stage is on the World Bank.

---

3 Eric S. Daffern, World Bank Policy in the Promotion of Energy-Producing Projects in Developing Countries (Reading, England, Apr. 7, 1987).

We initiated this study at the request of the United States Treasury Department, which instructs the U.S. representatives to the MDBs. The goal of the study is to establish the need for and to make recommendations concerning structural and institutional reform of the MDBs' energy lending priorities. We expect to use the final report to stimulate discussion on this important issue among key policymakers, such as bank officials, the Treasury Department, the U.S. Congress, and officials of other donor and borrowing countries. Because the document was requested by the U.S. government, we expect it to have a particularly high profile among policymakers.

Some of the preliminary recommendations we have been considering include the following:

- Substantial increases in the number of professional staff trained in end-use energy efficiency and conservation. In order to engage in lending in the areas of efficiency and conservation, it is apparent that the banks must have significantly improved technical capacity and adequate personnel to devote to project preparation. At present there is virtually none.

- A least-cost planning requirement that includes end-use efficiency and conservation. Experience has shown that efficiency and conservation invariably win when they compete with new supply on economic grounds. The banks should expand their least-cost methodologies to reflect the fact that a kilowatt saved is as good as a kilowatt generated, and environmentally preferable as well. For instance, a component reflecting adverse effects on climate should be included in the calculated cost of a new fossil fuel-fired power installation. Staff professionally trained in end-use efficiency should have the responsibility to review these studies and the power to question or reject their conclusions.

- Establishment of lending targets for efficiency and conservation. One of the principal impediments to undertaking energy efficiency and conservation projects is the lack of incentives for development bank staff to promote such projects and for borrowing country governments to propose them. For instance, the World Bank's controversial $500 million electric sector loan to Brazil, scheduled for consideration by the Bank's Board of Directors in January of next year, reportedly includes only $1 million--0.2% of the total--for efficiency improvements. The Bank should set aside a sum of money--beginning with at least 10% of the World Bank's $3.5 billion annual energy portfolio and increasing over time--strictly to finance end-use efficiency and conservation projects.

Just as important as--or perhaps more important than--the final product is the process by which it is produced. As in all of our work on the issue of environmental reform of the development banks, a close partnership with nongovernmental organizations (NGOs) in the Third World is essential. We anticipate sponsoring an in-depth review of successive stages of this project, beginning with a draft outline. It is already apparent that the issue is far more complex than can be dealt with in a presentation of this length or, indeed, by the staff NRDC has to devote to it. Organizations and individuals--and particularly those from developing countries--with experience and expertise in this field are warmly invited to contact us to participate in this process to assure both the quality and the legitimacy of the final product. Simultaneous with this broad review process, we will be informally consulting with development bank and U.S. government officials to assure the ultimate implementation of recommendations advocated in the document, which should be highly influential in encouraging positive change in this critical sector.

International Progress on the Montreal Protocol

Presented
to the

Second North Amercian Conference
on Preparing for Climate Change
...a Cooperative Approach

by
G. Victor Buxton
Environment Canada

Washington, D.C. December 6-8, 1988

Does the Montreal Protocol offer hope for the future? Indeed it does! Lets begin by reminding ourselves that international agreements do not bring about environmental change - people do. The role of the agreement is to provide the sense of direction and schedule for the required change. It does this by providing a clear enunciation of obligations which in turn assist in creating the required political will.

As most of you are aware, the Montreal Protocol puts in place an international process for controlling all ozone depleting substances and not just those currently earmarked for controls. It also provides the means for increasing the stringency or tightening the time frame for such controls.

What I would like to do today is to give you a very brief overview of the international progress we are making on ozone layer protection via the Montreal Protocol. I will do this by giving you a short report on the major meetings held in the Hague, 17-26 October 1988 and what the future has in store with respect to amending the Protocol. I'll do this by giving you a summary of two formal UNEP reports: one prepared by my colleague, Mr. Peter Usher, on the State of the Science; and, one I prepared myself for UNEP on the state of the technological aspects.

The Science and Stratospheric Chemistry

The purpose of the science meeting was to review current scientific information on the state of the ozone layer, on processes involved in its destruction and on the environmental effects of ozone layer depletion. Emphasis was given to new scientific information which had become available since the adoption of the Montreal Protocol (Sept.87); and, to recommend procedures for carrying out a review of the current scientific information which could be used, along with similar reviews on the technology, environmental and health effects and economics, to assess the efficacy of existing controls. With respect to the Antarctic ozone hole, the key findings were that the unique meteorology during the winter and spring over Antarctica establishes an isolated air mass (polar vortex) with temperatures sufficiently cold to form polar stratospheric clouds (PSCs) which are responsible for the observed perturbed chemical composition. Laboratory studies show that PSCs are capable of converting photochemically inactive chlorine species into active forms which are capable of destroying ozone entirely consistent with the high levels of active chlorine (ClO) observed within the polar vortex in 1986 and 1987. The observed strong relationship between ClO and ozone from the 1987 aircraft measurements indicates that these industrially derived chlorine species are primarily responsible for the observed decrease of ozone within the polar vortex.

The conclusion is that even if the Montreal Protocol with its present control measures was ratified by all nations, the Antarctic ozone hole would remain forever. This conclusion is based simply on the fact that even with a fully ratified Montreal Protocol the atmospheric abundance of chlorine will approximately double from today's level of about 3 ppbv during the next few decades. Assuming our current understanding of the role of man-made chlorine in producing the Antarctic ozone hole is correct, then the Antarctic ozone hole will not disappear until the atmospheric abundance of chlorine is reduced to the levels of the late 1960's or early 1970's of about 2ppbv. This would require a reduction of greater than 85% (close to a complete phase-out) of the emissions of the non-fully halogenated chlorine containing chemicals, such as methyl chloroform. Even with a complete cessation of the fully halogenated CFCs, and a cut-back in other gases such as meythl chloroform, it will take many decades for the atmospheric abundance of chlorine to decrease to 2ppbv. It should be noted that the Antarctic ozone hole could disappear if there were to be a significant increase in the temperature of the Antarctic stratosphere thus precluding the formation of the Polar Stratospheric Clouds (PSCs) which are required for the man-made chlorine to produce the Antarctic ozone hole.

With respect to global warming, all CFCs, including the proposed substitutes for CFC11 and CFC12, have infra-red absorption bands comparable in strength to CFC11 and CFC12. Since for a given flux of these gases the atmosphere concentrations are proportional to their atmospheric lifetimes, the atmospheric lifetimes provide a rough guide to their greenhouse potential. With global reductions currently called for in the Montreal Protocol, the contribution of CFC11 and CFC12 to the predicted global warming over the next century should be about 10% of the total warming.

The Effects of Ozone Depletion

The major anticipated effects can be summarised as follows:

- Decreases in the total atmospheric content of ozone will lead to increased penetration of solar UV-B radiation to the earth's surface. A change in the vertical distribution of ozone will contribute to the greenhouse effect, which in turn will lead to climate changes and a rise in sea level.

- With respect to direct effects of increased UV-B radiation on human health, there is concern for:

(a) A possible increase in infectious diseases;
(b) Increased incidence of cataracts.
(c) Increases in skin cancer. (Non-melanoma skin cancer is predicted to increase by 3% for every 1% decrease of ozone. The more dangerous malignant melanomas are also expected to increase.)

- About two-thirds of the plant species and cultivars investigated were sensitive to UV-B radiation. These plants showed decreased growth, especially of seedlings, and reduced yield and food quality. UV-B radiation appears to limit the growth of some plants even under present conditions, and increased UV-B may lead to decreased productivity. Enhanced UV-B radiation may also change the competitive balance within ecosystems.

- Increased UV-B has been shown to have a negative influence on aquatic organisms, such as phytoplankton, zooplankton, larvae of crabs and fish. Many of these small organisms are at the base of the marine-food web. Increased UV-B radiation will therefore have a negative influence on the productivity of fisheries.

- Identified effects of increased UV-B radiation on the non-biological world include: Increased severity of air pollution in urban and industrial areas; and damage to plastic materials outdoors.

- More research is needed to gain a more complete knowledge of many of these effects, especially in tropical and subtropical areas. Even with the present state of knowledge, these effects represent a real threat. Damage to the eyes and a possible increase of infectious diseases will be more serious for people in tropical and subtropical areas, where UV-B exposure is already much higher than elsewhere. Decreased food production will most greatly affect people in areas where shortages occur now, (mainly in developing countries). Skin cancer would primarily affect people with little protective pigment in their skins. Polar and circumpolar areas receive comparatively little UV-B radiation, but the increases already apparent are more drastic than elsewhere, and organisms in these regions may be least adapted to deal with this rapid increase in exposure.

The Workshop on Substitutes and Alternatives

The highlights and conclusions from this meeting were as follows:

- For the refrigeration, air conditioning and heat pumps sector, the solutions are emission reduction and equipment re-design using ozone safe, health safe substitutes without significant climate warming. HCF-123, FC-134a, FC-152a, and possibly FC-125 are promising substances in this respect. Auto air conditioners offer immediate recycling possibilities. Recycling CFCs from refrigeration equipment will require the creation of additional infrastructure.

- For the rigid foams sector , HCFC-123 and HCFC-141b can substitute for CFC-11, as can a blend of water and CFC-11. Toxicology testing for HCFC-123 and HCFC-141b is not complete. Other technologies are emerging (vacuum panels). There is no longer a need in packaging.

- For the flexible foams sector, the technology is pending for producing all grades without CFCs.

- For the electronics solvents sector, the reduction programme is concentrating on: the development of fluxes which do not require cleaning, aqueous cleaning technologies, alternate solvents (eg terpenes), and recycle and recovery. A 50% reduction in CFC-113 could be achieved by conservation measures alone.

- For the metal cleaning sector, alkaline or acid cleaning looks the most promising.

- For the dry cleaning sector, the only promising substitute is HCFC-123.

- For the aerosol sector, there is no technically valid reason for not moving away from CFC use now. However, there will be a continuing need for CFCs in certain medical uses such as asthma sprays.

- For the halons sector, only 4% and 7% respectively of annual consumption (Halon-1211 and Halon-1301) were used on fires with 12% and 23% respectively emitted in non-fire events such as equipment testing and training. The scope for emission reduction is enormous. Current programmes in this sector are focused on: eliminating unwanted discharges, alternatives to discharge testing, recovery and recycle, developing alternative agents, and managing the enormous "bank" (84% and 70% respectively) of halons currently in existing equipment.

The very preliminary conclusions that were drawn in the context of the Montreal Protocol were:

For controlled CFCs, it is not technically apparent that we can significantly hasten the phase-down schedule for the 20% first step in the Protocol (July 1993-June 1994). However, given the tremendous technical progress anticipated by 1994, it should be possible to reduce dramatically the extent of the second step and the time frame.

For Halons, little is known about 2402 but for 1301 and 1211, conservation measures alone should be able to reduce global consumption by more than 50% and new non-ozone depleting substances under development could permit almost total elimination within 5-10 years. Notwithstanding the above, it is extremely important to manage the bank properly.

Figure 1 illustrates the four expert panels that have been created and the decision framework that will give rise to amending the Protocol. (Article 6 of the Protocol calls upon the Parties to convene expert panels to assess the control measures provided for in the Protocol.) There will be four such panels. The first panel (Science) will spell out the existing state of the ozone layer and demise scenarios associated with various chlorine concentrations in the stratosphere. The second (Effects) will describe the human health and environmental effects (plants, aquatic, climate change, etc.) associated with increased levels of UVB. The third (Technology) will spell out the technical viability, as a function of time, of reducing our dependency on these chemicals; and the fourth, (Economics) will try to conduct both a cost/benefit analysis on the technical options and attempt to quantify the generic costs associated with ecosystem disruption, increases in global skin cancer, reduction in the global food supply, etc.

These four panels will prepare reports for presentation to an international multi-disciplinary committee of legal and technical experts appointed by the Parties. Since the Protocol will not enter into force (EIF) until January 1, 1989, we will not have "Parties" per se until after that date, so this Committee is shown on the sketch as "Ad Hoc". The Committee of the Parties will have the task of integrating the 4 reports into one report which will assess the adequacy of the existing control measures in the Protocol and make formal recommendations to the Parties. To assist in this difficult integration and policy making task, Dr. Tolba intends to convene a high level steering committee composed of himself, 6-8 Environment Ministers from key countries and the 4 panel chairpersons. The first formal meeting of the Parties (April 1989) will formalize this "Committee of the Parties".

Figure 1

MEETING OF THE PARTIES

HIGH LEVEL STEERING COMMITTEE

(AD HOC) COMMITTEE OF THE PARTIES

SCIENCE

WATSON (USA)

EFFECTS

VAN DER LEUN (N)

TEVINI (FRG)

TECHNOLOGY

BUXTON (CANADA)

B

ECONOMICS

EEC-?

A

B

A – ?
B – ANDERSEN (USA)

Figure 2 shows a time line sketch for the process and you will note that the process is now well under way.

Figure 2

TIME LINE FOR THE PROCESS

- 4 Review panels created	October 1988
- Montreal Protocol enters into force	January 1989
- Advisory Group meets to prepare documents for the first meeting of the Contracting Parties	February 1989
- First meeting of the Contracting Parties	April 1989
- Reports from the 4 review panels finalized	July/August 1989
- Committee of Parties integrates report of expert panels and formulate recommendations	September 1989
- Second meeting of the Contracting Parties to decide on amendments by consensus or vote as set out	April 1990
- Amendments enter into force automatically	October 1990

This brings me full circle. You have now seen how the Protocol framework itself can bring about the sense of direction and the schedule for change. However, as I said at the beginning, agreement doesn't bring change - people do. If we have a consensus of political will to move in the direction I have outlined, the Protocol will be amended in October 1990. I am hopeful that this science driven Treaty will demonstrate that the global community has come of age and is prepared to respond to this risk to future generations in an environmentally appropriate manner and now...not later.

TOWARD AN INTERNATIONAL CONVENTION FOR THE PROTECTION OF THE GLOBAL CLIMATE: FINANCIAL FRAMEWORK[1]

by

Wilfrid Bach

Center for Applied Climatology
and Environmental Studies
Climate and Energy Research Unit
Department of Geography
University of Münster
Robert-Koch-Str. 26
4400 Münster
Federal Republic of Germany

and

Hermann Scheer

Member of the German Parliament
Bundeshaus
53 Bonn 1
Federal Republic of Germany

ABSTRACT

The conceptual framework of a strategy for the protection of the global climate is presented. It is based on a business deal between the industrialized countries (ICs) and the developing countries (DCs). To defray larger costs from the impacts of the greenhouse gases - induced global climatic change, the ICs are willing to pay for an emission reduction in the DCs through a decrease in forest and soil destruction, reforestation, and the use of efficient energy technology and renewables. The financial scheme involves initially a uniform ¢ 0.5 /kWh CO_2-tax on fossil fuel use in OECD or EC countries amounting to \$ 200 bill. and \$ 50 bill., respectively. The revenues, to be put into an international fund, are earmarked for use by the DCs: i.e. 50 % for debt liquidation, and the other 50 % for forest and soil preservation, afforestation, agricultural reforms, improvement of the energy economy, family planning coupled with social security, etc.

[1] This is based on a concept worked out by EUROSOLAR, an Association for the Solar Energy Age, of which Hermann Scheer is chairman and Wilfrid Bach is a member of the board.

THE FOCUS

It can no longer be overlooked that, at the continuation of present trends, mankind is heading toward a major climatic and environmental disaster. It is, therefore, urgent to set up a comprehensive global strategy for the protection of climate which involves the reduction of the emission of CO_2 and other greenhouse gases into the atmosphere, puts an end to forest destruction, and initiates massive reforestation programs. This requires a framework of specific political agreements itemizing both the tasks of the international institutions and the diverse actions to be taken by individual countries. Above all, to be successful, it needs a workable financial framework. In this context emphasis is on forest policy and energy policy.

FOREST POLICY

Measures favoring forest preservation in the developing countries (DCs) and the reduction of pollution emission responsible for forest dieback in the industrialized countries (ICs) have the highest priority, because they are cheaper than reforestation programs. But the protection of degraded areas and afforestation on a massive scale are also very important. While it would be physically possible to remove from the atmosphere all of the currently emitted ca. 20 Gt of CO_2 by planting new forests and intensively managing existing forest land, such a program would presently be impractical, because of financial and political constraints. Nevertheless, reforestation and forest management could play a significant role.

Concomitant measures have to be taken, these include:

- the expansion of national and international forestry institutions with the ICs pledging considerable scientific and staff support;
- agricultural development programs dealing with soil protection, reduction of synthetic fertilizer use, and problems with cattle feedlots.

Reforestation and forest management can ease the CO_2-situation and thus help pave the way from the present wasteful energy economy to a new policy of more efficient energy use.

ENERGY POLICY

The new energy policy will have to focus on three points:

- a more efficient energy use in all conversion and end-use areas as well as a greater coupling of all energy conversion processes (e.g. cogeneration, integrated energy use concepts);

- a more expeditious deployment of the inexhaustible solar-based energy resources (e.g. solar radiation, wind and hydro power, biomass) in diverse but regionally and locally appropriate forms;

- a much greater effort to produce hydrogen efficiently and economically from renewable energy resources; this fuel can be easily transported and stored, and it allows to tap the enormous potential of the renewables.

In terms of implementation, the above represents a priority listing. More efficient energy use is the fastest way of reducing CO_2-emissions. Such measures have top priority in the ICs. Success in this area depends, however, to a large extent on a high political-administrative level, a highly developed economic infrastructure as well as a high level of education and information of the population. Also, in the DCs the more efficient energy use can be successful, if more systematic developmental and structural policies are pursued. Technical appliances have a low efficiency, and the way biomass is generally used is extremely wasteful. This means, for example, that the very poor spend annually three to ten times more energy in terms of firewood for cooking than those who have a modern stove running on electricity or gas. A more rapid deployment of more efficient energy use in the DCs could compensate a large part of the energy increase expected as a result of population growth. A successful introduction of more efficient energy use will win time and room for maneuvering necessary to build up a solar-based energy economy.

Therefore, all suitable technologies utilizing renewable energy resources must be simultaneously developed and deployed, especially in countries with a high availability of solar energy, as is the case in many DCs. However, it is only the production of hydrogen with renewable energy on a substantial scale that a drastic reduction in fossil fuel use can be brought about. The time gained by more rational energy use and the direct use of renewable energy, should be utilized for a worldwide development and deployment of cost-effective and efficient hydrogen production methods.

The use of nuclear power has also been suggested as a means to reduce the global trace gas - climatic threat. Its current contribution of ca. 5 % to global primary energy use is too small to have much of an effect. Nuclear power produces only electricity from centralized power plants. About 75 % of the world's population lives in rural areas with no electricity supply grid. There is, therefore, not much demand for this type of energy. Furthermore, based on the unsolved problems of nuclear waste disposal and the, on principle, unsolvable problems both of a large reactor accident and the proliferation of bomb-grade radioactive material, it is highly unlikely that a further development of nuclear power will be accepted of a magnitude that can make a noticeable contribution to the reduction of the CO_2/climatic risk. Additionally, each dollar invested in efficient energy use can avoid seven times as much CO_2-emission than investment of that dollar in nuclear power. Finally, a drastic expansion of nuclear power is incompatible with the expeditious deployment of renewable energy sources, because it absorbs almost all of a country's funds available for investment.

Without doubt, mobilization of the solar option will require massive investments of the kind only ICs will be able to afford. It is only through reallocation of the world's economic resources and a priority change in the use of the gross world product that there will still be a chance of effectively reducing the greenhouse effect. Questions of current cost-effectiveness in the narrow business operational sense can no longer be permitted to determine future energy policy.

CONCEPTUAL FRAMEWORK

The framework consists of internationally binding principles, proposals both for a financial framework and institutional measures, as well as further ideas for implementation.

The Principle

In view of the different causes and producers of the CO_2/climatic threat, the diverse conditions and opportunities for action as well as the different national capabilities, an "International Convention for the Protection of the Global Climate" would have to be based on a catalog of obligations that is equally effective but not of the same kind. This is the basic idea which all measures must follow, if they are to be more than just an appeal to the international community.

The binding measures must be affordable and quickly realizable. They require a global economic framework. They obtain their binding force by a resolution of the United Nations or some other international organization, supplemented by national law. Since international organizations have no direct means of imposing sanctions in case of a nation's none-compliance, implementation must be ensured by international public pressure and positive financial incentives.

The Financial Framework: An International CO_2-Tax

It is stipulated to impose a CO_2-tax on all ICs in accordance with their respective CO_2-emissions. This revenue would flow into an international fund. In the case that political efforts for global actions failed, the OECD countries and the member states of the Council of Europe should take the initiative. Although a global initiative would be very important, we cannot afford to wait any longer for a consensus to be reached on action that is needed now. Therefore, for a start, initiatives from a group of nations may be necessary, if it proved impossible to get global action.

Specifically, the polluter-pays principle is applied to individual nations. Each country may decide for itself whether and in what way it will pass on the CO_2-tax to individual polluters, and which domestic strategies are the most appropriate for CO_2-emission reduction. The greater the success of a country is in reducing CO_2 (e.g. through efficiency improvements and the use of CO_2-free energy sources), and absorbing CO_2 (e.g. by reforestation), the lower will be its tax payments. This will create a positive incentive for the main polluters to reduce the CO_2/climatic threat.

The revenue flowing into the international fund will be made available to the DCs to finance measures against the greenhouse effect and related problems as well as to reduce their debt. Conditions for eligibility are that funds must be earmarked for measures which reduce the greenhouse risk. It is already the preservation of forests that is rewarded with a positive financial incentive, while any further deforestation results in a stop of all payments. Measures going beyond the preservation of forests and leading to a net growth of CO_2-absorbing biomass, thereby counteracting the greenhouse effect, will merit additional financial assistance from the fund. Consequently, the polluter-pays principle applies also to the recipient countries, if they make a positive contribution to save the world's climate.

Moreover, fund money used for debt liquidation without any strings attached increases the DCs latitude for setting up a more independent political and economical framework for action. Based on these criteria, the following use of funds is conceivable:

- 50 % for liquidation of debt and payment of interests as well as for general measures to improve the infrastructure;

- 50 % earmarked for forest preservation, reforestation, agricultural reforms, soil preservation, investment in cost-effective energy efficiency improvements, and transfer of best available technology to tap the renewable energy resources.

If climate damage caused by CO_2 from fossil fuel burning (coal, oil, gas) were taxed with ¢ 1 per kWh of mean fossil energy content and paid by all countries, this would mean an annual revenue for the international fund of something of the order of $ 700 billion. This calculation is, however, more theoretical than practicable, since such tax payment by the DCs is hardly feasible, and taxation of countries outside the world monetary system would fail due to lack of foreign exchange.

More realistic, also in terms of development policy, would be to tax only the OECD countries amounting to annual revenues of the order of ca. $ 400 billion. If only the member states of the Council of Europe would take the appropriate initiative, the annual tax would still add up to about $ 100 billion. Additionally, we propose that countries, not contributing to the fund, be sanctioned with an import tax on their industrial products by member countries.

Due to the different carbon contents of fossil fuels, the tax must be related to a common CO_2-emission base as follows: One kWh from coal, oil, and gas would be taxed with ¢ 1.25, 1.00, and 0.67, respectively. Natural gas, being relatively environmentally benign, would thus be favored.

To get the funding scheme quickly off the ground, we propose to start with a tax of ¢ 0.5 per kWh for an initial five-year period. Considering only the OECD and the Council of Europe, this would still result in revenues of the order of $ 200 and $ 50 billion, respectively.

INSTITUTIONAL REGULATIONS

We recommend the creation of an "International Agency for the Protection of the Global Climate" within the framework of the United Nations with the purpose of coordinating, assessing, and evaluating the relevant measures to be carried out by such international organizations as FAO, UNEP, International Monetary Fund, World Bank, and Regional Development Banks. The Agency would assume responsibility for implementing the tasks connected with the conceptual framework. One or more international financial institutions would have to be linked to the Agency for administering the CO_2-tax fund and for carrying out both the debt liquidation and the financial assistance schemes. The Agency's tasks would also include establishing and maintaining scientific institutions with the purpose of monitoring and assessing the overall danger potential, and evaluating the effectiveness of measures taken against the greenhouse effect.

REFINEMENT OF THE CONCEPT

To refine the concept, additional studies and discussions are necessary. These include:

- The question of including the COMECON countries in the overall strategy. For some of them, liquidation of their international debts would be necessary to free funds for their political, economical and ecological reforms. Since they are contributing, however, significantly to CO_2-emission, they should actually have to comply with the international CO_2-tax. In a resolution adopted recently, the governments of the Warsaw Pact countries solomnly announced their willingness, in principle, to assume international responsibility and to carry the burden associated with preserving the environment. In view of a further refinement of this framework, it would be necessary to examine in what way the COMECON countries could assume responsibility, or to what extent they could be exempted from the CO_2-tax for initiatives of their own to reduce the greenhouse effect, or under what conditions they would be eligible for debt liquidation and financial aid.

- A synopsis and an evaluation of current national and international efforts to reduce the greenhouse effect, and to use existing experience to refine the implementation steps.

- The development of programs for a sensible use of the international fund's revenues. This should take into account that reasonable initiatives on the part of the recipient countries must be expected in the interest of building up an efficient and sustainable solar energy economy.

- The development of a list of specific criteria for imposing the CO_2-tax, for lowering the tax in accordance with reducing the CO_2-emissions, for distributing the funds, and for avoiding misuse of the implementation instruments.

- Plans both for establishing institutions responsible for implementing the framework, and for including existing banks, special organizations, as well as other international organizations.

- The conceptual inclusion of other problems of global dimension. Raising eventually the base tax to the one-cent level would generate a considerable amount of extra money which, in turn, could be used to contain population growth - for example by financing an extensive social security system.

CONCLUSIONS

An initiative taken on the basis of this concept could, within a short period of time, drastically reduce the debt burden of the DCs (currently some $ 1.2 trillion). Use of the funds from the CO_2-tax for active measures against the greenhouse effect would help create a framework that is of a financial dimension which could bring quick and effective results. A considerable part of the fund provided by the ICs would, in addition of having a positive environmental effect, offer them new latitude for action.

A restriction of international competitiveness among individual economies should not be expected, since this would be a new framework for everyone. Political steps would have to be taken, however, to prevent increased use of environmentally-damaging energy sources outside the ICs. The financial framework would also be appropiate for this situation, since such misbehavior could be sanctioned by reducing the payments.

This concept would promote the responsibility of all participating nations. The more effective the measures against the greenhouse effect, the lower would be the CO_2-tax, and the higher would be the financial aid. The latitude for specific national activities would remain. The urgently needed disencumbrance of the DCs could take place in an orderly way without risking the collapse of national banking systems. In many countries, prosperity and minimum social standards could become more secured and be brought more in line with human dignity. Moreover, in many regions of the world one could take precautions against societal and political crises instead of responding in an unsatisfactory and questionable way through expensive disaster relief programs or security-related investments.

The production and transfer both of the most efficient energy technology and the most appropriate renewable energy technology to the DCs would set in motion a mass-production process in the ICs. This would lower the overall costs and accelerate the market penetration, thereby reducing the CO_2-emissions, and hence the CO_2-tax. The end result could be a coupled system for an ecological reform of industrial society with interacting and self-enforcing measures.

Presentation

Effects of Global Warming on International Treaty Obligations Relating to Water Rights

James M. Strock
1001 North Vermont #514
Arlington, VA 22201
703-525-1299

Introductory

Given the importance of water supplies to human populations, any significant change in water availability--increases, decreases, or unpredictably wide fluctuations from historical norms--will have multifarious effects on a range of institutional arrangements. As a result, the legal and political challenges concomitant with various global warming scenarios are profound. An example is seen in the western United States, where water has always been in short supply, and the possibility of even greater scarcity could lead to major reallocations of legal and political authority. In particular, the need to ensure that water is targeted to those uses deemed by society to be most beneficial would likely lead to major assertions of national authority, overriding the traditional leadership role reserved to the states.[1]

When one throws in the additional factor of international water allocation arrangements, the foreseeable diminution of private rights, as well as local and state control over water law, becomes even more pronounced. By necessity, international treaty obligations override conflicting laws of national governments or their subdivisions.[2]

In considering the potential effects of global warming scenarios of international treaty obligations, it is useful to examine how the United States would be affected. The choice of the U.S. is not because of national chauvinism or a lack of interest in other nations. Rather, the American example is instructive because the U.S. is presently a party to several international treaties both as an upstream nation (as in treaties with Mexico and Canada) and as a downstream nation (with Canada). Thus the American example necessarily includes both situations which other nations will face in crafting transnational water allocation responsibilities to deal with the challenge of global warming. Further, because the division of water regulation authority in the U.S. is familiar, the U.S. example is accessibly illustrative of the wide-ranging effects of changing international water allocations which all nations will face.

United States-Mexico Water Relations

In the case of the Colorado River, the United States stands as an upstream nation in relation to Mexico. As an upstream nation, the U.S. historically relied upon the so-called "Harmon Doctrine," claiming absolute territorial sovereignty over the flows of the Colorado River within America's borders, not acknowledging Mexican claims to predictable flows. Building upon the 1922 Colorado River Compact (an interstate agreement allocating water among participant states, in this case providing that upper and lower basin states would contribute equally in supplying water flows to Mexico which might arise in the future), the United States entered into treaty negotiations with Mexico during the administration of Franklin Roosevelt, as part of the "Good Neighbor" policy.

The resulting 1944 treaty, which continues in effect today, allocates a guaranteed annual minimum flow of 1.5 million acre feet to Mexico, with contingencies in the event of drought. However, even without factoring in the possibility that global warming might dramatically decrease the flows of the Colorado, the effective functioning of the treaty is limited by its failure to resolve two key questions: the water quality to which Mexico is entitled, and overallocation of the water in the rapidly developing western United States. These matters are made more difficult because they are intimately interrelated.

At the time the treaty was concluded in 1944, the matter of water quality was not a focus of attention, both because of the then-contemporary lack of awareness of agricultural pollution and the assumption that excess flows would be of sufficient quantity to maintain safe levels of salinity. However, as American consumption increased, salinity levels rose dramatically, making the excess flows highly saline and therefore unusable for agricultural production in Mexico. In response to Mexican protests, the United States joined in framing Minute 242 of the International Boundary and Waters Commission. Minute 242 includes a ceiling on salinity levels in the flows reaching Mexico, and was implemented through federal salinity abatement measures, including state limitations.

In turn, however, federally aided salinity limitation measures provided the states with continued incentives to allocate water flows in their growing jurisdictions, with less concern about the effects of usage on downstream salinity levels. And, as development in the American west has exploded, the waters of the Colorado have become overallocated--to an extent that the U.S. may have great difficulty meeting its treaty obligations during the next decade. Thus, even without any allowance for the effects of global warming, the need to supply our guaranteed treaty levels may lead to federal override of state allocations in the near future.

Finally, a major flaw, from today's perspective, in the 1944 treaty is its omission of groundwater provisions. Although this has begun to be remedied, as in Minute 261 of the International Boundary Waters Commission, a comprehensive approach will be needed in the years ahead. If global warming scenarios lead to surface water scarcity, then groundwater supplies will be further depleted as the hydrologic cycle is altered.

Not surprisingly, the treaty did not address the possibility of climactic change, focusing solely on the familiar "extraordinary drought," which would allow for Mexican flows to be reduced proportionately to lower upstream flows. While there is at least modest precedent in American law for including climactic change as a factor in Supreme Court consideration of equitable allocation of waters among the American states,[3] the international legal consequences remain unclear.

The only firm conclusion that may be drawn is that federal regulation of domestic water supplies would continue to increase under these circumstances. Federal judicial or legislative action could resolve interstate water through application of "equitable apportionment" principles--and even intrastate disputes might be resolved at the state level in an analogous way.[4]

United States-Canada Water Relations

The Columbia River Treaty between the U.S. and Canada differs from the treaty with Mexico, most significantly here because the U.S. is both an upstream and downstream nation in this case. However, while it useful to consider the Canada-U.S. accord with a focus on the U.S. as a downstream nation, there does not appear to be a large body of scientific evidence pointing to shortages of supply from global warming in the northern region of the U.S. Thus there may not be action-forcing events altering that relationship, unlike the Mexican-American accord, which may well require readjustment whether or not global warming scenarios are borne out. If the United State comes to face a situation analogous to that of Mexico, that is, as a downstream nation facing shortages in anticipated flows from an upstream neighbor, this could add an additional complexity to America's computation of its national interest in participating in multinational attempts to fashion a comprehensive response to perceived global warming trends.

Conclusions

If, as is widely assumed, the U.S.-Mexico treaty will need to be changed to take into account global warming scenarios of increased water scarcity, certain results are foreseeable. One is that there will be tremendous domestic resistance in the United

States to any limits on states' ability to allocate flows within current limitations. There will be little incentive for the U.S. to ascribe decreased excess flows to global climactic changes rather than "extraordinary droughts" contemplated by the 1944 treaty. Under these circumstances, it is unrealistic to anticipate that the U.S. would, because of necessarily uncertain scientific forecasting, initiate a reexamination of the 1944 treaty.

However, because the Colorado River may well be overallocated in the near future even without additional scarcity resulting from global warming, the 1944 treaty may well be reexamined on more traditional grounds. Any such reexamination could be expected to include climate-related matters, at least in the context of new provisions for shortages however induced. It is, of course, impossible to predict how such a reexamination would be concluded. It is clear that new water development in the west would face opposition not only from those immediately benefiting from the status quo, but also from a broad array of groups opposed on environmental and budgetary grounds.

Further, to the extent that global warming scenarios do become an explicit factor of debate, it is to be expected that national security concerns may lead to new political legitimacy for the traditional doctrine of absolute territorial sovereignty. This may particularly occur in the case of Mexico, both because of the apparent American need for scarce water supplies and the alternative availability of petroleum, Mexico's most apparent "bargaining chip." Thus, while viewing global warming as a national security concern may be a useful construct in seeking to craft transnational agreements to prevent anthropogenic atmospheric destruction, such an approach may frustrate subsequent efforts to renegotiate international agreements intended to respond to changes in water flows resulting from global warming. If this potentiality is taken seriously, it provides yet another political argument for focusing on preemptive global action to prevent or mitigate the effects foreseen in some global warming scenarios.

1. See Strock, "Adjusting Water Allocation Law to Meet Water Quality and Availability Concerns in a Warming World," in Proceedings, First North American Conference on Preparing for Climate Change: A Cooperative Approach, 382 (1988).

2. See, e.g., United States Constitution, Article I, Sections 8 and 9; Sanitary District of Chicago v. United States, 266 U.S. 405 (1925).

3. See Washington v. Oregon, 297 U.S. 517, 527 (1936), Colorado v. New Mexico, 456 U.S. 176 (1982).

4. See Strock, supra note 1, at 385.

INTERNATIONAL LEGAL AND POLICY OPTIONS FOR DEALING WITH GLOBAL WARMING AND CLIMATIC CHANGE[1]

KILAPARTI RAMAKRISHNA
WOODS HOLE RESEARCH CENTER
WOODS HOLE, MA 02543, USA

The first and most significant pronouncement on international legal initiatives in dealing with global warming and climate change emanated from the Toronto Conference on the Changing Atmosphere: Implications for Global Security, June 1988. The Statement adopted at the end of the Conference recognized that no single international organization, country, industry, or individual can tackle this problem in isolation, and called on all to take specific actions to reduce the impending crises caused by the pollution of the atmosphere. The Statement also called upon governments "to work with urgency towards an action plan for the protection of the global atmosphere". The issue as framed reads as follows[2] -

> Continuing alteration of the global atmosphere threatens global security, the world economy, and the natural environment through:
>
> > Climate warming, rising sea-level, altered precipitation patterns and changed frequency of climatic extremes induced by the "heat trap" effects of greenhouse gases;
> >
> > Depletion of the ozone layer;
> >
> > Long range transport of toxic chemicals and acidifying substances.

The Prime Minister of Canada, who inaugurated the Conference, went so far as to set 1992 as the year by which either a fully worked out agreement is to be ready for signature or all the elements of such an agreement are to be in place.

A Workshop on Global Climatic Change held at Woods Hole during September 12-13, 1988 discussed various steps toward an international convention for stabilizing the composition of the atmosphere[3]. The participants recognized that progress will require discussion of the challenges among all the nations and much broader understanding of the necessity and universal advantages of action than exists currently. To address this and other issues the participants proposed that the General Assembly of the United Nations be encouraged to adopt a Declaration of Principles defining the problem and suggesting the pattern of solutions outlined at the Workshop. On the subject of international conventions and protocols the participants felt that any attempt at codification and progressive development of an international law of the atmosphere along the lines of the law of the sea, while desirable in the long run, is overly ambitious and is unlikely to be realized in the near future. It was even felt that such a move could slow down rather than facilitate negotiation of a Convention. Energies should be focussed on obtaining a convention aiming at the stabilization of the greenhouse gas composition of the atmosphere.

In December 1988 the General Assembly of the United Nations, at the initiative of Malta and supported by 19 other countries, deliberated the agenda item on the conservation of climate as part of the common heritage of mankind and adopted a resolution on the protection of global climate for present and future generations of mankind. The resolution recognizes that climate change is a common concern of mankind and requested immediate action leading to a comprehensive review and recommendations with respect to:

> the identification and possible strengthening of relevant existing international legal instruments having a bearing on climate;

elements for inclusion in a possible future international convention on climate.

About the same time as the General Assembly resolution was debated, the World Meteorological Organization and the United Nations Environment Programme have established an Intergovernmental Panel on Climate Change (IPCC). IPCC since then split its work into three major working groups. Working Groups I and II are concerned with science and socio-economic impacts of climatic change and Working Group III is looking at response strategies. International legal and policy options form part of this Group's priorities.

All this activity might lead one to think that awareness of the dangers of increases in greenhouse gases is of relatively recent origin. Several meetings in the past have discussed the potential effects of, as well as policies to prevent, mitigate, and adapt to global warming beginning with the World Climate Conference of 1979[4]. Other meetings of importance were held in Villach, Austria in 1980, 1983, 1985, and more recently in 1987.[5] The last mentioned meeting, continued at Bellagio, Italy in November 1987, and the Toronto and Woods Hole meetings were the only ones that devoted their attention to developing policy options for responding to climatic change.

Given the amount of news media coverage, and congressional hearings in the United States and elsewhere, and the resulting popular concern, it is reasonable to state that interest in the policy making aspects of the global warming issue is at a peak. Never before has the policy issue been so central to the global warming debate.[6]

All this activity in such a short time span, the time when scientific concerns crystallized and the time when the policy imperatives recommended, distinguishes this particular environmental concern with the gestation period taken up by other environmental concerns. Other distinguishing features of the various recommendations referred to above are the linkages made in this case with the urgent need of less developed nations for sustainable development and linkages with several other major concerns in the areas of balance of payments and foreign debt burdens etc. The only other time such a prominent role given to developing country economic concerns was at the time of the Stockholm Conference on the Human Environment in 1972. The questions that therefore need to be addressed in charting out appropriate international legal and policy options for the world are: a review of interest articulation of the developing countries in the international environmental movement in the past twenty years with a view to consolidate their position for what promises to be a period of intense international activity; and the appropriateness of discussing a framework convention for stabilizing the greenhouse gas composition of the atmosphere as opposed to working on an omnibus multilateral convention dealing with all aspects of atmospheric pollution.

The history of environmental management is similar to the history of the rules of the road at sea, industrial safety, and laws pertaining to human rights. The critical insights have come from catastrophes of various types. A classic example exists in nations' attempts to deal with marine pollution. A number of multilateral conventions brought into existence either under the auspices of the International Maritime Organization[7] or by the initiatives of countries sharing a region[8] suffered from this drawback. They were concluded in response to major casualties from marine pollution. Even the conventions entered into by the International Atomic Energy Agency on early warning following nuclear accidents that might have transboundary effects and a convention for providing for emergency assistance were responses to the accident at Chernobyl.

Beginning with the UNEP Regional Seas Conventions[9], a trend away from this practice has been developing. A more recent example would be the Vienna Convention for the Protection of the Ozone Layer and its Montreal Protocol. It is important to bear in mind that even these actions resulted after some considerable damage had taken place though it is not "felt" with the same real intensity. The success of the UNEP conventions is due to UNEP's ability to place marine pollution problems in the overall framework of environmental problems and approaching the problem with a comprehensive plan concerning not only the consequences but also the causes of environmental degradation of the region concerned.

Apart from this example, several multilateral legal instruments beginning with conventions on arms regulation provide useful insights on how the proposed exercise of stabilizing the composition of the greenhouse gases may proceed. Notably they are the 1963 treaty banning nuclear weapon tests in the atmosphere to the 1977 convention on the prohibition of military or any other hostile use of environmental modification techniques; the 1979 convention on long-range transboundary air pollution, with its Helsinki protocol on the reduction of sulphur emissions, and the 1988 Sofia protocol on the control of emissions of nitrogen oxides; the 1982 law of the sea convention; and finally the 1985 Vienna convention for the protection of the ozone layer with its Montreal protocol of 1987.

This experience provides us with a picture of the world community that can adopt appropriate measures to deal with any kind of problem depending on how it is presented. An attempt is currently made by the world community and interested institutions on a global scale to explore appropriate avenues in reaching a global agreement to stabilize the greenhouse gas composition of the atmosphere. It is important, particularly in seeking the support and participation of developing countries, to look at the causes of the current concentrations of greenhouse gases and the consequences of any regulatory mechanism for developing countries.

In recent times conservation groups were successful in a program known as "debt-for-nature swaps". In this the developing countries are given incentives to reverse the destruction of tropical forests when the conservation groups, with help from philanthropic foundations, buy Third World debt in return for commitments to preserve extensive tracts of forests. The proponents of the idea agree that this was not a solution to the debt crises facing the Third World, but believe that it does help in lessening some portion of the debt burden and helps as well with the broadly desirable objective of conserving nature.

There is a need to work with such ideas in accomplishing the CO_2 reductions. One suitable idea appears to be the appropriateness of the "bubble policy" introduced for the first time by the USEPA to encourage industries to control pollution more than they are required to. Also known as "emissions trading," the "bubble policy allows industry to be excused from pollution controls at one or more emission points in exchange for increasing controls at other emission points - as if all points were placed under an imaginary bubble. The emissions from each bubble must have an impact on air quality equal to or better than the impact of the original controls"[10]. The ensuing fight over emissions trading, according to a Report of the Conservation Foundation, illustrates both the benefits and the difficulties encountered[11].

Given the fact that the only way to prevent increased CO_2 levels is by stopping the combustion of fossil fuels and by halting further erosion in the mass of carbon stored in vegetation and soils, and given the fact that a large number of nations are trying to feed an ever increasing population with economies that are at present largely dependent on consuming fossil fuels for their energy supply, one possible solution seems to be to modify the bubble concept referred to above so as to manage the level of CO_2 in the atmosphere.

The scientific community is aware of global emissions of CO_2 from combustion of fossil fuels on an annual basis[12]. They are also aware of how much CO_2 is contributed by each country and by how much the emissions of CO_2 must be reduced to stabilize the CO_2 component of the greenhouse gases. The proposal is to recommend a quota system taking fully into account questions of equity, relative levels of development and the consequent need of the developing countries to attain desired levels of economic development. With this as the objective, the quota system will arrive at how much CO_2 can be released henceforth by each country. The current levels of development attained and aspired to obviously play an important role in assigning the quotas. If a country cannot reduce its CO_2 releases to be in compliance with the quota, it will have the option either to contribute to a Fund to be established for this purpose or, better still, to enter into an agreement with a country that is not releasing its fullest quota of CO_2

emissions. Such an agreement would provide on the one hand for afforestation, and on the other, for technology transfer to facilitate energy development away from its traditional dependence on fossil fuels. Without a doubt it would have considerable appeal to the developing countries, it addresses many of the challenges that the leaders of the environment movement have set for the industrialized nations, and, what is more, it even encourages participation by the developing nations in framework convention and specific protocols for dealing with greenhouse gases.

***

ENDNOTES

1. An abridged version of the paper presented at Session X on International Policy Options during February 23, 1989, at the International Conference on Global Warming and Climate Change: Perspectives from Developing Countries, Feb. 21-23, 1989, New Delhi, India.

2. These changes according to the Conference Statement will:

   [i]mperil human health and well-being; diminish global food security, through increased soil erosion and greater shifts and uncertainties in agricultural production, particularly for many vulnerable regions; change the distribution and seasonal availability of fresh water resources; increase political instability and the potential for international conflict; jeopardize prospects for sustainable development and reduction of poverty; accelerate extinction of animal and plant species upon which human survival depends; after yield, productivity and biological diversity of natural and managed ecosystems, particularly forests.

   To this scary scenario was added the warning that if rapid action is not taken now by the countries of the world, these problems will become progressively more serious, more difficult to reverse, and more costly to address.

   See the Toronto Conference Statement, July 5, 1988 version, pp: 2-3.

3. See Kilaparti Ramakrishna and George M. Woodwell, Summary of Workshop on Global Climatic Change: Steps Toward an International Convention Stabilizing the Composition of the Atmosphere (Document prepared at the end of the Workshop during September 13-14, 1988 at Woods Hole, Massachusetts).

4. See: World Meteorological Organization, World Climate Conference ; Declaration and Supporting Documents (WMO, Geneva, 1979).

5. For a survey of both scientific literature and some of these conferences, see: William Kellogg, " Mankind's Impact on Climate: The Evolution of an Awareness", Climatic Change, Vol.10 (1987), pp: 111-136.

6. Lynton Caldwell prescribed a three pronged approach for the debate to reach the current level of discussion. They are: a general and clearly articulated consensus in principle among the sciences regarding the nature and significance of atmospheric/climatic change, and what action could be taken to forestall it; a traumatizing event of such scope and magnitude that current social assumptions and attitudes would be destabilized beyond the reach of a technological fix; and the presence of an organized, and persuasive social movement committed to the achievement of an ecologically sustainable order, and willing to pay the costs required. A semblance of each of the above can be discerned in the current level of refinement. See Lynton Caldwell, "Societal Response to Climatic Change: Alternative Conjectures" (Indiana University, 1987, p. 6).

7. These conventions include: International Convention for the Prevention of Pollution of the Sea by Oil, 1954, with its amendments; International Convention Relating to Intervention on the High Seas in Cases of Oil Pollution Casualties, 1969, with its protocols; International Convention on Civil Liability for Oil Pollution Damage, 1969, with its protocols; International Convention on the Establishment of an International Fund for Compensation for Oil Pollution Damage, 1971, with its protocols; International Convention Relating to Civil Liability in the Field of Maritime Carriage of Nuclear Material, 1971; Convention on the Prevention of Marine Pollution by Dumping of Wastes and Other Matter, 1972; International Convention for the Prevention of Pollution from Ships, 1973.

8. These include: Agreement Concerning the International Commission for the Protection of the Rhine Against Pollution; Agreement for Cooperation Dealing with Pollution of the North Sea by Oil, 1969; Agreement Concerning Cooperation in Taking Measures Against Pollution of the Sea by Oil, 1971; Convention for the Prevention of Marine Pollution by Dumping from Ships and Aircrafts, 1972; Convention on the Protection of Marine Environment of the Baltic Sea Area, 1974; Convention for the Prevention of Marine Pollution from Land Based Resources, 1974; Convention for the Protection of the Rhine Against Chemical Pollution.

9. Conventions of some major significance with their various protocols are the following: Convention for the Protection of the Mediterranean Sea Against Pollution, 1976; Kuwait Regional Convention for Cooperation on the Protection of the Marine Environment from Pollution, 1978; Convention for the Cooperation in the Protection and Development of the Marine and Coastal Environment of the West and Central African Region, 1981; and Regional Convention for the Conservation of the Red Sea and the Gulf of Aden Environment, 1981.

10. Considerable controversy was generated over thorny implementation issues. They include: are trading proposals legal? what are their consequences for enforcement? will the proposals, in fact, produce equivalent or better environmental results than existing requirements? what administrative safeguards are necessary to preclude or reduce abuse? what is the appropriate balance between placing constraints on trading to limit abuse and making sure rules are flexible enough to yield the benefits trading should produce? See for this and an excellent review of the bubble policy, Richard A. Liroff, Reforming Air Pollution Regulation: The Toil and Trouble of EPA's Bubble (The Conservation Foundation, Washington D.C., 1986).

11. See ibid. The Report points out that bubbles have produced significant cost savings and have reduced pollution more than compliance with conventional regulatory requirements would have. A few bubbles have sped pollution abatement producing reductions in emissions faster than would have occurred in response to conventional control requirements. On the negative side, difficulties in changing regulatory policy demonstrated how reforms that seem quite sensible on their face can be quite hard to put into practice, and how reforms like the promise of so many government programs, do not always deliver all that they claim.

    See also Robert W. Crandall, Controlling Industrial Pollution: The Economics and politics of Clean Air (Brookings Institution, Washington D. C., 1983).

12. Rough estimates of the contribution of deforestation to atmospheric concentrations of CO2 indicate it to be significant, although the contributions from individual countries are poorly known at present. While these data are important, the quota system discussed in this paper has some merit for its consideration, even without the deforestation data, in establishing a meaningful quota system.

ASSESSING THE THREAT TO ANTIQUITIES POSED BY CLIMATE CHANGE, SEA LEVEL RISE AND AIR POLLUTION

by Dr. Hind Sadek and John C. Topping, Jr.*

Many of earth's most prized antiquities may be in grave danger over the next generation as a result of global climate change and local air pollution.

Air pollution, a growing concern in such cities as Cairo, Delhi, and Mexico City, has already caused serious damage to the Parthenon in Athens. Air pollution has also been implicated as a significant contributing factor to the rapid deterioration of a number of English cathedrals including Salisbury Cathedral.[1]

Climatic factors have been identified as a possible cause of the collapse in March 1989 of the 900 year old Pavia Tower in Pavia, Italy. The crumbling of this 255-foot tower killed two persons, injured fifteen and damaged an adjacent cathedral. Experts have reported that the tower's collapse may have been due either to natural decay in the mortar holding the bricks together or to a sudden sinking following a dry spell that lowered the water table.[2]

I. AN OVERVIEW OF THE THREAT

Added to the fairly visible threat of growing urban air pollution in major cultural centers of the developing world are the likely multiple stresses

---

1 Craig Whitney writes in a March 25, 1989 article in The New York Times, "Salisbury Cathedral's 404-foot spire, its limestone blocks cracked and worn by wind, rain, frost and poisons in the atmosphere to a thickness of only three inches in places, threatened to fall apart after more than 600 years, and is now swathed in 280 tons of steel scaffolding. At the dizzying height of the spindly structure, the once spiky Gothic shrines and decorations look worn to nubs, like the battlements of sand castles on the beach after a high wave." Whitney, "Crumbling English Cathedrals Go Abroad for Aid," op. cit., at p. 8.

2 "Italian Tower Collapse," The Washington Post, March 18, 1989.

***A Senior Fellow at the Climate Institute, Dr. Sadek is the Coordinator of the December 17-21, 1989 Cairo World Conference on Preparing for Climate Change and Director of the Institute's Antiquities Project. An Egyptian born archeologist and Harvard Ph.D, she has served as Director of the National Museum of Natural History of Iran prior to 1988. The President of the Climate Institute, Mr. Topping, is the senior author of the Clean Air Handbook and former Staff Director of the Office of Air and Radiation of the U.S. Environmental Protection Agency.**

accompanying rapid climate change. These include increased coastal erosion associated with sea level rise, enhanced risk of storm damage in coastal areas, changed and often increased salinity, and altered rainfall and drought patterns. Moreover, many of these factors would be operating in combination with urban air pollution and increased ultraviolet radiation resulting from a thinning of the stratospheric ozone layer.

With the nineteen eighties having produced the six warmest years of the past century, there are strong indications that the greenhouse warming long predicted by climatologists may well be underway. Concentrations of carbon dioxide, the most important greenhouse gas, have jumped from 315 parts per million by volume to about 350 parts over the past three decades, while concentrations of the other greenhouse gases such as methane and chlorofluorocarbons have risen more sharply. Virtually all climatologists expect an acceleration of this warming trend if the concentrations of greenhouse gases continue to rise.

Such a greenhouse effect induced global warming would be expected to trigger significant worldwide rises in sea level, largely as a result of a thermal expansion of the upper layers of the ocean and a melting of Northern Hemisphere glaciers. Mid-range estimates of global sea level rise over the next century are around one meter. About a one foot sea level rise has occurred over the last century around the U.S. Atlantic Coast, about half due to global sea level rise and the remainder to local subsidence. The National Academy of Engineering has projected a reasonable prospect that this rate of sea level rise may increase as much as fivefold over the next century.[3]

In many areas the local sea level rise will occur much more rapidly as natural and man-made subsidence amplifies the global rise associated with greenhouse warming. The mouth of the Nile, the Mississippi, and the Ghanges Brahmaputra, are all vulnerable to such rapid rise. Entire parishes of Louisiana are likely to be inundated as levee construction along the Mississippi River and industrial development within the coastal parishes have aggravated natural factors producing subsidence.

Similar circumstances are operating along the mouth of the Nile, a much more densely populated region. Subsequent to the construction in 1964 of the High Aswan Dam, eighty to a hundred million tons a year of sediment that would have washed down the Nile to replenish the mouth have been blocked from descending. As a result of this loss of sediment and local tectonic subsidence, the Nile delta has experienced spectacular erosion in the last 25 years. Many

---

[3] National Research Council, Responding to Changes in Sea Level: Engineering Implications, National Academy of Science Press, Washington, D.C., 1987.

areas along the delta's coast with the Mediterranean are eroding at rates in excess of one meter annually, and in some spots erosion has exceeded 100 meters a year.[4]

The coastal delta includes a number of ancient sites including the city of Alexandria. Although coastal erosion itself may imperil some sites, other related factors may be of similar consequence. As sea level rises the wedge of salt water underlying the delta's fresh ground water will be forced further and further inland. Besides contaminating ground water, this increased salinity may also accelerate deterioration of monuments. Rising sea levels also increase vulnerability to storm surge.[5]

In the absence of a successful international effect to curb greenhouse emissions and slow the rate of global sea level rise, Egypt can be expected to experience profoundly adverse effects. A one meter relative sea level rise, a likelihood during the first half of the 21st century would produce, it has been projected, a loss of about 12 to 15 percent of Egypt's arable land, an area containing about 16 percent of its population.[6] Although no systematic analysis exists of the vulnerability of Egyptian antiquities to climate change, such enormous coastal erosion would almost certainly have sweeping effects. In addition, the severe human dislocation associated with such sea level rise might further imperil such sites in ways we can not yet discern.

Meanwhile the more localized air pollution as Cairo's population increases is becoming a factor which could jeopardize some antiquities around Egypt's capital. Moreover, the degradation of materials associated with such pollution could be augmented by increased ultraviolet radiation associated with global warming.

Sea level rise and air pollution are likely to pose little threat to the antiquities of Upper Egypt, but significant shifts in weather circulation patterns, changing humidity and salinity are posing serious threats to important sites in Upper Egypt, including the temples of Luxor and Karnak and the tombs in the Valley of Kings, Queens and Nobles.

---

4 James Broadus, John Milliman, Steven Edwards, David Aubrey and Frank Gable, "Rising Sea Level and Damming of Rivers: Possible Effects in Egypt and Bangladesh," Effects of Changes in Stratospheric Ozone and Global Climate, Volume 4: Sea Level Rise, Washington, D.C., October 1986, p. 167.

5 Ibid, p. 166.

6 Ibid, p. 171.

II. ROLE OF THE CAIRO CONFERENCE IN CATALYZING INTEREST IN THIS ISSUE

A major focus of the World Conference on Preparing for Climate Change to be held December 17-21, 1989 in Cairo, Egypt will be on the threat posed to the cultural treasures of the world by climate change, sea level rise and air pollution. There are several reasons that the Cairo Conference may be an optimum place for creating public awareness of this problem.

First, Egypt is perhaps the nation most associated in the public mind with antiquities. An Egyptian setting should tend to reinforce interest in this issue, not only as it pertains to Egyptian antiquities, but also to cultural treasures in Greece, Italy, Thailand, India and other countries.

Second, the Egyptian government is providing high level support to this Conference. Egypt's First Lady, Mrs. Suzanne Mubarak, is serving as Honorary Conference Chairman. H.E. Dr. Atef Ebeid, Minister for Cabinet Affairs and Chairman of the Egyptian National Committee for the Conference, is Egypt's chief environmental official, and H.E. Dr. Farouk Hosni, Minister for Culture, will play an active role. These distinguished Egyptians all recognize the importance of highlighting the threat to the antiquities, a vital cultural and economic asset.

Third, unlike some developing countries that are warily viewing cooperative efforts to protect the earth's climate, Egypt has been at the lead of the international effort to protect the global atmosphere. The United Nations Environment Programme's Executive Director, Dr. Mostafa Tolba, an Egyptian, was the driving force behind the negotiation and speedy ratification of the Montreal Protocol to protect the ozone layer, and Egypt was one of the first nations to ratify the accord. Dr. Tolba's accomplishments will be honored at the Conference, thus recognizing that environmental leadership is not confined to North America or Europe.

III. AN ANTIQUITIES FOCUS AS A UNIFYING THREAD

With many antiquities located in vulnerable coastal regions or near increasingly polluted cities an early start is essential to safeguard such treasures. Such a strategy has a number of elements including identifying what is at risk and developing workable response strategies. Some such strategies are site specific, others may require regional cooperation such as abatement of air pollution and others such as protection of the climate may require global cooperation.

Yet an interest in preservation of a nation's cultural heritage is nearly universal. To the extent that nations of the world realize that much of their priceless cultural heritage may be at risk, they may be more likely to take the actions which will slow greenhouse emissions and mitigate damage to cultural and human values.

Cultural preservation efforts have a tremendous potential to draw broad based support and reinforce sentiments of national pride. The Statue of Liberty restoration effort in the United States drew several million contributors and created great public interest.

Although assessments of environmental threats to cultural treasures are still largely anecdotal, the Cairo Conference could develop some focus on this problem both in Egypt and elsewhere.

Venice, a city which has long coped with the water, is facing a problem of regular inundation of St. Mark's Square. Where will this city be a generation hence?

Ravenna, an Italian City which is the site of some of the world's greatest murals, is already witnessing inundation of some of this art located on the floor of cathedrals.

Thanks largely to man-made development pressures, Bangkok, Thailand, is sinking steadily each year and with it some of the most beautiful buildings may be headed to oblivion.

In two antiquities workshops, one focusing on Egyptian antiquities and the other on antiquities in other nations, the Cairo Conference will provide an early glimpse of the potential scope of this problem. The meeting should also provide an opportunity to catalyze interest on the part of a number of culturally conscious governments. If successful, the Cairo Conference should stimulate many participating governments to examine the threat that the deterioration of the earth's atmosphere poses to their heritage.

IV. THE ELEMENTS OF AN ANTIQUITIES EFFORT

The Cairo Conference will leverage antiquity concerns onto the global environmental agenda and in so doing could introduce a more unifying thrust. Its role in introducing this issue will largely be one of sensitizing opinion leaders to the existence of the problem. Cairo will also provide a scene for formal and informal discussions of how we proceed to assess the dimensions of this problem.

The Laboratory for Coastal Research of the University of Maryland, widely regarded as a leading center for research into implications of sea level rise for coastal areas, has agreed to undertake the assessment of the threat posed by sea level rise to some of the world's leading antiquities. In addition, EPA's Office of Air Quality Planning and Standards has provided some technical assistance on air pollution aspects. Its parent, the EPA Office of Air and Radiation, is an active Cairo Conference sponsor.

Already a number of distinquished individuals associated with scholarly and antiquities interests have agreed to help in planning the Cairo Conference.

Before the Cairo Conference we plan to convene a small focus group to sketch out initial priorities for an antiquities scoping effort expected to extend for up to two years. This meeting would also begin to develop some picture of the resources needed to carry out such studies and implement recommendations.

We anticipate that one or more public meetings would be held in 1990 in the Mediterranean region to explore the implications for cultural sites in that area. Follow-on meetings would also be anticipated in Egypt with the strong involvement of the concerned government agencies.

During the first few months several significant sites would be identified for a more intensive on-site followup. Cooperation would be elicited from cultural authorities and institutions in each of the countries in which sites are situated. By this summer we would hope to have a reasonable picture of the resources required in 1990 and 1991 to conduct such assessments.

Although the resources needed for this project remain uncertain, it is likely that they would be a tiny fraction of the funds committed annually to cultural preservation efforts. Such a study, if well implemented, could ensure that cultural preservation efforts are not negated by environmental factors which were not considered in project planning. Finally, by highlighting the prospect that humanity's cultural heritage may be imperiled by rapid climate change and air pollution, the project could provide an impetus toward regional and global action to address these problems.

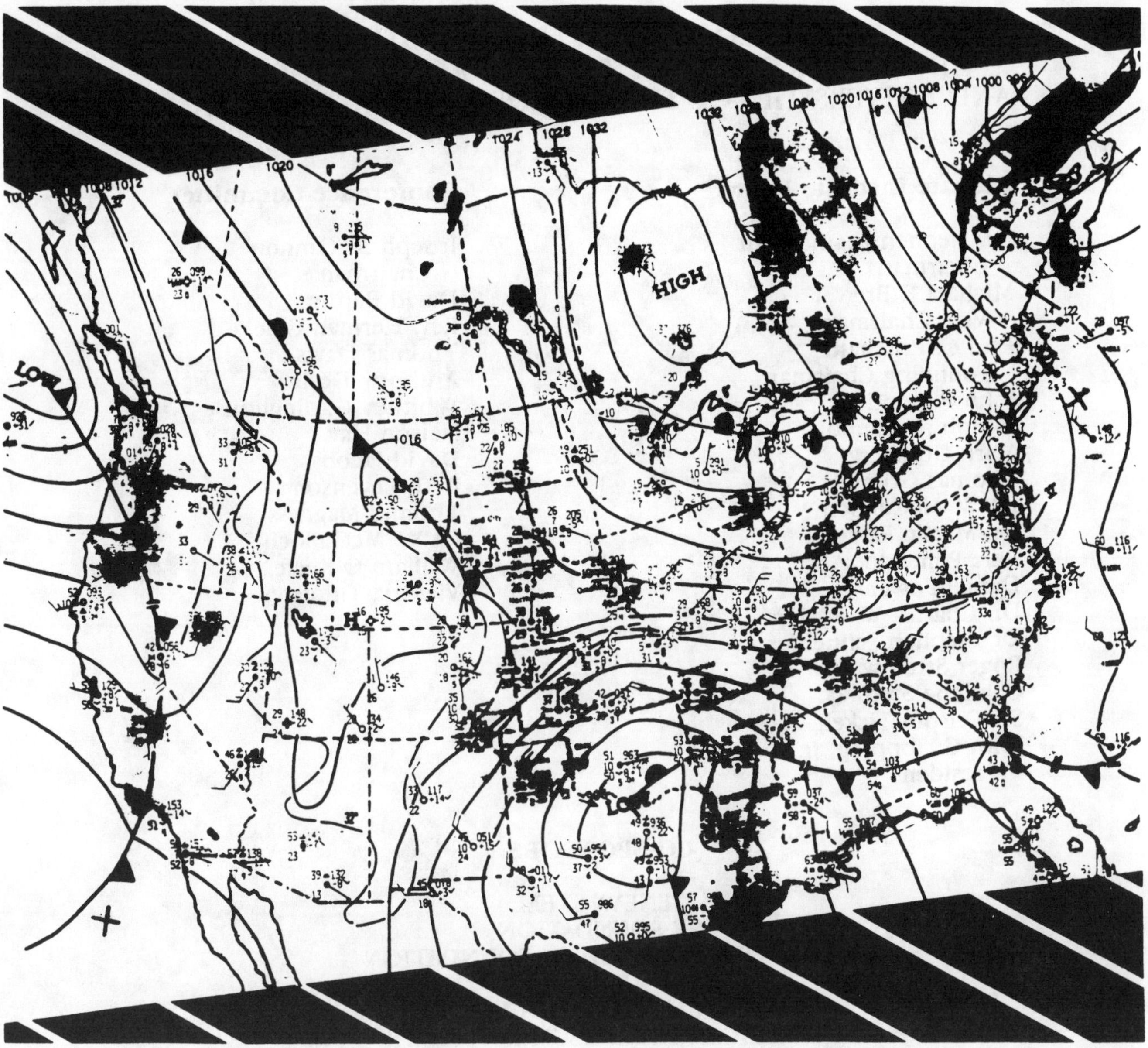

SECOND NORTH AMERICAN CONFERENCE ON PREPARING FOR CLIMATE CHANGE: A COOPERATIVE APPROACH

December 6–8, 1988

Mayflower Stouffer Hotel
1127 Connecticut Avenue, N.W.
Washington, D.C. 20036

CONVENED BY THE CLIMATE INSTITUTE IN COOPERATION WITH PARTICIPATING AGENCIES AND UNIVERSITIES OF THE CANADIAN CLIMATE PROGRAM

Climate Institute Board

Dr. Stephen Leatherman
  Chairman
Michael F. Brewer
  Vice Chairman
Paul C. Pritchard
  Founding Chairman
John P. Bond
Joseph A. Cannon
Mark Goldberg
Thomas Grumbly
J.A.W. McCulloch
Dr. Martin Parry
Rafe Pomerance
Daniel Power
Dr. Charles W. Powers
Dr. Stephen Schneider
Roger Strelow
Charlene Sturbitts
Sir Crispin Tickell
John C. Topping, Jr.
  President

Conference Committee

Joseph A. Cannon
  Chairman
David Bardin
Rev. Herman Cole
Thomas Davis
Anthony Desir
Whitney Garlinghouse
Nelson Hay
David Jacobs
Ruth Levenson
Thomas Magness
J.A.W. McCulloch
William O'Keefe
Victoria Tillotson

CO-SPONSORS

GENEVA STEEL
GE FOUNDATION
THE WILLIAM BINGHAM FOUNDATION
AT&T
AMERICAN GAS ASSOCIATION
AMERICAN PETROLEUM INSTITUTE
MOTOR VEHICLE MANUFACTURERS ASSOCIATION
EBASCO SERVICES, INC.
TEXACO, INC.
WORLD RESOURCES INSTITUTE
NATIONAL PARKS AND CONSERVATION ASSOCIATION
NATIONAL AUDUBON SOCIETY
WOODS HOLE RESEARCH CENTER
ATLANTIC CITY ELECTRIC COMPANY
PPG FOUNDATION
U.S. ENVIRONMENTAL PROTECTION AGENCY
U.S. DEPARTMENT OF ENERGY
NATIONAL CLIMATE PROGRAM OFFICE
NATIONAL OCEANIC & ATMOSPHERIC ADMINISTRATION
U.S. FOREST SERVICE
NATIONAL AERONAUTICS AND SPACE ADMINISTRATION
NATIONAL SCIENCE FOUNDATION

Conference Consultants
Hartnett & Associates

CONFERENCE AGENDA

First Day: Tuesday, December 6, 1988

Morning: *Scientific Overview*

8:30–11:30 (East/State)

Opening Remarks of Conference Chairman, Joseph Cannon, President, Geneva Steel

Scientific Keynote, Dr. Stephen Schneider, National Center for Atmospheric Research

"Likely Climate Changes in North America and the Caribbean," Dr. James Hansen, Goddard Institute for Space Studies

"Likely Sea Level Rise," Dr. Stephen Leatherman, University of Maryland and Chairman, Climate Institute

"Stratospheric Ozone Depletion," Dr. Robert Watson, NASA

"Health and Environmental Effects of Ozone Depletion and Climate Change," John Hoffman, U.S. EPA

Luncheon Panel: *12:15–1:45 (East/State)*

"Is Antarctic ozone hole lessening? Review of 1988 results. Is there an Arctic ozone hole?" Dr. Robert Watson, NASA, Dr. Alex Chisholm, Atmospheric Environment Service, Canada, Dr. Michael McElroy, Harvard University and Dr. F. Sherwood Rowland, University of California at Irvine

---

Afternoon: **Resolving Scientific Uncertainties**

1:45–3:30 p.m. *Simultaneous Panels*

A. Role of Climate Feedbacks in Global Warming *(East)*

Chairman, ***Dr. Stephen Schneider***

"Potential volume and rate of methane hydrate release," Dr. Gordon MacDonald, MITRE Corporation

"The contribution of soil feedbacks to global warming," Dr. George Woodwell, Woods Hole Research Center

"Feedback processes that may affect future concentrations of greenhouse gases," Dr. Daniel Lashof, U.S. EPA

"The role of clouds in enhancing or reducing greenhouse warming," Dr. Robert A. Schiffer, NASA

B. The Role of Oceans in Climate Change *(Massachusetts)*

Chairman, ***Dr. John Carey, President, American Oceanic Organization***

"CO2 uptake by the oceans: future prognosis," Dr. Taro Takahashi, Lamont-Doherty Geological Observatory

"How climate change might affect the Gulf Stream and other ocean currents," Dr. Robert Molinari, NOAA

"Understanding El Nino and long term climate variability over oceans," Henry Diaz, NOAA-ERL

"The cost of obtaining decent ocean data," John Bond, Climate Institute

C. Role of Agriculture in Altering Climate *(New Hampshire)*

Chairman, *Dr. Martin Parry, Climate Institute*

"Factors affecting biological methane production," Dr. C.C. Delwiche, University of California at Davis

"Greenhouse gas contribution of deforestation to create farmland," Dr. Richard Houghton, Woods Hole Research Center

"The potential role of increased animal productivity in reducing methane emissions," Dr. Henry Tyrrell, U.S. Department of Agriculture

"Reducing greenhouse gases through shifting staple production to woody plants," Philip Rutter, American Chestnut Foundation

D. Potential Impact of Global Warming on Public Health in North America *(Rhode Island)*

Chairman, *Dr. Janice Longstreth, ICF Clement*

"An overview of potential impacts of global warming on public health," Dr. Janice Longstreth, ICF Clement

"Changes in mortality and health from extreme heat or cold," Dr. Larry Kalkstein, University of Delaware

"Implications of climate change for natural hazard management," Dr. William Riebsame, University of Colorado

"Impact of global warming on spread of pathogens and pests," Dr. Andrew Dobson, University of Rochester

E. Potential Effects of Global Warming and Stratospheric Ozone Depletion on Ground Level Air Quality *(Pennsylvania)*

Chairman, *Thomas Grumbly, President, Clean Sites*

"Interactive effects of climate change, ground level ozone and UVB on natural systems," Dr. Chris Bernabo, Science and Policy Associates

"Implementation of future air quality control strategies to prepare for climate change," Gerald Emison, Director, EPA Office of Air Quality Planning and Standards

"Linkages between air pollution control strategies and climate change," Dr. William Moomaw, World Resources Institute

F. Status and Greenhouse Effect Implications of Renewable Energy and Conservation Technologies *(Georgia)*

Chairman, *Dr. William Chandler, Battelle Pacific Northwest Laboratories*

"Potential of sea thermal energy," J. Hilbert Anderson, President, Sea Solar Power, Inc.

"Potential energy uses and greenhouse implications of hydrogen," Peter Hoffmann, Editor, Hydrogen Letter

"Status of photovoltaic solar technology," Dr. Dan Arvizu, Sandia National Laboratory

"Potential of geothermal energy," Dr. Jefferson W. Tester, Massachusetts Institute of Technology

3:45–5:30 p.m. ***Simultaneous Panels***

A. Changes in Weather Circulation Patterns *(Massachusetts)*

Chairman, ***Jim McCulloch, Richmond Hill, Ontario, Canada, Climate Institute***

"Potential shifts in monsoon patterns associated with climate warming," Dr. Eugene Rasmusson, University of Maryland

"Potential changes in precipitation and in temperature variability," Dr. David Rind, GISS

"Implications of urbanization for local and regional temperatures in the United States," Dr. Arthur Viterito, George Washington University

B. Projections of Global Emission Trends *(New Hampshire)*

Chairman, ***Dr. Michael Barth, Senior Vice President, ICF***

"Base case emission scenarios," Dr. Daniel Lashof, U.S. EPA

"Effect of policy options on emission forecasts," Dennis Tirpak, U.S. EPA

"Uncertainties in energy models," Dr. Jae Edmonds, Battelle Pacific Northwest Laboratories

C. Addressing Impacts of Climate on Agriculture *(East)*

Chairman, ***Cynthia Rosenzweig, Columbia University***

"An overview of effects of climate on agriculture," Dr. Martin Parry, University of Birmingham, U.K.

"Likely effects of climate change on U.S. agriculture," Dr. Leon Hartwell Allen, Jr., University of Florida

"Likely impact of climate change on Canadian agriculture," Dr. Barry Smit, University of Guelph

"Translating climate impacts into legislative responses," James Cubie, U.S. Senate Agriculture Committee

D. Impact of Stratospheric Ozone Depletion on Human Health *(Rhode Island)*

Chairman, ***Dr. Charles W. Powers, Resources for Responsible Management***

"An overview of health risks from stratospheric ozone depletion," Dr. Janice Longstreth, ICF Clement

"Implications of UVB for skin cancer and eye disease," Dr. Andrea Kornhauser, Food and Drug Administration

"Stratospheric ozone depletion, sunlight and immune suppression: a new connection," Dr. Edward DeFabo, George Washington University Medical Center

E. Uncertainties in Our Understanding of Atmospheric Change *(Pennsylvania)*

Chairman, ***Rafe Pomerance, World Resources Institute***

"An overview of interrelationships among changes in the stratosphere, troposphere and biosphere," Dr. Michael McElroy, Harvard University

"Reducing uncertainties in projections of regional ecological impacts," Dr. Peter Jutro, U.S. EPA and Dr. Anthony Janetos

"The potential of increased methane concentrations to enhance depletion in polar stratospheric clouds," Dr. F. Sherwood Rowland, University of California at Irvine

F. Status and Greenhouse Implications of Nuclear Energy Technologies *(Georgia)*

Chairman, ***Dr. Gordon MacDonald, Vice President and Chief Scientist, MITRE Corporation***

"An overview of the greenhouse implications of various nuclear energy technologies," Mark Mills, President, Science Concepts

"Directions in advanced reactor technology," Dr. Michael Golay, Massachusetts Institute of Technology

"Status of aneutronic fusion energy," Dr. Bogdan Maglich, President, Advanced Research Corporation

---

7:00–10:00 p.m. AWARDS RECEPTION AND DINNER *(East/State)*

Chairman, ***Sir Crispin Tickell, Ambassador from the United Kingdom to the United Nations***

Remarks by Ambassador Tickell

Award Presentations, Dr. Stephen Schneider, Chairman, Awards Committee

Presentation to Dr. F. Sherwood Rowland for his outstanding scientific contributions to our understanding of the stratospheric ozone layer.

Presentation to Dr. J. Murray Mitchell, Jr. for his pioneering scientific research into past climate change and its potential implications for future greenhouse warming.

Presentation to Ambassador Richard Benedick for his outstanding work in 1988 in advancing understanding within the international/diplomatic community of the challenges posed by potential greenhouse warming and stratospheric ozone depletion.

Entertainment by Captain Sea Level, Balladeer of the Climate Community

---

Second Day: Wednesday, December 7, 1988

Morning 8:30–11:30 a.m. **Plenary: Agenda for the Next Decade** *(East/State)*

"Status of USA-USSR Global Climate Agreement," Dr. Alan Hecht, Director, U.S. National Climate Program Office

"International Space Year and Mission to Planet Earth," Dr. Lennard A. Fisk, Associate Ad-Administrator for Space Sciences and Applications, NASA

"Lessons from UNEP Protocol for a Greenhouse Convention," Ambassador Richard Benedick, The Conservation Foundation

"UNEP's role in addressing climate change," Dr. Noel Brown, Director, North American Operations, UNEP

"Deforestation in the Western Hemisphere," Prof. Eneas Salati, InterAmerican Development Bank

"A Canadian perspective on climate change," Dr. Kirk Dawson, Director General, Canadian Climate Centre

---

Luncheon: 12:15–1:30 p.m.

Chairman, ***Michael Brewer, Vice President, Dun & Bradstreet***

Remarks by Robert Grady, Office of the President-elect

Policy Keynote Speaker, ***Ambassador Stephen Lewis, Chair, Toronto Conference on the Changing Atmosphere***

ASSESSING REGIONAL IMPACTS

1:45–3:30 p.m. **Simultaneous Panels**

A. Potential Effects of Climate Change and Sea Level Rise in Caribbean

Chairman, ***Anthony Desir, Trinidad and Tobago***

"Likely sea level rise in the Caribbean," Frank Gable, Woods Hole Oceanographic Institution

"Implications for Caribbean society of climate change, sea level rise and shifts in storm patterns," Dr. Orman Granger, University of California at Berkeley

"Implications of climate change for coral reefs and islands, mangrove swamps and wetlands in the Caribbean," Dr. David Stoddard, University of California at Berkeley

B. Potential Impact of Changes in Climate in California *(Rhode Island)*

Chairman, ***Dr. Peter Gleick, Pacific Institute***

"Impact on water resources," Dr. Robert Buddemeier, Lawrence Livermore national Laboratory

"Potential effects on irrigated agriculture," Dr. Daniel Dudek, Environmental Defense Fund

"Impact of climate change on water resources in the San Francisco Bay," Philip Williams

C. Potential Impact of Climate Change and Sea Level Rise on Chesapeake Bay *(Pennsylvania)*

Chairman, ***Dr. Stephen Leatherman, Climate Institute***

"Marsh loss and shore erosion," Dr. Michael Kearney and Dr. J. Court Stevenson, University of Maryland

"Impact on aquatic animals and fisheries," Dr. Victor Kennedy, University of Maryland

"Implications of climate change for environmental protection strategies for the Chesapeake Bay," Dr. Alvin Waller, EPA, Region III

D. Likely Impact of Rapid Warming on Arctic *(New York)*

Chairman, *Dr. Fred Roots, Science Advisor, Department of Environment, Canada*

"Implications of expected change in ice cover," Dr. Claire Parkinson, NASA

"Effects on migratory birds and other wildlife," Dr. J.P. Myers, National Audubon Society

"An overview of potential effects of rapid warming on the Canadian Arctic," Dr. Stephen C. Lonergan, McMaster University

E. Potential Impact of Climate Change on Southern U.S. *(Georgia)*

Chairman, *Daniel Power, Nashville, Tennessee, Climate Institute*

"Implications of global climate change for TVA reservoir system," Dr. Barbara Miller, TVA Engineering Laboratory

"Potential impact of climate change on water quality in the Southern U.S.," Dr. Ellen Cooter, Oklahoma Climatological Survey

"Policy implications of potential climate change in Southern U.S.," Dr. Mark Meo, University of Oklahoma

---

3:45–5:30 p.m. *Simultaneous Panels*

ASSESSING IMPACTS ON MAJOR SECTORS

A. Impacts on Water-Resources *(East)*

Chairman, *Dr. Paul Waggoner, Chairman, AAAS Water Panel*

"The state of future water resource supply and demand in North America even without climate change," Dr. J.E. Schefter, U.S. Geological Survey

"Vulnerability of North American water systems to climate change," Dr. Peter Gleick, Pacific Institute

"Likely changes in evapotranspiration," Dr. Norman Rosenberg, Resources for the Future

"Likely changes in water quality related to climate change," Dr. Henry D. Jacoby, Massachusetts Institute of Technology

B. Impacts of Climate Change on Urban Planning *(Rhode Island)*

Chairman, *Neal R. Peirce*

"Likely effects of climate change on municipal infrastructure in Miami, New York and Cleveland," Dr. Ted Miller, Urban Institute

"Effects of climate change on coastal infrastructure," Jim Titus, U.S. EPA

"Emergency preparedness to address climate change," Dr. Sherry Oaks, Pennsylvania State University

"Perspectives of a civil engineer and planner," Daniel Power, Climate Institute

"Climate change and water resources infrastructure," Mark Mugler, Apogee Research

C. Impact of Climate Change on Fish and Wildlife *(Pennsylvania)*

Chairman, *Dr. Robert Peters, World Wildlife Fund*

"Potential implications of climate change for mammals," Dr. Dewey McLean, Virginia Polytechnic Institute and State University

"Potential impact of climate change on Great Lakes fisheries," Dr. Henry Regier, University of Toronto

"Potential impact of climate change on U.S. wildlife refuges," Robert Breckenridge, Idaho Falls Engineering Laboratory

"Implications of potential effects of climate change for Department of Interior wildlife and resource management," Dr. Indur Goklany, U.S. Department of Interior

D. Impact of Climate Change on Forests *(Georgia)*

Chairman, *Rev. Herman Cole, Chair, Adirondack Park Agency*

"An overview of the likely impact of climate change on North American forests," John Wells

"Climate change and U.S. forest markets," Dr. Larry Regens, University of Georgia

"Implications of climate change for Canadian forests," Dr. James B. Harrington, Canadian Forestry Service

"Climate change and forest fires," Dr. Michael A. Fosberg, U.S. Forest Service

E. Potential Impact of Stratospheric Ozone Depletion on Food Chain and Vegetation *(New York)*

Chairman, *John Hoffman, U.S. EPA*

"Impact on marine food chain," Dr. Robert Worrest, U.S. EPA

"Impact on crops," Dr. Alan Teramura, University of Maryland

"Impact on vegetation," Dr. Manfred Tevini, University of Karlsruhe (F.R. Germany) (Tentative)

Wednesday Evening

7:00–10:00 p.m. ***Dinners in private homes in the Washington area for conference participants***

Third Day: Thursday, December 8, 1988

Response Strategies

8:30–10:15 a.m. *Simultaneous Panels*

A. Prospects for CFC and Halon Substitutes *(Georgia)*

Chairman, *Dr. Stephen Andersen, U.S. EPA*

"Halon substitutions," Carl Jewell, Halon Research Institute

"Substitutes in mobile air conditioning," Simon Oulouhajian, Mobile Air Conditioning Society

"Incentives for substitution and efficiencies among CFCs and halons," Prof. Alan Miller, University of Delaware Law School

B. Preparing for Climate Change in the Great Lakes *(Pennsylvania)*

Chairman, *Dr. Stewart Cohen, Canadian Climate Centre*

"An overview of EPA studies of climate change impacts on Great Lakes region," Joel Smith, U.S. EPA

"A summary of U.S.-Canadian Great Lakes Climate Change Conference recommendations," Dr. William Bolhofer, U.S. National Climate Program Office

"Climate Change and the IJC," Dr. Murray Clamen, International Joint Commission, Canada

"Implications of changing Great Lakes levels for commercial shippers," Angus Laidlaw, Dominion Marine Association

C. Preparing for Climate Change in New England and Atlantic Canada *(Massachusetts)*

Chairman, *Greg Watson, Executive Director, Massachusetts Office of Science and Technology*

"Strategies to respond to climate change and sea level rise in Atlantic Canada," Dr. Peter Stokoe, Dalhousie University

"Upland erosion in New England," Dr. Graham Giese, Woods Hole Oceanographic Institution

"Strategies for Cape Cod to respond to sea level rise," Dr. Stephen Leatherman, Climate Institute

D. Factoring Climate Change into Corporate Planning *(Rhode Island)*

Chairman, *Roger Strelow, Vice President, General Electric Company*

"Implications of climate change for U.S. electrical demand," Ken Linder, ICF

"Implications of climate change for environmental engineering and construction industry," Dr. James J. Ferris or J.F. Silvey, Ebasco Services

"Implications of climate change for insurance industry," Dr. Donald Friedman, Travelers Insurance

"Implications of climate change for securities underwriting," Al Mediolo, Senior Vice President, Moody's

"Implications of climate change for investment planning," Ted Harris, Oxford Analytica

E. Energy Strategies to Restrict Emissions Growth *(East)*

Chairman, *Ted Williams, U.S. Department of Energy*

"Supply side strategies to reduce greenhouse emissions," Philip Jessup, Energy Probe, Toronto

"Government strategies to limit buildup of greenhouse gases," David J. Bardin, Esq., Arent, Fox, Kintner, Plotkin & Kahn

"Energy conservation as a strategy to reduce greenhouse emissions," Marc Ledbetter, American Council for an Energy Efficient Economy

10:30 a.m.–12:15 p.m. Simultaneous Panels

A. How to Protect the North American Coastline *(Massachusetts)*

Chairman, Thomas Magness, Ebasco Services

"Relative vulnerability of North American coasts to shore line erosion," Jim Titus, U.S. EPA

"Limitations of fortification and beach nourishment strategy," Dr. Stephen Leatherman

"Policy considerations in coastal protection," Lynne Edgerton, Esq., Natural Resources Defense Council

B. International Actions on the Ozone Layer and Greenhouse Warming *(East)*

Chairman, *Ambassador Richard Benedick, The Conservation Foundation*

"International progress on implementing the Montreal Protocol," Victor Buxton, Environment Canada

"A framework for a greenhouse convention," Dr. Wilfrid Bach, University of Munster, FRG

"Factors to consider in global negotiations," Dr. Konrad Von Moltke, Dartmouth College and The Conservation Foundation

C. Preservation of Tropical Forests *(Georgia)*

Chairman, *Dr. Norman Myers, Oxford, UK, World Wildlife Fund*

"Role of deforestation in regional climate change," Dr. Norman Myers, Oxford, U.K., World Wildlife Fund

"Strategies to finance retention of tropical forests," Dr. Ata Qureshi, Climate Institute

"Role of development assistance organizations in forest preservation," Prof. Eneas Salati, InterAmerican Development Bank

D. Developing Country Energy Strategies *(Rhode Island)*

Chairman, *Dr. Richard Morgenstern, Director, Office of Policy Analysis, U.S. EPA*

"An overview of the climate implications of energy strategies," Dr. Irving Mintzer, World Resources Institute

"The impact on global climate of energy strategies in Asia," Dr. William Chandler, Battelle Pacific Northwest Laboratory

"Making efficiency and conservation a priority in World Bank energy lending," David Wirth, Esq., Natural Resources Defense Council

E. Development of Legal and Institutional Mechanisms for Climate Cooperation *(Pennsylvania)*

Chairman, ***Dr. Alan Hecht, Director, U.S. National Climate Program Office***

"Next steps for climate cooperation," Dr. Kilaparti Ramakrishna, Woods Hole Research Center

"Mobilizing a global citizens movement," Dr. Hind Sadek, Climate Institute

"Impact of climate change on availability of Colorado River water in the U.S. and Mexico," Dr. Peter Gleick, Pacific Institute

"Legal mechanisms for water sharing in time of drought," James Strock, Esq.

12:45–2:30 p.m. Luncheon: Concluding Speaker

Senator Albert Gore, Jr. (Ballroom)

Award to Paul Pritchard, President, National Parks and Conservation Association

---

NOTES

Climate Institute Honors Rowland, Mitchell, and Benedick

At a dinner on December 6, the first night of the 3-day Second North American Conference on Preparing for Climate Change, the Climate Institute will present awards to three individuals who have made outstanding contributions to furthering knowledge of the global climate system and to controlling damage to the stratosphere.

Dr. F. Sherwood Rowland is recognized for scientific research advancing our understanding of the stratospheric ozone layer.

Dr. F. Sherwood Rowland.

In a paper published in the British journal *Nature* in 1974, F. Sherwood Rowland and his co-author Mario J. Molina first established a possible link between chlorofluorocarbons and eventual stratospheric ozone depletion. He has remained at the forefront of stratospheric research since then, playing a central role in NASA's Ozone Trends Panel. The March 1988 Panel Report established that global ozone levels have dropped appreciably in recent years. This finding culminated in international policy action in the Montreal Protocol in September 1987.

Fluorocarbon depletion of stratospheric ozone is one of the many interests of this scientist whose prodigious research activities include work on the chemical effects of nuclear transformations, radiation chemistry, tracer reactions, isotopes applied to geochemistry, and tropospheric as well as stratospheric chemistry of halocarbons, methanes and other hydrocarbons.

Dr. Rowland has been Daniel G. Aldrich Professor of Chemistry at the University of California, Irvine, since 1985. He also serves on the National Academy of Sciences Board of Environmental Studies and Toxicology, the Scientific Committee on Problems of the Environment of the International Council of Scientific Unions, and the Ozone Committee of the International Association of Meteorology and Atmospheric Physics, as well as being a member of the science advisory board of the Max Planck Institute of Cosmochemistry and Geochemistry in West Germany.

Dr. Rowland received his Ph.D. degree from the University of Chicago and has also taught at Princeton University and the University of Kansas, Lawrence.

Dr. J. Murray Mitchell, Jr., is cited as a pioneer researcher on climate change. He was the first scientist to study paleoclimatic data to explain large scale past climate change and also the first to look into data records for the importance of human impacts. He was in the vanguard in plotting climatic futures based on climate sensitivities and carbon dioxide burdens.

J. Murray Mitchell, Jr.

Except for occasional semester sabbaticals at academic institutions around the country, Dr. Mitchell has spent his entire career in research in the field of climatic change with the Federal Government. Starting with what was then called the U.S. Weather Bureau, he continued his work through various governmental transformations of that agency to what is now the National Environmental Satellite, Data and Information Services (NESDIS). Dr. Mitchell has recently retired from NESDIS.

Ambassador Richard Elliot Benedick is honored for his very significant role in drawing the diplomatic community into a realization of the implications of climate change and toward a willingness to consider international approaches to preparations for change.

Ambassador Richard E. Benedick.

Ambassador Benedick was principal U.S. negotiator for the Montreal Treaty concluded in 1987, a landmark international agreement limiting the dispersion of chlorofluorocarbons and halons into the atmosphere.

Last March, he chaired a symposium on the "Impact of Climate Change on the Third World: Implications for Economic Development and Financing," in Washington that was so successful it was repeated for members of permanent missions to the United Nations in New York in June. Nearly two dozen ambassadors expressed appreciation for this initiative, and papers from the symposia have entered the policy process of governments. The symposium was sponsored by the Climate Institute and co-sponsored by The Conservation Foundation.

Ambassador Benedick has also spoken at many meetings and conferences around the world, drawing lessons and implications from the successful Montreal negotiations for future actions among nations to ameliorate global atmospheric problems. He is now Senior Fellow at The Conservation Foundation.

This is the second year in which Climate Institute awards have been made. Last year the awards were made to Dr. Roger Revelle, University of California, San Diego, for scientific achievement, and to Senator John Chafee for his contributions to public understanding of climate problems.

CLIMATE ALERT

Winter 1988

Vol. 1, No. 4

WORLD RESOURCES INSTITUTE
A CENTER FOR POLICY RESEARCH

1709 New York Avenue, N.W., Washington, D.C. 20006, Telephone 202-638-6300

Introduction of DR. F. SHERWOOD ROWLAND University of California, Irvine

Rafe Pomerance
Senior Associate for Policy Affairs

I know Sherry Rowland. Sherry Rowland is a friend of mine. Steve Schneider, you are a great scientist too.

Sherry Rowland is a fine scientist and a great citizen of this planet. He is a world figure now not only because he and Mario Molina proposed, in 1974, that CFCs could destroy ozone and they were right.

Rather, Sherry Rowland is a world figure for what he has given to protect the ozone layer since 1974. In session after session, Rowland has maintained that chlorine would destroy ozone. The ups and downs of model results did not alter his fundamental concern. That should teach us what we should listen for from the scientific community: Don't worry too much about detailed forecasts of a problem -- watch the direction and consider the surprise.

Perhaps nothing characterizes Sherry Rowland's journey more than the stamina, vitality and skill he has maintained in the public arena. Some tried to ignore the warnings. But Sherry would not be stopped. He has stayed in the debate, appropriately cautious and yet appropriately concerned. His strength, too, has been his humility. He was always there to talk to those environmentalists who wanted to understand, to the media who wanted to educate, to political leaders who were living up to their responsibilities. He was always ready to talk despite the commute from California.

Scientists are citizens. Sherry saw his duty as a citizen of this country: that is to speak out. That is not easy for scientists in this field to do. He handled himself well in policy discussions. He did not need to give specifics for a solution. He heeded and did suggest the direction and allowed officials and the executive branch and, finally, the international community, to do their job. He understood, he educated, he led, he shared his knowledge.

Perhaps he became endeared to us most when, at an important Senate hearing chaired by Senator Chafee in June of 1986, the following exchange took place:

> Senator CHAFEE. Gentlemen, I am going to ask you a question. Suppose each of you were king and you had what you might call unlimited authority. What you you do about this problem? I will start with King Rowland.
>
> ROWLAND. If I were king, the first thing I would do is consult with the queen, who is sitting behind me and who has a very good view on what the sensible things to do in such cases are.

It gives me great pleasure to introduce Dr. Sherry Rowland.